Silicon Sensors and Actuators

Benedetto Vigna • Paolo Ferrari
Flavio Francesco Villa • Ernesto Lasalandra
Sarah Zerbini

Editors

Silicon Sensors and Actuators

The Feynman Roadmap

Editors
Benedetto Vigna
Analog MEMS & Sensors R&D
STMicroelectronics (Italy)
Cornaredo, Milano, Italy

Flavio Francesco Villa
ST Microelectronics, Analog MEMS
and Sensors Group, MEMS Technology
and Design R&D
Agrate Brianza
Monza Brianza, Italy

Sarah Zerbini
Analog, MEMS and Sensors Group
STMicroelectronics (Italy)
Cornaredo, Milano, Italy

Paolo Ferrari
ST Microelectronics, Analog MEMS
and Sensors Group, MEMS Technology
and Design R&D
Agrate Brianza
Monza Brianza, Italy

Ernesto Lasalandra
Analog, MEMS and Sensors Group
STMicroelectronics (Italy)
Cornaredo, Milano, Italy

ISBN 978-3-030-80137-3   ISBN 978-3-030-80135-9   (eBook)
https://doi.org/10.1007/978-3-030-80135-9

This Springer imprint is published by the registered company Springer Nature Switzerland AG
The registered company address is: Gewerbestrasse 11, 6330 Cham, Switzerland

Foreword

I have spent my entire professional career in the semiconductor industry, and I had the pleasure and the luck to see in first person two important turning points of this industry. In the 1960s, the technological community started to integrate successfully on silicon many transistors aiming to mimic the function of the brain, the memory, the nervous system, and the muscle of human beings. In the 1990s, instead, new components, such as variable capacitors and variable resistors, became the core of miniaturized silicon transducers, either sensors or actuators, and we started to integrate into silicon the five senses of human beings. It has been for sure a big revolution.

In the last 20 years, the pace of transducer development accelerated thanks to the progresses in microfabrication techniques and successful high-volume market applications. Nowadays, transducers are all around us: in cars, in smartphones, in factories, in printers, in tablets, in satellites, in drones, in smart speakers, in watches, and even in medical patches and shoes.

There is no doubt we are living in interesting times, with many challenges opportunities in front of us. Internet of Things, artificial intelligence, and 5G networks will boost the GDP of many countries and will enable new business models. I really hope these new technologies will help reduce the negative impact of humankind on nature and will also help reduce the gap between the richest and poorest countries in the world.

Within the frame of this new blurred world, micromachined silicon sensors will play even a more important role than today. Several parameters, such as vibrations, sound, atmospheric pressure, pollution level, will reach the digital world without the need of people typing on a keyboard or moving a mouse or touching an advanced display.

The most successful transducers in the market are not exploiting complex quantum physics effects. Their operation mode is relatively easy to be understood. But under this apparent simplicity, two big challenges reside. On one side, being multidisciplinary devices, they require the mastering of many disciplines (physics, engineering, electronics, and materials science) for their proper design. On the other side, there are many secrets and details to be tailored during their qualification and their production so to reach the high-yield high-reliability low-cost target. In my career, I have been contacted by many professors and startups and I read many papers about innovative transducers. But only few of them reached the market successfully, because they tackled properly all the dimensions, beyond the intrinsic beauty of a new silicon structure.

Physics and technology of silicon-based transducers are rather complex, and they are treated in numerous publications scattered throughout the literature. Therefore, a clear need exists for a book that thoroughly and systematically reviews the present basic knowledge on these devices. My enthusiastic welcome, therefore, goes to this enlightening book, *Silicon Sensors and Actuators,* written with the contributions of more than eighty authors with different backgrounds and education (technologists, physicists, and electronic, mechanical, and biotech engineers). At the origin of this book and its scientific contents, there is a fantastic adventure, which began about 25 years ago in STMicroelectronics, a mature hi-tech semiconductor multibillion-dollar company. All these people, with the support of whole organization of STMicroelectronics and the one of Academic and research centers all over the world, have been able to conceive, design, qualify, produce, and sell more than 20 billion transducers to many customers big and small.

This book provides a complete and up-to-date overview of these devices, including industrialization's biggest challenges related to reliability, packaging, and engineering. I believe that students, researchers, or engineers involved in silicon-based sensor and actuator research and development will find a wealth of useful information in this book, thanks to the proven track of record of the authors. The reader will be able to acquire a solid theoretical and practical background that will allow them to analyze the key performance aspects of these devices, critically judge a fabrication process, and conceive and design new ones for future applications. This book and its great collection of achievements represents a milestone for technicians and passionate supporters of the field. I believe it can help to stimulate the fantasy and creativity of the readers so as to generate new devices that can enhance the brilliant thinking nature of Homo Sapiens, such as we are.

I had the opportunity and the pleasure to work with some of these highly talented people. Twenty-five years are gone, but without any doubt the challenges of these transducers' development and industrialization helped all of us stay young in spirit.

I really hope you will enjoy reading this book.

Bruno Murari
ST Technical advisor and former director
of Divisional Research Development
Center – Cornaredo (Mi), Italy

The original version of this book was revised: The author name "Paolo Ferrari" has been changed to "Paolo Ferrarini" in Chapters 4 and 10. The correction to this book is available at https://doi.org/10.1007/978-3-030-80135-9_28

Introduction

Fourteen was the number of the Third Industrial Revolution, also known as the Digital Revolution, and Fourteen is the number of the Fourth Industrial Revolution. Fourteen is the number of electrons around a nucleus composed by 14 protons and 14 neutrons. These elementary particles all together compose the beautiful atom of silicon, discovered in 1824 by the Swedish chemist Berzelius. Today, without silicon, we would not have any of the key blocks of the Digital Revolution (computers, mobile phones, the World Wide Web) that nowadays we all take for granted; we also couldn't think about the present and future waves of artificial intelligence.

In the second half of the twentieth century, the silicon transistor, invented in 1947, boosted the pace of innovation, and consequently the worldwide gross domestic product was as never seen before in the history of mankind. Silicon started to be used for its semiconductor properties in analog and digital circuits, in discrete and integrated chips, and it drove the third industrial revolution of the 1970s. The first operational amplifier, the first microcontroller, the first memory, the first analog to digital converter were all using silicon as their base material. Nowadays, the semiconductor industry has a value of about 400B$ and it employs several million people around the world. As often happens during the progress of mankind, military applications drove the development of this strategic industry, and at the time in which this book is being written, it is still creating tense geopolitical frictions between the USA and China.

What triggered the Silicon Revolution was a simple silicon transistor, which was realized on 1" wafer, replacing the mature and reliable vacuum tube thanks to its semiconductor properties. Moreover, during the past 60 years, lithographic pitch reduction has been driving and aligning all the players of this industry along the guidelines of the well-known Moore's Law.

In parallel, in the last four decades, in a few research laboratories scattered throughout the USA and Europe and in a handful of almost-obsolete manufacturing plants, a few visionary and brave pioneers have been exploiting other physical properties of silicon (see Part I of this book). These people were the actor of what I like to call the " Secret MEMS Revolution." These researchers didn't appreciate deep submicron technologies running in 12" high capital-intensive wafer fabs, they didn't fall in love with Moore's Law. Inspired by the motivating words of the Noble Prize winner R. P. Feynman's legendary talk ("There's Plenty of Room at the Bottom," 1959), the ambition of those pioneers was very simple: to manufacture low-cost energy-efficient miniaturized sensors and actuators on silicon wafers to help as many people as possible. Later, those miniaturized devices became more known as micro-electromechanical Systems (MEMS): millimeter-sized systems where not only electrons are moving, but also fluids, cantilevers, and membranes. The manufacturing processes required to realize such devices are well described in Part II of this book.

All those explorers, out of the mainstream silicon technology development roadmap and with limited support by the Semiconductor Industry Association, have gone through many theoretical and practical challenges to bring their ideas to the market. Only a few innovative companies with brave and resilient teams, who dared to challenge themselves with a clear vision that beyond Moore's Law there is another business worth world, have been successful. STMicroelectronics is among the few successful semiconductor companies thanks to its early investment in 8" MEMS manufacturing line in 2005 and to a great team whose words you can read in this book.

Tool making has always differentiated our species from all others on Earth. Four hundred thousand years ago, Homo Sapiens carved aerodynamically shaped wooden spears to kill animals to feed themselves properly and all the members of their tribe. Since the Cognitive Revolution thirty-five thousand years ago, Homo Sapiens have been able to imagine a new world and influence it. Homo Sapiens are equipped with five senses (smell, touch, sight, taste, hearing), two couples of actuators (arms and legs), and, most importantly, the best brain in nature able to integrate the signals gathered by sensors and then moving the innate actuators. Over time, our ancestors realized that they needed more precise sensors and more efficient actuators to better master the world around them, and thus they started to use their brains to conceive, develop, and manufacture sensors and actuators to augment the limited capabilities of their body.

Time has been the first variable that mankind has been interested to measure. In the ancient world, time measurement was important for many different purposes: from knowing the exact hour during which to hold religious rites to the time slot allocated to the defense lawyers in ancient Rome's courts and then to the paid time slot allocated to prostitutes' customers in many brothels all over the world. The Egyptians used large obelisks to track the movements of the sun and to measure the passing of time; they also developed water clocks, later used by both the Greeks (a.k.a. clepsydrae) and the Chinese. With the exception of a few other instruments (like the compass invented in China, used for divination first and later for navigational orienteering around 1050 AD), we will need to wait for the Scientific Revolution of the seventeenth century, whose foundational Galilean scientific method required the development of many more instruments to challenge the dogmatism of the pre-modern era. Barometers, accelerometers, thermometers, and other bulky and expensive sensors were all invented starting from the seventeenth century. Their widespread use has been limited by their high price, also linked to the amount of raw material used to realize those sensors. Only in the last 25 years, a strong boost to the massive adoption of sensors came from the automotive market (airbags, . . .) and, later, from the consumer (Nintendo Wii Console) and personal electronics market (computers, mobile phones, watches, etc.), thanks to the sensors' increased reliability, optimized power consumption, miniaturized size, and, most importantly, more affordable cost (see Parts III, VI, and VII of this book).

Almost eight thousand years ago, Homo Sapiens started using cows for tilling the land for more efficient farming, and later began riding horses to move more quickly from one point to another. Much later, in the eighteenth and the nineteenth centuries, J. Watt's steam and the electromagnetic fields of T. Edison and N. Tesla, respectively, ignited the First and the Second Industrial Revolution in the Old Continent. All these innovations were meant to offset the natural limit of our legs and arms and sustain the economic and demographic growth of mankind thanks to an increase in productivity.

Today, these sensors and actuators surround us, and we interact with them daily.

We can find several types of sensors in cars, smartphones, pacemakers, drones, smart speakers, washing machines, and many other equipment. Their uses are very widespread. They can help us interact in an easier way with complex digital devices or they can make our cars greener, smarter, and safer. These sensors are all made in silicon, and a detailed description of these sensors can be found in Part III of this book. The sensors that we find all around us are much smaller than their macroscopic counterpart of the past. As an example, let's consider the example of the gyroscope, a sensor able to measure angular rates of the system where it is mounted. We can find it in all medium and higher-end cars and, since 2010, also in medium-high-end smartphones of several brands. The silicon gyroscope occupies a volume of few cubic millimeters, it has a weight of few milligrams, and it is much smaller than the Foucault's pendulum (year 1851) used to measure the Earth's rotation, thanks to a suspended 28-kilogram brass-coated lead bob attached to a 67-meter long wire!

Today, we are used to printing many documents easily, remaining in the comfort of our homes and offices. These printers are much smaller and cheaper than the original press machine of Gutenberg (year 1453). That machine was about three cubic meters with a weight about 200 kg ! Today, ink-jet printers are much smaller thanks to a micromachined thermal or piezoelectric actuator able to eject accurately and quickly picoliter-size droplets of ink. (see Part IV).

This field of sensors and actuators is so diverse and multidisciplinary that it is difficult for any single person to follow up all its activity. Thus, I asked my colleagues, expert in sensors and actuators, to join me in writing all the relevant and specific topics that concern this vast field. The result is this new book on silicon sensors and actuators.

This book is intended for practicing engineers, scientists, and advanced graduate students who seek a broader understanding on important subjects regarding the micro-sensor and micro-actuators field. The topics in this book are arranged in logical order in the form of eight parts. Besides this introduction and a part related to the silicon properties, five other important areas are covered: micromachining technology, device modeling and required circuitry, assembly and calibration techniques, reliability tests, and present and future device applications.

The first part provides a good overview of the silicon properties. The second part describes the different micromachining technologies and the varied materials used to realize sensors and actuators. Parts III and IV describe in detail the theory and working mode of the transducer element of different type of sensors and actuators, while Part V addresses the challenges of the related electronic circuitry. Part VI focuses on the importance of assembly and calibration on the performances and high-volume manufacturability of MEMS, while the seventh part addresses reliability, a very important, but often-forgotten topic. In the last part, we take a quick glance at potential future applications of sensors and actuators.

In this book, we purposely decided not to address the topic of CMOS image sensors, since they require 12" factories and, like microprocessors and memory chips, follow more closely the Moore's law.

Most of the 80 writers of this book are coming from STMicroelectronics. We also received valuable contributions from colleagues at Politecnico di Milano, and from uSound and Polight, two European startups with which we cooperate. The reader of this book will have the pleasure to see all the theoretical and industrial challenges explained in detail by a very talented team which has been able to scale up the production of several MEMS products, from few low-yield prototypes to high-volume high-yield production scale. Together, this team acquired in 25 years thousands of patents and thousands of years of experience in MEMS, through successes and, most importantly, through failures. This is the team behind the twenty billion units of silicon sensors and actuators deployed in the market in the last two decades. This team has been able to grow such business thanks to strong teamwork, high talent, and obviously some luck. "Audaces Fortuna Iuvat," old Romans were saying.

Each part of this book is self-contained, and readers interested in a subject will be able to find the needed information easily. It is my sincere wish that the combination of the variety and depth of the topics and industrial experience shared in this book will make it a valuable reference as well as a useful teaching text.

Benedetto Vigna

Contents

Part I
Silicon as Sensor Material

Chapter 1
Silicon Properties and Crystal Growth

Flavio Francesco Villa

1.1 Properties of Silicon

Silicon is an abundant element found in the Earth's crust in various compounds. Silica, or silicon dioxide, is the most common starting raw material for purified silicon for semiconductor and sensor applications, and the Siemens process is the most used in semiconductor-grade silicon production.

The advent of solid-state electronics dates from the invention of the transistor by Bardeen, Brattain, and Shockley [1]. During the early 1950s, the electronic components were made from germanium, a material whose melting point is lower in respect to the silicon one. This characteristic allows to grow the germanium ingot with greater ease in respect to the silicon one; however, its narrow bandgap (0.66 eV) limits the operation of germanium-based devices to temperatures of roughly 90 °C due to the major leakage currents observed at higher temperatures. On the contrary, silicon electronic devices are capable of operating at up to 200 °C, thanks to wider bandgap of silicon (1.12 eV). But there is a more serious problem than the narrow bandgap: germanium does not readily provide a stable passivation layer on the surface. For example, germanium dioxide (GeO_2) is water soluble, dissociates at approximately 800 °C, and is a poor electrical insulator. Silicon, unlike germanium, has a remarkable synergy with its oxide, silicon dioxide (SiO_2). By simply heating silicon in an oxygen atmosphere, a high dielectric strength and electrically insulating silicon dioxide is inexpensively formed. This oxide layer is chemically and mechanically very stable, effectively passivates the surface states of underlying silicon, form an effective diffusion barrier for the commonly used dopant species, and can be easily preferentially etched from the silicon, and vice

F. F. Villa (✉)
ST Microelectronics, Analog MEMS and Sensors Group, MEMS Technology and Design R&D, Agrate Brianza, Monza Brianza, Italy
e-mail: flavio.villa@st.com

© Springer Nature Switzerland AG 2022
B. Vigna et al. (eds.), *Silicon Sensors and Actuators*,
https://doi.org/10.1007/978-3-030-80135-9_1

versa, with high selectivity. Other advantages of silicon are that it is completely nontoxic, and that silica (SiO_2), the raw material from which silicon is obtained, constitute together with oxygen, about 75% of the Earth's crust. This implies that the silica is available in plentiful supply to the semiconductor industry. Moreover, electronic grade silicon can be obtained at less then one-tenth the cost of germanium. All these advantages have caused silicon to almost replace germanium completely in the semiconductor industry.

Silicon is one of the most studied elements in the periodic table and many of its physical properties have been measured. A more comprehensive treatise on basic silicon properties can be found, for instance, in a handbook edited by Hull [2] or in Landolt-Börnstein [3, 4].

1.1.1 Crystal Planes and Orientation

Silicon crystallizes into a diamond cubic crystal structure (Fig. 1.1) in which the atoms are covalently bonded. The unit cell contains eight atoms, and the atoms follow a face centered cubic (fcc) Bravais lattice. The unit cell length (**as** in Fig. 1.1) at room temperature is 0.5431 nm. This value is one of the most precisely known among elements, since silicon crystal can be grown almost perfectly, and the lattice parameter can be measured precisely [5] with an uncertainty of about $3\text{–}6 \cdot 10^{-8}$.

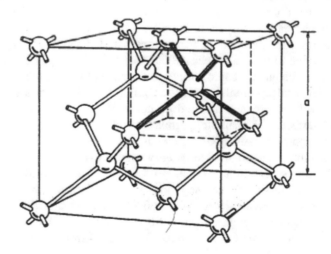

Fig. 1.1 The silicon lattice unit cell

1.1.1.1   Miller Index System

A convenient way to describe atomic planes and directions in the crystal lattice is to use Miller indexes. When the lattice axes are orthogonal and the lattice parameters in all directions x, y, z are identical as for the silicon lattice, the Miller notation is easy to use. In fact, let us consider, as in Fig. 1.2, an orthogonal coordinate system, with axes x, y, and z, with equal unit vector size, **a**. The plane intercepts, Fig. 1.2a, the x axis at a distance **a** from the origin and is parallel to the y- and z-axes. The intercept points in each axis are thus **a**, ∞, ∞. Since this is an uncomfortable notation, Miller indexes (*hkl*) are constructed in such a way that reciprocals of the intersect points are taken. Thus, in the case of Fig. 1.2a, the Miller indexes (*hkl*) are **a**/**a**, **a**/∞, **a**/∞ or (100). A plane, Fig. 1.2b, intersecting axis x at **a**, axis y at **a** and z axis at ∞, respectively, is (110). A plane, Fig. 1.2c, intersecting axis x at **a**, axis y at **a** and z axis at **a** is (111).

Some other conventions are also followed, which are reported below.

1. (hkl) identifies a plan.
2. {*hkl*} identifies all planes in the same family, and in a cubic lattice it means that {khl} includes all eight identical plans.
3. [*hkl*] identifies a direction. The direction is perpendicular to the plane (*hkl*).
4. <*hkl*> identifies a family of directions, and again in cubic lattice it includes all identical directions.

Those who would like to have a comprehensive picture of Miller indexes should consult, for instance, [6].

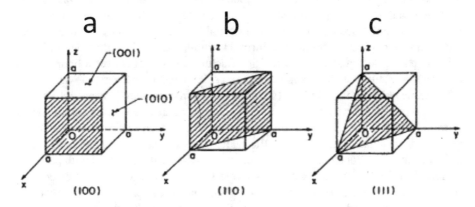

Fig. 1.2 Miller indexes

1.1.2 Electronic and Mechanical Characteristics

Semiconductors, such as silicon, are solid materials that have electrical conductivities in between those of conductors and those of insulators. The physical reason causing a material to behave as a conductor, semiconductor, or insulator lies in the availability, or lack thereof, of free current carriers in the material. Semiconductors are characterized by the narrow bandgap between the valence bands, occupied by electrons, and the conduction band, in which electrons move freely according to applied electrical fields. Intrinsic (i.e., pure) semiconductors act as insulators at room temperatures, but their behavior changes dramatically with temperature, and, more to the point, with small impurities present in the crystal. Very small amounts of electrically active impurities can totally alter the electrical properties of semiconductors such as silicon. This is because the electrically active impurities either easily donate valence electrons (donors) or accept them, creating holes (acceptors). These electrons or holes are free (i.e., not bound to individual atoms). Their movement due to applied electrical fields carries electrical currents, giving rise to the term charge carriers used to denote them. The electrical properties of semiconductor materials such as single-crystal silicon is thus defined by the impurity concentrations present in the silicon lattice. Impurities are introduced into the starting materials during crystal growth and modified during device processing by additional doping of the silicon material with electrically active impurities. In intentional doping of silicon, impurity atoms from group III (acceptors) and group V (donors) are used. Techniques used include both very traditional methods, such as deep diffusions of dopants, which have been abandoned in mainstream semiconductor processes, and current standard techniques such as ion implantation and epitaxial deposition. Silicon is a semiconductor whose resistivity can be adjusted by doping from sub-mΩ cm to several kΩ cm; it is quite inert in a normal environment, hard, transparent in an infrared region, and elastic at room temperature with no plastic deformation and with high fracture strength. However, silicon wafers do break-sometimes without apparent cause; silicon wafers and parts of wafers may 'also easily chip'. In fact, defects (scratches, dents, mechanical damage, etc.) on the surface or periphery of the silicon wafer can cause premature cracking under stress along cleavage planes, because of the notch effect. Silicon wafers having a rough surface or residual damage left on the wafer surface (cut wafer, lapped wafer, or ground wafer without stress relief) are more prone to break compared with polished wafers. The ideal, "strong" wafer is double side polished, and the wafer edge is polished and round, also. The potential of silicon as a micromechanical material was widely described by Kurt E. Petersen [7]. This review shows the advantages of employing silicon as a mechanical material, the relevant mechanical characteristics of silicon, and the processing techniques which are specific to micromechanical structures. The basis of micromechanics is that silicon, in conjunction with its conventional role as an electronic material, and taking advantage of an already advanced microfabrication technology, can also be exploited as a high-precision, high-strength, high reliability mechanical material, especially applicable wherever

Table 1.1 Silicon physical properties

Atomic number of Si	14
Crystal structure	Diamond
Lattice constant	0.5431 nm
Si atoms	5×10^{22} atoms.Cm^{-3}
Melting point	1687 K
Boiling point	3538 K
Specific density	2.329 g.cm^{-3} at 298 K
Specific density (liquid)	2.57 g.cm^{-3}
Specific heat capacity	19.79 J.Mol^{-1} K^{-1}
Heat of fusion	50.21 kJ/mol
Heat of vaporization	383 kJ/Mol
Molar heat capacity	19.789 J/(Mol·K)
Speed of sound	8433 m.s^{-1}
Volumetric compression coefficient	1.02×10^{-8} KPa^{-1}
Index of refraction f(T, λ)	~ 3.54 λ = 1.1 μm
	~ 3.48 λ = 2 μm
	At room temperature

Table 1.2 Silicon mechanical and thermal properties

	Si	C	SiC	SiN	Fe	W	Steel	Mo	Al
Yield Strength (GPa)	7	53	21	14	12.6	4	2.1	2.1	0.17
Knoop Hardness (Kg/mm2)	850	7000	2480	3486	400	485	660	275	130
Young Modulus (100 GPa)	1.9	10.3	7	3.8	1.96	4.1	2	3.43	0.7
Density (g/cm3)	2.3	3.5	3.2	3.1	7.8	19.3	7.9	10.3	2.7
Thermal Conductivity (W/cm K)	1.57	20	3.5	0.19	0.8	1.78	0.32	1.38	2.36
Thermal Expansion (ppm/K)	2.33	1	3.3	0.8	12	4.5	17.3	5	25

miniaturized mechanical devices and components must be integrated or interfaced with electronics. Tables 1.1 and 1.2 list some basic properties of silicon. Some of the parameters depend on doping level as well as temperature.

Silicon is almost an ideal structural material: It has about the same Young's modulus as steel, but is as light as aluminum, has low thermal expansion (Its thermal expansion coefficient is about eight times smaller than that of steel and is more than ten times smaller than that of aluminum), a high speed of sound and very low intrinsic mechanical losses. Silicon shows virtually no mechanical hysteresis. It is thus an ideal candidate material for MEMS, Sensors, and Actuators.

The performance of Micro Electronic and Mechanical Systems (MEMS) strongly depends on the mechanical properties of materials used. The evaluation of the mechanical properties of MEMS materials is indispensable for designing MEMS devices. Accurate values of mechanical properties (elastic properties, internal stress, strength, fatigue) are necessary for obtaining the optimum performances. For an example, elastic properties are necessary in prediction of the amount of deflection from an applied force and material strength sets device operational limits. Also, in view of reliability and lifetime requirements, mechanical characterization of MEMS materials becomes increasingly important. Reliability, accuracy, and repeatability of evaluation methods also became an issue.

1.1.2.1 Elastic Properties

Elastic properties are directly related to the device performance. Young's modulus and Poisson's ratio are basic elastic properties that govern the mechanical behavior. Since two independent mechanical properties are necessary for full definition of mechanical properties of MEMS materials, their properties can be accurately determined by measuring Young's modulus and Poisson's ratio. Young's modulus (E) is a measure of a material stiffness. It is the slope of the linear part of stress-strain (ε-σ) curve of a material (Fig. 1.3). Poisson's ratio is a measure of lateral expansion or contraction of a material when subjected to an axial stress within the elastic region. Young's modulus, Poisson's ratio, and shear modulus are isotropic on silicon (111), whereas the variations on silicon (100) and (110) are quite significant.

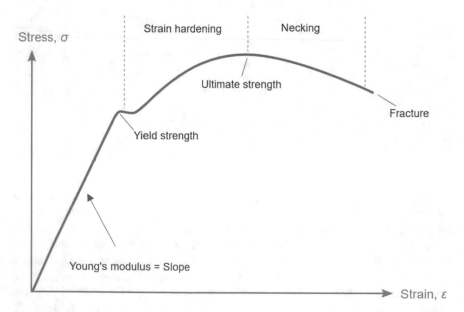

Fig. 1.3 Stress-Strain curve in ductile materials

In engineering and materials science, a stress–strain curve [8] for a material gives the relationship between stress σ (*a force applied over area-for uniaxial stress*) and strain ε (*a measure of the deformation of the material*). It is obtained by gradually applying load to a test coupon and measuring the deformation, from which the stress and strain can be determined. These curves reveal many of the properties of a material, such as the Young's modulus, the yield strength, and the ultimate tensile strength. The first stage is the linear elastic region. The stress is proportional to the strain, that is, obeys the general Hooke's law ($\sigma = \varepsilon\,E$), and the slope is Young's modulus or modulus of elasticity E. In this region, the material undergoes only elastic deformation. The end of the stage is the initiation point of plastic deformation. The stress component of this point is defined as yield strength. The yield strength or yield stress is a material property and is the stress corresponding to the yield point at which the material begins to deform plastically. The yield strength is often used to determine the maximum allowable load in a mechanical component, since it represents the upper limit to forces that can be applied without producing permanent deformation; for Silicon (annealed) the yield strength range is about 5000–9000 MPa [9].

1.1.2.2 Strength

The strength of a material determines how much force can be applied to a MEMS device. It needs to be evaluated to assure reliability of MEMS devices. Strength depends on the geometry, loading conditions, as well as on material properties. As the useful measure for brittle materials, the fracture strength is defined as the normal stress at the beginning of fracture. The flexural strength is a measure of the ultimate strength of a specified beam in bending and it is related to specimen's size and shape. For inelastic materials, the yield strength is defined as a specific limiting deviation from initial linearity. Silicon is a hard, brittle material, and at room temperature under stress, silicon single crystal elongates elastically until fracture stress appears without significant plastic deformation (Fig. 1.4). It is well known that while the amorphous and polycrystalline brittle solids behave isotropically, brittle single crystals manifest mostly anisotropic behavior. For the silicon single crystal, the fracture anisotropy has been reported [10] with reference to the low index plane, such as {001}, {110}, and {111}; silicon has two principal cleavage planes: {111} planes, usually the easy cleavage planes, and {110} planes [11–13]. On the contrary Polysilicon, an aggregation of pure silicon crystals with random orientations deposited on the top of silicon substrates, is even stronger than single silicon crystals; Polysilicon is the most frequently used MEMS material.

1.1.2.3 Fatigue

MEMS devices are often exposed to cyclic or constant stress for a long-time during operation. Such operational conditions may induce fatigue. Fatigue may be observed

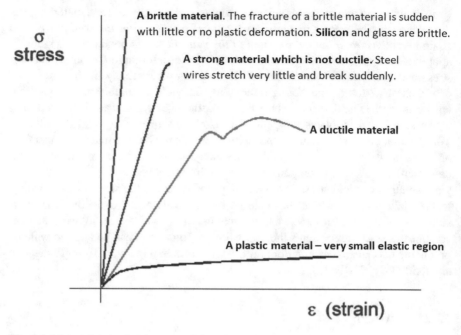

Fig. 1.4 Materials behaviors in stress-strain curve

as change in elastic constants and plastic deformation leading to sensitivity changes and offset drift in MEMS devices. It may also be observed as the strength decreases that may lead to fracture and consequentially failure of the device. Fatigue behavior of a MEMS device also depends on its size, surface effects, effect of the environment such as humidity and temperature, resonant frequencies, etc. To realize a highly reliable MEMS device, a detailed analysis of the fatigue behavior must be performed using accelerated life test method as well as life prediction method.

1.2 Starting Materials

The preparation of silicon single-crystal substrates, as it well known, is the first step in the long and complex process of device and circuits fabrication. Silicon is the second most abundant element, after oxygen, in the earth's crust, in the form of silica (SiO_2) and silicates and is readily extracted from silica-rich sands. In nature, it typically contains large amounts of impurities and so the initial stages of the manufacturing process are concerned with reduction and purification of the silica sand to produce Metallurgical Grade polysilicon (MG), typically greater than 98% pure silicon. This is then further refined to produce Electronic semiconductor Grade polysilicon (EG), which can be used as basic material for crystal growth to produce single crystal silicon substrate suitable for device and circuits manufacturing.

1.2.1 Metallurgical-Grade Silicon

Metallurgical Grade polysilicon (MG) is produced through the reduction of silica by mixing it with carbon in the form of coal, coke, or wood chips and heating the mixture to high temperatures (more than 2000 °C) in a submerged electrode arc furnace (Fig. 1.5). The reaction that takes place is the following:

$$SiO_2 + 2C \rightarrow Si + 2CO$$

Silicon produced by this method typically contains ~2% impurities as Fe and Al and other impurities, including B and P.

1.2.2 Electronic-Grade (EG) Polysilicon

The metallurgical grade silicon is converted into trichlorosilane ($SiHCl_3$) in a fluidized bed by reaction with anhydrous hydrogen chloride:

$$Si\,(MG) + 3HCl \rightarrow SiHCl_3 + H_2$$

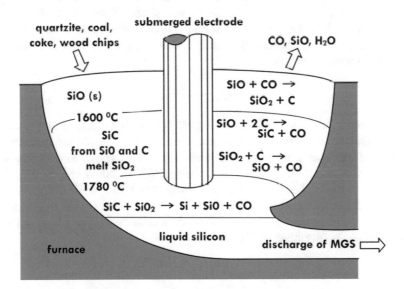

Fig. 1.5 Submerge-electrode arc furnace for Metallurgical-Grade Silicon (MGS) production. The temperature is maintained above the melting point of silicon so that the molten semiconductor is removed from the bottom

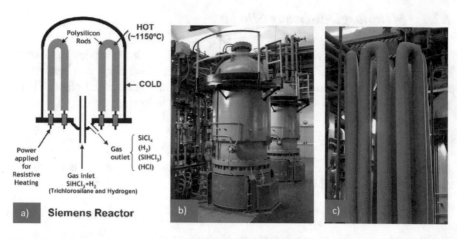

Fig. 1.6 (a) Schematic sketch of reactor vessel. (b) External view. (c) As grown polysilicon rods after reactor run (Current generation reactors have many rods)

During this reaction, which takes place at about 300 °C, various impurities such as Fe, Al, and B are removed by converting them into their halides ($FeCl_3$, $AlCl_3$, and BCl_3, respectively), and byproducts such as $SiCl_4$ and H_2 are also produced. Subsequently the trichlorosilane is fractionally distilled from the impurity halides to obtain a very high purity compound (impurities are in the low parts per billion (ppb) range or less) a necessary requirement for production of semiconductor devices. The liquid TCS is then converted to solid polysilicon by the Siemens process. In this process, the triclorosilane is thermally decomposed into silicon and HCl in the presence of hydrogen as follows:

$$SiHCl_3 + H_2 \rightarrow Si\,(EG) + 3HCl$$

The decomposition of $SiHCl_3$ (1000–1200 °C) is accomplished in a reactor in which a thin rod of silicon known as a *slim rod* is available for the deposition of the silicon formed as a result of above reaction. Other intermediate compounds such as silicon tetrachloride ($SiCl_4$) and silane (SiH_4) can also be used. Figure 1.6a) is a schematic sketch of the reaction vessel in which the decomposition of $SiHCl_3$ takes place. The *slim rod* has to be of a purity level comparable to that of the deposited silicon. About 80% of the world's polisilicon is produced using the Siemen's process developed in the 1950s [14].

1.3 Growth of Single Crystal Silicon

The growth of single crystal of silicon from high-purity polysilicon represents the first significant step in the manufacture of a highly perfect semiconductor material.

The industry standard for production of monocrystalline silicon for semiconductors is the Czochralski (CZ) method.

Historical Development of Single Crystal Growth – Czochralski's Creative Mistake.

The method currently used to grow a Single Crystal was discovered in 1916 [15] by Jan Czochralski (Fig. 1.7), a Polish chemist, while investigating the rate of crystallization of metals. The method was developed as a result of an accident and through Czochralski's careful observation [16]. One evening he left aside a crucible with molten tin and returned to writing notes into the laboratory notebook. At some moment, lost in thought, instead of dipping his pen in the inkpot he dipped it in the crucible. Alarmed, he quickly lifted his arm and noted that a long, thin thread was hanging onto the pen, which later proved to be a *single* crystal. They were not single crystals in today's understanding, though, they were "threads" of such metals as zinc, tin, and lead. At the time, there was no need whatsoever for monocrystalline materials, as all materials industrially exploited at that time were polycrystalline. It is worth noting that up to the Second World War, scientists were mainly interested in the properties of metals and their alloys, so mainly such crystals were grown at that time. After invention of germanium-based transistor in 1947, G.K. Teal and J.B. Little from Bell Laboratory used the Czochralski method to obtain germanium single crystals [17, 18]. Later, Teal and Buchler grew CZ–Si using the same technique, but they could only grow dislocated single crystals of specified orientation [19]. The first demonstration of dislocation-free CZ silicon crystal growth was demonstrated in 1959 by Dash, [20] using a modified seeding technique. Growth of silicon crystals by the CZ method have been widely studied

Fig. 1.7 Jan Czochralski, around 1907. (Photograph by kind permission of Prof. P.E. Tomaszewski)

over the course of the following five decades and significant progress has been made. Dislocation-free, high-purity silicon crystal up to 450 mm in diameter is now possible on a commercial scale.

1.3.1 The Czochralski Crystal Pulling Method

Schematic view of typical Czochralski silicon crystal growing system, today in use, is represented in Fig. 1.8A. It consists of a silica crucible (which is often called quartz crucible), a graphite susceptor around the silica crucible, a cylindrical heater, and a heat shield around the heater. As the silica crucible contains the melt, it is the most important component of the growth apparatus. The crucible material must be chemically unreactive with molten silicon. Also, the material must have high melting point, thermal stability, and hardness. The materials for crucible, which satisfy these properties, are silicon nitride (Si_3N_4) and fused silica (SiO_2). The latter is in exclusive use nowadays. The graphite susceptor, around the silica crucible, is mandatory because the high temperature causes the crucible to soften, and a proper mechanical support is therefore needed. The susceptor also helps distribute heat around the silica crucible a little bit more uniformly. Graphite is the material of choice because of its high-temperature properties. The graphite must be pure to prevent contamination of the crystal from impurities that could be volatilized from the graphite at the temperature involved. The susceptor rests on a pedestal

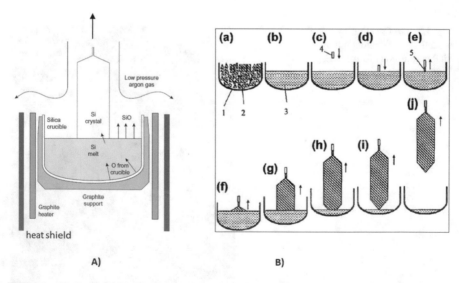

Fig. 1.8 (**A**) Czochralski silicon crystal growing system. (**B**) Czochralski Process steps: (a) The polycrystalline feedstock (1) is put inside a crucible (2); (b) it is melted (3); (c, d) Seeding procedure: The seed crystal (4) is dipped into the melt, followed (e) by Dash necking (5), (f) shouldering, (g) cylindrical growth, (h) growth of end cone, (i) lift off, (j) cooling down and removing of the crystal

whose shaft is connected to a motor that provides rotation. The heater is usually connected to two or four electrodes at its lower edge, typically by supporting elements also made of graphite. The electrodes deliver the required power, which is at least tens of kilowatts, but often well exceeds 100 kW. The heater is normally of picket type, which means that it has vertical slits cut into it in a manner that forces the electric current to flow up and down, in opposite directions in neighboring pickets. The voltage is quite low, and amperage is high, mainly because of the high electrical conductance of the heater, but also because of the poor electrical insulation properties of the inert gas at the low pressure and high temperature normally used in silicon CZ growth. The heat shield around the heater drastically reduces the power consumption, and it also helps create a more controlled temperature distribution around the susceptor.

At the beginning of the process, Fig. 1.8B, the feed material is put into a cylindrically shaped crucible and melted by resistance or radio-frequency heaters to one temperature that is slightly more than the silicon melting point of 1420 °C. A small single-crystal rod of silicon called a *seed crystal* is then dipped into the silicon melt. The conduction of heat up the seed crystal will produce a reduction in the temperature of the melt in contact with the seed crystal to slightly below the silicon melting point. The silicon will therefore freeze onto the end of the seed crystal, and as the seed crystal is slowly pulled up out of the melt it will pull up with it a solidified mass of silicon that will be a crystallographic continuation of the seed crystal. Both the seed crystal and the crucible are rotated but in opposite directions during the crystal pulling process to produce crystalline ingots of circular cross section. During the silicon growth process the shape of the crystal, especially the diameter, is controlled by carefully adjusting the heating power, the pulling rate, and the rotation rate of the crystal. In the initial stages, the pull rate is quite high, and the growing crystal is only about 3–5 mm in diameter. This narrow portion of the crystal is called the "neck" and was first used by Dash [20, 21] for producing a dislocation-free crystal and is standard practice in the industry today. In fact, in (111) and (100) oriented Si seed crystals, dislocations introduced, due to the thermal stress generated between seed (*cold*) and the melt (*hot*), will propagate obliquely to the growth direction, and will terminate on the sides of the neck rather than propagating down into the body of the growing crystal, provided the neck length to diameter ratio is sufficiently large. This was an important breakthrough for yield and reliability of semiconductor device fabrication processes, since dislocations can be killer defects in diode and transistor fabrication. Once the neck is several centimeters long, the pull rate is slowed allowing the crystal diameter to increase to the desired dimension. At the end of crystal growth, the diameter of the crystal is tapered to form a conical tail that minimizes the number of dislocations formed by the thermal shock of withdrawing the crystal from the melt and allows these dislocations to propagate out of the sides of the cone rather than into the body of the crystal, maintaining a dislocation free crystal. Silicon CZ growth takes place under a continuous flow of inert gas. A schematic picture of the gas flow pattern is shown in Fig. 1.8(A) Considering the high temperature and the reactivity of silicon melt, noble gases are the only ones allowed. Argon is the gas of choice because of its much lower

Fig. 1.9 (**a**) A single crystal of silicon, 300 mm in diameter, 2 m long, and weighing 265 kg drawn using the CZ process; photograph reprinted here with permission of Wacker-Chemie – from Ref. [16]. (**b**) Increase in areas of CZ silicon crystal sections over the years

cost than that of other noble gases. Argon has also the advantage of poor thermal conductivity compared with, for example, helium, and this feature facilitates the effort to insulate hot areas around the melt from the water-cooled vacuum-chamber walls. Oxygen is dissolved from the crucible, and the purge gas removes volatile silicon monoxide from the vicinity of the melt and the crystal. Only a small fraction of total dissolved oxygen ends up in the growing crystal. Figure 1.9a) displays a cooled silicon single crystal that was produced by the CZ method. At the top of the single crystal is situated the thin seed which at the start of drawing is dipped into a crucible containing molten silicon. In Fig. 1.9b) is shown the increase in areas of CZ silicon crystal sections over the years.

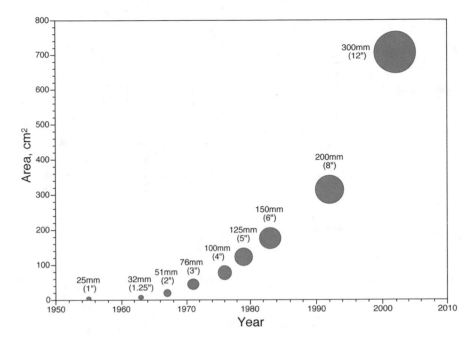

Fig. 1.9 (continued)

About *Seed Crystal*

The first silicon single crystals were obtained in the fifties of the last century, starting from polycrystalline germs. The technique used, proposed by *Stockbarger,* was to grow preferentially a "polycrystalline grain" over the others. In this method, the substance, in our case the polycrystalline silicon, is melted in a crucible, Fig. 1.10a), where there is a temperature gradient. The crucible is then lowered slowly, towards the lower temperature zone, so that the crystallization starts from its bottom, which is shaped like a cone. In a first time, several crystalline or disoriented polysilicon grains are formed, but continuing the crystallization (and the lowering of the crucible) one of them will have the upper hand over the others and affect growth of the whole rest of the melted, obtain a single crystal with the same orientation of the initial "grain". Things can be facilitated by shaping the crucible tip as shown in Fig. 1.10b). This shaping also has the advantage of preventing dislocations in the growing single crystal. Once the first crystal was made available, subsequent germs were obtained directly by cutting the crystal. Today's industrial germ-building process reproduces itself with a simpler cloning process; from a single crystal the new germs are cut out for many other single crystals.

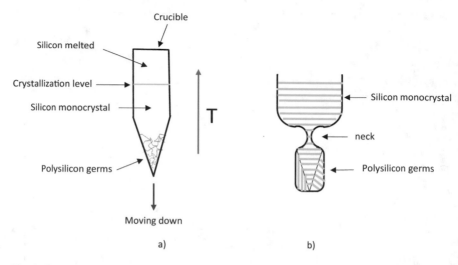

Fig. 1.10 Technique used to grow the first silicon seed crystal

1.3.2   Impurities in Czochralski Silicon

The CZ silicon crystals contain small amounts of impurities of which some are essential, in controlled amounts, for the quality of the crystal and wafers as oxygen, while other impurities are harmful if their concentration is too high and, thus, the latter elements must be avoided. The harmful impurities in CZ silicon crystals include transition-metal impurities (e.g., Fe, Cu, and Ni) [22–25], and alkali and alkali earth metals (e.g., Na). These impurities, as well as accidental dopants, may originate from the starting materials (polysilicon, recycled silicon, if used, and dopants), silica crucible, and materials and contamination of the furnace and the hot zone [26]. Carbon, (typically less than 0.5 ppma) introduced from the graphite parts used in the crystal puller hot zone, is electrically inactive in silicon, while transition-metal impurities may form deep impurity levels within the band gap between the conduction and valence bands, behaving as recombination centers or as unwanted acceptors and/or donors, or forming complexes with dopants under certain conditions. The transition metals, or their precipitates, may increase the recombination lifetimes and cause leakage currents in the junctions of the devices, quality problems with gate oxides in metal-oxide-semiconductor structures, or even a shift of the resistivity. For example, iron can be very effective in decreasing the recombination lifetime, especially in p-type silicon [27], while copper is considered to have a greater impact on the lifetime of n-type wafers than on p-type wafers. Oxygen originating from the silica crucible is one of the most characteristic impurities in CZ silicon. A suitable oxygen concentration is among the main quality factors to be controlled in the crystal-growing process. Oxygen in interstitials gives the wafers mechanical strength against thermal or mechanical stress during heat treatments, decreasing the risk of slip dislocation formation and wafer warping

[28, 29]. Furthermore, oxygen forms precipitates [30–32], which are often used for impurity gettering in device processing, where gettering is a method to remove harmful metal impurities from the device region of the wafer [33]. Also, too high an oxygen concentration can be harmful if it induces too strong precipitation of oxygen and formation of defects. In fact, oxide precipitates formed in the active device region near the wafer surface can cause the device to fail but those away from the surface, in the bulk of the wafer, can provide a benefit by trapping metal impurities that may contaminate a wafer during processing. The process of trapping impurities in this way is called *internal gettering*. For gettering to take place with maximum advantage, it is necessary to ensure that oxygen precipitates form only in the wafer bulk and not near the surface. This can be achieved by employing silicon wafers with tightly controlled oxygen levels (typically within the range ~8–15 ppma depending upon the specific application) and subjecting them to a thermal cycle at sufficiently high temperature and length of time to out diffuse oxygen near the wafer surface such that the concentration is below the critical supersaturation level to form precipitates. Under these conditions, during the remaining thermal processes of the device manufacturing, the oxygen near the wafer surface remains in solution, while in the supersaturated wafer bulk it precipitates forming gettering centers [34]. The near-surface region without precipitates is known as the *denuded zone* (DZ).

1.3.3   Defects in Silicon Crystals

Although single crystalline silicon made with CZ growing method is almost perfect contain anyway some defects. Furthermore, during the processing of silicon wafers for implementing the devices other defects are formed. These defects can be classified as:

(a) Point defects: vacancies and self-interstitials or their agglomerates
(b) Linear defects: dislocations
(c) Planar defects: stacking faults
(d) Volume defects: precipitates, COP (Crystal Originated Particles)

Short descriptions of the typical defects in CZ silicon are given in the list below.

Vacancies and self-interstitials: They are generating during silicon crystals grown from the melt because of their low diffusivity; in such a way the vacancies and self-interstitials are "frozen in" the crystal during the cooling phase. Vacancies and interstitials have equilibrium concentrations that depend on temperature and they play a very important role in the kinetics of diffusion and oxidation in the technical processes of silicon. In fact, the diffusion of many dopants depends on the vacancy concentration, as does the oxidation rate. Reviews of Falster et al. [35, 36] give an overview of intrinsic point defects in CZ silicon.

Linear defects are called dislocations. In silicon single crystals grown by CZ or in wafers cut from these crystals, the dislocations density is practically zero. During the processing of the silicon wafer at higher temperatures, the applied stress may generate dislocations. The origin of these stresses can be thermal gradients during temperature transients in heat treatment steps, or mismatch stresses caused by thin films or heavily doped areas in combination by high temperatures. Movement of dislocations through the crystal lattice causes permanent deformation. Dislocations in silicon move on {111} planes along [110] directions. If the deformation is heavy, the wafer warpage may increase. In general, dislocations in silicon wafers have adverse effects on device performance. Dislocations attract impurities, and the result can be that moving dislocations leave precipitate colonies behind.

Planar defects are called stacking faults. In silicon, stacking faults are along (111) atomic planes and they are surrounded by partial dislocation. A silicon lattice may contain an extra atomic place; this stacking fault is called extrinsic. If an atomic plane is missing, the stacking fault is intrinsic. Most stacking faults in silicon is of the extrinsic type. Extrinsic stacking faults are formed by excess silicon interstitials; oxidation-induced stacking faults or stacking faults around growing oxide precipitates are typical examples. A large oxygen precipitate may generate stacking fault(s) to reduce misfit stresses. Wet oxidation of silicon wafer also may induce stacking faults. Figure 1.11a) displays surface Stacking Faults (SF) on CZ wafer after Secco D' Aragona chemical etch (3 min); Fig. 1.11b) shows bulk Stacking Faults and dislocations on CZ after 48 h dry oxidation at 1100 °C (3 min Secco, 500x). The lattice strain caused by Oxidation Stacking Faults (OSF) is strongest at those positions where the partial dislocations enter the surface. This leads to the formation of stacking fault etch pits in the form of "dog-bones".

Volume defects in silicon are precipitates. They can be coherent (the lattice of the precipitate is aligned with the host silicon lattice), or non-coherent, when there

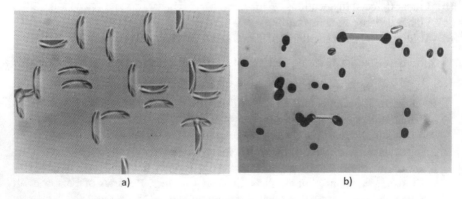

a)   b)

Fig. 1.11 (**a**) Surface stacking faults (SF) on CZ (3 min Secco). (**b**) Bulk SF and dislocations on CZ (3 min Secco)

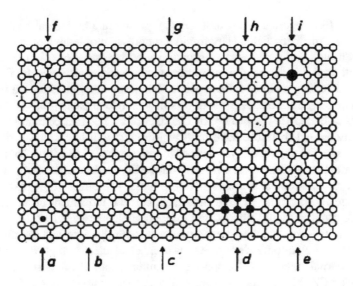

Fig. 1.12 Various crystal defects in a simple cubic lattice (**a**) Interstitial impurity atom, (**b**) edge dislocation, (**c**) self-interstitial, (**d**) coherent precipitate of substitutional impurity atoms, (**e**) small dislocation loop formed by agglomeration of self-interstitials, (**f**) substitutional impurity atom widening the lattice (tensile strain), (**g**) vacancy, (**h**) small dislocation loop formed by agglomeration of vacancies, (**i**)substitutional impurity atom compressing the lattice (compressive strain)

is a mismatch in lattice alignment. Small precipitates are often coherent, and when they grow, they turn first semi coherent, and then non-coherent. Precipitates are normally compounds of silicon: silicates or silicon dioxide. COP (Crystal Originated Particles), a cluster of vacancies, is also a volume defect. The size and density of this defect depends on the crystal growth process; the average size can be from tens of nanometer to less than 200 nm. COP has mostly an effect gate oxide quality in CMOS devices. Because the COP originates from crystal growth process involving melt, epitaxial layers are free from COPs.

A schematically representation of various crystal defects in a simple cubic lattice (silicon) is shown in Fig. 1.12.

1.4 New Crystal Growth Methods

1.4.1 Czochralski Growth with an Applied Magnetic Field (MCZ)

The use of magnetic fields for the growth of semiconductor crystals have already been considered many decades ago. As early as in 1966, Chedzey et al. [37] and

Utech et al. [38] reported about InSb crystals grown in a horizontal boat under the influence of a magnetic field. They found a suppression of temperature fluctuations in the InSb melt and a decrease of growth variations (striations) in the crystal. In 1970, Witt et al. [39] applied a static transverse (horizontal) magnetic field to the Czochralski (CZ) growth of InSb crystals. Ten years later, in 1980, the transverse field was also used for the CZ growth of silicon single crystals [40, 41]. Originally MCZ was intended for the growth of CZ silicon crystals that contain low oxygen concentrations and therefore have high resistivities with low radial variations. In other words, MCZ silicon was introduced to replace the FZ (Floating Zone method) silicon almost exclusively used for power device fabrication. Since then, the method has received considerable attention over the years. The effect of the magnetic field is explained by its ability to fluid flow damping; the movement of an electrically conducting fluid, as liquid silicon, in a magnetic field causes, due to Lorentz force, an electromotive force that is perpendicular to both the field and the direction of movement. This electromotive force tends to create electric currents that, interacting with the external magnetic field, oppose the fluid motion. The field strengths typical for magnetic CZ (MCZ) processes are sufficient to slow down various flow patterns very significantly. This has implications to a range of crystal properties such as oxygen and dopant distributions; but they also tend to reduce crucible wear, thus extending the lifetime of the crucible, and reduce thermal fluctuations in the melt, which, together with reduced crucible wear, results in better dislocations – free growth yields. After the MCZ technology had been established as a standard method for low-oxygen CZ material, it was found that magnetic fields were also useful for the growth of advanced CZ crystals, e.g., the so-called "perfect", "pure", or "ultimate" silicon [42–44]. Silicon crystals of this type are grown under conditions which avoid large agglomerates of intrinsic point defects, i.e., L pits (Si interstitial aggregates) and voids (vacancy aggregates). Below a critical size, these defects are no longer harmful to device functioning and, hence, can be tolerated. Today, a substantial part of the CZ pullers is equipped with a transverse magnetic field of 3000–5000 G, which has been shown to be most suitable to produce this advanced material.

1.4.2   *Continuous Czochralski Method (CCZ)*

Crystal production costs depend to a large extent on the cost of materials, in particular the cost of those used for quartz crucibles. In the conventional CZ process, called a *batch process*, a crystal is pulled from a single crucible charge, and the quartz crucible is used only once and is then discarded. This is because the small amount of remaining silicon cracks the crucible as it cools from a high temperature during each growth run. Today, supply for large silicon wafers has driven the growth of silicon crystals from 200 to 300 mm in diameter. With the increasing silicon ingot sizes, melt volume has grown dramatically, exacerbating the costs and problems related to oxygen control and resistivity in the silicon monocrystal. A proposed

Fig. 1.13 The system for Continuous Czochralski growth (CCZ); a quartz baffle ("black" in figure) is required to prevent the melt turbulence caused by feeding in the solid material around the growth interface

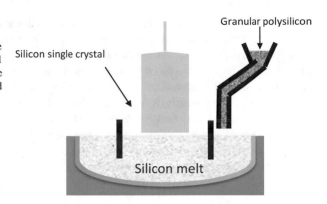

strategy to resolve these issues consists of continuously adding feed as the crystal is grown and thereby maintain the melt at a constant volume. According to this method, named Continuous Czochralski Method (CCZ), the silicon monocrystal can be pulled while the amount of the silicon melt in the crucible remains constant. Therefore, the oxygen concentration and resistivity of the silicon monocrystal, which depend on the amount of the silicon melt in the crucible, can be constantly maintained. This makes it possible to considerably improve the manufacturing yield of the silicon monocrystal, which in turn allows reduction in manufacturing costs and provides an ideal environment for silicon crystal growth. Continuous charging is commonly performed by polysilicon feeding, as shown in Fig.1.13 This system consists of a hopper for storing the polysilicon raw material and a vibratory feeder that transfers the polysilicon to the crucible. In the crucible that contains the silicon melt, a quartz baffle is required to prevent the melt turbulence caused by feeding in the solid material around the growth interface. The CCZ method certainly solves most of the problems related to inhomogeneities in crystal grown by the conventional CZ method. Moreover, the combination of MCZ and CCZ (the Magnetic-field-applied Continuous CZ [MCCZ] method) is expected to provide the ultimate crystal growth method, giving ideal silicon crystals for a wide variety of microelectronic applications [45, 46].

1.4.3 Neckingless Growth Method

As mentioned in 1.3.1, Dash's necking process, able to eliminate grown-in dislocations, has been the industry standard for more than 40 years. However, recent demands for large crystal diameters (greater than 300 mm, weighing over 300 kg) have resulted in the need for larger diameter necks that do not introduce dislocations into the growing crystal, since a thin neck 3–5 mm in diameter cannot support such large crystals' weight. It is estimated that large diameter necks 12 mm in diameter can support CZ crystals as heavy as 2000 kg [47]. Such a large size of the neck, however, is unable to block the propagation of dislocations into the body of the growing crystal. Fortunately, it was found that the boron concentration limit in the seed for no dislocation formation due to thermal shock was 1×10^{18} atoms/cm^3, and the maximum admissible discrepancy of boron concentration in the seed and in the crystal was 7×10^{18} atoms/cm^3 for no dislocation formation in the crystal due to the lattice misfit. Consequently, it is revealed that dislocation-free CZ-Si crystals can be grown from undoped Si melt without the Dash-necking process when a seed with a suitable boron concentration is used. The mechanism by which dislocations are not incorporated into the growing crystal has been primarily attributed to hardening effect of the heavy doping of boron in the silicon.

1.5   Silicon Wafers Preparation and Properties

After the silicon crystal has cooled, it is removed from the crystal puller for machining. Silicon crystals, or ingots in the terminology, commonly in use are typically up to 2 m in length, and they have a small over-diameter to eliminate yield loss in final wafers due to diameter fluctuation and small deviations from round shape during the growth. Therefore, crystals are typically grown a few millimeters oversize on diameter and ground down to the required size prior to slicing. Initially the shoulder and tail portions are removed with a diamond blade cropping saw and recycled. The body of the crystal is then sawed into lengths that can be accommodated by the subsequent slicing operation and prepared for grinding to the required diameter. In the days of smaller wafers, a flat was ground along one side to indicate crystallographic direction as determined by X-ray diffraction, but later, on larger wafers and in the interests of preserving surface area for saleable devices, the flat was replaced by a notch (see figure on the right side).

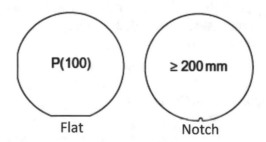

P(100) $\geq$ 200 mm

Flat   Notch

1.5.1   Silicon Wafers Manufacturing Process

The **wafer production process** (Fig. 1.14) starts with slicing of the crystal ingot. The ingot to be sliced into wafers needs a support to hold cut wafers; therefore, a ceramic

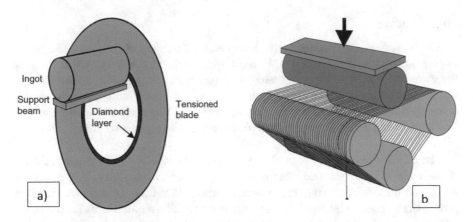

Fig. 1.14 (**a**) Internal Diameter (ID) saw machine; (**b**) Wire saw machine

or graphite beam is glued onto the piece of ingot. Two types of slicing methods have been used in the silicon wafer industry: the internal diameter (ID) saw, Fig. 1.14a, and the wire saw, Fig. 1.14b. The ID saw uses a thin annular blade with a diamond bonded region on the inside edge of the annulus. ID saws cut only one wafer at a time, taking a few minutes to slice each wafer from the crystal. Today, wafers (and from 200 mm up, extensively) are increasingly cut in wire-cutting machines because of productivity reasons. ID cutting, however, retains the flexibility of manufacturing smaller lots, but with higher cost.

Wire saw slicing of silicon is achieved predominantly by Free Abrasive Machining (FAM) using a slurry comprising silicon carbide (SiC) grit suspended in either oil or ethylene glycol. The wire acts to transport slurry to the silicon ingot where the grit becomes trapped between the tensioned wire and the ingot creating what is known as a rolling and indenting cutting mechanism [48–50]. Recently, wire cutting using diamond coated wire has been introduced. The method was introduced first to solar grade silicon wafer manufacturing and is now becoming more common in semiconductor grade wafer manufacturing. By using diamond wire, it is possible to use water as a coolant, creating savings in waste treatment and as-sawn wafer surface quality is improved [51, 52]. However, reports indicate that diamond wire wafers in as-sawn state are more prone to fracturing [53]. Ideally, the slicing process would produce wafers with a polished surface perfectly flat and clean ready for device fabrication. In practice, cut wafers have significant crystallographic damage induced by the slicing process. They are typically slightly warped, of nonuniform thickness with rough, contaminated surfaces, and a square edge profile which is easily chipped. Independent of the cutting method, wafers are cleaned after cutting, and the supporting beam piece is removed. In addition to cleaning, a slight etch may be used to reduce surface damage to make the wafer mechanically stronger, and reduce the stress-related warp. The wafering process sequence after sawing is designed to produce the ideal wafer at the lowest cost and includes the steps of *Edge Grinding, Lapping and Etching, Polishing and Cleaning*.

Edge Grinding

Silicon wafers after cutting have sharp edges, and they chip easily. The wafer edge is shaped to remove sharp, brittle edges; rounded edges minimize the risk for chipping. The edge shaping operation makes the wafer perfectly round (off-cut wafers are oval shaped after slicing), the diameter is adjusted, and orientation flat(s) or notch is dimensioned or made. This operation is done normally with a profiled diamond wheel, and the edge profile follows the shape given by SEMI M1-0414 [54].

The SEMI standard profile is often not suitable or ideal for MEMS applications. If the wafer is to be thinned after processing, the wafer edge may, as a result, be sharp and thus brittle using the standard profile; therefore, an asymmetric edge profile can be used. Also, wafer bonding may set specific requirements for the wafer edge shape; typically, for bonding applications, a blunter profile is recommended to achieve a good bond up to wafer edge. In Fig. 1.15 are shown some examples of the edge profiles. Profiles (a) and (b) are conventional. Profiles (c) and (d) are optimized for wafer thinning after processing. Profile (d) is also suitable for various bonding applications.

Lapping and Etching

Cut wafers from a saw typically have a higher TTV (Total Thickness Variation: the absolute variation of wafer thickness relative to a reference plane) than that required for state-of-the-art lithographic processes used by device manufacturers. This state is corrected by subjecting the wafers to a grinding (fixed abrasive) or lapping (loose abrasive) process. The macroscopic flatness of wafers is typically determined by this lapping or grinding process. Subsequent chemical etching removes the crystallographic damage that these processes produce.

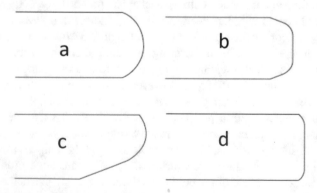

Fig. 1.15 Examples of the wafer edge profiles

Polishing

With ever smaller device geometries, the wafer surface needs to be highly planar on even the smallest length scales, especially for critical photolithographic processes. The development of chemical mechanical polishing (CMP) process by Walsh in the early 1970s has been a critical enabler in manufacturing of polished wafers without micro scratches or surface damage [55, 56]. CMP uses an abrasive and corrosive chemical slurry (commonly a colloid) in conjunction with a polishing pad and retaining ring, typically of a greater diameter than the wafer. The pad and wafer are pressed together by a dynamic polishing head and held in place by a plastic retaining ring. The dynamic polishing head is rotated with different axes of rotation (i.e., not concentric). This removes material and tends to even out any irregular topography, making the wafer flat or planar.

Cleaning

A clean silicon surface is essential to device performance. Contaminants can cause leakage, low breakdown voltage, nonuniform oxide growth, lithography errors, and many other problems for device manufacturers. The process developed by Kern and Puotinen, which came to be called "RCA clean", in 1970 remains the basis for the processes currently used in the industry [57]. The key phases in cleaning a silicon surface are:

- Removal of insoluble organic contaminants with a 5: 1: 1 H_2O: H_2O_2: NH_4OH solution
- Stripping of a thin silicon dioxide layer using a diluted 50: 1 H_2O: HF solution
- Removal of ionic and heavy metal contaminants using a solution of 6: 1: 1 H_2O: H_2O_2: HCl
- Passivation of the highly reactive bare, clean, silicon surface in H_2O

These steps are usually carried out in a high throughput automated wet bench, in which a robotic carrier lifts a cassette loaded with wafers from one tank to the next. The series of tanks contain chemical solutions or deionized water (DIW). Clean wafers are then packed in clean shipping boxes. Packaging technology has also developed over time with improved materials and designs, which add fewer particles during shipping. In Fig. 1.16 are shown silicon crystal ingots and silicon wafers at the end of the manufacturing process. In **silicon-based MEMS and sensors,** the silicon wafer provides both the substrate and the material for the active device layer. Additionally, polished silicon wafers are often bonded on the processed wafers to form hermetically enclosed packages. Mainly three types of wafers are used: polished bulk, epitaxial, and silicon-on-insulator (SOI) wafers. If device processing is performed from the front side only, single-side polished wafers are enough. It is quite common, though, to process the wafer from both sides; in these cases, double-side polished wafers are used.

Fig. 1.16 Silicon crystal ingots and silicon wafers

1.5.2 Standard Measurements of Polished Wafers

Typical standard measurements are: (1) oxygen and carbon concentration; (2) metal concentration; (3) resistivity; (4) wafer geometry; and (5) particles. In the following, a brief description of each measurement is given.

– Oxygen and Carbon Concentration

Oxygen is in interstitial sites in silicon lattice, and in CZ silicon it is super-saturated at normal processing temperatures. The typical achievable range in CZ growth process is approximately 6–18 ppma, or $3–9 \times 10^{17}$ atoms/cm^3. Precision measurements are made using double-side polished, or thick samples cut from grown crystals with Fourier transform infrared (FTIR) spectrometer. Interstitial oxygen absorbs at 1107 cm^{-1} wave number (9.03 μm wavelength IR-radiation). IR spectroscopy is the only technique that can discriminate between interstitially dissolved oxygen and oxygen in complexes or precipitates. Other techniques such as SIMS (Secondary-Ion Mass Spectrometry), CPAA (Charged Particle Activation Analysis), photon activation analysis, X-ray diffraction, and gas fusion analysis are used to detect the total oxygen concentration in the bulk. Carbon lies in substitutional lattice sites, and the concentration is typically less than 0.3 ppma or less than 1.5×10^{16} atoms/cm^3. Carbon absorbs at 605 cm^{-1} wave number, or 16.53 μm wavelength, and at room temperature a phonon band has a strong absorption at 610 cm^{-1}.

– *Metal concentration*

Bulk metal concentrations in silicon wafers are generally very low, and standard practice is to follow only bulk iron level through minority carrier lifetime measurements in lightly boron-doped silicon wafers.

A common technique in use is the μPCD (photoconductivity decay) method based on photo-induced minority carrier lifetime, according to SEMI MF1535-0707 standard [58]. Carriers are generated by light pulses, and the decay of the carrier density is measured by microwave reflection. Iron concentration can be calculated by comparing minority carrier lifetimes before and after iron-boron pair dissociation in p-type material.

– *Resistivity*

Resistivity of the wafer is measured with a contact method using a four-point probe according to SEMI

MF43-0705 and MF84-0312 [59, 60]. As contact methods damage and thus destroy the wafer surface, these methods are used typically for monitoring purposes only with monitor wafers. Polished prime wafers are measured with noncontact methods based on eddy current measurement, according to SEMI MF673-1105 [61]. Noncontact measurements have an accuracy and repeatability of less than 2–3% once properly calibrated and are used in volume production environment. The method is not able to resolve small-scale variations, like dopant striations. If small-scale resistivity variations need to be measured, a spreading resistance (SPR) measurement is done according to SEMI MF525-0312 [62].

– *Wafer geometry*

Wafer measurements for thickness, total thickness variation (TTV), and shape (warp, bow) are done with a noncontact capacitive method according to SEMI MF 1530-0707 [63], and commonly, 100% of wafers are measured. Here in the following definitions for TTV, Bow and Warp are given (Fig. 1.17).

TTV: the difference between the maximum and minimum values of wafer thickness; **Bow:** The deviation of the center point of the median surface of a free, unclamped wafer from the median surface reference plane established by three points equally spaced on a circle with a diameter a specified amount less than the nominal diameter of the wafer. **Warp:** the difference between the maximum and minimum values of the median surface from an established reference plane.

– *Particles*

Particles can be controlled visually under bright collimated light according to SEMI MF523-1107 [64]. Small (<0.3 μm) particles can be seen reliably only with dedicated laser scanners [65]. Automated laser scanners allow measurement of every wafer. Care must be taken when measuring small particles in the 0.1 to approximately 0.16 μm range not to mix particles with crystal originated particles (COPs) born from crystal growth. COPs are small vacancy agglomerates that are harmful in certain CMOS processes. For a complete list of measurements, the reader should consult reference [66].

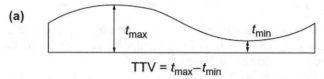

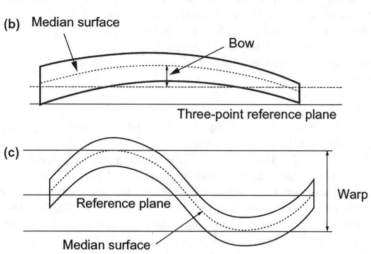

Fig. 1.17 Basic deformations of silicon wafers (overdone): thickness variation (**a**), bow (**b**), and warp (**c**)

Bibliography

1. Shockley, W. (1949). The theory of p-n junctions in semiconductor and p-n junction transistors. *Bell System Technical Journal, 28*, 435.
2. Hull, R. (Ed.). (1999). *Properties of crystalline silicon*. Inspec.
3. Landolt-Börnstein Group III. (2002). *Condensed matter*, volume 41A2a and volume 41A1b. Springer.
4. Hull, R. (Ed.). (1999). *Properties of crystalline silicon*. Inspec.
5. Martin, J., et al. (1998). The silicon lattice parameter—An invariant quantity of nature? *Metrologia, 35*, 811–817.
6. Kittel, C. *Introduction to solid state physics*. Wiley.
7. Petersen, K. E. Silicon as a mechanical material. Proceeding of the IEEE, vol. 70, NO. 5, MAY 1982.
8. Courtney, T. H. (1990). *Mechanical behavior of materials*. McGraw-Hill.
9. Howatson, A. M., Lund, P. G., & Todd, J. D. *Engineering tables and data*, p. 41.
10. Ebrahimi, F. (1995). *Scripta Metallurgica, 32*, 1507.
11. George, A., & Michot, G. (1993). *Materials Science and Engineering A, 164*, 118.
12. Li, X., Kasai, T., Nakao, S., Ando, T., Shikida, M., Sato, K., et al. (2005). *Sensors and Actuators A, 117*, 143.
13. Michot, G. (1988). *Crystal Properties and Preparation, 17_18*, 5.
14. Bischoff, F. Apparatus for vapor deposition of silicon. U.S. Patent 3 335 697, Aug. 15, 1967, (Original German priority date May 18, 1954).

15. Czochralski, J. (1918). Ein neues Verfahren zur Messung des Kristallisationsgeschwindigkeit der Metalle [A new method for the measurement of the crystallization rate of metals]. *Zeitschrift für Physikalische Chemie, 92*, 219–221. The paper was received by the editorial board on August 19, 1916 and was published in 1918, with a two-year delay. In the scientific literature, the year 1916 was adopted as the date of elaboration of the method.
16. Evers, J., Klüfers, P., Staudigl, R., & Stallhofer, P. (2003). Czochralski's creative mistake: A milestone on the way to the gigabit era. *Angewandte Chemie, 42*, 5684–5698.
17. Little, J. B., & Teal, G. K. Production of germanium rods having longitudinal crystal boundaries. U.S. Patent 2 683 676, issued Jul. 13, 1954, filed Jan. 13, 1950.
18. Teal, G. K., Sparks, M., & Buehler, E. (1951). Growth of germanium single crystals containing p-n junctions. *Physics Review, 81*, 637–647.
19. Teal, G. K., Sparks, M., & Buehler, E. (1952, August). Single-crystal germanium. *Proceedings of the IRE, 40*, 906–909.
20. Dash, W. C. (1959, April). Growth of silicon crystals free from dislocations. *Journal of Applied Physics, 30*(4), 459–474.
21. Dash, W. C. Method of growing dislocation-free semiconductor crystals. U.S. Patent 3 135 585, Jan. 2, 1964.
22. Davis, J. R., Rohatgi, A., Hopkins, R. H., Blais, P. D., Rai-Choudhury, P., McCormick, J. R., et al. (1980). Impurities in silicon solar cells. *IEEE Transactions on Electronic Devices, ED-27*(4), 677–687.
23. Graff, K. (2000). *Metal impurities in silicon-device fabrication* (Springer series in materials science 24) (2nd ed.). Springer.
24. Istratov, A. A., Hieslmair, H., & Weber, E. R. (2000). Iron contamination in silicon technology. *Applied Physics A: Materials Science & Processing, 70*(5), 489–534.
25. Istratov, A. A., & Weber, E. R. (2002). Physics of copper in silicon. *Journal of the Electrochemical Society, 149*(1), G21–G30.
26. Harada, K., Tanaka, H., Matsubara, J., Shimanuki, Y., & Furuya, H. (1995). Origins of metal impurities in single-crystal Czochralski silicon. *Journal of Crystal Growth, 154*(1), 47–53.
27. Schroder, D. K. (1997). Carrier lifetimes in silicon. *IEEE Transactions on Electronic Devices, 44*(1), 160–170.
28. Sumino, K., & Yonenaga, I. (1994). Oxygen effect on mechanical properties. In F. Shimura (Ed.), *Oxygen in silicon, semiconductors and semimetals* (Vol. 42, pp. 449–511). Academic.
29. Yonenaga, I. (2001). Dislocation behavior in heavily impurity doped Si. *Scripta Materialia, 45*(11), 1267–1272.
30. Borghesi, A., Pivac, B., Sassella, A., & Stella, A. (1995). Oxygen precipitation in silicon. *Journal of Applied Physics, 77*(9), 4169–4244.
31. Takeno, H., Otogawa, T., & Kitagawara, Y. (1997). Practical computer simulation technique to predict oxygen precipitation behavior in Czochralski silicon wafers for various thermal processes. *Journal of the Electrochemical Society, 144*(12), 4340–4345.
32. Sueoka, K. (2006). Oxygen precipitation in lightly and heavily doped Czochralski silicon. In C. L. Claeys, R. Falster, M. Watanabe, & P. Stallhofer (Eds.), *High purity silicon 9* (pp. 71–87). The Electrochemical Society.
33. Myers, S. M., Seibt, M., & Schröter, W. (2000). Mechanisms of transition metal gettering in silicon. *Journal of Applied Physics, 88*(7), 3795–3819.
34. Villa, F. F., & Paciaroni, L. (1989). Effect of substrate oxygen content on smart power ICS yield. Intrinsic gettering effectiveness and denuded zone calculation. In M. A. Shibib (Ed.), *Symposium on high voltage and smart power ICs* (p. 228). The Electrochemical Society.
35. Falster, R., Voronkov, V. V., & Quast, F. (2000). On the properties of the intrinsic point defects in silicon: A perspective from crystal growth and wafer processing. *Physica Status Solidi B: Basic Solid State Physics, 222*, 219–244.
36. Falster, R., & Voronkov, V. V. (2000). The engineering of intrinsic point defects in silicon wafers and crystals. *Materials Science and Engineering, B73*, 87–94.
37. Chedzey, H. A., & Hurle, D. T. (1966). *Nature, 239*, 933.
38. Utech, H. P., & Flemings, M. C. (1966). *Journal of Applied Physics, 37*, 2021.

39. Witt, A. F., Herrman, C. J., & Gatos, H. C. (1970). *Journal of Materials Science, 5*, 822.
40. Hoshi, K., Suzuki, T., Okubo, Y., & Isawa, N. (1980). Abstract 324. *The Electrochemical Society Extended Abstracts, 80–1*, 811.
41. Suzuki, T., Isawa, N., Okubo, Y., & Hoshi, K.. (1981). *Semiconductor Silicon* 1981 (H. R. Huff, R. J. Kriegler, & Y. Takeishi, Eds., p. 90). The Electrochemical Society.
42. Takasu, S., Homma, K., Toji, E., Kashima, K., Ohwa, M., & Takahashi, S.. (1988). *PESC'88 Record*, vol. April 1988, p. 1339.
43. Eidenzon, A. M., & Puzanov, N. I. (1997). *Inorganic Materials, 33*(3), 272.
44. Park, J. G., & Chung, H. K.. *Silicon Wafer Symposium SEMICON West 99* (SEMI 1999, D–1).
45. Shimura, F. (1988). *Semiconductor silicon crystal technology*. Academic.
46. Arai, Y., Kida, M., Ono, N., Abe, K., Machida, N., Futuya, H., & Sahira, K. (1994). *Semiconductor silicon* (p. 180). The Electrochemical Society.
47. Kim, K. M., & Smetana, P. (1989). *Journal of Crystal Growth, 100*, 527.
48. Li, J., Kao, I., & Prasad, V. (1998, June). Modeling stresses of contacts in wire saw slicing of polycrystalline and crystalline ingots: Application to silicon wafer production. *The Journal of Electronic Packaging, 120*, 123–131.
49. Hauser, C., & Nasch, P. M. (2004). Advanced slicing techniques for single crystals. In H. J. Scheel & T. Fukuda (Eds.), *Crystal growth technology*. Wiley.
50. Hauser, C. (1988). Device for wire sawing provided with a system for directing wire permitting use of spools of wire of very great length. U.S. Patent 5 829 424, 1998.
51. Bye, J.-I., Jensen, S. A., Aalen, F., Rohr, C., Nielsen, Ø., Gäumann, B., et al. Silicon slicing with diamond wire for commercial production of PV wafers, 24th European Photovoltaic Solar Energy Conference 2009, Hamburg, Germany, p. 1269.
52. Bye, J.-I., Norheim, L., Holme, B., Nielsen, Ø., Steinsvik, S., Jensen, S. A., et al. Industrialized diamond wire wafer slicing for high efficiency solar cells, 26th European Photovoltaic Solar Energy Conference and Exhibition 2011, Hamburg, Germany, p. 956.
53. Bidiville, A., Heiber, J., Wasmer, K., Habegger, S., & Assi, F.. Diamond wire wafering_ wafer morphology comparison to slurry sawn wafers, 25th European photovoltaic solar energy conference and exhibition, 2010, Valencia, Spain, p. 1673.
54. SEMI M1-0414, Specifications for Polished Monocrystalline Silicon Wafers, Semiconductor Equipment and Materials International 2014, www.semi.org
55. Walsh, R. J. Apparatus for processing semiconductor wafers. U.S. Patent 3 964 957, June 22, 1976.
56. Walsh, R. J. Process for chemical-mechanical polishing of III-V semiconductor materials. U. S. Patent 3 979 239, Sept. 7, 1976.
57. Kern, W., & Puotinen, D. A. (1970). Cleaning solutions based on hydrogen peroxide for use in silicon semiconductor technology. *RCA Review, 31*, 187–206.
58. SEMI MF1535-0707 test method for carrier recombination lifetime in silicon wafers by non-contact measurement of photoconductive decay by microwave reflectance, semiconductor equipment and materials international 2014.
59. SEMI MF43-0705 test methods for resistivity of semiconductor materials, semiconductor equipment and materials international 2014.
60. SEMI MF84-0312 test method for measuring resistivity of silicon wafers with an in-line four-point probe, semiconductor equipment and materials international 2014.
61. SEMI MF673-1105 test method for measuring resistivity of semiconductor wafers or sheet resistance of semiconductor films with a noncontact Eddy-current gauge, semiconductor equipment and materials international 2014., www.semi.org
62. SEMI MF525-0312 test method for measuring resistivity of silicon wafers using spreading resistance probe, semiconductor equipment and materials international 2014.
63. SEMI MF1530-0707 test method for measuring flatness, thickness, and Total thickness variation on silicon wafers by automated non-contact scanning, semiconductor equipment and materials international 2014.
64. SEMI MF523-1107 practice for unaided visual inspection of polished silicon wafer surfaces, semiconductor equipment and materials international 2014., www.semi.org

65. SEMI M35-1107 guide for developing specifications for silicon wafer surface features detected by automated inspection, semiconductor equipment and materials international 2014.
66. International SEMI Standards, semiconductor equipment and materials international, November 2014.

Part II
MEMS Processes

Chapter 2
Epitaxy

Roberto Campedelli and Igor Varisco

2.1 Principle of Epitaxy

Epitaxy (from the Greek roots epi, meaning "above", and taxis, meaning "an ordered manner") is a term applied to processes used to grow a single thin crystalline layer on a single crystalline substrate. If the substrate and thin film are of the same material, the process is known as homoepitaxy. If the substrate and thin film are of different materials, the process is known as heteroepitaxy.

The substrate acts as a seed for the growth and the epitaxial layer maintains the same crystallographic orientation.

In the case of two different materials, the lattice mismatch becomes a key issue generating defects as dislocations. The strain due to lattice mismatch can be accommodated elastically if the layer thickness is lower than a critical value [4].

Chemical Vapor Deposition (CVD) is the most widely employed means to deposit semiconductors and dielectric materials for MEMS, Microfluidic devices, and Integrated Circuits. These materials include those commonly used in the IC industry such as polycrystalline silicon, epitaxial silicon, silicon–germanium, as well as other semiconductors, such as gallium arsenide, indium phosphide, and silicon carbide. Chemical vapor deposition can be achieved from solid phase, liquid phase, or vapor phase. Epitaxy is not limited to CVD, in fact, methods based on Physical Vapor Deposition (PVD) are also commonly used to grow binary, ternary, and quaternary III-V compound semiconductors. Molecular Beam Epitaxy (MBE) can also be used to grow epitaxial Si, SiGe, and even cubic structure SiC (3C-SiC) films.

R. Campedelli (✉) · I. Varisco
ST Microelectronics, Analog MEMS and Sensors Group, MEMS Technology and Design R&D,
Agrate Brianza, Monza Brianza, Italy
e-mail: roberto.campedelli@st.com; igor.varisco@st.com

© Springer Nature Switzerland AG 2022
B. Vigna et al. (eds.), *Silicon Sensors and Actuators*,
https://doi.org/10.1007/978-3-030-80135-9_2

Vapor-Phase Epitaxy (VPE), a special form of CVD, is the most used method for silicon epitaxial growth, allowing a very high quality of the grown layers with high growth rate (up to several microns per minute) and is based on the transport in the form of volatile compounds of silicon precursors like chlorosilanes (SiH_2Cl_2, $SiHCl_3$, $SiCl_4$) or Silane (SiH_4) on the substrate surface where they will chemically react to form an epitaxial layer.

As shown in Fig. 2.1, heterogeneous Chemical Vapor Deposition reaction involves the following key steps:

1- Introduction of precursors in the deposition region.
2- Transport of reactants from gas stream through the boundary layer to the wafer surface.
3- Surface processes that include adsorption and dissociation of precursors into reactants species.
4- Surface diffusion of reactants to reaction sites; decomposition or reaction, migration, and attachment at kink/ledge sites.
5- Desorption of reaction byproducts from the substrate surface.
6- Transfer and removal of residual reactants and by products from the substrate surface.

Two different regimes based on temperature are considered for epitaxial processes: mass transfer-controlled and reaction-controlled regime. In Fig. 2.2, a typical Arrhenius plot of growth rate as a function of inverse temperature for different silicon precursors is shown.

At relatively low temperatures, the process is in the reaction-controlled regime and surface reactions govern the process. In this regime, the growth rate is much more sensitive to temperature. (region A).

At relatively high temperatures, the process is in the mass transfer-controlled regime and the process depends by the flux of reactants species through the

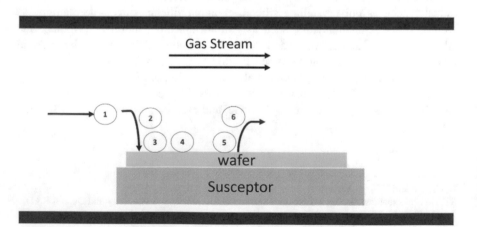

Fig. 2.1 Heterogeneous reaction model for a Chemical Vapor Deposition. Growth mechanism

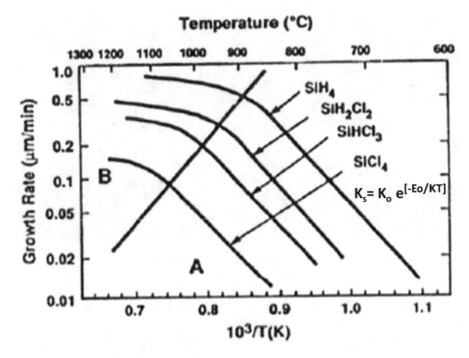

Fig. 2.2 Area A = Surface reaction kinetics limited (low temperature); Area B = Mass transport limited (high temperature)

boundary layer. Proper chamber design assures a constant boundary layer thickness at the wafer surface; therefore, the growth rate varies relatively slowly with increasing temperature because the mass transfer coefficient is relatively constant with temperature. (region B).

Epitaxial process recipes are generally designed to operate well within either the mass transport-controlled regime or the reaction-controlled regime as determined primarily by the temperature range required by the process to produce the desired film.

At moderate temperatures, both fluxes influence the deposition rate and thus control of the process is challenging. If a single crystalline film is desired, a high-deposition temperature is necessary to initiate epitaxial growth and thus a process in the mass transfer-controlled regime is selected. Likewise, deposition of polysilicon requires a much lower substrate temperature; therefore, a process in the reaction-controlled regime is selected. Single crystal growth occurs when the reactant species adsorbed on the silicon surface (adatoms) diffuse towards thermodynamically more favorable sites (kink sites) [5] as shown in Fig. 2.3.

At high temperature, with increasing the surface migration, the crystal growth is promoted.

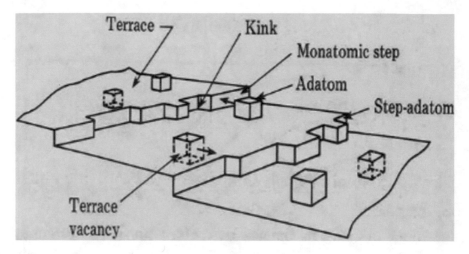

Fig. 2.3 Schematic illustration of epitaxial growth with adatoms

At high growth rate, in the initial stage of the growth, atoms aggregate to form a 3D island helping the polysilicon nucleation.

A more comprehensive treatise on the fundamentals of epitaxial deposition processes can be found in several published articles, books, and proceedings concerning silicon epitaxy, and related topics can be used as general references for the whole chapter [1, 3, 6].

2.2   Epitaxial Reactors

Cold-wall epitaxial reactors are classified according to the reaction chamber design, to the type of susceptor, and to the type of heating system applied to provide the thermal energy to the reaction.

Horizontal reactors for single or multiple wafers processing can be considered a landmark in the field of the reactors commonly used for epitaxial depositions on 200 mm substrates, can operate under cold wall conditions and offer high throughput, high defectiveness control, low costs of ownership, and a good control of thickness and resistivity of the film. Schematic design of these types of reactors is shown in Fig. 2.4.

A cylinder reactor, namely, a barrel reactor, is a vertical reactor mainly used for wafers' diameters up to 150 mm, where the susceptor contained in an externally cooled quartz bell is a truncated pyramid that may accept one or more rows of wafers into cavities located on the pyramid faces. Schematic design of these types of reactors is shown in Fig. 2.5. Barrel reactors have the advantage of processing many wafers per batch, providing an excellent control of the grown layer in terms of thickness uniformity and auto doping.

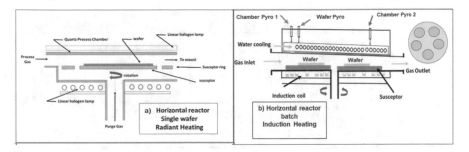

Fig. 2.4 Horizontal Epitaxial reactors: a) Single wafer, radiant heating; b) batch reactor, induction heating

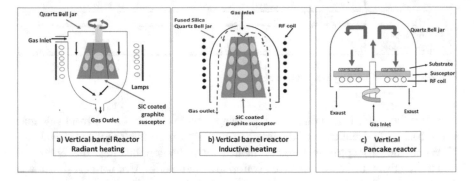

Fig. 2.5 Barrel and pancake Epitaxial vertical reactors with radiant or induction heating

The substrate heating method determines the basic characteristics of the growth process. Two types of heating are commercially employed: Radiant heating and Inductive heating.

In the first case, high-energy IR halogen lamps positioned outside the process chamber are used, while in the second case, a graphite susceptor covered with silicon carbide (SiC) is heated by RF (20–200 kHz) for inductive coupling of the susceptor with an inductor positioned outside and under the reaction chamber (coil).

Clear-fused quartz is used in cold-wall reactors for its transmittivity in the wavelength range of the halogen lamps or RF induction heating. This allows to keep the quartz chamber cold enough, minimizing coating.

Typically, a gold plate reflector is placed behind the lamps array to enhance radiation towards the wafer.

Atmospheric or reduced pressure processes can be used on horizontal single wafer reactors, while batch reactors are commonly limited to atmospheric processes.

The heating method affects the relative temperature between wafer and susceptor and, consequently, the thermal stress of wafer and mass transfer of silicon at the back as summarized in Fig. 2.6.

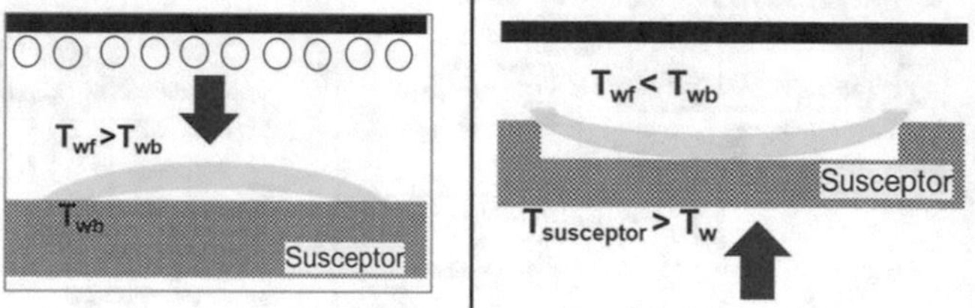

RADIANT HEATING	INDUCTION HEATING
$T_{wf} > T_{wb}$ T_{wb} Susceptor	$T_{wf} < T_{wb}$ Susceptor $T_{susceptor} > T_w$

Energy Transfer	From Front wafer to back side	From Back wafer to front sideside
Wafer Temperature	T front > T back	Tback > T front
Wafer Bow	Positive	Negative
Slip lines	low	High
Backside coating (Mass Transfer)	Lower	Higher

Fig. 2.6 Impact of heating method on substrate

When the wafer is radiant heated, its temperature is higher on the front side than at the back as consequence of the energy transferred from the lamps to the wafer.

When the susceptor is directly heated by induction heating, the energy is transferred from the susceptor to the wafer, so the temperature of wafer results lower than the susceptor.

In batch reactors using induction heating, spherical profile into the susceptor's pockets are optimized according to the wafer thickness to minimize wafer bow and thermal stress caused by the temperature gradient between the front and back side of the wafers.

Protection of the back side of the wafer is particularly needed to prevent autodoping.

ASM-Epsilon reactors [7] and Centura Epi reactors [8] are widely used for single wafer epitaxial processes for 200 mm substrates.

Process gases flow horizontally through an air-cooled quartz process chamber from lateral inlet injectors to the exhaust assuring a laminar flow across the wafer and through the chamber. Wafer heating capability is provided by radiant lamp array. In ASM reactor, the upper linear array of tungsten–halogen lamps heat the top side of the wafer while the lower one heats the bottom side. Four additional lamps positioned under the susceptor provide more energy to heat the susceptor. Radiation from the wafer edge is reduced by a silicon carbide guard ring. A schematic of this reactor is shown in Fig. 2.7.

Temperature is controlled by four thermocouples not in contact with the wafer at four locations on the susceptor ring (side, front, and rear) and susceptor center. Thermocouple readings are used in Proportional–Integral–Derivative (PID) control loops for temperature control. The reactor can operate at both atmospheric and reduced pressure (e.g., 20 Torr).

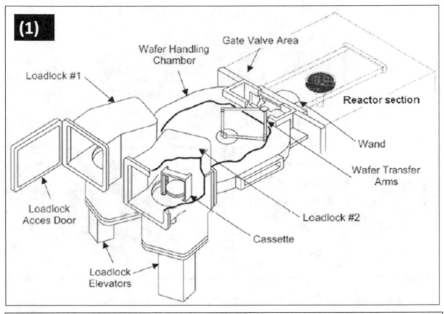

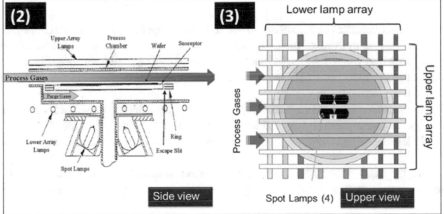

Fig. 2.7 Schematic Illustration of an ASM Epsilon reactor: (1) Process chamber, transfer station and Load locks; (2) Side view: (3) Schematic of lamps array

Both the reactors are equipped with a rotating susceptor to increase the radial uniformity of the deposited film. Transfer module isolates the process chamber from load locks at each loading or unloading of wafers.

Epi Centura is a single-wafer cluster tool to which up to three separate atmospheric or reduced pressure epi chambers can be attached. High throughput level is assured when all the chambers operate at the same time at atmospheric or reduced pressure. Centura Platform has independent cooldown and load locks.

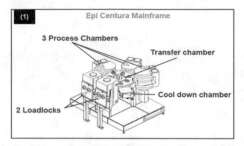

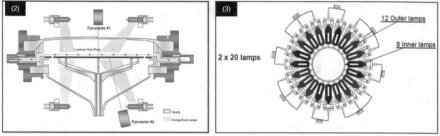

Fig. 2.8 Single wafer Epi Centura Reactor Chamber and radial heating array overview

The epi chamber is basically composed of two (upper and lower) quartz domes.

The top and bottom of the chamber, is heated by radiant lamp modules divided into inner and outer rings of lamps. Chamber temperature is monitored via two optical pyrometers, one near the bottom of the susceptor and one on the center of the wafer's surface.

Temperature set points are maintained by PID controllers that use feedback from the optical pyrometers to vary the power delivered to the lamp modules.

Schematic of Epi Centura chamber, including radial heating elements, are shown in Fig. 2.8.

On the market, horizontal epitaxial batch reactors are also available. In batch epitaxial reactor like PE3061 [9], up to five wafers of 200 mm can be loaded for each batch.

This type of reactor is limited to atmospheric processes and assures wafer quality close to single wafer reactor and low cost of ownership for thick epitaxial deposition processes (> 15 μm) for MEMS and discrete technology like Power MOS, Insulated Gate Bipolar Transistor (IGBT). Thermal stress needs to be carefully optimized to reduce slip lines and dedicated susceptors are required for processing different thickness of wafers. A schematic of this reactor is shown in Fig. 2.9.

Medium Radio-frequency induction heating is used as heating methods of graphite silicon carbide coated susceptor. A water -cooled copper induction coil serves as primary winding of a transformer. The graphite susceptor serves in effect as a single-turns secondary winding. The voltage induced in the susceptor produces a circulating eddy current which, as a result of heating, produced by I^2 power loss, raises the temperature to the required values.

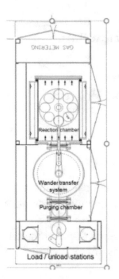
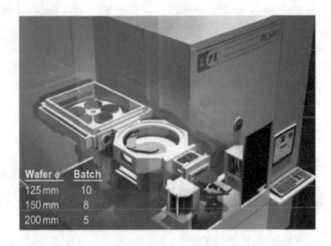

Fig. 2.9 Horizontal batch reactor PE3061 overview

2.2.1   Future Requirements for MEMS

The growing demand for ever-thicker epitaxial or polycrystalline silicon layers with a tight thickness and doping uniformity control for new advanced power devices and for MEMS applications, is pushing the suppliers to upgrade the current 200 mm single wafer epitaxial reactors primarily used till now for thin epitaxial growth (< 15 μm).

Re-design must address all the challenges of growing high-quality thick films (20–100 μm) in a single wafer reactor exceeding batch reactor performance-like process throughput, cost of ownership, and maintaining the same or better on-wafer performances. Enhanced growth rate up to 6 μm/min is required to significantly reduce cost of ownership. Thick epitaxial layer must be grown up to 100 μm in a single pass deposition without intermediate chamber cleaning step.

Due to the longer deposition times involved, real time and closed-loop control of quartz chamber temperature must be implemented to reduce coating buildup, degradation of the quartz chamber quality, particles generation and process drift. Quartz chamber and cooling system must be adequately designed.

Uniformity of thickness and dopant concentration across the wafer surface and between wafer and wafer are most critical parameters to achieve and control. Wafer sticking, wafer placement on the susceptor, thermal stress (slip lines), backside growth, temperature, and precursor flow stability must be carefully controlled when thicker films are grown to guarantee key parameters as in wafer thickness and dopant concentration uniformity, Total Thickness Variation (TTV), and site flatness.

Furthermore, as regards the growth temperature, it is necessary to improve its uniformity on the wafer, its stability during the growth cycle, and its reproducibility to guarantee better control of wafer bow and residual stress of the layer, when polycrystalline silicon membranes for micro actuator devices are realized.

2.3 Epitaxial Silicon Processes: Applications and Process Generalities

2.3.1 Applications

With the epitaxial process is possible to obtain materials with chemical and physical properties better than the starting substrate and consequently, semiconductor devices with improved performances.

Silicon Epitaxy remain the easiest known method for controlling doping concentration level, doping profile, and thickness uniformity. Examples of epitaxial growth on silicon substrates are reported in Fig. 2.10.

Lightly doped epitaxial layer grown on top of heavily doped substrates or heavily doped buried layer patterns were used in bipolar devices to minimize collector resistance and increase collector-substrate breakdown voltage [10, 11] while in CMOS technologies, to minimize latch up effects [12]. P/P+ epitaxial substrates show also good internal gettering properties even in low thermal budget processes and very good robustness to slip lines generated by thermal stress. When no patterned structures are defined on the starting materials, the epitaxial growth process is directly manufactured by silicon vendors. Conversely, when structures like N+ buried layer or P+ isolation are required, the epitaxial layer is grown into the manufacturing line.

Surface defects induced by polishing and grown-in defects as vacancies clusters (Crystal Originated Particles or COPs) associated to the Czochralski crystal growth method, can be suppressed introducing an epitaxial layer with a beneficial effect on gate oxide electrical breakdown in CMOS technologies.

Oxygen precipitate-free zone is also created in the preliminary H_2 bake step of the epitaxial process used to remove the native oxide on the wafer surface.

Epitaxial processes are not commonly employed in Si-based surface micromachining technologies. Epitaxy can be used in MEMS devices like thermal inkjet print heads where the monolithic integration with CMOS devices is applied.

Silicon epitaxial layers can be also used as etch stops in micromachining technologies to form Bonded Etch-Back Silicon on insulator (BESOI) structures [13].

When single crystal silicon membranes are required for MEMS applications like micro-μactuators, capacitive and Piezo Micromachined Ultrasonic Transducers, silicon-on-insulator substrates are typically employed. Active layer acts as structural layer while buried oxide may also be used as a sacrificial layer. Thin membranes in the thickness range 0.4–1.5 μm and with excellent thickness uniformities are obtained using Smartcut SOI technology [7, 14]. Epitaxial process on Smartcut SOI

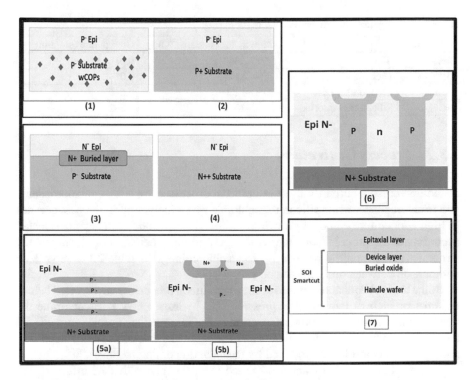

Fig. 2.10 Summary of different conventional Epitaxial growth: (1) epitaxy on CZ substrates for Crystal Originated Particles free substrate; (2)and (4) N- or P- epitaxy on highly doped substrates to minimize latch up in CMOS applications; (3) Epitaxial growth on patterned N+ buried layers; (5a) multiple N- epitaxial steps on P doped area to obtain a thick epitaxial layer and after final annealing to build a P- deep layer for superjunction MOSFET; (6) Etched thick N- Epi successively filled with P doped Epitaxy for superjunction power MOSFET; (7) Epitaxial growth on SmartCut SOI to target device layer thickness

substrates is added to obtain thicker membranes with a tight control of the thickness uniformity (percentage standard deviation below 1.5%).

Other materials like GaN or Cubic silicon carbide (3C-SiC) can be grown on silicon or SOI substrates by epitaxy for High voltage power and high frequency devices, electronic and biomedical applications [15–17].

Epitaxial silicon process is also used for specific MEMS devices like micro-machined piezo-resistive pressure sensors. By combining an array of trench structures, epitaxial silicon growth and hydrogen annealing process, monolithic crystalline silicon (c-Si) membranes can be realized. In the case study reported in paragraph 5.3.3, this specific application will be examined in depth.

Table 2.1 Comparison of Silicon precursors for epitaxy

Si Precursor		Growth Temperature (°C)	Growth rate (um/min)
SiH_4	Silane	500–1000	0.01–0.8
SiH_2Cl_2	Dichlorosilane	750–1100	0.05–2
$SiHCl_3$	Trichlorosilane	1050–1200	0.6–6
$SiCl_4$	Silicon tetrachloride	1100–1270	0.6–2

2.3.2 Process Generalities

Silicon precursors currently used in epitaxial processes are silane, dichlorosilane, and trichlorosilane, while silicon tetrachloride was widely used in the past decades in high temperature epitaxial growth processes (1150–1300 °C). They are selected according to the process temperature, the growth rate, specific process concern, the structures present on the wafer prior to starting epitaxy. In Table 2.1 is reported a comparison between different silicon precursors.

Some basic chemical reactions for the main silicon precursors are reported below. Parameters such as temperature, pressure, and hydrogen flow rate determine the mechanisms of decomposition, adsorption of the intermediate species formed. All the reactions involving chlorosilanes and hydrogen are reversible, and silicon film and hydrogen chloride are produced by the reduction of chlorosilanes. [18].

The reactant molecules are transported from gas phase to the surface of the silicon wafer and by thermal decomposition at a certain rate, $SiCl_2$ * species previously adsorbed are attacked by the hydrogen atoms to reduce $SiCl_2$* to Si and release HCl gas.

- $SiCl_4 + 2H_2 \longleftrightarrow Si_{(s)} + 4HCl$ (at very high temperature)

  $SiCl_4 + H_2 \longleftrightarrow SiHCl_3 + HCl$
  $SiHCl_3 \longleftrightarrow SiCl_2* + HCl$ (Decomposition and adsorption)

- $2SiHCl_3 \longleftrightarrow Si_{(s)} + SiCl_4 + 2HCl$

  $SiHCl_3 + H_2 \longleftrightarrow Si_{(s)} + 3HCl$
  $SiHCl_3 + H_2 \longleftrightarrow SiH_2Cl_2 + HCl$
  $SiH_2Cl_2 + H_2 \longleftrightarrow Si_{(s)} + 2HCl$ (the reverse reaction is Si etching)
  $SiH_2Cl_2 \longleftrightarrow SiCl_2* + H_2$ (Decomposition and adsorption)
  $SiCl_2* + H_2 \longleftrightarrow Si_{(s)} + 2HCl$ (the reverse reaction is Si etching)

The reactions of silane as silicon precursor are shown. The decomposition of silane occurs by pyrolysis, but intermediate species react at the surface much more readily than silane, contributing to higher deposition rates.

- $SiH_4 \rightarrow Si_{(s)} + 2H_2$

  $SiH_4 \longleftrightarrow SiH_2* + H_2$
  $SiH_2* + SiH_4 \longleftrightarrow Si_2H_4* + H_2$

$$Si_2H_4 * \longleftrightarrow Si_2H_2*+ H_2$$
$$Si_2H_2 * \longleftrightarrow Si_{2(s)} + H_2$$
$$SiH_4 \rightarrow Si_2(s) + H_2$$

Si_2, Si_3 and greater Si multiples, are greatly suppressed by addition of hydrogen. Increase in gas phase nucleation particulates follows increase of Si_2, Si_3 density or an increase in temperature or pressure.

Trichlorosilane is preferentially used for high temperature processes and thick epitaxy at atmospheric pressure; however, for thin layers, low pressure can be also used. Dichlorosilane, at reduced pressure, is mainly employed for relatively thin epitaxial layers ($< 10\,\mu$m) or for a better control of arsenic autodoping. For Selective Epitaxial Growth (SEG) [19] only in exposed silicon areas, dichlorosilane and hydrochloric acid are commonly used on wafers patterned with silicon dioxide or silicon nitride [19]. Chlorine-free precursor, like silane, is used for low-temperature epitaxial processes (lower than 1000 °C) when a low transition region or no pattern shift is requested. Furthermore, silane is also used as seed layer on wafers with oxide for low temperature polycrystalline silicon processes at atmospheric or reduced pressure.

Typical pressure applicable for conventional vapor phase epitaxy are in the range 20 Torr–200 Torr.

In horizontal reactors, process temperature is usually in the range 800–1190 °C with radiant heated reactors and 900–1160 °C with induction heated reactors.

Hydrides like diborane, phosphine, or arsine are used for P type or N-type doping gases.

Typical dopant concentration is from 20 ppm to 300 ppm in hydrogen and additional dilution is performed before entering in the mainstream. In MEMS epitaxial processes, when low resistivities are required, phosphine or diborane, with a concentration of up to 1% in hydrogen, could be used.

The dopant is incorporated into the grown layer in substitutional lattice sites. The lower achievable resistivity is related to the solid solubility in silicon of the dopant and to the temperature. Achievable range of resistivity are 0.005–50 Ω•cm for N-type and 0.01–150 Ω•cm for P type epitaxy.

Resistivities in the range 1–50 Ω•cm are used in CMOS or smart power processes and lower than 15 mΩ•cm for specific MEMS technologies.

Epitaxial layers in the range 1.5–70 μm are used for inertial sensors and up to 100 μm for some specific applications.

Typical tolerance of the epitaxial process manufactured on 200 mm substrate is +/− 5% for the thickness and +/− 5% for resistivity. For some specific MEMS applications where the thickness has a significant impact on the performances of the devices (bow of membrane, resonant frequency, sensitivity) the tolerance requirements are below 2%.

An epitaxial recipe can be summarized in the following main steps:

− Chamber etching.
− Load wafer.
− Purging.

- Ramp up.
- Bake.
- Deposition.
- Ramp down.
- Unload.

Hydrochloric acid (40–50% in hydrogen) at high temperature is used for quartz chamber and silicon carbide susceptor cleaning after each wafer or batch deposition to guarantee an accurate control of the defectiveness and contamination and good reproducibility of thickness and resistivity.

After wafer loading and purging, native oxide is removed from the wafers with a hydrogen step at high temperature with a time of 10–60s (bake step) before starting the growth. This step assures a good quality of the grown layer without any defects.

Tight control of temperature uniformity on wafer is required to reduce thermal stress and avoid generation of dislocation like slip lines.

2.3.3 High Temperature Hydrogen Annealing

Thermal treatments in hydrogen atmosphere have been extensively used in the IC industry as in situ surface preparation and to remove native oxide before silicon epitaxy [20].

Since the early 2000s, high temperature hydrogen annealing processes in conventional epitaxial reactors or in specifically designed vertical furnaces have been employed to improve electrical characteristics of the thin gate oxide grown on Czochralski Si wafers (CZ) [21, 22].

Silicon crystals prepared using CZ growth method and high pulling rates result in agglomeration of vacancies and formation of Crystal Originated Particles (COPs) [23] that have a detrimental impact on thin gate oxide integrity.

COPs have shape of octahedral voids with sizes ranging from 50 nm to 150 nm and hydrogen annealing has a significant effect on reducing near-surface COPs defects density and size.

Silicon migration by hydrogen annealing and transformation of microstructure has been applied to various fabrication. Many studies have been also focused on the transformation of deep trenches [24] for different applications, like trench corner rounding, deep trench sidewall smoothing, and scallop's suppression generated during trench definition by Deep Reactive Ion Etching (Bosch Process), in mono silicon or Silicon on insulator substrates.

Sidewall surface smoothing drastically improves characteristics of the thin oxide grown in the deep trench capacitor. The shape of the trench can be controlled by the various parameters of the hydrogen annealing, like temperature, time, and pressure. Trenches, with Aspect Ratio 15:1 realized with DRIE Bosch process, are shown in Fig. 2.11. As example, scallops with nonuniform size along the trench's depth are

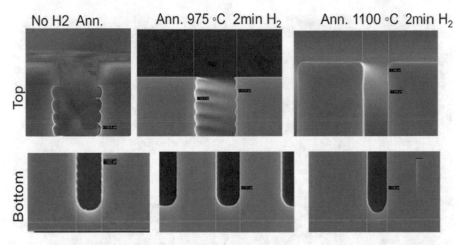

Fig. 2.11 SEM images of the initial shape of 30 μm deep trench (AR 15:1 after Bosch DRIE and after hydrogen annealing at atmospheric pressure for 2 min at 975 °C and 1100 °C)

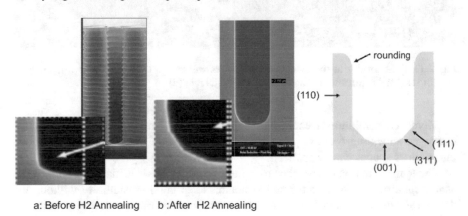

Fig. 2.12 Trench sidewall before (**a**) and after 1100 °C annealing H2 100% 2 min. (**b**). At bottom of trench (111) and (311) planes are highlighted

completely suppressed and trench rounded, increasing the annealing temperature in hydrogen atmosphere.

During high temperature hydrogen annealing, migration of silicon atoms is promoted transforming silicon structures like trenches. Desorption and surface diffusion of silicon atoms are the dominant factors for the migration mechanism and shape evolution, while the driving force of the transformation is the surface energy minimization [25]. Surface atoms migrate to reduce the number of dangling bonds and lowering the total surface energy. So, the transformation starts from the part that the radius of the curvature is the smallest, such as the corner of the rectangular shape and especially the corner at the bottom of the trench. Thus, the sharp corner of the deep trench can be selectively rounded. High density crystallographic planes as (111) and (311) are generated as shown in Fig. 2.12.

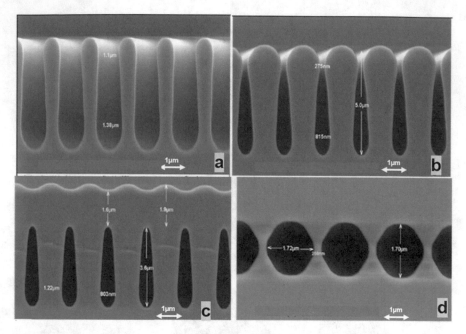

Fig. 2.13 SEM images of the evolution of trenches with AR 6:1 after H2 bake (**a**); Epitaxial growth 15s (**b**); 45s (**c**); after H2 annealing 1190 °C 20 min (**d**)

Moreover, combining trenches with different aspect ratio, thickness, pitch, and hydrogen annealing process parameters, micro-structure transformation of silicon into various shapes can be realized.

The transformation of trenches to spherical shaped voids is mainly driven by the aspect ratio of trench. The formed structures keep good crystallinity without any defects and their surfaces have a lower roughness comparable with that of bulk.

In Fig. 2.13, the evolution of a microstructure realized with hydrogen annealing and epitaxial growth is shown as an example.

Silicon migration is enhanced at high temperature and low pressure. Increasing pressure, the number of the collided hydrogen atoms at the surface increase minimizing the migration of silicon. Aspect ratio of the trenches and their layout play an important role in defining size and shape of the transformed microstructures.

High temperature hydrogen annealing processes for silicon migration can be performed in any epitaxial reactors or in dedicated furnaces fully equipped with appropriate safety systems. For high temperature annealing of up to 1200 °C, atmospheric reactors or furnaces are used, while for reduced pressure process, lower temperature is employed. Moisture content in the environment is extremely important to keep under control; few tens ppm of oxygen can increase surface roughness during high temperature cycles.

2.3.4 Case Study: Single Crystal Silicon Membrane for Piezoresistive Pressure Sensors

Thin silicon membranes are currently widely used to realize piezoresistive pressure sensors. A MEMS technology named VENSEN developed in STMicroelectronics in 1997 has been applied in the early 2000s to manufacture single crystal silicon membrane for piezoresistive sensing of the environmental pressure. The name of this process derives from similarities between the house-building technique used in the city of Venice (on pile-works) and the process itself. By combining in single crystal silicon an array of pillars structures realized by deep reactive ion etching (DRIE), to which follows an epitaxial silicon growth and silicon migration by hydrogen annealing, vacuum sealed cavities can be obtained; the entire structure is a monolithic crystalline silicon (c-Si) and is called, by some authors, "Silicon on nothing" [26, 27].

This process allows the fabrication of micromachined devices with an excellent long-term stability and reliability, where under vacuum cavities are achieved without hermetical sealing process, such as anodic or direct bonding technologies.

An extensive treatise concerning the realization of piezo resistive pressure sensors is reported in chapter 16.

These thin membranes have a square shape with side width 200–800 μm and a thickness from 5 to 60 μm according to the performance required by the device; the gap (empty space under the membrane) is in the range 1.5–4 μm. Membranes bow is in the range of 0.1–2 μm and strictly related to their dimensions and thicknesses.

In this section, the realization of monolithic crystalline silicon (c-Si) membrane by epitaxy followed by hydrogen annealing is presented.

The main technological steps that are involved in the realization of a silicon crystal membrane are hereafter summarized and displayed in Fig. 2.14:

(a) Mask Pillar.
(b) DRIE silicon etching.
(c) Pillar cleaning (polymer removal).
(d) Pre epitaxial growth cleaning.
(e) Epitaxial Growth.
(f) Cleaning.
(g) Hydrogen Annealing.

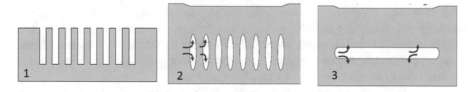

Fig. 2.14 Schematic of membrane formation: (1) Pillar definition;(2) Epitaxial Growth; (3) Silicon Migration by high temperature H2 annealing

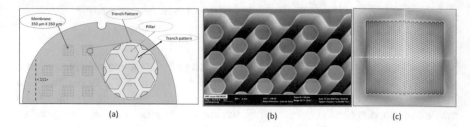

Fig. 2.15 Pillar array: (**a**) mask pillar layout; (**b**) SEM image of pillars after Deep silicon etching; (**c**) shape of membrane before Epitaxy

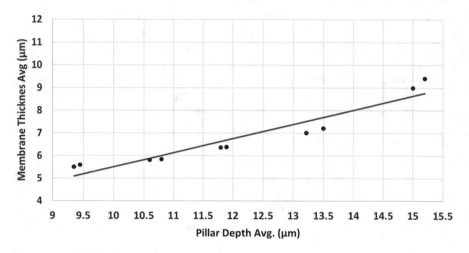

Fig. 2.16 Dependence of Pillar depth on final membrane thickness. Epitaxial process at reduced pressure

Starting material is a 200 mm substrate grown by Magnetic Czochralski (MCZ) method for low interstitial oxygen content, N-type doping in the range 1E14–1eE16 atoms.cm^{-3}, (100) orientation, with device dicing along <110> direction.

Mask pillar: An array of pillars defines the final size of the suspended silicon membrane. Pillars' size is designed to assure a uniform sealing and allow a complete pillars' migration during H2 migration. (Fig. 2.15). Square Membranes with sides in the range of 100–800 μm has been characterized.

Trench Etch: Pillars are defined by Deep reactive ion etching of silicon using an inductively coupled plasma source, the Bosch process [28]. The balance between etching and sidewall passivation allows to obtain well-controlled pillars' profile. Tight control of trench depth and width is needed to guarantee a uniform membrane's formation, the required thickness, and the correct cavity shape (empty space or gap). Aspect ratio from 10:1 to 15:1 is usually set.

The strong dependence of the trench depth with the final thickness of the membrane after the hydrogen annealing is shown in Fig. 2.16. Deeper trenches provide thicker membranes.

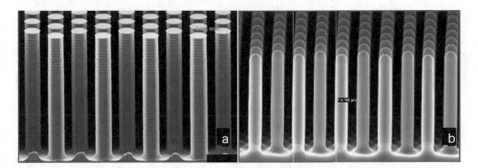

Fig. 2.17 Pillars' transformation: (**a**) after DRIE; (**b**) after hydrogen bake step during epitaxial process

Proper process cleaning sequences need to be introduced to efficiently remove polymers generated during the DRIE Bosch process, remove metallic contamination and prepare the surface just before the epitaxial process. A thin hydrophilic carbon free oxide is formed [29, 30] and successively in situ removed by a H_2 pre-bake at high temperature during in the first stage of the epitaxial process.

With the epitaxial growth starts the modification of the shape of the trenches and the sealing of the array of pillars. The required deposition time for complete sealing is related to epitaxial parameter like temperature, growth rate, pressure, and to pillar's geometrical dimensions and spacing.

During H_2 pre-bake at high temperature, silicon migration starts from the surfaces where the radius of curvature is the smallest driven by the surface energy minimization. Corners are consequently rounded, and sidewall smoothed as already mentioned in paragraph 5.3. Figure 2.17 shows the first stage of the pillar transformation with hydrogen bake.

Proceeding with the growth, the structure is sealed, the transformation of the underlying cavity continues to evolve sustained by the hydrogen flow used as carrier, by the hydrogen trapped inside the cavities and by the temperature.

Silicon migration inside the sealed structures is temperature- and time-dependent. Moreover, combined mechanism of convective and conductive heat transfers to the wafer mainly due to the different methods of heating and gas flow rate, play an important role in silicon migration.

Relevant differences have been found between reactors with induction heating or with radiant heating as well as vertical furnaces.

The selected temperature and pressure for the epitaxial process establish in agreement with the ideal gas law the final residual pressure inside the formed cavities. For instance, as reported in Table 2.2, growth at atmospheric pressure in the temperature range 1050–1185 °C allows to achieve a final pressure of 200–255 mbar, while at reduced process pressure of 20–60 Torr, a final residual pressure of 5–18 mbar will be reached.

Epitaxial sealing process is usually performed at reduced pressure to improve sensitivity of the devices. The lower pressure achievable in a single wafer epitaxial

Table 2.2 Final pressure inside cavities for different epitaxial process parameters

Process	Pressure (Torr)	Temperature (°C)	Final Pressure (mbar)
ATM	760	1185	205
ATM	760	1050	226
RP	20	1150	5.5
RP	20	1050	5.95
RP	60	1050	17.6
RP	60	1150	16.6

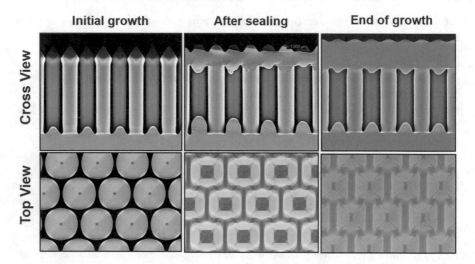

Fig. 2.18 Membrane formation and evolution at different step of the epitaxial growth at reduced pressure. SEM Top view and Cross Section

reactor is 20 Torr and deposition temperature is in the temperature range of 1050 to 1150 °C using Trichlorosilane as silicon precursor.

When atmospheric pressure process is used, temperature can be increased up to 1190 °C.

The transformation of the pillar's array during the epitaxial process at reduced pressure is shown in Fig. 2.18.

The pressure during the epitaxial growth determines the features of the structure as membrane thickness, empty space size (cavity), and the step height difference between membrane and outer membrane.

At atmospheric pressure, the larger amount of material is preferentially deposited above the substrate surface and on the pillar sides rather than on the top. Thicker membranes and thinner cavities are formed with a large (>2.5 um) step height at the membrane edge. At low pressure, the preferential growth on top of pillars leads to increased cavity size and to step height substantially reduced to less than 0.5 um. The low-pressure process allows to reduce step at the membrane edge favoring the

Fig. 2.19 Evolution of membrane formation during H2 annealing with time

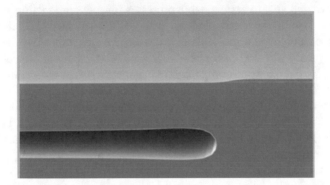

Fig. 2.20 Membrane after epitaxial growth and high temperature hydrogen annealing

following lithographic process steps; the increased cavity height provides increased margin to membrane deflection when net pressure acts on the device.

After sealing, the buried pillars are completely rearranged by silicon migration in hydrogen atmosphere. Silicon is redistributed between the membrane and the substrate determining final thickness of the membrane and size of the cavity (gap).

Additional high temperature annealing at atmospheric pressure in hydrogen atmosphere is performed after epitaxial process. A 100% H_2 annealing in the range temperature 1150–1200 °C are commonly employed. In Fig. 2.19 is shown how membrane formation proceeds with time by silicon migration and pillars thinning during annealing.

A cross section of the membrane at the end of the process is shown in Fig. 2.20.

2.4 Applications and Process Generalities of Thick Epitaxial Polycrystalline Growth

2.4.1 Introduction and Application

To realize MEMS devices, different silicon micromachining technologies can be used: bulk micromachining and surface micromachining. In bulk micromachining technologies, anisotropic silicon etching (e.g., KOH, TMAH) processes are mainly employed. In surface micromachining [31] the substrate is mainly a mechanical support and the mechanical structural layers realized above the silicon substrates are released by the removal of a sacrificial silicon oxide layer.

At the beginning of 1980s, researchers of Berkeley's University, California, realized the first polycrystalline silicon microstructures manufactured using surface micromachining technology. [32].

Polycrystalline silicon is the most common structural material in MEMS technologies, and it is used for a huge variety of applications. The deposition technology is well known and able to assure stable processes for mass production.

For both CMOS, Discrete devices, and MEMS technologies, thin films of amorphous or polycrystalline silicon (polysilicon) are commonly deposited by LPCVD. Typically for 200 mm wafer processes, vertical furnaces are used. Film is deposited by gas phase decomposition of silane (SiH4) at temperatures ranging from 525 to 630 °C and pressures from 200 to 800 mTorr [31, 33–36].

Different methods for Polysilicon doping have been developed to produce conductive layer: by Solid source diffusion, by liquid source as phosphorous oxychloride ($POCl_3$) [37], by ion implantation or by in situ doping during LPCVD process deposition.

The first two methods were diffusely employed in 100 mm and 150 mm wafer processing. Ion implantation requires an additional annealing to electrically activate the implanted dose. In situ doping is obtained by incorporation of dopant gas like diborane (B_2H_6), arsine (ASH_3), or phosphine (PH_3) during LPCVD process. State of the art LPCVD vertical furnaces for 200 mm substrates configured with multi-injectors to compensate depletion of gases along the tube, can provide polysilicon film with good uniform doping, very tight control of thickness and residual stress.

A more comprehensive treatise on polysilicon deposition processes and electrical and mechanical properties can be found in Chap. 3 dedicated to thin films deposition techniques.

Amorphous or polycrystalline silicon film by LPCVD is widely used for different applications:

- Buried interconnections (for inertial sensors).
- Filling of deep trenches in Through silicon Vias technologies.
- Seed layer to promote on oxided wafers growth of polycrystalline silicon by epitaxy (EPIPoly).
- In Microfluidics, polysilicon can be used for making channel walls and for sealing etched channel structures (Lab on Chip).
- Polysilicon resistors.
- Realization of thin membranes with strict requirements in terms of thickness and residual stress (acoustic sensing membranes for MEMS microphones).

The transition from amorphous to crystalline typically occurs around 570 °C and even at lower temperatures by increasing the phosphine doping concentration. The crystalline orientation of the polysilicon grains is also dependent on temperature; at 620 °C, grains are preferentially {110} oriented, the grains are large and have a columnar structure aligned perpendicular to the plane of the substrate.

Residual stress in polysilicon layer by LPCVD can be tuned and controlled using specific deposition process conditions. Layers with high compressive or tensile residual stress can be obtained. After deposition, residual stress can be fine-tuned and reduced with an additional thermal annealing process. Typically, compressive film in the order of 300–380 MPa and deposited by LPCVD at 600–625 °C, after high temperature annealing, can reach up to − 10 MPa, while film with tensile stress film in the order of 300–400 MPa deposited at 550–590 °C; stress can be reduced to neutral stress or turn to compressive values after high temperature annealing.

LPCVD polycrystalline silicon deposition is a slow process in terms of deposition rate and, when in situ doped with phosphorous, its growth rate is further reduced. Growth rate from 12 nm/min for polycrystalline silicon deposited at 620 °C is reduced to 4 nm/min with a doping concentration of 1×10^{19} cm^{-3} and to less than 1 nm for amorphous silicon doped up to 2.5×10^{20} cm^{-3} at 525 °C. In terms of process throughput, for example, an undoped polysilicon layer of 2 um requires a deposition time of 7 h for batch and at least 15 h if doped.

Consequently, polysilicon depositions by LPCVD for thick structural layers (in the range 2–5 μm for application like micro actuators and in the range 15–35 μm for inertial sensors) cannot be proposed for the low throughput of the process. Moreover, the high stress gradient and residual stress values obtained by LPCVD deposition, are not always suitable for inertial sensor devices.

The use of conventional horizontal single wafer or batch epitaxial reactors allows the growth of both relatively thin layers (1–4 μm) and thicker layer of 4–60 μm with acceptable growth rate and reasonable process throughput.

The term EPIPpoly is applied to a polycrystalline silicon grown at high temperature in an epitaxial reactor.

Polycrystalline silicon film by epitaxy has proven a valid and low-cost alternative to the use of Silicon on Insulator (SOI) substrates in MEMS manufacturing technologies. It is mainly employed as

- structural layer for inertial sensors (gyroscopes and accelerometers). Typical thicknesses are in the range 15–70 μm.
- thin membranes for MEMS PMUTs (Piezoelectric Micro machined Ultrasound Transducer), loudspeaker and piezo inkjet μactuators. Typical thicknesses are in the range 1.5–15 μm.
- structural layer combined with SOI substrates for the realization of μmirror devices using MOEMS technology (Micro-Opto-Electro-Mechanical Systems). Typical thicknesses are 40–60 μm,
- cap sealing of released structures at controlled pressure (e.g., resonators, capacitive pressure sensors, micro vacuum tubes) [38].

2.4.2 Process Generalities

EPIPpoly process have been mainly developed in atmospheric epitaxial reactors using trichlorosilane chemistry with a deposition rate as high as 5 μm/min at temperatures in the range 1050–1190 °C. Silicon precursor like dichlorosilane can be also used limiting the deposition temperature to 1100 °C.

A thin layer of polysilicon, namely, a seed layer is previously deposited using SiH$_4$ as silicon precursor at lower temperature. Chlorosilanes precursor cannot be directly employed because the supplemented etching action of hydrochloric acid produced from the source gas reaction limits the polysilicon nucleation on the oxide layer favoring only random agglomerates into polysilicon islands.

This layer, instead, ensures a uniform nucleation and growth of EPIPoly on oxide-coated wafers.

Both the seed layer and the EPIPpoly growth process are mass transport limited in the typical range of high temperature employed and consequently the growth rate is relatively stable with temperature.

Conductive structural layers in the order of 5–50 mΩ•cm (N-type) are required to assure an electrical connection between the layer and buried interconnection in inertial sensor. Conversely, intrinsic polysilicon layers can be used as membranes in several μactuator devices.

Different methods can be applied to reach the required level of doping. Thin seed layer with intrinsic resistivity can be doped by ion implantation followed by an annealing for the full activation of the dopant.

Seed and thick EPIPoly layer may also be doped successively in a diffusion furnace by liquid sources such as phosphorous oxychloride (POCl$_3$) [39] and treated at high temperature in inert atmosphere to diffuse the dopant along the entire thickness of the layer.

On 200 mm substrates, the EPIPpoly process is mainly in situ doped and all the layers entirely integrated in the epitaxial reactors.

In Fig. 2.21, the resistivity profile along the thickness of the EPIPpoly is shown and compared with the same process grown on mono silicon crystal. Resistivity on polysilicon is two times higher than the resistivity on mono silicon crystal. The significant reduction of conductivity of polysilicon is mainly caused by dopant segregation and carrier trapping at grain boundaries of the film.

The main precursors, the typical temperature, and growth rate used for the EPIPpoly processes (seed layer and high temperature poly are summarized in Table 2.3.

Polycrystalline seed layer, mainly in situ doped, can be deposited by LPCVD, or directly grown at lower temperature in the epitaxial reactor (integrated process).

In the first case, the poly seed layer will be defined by lithography and etched from the region of the wafers where the oxide has been already removed (usually where alignment marks are defined).

The result will be a uniform growth of EPIPoly on the oxide and a uniform growth of mono crystalline silicon in the region where the oxide has been removed.

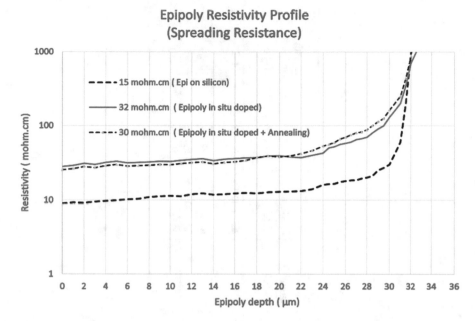

Fig. 2.21 Resistivity profile by Spreading resistance (SPR) of in situ doped EPIPoly layer

In the transition regions, mono-crystalline and poly-crystalline growth domains merge forming a distinct front at 54.7 ° angle, which is the angle between the (100) and (111) crystalline silicon planes. Figure 2.22 show as example a SEM cross-section where is clearly visible the front of growth between monocrystalline and polycrystalline Silicon.

Temperature profiles of EPIPoly recipes at atmospheric pressure in single wafer reactor (radiant heating) for film thickness < 20 um (a) or thicker thicknesses (b) are shown in Fig. 2.23.

The main steps, such as chamber etching, bake, seed layer deposition, and poly temperature deposition, are highlighted. Fast temperature ramps in the order of 120–180 °C/min and growth rate in the range of 3–5 μm/min are applied. Loading and unloading temperature is around 700–800 °C.

In single wafer reactor, when high temperature deposition is required, an intermediate chamber cleaning step is carried out at half of the deposition with film thicker than 20 μm. In this way, excessive coating buildup, degradation of the quartz chamber quality and particles generation can be reduced.

Temperature deposition is selected to meet the final residual stress requirements and to avoid any pattern distortion of alignment marks required for next lithographic steps. When no alignment marks in silicon are present, or different alignment approaches are used, EPIPoly temperature deposition is significantly reduced as for the realization of thin EPIPoly membranes.

Table 2.3 Silicon precursors and typical process parameters for EPIPoly process

Layer	CVD	Precursor	Thickness (μm)	Growth Temperature (°C)	Growth rate (um/min)	Pressure	Dopant Type	Resistivity (mΩ•cm)
Integrated seed	VPE	Silane	0.3–1.2	800–980	0.1–0.4	ATM	N type PH3	10–50
Seed	LPCVD	Silane	0.45–1.2	620	0.012	200–800 Torr	N type PH3	0.4–2
Epipoly	VPE	Dichlorosilane	2	1080–1100	2	ATM 60–200 Torr	N type PH3	10–50
Epipoly	VPE	Trichlorosilane	1.5–60	1080–1190	2–5	ATM	N type PH3	10–50

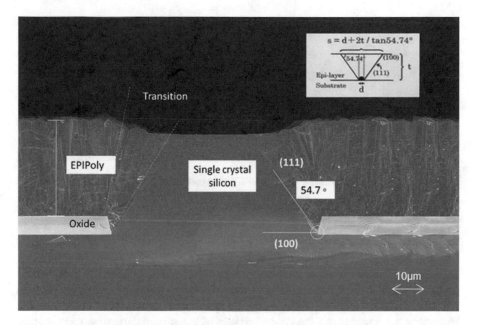

Fig. 2.22 SEM cross section of Epitaxial growth in the transition region with mono-poly interface at 54.7 °

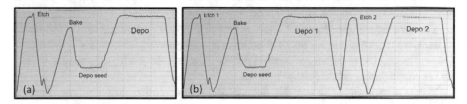

Fig. 2.23 Typical high temperature EPIPoly recipes in single wafer reactor for film thickness < 20 um (**a**) or higher thicknesses (**b**). An intermediate chamber etch is carried out to reduce excessive buildup of silicon in the chamber

Temperature profile of EPIPoly process in a batch reactor with induction heating (LPE PE3061) is shown in Fig. 2.24. Typical ramp up temperature is 25–30 °C/min and growth rate in the range 1.8–2.2 μm/min. Wafers are loaded at relatively low temperature.

2.4.3 Mechanical Properties of Thick Polycrystalline Silicon

Mechanical properties of polysilicon are excellent. Young's modulus and rupture resistance of polycrystalline silicon are comparable to materials like steel. Its density is less than aluminum's. Silicon has high thermal conductivity and small thermal

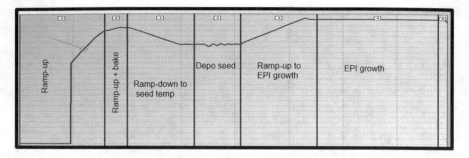

Fig. 2.24 Typical epitaxial recipe for thick EPIPoly on batch reactors with induction heating

expansion coefficient. The excellent thermal properties make it a very good material for high temperature applications.

Polysilicon has properties like single crystal silicon, shows excellent adhesion and chemical resistance, and by means of doping is possible to modify its electrical properties. Since polysilicon is an aggregate of small single crystal domains (grains) whose orientation changes with respect to each other, its properties depend on the behavior of the grains, on the characteristics of grain boundaries, on their shapes, and preferential crystal orientation.

Process parameters, such as temperature, pressure, doping level, and growth rate, affect the final properties of the material.

The measure of these properties is a very difficult and challenging task in terms of measurement techniques and methodologies. Significant differences in the elastic properties and the nominal strength of polysilicon were found by several researchers and reported in literature [40, 41].

Young's modulus (E) can be determined using several test devices. Among others: tension tests, bending of cantilever, resonant devices, bulge tests, buckling tests. The elastic modulus ranges from 132 to 174 GPa, and the fracture strength from 1.2 GPa to 2.7 GPa.

In Polysilicon layer, commonly used as cantilever or membrane, residual stress and stress gradient of the film are fundamental for the prediction of elastic, static, and dynamic behavior of structures with multiple degrees of freedom such as inertial devices. The residual stress is the result of the intrinsic stress of the deposited layer and thermal stress due to the Thermal Expansion Coefficient (CTE) difference between the different materials which is proportional to the temperature. [42].

Residual stress, can be easily calculated from the measurement of the substrate curvature (bow) before and after the film deposition and applying the Stoney equation [43]:

$$\sigma = \frac{1}{3} \cdot \frac{E_s}{1 - v_s} \cdot \frac{t_s^2}{t_f} \cdot \frac{\delta}{r^2}$$

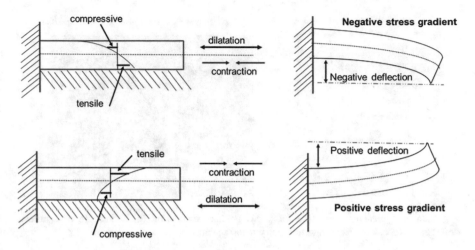

Fig. 2.25 Schematic of bended cantilevers with positive and negative stress gradient

where σ is the layer residual stress, Es is Silicon Young Modulus, vs is Silicon Poisson ratio, t_s is the wafer thickness, r is the wafer radius, t_f is the thickness of the deposited layer, δ is the wafer bow difference before and after deposition.

Stress gradient is the result of a non-homogeneous stress profile along the thickness of the film. It determines a deformation of the structures upwards or downwards once the structures are released.

In Fig. 2.25, a schematic representation of a free cantilever anchored to a wall is shown. If the structure is bent upwards, this means a positive stress gradient, while it bends downwards in the case of a negative stress gradient.

The stress gradient of an EPIPoly structural layer can be determined using appropriate test structures with cantilevers of different length. After release, the deflection of the cantilevers is measured using an optical interferometer as shown in Fig. 2.26.

The Δz deflection is correlated to the stress gradient (Γ) with the following relation:

$$\Gamma = \frac{2E}{l^2}\Delta z$$

and where Γ is the stress gradient, E the EPIPOLY Young Modulus, l the cantilever length and Δz the cantilever deflection.

Specific test structures fully compatible with the device process flow allow to determine the stress gradient by electrical measurements.

As shown in Fig. 2.27, two EPIPoly anchored structures with different length (l_1, l_2) and same width (w) are electrostatically actuated towards a reference poly backplate. The stress gradient is obtained from the pull-in voltage ratio of the

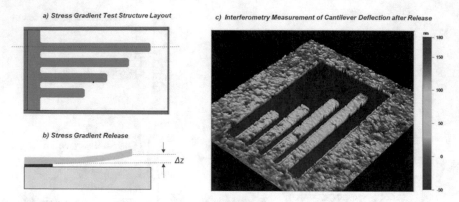

Fig. 2.26 A typical test structure for stress gradient evaluation (a) and image of cantilever deflection after release using interferometry measurement

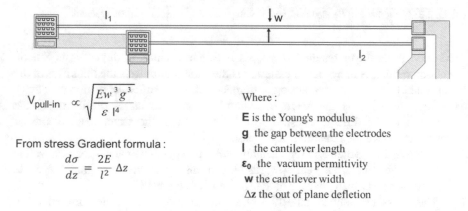

$$V_{\text{pull-in}} \;\propto\; \sqrt{\frac{E w^3 g^3}{\varepsilon\, l^4}}$$

From stress Gradient formula :

$$\frac{d\sigma}{dz} = \frac{2E}{l^2}\,\Delta z$$

Where :

E is the Young's modulus
g the gap between the electrodes
l the cantilever length
ε_0 the vacuum permittivity
w the cantilever width
Δz the out of plane defletion

Fig. 2.27 Test structure layout for stress gradient measurements

two cantilevers ($V_{\text{pull-in}} \; \alpha \; \Delta h$) and using the stress gradient formula ($\Delta h \; \alpha$ *Stress Gradient*).

As previously mentioned, residual Stress and stress gradient of the structural layer are important parameters and must be monitored for process control.

Factors influencing stress and stress gradient are the deposition temperature, the level of doping, the deposition rate, and the thickness of layer. Final stress of the structural layer is also influenced by thermal treatment when change of morphology and dopant distribution occur. Moreover, oxidation processes and an oxygen diffusion inside the polycrystalline layer determine a variation of the stress gradient.

Final residual stress of EPIPoly is strongly impacted by the growing temperature, the thickness as well as the type of epitaxial reactor, its thermal budget, and heating method.

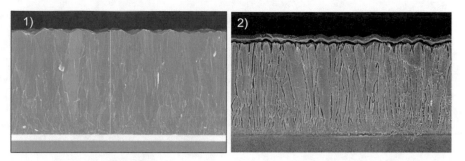

Fig. 2.28 SEM Cross section of EPIPoly layer; as grown (1) and after preferential etching (2). Columnar structure is highlighted

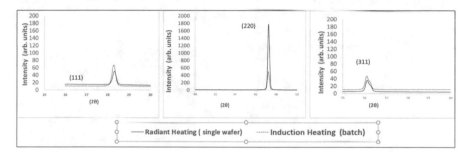

Fig. 2.29 XRD main peaks intensity of a 30 μm thick polysilicon layer grown on different reactors: Single wafer reactor (radiant heating) and batch reactor (induction heating)

The stress of the EPIPoly is influenced by grain structure. In the typical range of temperature employed to grow this layer (1120–1190 °C), the EPIPoly has always large polycrystalline grains with a columnar structure as shown in Fig. 2.28.

In high temperature growth of columnar structures, the initial heterogeneous nucleation at the interface exhibits slightly smaller average grain size with a high compressive stress. Increasing the film thickness, the columnar morphology develops with grains always larger in size and consequently, the stress will be less compressive.

The structural evaluation of EPIPoly films grown on different reactors has been performed by XRD. The EPIPoly processed in a single wafer reactor and in a batch reactor at high temperature (higher than 1150 °C) have very similar diffraction patterns, although a slightly different preferential orientation is present. EPIPoly samples have a strong [110] orientation out of plane. In single wafer reactor, EPIPOLY Si film is much more preferentially oriented than batch reactors (induction heating). A difference of the relative intensities of the diffracted maxima is also present, with higher (220) and lower (111) and (311).

The corresponding XRD (111), (220), and (311) peak intensity of 30 μm EPIPoly is reported in Fig. 2.29.

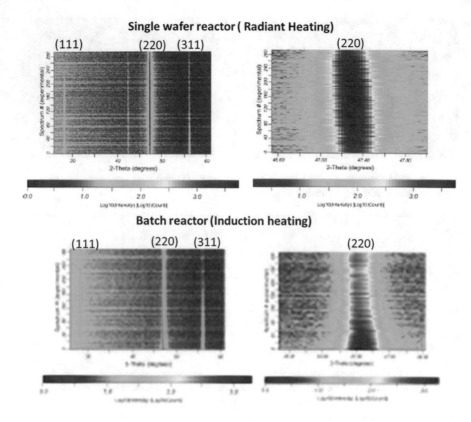

Fig. 2.30 Bragg Brentano plot (Left) and detail of the 220 peaks (Right) for thick EPIPoly samples processed in single wafer or batch epitaxial reactors

Figure 2.30 shows the results obtained analyzing samples processed in two different epitaxial reactors in Bragg Brentano configuration. The variation of intensity of each diffracted line is correlated with the different texturing of the EPIPoly layers. Bragg Brentano plot and a magnification of the (220) peak is displayed.

Visually, the (220) intensity is higher than the others and the samples processed on a single wafer reactor confirm to be much more preferentially (220) oriented.

These results, have been confirmed also for thinner EPIPoly films (7–15 μm) grown in the range temperature 1150–1200 °C. In EPIPoly with low temperature LPCVD seed layer, (311) and (111) peaks intensity is higher compared with epi seed layer. EPIPoly without any seed has the lower relative intensity of (111) and (311) planes (Data not reported here).

In Fig. 2.31, the dependence of residual stress with EPIPoly thickness for single wafer reactor (radiant heating) and batch reactor (induction heating) is shown. For both the reactors, increasing the EPIPoly thickness in the range 20–66 μm, the tensile residual stress increases.

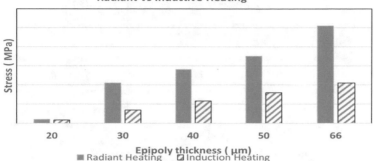

Fig. 2.31 Dependence of residual stress with EPIPoly thickness for single wafer reactor (radiant heating) and batch reactor (induction heating)

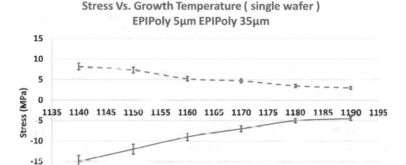

Fig. 2.32 Dependence of stress with growth temperature for thin layer 5 μm) and thick EPIPoly (35 μm) in single wafer radiant heated reactor

For the realization of MEMS devices, low stress EPIPoly structural layers in a wide range of thicknesses are used. In micro actuators, thin EPIPoly membranes are in the range 1–15 μm, while for inertial sensor, comb fingers structure in the thickness range 15–60 μm are required. In Fig. 2.32, the residual stress of thin EPIPoly layer (5 μm) and thick layer (35 μm) is shown. Thin layers always exhibit a compressive residual stress also at high growth temperature. In 35 μm EPIPoly layer, the stress, as previously mentioned, is always tensile and slightly dependent with temperature. Estimated stress vs. temperature sensitivity is +0.14 MPa/°C for thick EPIPoly and 0.25 MPa/°C for thinner layers.

The growth rate and consequently the process time contribution to the final residual stress is negligible for thick EPIPoly layers. When high growth rate is applied in thin film deposition, the compressive stress increases toward much more compressive values, as shown in Fig. 2.33.

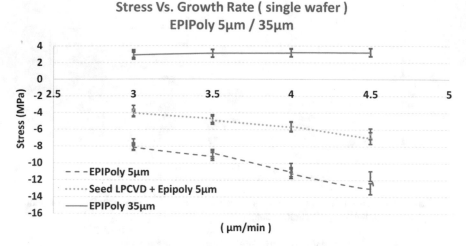

Fig. 2.33 Dependence of stress vs. growth rate in a single wafer reactor for thin and thick EPIPoly layers

2.4.4 Case Study: Displacement Optimization in Single Axis Accelerometer

Thermal treatments done after the EPIPoly growth can have a relevant effect on the EPIPoly structural layer. The combination of High temperature and oxygen atmosphere determine a large change in stress and stress gradient. The oxygen diffusion and incorporation inside the EPIPoly and at grain boundary develop a compressive stress with a gradient correlated with the oxygen concentration profile in the film. This effect is not significant at temperature lower than 1000 °C, but is relevant at higher temperature with an increase of the overall compressive stress and a shift of the stress gradient to negative values[[44].The possibility to fine tune the final stress and the stress gradient of the structural layer in different surface micromachined MEMS devices as accelerometers, gyroscopes, and in membranes for micro actuators, is of relevant interest and commonly applied to optimize the performances of the products.

By increasing the negative stress gradient, it is possible to modify the deflection of the released structures upwards, while increasing the positive stress gradient the structures will be deflected downwards. The case study shown in the following paragraph describe the optimization of the annealing to adjust stress and stress gradient in specific applications.

New MEMS products and applications involve the development of thick silicon accelerometers with dedicated specification in terms of sensitivity, noise, and power consumption. 60 μm EPI thickness is mandatory to realize sensors with high sensitivity and tight range for min/max sensitivity variation (0.2–0.7 pF/g) and with very low Brownian noise→ 10 ng/$\sqrt{}$Hz.

Single axis(Y) accelerometer (4 × 4 mm2) with a target in sensitivity of 2 pF/g has been developed with an EPIPpoly structural layer of 60 μm after CMP.

The EPIPoly film has been grown in batch reactor with induction heating at high temperature and atmospheric pressure in a single deposition. The seed layer grown using silane at low temperature, is integrated in the EPIPoly recipe where Trichlorosilane (SiHCl3 or TCS) is used as silicon precursor. Growth rate was 0.2 um/min for seed layer and 1.8–2.3 μm for the EPIPOLY. In situ doping with phosphorus equivalent to a resistivity in the order of 0.015–0.030 Ω•cm was required for both the layers.

Residual EPI-Poly stress is ranging between 2 and 5 MPa (tensile) with a positive stress gradient of +0.1 MPa/μm.

Surface roughness of the deposited layer is very high. Root mean square (RMS) is ≥900 nm. After a Chemical Mechanical Polishing process (CMP), surface roughness is reduced to about 5 nm.

EPIPoly sensing mobile masses have shown in the first prototypes, at the end of the process a concave shape with a minimum point at the center structure and low yield. In Fig. 2.34, X and Y profile of the structure is shown with a displacement of the structure of −1.2 μm. EPI-Poly mobile mass deformation is related to the positive stress gradient of the material (+0.1 MPa/μm) and structure design.

Correction of the final shape has been addressed optimizing temperature and atmosphere (oxygen/nitrogen ratio) of the EPIPoly annealing performed after CMP.

Displacement variation (in μm) between the EPIPoly fixed structure (used as a reference plane) and the EPIPoly mobile structure in function of the annealing was evaluated using an optical profiler.

A summary of the results as displacement of the structure in the temperature range 875–1050 °C and with a different oxygen content (0–10%) is reported in Fig. 2.35. Annealing time was kept constant at 90 min.

The residual stress becomes more compressive, increasing temperature annealing. The stress gradient, here not reported, shift towards more negative values. [34].

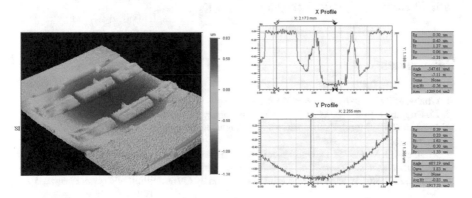

Fig. 2.34 X and Y profile after structure release using optical profiler (WYKO)

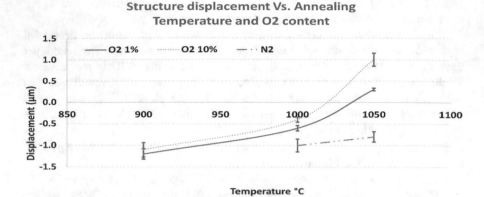

Fig. 2.35 Displacement of structures at different annealing temperature and content of oxygen

Fig. 2.36 Optical images of the structure after different thermal treatment

Only at temperature higher than 1000 °C, the oxygen starts to diffuse inside EPIPoly. Segregation and oxidation along the grain boundary leads to lattice expansion and to develop compressive stress.

A summary of the results as bending of the structure after annealing at 1050° C with different O_2/N_2 ratio is shown in Fig. 2.36.

References

1. Chang, & Sze. (1996). *Wolf and Tauber*.
2. D. Crippa, D. Rode, M. Masi (2001) *Silicon epitaxy, semiconductors and semimetals*.
3. Markku Tilli, Mervi Paulasto-Krockel. *Handbook of silicon based MEMS materials and technologies*.
4. Matthews, J. W. (1975). Defects associated with the accommodation of misfit between crystals. *Journal of Vacuum Science and Technology, 12*, 126.
5. Bloem, J. (1980). Nucleation and growth of silicon by CVD. *Journal of Crystal Growth, 50*(3), 581–604.
6. Crippa, D., Rode, D., & Masi, M. (2001). *Silicon epitaxy, semiconductors and semimetals* (Vol. 72). Academic Press, Now Elsevier. ISBN 0-12-752181-X.

7. ASM America, Phoenix AZ; www.asm.com
8. Applied Materials, Inc., Santa Clara, CA; www.amat.com
9. PE-3061 Product Brochure, LPE S.p.A., Via Falzarego, 8-20021 Baranzate (MI) Italy, www.lpe-epi.com
10. Grove, A. S. (1967). *Physics and technology of semiconductor devices*. Wiley.
11. SZE, S. M.; DEVICES, "Semiconductor Physics and technology", (1985).
12. Sze, S. M. E. (1981). Physics of semiconductor devices. In C. J. Hu et al. (Eds.), *A self-aligned 1 μm CMOS Technology for VLSI* (2nd ed.). Wiley.
13. Celler, G. K., & Cristoloveanu. (2003). Frontiers of silicon-on-insulator. *Journal of Applied Physics, 93*(9), 4955–4978.
14. Aspar, B., Bruel, M., Moriceau, H., et al. (1997). *Microelectronic Engineering, 36*(1–4), 233–240.
15. Reyes, M., Frewin, C., Ward, P. J., & Saddow, S. E. (2013). 3C-SiC on Si hetero-epitaxial growth for electronic and biomedical applications. *ECS Transactions, 58*, 119–126.
16. Ranjan, K. Yadav, Yogendra Gomes, Umesh Rathi, Servin Biswas, Dhrubes. "A strategic review on growth of GaN on silicon substrate for high power high frequency microwave & millimeter wave switch devices" (2012).
17. Semond, F. (2015). Epitaxial challenges of GaN on silicon. *MRS Bulletin, 40*(5), 412–417. https://doi.org/10.1557/mrs.2015.96
18. Kommu, R. S., Wilson, G. M., & Khomami, B. (2000). A theoretical/experimental study of silicon epitaxy in horizontal single-wafer chemical vapor deposition. *Journal of the Electrochemical Society, 147*(4), 1538.
19. Goulding, M. R. (1991). The selective epitaxial growth of silicon. *Le Journal de Physique IV, 2*(C2), C2-745–C2-778.
20. Morishima, K., et al. (1991). Japan Patent No. JP03123027.
21. Matsushita, Y., Samata, S., Miyashita, M., & Kubota, H. Improvement in thin oxide quality by hydrogen annealed wafer. In *Proceedings of 1994 IEEE international electron devices meeting*.
22. Gräf, D., Lambert, U., Brohl, M., Ehlert, A., Wahlich, R., & Wagner, P. (1995). Improvement of Czochralski silicon wafers by high temperature annealing. *Journal of The Electrochemical Society, 142*(9).
23. Voronkov, V. V., et al. (1997). *Proceedings of the Electrochemical Society, 97–22*, 4.
24. Kuribayashi, H., Hiruta, R., Shimizu, R., Sudoh, K., & Iwasaki, H. (2003). Shape transformation of silicon trenches during hydrogen annealing. *Journal of Vacuum Science & Technology A, 21*(1279).
25. Sato, T., Mitsutake, K., Mizushima, I., & Tsunashima, Y. (2000). Micro-structure transformation of silicon: A newly developed transformation technology of patterning silicon surfaces using the surface migration of silicon atoms by hydrogen annealing. *Japanese Journal of Applied Physics, 39*, 5033–5038.
26. Mizushima, I., Sato, T., Taniguchi, S., & Tsunashima, Y. (2003). Empty-space-in-silicon technique for fabricating a silicon-on-nothing structure. *Science & Technology A, 21*(4), 1279–1283.
27. Wong, Y.-P., Bregman, J., & Solgaard, O. (2017). Monolithic silicon-on-nothing photonic crystal pressure sensor. In *19th international conference on solid-state sensors, actuators and microsystems*.
28. Laermer, F. & Schilp. (1996). *Method of Anisotropically etching silicon*. U.S. Patent No 5,501,893.
29. Kern, W. (1970). Cleaning solution based on hydrogen peroxide for use in silicon semiconductor technology. *RCA Review, 31*(187–205).
30. Ishizaka, A., & Shiraki, Y. (1986). Low temperature surface cleaning of silicon and its application to silicon MBE. *Journal of the Electrochemical Society, 133*(4), 666.
31. Bustillo, J. M., Howe, R. T., & Muller, R. S. (1998). Surface micromachining for microelectromechanical systems. *Proceedings of IEEE, 86*(8), 1552–1574.
32. Madou, M. (1997). *Fundamentals of microfabrication*. CRC Press.

33. Zhou, H.-W., Kharas, B. G., & Gouma, P. I. (2003). Microstructure of thick polys-crystalline silicon films for MEMS applications. *Sensors and Actuators A: Physical, 104*(1), 1–5.
34. Michelutti, L., Chovet, A., Stoemenos, J., Terrot, J. M., & Ionescu, A. M. (2001). Poly-cristalline silicon thin filmsfor microsystems: correlation between technological parameters, film structure and electrical properties. *Journal of Thin Solid Films, 401*, 235–242.
35. French, P. J. (2002). Polysilicon: A versatile material for microsystems. *Sensors and Actuators A: Physical, 99*(1), 3–12.
36. Dana, S. S., Anderle, M., Rubloff, G. W., & Covic, A. (1993). CVD growth of rough-morphology silicon films over a broad temperature range. *Applied Physics Letters, 63*(10), 1387–1389.
37. Merck, Process guidelines for using phosphorous oxychloride as N-type silicon dopant, https://www.merckgroup.com
38. Ng, E. J., et al. (2012). Ultra-stable epitaxial polysilicon reasonators. *Solid State Sensors, Actuators, and Microsystems Workshop*, 271–274.
39. Elbrecht, L., Catanescu, R., Zacheja, J., & Binder, J. (1997). Highly phosphorus-doped polysilicon films with low tensile stress for surface micromachining using POCl3 diffusion doping. *Sensors and Actuators A: Physical, 61*(1–3), 374–378, ISSN 0924-4247.
40. Suo, Z. (2001). Fracture in thin films. *Encyclopedia of Materials: Science and Technology, 3290*.
41. Corigliano, A., Cacchione, F., & Zerbini, S. Chapter 13: Micro and nano mechanical testing of materials and devices. In F. Yang & J. C. M. Li (Eds.), *Mechanical characterization of low-dimensional structures through on-chip tests*.
42. Transducers. (1995). Highly sensitive internal film stress measurement by an improved micromachined indicator Structure. In *Proceedings of 8th International Conference on Solid-State Sensors and Actuators (Transducers '95)* (pp. 84–87).
43. Chou, T.-L., Yang, S.-Y., & Chiang, K.-N. (2011). Overview and applicability of residual stress estimation of film–substrate structure. *Thin Solid Films, 519*(22), 7883–7894.
44. Matthias, F., et al. (1997). Comprehensive study of processing parameters influencing the stress and stress gradient of thick polysilicon layers. *Micromachining and Microfabrication Process Technology III. International Society for Optics and Photonics*, 130–141.

Chapter 3
Thin Film Deposition

Roberto Campedelli, Luca Lamagna, Silvia Nicoli, and Andrea Nomellini

3.1 Low Pressure Chemical Vapor Deposition (LPCVD)

Low Pressure Chemical Vapor Deposition (LPCVD) is a chemical deposition technique largely used in the fabrication of several semiconductor technologies [1] and, in this respect, MEMS are no exception [2]. This section of the chapter is dedicated to the general aspects of LPCVD and describes the deposition of polysilicon and silicon nitride films. Case studies that illustrate the application of LPCVD films in MEMS device are also included.

3.1.1 Process Generalities and Equipment

The advantages of LPCVD for MEMS fabrication process resides in the ability to deposit high quality thin films [1–3] in a wide range of thickness (from hundreds of angstroms to few microns) and with tunable material properties (resistivity, residual stress) maintaining high process repeatability. Low deposition rate (from 1 nm/min to 20 nm/min) is balanced by processing up to 100–150 wafers simultaneously, thus resulting in acceptable throughput. LPCVD technique is based on the thermal decomposition of inorganic precursor to form the desired material. Amongst the most used precursors there are Silane (SiH_4), Disilane (Si_2H_6), Ammonia (NH_3) for silicon and silicon nitride deposition, phosphine (PH_3) and diborane (B_2H_6) for doped polysilicon; Tetraethyl orthosilicate (TEOS) is used for silicon dioxide.

R. Campedelli · L. Lamagna (✉) · S. Nicoli · A. Nomellini
ST Microelectronics, Analog MEMS and Sensors Group, MEMS Technology and Design R&D,
Agrate Brianza, Monza Brianza, Italy
e-mail: roberto.campedelli@st.com; luca.lamagna@st.com; silvia.nicoli@st.com;
andrea.nomellini@st.com

© Springer Nature Switzerland AG 2022
B. Vigna et al. (eds.), *Silicon Sensors and Actuators*,
https://doi.org/10.1007/978-3-030-80135-9_3

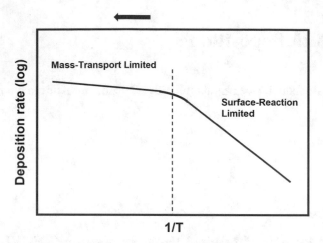

Fig. 3.1 Mass-transport and surface-reaction Limited predominance depends on the deposition temperature [1]

Precursors and carrier gases are continuously introduced in the hot wall reactor while unreacted gases and reaction byproducts are removed. The deposition process is influenced by the kinetic factors of the used gases and by the thermodynamics of the decomposition reaction which are mainly temperature and pressure. Gases diffuse inside the reactors via forced convection and then diffuse towards the wafer surface through the so-called boundary layer. Here the gases are adsorbed on the wafer where they bond to the surface, desorbing the reaction product that is mainly H_2. LPCVD processes are usually carried on between 500 °C and 700 °C, in pressure conditions of a few Torr to tens of mTorr. In this range of temperature and pressure, the reaction is surface reaction limited [1, 3], instead of mass transport limited [Fig. 3.1]. The reaction-limiting condition is crucial to obtain uniform film deposition along the wafers' diameter, allowing simultaneous process of multiple closely spaced wafers in a hot wall tube type reactor.

A typical LPCVD reactor, more commonly called vertical furnace, is schematically represented in [Fig. 3.2]. The reactor consists of a long vertical quartz tube which can accommodate up to 150 wafers. Alternatively, horizontal furnaces are still used in many fabs, despite their belonging to an old technological platform for this process equipment. The wafers are loaded inside a dedicated carrier, called boat, usually made in fused quartz or silicon carbide. The substrates are loaded horizontally on the boat, with a spacing of a few millimeters: the specific pitch vary based on the furnace's original design. The boat is placed inside the furnace tube thus closing the reactor environment at the same time. The system is heated using large resistivity heater elements positioned around the reactor tube. Temperature is a key parameter, it is controlled independently in different areas along the reactor tube length, usually up to 5 different zones. This allows the LPCVD process to have precise temperature control, in the range of ±1 °C, ensuring process uniformity and repeatability on all the processed wafers. Precursor's gases are introduced inside the

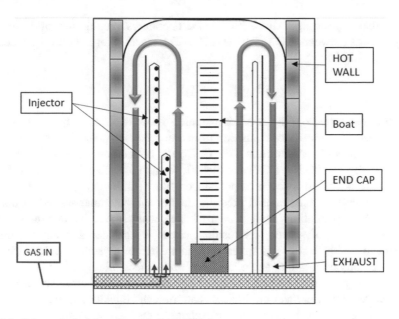

Fig. 3.2 Schematic drawing of a vertical LPCVD furnace

furnace via a gas manifold controlled by mass flowmeters. Manifolds are dedicated to specific precursors, with dedicated calibrated flow meter to ensure precise control of the species' partial pressure during the process. Process gases enter the reactors from the flange positioned at the bottom or top of the tube. More precise gas distribution can be achieved using injectors tubes connected to the main flange. This solution allows a more evenly disposal of the gases along the tube height; injectors designed are usually based on the LPCVD specific process and furnace hardware. A vacuum system is attached to the end cap opposite the gas injection flanges in order to maintain the desired deposition pressure. The typical vacuum system consists of a large, high-throughput rotary vane pump attached to an end cap through a vacuum line that contains both a pressure control valve and a vacuum isolation valve.

3.1.2 Polysilicon

3.1.2.1 Material Properties and Applications

Polycrystalline silicon, commonly abbreviated as "polysilicon", is extensively used in MEMS devices [1, 2, 4], as structural layer, to fabricate moving membranes or for electrical applications such as interconnections and TSV structures [5, 6]. MEMS applications require polycrystalline silicon with a thickness up to few microns; therefore, LPCVD is the more suitable deposition technique of choice. Alternatively, thicker layers can be grown in an epitaxial reactor. Polysilicon is deposited by thermal decomposition of silane (SiH_4) at temperatures ranging from 560 to 680 °C and pressures between 200 and 1000 mTorr. The reaction formula for the precursor decomposition is:

$$SiH_4(g) \rightarrow Si(s) + 2H_2(g)$$

Processes carried out at lower temperatures result in silicon deposited in amorphous phase; indeed, silicon transition from amorphous to crystalline phase in most LPCVD process happens around 570 °C. Crystal orientation is preferentially (111) and (311) for T<600 °C, (220) for T>600 °C, (100) at T>650 °C [7, 8]. Lower temperature depositions result in polysilicon film with bigger grain size, while at higher temperature, the material presents smaller and columnar-shaped grains. For many MEMS applications such as microphones, where the active moving structure is fabricated in polysilicon, its residual stress is considered a key parameter. Polysilicon residual stress is inherently bound to its crystalline microstructure, crystallography orientation, and grains size. From a technological point of view, the main process parameters which influence polysilicon residual stress are deposition temperature, pressure, layer doping concentration and thickness [9–12]. As general trend, low deposition temperature amorphous films have a compressive residual stress, same as polysilicon deposited at high temperature ($\geq$600 °C). Tensile stress can be obtained for layers deposited around the amorphous/polycrystalline transition phase (560–590 °C).

Conductive polysilicon layers are achieved using in-situ doping [13, 14]. During the deposition, phosphine (PH_3) or diborane (B_2H_6) are added to the main precursor silane, thus resulting in n- or p- evenly doped film. Incorporated phosphorus or boron are partially electrically activated if deposition is done at higher temperatures; however, complete activation is done by performing an additional thermal annealing at temperatures above 950 °C. Doping has an influence on deposition rate since boron slightly increases it, while phosphorus inhibits the silicon nucleation, thus resulting in slower depositions [15, 16]. On the other hand, phosphorus doping favors grain nucleation, lowering the amorphous/polycrystalline transition temperature around 560–570 °C, resulting in a bigger grain size if compared to undoped films. Phosphorus doped polysilicon deposited around 540–580 °C shows high tensile residual stress, which can be useful for many applications like sensing membranes or moving cantilevers [Fig. 3.3].

Another approach to fabricate conductive polysilicon films is to start depositing an undoped layer; doping is then achieved by using ion beam implantation or a solid source of dopant. A very common n-type dopant source is phosphoryl chloride ($POCl_3$). In this case, a doped phosphorus glass is deposited on top of the undoped polysilicon layer in dedicated furnaces. Afterwards, high temperature annealing is performed to promote the P diffusion from the $POCl_3$ glass throughout the polysilicon film thickness.

Another interesting use of polysilicon for MEMS fabrication is the so-called permeable poly. It consists of a thin polysilicon layer which, when deposited in tight specific pressure and temperature conditions, has a peculiar property. The

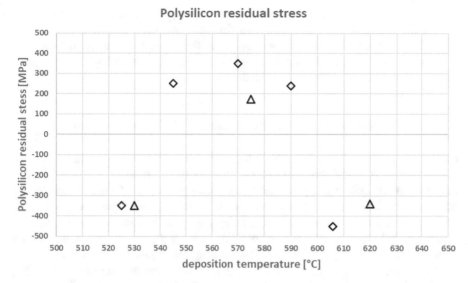

Fig. 3.3 Typical residual stress values for 1.0um thick polysilicon. Triangles represent undoped poly; diamonds are for n-doped layers (phosphorus, 1E19 – 1E20 atm/cm^3)

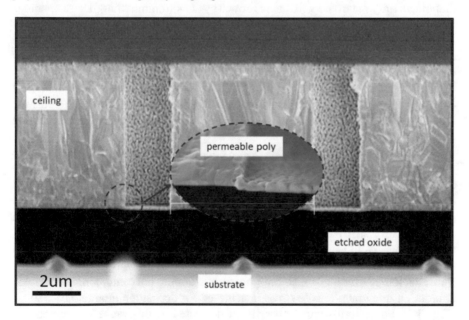

Fig. 3.4 Permeable poly can be used to fabricate buried cavity by means of oxide HF release

grains are not yet completely formed and coalesced, meaning there are holes (or pores) in between the grain borders [Fig. 3.4]. The polysilicon film maintains its structural integrity but its porosity at the grain border makes it permeable to HF vapor. This property allows to use the permeable poly to create buried cavity by means of depositing and patterning the permeable poly on a sacrificial oxide layer and etch it with an HF-based chemistry. After the oxide etch, the ceiling and the porous part that remains after the release process might be sealed depositing another film of polysilicon or alternatively a dielectric.

3.1.2.2 Case Study: Polysilicon Membrane for MEMS Microphone

Polysilicon plays an important role in the fabrication of MEMS capacitive acoustic microphones. A microphone is an electro-acoustic transducer realized for the detection of airborne sound pressure. Since sound pressure levels are several orders of magnitude lower than atmospheric pressure, microphones need very sensitive and thin membranes in comparison with diaphragm for traditional pressure sensors. The functional elements of a condenser microphone include a moving plate, a back-plate, an air gap, and a rear volume. The sensing membrane can be realized using a LPCVD polysilicon layer, which is able to move upon the transmission of the air vibrations that are associated to the sound. The back plate (the reference plate that acts as a counter electrode for the membrane movements) is designed as a rigid diaphragm fully patterned with holes formed by anisotropic etching. These openings let the air flow between the air gap and the rear volume, thus reducing the viscous resistance. Because of the movement, the capacitance between the membrane and another electrode is varied and it is then converted into an electrical signal. The back chamber allows the membrane movement and its excursion along the vertical axis. Figure 3.5a shows a SEM cross section with a close-up of the polysilicon membrane of a MEMS capacitive microphone. Membrane can be anchored to the substrate along all its perimeter; alternatively, another design could present the membrane anchored to the substrate only in some selected points.

Polysilicon sensing membrane acts as a drumhead and resonate with the acoustic waves. Accordingly, it requires by design a certain rigidity which is directly determined by the layer thickness and its residual stress, which, for instance, and depending on the device design, should be slightly tensile. Polysilicon residual stress is a key parameter for the correct functioning of the MEMS microphone. An excessively stiff membrane would not move properly, while on the contrary, a relaxed one would not vibrate properly [17, 18]. Tensile layer can be obtained with phosphorus doped polysilicon deposited around the amorphous/polycrystalline transition temperature. For medium phosphorus doping concentration, 1E19–1E20 atm/cm^3, a maximum tensile stress is achieved for deposition around 570–580 °C [Fig. 3.3]. Phosphorus-doped polysilicon deposited in this range of temperature presents grains of large size compared to polysilicon deposited at >600 °C (Fig. 3.5b and Fig. 3.5c) oriented preferentially along the (111) plane compared to (220) and (311). Depending on the exact deposition temperature and layer thickness, silicon can be not completely crystallized, with grains distributed inside amorphous areas. However, after a thermal treatment above 950 °C, it completely recrystallizes with the crystallographic properties described above.

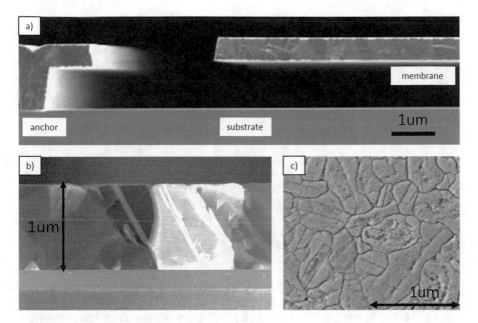

Fig. 3.5 SEM images of polysilicon layer used as moving membrane in a MEMS microphone. (**a**) is a cross section of the membrane, it is also recognizable as the anchor to the substrate; (**b**) is a TEM cross section, big grains microstructure is evident. (**c**) is a SEM top view where the poly grain boundaries are highlighted

Figure 3.6 shows the (111) preferential orientation of an n-doped (phosphorus, ~5E19 atm/cm^3) polysilicon layer deposited at 570 °C and then annealed at 1070 °C. The intensity peaks for the crystallographic spectra were obtained by XRD measurements performed in Bragg Brentano configuration.

Preferential orientation of the grains is summarized in Table 3.1.

Residual stress of the as deposited polysilicon layer is typically too high to be directly employed in MEMS devices. Target stress values for those applications are usually in the 0–100 MPa range. In this respect, the solution is to perform a thermal annealing in inert atmosphere on the wafer after the polysilicon layer deposition. With thermal treatments in the range of 950–1100 °C, the layer completely crystallizes and release some of its initial stress, thus shifting the residual stress value toward in the slightly tensile regime (0–40 MPa). This is a particularity of phosphorus-doped layers. Differently, undoped polysilicon layers usually present highly compressive stress values as deposited and then a shift towards neutral values (between −40 and −10 MPa) upon high temperature annealing [13]. It is important to underline that performing an annealing in the process flow directly after the polysilicon deposition could result in a compressive layer even when starting stress is highly tensile. In order to maintain the tensile stress, while lowering an overall residual stress from a starting point of hundreds of MPa, it is important to prevent the phosphorous outgassing from the film upon annealing. One possible

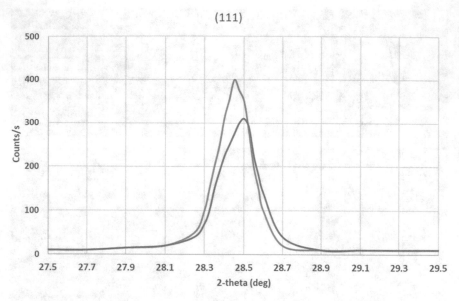

Fig. 3.6 XRD peaks intensity of a 1 μm thick phosphorus-doped polysilicon layer deposited at 570 °C (orange line), and after thermal annealing at 1070 °C (blue lines)

Table 3.1 Table reports relative peak intensity compared to silicon powder values (random orientation); preferential orientation along (111) plane is observed

XRD Bragg Brentano peaks intensity	Si Powder random orientation	Layer as deposited	Layer after annealing
Intensity (111)/Intensity (220)	1.8	7.7	8.4
Intensity (111)/Intensity (311)	3.3	7.7	8.5

solution to address this item is to deposit a thin oxide cap layer on top of the polysilicon film juts before the thermal treatment. This technical solution prevents the phosphorus outgassing at high temperature and maintains the phosphorous concentration in the layer. Figure 3.7 shows an example of polysilicon residual stress trend versus annealing temperature of 1.0 μm thick layer with an as deposited stress of ∼350 MPa. With an adequate tuning of the annealing temperature, annealing duration, and oxide cap thickness, it might be theoretically possible to target any residual stress value for a polysilicon film. It is worth noticing that oxygen impurities can also play an important role during polysilicon deposition or annealing since they can jeopardize the residual stress tuning. In fact, oxygen tends to be gettered by polysilicon accumulating at the grains borders and changing the overall layer stress [1]. For many MEMS application also, the layer stress gradient is a very important parameter that must be monitored during the fabrication process of a functional polysilicon layer. The stress gradient represents the variation of the polysilicon residual stress along the layer thickness. For fully anchored membranes, it has almost no impact on the device functionality. On the contrary, for applications

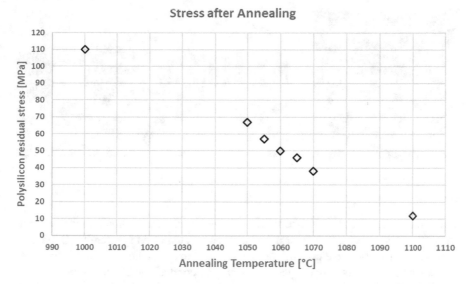

Fig. 3.7 Residual stress of a 1 μm thick, phosphorus-doped polysilicon layer deposited at 570 °C, after thermal annealing at different Temperatures (annealing time 30′). As deposited residual stress was 370 MPa

where the polysilicon is anchored only in few points, the stress gradient directly influences how the structure is bent upon the release process. Stress gradient values strongly vary with annealing temperature and time: longer annealing and higher temperature result in polysilicon with negative stress gradient values, while with lower temperature and short duration of the thermal treatment, the layer presents a positive stress gradient.

3.1.2.3   Case Study: Polysilicon Interconnection for Inertial Sensors

Inertial sensors products are one of the most diffused mass market application of MEMS technology. An inertial sensor is a device capable of sensing and communicating the movement via electric signal. Accelerometers can sense accelerations by means of electrical capacitance variations of capacitors sensitive to external forces. Gyroscopes, instead, can sense rotations along the axis, using a more complicated design which revolves around the Coriolis force. Capacitors are formed by one static plate and one moving plate, separated by air (Fig. 3.8). The released plate can move accordingly with the external accelerations, its relative position to the fixed plate produces a change in the capacitance. This electrical signal is collected and then communicated externally to the application-specific integrated circuit chip (ASIC). Plates are polysilicon masses and, depending on the architecture, they can be of variable thickness (10–60 μm). These polysilicon thick layers are not commonly deposited by LPCVD but, instead, they are grown inside epitaxial reactor at higher

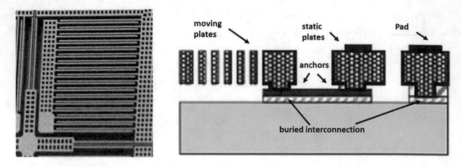

Fig. 3.8 Schematic cross section of a MEMS accelerometer

temperature (i.e., above 1000 °C). The moving masses are anchored to the substrate in the same way a cantilever or a spring is connected at one extremity. Other than mechanical function, the polysilicon masses need to act as the plates of a condenser, so they must have also an adequate resistivity to conduct the electrical signal.

Multiple capacitors made of parallel static/moving plates are present in a single die, in order to improve the electrical signal collected. The moving plates cannot be electrically connected with a standard metallization on top of the masses. One possible solution is to gather the electrical signal from all the plates at the same time and collect it in one metallic pad. This can be done using buried interconnection, which collect the electrical signals from all the anchors and bring it out of the MEMS moving body where the signal is then collected. This structure is like an underground railroad for the electrical signal, which tracks start from the anchors and travel below the suspended masses to reach the pads. The electrical interconnection can be fabricated in polysilicon and grown by LPCVD to benefit from a good thickness control and low resistivity of the film. Moreover, to avoid the sticking of the moving masses that can touch during their movements the buried interconnections, this polysilicon layer is deposited with a controlled surface roughness in order to reduce the contact area and the related adhesion forces that can promote undesired stiction [19]. Nevertheless, in LPCVD technology, a polysilicon layer with both low resistivity and high roughness is not a very common task. Low resistivity polysilicon is usually achieved by depositing a highly doped amorphous layer at T<550 °C and then crystallized with an annealing. However, this leads to a layer with low surface roughness. On the other hand, to achieve high roughness, a fully crystalline polysilicon is generally deposited at high temperature [20, 21]. Figure 3.9 reports RMS (Root Mean Square) roughness of phosphorus doped polysilicon deposited at different temperatures, measured with AFM technique. Roughness increases with deposition temperature, amorphous layer recrystallized by high temperature annealing results in RMS<10, still much lower than polysilicon deposited at T>600 °C. It is important to notice that roughness also varies greatly with layer thickness (Fig. 3.9). A polysilicon layer deposited at 610 °C of 200/600 nm thickness results in much lower roughness compared with the same layer 900 nm thick.

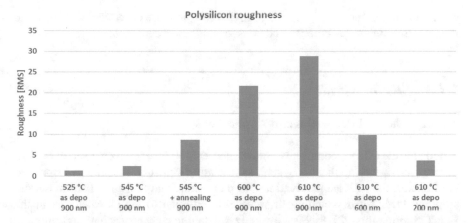

Fig. 3.9 Typical surface roughness for polysilicon layers deposited at different temperatures and thickness. Polysilicon is n-doped (phosphorus, 1E20 at/cm^3)

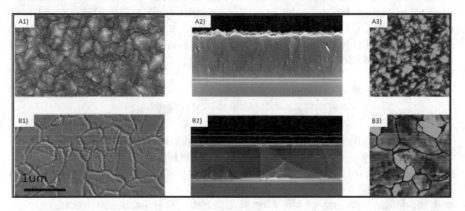

Fig. 3.10 SEM top view (A1), SEM cross section (A2) and AFM image (A3) of a 900 nm thick polysilicon layer deposited at 610 °C. B1, B2, and B3 are the same images for a polysilicon layer deposited at 545 °C + annealing at 950 °C. Both layers are n-doped (phosphorus, 1E20 atm/cm^3)

Figure 3.10 shows SEM images and AFM maps of a polysilicon layer with high roughness deposited at 610 °C, and of a layer deposited at 545 °C and then recrystallized with an annealing. Grains size and shape are appreciably different between the two layers. High temperature poly has smaller grains arranged in a columnar shape, while low temperature recrystallized poly has bigger, more squared grains. For this application, it is required to maintain low resistivity, namely, 1 mΩ•cm or less, even though it is not trivial to achieve it for poly films deposited at T>600 °C. Indeed, at this temperature, mass-transport limited regime overcomes surface-limited regime in the CVD kinetics, leading to fast reaction of silane and phosphine on the wafers in the reactor tube. This could cause depletion of gases (mainly phosphine) along the tube length, making it difficult to then achieve a good resistivity uniformity. A solution to address this issue is using injectors to distribute evenly the gases and, in addition, working at low pressure to reduce the deposition rate by inhibiting gases kinetics and consequently their depletion.

3.1.3 Silicon Nitride

3.1.3.1 Material Properties and Applications

Silicon nitride (Si_3N_4, commonly abbreviated as SiN) is extensively used in MEMS devices for various applications, such as insulating structural material in suspended membranes, final device passivation and protection from moisture, hard mask and stopping layer for dry etching or release, in addition to the classical use in the LOCOS formation [22]. SiN is extremely resistant to etchants such as concentrated HF, thus making it an appealing material for specific MEMS applications where oxides are used as temporary sacrificial layers. Silicon nitride is also frequently used as insulating layer thanks to its high resistivity of $\sim 10^{14}$ $\Omega \cdot$ cm and high dielectric strength. Very high-quality thin SiN films can be grown using LPCVD by the reaction of dichlorosilane (DCS, $SiCl_2H_2$) and ammonia (NH_3) at temperatures between 700 and 850 °C and pressures in the range of 100–300 mTorr. The reaction is:

$$3SiCl_2H_2 + 4NH_3 \rightarrow Si_3N_4 + 6HCl + 6H_2$$

Due to the high energy that is needed to decompose ammonia molecules, during deposition, the gas flow for ammonia is usually higher than DCS flow. Silicon nitride layer deposited on silicon substrate has typically high tensile stress. This intrinsic stress is caused by the discrepancy between the SiN layer volume units and the lattice constant of the silicon substrate. This mismatch forces the SiN structural units to extend vertically to match horizontally the silicon atom structure and thus causing high tensile stress in the order of 1 GPa or even more. This aspect limits the maximum thickness of SiN layer to few hundreds of nm because higher thicknesses may cause cracks in the film due to an excessive strain. Changing the SiN stoichiometry by depositing a silicon rich layer is used as a process-tuning strategy to change the material residual stress [23–25]. Silicon-rich SiN has lower stress values compared to the stoichiometric one: the stress can be varied from tensile to compressive. Stress in Si_xN_y film decreases with increasing DCS/NH_3 gas ration during deposition. Silicon-rich SiN is commonly also called Low-Stress silicon nitride (LSSIN). The clear advantage of LSSIN, once the risk of an excessive strain has been addressed, is the possibility to deposit higher film thicknesses for specific applications, up to a few microns. Other than lower stress, silicon rich SiN also presents lower etch rate in HF [Table 3.2]. Main drawbacks associated to the LSSIN are its worst performances as insulator, such as less resistance to leakage current, and the more challenging deposition process in terms of thickness and material properties reproducibility along the batch.

Table 3.2 Properties of LPCVD silicon nitride

	Standard Si_3N_4	Low stress Si_3N_4
Crystallographic phase	Amorphous	
DCS/NH3 gas ratio	1:3.5	>1:3.5
Residual stress	+1 GPa	From −200 MPa to +1 GPa
Max thickness	<0.3 μm	>3.0 μm
Etch rate in HF 40%, 25 °C	~100 Å/min	<100 Å/min
Electrical characteristics	Good insulating properties	Bigger leakage compared to std. nitride, increase with DCS/NH$_3$ ratio
Other characteristics	Biocompatible material, barrier to moisture and Na$^+$, and resistance to KOH and TMAH chemicals	

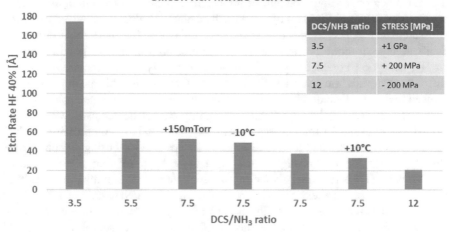

Fig. 3.11 Silicon nitride etch rate in HF lowers with higher DCS/NH$_3$ ratio

LSSIN properties such as stress and etch rate in HF solutions strongly depend on the DCS/NH$_3$ ratio, and in second order from deposition temperature and pressure. Figure 3.11 reports etch rates value in HF 40% for different nitrides varying the DCS/NH$_3$ ratio, from stoichiometry Si_3N_4 (1:3.5) to silicon rich SiN (1:12). Residual stress drastically decreases with DCS/NH$_3$ increase, from 1 GPa stoichiometric film to +200 MPa (1:7 ratio) and −200 MPa (1:12 ratio) for silicon rich layers.

3.1.3.2 Case Study: Low Stress Silicon Nitride for MEMS Microphone

MEMS capacitive microphones architecture requires in addition to a sensing membrane, a rigid backplate (Fig. 3.12). The backplate acts as the reference capacitor plate that evaluates electrically the movements of the sensing membrane.

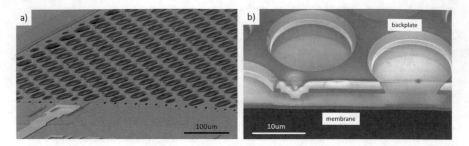

Fig. 3.12 SEM top view (**a**) and cross section (**b**) of a capacitive MEMS microphone. The holed backplate is recognizable above the membrane

The reference plate should be almost insensible to pressure variation in the audible range (from 20 Hz to 20 kHz) because the backplate should not be moved by sound pressure, since it acts as the reference capacitor plate. The main requirements to fabricate this structure are tensile residual stress, high thickness, good resistance to HF, and good conductivity [18]. In order to satisfy all the requirements, two different materials are used for the backplate fabrication. Thin polysilicon layer is used to grant adequate electrical conductivity, the film is n-doped (phosphorus) to target 1 mΩ•cm resistivity. Backplate rigidity cannot be achieved with a thick polysilicon because for thicknesses >1 μm poly tends to have compressive residual stress. High thickness and tensile residual stress are provided using silicon nitride. Stoichiometric Si_3N_4 has high stress but cannot be used with high thickness; therefore, LSSIN is the ideal choice for this application. Several microns of LSSIN with adequate tensile stress can be deposited using a DCS/NH$_3$ ratio of 1:7 without any film crack that could affect the backplate integrity. Silicon nitride grants the essential HF protection to resist the long sacrificial oxides release. Moreover, LSSIN also act as good electrical isolating layer to avoid parasitic currents in the capacitor.

3.2 Plasma Enhanced Chemical Vapor Deposition (PECVD)

Plasma Enhanced Chemical Vapor Deposition (PECVD) is a deposition technique that belongs to the large family of CVD processes. PECVD occurs when a plasma is created in a pressure-controlled environment to assist the film deposition while the chemical precursors are introduced in vapor phase into the heated reaction chamber. This technique is widely employed in microelectronics and semiconductor technology to deposit films with tailored properties for the realization of all the devices; in this respect, it can be considered a key deposition technique also for MEMS.

This section of the chapter begins with an overview of the general principles of PECVD and continues describing the properties of dielectric materials deposited by PECVD for MEMS fabrication. Selected case studies that illustrate aspect and applications of PECVD films in MEMS devices are also included.

3.2.1 Principles and State of the Art/Methodology/Equipment

In CVD, the film deposition takes place during a thermo-chemical reaction that occurs on the substrate surface and involves precursor molecules that are provided in vapor phase. Temperature, gas flows, and pressure are the main parameters that are influencing the deposition process. As already described in the previous section, CVD reactions generally occur at high deposition temperatures to provide the energy needed to promote molecules dissociation. Differently, in the PECVD depositions, the plasma is used to break the molecules and thus promote the deposition mechanism.

The main advantages of PECVD compared to conventional CVD processes is the use of relatively low process temperature (100 °C – 400 °C). The low thermal budget becomes a very important aspect for MEMS applications where the fabrication flow might present temperature sensitive features such as bonded wafers or functional materials that cannot withstand high temperatures processing. The pressure employed during the deposition is in the range of 1–10 Torr depending also on the reactor characteristics.

Moreover, PECVD can afford deposition of dielectrics with significantly high deposition rates that are typically in the range of $0.3 \, \mu m/min$–$1 \, \mu m/min$, but it can reach even few $\mu m/min$ in very high throughput configurations. The film properties are easily tunable through the process parameters to address any specific device characteristics and the related requirements in terms of thickness, refractive index, and stress.

In PECVD, the plasma is generated by electrical discharge between two electrodes using radio frequency (RF) generators. Plasma is a partially ionized gas consisting of equal numbers of positive and negative charges, and different numbers of neutral molecules. In a plasma, electrons disassociate molecules and generate free radicals that are chemically very reactive and thus increase chemical reaction rate in PECVD processes [26]. The reactant gas is partially ionized by electrons and the radicals that are generated react together, diffuse to the substrate, and form the film on its surface, thus resulting in a material deposition. Reactors can be equipped with two RF generators to benefit from a dual-frequency approach and improve the control on the properties of the deposited materials. Indeed, the use of a combination of high frequency (HF – i.e., 13.56 MHz) and low frequency (LF – i.e., 375 kHz) generators enable a direct control on the stress and the density of the film, thus providing the opportunity to obtain a wide range of different technological solutions that would not be available with a standard CVD deposition.

A capacitive plasma source, typically used in a PECVD reactor, is shown in Fig. 3.13.

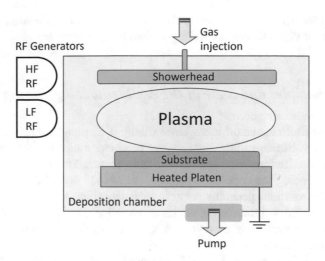

Fig. 3.13 Sketch of a PECVD process chamber

The two parallel electrodes are the two main components of one PECVD chamber: the showerhead and the platen, where the wafers are placed during the deposition. The showerhead and platen are two critical parts for the control of plasma deposition in any PECVD chamber. Indeed, the showerhead design affects the uniform distribution of gas into the chamber and thus directly influences the uniformity of the deposition process. On the other hand, the platen is involved in the heating of the substrate and at the same time is the housing of the wafer while the film deposition takes place. The combination of showerhead and platen design and the selection of the process parameters define the geometry and therefore the final shape of the plasma that is responsible for any PECVD deposition process [27]. The evolution of PECVD reactors design can afford a thickness non-uniformity below 1%, thus allowing to deposit films with an excellent control.

Nowadays there are many commercial PECVD reactors available on the market that have different hardware configurations that have been developed so far. Chambers can be lamp heated or electrically heated. In lamp heated deposition chambers, the external heat that supplies energy during the deposition is provided by infrared radiation emitted from lamps. In this configuration, the lamps heat the susceptor through quartz windows. A closed loop temperature control sub-system senses the temperature of the susceptor using thermocouples and regulates the temperature by varying the intensity of the lamps [28]. In an alternative configuration of the chamber, it is used an electric heater to heat the susceptor, and therefore the wafer. In both designs, the chamber body is generally made of aluminum and it is kept at a constant temperature.

Another hardware configuration consists in the use of a resistive heater coupled with a grounded platen electrode. The use of an air-cooled platen body maintains stable temperatures during high power process, it is suitable for fast cool down for the periodic module maintenance. The showerhead and the platen are typically made of aluminum or anodized aluminum, while the chamber shields are made of ceramics. These PECVD chambers can use either a direct or a remote approach for the cleaning processes of the chamber. Using a remote plasma turned out to be beneficial for all the chamber components, thus reducing the risk of damaging hardware parts due to ion bombardment such as occurs during the in-situ plasma cleaning. In more advanced PECVD reactors, there are available also alternative configurations such as the use of twin deposition chambers, remote plasma source (RPS) for the chamber cleaning and other hardware options that have been introduced targeting very high throughput for mass production applications [29].

A general PECVD process is not only represented by the actual deposition recipe but, more generally, it can be described referring to three phases. There is an initial pre-deposition that is performed without any wafer into the chamber and that has the aim to condition the chamber. Then the actual deposition takes place. This process starts with a first step of stabilization of the gas flows and pressure; afterwards, the wafer is introduced into the chamber and positioned on the platen and then the plasma strike occurs thus starting the deposition. The process ends with a final sequence of steps of pump and purge to evacuate all the process gases. After completing the deposition and removing the wafer from the chamber, the third phase is represented by a cleaning process that is run to remove all the accumulated deposition from the chamber sidewalls and all the other parts that have been exposed to the plasma. The cleaning step, which is an etch process, can be run on every wafer or after a variable number of wafers, based on the chamber and process types and depending on the deposited thickness. The built-up of deposition inside the chamber or a not consistent and repeatable cleaning approach can cause a change of the chamber impedance. Consequently, this would impact the RF coupling into the chamber thus causing an incorrect plasma formation, plasma instability, and a drift of the film properties. The cleaning step plays an important role and can be performed either by in situ cleaning based on the use of C_3F_8 and O_2 or using more advanced RPS systems with NF_3 and Ar plasma. The latter approach is faster, less aggressive for the chamber components, and therefore guarantees higher productivity associated to longer lifetimes for the hardware parts.

Amongst the family of the PECVD processes, there is the High-Density Plasma (HDP) CVD that is based on the use of high-density plasma (10^{11}–10^{12} ions/cm^3) created by inductively coupling approach. The high density of the ions enables to combine a sputtering component, using Ar, with the deposition component based on SiH_4 and O_2. This combination of deposition and in situ sputtering phases into the same process gives to the HDP-CVD its exceptional gap-filling characteristics without generating any undesired voids in the film (Fig. 3.14). HDP-CVD has been developed for the most advanced gap-filling requirements in advanced semiconductors and is used as filling deposition strategy for dimension <0.35 μm [30].

PECVD is widely employed in semiconductor industry as one of the reference techniques for the deposition of dielectrics that can be applied in shallow trench isolation, sidewall spacer, pre-metal and inter-metal dielectrics, anti-reflection coat-

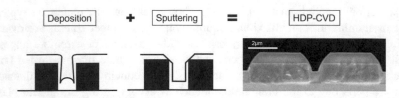

Fig. 3.14 HDP-CVD approach: deposition and sputter sequence

Table 3.3 Summary of PECVD films, deposition temperature range, and their application in MEMS devices such as sensors and actuators

PECVD film	Deposition Temperature (°C)	MEMS application
SiO₂ TEOS	100–400	Sacrificial layer, structural dielectric, electrical insulation, optical layer
SiO₂ USG	100–400	Passivation, structural dielectric, mechanical protection, electrical insulation
SiN	200–400	Passivation, structural dielectric
NH₃ free SiN	150–350	Passivation, structural dielectric
SiON	150–400	Passivation, optical layer
SiC	200–400	Mechanical protection, passivation
a-Si	100–400	Protective or sacrificial layers

ing and passivation applications. Nonetheless, PECVD processes are of paramount importance in MEMS application for their high deposition rates compared to the conventional CVD methods having at the same time the possibility to directly tune film properties such as the stress that play a very important role in MEMS devices such as sensors and actuators. Table 3.3 above summarizes the main applications of PECVD depositions in MEMS technology.

In MEMS devices such as the inertial sensors, the sacrificial dielectric plays the important role to act as mechanical spacer during the fabrication flow; they must be stable and withstand the whole thermal budget. Sacrificial films are SiO_2 layers deposited under structures that will become free to move only after the MEMS final release process step.

The dielectric films are commonly employed for electrical insulation of conductive structures and they are inserted as electrical spacer between active electrodes present inside the structure of device. PECVD films are employed for the final passivation of the active stack of the sensors or actuators. In this respect, SiN is the layer that provides the best performances in terms of resistance upon moisture exposure, demonstrating a very good stability of stress and film hermeticity along time. The low deposition temperatures of PECVD represents a key aspect for the use of this technique in MEMS fabrication because it can easily satisfy low thermal budget requirements that are characteristic for some specific cases [31]. For instance, during the deposition of the passivation for ferromagnetic materials, the temperature must be kept low enough to prevent the degradation of the magnetic properties. In this case, a combination of silicon oxide and nitride films deposited at 175 °C is employed as passivation solution. Furthermore, SiN films can be used to passivate and prevent the degradation of NiFe layers acting as a barrier to O_2 diffusion with respect to TEOS deposition.

In general, low deposition temperatures are used in all the process flows that require the control of the thermal budget under certain temperatures. Bonded wafers, depending on the bonding technology solution, can become unstable at temperatures higher than 200 °C. In this case, dielectric films are generally deposited using the low temperature PECVD processes. Those films can be used also to enhance adhesion between different materials; for example, thin silane-based layers might be used to promote the adhesion of thick TEOS films with noble metallic surfaces such as Pt or Au that are employed to realize the electrical pads of the devices.

A very important role is played by the PECVD films, deposited with a tailored stress, in MEMS structures, where the design of the device is based on the realization of a multilayer stack where every film must provide a specific contribution to the overall mechanical behavior. High stress PECVD SiN films represent the best solution to act as passivation layer, but also at the same time as mechanical component of the device. The principal dielectric thin films deposited with PECVD technology can be divided into Silicon Oxide and Silicon Nitrides, then there are other materials as Silicon Carbides (SiC) or Amorphous Silicon (a-Si). In the next section, these materials will be briefly described also using dedicated case studies that illustrate the use of the PECVD films in MEMS devices.

3.2.2 Silicon Oxides

3.2.2.1 Material Properties and Applications

There are two silicon precursors that are used for the PECVD oxide deposition: silane (SiH_4), which is a gas, and TEOS tetraethyl orthosilicate $Si(OC_2H_5)_4$, which is a liquid at room temperature. Both processes produce very high quality SiO_2 films; the two solutions can fit very well in specific applications and requirements. In MEMS technology, the SiO_2 films deposited using $Si(OC_2H_5)_4$ are commonly defined TEOS layers, while the SiO_2 films deposited using SiH_4 can be defined as SiO or Undoped Silicon Glass (USG). It is known that the conformality of the TEOS films is superior compared with that of the silane-based processes and this is due to the different nature of the two silicon sources.

The PECVD SiO_2 can be doped directly during the deposition with boron and phosphorous. The addition of phosphorous forms phosphosilicate glass films (generally called PSG). This results in a reduced stress in the film, which is associated to an improved step coverage [32].

The reaction for SiO_2 deposition using TEOS is:

$$Si(OC_2H_5)_4 + O_2 \xrightarrow{\text{plasma\&heat}} SiO_2 + \text{other volatiles}$$

Table 3.4 typical TEOS and
USG thickness and stress
range in MEMS

	Thickness range	Stress [MPa]
TEOS	50 nm – up to 2 μm	−200/+150
USG	100 nm – up to 3 μm	−80/+20

The addition of oxidant gas such as oxygen or O_3 to TEOS plasma improves the film quality resulting in a deposition with a low content of impurities, a smooth surface roughness, better conformality, and step coverage. TEOS is a high-quality silicon dioxide film which can be deposited with high deposition rate and very good conformality also at low deposition temperature [33]. TEOS layers are extremely useful, low-cost, versatile, and therefore they are used in almost all the fabrication flows of MEMS devices.

The silane-based oxide, the SiO or USG films, are deposited using SiH_4 in combination with N_2O as oxygen source; Ar and N_2 are also added to the plasma.

The reaction for SiO_2 deposition using SiH_4 is:

$$SiH_4 + 4N_2O \xrightarrow{\text{plasma\&heat}} SiO_2 + 4N_2 + 2H_2 + O_2$$

USG films are commonly employed in the passivation stack of MEMS devices or as structural dielectric layer with a controlled low stress.

Table 3.4 summarizes the range for thickness and stress of TEOS and USG films deposited by PECVD for common MEMS applications.

3.2.2.2 Case Study: Warpage Compensation in MEMS Fabrication

During MEMS fabrication, a serious issue could be represented by the uncontrolled wafer warpage that makes it problematic to continue in the process flow performing the following process steps. The uncontrolled wafer bow is due to differences in the coefficient of expansion of the thin films and substrate, thus resulting in residual stress during the cool-down from the deposition temperature. This effect can be the result of the superposition of several process steps and represents a major issue during MEMS fabrication. These residual stresses represent stored energy that can be released afterwards both with time and during following process steps. The total film stress is a function of film thickness, elastic modulus, and of the morphology and the density of the material. Furthermore, the stress may not be uniform throughout the film thickness or along the wafer diameter. Films under compression will try to expand resulting in a positive bow. The film will bend the substrate with the film being on the convex side. If the film has a tensile stress and a negative bow, the film will try to contract and bend the substrate in the opposite direction.

The difference in wafer bow is illustrated in Fig. 3.15.

Fig. 3.15 Compressive thin film generates a positive bow, while tensile thin film a negative bow

Table 3.5 Thickness and stress of different TEOS layer

TEOS type	Deposition temperature [° C]	Layer thickness [μm]	Layer stress [MPa]
(A) Low stress	175	1	−15
(B) Compressive	175	1.2	−145
(C) Thick and compressive	175	1.8	−145

The most common technique for measuring the total film stress is by evaluating the deflection (bowing) of a thin substrate on which the film has been deposited. By knowing the mechanical properties of the substrate and film material, the film thickness, and the deflection, the film stress can be calculated [34]. The stress, then, can be calculated using Stoney's equation.

$$\sigma = \frac{E}{1-\nu} \frac{t_s^2}{6t_f} \left(\frac{1}{R} - \frac{1}{R_0} \right)$$

Where $E/1-\nu$ is the biaxial modulus of the substrate, in our case silicon = 1.805 E12, t_s and t_f are the thicknesses of the substrate and film, respectively, and R_0 and R are the radii of curvature of the substrate before and after film deposition.

However, there is the possibility to recover and tune the wafer warpage by depositing different PECVD layer with engineered film properties such as thickness and stress. Table 3.5 summarizes some TEOS films that have been considered to perform warpage compensation.

The selection of TEOS layer can be combined with different wafer warpages acting as compensation layers. This application is very important because the specific deposition of a compensation TEOS film can balance the warpage of the wafer, thus making them flatter and therefore processable in the following steps. A certain warpage could be balanced with the deposition of a proper compensation film. A key aspect of this application is that TEOS can be deposited on both the front side and the back side of a wafer, thus compensating warpage in both directions. For example, if the wafer has initially a convex shape, there are two possible approaches to recover the warpage: a compressive layer can be deposited on the back side of the wafer, or alternatively, a tensile layer can be used on the front side for compensation.

The present MEMS case study considers warpage compensation that involves a 400 μm thick substrate which is stressed by a thick (about 100 μm) polysilicon epitaxial growth on the wafer backside. After the growth of the EPI layer, the wafer presents an unstable and very high bow above 200 µm, which makes it impossible to perform the next process steps. The solution adopted was to employ PECVD TEOS deposition to compensate the excessive wafer bow. Recovering the warpage by depositing the TEOS directly on the EPI surface had also the advantage to protect this layer from the following phases of the process flow.

Figure 3.16 shows the warpage recovery after the different TEOS depositions with the film's characteristics described in the Table 3.5.

A linear behavior is observed between the bow measured before and after the TEOS deposition, thus enabling to calculate an empirical formula to predict, according to the initial bow of the wafer and the type of deposition, the final wafer bow. The experimental results achieved for different cases agreed with the expected trends. Nevertheless, it turned out that there is also an important substrate effect that must be carefully considered. Indeed, in MEMS technologies, thin silicon substrates down to 400 µm can be used and this could directly affect the warpage compensation strategy.

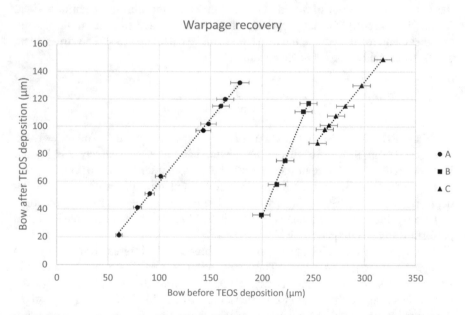

Fig. 3.16 Bow evolution after different types of TEOS deposition, starting with different value of wafer bow before deposition

3.2.3 Silicon Nitrides

3.2.3.1 Material Properties and Applications

Silicon nitride films deposited by PECVD can be carefully engineered in all its physical and chemical properties by tuning the deposition parameters, such as gas ratio, pressure, plasma density, and temperature. Depending on the specific application, it might be more critical to tune the refractive index or the film stress. Nevertheless, all the PECVD silicon nitrides do not have a stoichiometric composition, since in those films the Si/N ratio is generally about 1.12–1.15 compared to the 0.75 that is calculated for the stoichiometric Si_3N_4 film.

Precursors employed in PECVD processes are ammonia (NH_3), silane (SiH_4), and N_2, the reaction that takes place depending on the plasma chemistry are reported below:

$$SiH_4 + N_2 + NH_3 \xrightarrow{\text{plasma\&heat}} SiN_xH_y + H_2$$

$$SiH_4 + N_2 \xrightarrow{\text{plasma\&heat}} SiN_xH_y + H_2$$

NH_3-based SiN is the most common film amongst PECVD nitrides; there is a lot of knowledge on these processes, and this results in a very good control of those films. However, for some applications, low hydrogen content turned out to be a critical aspect and therefore the NH_3-free SiN has been developed. This SiN has a lower hydrogen content, lower wet etch rate, and is characterized by a better stability at lower deposition temperatures.

Typically, the hydrogen content of SiN films is measured with Fourier transform infra-red (FTIR) spectroscopy. In Fig. 3.17 there is a comparison of standard SiN vs. NH_3-free SiN.

As shown by the superposition of the spectra, the Si-N peak is positioned at 850 cm^{-1}, the Si-H peak at 2200 cm^{-1} and the N-H peak at 3300 cm^{-1}. In the band of interest (N–H stretching), large hydrogen concentration has important impacts. As the number of N–H bonds in the layers increases, the contributions from N–H•••N vibrational absorption increases as well. This bonding type takes place between the hydrogen atoms in the N–H bonds and lone pair electrons of nearby N atoms [35]. Therefore, from the FTIR spectra, it is possible to observe that SiN deposited using an NH_3-free plasma has low hydrogen content, since the N-H peaks is absent.

In the framework of optical applications, for instance, when the light is transmitted through a SiN layer, H concentration is very important because it is directly related to propagation losses. Minimizing the hydrogen content in the material consequently minimizes those losses and therefore the overall power consumption of the device during its operation.

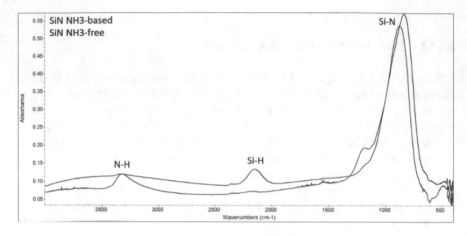

Fig. 3.17 Comparison of SiN NH_3-based vs. NH_3-free PECVD SiN

Table 3.6 Chemistry composition for different Silicon Nitride

SixNx	Deposition Temperature [°C]	Deposition Pressure [Torr]	Si at%	N at%	RI
LPCVD Si_3N_4	780	0.2	57.1 ± 0.2	42.9 ± 0.2	2
PECVD SiN NH_3-based	400	5	42 ± 0.2	58 ± 0.2	1.9
PECVD SiN NH_3-free	350	2.5	53 ± 0.4	47 ± 0.4	2.3

The requirements for low H% in the films are also the reason in trying to deposit films that have compositions close to the ideal (stoichiometric) Si_3N_4. If there is no excess of Si or N, hydrogen cannot bond into the film.

In terms of chemistry composition, it can be observed from Table 3.6 that ammonia-free SiN is close to a stoichiometric silicon nitride, while the ammonia-based SiN has a lower Si concentration. This explains the difference in the refractive index of the layers.

Silicon nitride is widely used as the solution for the passivation in MEMS, microelectronics, and optoelectronic industries for its excellent moisture diffusion barrier properties. The performance of MEMS device is strictly limited by fabrication issue such as the adverse effect of residual stress. In this respect, there are many applications where a precise stress control is required. Compressive stress may cause undesired buckling and bending of released micro machined structures: moreover, cracking may occur if the tensile stress is too high [36].

3.2.3.2 Case Study: High Stress Silicon Nitride for MEMS Actuators

During the fabrication of MEMS actuators, a key aspect is played by the thickness and the stress of each layer that is composing the final structure of the device. Actuators, such as inkjet printhead, autofocus, speakers, and micromirrors, are MEMS devices that, upon a proper actuation that can be provided from an external or an internal source, are able to move some of their parts significantly with an excursion of several μm. In this respect, the mechanical properties of each layer must be carefully addressed during the fabrication so that the design requirements can be fully satisfied, thus targeting the correct functionality of the device. This is the case of the silicon nitride passivation layer that is used as the final sealing of MEMS piezoelectric-actuated autofocus and speakers.

The description of those devices and the insights about their structure and their operations can be found in Chaps. 20 and 21 of this book.

This case study discusses the engineering of high stress PECVD SiN film. First, this layer must act as a passivation solution for the device. Therefore, it must be deposited with a controlled thickness and it should have a very good conformality on top of all the different geometries and topographies. This aspect is very well accomplished by PECVD SiN films. On the other hand, at the same time, the SiN layer acts as an important player during the device actuation. Indeed, SiN stress and elastic modulus are directly affecting the mechanical response of the whole actuator. The SiN film's mechanical properties take part in the definition of the bending of the MEMS after the release, when the actuator is resting free and suspended after the removal of the sacrificial layer that was keeping it frozen and stable. Moreover, during the actuation, the stress modulation of the SiN layer can directly affect the final performances in terms of excursion.

In this respect, the modeling and the simulations have demonstrated that the stress of silicon nitride must be carefully tuned: the use of a high compressive SiN has been required by design for piezoelectric autofocus and piezoelectric speakers. For this specific application, a PECVD process for the deposition of a SiN layer was developed at 350 °C obtaining an NH_3-free high compressive stress SiN. The stress target was set at -350 MPa with a required control on stress variability to be within the ± 30 MPa. In terms of process parameters, it is worth highlighting that this process is controlled by a single RF frequency; therefore, stress tuning is controlled by the high frequency RF power modulation. To achieve a high compressive stress, this HF power is set at high value in the range of 1800–3000 W.

The thickness target, depending on the application, is in the 300–600 nm range.

A design of experiment study was carried out to achieve the high compressive stress (-350 MPa). The effect of the PECVD process parameters on the deposition was thoroughly investigated. The dependence of thickness, stress, and RI on deposition power on variation of the pressure, HF RF power, and SiH_4 flow are shown in Table 3.7.

Figure 3.18 shows a SEM cross section image of the 300 nm thick high compressive SiN films deposited as passivation layer for the piezoelectric stack. The

Table 3.7 Summary of the influences of process parameters on the SiN film properties. (Stress decrease means that the stress is becoming more compressive)

	Thickness	Stress	RI
Pressure	↓↓	↓↓	↓↓
HF power	↑	↓↓	↓↓
Spacing	↓	↓↓	↓↓
N_2	=	↓	=
SiH_4	↑↑	↑↑	↑

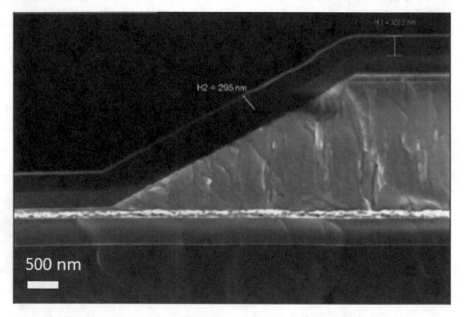

Fig. 3.18 SEM cross image of 300 nm thick high compressive SiN films deposited on top of the piezoelectric stack

film covers with very good thickness uniformity and conformality the PZT stack that is the active core of the piezoelectric device.

In terms of stress repeatability, the PECVD processes showed a good film stability as requested by the design specifications for the MEMS products.

This case study is an example of how the PECVD deposition technique can address specific requirements in terms of film stress, thus supporting in a resolutive way the fabrication of a MEMS device.

3.2.3.3 Others (a-Si, SiC)

PECVD can be used to deposit also different films with respect to the most common silicon oxide and silicon nitride. Examples of alternative materials are amorphous silicon (a-Si) or Silicon Carbide (SiC). Amorphous silicon is the non-crystalline allotropic form of silicon deposited in thin films using pure SiH_4 or combining it with H_2, He, or Ar.

The reaction for the formation of this material is the following:

$$Si H_x + Ar \xrightarrow{\text{plasma\&heat}} aSi + 2H_2 + Ar$$

a-Si can be used in MEMS technology for different applications such as protective or sacrificial layers. In general, the films deposited by PECVD present good step coverage and an interesting feature such as the resistance in concentrated HF solutions. Those properties are associated to high deposition rates compared to the ones achieved by LPCVD or PVD.

It is important to control the residual stress in a-Si films; indeed, one of the main issues is the generation of hillocks. This problem is due to the large compressive stress that generally characterizes the a-Si layer, it is also attributed to the substrate-dependent amorphous film formation [37–39].

For PECVD of SiC films, the precursors typically employed are SiH_4 and CH_4 with the following reactions:

$$Si H_4 + C H_4 \xrightarrow{\text{plasma\&heat}} SiC + 4H_2$$

The properties of SiC films deposited by PECVD are heavily dependent on process parameters; temperature plays a key role in governing the amount of hydrogen that may get incorporated into the films. PECVD SiC films are used as a structural material in micromachining processing and as a material employed for surface protection.

PECVD SiC films are resistant to HF, this is an important requirement for MEMS technologies where release processes are often present. In this respect, SiC can be used as protective layers in many MEMS products, like gyroscopes, accelerometers, and combo [40–42].

References

1. Kamins, T. (2012). *Polycrystalline silicon for integrated circuits and displays* (pp. 245–315). Springer Science & Business Media.
2. Stoffel, A., Kovacs, A., Kronast, W., & Muller, B. (1196). *Journal of Micromechanics and Microengineering, 6*, 1–13.
3. French, P. J. (2002). *Sensors and Actuators A, 99*, 3–12.
4. Ghodssi, R., & Lin, P. (2011). *MEMS materials and processes handbook. Vol. 1* (pp. 45–88). Springer Science & Business Media.
5. Dixit, P., & Henttinen, K. (2015). Via technologies for MEMS. In *Handbook of silicon based MEMS materials and technologies* (pp. 694–712). William Andrew Publishing.
6. D. Henry, X. Baillin, V. Lapras, MH. Vaudaine, JM. Quemper, N. Sillon, B. Dunne, C. Hernandez, E. Vigier-Blanc, 2007 Proceedings 57th electronic components and technology conference. IEEE, 830–835 (2007).

7. Lee, E. G., & Rha, S. K. (1993). *Journal of Materials Science, 28*(23), 6279–6284.
8. Maier-Schneider, D., Koprululu, A., Ballhausen Holm, S., & Obermeier, E. (1996). *Journal of Micromechanics and Microengineering, 6*, 436–446.
9. Maier-Schneider, D., Maibach, J., & Obermeier, E. (1995). *Journal of Micromechanics and Microengineering, 5*(2), 121.
10. D. G. Oei, S. L. McCarthy, MRS Online Proceedings Library Archive, 276 (1992).
11. Temple-Boyer, P., Imbernon, E., Rousset, B., & Scheid, E. (1998). *MRS Online Proceedings Library Archive, 518.*
12. Yang, J., Kahn, H., He, A., & Phillips, S. M. (2000). *Journal of Microelectromechanical Systems, 9*(4), 485–494.
13. Biebl, M., Mulhern, G. T., & Howe, R. T. (1995). Proceedings of the international solid-state sensors and actuators conference-TRANSDUCERS'95. *IEEE, 1*, 198–201.
14. McMahon, J. J., Melzak, J. M., Zorman, C. A., Chung, J., & Mehregany, M. (1999). *MRS Online Proceedings Library Archive, 605.*
15. Krulevitch, P., Johnson, G. C., & Howe, R. T. (1992). *MRS Online Proceedings Library Archive, 276.*
16. Mulder, J. G., Eppenga, P., & Hendriks, M. (1990). *Journal of the Electrochemical Society, 137*(1), 273.
17. Dehé, A., Wurzer, M., Füldner, M., & Krumbein, U. (2013, 2013). Proceedings of the European solid-state device research conference (ESSDERC). *IEEE*, 292–295.
18. Torkkeli, A., Rusanen, O., Saarilahti, J., Seppa, H., Sipola, H., & Hietanen, J. (2000). *Sensors and Actuators A: Physical, 85*(1–3), 116–123.
19. Maboudian, R., & Howe, R. T. (1997). *Journal of Vacuum Science & Technology B, 15*(1).
20. Dana, S. S., Anderle, M., Rubloff, G. W., & Acovic, A. (1993). *Applied Physics Letters, 63*(10), 1387–1389.
21. Sciuto, M., Papalino, L., Gagliano, C., Padalino, M., Coccorese, C., Mello, D., Renna, G., & Franco, G. (2005). *Crystal Research and Technology: Journal of Experimental and Industrial Crystallography, 40*(10–11), 955–957.
22. Kaloyeros, A. E., Jove, F. A., Goff, J., & Arkles, B. (2017). *ECS Journal of Solid State Science and Technology, 6*(10), P691.
23. Gardeniers, J. G. E., Tilmans, H. A. C., & Visser, C. C. G. (1996). *Journal of Vacuum Science & Technology A: Vacuum, Surfaces, and Films, 14*(5), 2879–2892.
24. Temple-Boyer, P., Rossi, C., Saint-Etienne, E., & Scheid, E. (1998). *Journal of Vacuum Science & Technology A: Vacuum, Surfaces, and Films, 16*(4), 2003–2007.
25. Toivola, Y., Thurn, J., Cook, R. F., Cibuzar, G., & Toivola, K. R. (2003). *Journal of Applied Physics, 94*(10), 6915–6922.
26. Ghodssi, R., & Lin, P. (2011). *MEMS materials and processes handbook.* Springer.
27. Porada, O. K., Ivashchenko, V. I., Ivashenko, L. A., Kozak, A. O., & Sytikov, O. O. (2019). *Journal of Supplementary Materials, 41*(1), 32–37.
28. Wang, N., David, N., et al. (1990). Process for PECVD of silicon oxide using TEOS decomposition, U.S. Patent No 4,892,753.
29. Lakshmanan, A., et al. (2008). *Overall defect reduction for PECVD films*, U.S. Patent Application No 11,508,545.
30. Nguyen, S. V. (1999). *IBM Journal of Research and Development, 43*(1.2), 109–126.
31. Jyrki, K., Hannu, K., Martti, B., Riikka, P., Mari, L., Panu, P., Jaakko, S., Heini, R., & Anna, R. (2010). Low-temperature processes for MEMS device fabrication. In *Advanced materials and Technologies for Micro/Nano-devices, sensors and actuators.* Springer.
32. Markku, T., Mervi, P. K., Matthias, P., Horst, T., Teruaki, M., & Veikko, L. (2015). *Handbook of silicon based MEMS materials and technologies.* Elsevier.
33. Abbasi-Firouzjah, M., Hosseini, S. I., Shariat, M., & Shokri, B. (2013). *Journal of Non-Crystalline Solids, 368*, 86–92.
34. Donald, M. M. (2001). *Vacuum technology & coating* (pp. 22–23).
35. Ay, F., & Aydinli, A. (2004). *Optical Materials, 26*, 33–46.

36. Tarraf, A., Daleiden, J., Irmer, S., Prasai, D., & Hillmer, H. (2004). *Journal of Micromechanics and Microengineering, 14*, 317–323.
37. Chung, C. K., Tsai, M. Q., Tsai, P. H., Lee, C., & Micromech, J. (2005). *Journal of Micromechanics and Microengineering, 15*, 136–142.
38. Jeyakumar, R., & Verma, A. (2012). *Materials Express, 2*(3), 177–196.
39. Street, R. A. (1991). *Hydrogenated amorphous silicon* (Cambridge solid state science series). Cambridge University Press.
40. Pasqualina, S. M. (2000). *Sensors and Actuators A, Physical, 82*(1–3), 210–218.
41. Izhevskyi, V. A., et al. (2000). *Cerâmica, 46*(297), 0366–6913.
42. Tong, L., Mehregany, M., & Tang, W. C. (1993). Proceedings IEEE micro electro mechanical systems. *IEEE*, 242–247.

Chapter 4
Thin Films Characterization and Metrology

Paolo Ferrarini, Luca Lamagna, and Francesco Daniele Revello

4.1 Films Characterization

4.1.1 Chemical Physical Characterization

Metrology and materials characterization are key activities in the development and fabrication of MEMS sensors and actuators. Chemical and physical techniques are employed since the early stages of process and materials characterization since they belong to the group of fundamental and mandatory analysis for the wide selection of materials employed in MEMS. In the following sections, we present some techniques employed in both thin and thick films regime for the characterization and monitoring of thickness, optical properties, and process performances. Structural analyses address crystallinity and morphology of the different materials and are nowadays even directly inserted as process control tools in many fabrication lines. Due to the extremely large variety of chemical elements used in the framework of MEMS devices, it is important to address by means of specific characterization also the metallic contamination to prevent any cross-contamination phenomena.

The original version of this chapter was revised: The author name "Paolo Ferrari" has been changed to "Paolo Ferrarini". The correction to this chapter is available at https://doi.org/10.1007/978-3-030-80135-9_28

P. Ferrarini (✉) · L. Lamagna · F. D. Revello
ST Microelectronics, Analog MEMS and Sensors Group, MEMS Technology and Design R&D, Agrate Brianza, Monza Brianza, Italy
e-mail: paolo.ferrarini@st.com; luca.lamagna@st.com; francesco.revello@st.com

4.1.1.1 Optical Properties

Spectroscopic Ellipsometry

Spectroscopic Ellipsometry (SE) is an optical technique which is widely used by the semiconductor industry to perform the accurate monitoring and control of thin films deposition steps which are present within the manufacturing process flows. In fact, it allows checking, in a very fast and non-destructive way, thickness and optical properties of the materials. Upon the analysis of the change of polarization of light, which is reflected off a sample, ellipsometry can yield information about layers in a very large thickness range, from microns down to fraction of monolayers. Ellipsometry can provide the optical constants that can be directly correlated to many properties such as morphology, crystallinity, chemical composition, and electrical properties. This technique is generally employed to characterize thickness for single film or complex multilayer stacks with an excellent accuracy and can be applied on various substrate materials. The output of the measurements can be thickness, the $n(\lambda)$ and $k(\lambda)$ dispersions, the optical gap, and the roughness of the film analyzed. SE can operate over a wide photon energy range, thus covering spectral range in the infrared, visible, or ultraviolet spectral region. SE analysis can provide films thickness very quickly; therefore, it turns out to be an excellent tool for developing and optimizing any kind of deposition processes. In this respect, SE has emerged in the last years also as a powerful choice to perform in situ study of film growth [1–5].

Ellipsometry is based on the analysis of the polarization state of a beam of light that is emitted by a lamp, linearly polarized by a polarizer, and then falls onto a sample. After reflection, the beam light passes a second polarizer, which is called analyzer, and falls into the detector (Fig. 4.1). For an SE measurement, the angle of incidence is equal to the angle of reflection and can be optimized as a function of the substrate. The polarization change of the light is determined only by the properties of the films such as thickness and optical constants of the materials. The light beam is polarized parallel or perpendicular to the plane of incidence; upon light detection, SE measures two parameters, which are conventionally denoted by Ψ and Δ. The polarization state of the light incident upon the sample is decomposed into the parallel (R_s) and perpendicular (R_p) components. SE measures the ratio ρ of those components and describe their interaction and the change in the light polarization by using the fundamental equations of ellipsometry reported below.

$$\rho = \frac{R_p}{R_s} = \tan \Psi \, e^{i\Delta}$$

It is important to recall that SE is an indirect method. The measured components cannot be converted directly into the thickness and the optical constants of the sample. A model analysis must be carefully performed to obtain the outputs of the measurement. The SE data modelling and analysis is based on the use of the correct

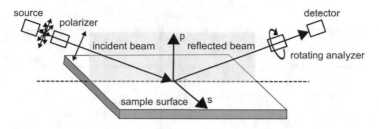

Fig. 4.1 Structure of a spectroscopic ellipsometry

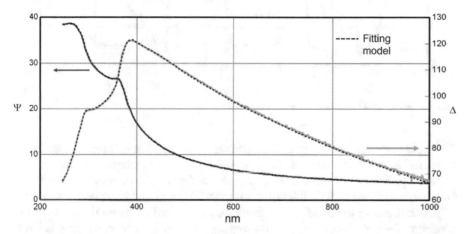

Fig. 4.2 Ψ and Δ experimental SE spectra (red and green) plotted along the spectral range together with the superposition of the data generated by the SE model (black dots) for a 10 nm thick film

optical constants and thickness parameters of all the individual layers of the sample, ordered in the correct sequence.

The sample is assumed to be composed of a certain number of discrete and defined layers that are optically homogeneous, isotropic, and must not fully absorb the light. By using an iterative procedure, unknown optical constants and thickness parameters are varied, and Ψ and Δ values are calculated. The calculated Ψ and Δ values, which match the experimental data, provide the optical constants and thickness parameters of the sample (Fig. 4.2). To quantify how well the data generated by the optical model fit with the experimental measured spectra, the concept of Mean Squared Error (MSE) value is introduced. The lower the MSE value, the better the agreement between the experimental data and model generated SE data, thus validating the results in terms of thickness and optical constant of the measured sample.

Ellipsometry is typically used for films whose thickness ranges from sub-nanometers to a few microns and it is therefore a powerful technique that can address many needs in the framework of MEMS sensors and actuators fabrication and their relative metrology issues. This technique is one of the most common to monitor

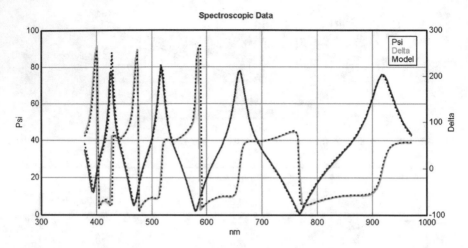

Fig. 4.3 SE spectra of a 1 μm thick dielectric film

dielectric films thickness and optical constants on both test and product wafers in many semiconductor process areas such as deposition, lithography, and dry etch.

One application of SE characterization, in the very thin film regime, is the antiwetting coating (AWC) thickness analysis. This material is deposited by means of physical evaporation and the molecules are deposited targeting the formation of coatings of about 10–20 nm. The antiwetting functionalization can play a key role on the surface of several MEMS microfluidic devices. SE can be used to characterize the thickness and the optical constants of those extremely thin films since it has been proven to be very sensitive to films even down to monolayer thickness.

Alternatively, SE is a well-known and very reliable technique that is widely used in the MEMS process line to characterize dielectric and polymeric films that are in the microns thickness regime going from few μm up to 40 μm (Fig. 4.3). The advantage of SE is always strictly connected to its fast speed, non-destructive character, and extreme flexibility in model creation and data analysis.

Reflectometry

A very common technique that is employed as an alternative to spectroscopic ellipsometry is the broadband spectrophotometry. This characterization technique works by obtaining reflectance and transmittance of the light over a wide wavelength range. The collected data can provide accurate information on the thickness, optical constants, and energy bandgap of films of many different materials. Similarly to ellipsometry, all the information cannot be obtained directly from the experimental spectra, but must be deduced from the analysis of reflectance, transmittance, or the phase shift of the polarized light.

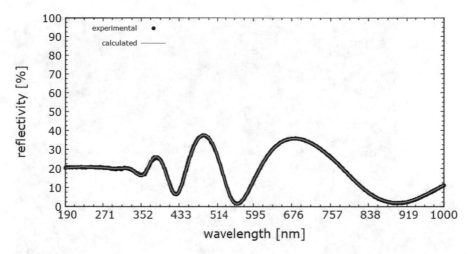

Fig. 4.4 Reflectometry spectra plotted along the spectral range for a 3000 Å thick PECVD silicon nitride film

Reflectometry uses the reflection of light to determine the optical properties (n and k) and thicknesses of ultra-thin (less than a few nm) and thick films. It can be applied also to more complex structures such as multilayers, patterned array of trenches and holes.

The data analysis is based on the Forouhi-Bloomer (F-B) dispersion equations that are used to determine thicknesses and n and k dispersion of the films. The measured (experimental) and the theoretical (calculated) reflectance spectra are compared along the experimental wavelength range (Fig. 4.4). The Forouhi-Bloomer dispersion equations for refractive index and extinction coefficient constitute a physical model that is applicable to a wide range of semiconductor and dielectric materials [6, 7].

Both spectroscopic ellipsometry, presented in the previous section, and reflectometry are non-contact optical techniques, and both require modeling to obtain a result. A reflectometer, however, measures an intensity ratio of light, whereas spectroscopic ellipsometry measures the change in the polarization state of light. A general comparison between the techniques would suggest that reflectometry is not sensitive to small changes in thin film thickness, so it is generally used on thicker (> 100 nm) samples, whereas ellipsometry is very sensitive to ultrathin films.

Reflectometry is largely used by semiconductor industries in their process lines and is suitable for various MEMS applications. Indeed, it provides a complete characterization of thickness and optical properties of films going from the more common and standard dielectrics to thick or even complex materials such as piezoelectric AlN or PZT films.

Alternatively, reflectometry can be a powerful solution if used to characterize much more complex structure that are often present in MEMS devices such as Si deep trench arrays.

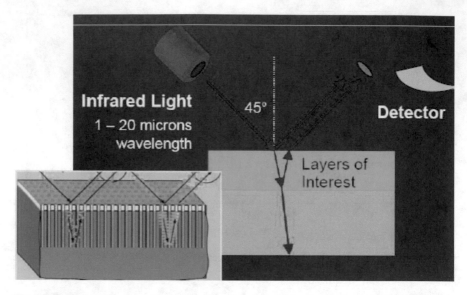

Fig. 4.5 Reflectometry measurements employed to characterize the depth of trench structures

Figure 4.5 shows that the technique allows the determination of the trench top width and depth by means of a dedicated modeling; measurements can be performed with a very high throughput and can address trench depths of 30 um. The tuning of the model depends on density and aspect ratio of the structures; however, there is a lot of flexibility and this fits very well with the requirements of many MEMS applications.

Fourier-Transform InfraRed Spectroscopy (FTIR)

Fourier-Transform InfraRed spectroscopy (FTIR) is a technique used to obtain an infrared spectrum of absorption in materials science and it is commonly employed for the characterization of thin films in semiconductor technology. The term Fourier-Transform Infrared spectroscopy originates from the fact that a Fourier transform is required to convert the raw data into the spectrum that is used to characterize the material that composes the sample under analysis. Infrared spectroscopy offers a metrology approach, complementary to other optical techniques such as ellipsometry and reflectometry, that provides excellent sensitivity to layer composition, including chemical bond densities through their vibrational mode intensities. Infrared spectroscopy is a powerful non-destructive process control tool that is present in many process areas of semiconductor production lines [8–11].

Samples of thin films such as oxides or nitrides can be characterized by using FTIR analysis. FTIR spectra are acquired in a certain spectral range between 600 cm^{-1} and 5000 cm^{-1}. The collected experimental spectra are referenced to a

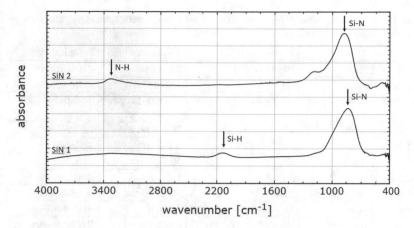

Fig. 4.6 Direct comparison of FTIR spectra acquired for two silicon nitride films (3000 Å thick)

baseline spectrum acquired on the reference Si substrates, and algorithms can be applied to analyze the data to determine the main absorption peaks.

FTIR analysis can be employed for the characterization of stoichiometry and it can act as an accurate tool to compare the chemical composition of thin films. Figure 4.6 shows the FTIR spectra acquired to compare the stoichiometry for two films deposited by PECVD (3000 Å Si_3N_4).

The alignment and the position at the same wave numbers of the characteristic peaks of a material can provide a direct insight on film structure and chemical composition (i.e., presence of specific chemical bindings to address film contamination).

4.1.1.2   Thickness

MEMS device fabrication requires the use of many different layers resulting in a high number of materials employed and a wide thickness range to be controlled varying from tens of angstrom to hundreds of microns. This huge variety implies the use of different measurement techniques and many different instruments for thickness inline process control. Most of them are based on reflectivity, polarimetry, and interferometry principles using UV-visible, Infrared, or X-ray radiation. Furthermore, because most of the techniques can accomplish the same measurement task, the choice to use one in particular depends on many factors and is often a tradeoff among accuracy, spatial resolution, throughput, and cost of ownership.

Table 4.1 summarizes the main techniques used in STMicroelectronics for the thickness characterization and inline control of thin and thick films and patterned features.

Case Study The control of PZT seed layer thickness is of paramount importance for the correct growth of (100) preferential crystal orientation. Small variation with

Table 4.1 Most used films types form MEMS fabrication with typical thickness range and associated measurement technique

Layer	Type	Film	Thickness range	Technique
Full sheet	glass	BPSG	0.6–18 μm	Reflectometry
Full sheet	metal	Ge	100–350 nm	Reflectometry
Full sheet	metal	Al, Ti, Au, Pt,TiW, Pd	100 nm–1.0 μm	Resistivity, XRF
Full sheet	nitride	LPCVD LS-SIN	200–1000 nm	Reflectometry
Full sheet	nitride	PECVD SIN	0.1–1 μm	Reflectometry
Full sheet	oxide	ITO	50–200 nm	Reflectometry
Full sheet	piezoelectric	AlN	1.0 μm	Reflectometry
Full sheet	piezoelectric	PZT	50–2000 nm	reflectometry
Full sheet	polymer	Polyimide	2–8 μm	Reflectometry
Full sheet	polymer	Polyimide	5–22 μm	Reflectometry
Full sheet	polymer	Temporary bonding polymer	20–30 μm	Reflectometry
Full sheet	Si functional	LPCVD Si-Poly (Doped/ undoped)	0.2–2 μm	Spectrometry (Poly-Si, A-Si)
Full sheet	Si oxides	Thermal, LPCVD, PECVD	10–100 nm	Ellipsometry
Full sheet	Si oxides	Thermal, LPCVD, PECVD	1–3 μm	Reflectometry
Full sheet	Si Structural	Epipoly after CMP	15–100 μm	FTIR
Full sheet	Si Structural	Silicon – grinding	300–1300 μm	Interferometry
Full sheet	Si Structural	Epipoly	0.3–110 μm	FTIR
Patterned	metal ECD	Au	1–10 μm	Mechanical profilometer
Patterned	metal ECD	Cu	3–30 μm	Mechanical profilometer
Patterned	metal ECD	Ni	1–5 μm	Mechanical profilometer
Patterned	metal ECD	Ni-Fe alloy	10 μm	Mechanical profilometer
Patterned	oxide	Getter	500–700 nm	Mechanical profilometer
Patterned	polymer	Anti-wetting coating	2–10 nm	XRR, Ellipsometry

respect to the thickness target could lead to the origin of misorientation varying film stress and piezoelectric properties [12–14].

PZT seed is measured before and after rapid thermal annealing with reflectometry technique using light in the range of visible spectra. Before the annealing, the film has a xerogel amorphous nature due to the deposition technique exploiting spin coating of sol-gel solution. At over 500 °C, the film starts crystallizing, passing from a nano-crystalline pyrochlore phase to the final perovskite phase. Local lead deficiency, caused by the contemporary evaporation and diffusion of lead through the Pt bottom electrode, give rise to unwanted pyrochlore phase mixed to the

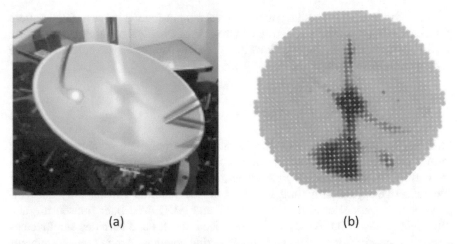

Fig. 4.7 (**a**) Wafers with PZT seed layer grain inhomogeneity at macro inspection and (**b**) GOF map resulting by reflectometry measurement

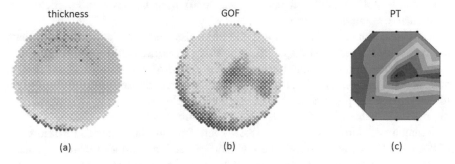

Fig. 4.8 (**a**) Thickness, (**b**) GOF, and (**c**) parameter testing maps of PZT seed layer with inhomogeneous grain

PZT perovskite grains. Other process variations during the deposition\annealing steps could result in seed layer morphology inhomogeneity, well recognizable in collimated light at optical macro inspection (see Fig. 4.7a). In this case the inhomogeneity is ascribable to grain size variation and a different ratio between perovskite and pyrochlore phase.

It is worth noting how the same pattern corresponds to the Goodness Of Fit (GOF) map of reflectance thickness measurements after annealing (see Fig. 4.7b). The models used to fit reflectance spectra and get information about film thickness are customized for the film before and after crystallization fixing n & k optical constant. The pattern found can be explained presuming a local variation of the optical constant in the film [15], hypothesis confirmed looking the maps in Fig. 4.8 reporting thickness, GOF, and electrical values coming from Parametric Testing (PT).

While thickness has a radial pattern, typical of spin coating process, GOF pattern points out a non-uniformity found also in the case of electrical measure at Parametric Testing. Further analysis showed how this electrical parameter was correlated to the crystal orientation of the film.

4.1.1.3 Micro-structure Analysis

Most of the film used in MEMS fabrication are crystalline or amorphous. The former class comprises metals or metalloid (Au, Al, Cu, Pt, Pd, Ti, WTi, Ge) used as electrical connection, electrodes, adhesion layer, barrier, or as bonding and reflective layers; structural layers as monocrystalline silicon, poly and epi-poly silicon; piezoelectric thin films (PZT and AlN) used both for sensing and actuation; oxides (ITO, Al_2O_3, TiO_2) or silicon nitrides and carbides employed as electrodes, passivation, barrier and anti-reflecting coatings. Among the amorphous, we can find different kinds of silicon oxides used mainly as passivation or release layer, doped Borophosphosilicate (BPSG) and polymers (benzocyclobutadiene, polyimides, polyamides, and epoxy compound) used as passivation, protective, permanent or temporary bonding.

This list is not exhaustive and just cites the most used materials, but many others are employed for the most disparate applications in MEMS field.

It is important to stress that the mechanical, electrical, piezoelectrical, insulating, and optical properties of these thin films strongly depend on the deposition technique and the process parameters used and is closely connected to the material microstructure that often differs from those of bulk materials.

Scanning Electron Microscopy (SEM) and Transmission Electron Microscopy (TEM) techniques are widely used for characterization (seldom for inline process control due to the need of lamellae and wafer breakage for cross-section analysis) both for imaging purpose and for micro-structure analysis as well as can provide information about chemical composition especially if couplet with Energy-Dispersive X-ray (EDX) spectroscopy techniques.

Spectroscopic techniques are instead widely used both for characterization and inline process control due to their nondestructive nature: FTIR, Raman, and XRD techniques can provide information about degree of polymerization, crystal structure, and defects.

X-Ray Diffractometry

All techniques based on the diffraction of X-ray radiation [16, 17] exploit the principle that electromagnetic waves with wavelength in the order of interatomic or intermolecular distances, when reflected by parallel crystal planes of crystalline materials, can undergo constructive or destructive interference depending on the angle of incidence. This principle is enunciated by Bragg's law as follows:

$$n \, \lambda = 2 \, d_{hkl} \sin (\theta)$$

where n is a positive integer, λ is the radiation wavelength, d is the spacing between crystal planes and θ is the angle that the outgoing reflected beam forms with the crystal plane. The d_{hkl} spacing depends on the indices h, k, l determined by the shape of the unit cell of the crystal. Therefore, all the possible θ values where we can have reflections are determined by the unit cell dimensions. The intensities of the reflected rays are determined by the distribution of the electrons in the unit cell that in turn depends on atoms type and position in the lattice. Planes passing through areas with high electron density will reflect strongly, while planes with low electron density will give weak intensities.

The typical XRD instrumental setup consists mainly of three elements:

1. *Source,* formed by a tube (generating the X-ray by means of the collision of accelerated electrons, emitted by a hot metallic filament anode, with a metallic cathode, usually copper or molybdenum), the optics (as filters or monochromators) and the slits regulating the beam divergence
2. *Goniometer,* which is the mechanical setup able to span through the different scan angles defined as in Fig. 4.9
3. *Detector,* formed by the optics, the slits, and a sensor able to collect the diffracted light

The most used XRD scans employed in thin film analysis are:

1. $2\theta\backslash\omega$ use the so-called Bragg-Brentano geometry, which is a symmetric scan fixing the ω:2θ angles ratio equal to 1:2. Keeping c angle equal to 0 degree, this mode results sensitive only for those crystal planes laying parallel to the sample surface. This mode is mainly used to check for sample texture, in plane crystallite dimension and stress analysis.
2. ω, also called rocking curves scan, scans through w angle keeping 2θ fixed. This mode is used to check for crystallite misalignment and mosaicity.

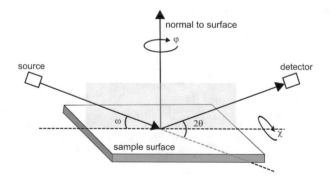

Fig. 4.9 Rotational degree of freedom used to sample scanning in XRD measurements

3. *Pole figure* consists of a series of scans through f angle, varying c for fixed ω/2θ value giving diffraction for a specific crystal plane. This kind of scan allows to build a spatial map of crystalline orientation.
4. *GIXRD* is performed fixing ω angle at grazing incidence with respect to the sample surface and varying 2θ angle. This geometrical configuration provides information coming only from the sample surface.

XRD measurement output is a diffractogram in which the detector intensity is plotted as a function of the angle scanned. Parameters coming from spectra analysis, as peaks position, intensity, and Full Width Half Maximum (FWHM) are used to get information about sample crystal structure, texture, mosaicity, and stress.

Case Study ITO, a mixed oxide of indium and tin [18, 19], widely used in optoelectronics for its characteristic to be conductive and transparent to visible light, is used as electrode in MEMS devices whenever the integration of other metal electrodes is made difficult or even impossible due to integration constraints. In this study, aimed to explore a wide deposition temperature process window and post deposition annealing effects, two ITO films with a nominal thickness of 100 nm have been deposited on a bare silicon wafer by PVD at 30 °C and 300 °C, respectively. Diffractogram in figure Fig. 4.10c shows that the sample deposited at 300 °C, referred as the Process of Reference (POR), is polycrystalline, while the sample deposited at 30 °C is completely amorphous Fig. 4.10a.

SEM images in Fig. 4.11 point out the different morphologies between the two samples: the crystallized film presents a rough surface studded with fine grains while the amorphous film is smoother.

Moreover, sheet resistance measurements reveal that the crystallized sample has a lower resistance (30 ohm/sq) compared to the amorphous film (380 ohm/sq).

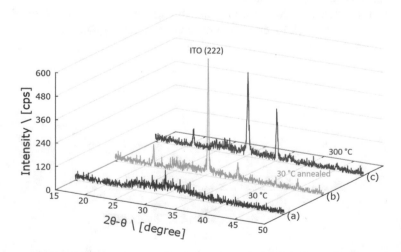

Fig. 4.10 ITO diffractogram curves for film deposited at (**a**) 30 °C, (**b**) 30 °C annealed and (**C**) 300 °C

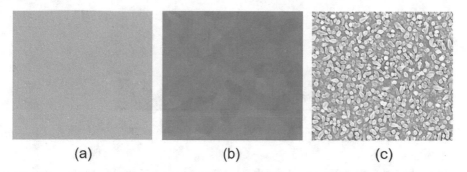

Fig. 4.11 ITO morphology for film deposited at (**a**) 30 °C, (**b**) 30 °C annealed and (**c**) 300 °C

The latter has been annealed for 300 s at 300 °C in Ar atmosphere to induce crystallization. After this treatment, both film morphology and sheet resistance change with a noteworthy difference between the center and the edge of the wafer. Diffractograms in Fig. 4.10b confirm the occurrence of film crystallization and the center-to-edge non-uniformity. This non-uniformity is emphasized by plotting the intensity of the more prominent ITO peak (222) versus the wafer radius. It is worth noting how the sheet resistance decreases in the same way going from the edge of the wafer to the center. Indeed, a linear correlation has been found between the sheet resistance and the degree of crystallinity evaluated as the intensity of the (222) peak.

4.1.1.4 Contamination Control

Several techniques have been employed for the direct detection of metallic contamination on silicon surfaces. Secondary Ion Mass Spectroscopy (SIMS) and Rutherford Backscattering Spectrometry (RBS) have been applied together with other analytical techniques such as Surface Photovoltage (SP) or Inductively Coupled Plasma-Mass Spectrometry (ICP-MS). Additional techniques have been developed and optimized by the semiconductor industry to support mass production of microelectronic devices with an extremely accurate control of metallic cross-contamination issues.

Total Reflection X-ray Fluorescence

Total Reflection X-ray Fluorescence (TXRF) has been largely demonstrated to be an excellent method of identifying and quantifying trace levels of metals on silicon surfaces. TXRF detection limits are on the order of 10^9–10^{10} ~ atoms/cm^2 for transition metals such as Fe and Cu, thus providing a very accurate control of metallic contaminations [20].

In TXRF, monochromatic X-rays impinge on an optically flat sample surface at an angle below the angle for total external reflection and excite only the top few

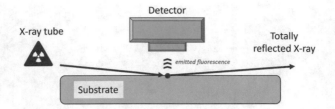

Fig. 4.12 schematic view of the TXRF stage, X-ray source, and detector configuration

atomic layers (about 3–8 nm in depth). The fluorescence X-rays from these top few monolayers emit in many directions, and a detector located perpendicular and close to the sample surface collects the emitted fluorescence X-rays and analyzes them according to energy (Fig. 4.12). The analysis of the spectra collected during the measurements allow a precise elemental identification.

TXRF software's execute from qualitative analysis to quantitative analysis automatically; after the acquisition of the fluorescence spectra, a precise fit is performed that allows peaks identification and the related identification of chemical elements on the wafers' surface.

Typically, using the TXRF configurations available on commercial tools, it is possible to identify:

1. Light elements: Na, Mg, Al
2. Transition metals: S, Cl, K, Ca, Ti, Cr, Mn, Fe, Co, Ni, Cu, Zn
3. Heavy elements: Au, Pb, Zr, Pt

Results are typically presented also in elemental maps that display the distribution and the intensity of the contaminants on the measured wafers (Fig. 4.13).

Contaminations control in MEMS process equipment is a mandatory activity due to the variety of chemical elements that are involved in the device fabrication flows. Differently from the conventional semiconductor industry, MEMS technology is often associated to the presence of unconventional chemical elements and there are many examples of those needs.

Wafer to wafer bonding is based on the use of Au, Pb, Al, and Ge depending on the specific approach that is employed; the fabrication of piezoelectric films requires the use of Pb, Zr, and Ti. All the aforementioned chemical elements, although necessary for the realization of the specific technological step, must not be spread in an uncontrolled way into a production line. Therefore, a tight contamination control is mandatory in each process area thus preventing cross-contamination phenomena.

4.1.2 Mechanical Characterizations

The knowledge of certain mechanical properties' values such as film stress and elastic modulus is very important for MEMS design and especially for the cor-

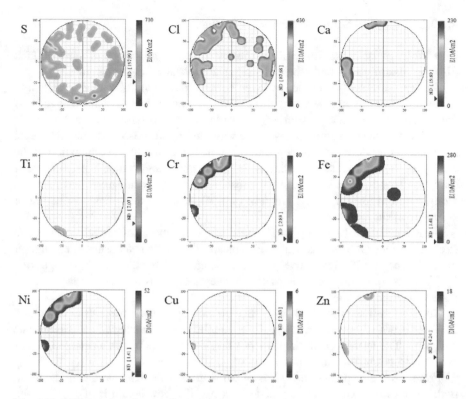

Fig. 4.13 TXRF elemental maps displaying the presence of metallic contaminants on the wafer

rect modeling of mechanical parts in Finite Element Methods (FEMs), routinely used by MEMS designers during the conception or optimization of new devices. Although most of the main commercial FEM softwares include wide material property libraries, frequently those values are related to macroscopic bulky materials while MEMS devices are built-up of many microscopic thin layers. Depending on thickness, deposition method, and the underneath substrate, films' mechanical properties can substantially vary with respect to the same bulk material. The use of approximate values can lead to erroneous simulation results or, even worse, the lack of information for these parameters makes the use of FEM unpractical due to a too high number of variables degree of freedom. Many indirect measurement methods, such as X-ray diffraction laser acoustic wave propagation, and direct, such as indentation or wafer/structures deflection, are used to get information about these parameters at local or whole film scale. Each technique has its own advantages and disadvantages as well as limitations mainly due to film nature and lack of knowledge about other film properties.

4.1.2.1 Film Stress and Wafer Bow/Warpage

Bow/Warpage wafer metrology allow the determination of stresses of thin and thick films on silicon wafers and other substrate materials at room temperature.

The measurement is based on laser optical triangulation, a dual wavelength technology permits to select 650 and 780 nm, and wavelength choice depends on the material's reflectivity optimization. The laser beam is moved above the sample along an axis; the curvature radius of the wafer along the axis is measured by a line scan before and after the film deposition and hence allows the determination of the change of curvature due to the deposition. This curvature change is calculated for total scan length, global stress, and for local scan length, local stress. Several scans along different axes allow the creation of a 3D image of the stress distribution or curvature of the wafer.

Scan length can be set up to 194 mm to cover whole wafer with 3 mm edge exclusion. For 2D/3D stress mapping, typically 4 to 6 diameter scans at an angular interval of 45° or 30°, respectively. During the measurement, the wafer is supported by 3 pins to guarantee wafer planarity. An integrated microbalance can measure the wafer substrate thickness.

The systems used for this characterization measure the wafer radius of curvature: the laser beam is moved across the wafer and the beam is deflected according to its curvature. The system determines the position z of the beam with a patented micro-positioning photodetector as a function of position x.

From these data points, the radius of curvature is calculated; knowing the beam path, the radius of curvature can be determined. The bow height H of a wafer with diameter D can be calculated and once radius has been calculated, the stress is then computed using Stoney's formula:

$$\sigma = \frac{E}{1-\nu} \frac{t_s^2}{6t_f} \left(\frac{1}{R} - \frac{1}{R_0} \right)$$

where $E/1$-ν is the biaxial modulus of the substrate, in our case, silicon $= 1.805\ 10^{12}$ t_s and t_f are the thicknesses of the substrate and film, respectively, and R_0 and R are the radii of curvature of the substrate before and after film deposition. The stress profile can be plotted along the wafer diameters thus characterizing the local stress distribution on the wafers (Fig. 4.14). With this analysis it is possible to address an extremely interesting aspect of the mechanical properties of the films such as the stress evolution at the edge of the wafer.

Wafer stress distribution can be also plotted in 3D pictures (Fig. 4.15).

4.1.2.2 Elastic Modulus and Hardness by Indentation

Indentation is a very versatile technique able to gather information about mechanical properties [21] as elastic modulus, hardness, creep, film adhesion, and scratch

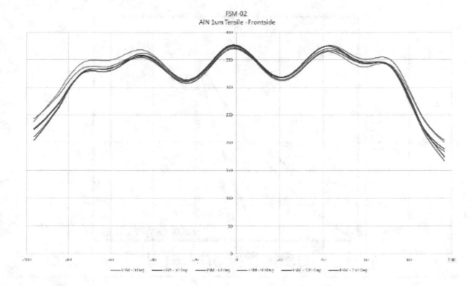

Fig. 4.14 Stress distribution along 6 different diameters for a 1 μm thick AlN film deposited by reactive PVD

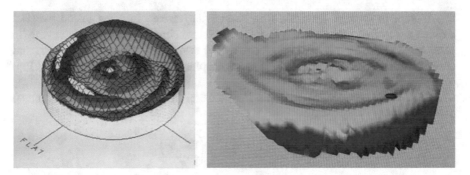

Fig. 4.15 3D map of the stress distribution for an AlN film deposited by reactive PVD

resistance as well electro-mechanical. Exploiting loading forces ranging from 10^{-3} N to 10 N nano and ultra-nano indentation can test both thick and thin film (up to hundreds Å) in a very wide range of hardness and elastic modulus. Coupled with scanning probe microscopy technique, indentation measurements can also give qualitative or semi-quantitative results regarding film fracture fragility and adhesion strength or with special modules add-on to follow the evolution of mechanical properties applying thermal budgets or change the environmental conditions. Furthermore, using special conductive tip and transducer, the electrical resistance of a contact can be measured during the indentation as well as the electrical response of a piezoelectric material subject to a loading force. In an indentation measure, a hard tip of know geometrical shape and elastic modulus (usually a Berkovich diamond

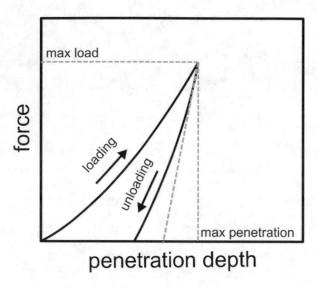

Fig. 4.16 Indentation curve

tip) is kept in contact with the film surface and driven into the tested material under a controlled force load. A transducer system simultaneously registers the tip penetration depth until a maximum load value is reached, then the tip is retracted drawing the unloading part of the loading-displacement curve as reported in Fig. 4.16.

The reduced indentation modulus E_r and material hardness H can be calculated using equation:

$$\frac{1}{E_r} = \frac{\left(1 - v_i^2\right)}{E_i} + \frac{\left(1 - v_s^2\right)}{E_s}$$

where v_i and v_s are the Poisson ratio of the indenter and sample, respectively, while E_i is the indenter elastic modulus obtained from the measure of a standard with known elastic modulus.

The technique limitations on the measure of the elastic modulus arise when the material undergoes permanent plastic deformation, giving rise to material pile-up or in the presence of cracks. High roughness or very thin film can also lead to huge error in the estimation of this parameter. The rules of thumb of indentation permits average roughness Ra < 1/5 h_{max} and h_{max} < 1/10 fth (film thickness). The last rule avoids any interference coming from taking into account for the substrate stiffness. Modern instruments overcome this issue through dynamic analysis technique [22] allowing the measure of very thin films.

Case Study Ultra-nanoindentation has been used to test the mechanical properties of two samples of 20 mm Borophosphosilicate glass (BPSG) films deposited with

Atmospheric Pressure Chemical Vapor Deposition (APCVD) and Sub-atmospheric chemical vapor deposition (SACVD). The nominal concentration of B and P are the same for the two samples ($B = 6.6 \pm 0.4$ wt % $P = 3.6 \pm 0.2$ wt %) while, to get the final thickness, films are deposited/annealed in multiple steps: 2 and 4 times for Sub-Atmospheric Chemical Vapor Deposition (SACVD) and Atmospheric Pressure Chemical Vapor Deposition (APCVD) techniques, respectively. The hygroscopic nature of BPSG material and the change of its chemical-physical properties due to water absorption through the film surface is thoroughly reported in literature [23]. For such thick films, water absorption is limited through the whole film thickness range, resulting in different properties of the film/air interface with respect to the bulk. This gradient can be measured using increasing loading forces that probes different tip penetration depths. Figure 4.17a shows the elastic modulus as a function of penetration depth pointing out how that property depends on the distance from the surface. Elastic modulus is lower for APCVD process and for both the techniques follows a trend presenting a minimum located at different penetration depths: at about 180 nm for APCVD and 250 nm for SACVD, respectively. Time-of-Flight Secondary Ion Mass Spectroscopy (TOF-SIMS) profiles analysis not reported here confirms a different concentration of B and P elements in the samples justifying the gap on the average properties values, while H profile, showed in Fig. 4.17b, could explain the shift in the minimum for elastic modulus curves. As can be noticed, H signal intensity, linked to the presence of absorbed water into the film, decrease until a limit value for different values of the sputtering times indicating that film deposited by APCVD absorb water in a deeper region with respect to SACVD, giving rise to local variation of the elastic modulus.

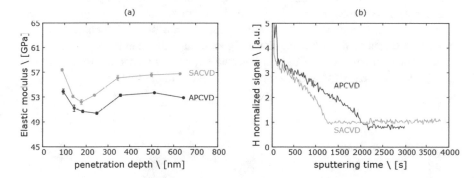

Fig. 4.17 (a) Variation of Elastic modulus and (b) TOF-SIMS Hydrogen signal as a function penetration depth for BPSG sample deposited with APCDV (dark blue) and SACVD (bright blue)

4.1.3   Dimensional Characterization

Most of the MEMS devices are characterized by complex structures, such as springs, metal coils, suspended membranes, and cavities which generally belong to the core of the sensors or the actuators. The design and the fabrication of these structures are complex because they have to take into account not only the physical properties of the layers that compose the device (i.e., thickness, stress, elastic modulus) but also geometrical aspects such as the shape factors of the three-dimensional features present.

To guarantee that all the design requirements are satisfied during the front-end manufacturing process, specific control measurements need to be performed both in two (2D) and three (3D) dimensions. In addition, the dimensional control of MEMS structures often requires a dedicated handling of the equipment to manage wafers that are bowed, with a very thin thickness or that are patterned with holes. Even more difficult is the case when the structures to be measured are hidden, as often occurs when two or more wafers are bonded together to form the functional MEMS. In this case, customized solutions in terms of metrology tools capability or design must be taken to overcome the problem. As an example, a typical MEMS application is reported in Fig. 4.18, where a dedicated window must be opened on the test dice to measure the back side cavity dimensions from the front side of the wafer.

For typical ULSI IC, semiconductor industries, following Moore's Laws, reduced year to year the dimension of the structures requiring the use of new advanced technologies for characterization and critical dimension (CD) measurement. Nanometer and Ångström have become the standard dimensional units and metrological SEM (Scanning Electron Microscope) and TEM (Transmission electron microscopy) are the common techniques used for this scope [24]. MEMS technologies are not dependent on Moore's Law and many new devices, particularly the actuators, have big structures that can hardly or cannot be measured with metrological SEM. In this context, optical inspection tools are used, allowing higher flexibility and a wider range of measurable features.

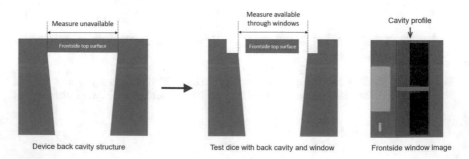

Fig. 4.18 Back cavity measurement example on test dice

Optical metrological tools are equipped with an illumination source and a camera for image capture. The illumination source can generate light in different wavelength; visible (VIS), ultraviolet (UV), or infrared (IR) wavelength is used depending on the application. Top surface measurements are taken with visible light, whereas hidden structures are measured with infrared light. IR-based characterization has some limitations due to materials properties; infrared light cannot pass through doped substrates or metal layers. For this reason, IR tools are equipped with two different IR illumination sources, one on top of the wafers and the other at the bottom to have the possibility to use transmitted light, reflected light, or a combination of both. Application performed with transmitted light (for example, bonding overlay measure) needs a dedicated edge grip ring chuck that allows the light to go directly from the source to the detector camera, passing through the wafer, with no obstacle in between which can overshadow the signal. New optical metrology tools have the possibility to work with UV illumination, this allows to take advantage of fluorescence phenomenon that enhance the contrast of organic materials, like polymer dry film, widely used in MEMS manufacturing. The measure of this film is extremely difficult because they usually are transparent and with visible light there is not enough contrast for profile detection of the structures. The last tool generation are also equipped with LED (light emitting diodes) illumination source. Compared to old generation tungsten-halogen bulb lamps, LED technology improves lamp lifetime, enhances light stability, and allows the user to choose a dedicated wavelength that helps to increase the image contrast and profile detection. On optical tools, the resolution of the image depends on the pixel size that is correlated to the power of magnification and the technology of the camera as reported in Table 4.2.

Increasing the magnification, the field of view decreases together with the depth of focus. It follows that the accuracy of a measure is correlated to the best magnification that allows the visualization of the entire structure inside the field of view of the objective.

Case Study Resonant micro-mirrors' functionality is strictly dependent on silicon springs that are connecting the structural part, with the metallic mirror, to the whole device frame. In particular, the mirror resonance frequency is correlated to the spring geometry and shape. The thickness of the spring is generally defined by the thickness of the active layer of the SOI (Silicon On Insulator) substrates that

Table 4.2 Pixel resolution, depth of focus, and field of view of different optical objectives

Objective	1X	2X	3X	5X	10X	20X	50X
Pixel size (μm)	9.5	4.75	3.02	1.9	1	0.5	0.2
Depth of focus (μm)	774	193	167	31	8	2	0.9
Field of view (mm × mm)	16.1 × 12.2	8 × 6.1	5.12 × 3.8	3.2 × 2.4	1.6 × 1.2	0.8 × 0.6	0.18 × 0.13

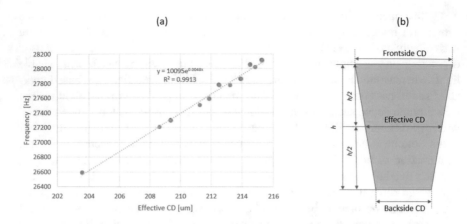

Fig. 4.19 (**a**) Correlation of resonation frequency and spring CD and (**b**) representation of effective CD on spring section

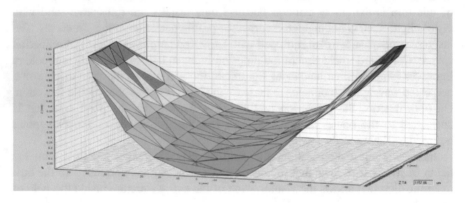

Fig. 4.20 Contour Z map of μ-mirror wafer on contactless chuck

are used for micromirrors. On the other hand, the CD variation induced by deep silicon etch process that fabricates the springs must be addressed during the device fabrication with a dedicated metrology. Figure 4.19 shows the linear dependency of the resonance frequency from the experimental spring CD measured on the devices.

With the aim to have an in-line process control during front-end manufacturing phase, both front and back side measurements have been collected for the same spring feature. Measurement tools equipped with a contactless chuck and double front/back side optics or special handlings able to flip the wafers preserving wafers integrity and avoiding scratches are used for this purpose. At the end of process flow, micro-mirror wafers can reach thickness less than 190 μm leading to warpage more than 1 mm when loaded on contact-less chuck, as showed in Fig. 4.20; to compensate this Z tilt, the metrological tool must have extremely accurate autofocus system.

4.1.3.1   3D Measures

To keep under control the behavior of a MEMS device, a full characterization is required, including not only electrical test but also 3D measures to verify if all the geometries respect the specifications in terms of depth, height, profile shape, thickness, and deflection. These measurements are usually performed at the end of process flow when the structures are released. Methods to perform 3D measurements can be divided in two main groups: destructive and nondestructive. The destructive one consists in cross section of the sample which reveal the profile of the structure to be measured: this is typically obtained with the combination of FIB (focused ion beam) and SEM/TEM techniques. These techniques are accurate and allow to have a full characterization of the structure, but the sample measured is "destroyed" and cannot be anymore used and correlated with final test. Furthermore, they give information only regarding specific points on the wafer and are time-consuming so these techniques are not suitable for mass production control. The nondestructive methods allow to take the measurements directly on the sample without jeopardizing the functionality of the device. Different techniques can be used, the choice depending on the kind of measure to be done (profile, thickness, depth), the architecture of the structure, the material composition, the desired accuracy, and the throughput. For instance, a mechanical profilometer can be used for quick and fast measure of bump height, X-ray for tilt measure of deep-etched gratings [25], confocal laser for metal coil height measure, interferometer can be used for TSV (Through Silicon Vias) depth measure [26], or for bonding glue thickness measure between two bonded wafers. Both methods, destructive and nondestructive, can be combined for initial characterizations and correlation between different techniques; this is useful to understand which is the best that fits the requirements of accuracy and throughput. For example, Fig. 4.21 shows the evaluation of the coil height by a SEM cross section in single point (a) compared

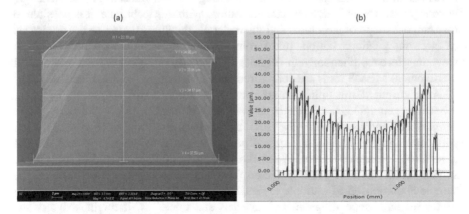

Fig. 4.21 (a) Example of coil profile evaluation by means of SEM cross section in a single point and (b) confocal laser signal of an entire coil

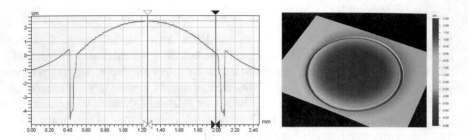

Fig. 4.22 2D profile and 3D representation of the same optical membrane structure

with the measure of the whole coil with a confocal laser (b); the first one gives a detailed and perfectly accurate characterization of single point of the structure, while the confocal laser profile provides valuable information about uniformity of the entire structure.

The output of these measurements is typically represented as a section of the structure, but some tools are also able to reproduce a full 3D image of the structure. For instance, Fig. 4.22 shows the 2D profile and the 3D representation of the same circular suspended membrane.

Advanced tools allow also to combine 3D measure with electrical stimulation of the device giving a real-time image of a structure during its actuation. This application is particularly useful for obtaining more information regarding the functionality of the device, for example the maximum displacement covered by a PZT membrane.

Interferometry

Due to its high flexibility, resolution, and throughput, the interferometric technique is widely used for MEMS 3D characterization. With this technique, it is possible to measure the shape of structures and the materials thickness thanks to the properties of electromagnetic radiation. Interferometry is a measurement method based on wave interference phenomenon (usually light, radio, or sound waves) and typical metrological MEMS tools are configured with a Michelson's optical Scheme [27].

As represented, this optical interferometry is designed with a light source where a beam is generated, the beam is subsequently divided in two equal parts, *beam sample (BS)* and *beam reference (BR)*, by a half-silvered mirror (Fig. 4.23). Each part continuing with different optical path, BS is sent to the sample and the BR is sent to a reference mirror. When the first beam of light arrived at the sample surface, some of the light is reflected (Ir_1) and some is transmitted (It_1). If the sample is composed of multiple layers, the transmitted light can pass through the first layer and, depending on the materials and thickness, it might arrive at the interface with the second layer where again some light is reflected (Ir_2) and some light is transmitted (It_2) through the second layer. This phenomenon will continue through the sample until the light is fully absorbed by the layers [28].

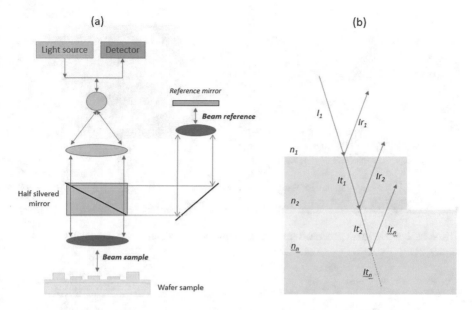

Fig. 4.23 (**a**) Schematic representation of interferometric system and (**b**) representation of beam behavior on sample surface

The intensity of reflected light is described with the Fresnel Equation:

$$\frac{Ir_1}{I_1} = \left(\frac{n_1 - n_2}{n_1 + n_2} \right)^2$$

where n is the index of refraction of each material.

The reflected beams from the samples and the beam from the reference mirror are finally recombined. The result of all the interference is a sinusoidal spectrum and the intensity variation, which depends on the path difference, is measured with a detector where an interferogram is generated. The interferogram is then put through a Fast Fourier Transform (FFT), which translates the information from the Frequency domain to the Time domain, giving us a plot of Intensity vs. Optical Thickness as represented in Fig. 4.24.

It is important to understand that the measurement is an optical thickness (T_O). To get the real thickness value (T_r) we must take into consideration the refractive index, n, as indicated in the equation below:

$$T_r = \frac{T_O}{n}$$

By this equation, for surface topography measure, the optical thickness corresponds to the real thickness, in this case the refractive index is 1 because we are measuring the thickness of the air on top of the wafer. For all other applications, it is extremely important that the value of n is known and used correctly. In this way,

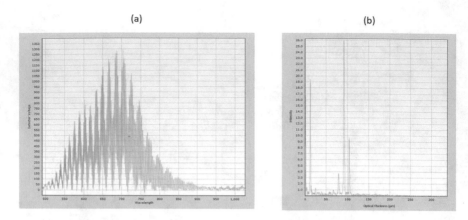

Fig. 4.24 Interferogram generated from (**a**) detector and (**b**) plot after Fast Fourier Transform

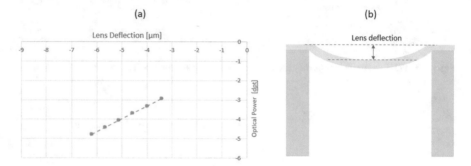

Fig. 4.25 (**a**) Correlation between Lens deflection and Optical Power and (**b**) representation of membrane lens deflection

it is possible to characterize both thickness and distance of surface layers in a single point, by the multiple acquisition along a desired line, it is possible to get the profile of the entire structure.

Case Study In a MEMS actuator fabricated for the autofocus optical application, there are many specifications that must be controlled on the final device to assess and guarantee the optical properties of the device. One of these is Optical Power (expressed in diopters) of the lens; this value is equal to the reciprocal of the focal length ($^1/_m$) that is directly correlated to the radius of curvature and the deflection of the lens on finished wafers as represented in Fig. 4.25.

The lenses have 1.5 mm diameter and are composed by Boron Phosphorous Silicon Glass (BPSG) suspended membrane and the risk to break them during the loading of the wafer is extremely high. The handling system must be configured for carefully managing the wafer, the vacuum can be done only close to the edge and the airflow inside the tool adjusted to avoid membrane vibration during the measure. To guarantee optical properties, these lenses are composed of a transparent layer; this

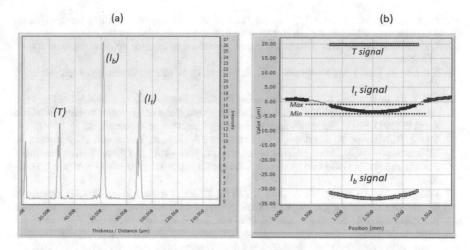

Fig. 4.26 (a) Interferometric signal in the middle of lens and (b) output signal of full acquisition across lens diameter

restricts the field of options and can be used for the deflection measure: i.e., confocal laser pass through the membrane and the signal does not return to the detector; besides, a mechanical profilometer does not have the right accuracy to perform this measure. In this context, the interferometric technique was found as the only suitable way to get lens deflection measure. As represented in panel (a) of Fig. 4.26 the interferometric signal of the membrane is composed of three different peaks: the membrane thickness (T), the interfaces between air with membrane top surface (I_t), and the interface between membrane bottom surface with the air below (I_b). For real thickness calculation, the signal value T must be corrected with the refractive index of the membrane, while the other two signals do not need any correction because are generated by an interface with the air. Picture b shows the full line acquisition across the lens diameter with the corrected signals, this allows with only one measure to calculate both profile and thickness of the entire membrane; lens deflection is finally obtained from the delta between minimum and maximum values of the I_t signal on top of lens.

References

1. Tompkins, H. G., & Irene, E. A. (2005). *Handbook of Ellipsometry*. W. Andrews Inc..
2. Tompkins, H. G. (1993). *A user's guide to Ellipsometry*. Dover Publications.
3. Hilfiker, J. N., Bungay, C. L., Synowicki, R. A., Tiwald, T. E., Herzinger, C. M., Johs, B., Pribil, G. K., & Woollam, J. A. (2003). *Journal of Vacuum Science and Technology, 21*, 1103.
4. Vedam, K. (1998). *Thin Solid Films, 313*, 1.
5. Langereis, E., Heil, S. B. S., Knoops, H. C. M., Keuning, W., van de Sanden, M. C. M., & Kessels, W. M. M. (2009). In situ spectroscopic ellipsometry as a versatile tool for studying atomic layer deposition. *Journal of Physics D: Applied Physics, 42*, 073001.
6. Forouhi, A. R., & Bloomer, I. (1986). Optical dispersions relations for amorphous semiconductors and amorphous dielectrics. *Physical Review B, 34*(10), 7018.

7. Forouhi, A. R., & Bloomer, I. (1988 Jul 15). Optical properties of crystalline semiconductors and dielectrics. *Physical Review B: Condensed Matter, 38*(3), 1865–1874.
8. Chabal, Y. J. (1988, May). Surface infrared spectroscopy. *Surface Science Reports, 8*(5–7), 211–357.
9. Kelly, M. J., Han, J. H., Musgrave, C. B., & Parsons, G. N. (2005). *Chemistry of Materials, 17,* 5305–5314.
10. Rai, V. R., & Agarwal, S. (2009). *Journal of Physical Chemistry C, 113,* 12962–19625.
11. Ay, F., & Aydinli, A. (2004, June). Comparative investigation of hydrogen bonding in silicon based PECVD grown dielectrics for optical waveguides. *Optical Materials, 26*(1), 33–46.
12. Muralt, P., Maeder, T., Sagalowicz, L., Hiboux, S., Scalese, S., Naumovic, D., Agostino, R. G., Xanthopoulos, N., Mathieu, H. J., Patthey, L., & Bullock, E. L. (1998). Texture control of PbTiO3 and Pb(Zr,Ti)O3 thin films with TiO2 seeding. *Journal of Applied Physics, 83,* 3835–3841.
13. Wang, L., Yu, J., Wang, Y., & Gao, J. (2008). Effect of excess Pb in PbTiO3 precursors on ferroelectric and fatigue property of sol–gel derived PbTiO3/PbZr0.3Ti0.7O3/PbTiO3thin films. *Journal of Materials Science: Materials in Electronics, 19,* 1191–1196.
14. Chen, S.-Y. (1998). Texture development, microstructure evolution, and crystallization of chemically derived PZT thin films. *Journal of the American Ceramic Society, 81,* 97–105.
15. de la Cruz, J. P., Joanni, E., Vilarinho, P. M., & Kholkin, A. L. (2010). Thickness effect on the dielectric, ferroelectric, and piezoelectric properties of ferroelectric lead zirconate titanate thin films. *Journal of Applied Physics, 108,* 1–8.
16. Birkholtz, M. (2006). *Thin film analysis by X-ray scattering.* Wiley-VCH Verlag GmbH & KGaA, Weinheim.
17. Waseda, Y., Matsubara, E., & Shinoda, K. (2011). *X-ray diffraction crystallography: Introduction, examples and solved problems.* Springer.
18. Synowicki, R. A. (1998). Spectroscopic ellipsometry characterization of indium tin oxide film microstructure and optical constants. *Thin Solid Films, 313–314,* 394–397.
19. Rogozin, A. I., Vinnichenko, M. V., Kolitsch, A., & Moller, W. (2004). Effect of deposition parameters on properties of ITO films prepared by reactive middle frequency pulsed dual magnetron sputtering. *Journal of Vacuum Science and Technology, 22*(2), 349–355.
20. Stoev, K. N., & Sakurai, K. (1999). Review on grazing incidence X-ray spectrometry and reflectometry. *Spectrochimica Acta Part B, 54,* 41–82.
21. Pharr, G. M. (1998). Measurement of mechanical properties by ultra-low load indentation. *Materials Science and Engineering, A253,* 151–159.
22. Olver, W. C., & Phar, G. M. (1992). An improved technique for determining hardness and elastic-modulus using load and displacement sensing indentation experiments. *Journal of Materials Research, 7,* 1564–1583.
23. Thorsness, A. G., & Muscat, A. J. (2003). Moisture absorption and reaction in BPSG thin films. *Journal of the Electrochemical Society, 150*(12).
24. Orji, N., Badaroglu, M., Barnes, B., Beitia, C., Bunday, B., Celano, U., Kline, R., Neisser, M., Obeng, Y., & Vladár, A. (2018). Metrology for the next generation of semiconductor devices. *Nature Electronics, 1.* https://doi.org/10.1038/s41928-018-0150-9
25. Song, J., Heilmann, R., Bruccoleri, A., & Schattenburg, M. (2019). Characterizing profile tilt of nanoscale deep-etched gratings via x-ray diffraction. *Journal of Vacuum Science & Technology B., 37,* 062917. https://doi.org/10.1116/1.5119713
26. Jin, J., Kim, J. W., Kang, C.-S., Kim, J.-A., & Lee, S. (2012). Precision depth measurement of through silicon vias (TSVs) on 3D semiconductor packaging process. *Optics Express, 20,* 5011–5016.
27. Hariharan, P. (2007). *Basics of interferometry* (2nd ed.). Elsevier.
28. Born, M., & Wolf, E. (1999). *Principles of Optics* (7th expanded ed.). Cambridge University Press.

Chapter 5
Deep Silicon Etch

Anna Alessandri, Filippo D'Ercoli, Pietro Petruzza, and Alessandra Sciutti

5.1 Silicon Processing

5.1.1 Deep Silicon Etch Evolution and Bosch Process

The standard plasma etching techniques are not able to define the silicon of the new MEMS devices. Both the gas mixtures (Cl, HBr) and the hardware usually adopted to realize the IC (Integrated Circuits) polysilicon gates are not adequate to pattern high aspect ratio designs. The processes are good in terms of anisotropy on IC scale, but they are slow and degrade performances when increasing the depth and the horizontal size of the structures. Fluorinated gases are the best candidate to develop innovative processes because of their high etching speed on silicon, especially sulfur hexafluoride (SF_6), but unfortunately their action is isotropic and exothermic, i.e., they spontaneously react at equal rates in all directions developing heat.

The first attempt to protect the sidewalls from the erosion is known as cryogenic dry etch [1] (1988): this approach consisted in adding oxygen to the fluorine gas mixture to allow the formation of silicon oxyfluoride (SiO_xF_y) as passivating layer and lowering the wafer's temperature into the cryogenic range (around $-100\ °C$) to slow down fluorine aggressivity and enhance oxide film formation. Even though profiles were anisotropic, sidewall morphology was good and etch speed was ten times greater than the standard one (5 μm/min vs. 500 nm/min), cryogenic etch did not really take off due to many drawbacks. The process was extremely sensitive to cryogenic conditions and oxygen percentage, photoresist materials were subjected to cracks and delamination, but above all, the special hardware needed to reach such

A. Alessandri (✉) · F. D'Ercoli · P. Petruzza · A. Sciutti
ST Microelectronics, Analog MEMS and Sensors Group, MEMS Technology and Design R&D, Agrate Brianza, Monza Brianza, Italy
e-mail: anna.alessandri@st.com; filippo.dercoli@st.com; pietro.petruzza@st.com; alessandra.sciutti@st.com

© Springer Nature Switzerland AG 2022
B. Vigna et al. (eds.), *Silicon Sensors and Actuators*,
https://doi.org/10.1007/978-3-030-80135-9_5

low temperatures was very expensive. Therefore, we had to wait until 1994 to break the ice, which during that time, was freezing MEMS development.

In 1994, Dr. Franz Laermer and Andrea Schilp from "Robert Bosch GmbH" developed a new process to help manufacturing their MEMS devices. This technique patented and licensed by Bosch company is frequently called TMDE (Time Multiplexed Deep Etching), but it is mostly known as the "Bosch process" [2]. It is a revolutionary and powerful technique because it combines the effectiveness of a fluorine-rich plasma on silicon surfaces with the etch-inhibitory property of Teflon-like films generated by fluorocarbon discharges without needing to reach cryogenic temperatures.

While the cryogenic approach is a single step or steady state process as in standard IC etching, i.e., the oxide passivation and the silicon removal occur at the same time in the same plasma, in the Bosch process, passivation and etch are separated steps: two plasmas are generated, alternatively switching from one to another. Passivation is provided by heavily polymerizing chemistries, generally C_4F_8 (deposition step), while silicon etch is still due to the chemical reaction of SF_6 (etch step). Due to the switching, the resulting silicon sidewalls are made of a series of small isotropic etch bubbles, the so-called *scallops,* giving a typical sidewall roughness which can be tuned and controlled by a proper process balancing.

The demand for higher precision and smaller sensors required even stricter control of lateral dimensions, so the early two-step approach evolved to a three-phase sequence, in order to further reduce the isotropic part of the process [3]. As explained in Fig. 5.1b, a physical etch step characterized by high verticality of action is inserted between the deposition and the chemical etch steps. Its purpose is to remove the polymeric film mainly from the bottom of the etched structure, affecting as little as possible the same protecting layer on the sidewalls.

The method proved to be extremely effective in realizing high Aspect Ratio structures (AR > 50:1), even "through the wafer" etching or wafer dicing. The

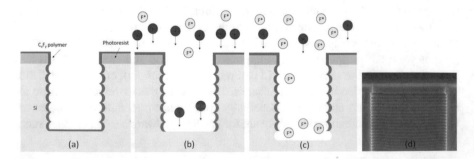

Fig. 5.1 Scheme of the Bosch process: the process is a repeating sequence of loops, each loop is made up of three phases or steps. (**a**) Passivation step: the protecting layer is deposited on all the surfaces, covering also the mask; (**b**) Passivation removal step: the protecting layer is preferentially removed from the bottom of the structure; (**c**) Etch step: silicon is removed from the bottom and the isotropic nature of this step is revealed through the sidewall roughness, i.e., scallop formation. Each scallop is equivalent to one complete loop (**d**)

polymer deposition ensures the anisotropy required, while maintaining high mask selectivity and a fast etch rate (up to 30 μm/min). Moreover, standard masks can be applied, such as hard masks (nitride, oxide) and photoresist as well, thanks to the standard temperature conditions. The hardware needed for Bosch process is simpler and cheaper than the cryogenic one, but more sophisticated than standard ICP-RIE (Ion Coupled Plasma – Reactive Ion Etching) tools. During the years, Bosch process proved to be able to meet the cost and quality production standards. Nowadays, it is the dominant technology for dry deep silicon etch in all the volume applications in MEMS field.

5.1.2 Bosch Process Terminology

Discussing about Bosch process implies becoming familiar with a specific terminology which identifies and describes the main geometrical and morphological features of a patterned structure. Some terms are shared with standard dry etch, others are peculiar of the technology itself, such as scallops, black silicon, and Aspect Ratio Dependent Etching (ARDE).

Black silicon (Fig. 5.2) refers to areas of incomplete silicon etch and it is the result of insufficient passivation removal due to an improper recipe balancing or surface contamination or mask residues. The unwanted effect generates a needle-shaped surface at the bottom of the structure (another term is grass) characterized by very low reflectivity; that is where its name comes from.

Scallops, as already defined in Fig. 5.1, are the typical indentations created on the silicon sidewall during the chemical part of the SF_6 etching reaction. They determine the sidewall roughness. The smaller the scallop, the better the anisotropy; the best trade-off consists in minimizing scallop dimension without lowering the etching speed. The search for the minimum scallop, however, can lead to *sidewall*

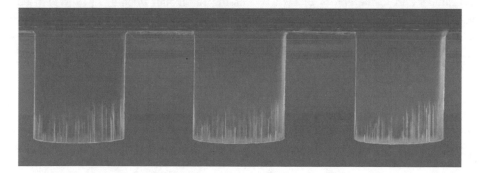

Fig. 5.2 SEM picture of Black Silicon

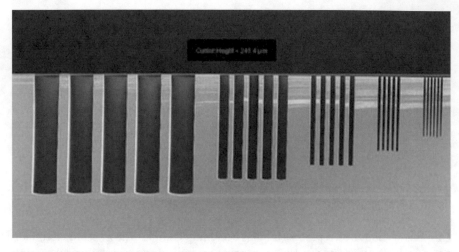

Fig. 5.3 SEM picture of ARDE or RIE lag effect

striations (see Table 5.1) and eventually to *black silicon* occurrence. Depending on the application, if the scallop minimization during the patterning is not enough or if it is not feasible, a subsequent *scallop erasing* is needed. It is a dedicated step which acts on the silicon sidewalls already defined through the Bosch process.

While etching high Aspect Ratio structures, it is observed that the etch rate depends on the mask opening. This effect is known as *ARDE (Aspect Ratio Dependent Etching)* or *RIE lag*, typical of Reactive Ion Etching tools, (Fig. 5.3). The lag is quantified by the following formula:

$$\text{ARDE or RIE lag} = (\text{depth@max width}) / (\text{depth@min width})$$

The effect is due to ion flux differences between low and high Aspect Ratio features and to the ability of etch processing in smaller openings [4]. When the lower etch rate occurs in the larger feature, the effect is called *inverse RIE Lag* [5].

Table 5.1 summarizes the terminology related to the Bosch process, just to better understand the technical challenges given by the process itself and by the applications that exploit it.

5.1.3 Process Parameters and Trends

The Bosch process is composed of three alternating plasmas, one for deposition step and two for etch steps. The three phases form the so-called *loop* which is repeated until the required depth is reached.

Profile control is a delicate balance between the etch and deposition components [6]. Due to the phase switching, there are many surface interactions during the

Table 5.1 Summary of process terminology related to Bosch process

Terminology	Definition	Picture
Critical Dimension (CD)	Nominal width of a structure to be etched, normally given by design through mask.	
CD loss Top-bottom	Difference between top and bottom width: CD loss = (WTop−WBottom)/2	
Aspect ratio (AR)	Ratio between the etch depth and the structure width: AR = depth/width	
Scallop size	Scallop width and height: They give an estimate of the sidewall roughness and it is mainly ruled by depc/etch balance.	

(continued)

Table 5.1 (continued)

Terminology	Definition	Picture
Top undercut	Difference measured from the mask edge to silicon sidewall (top "bite" larger than scallops). It is linked to insufficient passivation protection.	
Profile	POSITIVE or TAPERED ≤90°. (WTop > WBottom) VERTICAL = 90°. Perfect anisotropy NEGATIVE or RE-ENTRANT ≥90°. (WTop < WBottom)	
Bowing effect	Profile distortion due to a localized widening Wmax. It is a result of poor ion directionality or secondary ion etching from the bottom of the structure.	

Sidewall striations	Vertical roughness along the sidewall. It is linked to insufficient passivation removal. The "curtaining" effect degenerates with the occurrence of shards.
Sidewall breakdown	Defects due to over-etch on the sidewalls. It is due to insufficient passivation protection or excessive mask erosion giving random scallop damage (holes).
Notching	Undesired silicon etching at the interface of an insulating etch stop material. It is due to charge accumulation at the interface and it is measured as lateral distance from the vertical sidewall.
Tilting	Profile deviation from perpendicularity giving structures which are not straight. It is due to steering of ions due to non-uniform plasma.

PHASE 1	C_4F_8 deposition step	
	under electron-impact dissociation	$e + CF_4 \Longrightarrow CF_x^+ + CF_x^* + F^* + e$
	polymeric passivation layer creation	$nCF_x^* \Longrightarrow nCF_2 \text{ (ads)} \Longrightarrow nCF_2 \text{ (Teflon-like film)}$
PHASE 2	Removal of passivation layer step	
	thanks to ion energy	$nCF_2 \text{ (f)} + F^* \Longrightarrow CF_x \text{ (ads)} \Longrightarrow CF_x \text{ (g)}$
PHASE 3	SF_6 chemical step	
	under electron-impact dissociation	$e + SF_6 \Longrightarrow S_xFy^+ + S_xF_y^* + F^* + e$
	free fluorine reacts with Si	$Si + F^* \Longrightarrow Si\text{-}nF$
	thanks to ion energy	$Si\text{-}nF \Longrightarrow SiF_x \text{ (ads) and } SiF_x \text{ (ads)} \Longrightarrow SiF_x \text{ (g)}$

overall process and, in principle, any of these takes part in controlling the etch-rate and the final profile (Fig. 5.4).

Unlike standard plasma etching techniques, process parameters involved during the Bosch process are related to three plasmas instead of only one. As a consequence, the process trend space is more complicated and depends also on the relative duration of each step. Changes on one parameter usually affect the other ones, therefore experience on process development on a given system helps to handle their links and their sensitivity. System design and hardware can also limit or enhance the process sensitivity to some parameters. A global dependence trend is listed in Table 5.2, giving the main knobs only. In some cases, it may work differently depending on the exact process parameter level and applications.

5.2 DRIE Technological Challenges

Challenges for equipment and process development are given by device designer's requests to realize complex silicon patterns with high Aspect Ratio or to define high load cavities.

The in-plane dimension widths involved can be very different, from <1 μm up to millimeters, and depths can range from some microns up to 725 μm (full thickness of a standard silicon wafer). The goal is always to reach the highest anisotropy (tight CD control) with the smoothest sidewalls (absence of defectiveness/roughness/small scallops) without any tilt and any notching (to minimize the impact on device functionality).

During the years, the feature widths have been shrunk to increase the number of dies on wafer, with the consequent rising in aspect ratio. In parallel, the MEMS portfolio has also grown, including nowadays products with more elaborated architectures. Silicon micro-machining must guarantee its performances not only on standard bulk and SOI wafers but more often on a large variety of different

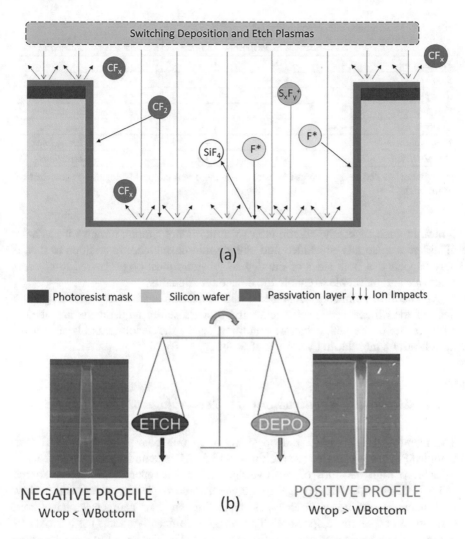

Fig. 5.4 (**a**) Many surface interactions coexist during the Bosch process; some of them are linked to the depo step, others to the etch step itself. (**b**) The balancing between etch and depo results in different profile angles: negative or re-entrant when the etch component exceeds, positive or tapered when passivation is stronger

substrates. For examples, patterns can be realized directly on thinner wafers (handle wafers down to 300 µm thickness) and sometimes the etch process should stop on big and suspended membranes (optical/acoustic products). If the silicon layer to be defined is not manageable because it is too thin, the layer itself is temporarily mounted on carrier wafers, which can be made of silicon or even of glass to become manufacturable. It happens also that the device itself is a stack of multiple wafers bonded to each other, with hardware limitations setting the upper thickness limit to

Table 5.2 Main silicon etch process trends

	Etch-rate	Selectivity	Profile	Scallop size	Sidewall
Pressure +	+	+	−v	+	Breakdown
Source power +	+	~ −	~ −v	+	~
Bias power +	+	−	−v	+	Breakdown
SF6 flow +	+	−	−v	+	Breakdown
C4F8 flow +	~ −	+	+v	−	Striations
Temperature +	+	−	−v	+	Breakdown
Depo/etch ratio +	−	+	+v	−	Striations

+: increase; −: decrease; ~: almost constant, +v: more vertical profile angle, −v: less vertical profile angle

1 mm. In such cases, the silicon etch substrate is not simple, because it has a lot of different materials embedded and already micromachined. In addition to this, it may happen that both sides of the device-substrate must be processed to connect front side and back side to enable the device functionality.

Architecture complexity represents one more challenge on top of the specific ones of the silicon etch itself: generally, process set-up adjustments are needed with respect to the bulk substrate condition due to the handling and heat transfer modifications introduced by a stack of wafers.

5.2.1 Hardware Requirements vs. Technological Challenges

The mainframe hardware for deep silicon etch is a conventional ICP-RIE (Ion Coupled Plasma – Reactive Ion Etching): a 13.56 MHz radio frequency ICP reactor, to achieve high densities of reactive species, and a separate RF power source (13.56 MHz or 400KHz) applied to the wafer to decouple ion current and energy. ICP antennas can be of three types depending on their geometry: planar coil, cylindrical coil, dome-shaped coil. The design significantly affects the behavior of magnetic field, hence plasma ionization uniformity [7]. The key difference that turns a RIE reactor in a DRIE (Deep Reactive Ion Etching) reactor is the possibility to enable faster etch rates and to manage plasma switching down to the time of 0.1 s. These performances have been achieved through heavy hardware changes (Fig. 5.5).

The first change is that gases are introduced directly into the top of the plasma chamber to minimize the delay in gas delivery time. Mass Flow Controllers (MFCs) are digital, faster, and with higher scale to maximize the gas quantity available to enhance etch rates. The operating pressure range could be very high, from 30 to 250 mTorr. Pumping systems are challenged by the amount of polymer evacuated from the chamber, hence large capacity Turbo Molecular Pump (TMP) are needed to allow high flow rates of etching gas with low chamber pressure. RF sources are more powerful (up to 5 kW full range) and are equipped with fast-response matching network to deliver controlled power into highly diverse and dynamically changing

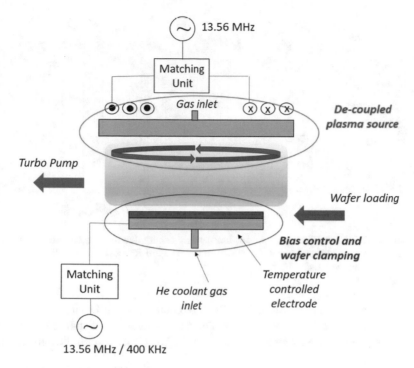

Fig. 5.5 Schematic view of an ICP DRIE reactor

plasma loads. Improved cooling systems have been developed because substrate temperature plays an important role in tuning profile angles and roughness due to the polymer condensation sensitivity (-10 to $30\,°C$ is the typical range). Generally, the heating load coming from plasma is significant: etch-rates are high, silicon exposed areas are large, and wafers can be composite, which worsens the heating transfer and makes temperature stability more critical. A parallel evolution of process control software becomes mandatory due to the lowering of single plasma step timescale and the need to minimize communication delays between the different hardware parts.

Separate mention must be made for hardware development related to notching and tilting issues and the software related to counteract high aspect ratio effect.

Notching (see Fig. 5.6) corresponds to undesired silicon etching at the interface of an insulating etch-stop material. It occurs whenever the silicon etch process lands on a dielectric (oxide, nitride, polyamide, tape, glass), i.e., with SOI wafers or composite wafers as substrates. It is due to charge build-up on the landing layer, causing incident ions to be deflected to the corners of the trench resulting in local SF_6 concentration, sidewall passivation breaking down, and lateral etch-rate increasing. Notching is measured as the lateral distance etched from the silicon sidewall. The most efficient way to get rid of this effect is to allow the charge to dissipate and this can be done by modifying the bias generator hardware to obtain a

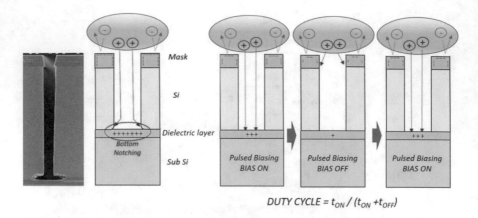

DUTY CYCLE = $t_{ON} / (t_{ON} + t_{OFF})$

Fig. 5.6 Notching formation mechanism: ion bending occurs due to charge build-up on buried oxide layer, so an additional pulsing is added to the standard RF power supply to allow charge dissipation during the BIAS OFF phase

pulsed biasing, i.e., alternating Bias ON to Bias OFF phases. During the Bias OFF phase, the dielectric discharge occurs. The new parameters in charge of notching effect control are the pulsing Frequency and the DC (Duty Cycle). Power set-point must be adapted [8].

Tilting corresponds to structure deviation from perpendicularity, and it is caused by plasma ion steering. Profile tilting heavily affects device functionality, modifies the functionality of gyroscopes, and threatens the communication between front side and back side structures in microfluidic devices (Fig. 5.7c). Two factors can rule ion injection angle: non-uniformity of plasma and the transition effect at wafer edge (Fig. 5.7a). Generally, the higher the etch-rate, the worst the profile tilting, because optimizing uniformity at high etch-rate regime is more difficult. On highly sensitive devices with very tight tilt specifications, the deviation may not be detectable by morphological analysis, as in Fig. 5.7b, but it is evaluable only by electrical measurements. To recover tilting, the process tuning approach is a minor knob compared to proper hardware design, such as changing chamber source geometry, introducing additional generators, accurately selecting ESC (Electro-Static Chuck) materials and layout.

ARDE (Aspect Ratio Dependent Etching) or RIE lag occurrence corresponds to the fact that etch depth is directly linked to the structure width: the larger the width, the deeper the trench. It is explained by gas transport and ion bombardment reduction in narrow deep trenches, and it is a complication for MEMS device fabrication, where the feature widths and aspect ratios involved are significant and different. To manage it, tool suppliers developed software packages that are able to adjust plasma parameters in real time in order to counteract the etching condition changes due to the depth proceeding, i.e., "ramping parameter" software option [9]. Also process modifications may be required to maximize the deposition at the

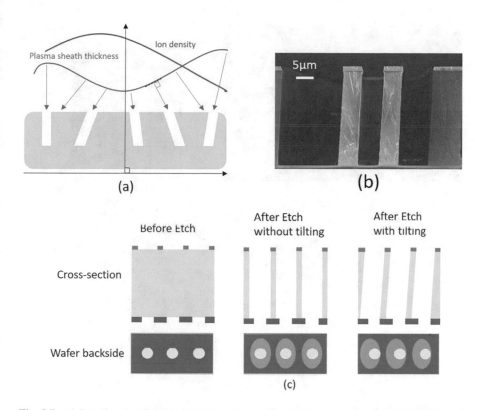

Fig. 5.7 (**a**) Ion directionality is affected by plasma sheath not-uniformity (**b**) SEM picture of steered structures (**c**) If tilting is present, the matching between front side and back side structures is no more guaranteed and their communication is lost

bottom of large features and to reduce their etch-rates. However, it is not an easy task, and it depends also on the hardware (Fig. 5.8).

5.3 Technological Challenges in MEMS Applications

5.3.1 Multi-Features Mask Devices: Accelerometers and Gyroscopes

MEMS gyroscopes and accelerometers are common sensors found in many consumer products, like smartphones and smartwatches, as well as in automotive, industrial, and medical applications. They are made by movable elements realized out of silicon defined by high aspect ratio Deep Reactive Ion Etching (DRIE) [9]. To

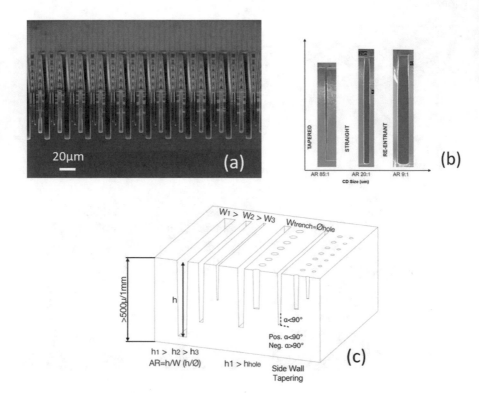

Fig. 5.8 (**a, b**) SEM picture of different width structures: not only depth, but also profile is affected by ARDE (**c**) At the same width, shape makes a difference: holes behave differently from trenches [10]

fabricate them, different manufacturing methods can be used. This paragraph will refer to ST ThELMA platform (Thick Epitaxial Layer for Micro-gyroscopes and Accelerometers, see Chap. 13 of the present book for further details). The sensitive movable masses are made of thick epipoly silicon layer grown on an oxide layer with a thickness range from 10 to 60 μm. The geometries of the MEMS mechanical elements are defined in the epipoly silicon with a mask (normally photoresist, called Trench mask) through a DRIE Bosch process landing on the oxide layer (Fig. 5.9) and performed on Inductively Coupled Plasma (ICP) DRIE reactors with high plasma density (Fig. 5.5).

Inertial sensor's trench mask presents many features of different dimensions spanning from less than 1 μm up to 30 μm, which must be etched simultaneously landing on the same insulation layer. This multi-features mask is leading to different issues to be faced with during the setup of the trench etch process, such as RIE lag and notching. Moreover, general aspects of the Bosch etch process, like Critical Dimension (CD) loss and profile control, need to be carefully evaluated (see Table 5.1 for definition).

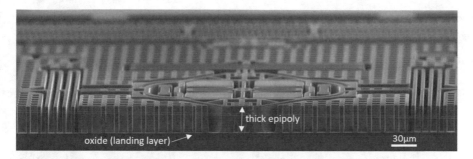

Fig. 5.9 SEM cross section of a gyro device fabricated using ST ThELMA platform. The thick epipoly layer grown on the oxide layer is defined through the DRIE Bosch process. Gyro springs and holes are visible.

In this paragraph the technological challenges of deep silicon etch process related to the realization of MEMS inertial sensors will be presented.

5.3.1.1 CD Loss and CD Control

The geometries fabricated by anisotropic etching are normally affected by a dimensional loss with respect to the nominal mask layout [11]. The difference between mask dimension and actual silicon geometry is commonly called CD loss.

The exact determination of the CD loss value is crucial in inertial MEMS design since it directly affects all the electromechanical parameters of the sensor. In gyroscopes and resonators, for example, CD loss is inversely proportional to the resonance drive frequency of the device which defines its whole behavior, i.e., lower CD loss corresponds to higher frequency and vice versa [12].

The CD loss effect is commonly considered in the modelling phase applying a homogeneous and average resize of the mask that defines the structural layer, but physical and electrical analysis, as described also in [13], have clearly shown that the dimensional loss directly depends on the local geometry exposed to the plasma etching and can change in different locations of the same chip. Not considering this effect could easily lead to rude error in the modelling phase [14]: for example, in the case of an accelerometer, a different CD loss between the mechanical spring and the electrode gap generates an error in the sensitivity of the device. If the different CD loss affects the distance between the movable mass and the mechanical stopper, this will cause a change in the mechanical full scale or a weakness with regards to stiction robustness.

For the above reasons, CD loss and CD control are technological challenges to be considered when setting up silicon etch processes for both accelerometers and gyroscopes.

Choosing the correct balance among different parameters in the process recipe helps to reach the goal of achieving lowest CD loss and lowest CD spread within wafer and wafer to wafer to match as closely as possible design values. Figure 5.10

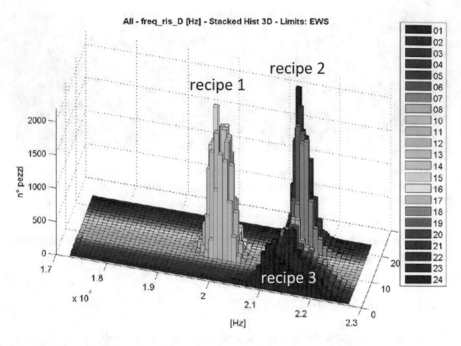

Fig. 5.10 Frequency distributions measured on gyro wafers etched with different process recipes. Recipe 2 shows a higher average frequency, i.e., lower CD loss, compared to other recipes. Recipe 3 shows a wider frequency spread compared to other recipes

shows how different process recipes lead to different frequency distributions which are strictly related to CD loss values.

As described in [15], the effect of using different etching equipment can also show a significant impact on CD loss, highlighting the need of different mask design customization for each process tool (Fig. 5.11).

From a morphological point of view, given a specific device, a preliminary analysis of the average CD loss after etching can be carried out in order to have a rough idea of the process output, i.e., if the CD loss is very high or very low. However, morphological analysis is very punctual and, for a precise and complete evaluation of CD loss on inertial sensors, electrical test is always needed.

5.3.1.2 ARDE or RIE lag

Inertial sensors normally present masks with different feature dimensions to be etched simultaneously and the Aspect Ratio (see Table 5.1 for definition) can reach up to 40:1. All the CDs have to be etched and to reach the oxide landing layer, so ARDE or RIE lag [16] is one of the technological challenges to be faced when setting up silicon dry etch processes for accelerometers and gyroscopes (Fig. 5.12).

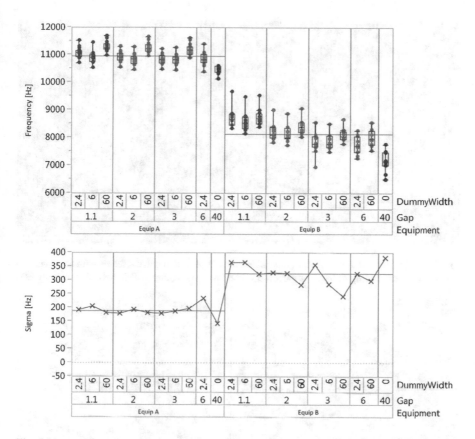

Fig. 5.11 Average value and standard deviation of the frequency obtained from the electrical measurements of wafers etched on different equipment. Equip A shows a significantly higher frequency (i.e., lower CD loss) and lower sigma compared to Equip B [15]

As a consequence of the ARDE effect, the bigger the mask CD, the sooner the feature will land on the oxide. Hence, the Bosch process must be set up to make sure that all the device features are completely open, even the most critical ones. In ST ThELMA technology platform, normally the most critical features to be opened due to RIE lag effect are the smallest holes designed on the device mass for the structures' release, generally carried out by hydrofluoric acid to remove sacrificial oxides.

If the ICP reactor is equipped with a Spectrometer, it is possible to follow the subsequent structure's opening, monitoring the reactants and products gases present in the chamber through optical emission spectroscopy. The optical signal from the plasma is translated into an electronic signal which, once reprocessed, gives the so-called EndPoint trace. This trace can be analyzed through different algorithms to stop the etching process when needed with the so-called EndPoint Detection (EPD) [17]. In Fig. 5.13, a typical endpoint trace of a gyro device shows that at the EPD

Fig. 5.12 ARDE or RIE lag effect in inertial sensor layout: etch depth decreases with narrower CD gap sizes corresponding to higher aspect ratios and lower etch rates (SEM cross section analysis)

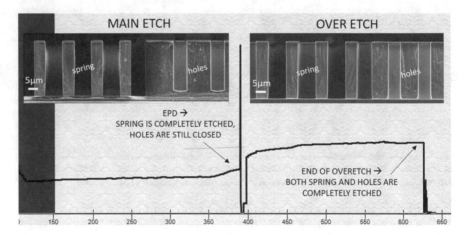

Fig. 5.13 Endpoint signal from a SOI gyro device during the etch process: Due to RIE lag, at endpoint detection, springs are completely etched while holes are still closed

the springs are completely etched and landed on oxide, while the holes are still to be completed. Adding a certain percentage of over etch, the holes also finally completely land on the oxide, but the springs will get higher etch time than needed.

Due to RIE lag, the process characterization on a specific new inertial device should start with etching one wafer, using a selected recipe with a certain number of loops (i.e., time when EPD occurs) to check which are the most critical areas. After that, SEM cross section should be performed to calculate residual silicon in the most critical features (Fig. 5.14).

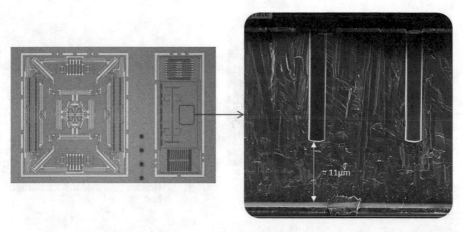

Fig. 5.14 A combo device (gyroscope + accelerometer) whose most critical features are the accelerometer mass holes (left). The SEM cross section at EPD shows that in the critical areas residual silicon is 11 μm (right)

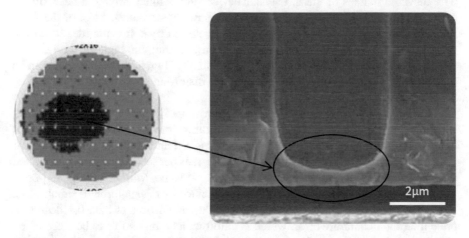

Fig. 5.15 Final electrical yield pattern of a wafer composed by accelerometer devices. Black areas in the picture on the left indicate non-yielding die due to structures not completely open as shown in the SEM picture on the right, while grey areas are good dies

With the analysis at endpoint detection, the correct percentage of over etch can be set and a complete etch on one wafer can be carried out to verify that all the features are fully landed on the insulation layer.

When the etch process is not correctly set up, some features may remain closed with a pattern related to recipe uniformity and therefore the vapor HF will not release the mass correctly. As consequence of this, the final yield will be affected as the dice turn out to be blocked at electrical testing as shown in Fig. 5.15.

 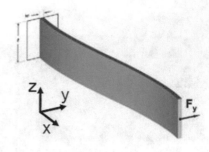

Fig. 5.16 Example of tilted structures giving a 2D structure distortion (left) and a 3D distortion (right) (SEM cross section analysis)

5.3.1.3 Profile Control and Tilt Effects

The movable masses of inertial sensor devices are defined through a deep silicon etch process which is supposed to give very straight structures. Most of the time, however, the profile of the trenches can be distorted by some non-ideality of the plasma and of the reactor itself (see Fig. 5.7a), causing an angle of the trenches which is translated in device fabrication non-ideality (Fig. 5.16, left). The tilt can be also non-uniform along the beam, giving a 3D distortion of the structure itself (Fig. 5.16, right).

These manufacturing imperfections in gyroscopes produce a mechanical coupling between the drive and sense mode vibratory modes. Excessive coupling is called quadrature error and typically leads to high yield loss [18]. Quadrature error values are normally electrically measured in degrees per second (dps) and are plotted in a wafer map with different colors based on the value. Minimal accepted q-error values are specifically defined for each gyro device by design through simulations. For some gyroscopes, for example, wall angle should be less than 0.1 degree to have minimal quadrature error, which is unfortunately impossible to be detected by physical analysis (i.e., SEM cross sections), but only by electrical measurements.

To avoid or minimize quadrature error, tilt control should be achieved balancing process parameters and choosing the correct hardware. Some process parameters are affecting tilting more than others [19]. Figure 5.17 is a representation on how the q-error yield loss is affected by pressure and source while other parameters, such as bias, temperature, gas flows, are not changing the final tilt. The different contribution of each parameter is not visible by morphological cross sections, but only by electrical measurements.

Figure 5.18 shows the effect of the reactor's hardware on the q-error yield loss: by increasing or decreasing the diameter of the plasma confinement ring in the chamber, it is possible to overturn the plasma profile tilting asymmetry, and therefore the initial quadrature pattern.

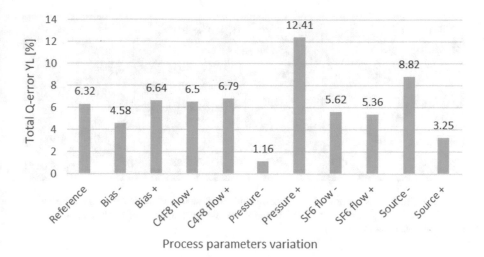

Fig. 5.17 Q-error yield loss on a gyro device with parametric process parameters variation. Data show that pressure and source are the main knobs for tilt reduction, while all the other parameters are aligned with reference

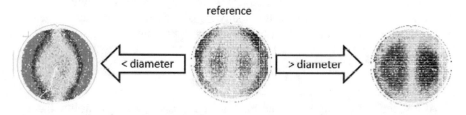

Fig. 5.18 Q-error yield pattern, black and grey areas indicate non-yielding die, whiter areas are good dies. Changing the diameter of the plasma confinement ring, the profile asymmetry is changing and therefore also the quadrature pattern on wafer with respect to reference

5.3.1.4   Notching on SOI

The moving masses of inertial sensor devices in ST ThELMA platform are defined through a deep silicon etch process with oxide as a stopping layer. When the etch process ends on an insulating layer, the oxide provides the accumulation of positive ions of the plasma at the bottom of the trench causing a deflection of other positive charges towards trench. This phenomenon is known as notching (see Fig. 5.6). The accumulation of charges is higher in features with smaller dimensions because ions can hardly exit from narrower holes [20]. To control this notching effect, a pulsing function with a square wave is used for triggering the bias generator set point. This helps to discharge the insulating layer and to reduce the notching effect [21].

Inertial sensors normally present masks with different CD to be etched at the same time to reach the same oxide layer, so notching is one of the technological challenges to be faced since the different dimensions of features will cause different

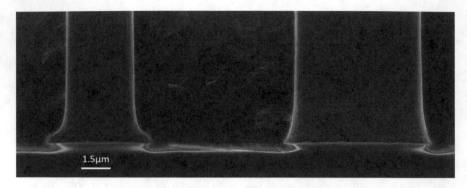

Fig. 5.19 SEM cross section of two features in inertial sensor with different CD size showing therefore different notching. The smaller one shows higher notching (left), while the bigger one shows smaller notching (right)

notching (Fig. 5.19). It is impossible to get rid of notching effect in a multi-feature mask like inertial sensor's one, but its minimization can be achieved by finding the correct balance among all process parameters.

Among all process parameters, square wave Duty Cycle (DC) and Frequency play a big role, because they directly modify the pulsing function which triggers the bias generator. The equation of Duty Cycle is the following:

$$DC = \frac{t_{on}}{t_{on} + t_{off}}$$

At first sight, reducing DC could mean reducing t_{on}, lowering therefore the time during which the ions remain at the bottom of the trench (Fig. 5.20). This could seem the main parameter to reduce notching, but cross-contamination of different effects, sometimes also unexpected, can appear.

Secondary effects on notching are also produced by process pressure and Bosch steps time modification. Increasing the etch step time of the Bosch phases, the notching increases (Fig. 5.21); the same trend happens lowering the pressure (Fig. 5.22).

Another fact that influences notching formation is the stack composition of the landing layer of the etch process: when under the insulator some other layers are present, i.e., poly layer under oxide, different notching entity can be detected due to local different accumulation of plasma ions (see Fig. 5.23).

Notching minimization is a fragile balance among many process parameters, which should be tuned properly for a specific device layout and a specific landing layer stack.

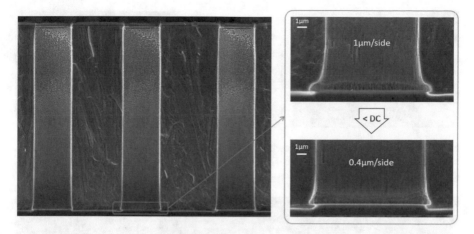

Fig. 5.20 Effect of square wave DC (Duty Cycle) on the notching. Decreasing the DC, the notching decreases from 1 to 0.4 μm/side (SEM cross section inspection on gyro holes)

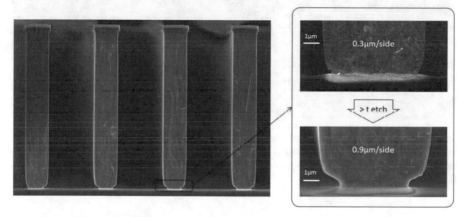

Fig. 5.21 Effect of etch step time (t_{etch}) on the notching. Increasing the etch step time in the Bosch process, the notching increases from 0.3 to 0.9 μm/side (SEM cross section inspection on gyro springs)

5.3.2 Complex Architectures: Micromirrors, Pressure Sensor, Microphone

In paragraph 5.3.1, technological challenges related to multi-features mask processing on single wafers are discussed. In this paragraph, issues and requirements linked to more peculiar wafer's architectures will be presented.

The evolution of MEMS devices leads to the conception of more complex structures [22] to achieve a higher level of integration. This brings up novel families

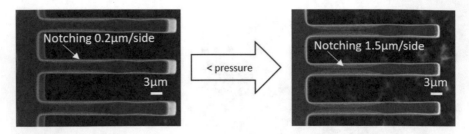

Fig. 5.22 Effect of pressure on the notching (view on the movable mass's back side after release, notching is the white area on the comb's back side). Decreasing the pressure of the Bosch process, the notching increases from 0.2 to 1.5 µm/side (SEM inspection on gyro combs back side)

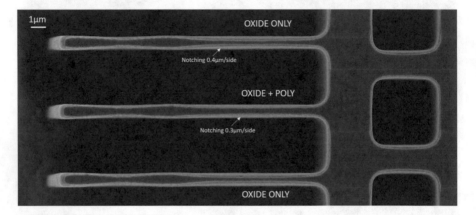

Fig. 5.23 Effect of landing layer stack composition on the notching (view on the movable mass's back side after release, notching is the white area on the comb's back side). On comb landing on oxide only, the notching is 0.4 µm/side; on combs landing on oxide + poly, the notching is 0.3 µm/side

of products and applications having challenging characteristics and requirements such as:

– Two or more wafers bonded together
– Very deep trenches (>100 µm)
– Very high exposed areas (up to 80%)

All these aspects generate a whole new class of challenges as far as the etch process is concerned, occurring alone, or mixed up together, such as:

– Landing on suspended thin layers or even vacuum
– Polymerization control due to multi-wafers architecture
– Different heat conduction
– Profile and tilt control to match previously defined structures
– ARDE and selectivity control related to stacked wafers

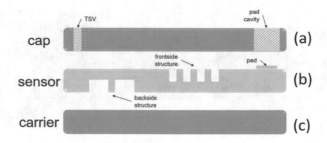

Fig. 5.24   Sketch of multi-wafers stacking

In this paragraph, all problems listed above will be described and discussed along with practical examples. Furthermore, some classes of products will be presented as case studies to illustrate all aspects more clearly.

5.3.2.1   Bonded Wafers

A way to obtain new system architectures is stacking two or more silicon wafers, bonded together temporarily or permanently with a suitable bonding technique.

Usually, the wafers can be identified as:

– Sensor wafer, which bears the actual MEMS (Fig. 5.24b).
– Carrier or handle wafer, which facilitates the handling of the sensor wafer (Fig. 5.24c)
– Cap wafer, which covers and protects the sensor wafer as well as implementing some accessory functions (Fig. 5.24a)

Thus prepared, the wafers can be further etched to create pad cavities or to open Through-Silicon Vias (TSV) or to define backside structures. This approach will be further detailed in Chaps. 16 and 18 of the present book.

In the following paragraphs, two classes of products which showed some of the mentioned problems will be discussed: micromirrors and pressure sensors.

Micromirrors (see Chap. 18 of the present book for further details) are a family of actuators devised for augmented reality (AR), virtual reality (VR), and mixed reality (MR) purposes. They vary in size, shape, and working principle, but they all need to be able to oscillate during their actuation. Hence, they require the definition of both front side and back side structures, in many cases using a carrier wafer to perform one etch without damaging the previous ones. In this configuration, the output of the etch process can be significantly affected.

If we consider two different micromirrors which are conceived for different purposes, their difference in process flows and in design lead to very different requirements. Micromirror 1 is a fast scan resonant mirror, it has very large and deep trenches and its actuation is bulk piezoelectric (Fig. 5.25, left). Micromirror 2

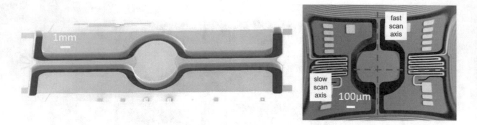

Fig. 5.25 Comparison between two different micromirrors. Both SEM pictures are taken from the front end. Micromirror 1 (left) is a bulk piezoelectric mirror, Micromirror 2 (right) is electromagnetic.

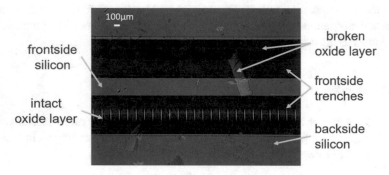

Fig. 5.26 Post etch optical image of Micromirror 1 back side. The oxide layer is broken and its shards are spread on the wafer. Both rotor and back side structures are visible

is a double axis, electromagnetically actuated mirror and it is smaller in dimensions (Fig. 5.25, right).

Micromirror 1 illustrates a case of landing on thin layers. It requires a carrier wafer to carry out the backside silicon etch process landing on a 1 μm oxide as a stopping layer. After the backside etch, the oxide layer is suspended in the air over the trenches previously realized on the other side of the wafer. Then, due to its intrinsic stress and the large dimension of the trench, it can break into pieces and its shards spreading on the wafer leading to high defectiveness (Fig. 5.26).

To prevent the oxide from breaking, one possibility is to perform a temporary bonding and fill the cavity with a large quantity of bonding polymer, so that the oxide layer is no longer suspended after backside etch (Fig. 5.27, left). But too much polymer can apply significant pressure over the backside, leading to cracks in the silicon substrate following the contour of the front end trench, as shown in Fig. 5.27 right. Therefore, a subtle tradeoff between the two situations is needed to achieve optimal process results.

On the contrary, Micromirror 2 follows a different process flow and the front side rotor structures are defined in the final part of the fabrication process, where a carrier wafer is needed as well. During the bonded wafers processing, the sensor wafer is exposed to the hot plasma, while the carrier wafer is in contact with the cooled

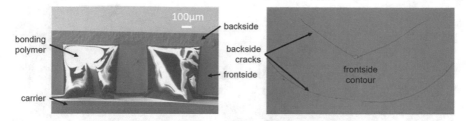

Fig. 5.27 On the left, front side trenches filled with polymer. On the right, cracks in back side silicon following the contour of front side trench (see Fig. 5.25, left for reference)

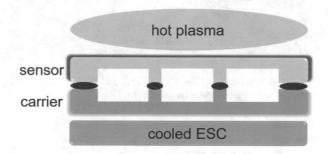

Fig. 5.28 Sketch of the processing of bonded wafer. Large cavities and few contact points between carrier and sensor wafers prevent effective heat conduction and therefore sensor wafer cooling

ESC (Electro-Static Chuck). The fact that the etched wafer is not directly chilled can lead to potential overheating problems, depending on both the process flow and the sensor-to-carrier bonding condition. They are bonded together via glue printing technique, meaning that the contact points are few. Therefore, a large empty gap is formed between cap cavities and sensor backside structures, preventing effective heat conduction (see Fig. 5.28). Therefore, the heat build-up in the sensor wafer can lead to resist burning and silicon defectiveness (see Fig. 5.29).

Another challenge related to bonded wafers and large cavities concerns polymer passivation. As already described in Fig. 5.1, in a Bosch process, the passivation step consists in the deposition of a polymeric layer (carbon-fluorine based) in order to protect the already-etched silicon from the ongoing process. Usually this polymer is highly conformal, but when a large cavity is opened, the vertical conformality is hindered because the same amount of polymer must be distributed on a larger surface. Therefore, the passivation layer gets thinner (Fig. 5.30) and the sidewall quality is directly affected (Fig. 5.31). Due to this, a recalibration of depo/etch ratio is necessary to preserve a satisfying etch quality.

A last challenge related to bonded wafer architecture in micromirrors is due to ARDE or RIE lag effect (see Fig. 5.8). As already said in previous paragraphs, the simultaneous presence of narrow and large structures usually implies a significantly prolonged over-etch during the silicon etch process. In this case, it is possible that the layout of the processed wafer might be transferred onto the carrier wafer.

Fig. 5.29 Effects of heat build-up. Optical image (left) and SEM image (top right) of resist burnings; SEM image of etch defectiveness (bottom right)

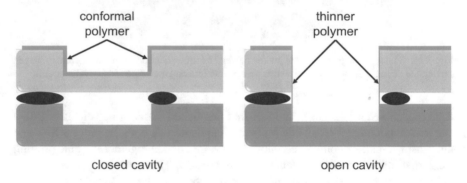

Fig. 5.30 Sketch of the issue related to large cavities opening. While the cavity is still closed (left), the conformity of the polymer is guaranteed. Once the cavity is opened, the passivation layer gets thinner (right)

In Fig. 5.32, an example of this effect can be seen through some fancy interference fringes. Of course, this event is highly undesirable for many reasons, therefore measures to prevent it must be taken both from design and process integration point of view.

Analogous problems occur also in pressure sensor's architecture. Pressure sensors are common devices found in many consumer products, like smartphones and smartwatches (see Chap. 16 of the present book for further details). Here, a cap wafer is needed to protect the sensing structures as well as to carry out two other important functions. The first one is to allow the electrical contacting to the pads of the sensor through specific cavities; the second one is to put in contact the sensing structure with the environment by means of chimney-like openings that go through all the cap wafer (Fig. 5.33).

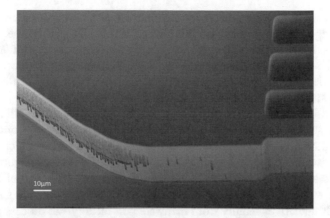

Fig. 5.31 Damaging on trench sidewall due to insufficient passivation

Fig. 5.32 On the left, example of ARDE lag causing layout transfer onto the carrier wafer. On the right, detail of interference fringes formed on the carrier wafers as a consequence of layout transfer

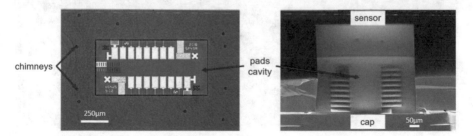

Fig. 5.33 Panoramic view of Pressure sensor's cap wafer structures: optical (left) and SEM (right) images of pad cavity and chimneys

Since any hardware for silicon dry etch intrinsically introduces a certain degree of center/edge non-uniformity, peculiar attention must be paid to the opening of pad cavities. In the region where the etch rate is faster, the cavity will open earlier, exposing the underlying structures to the etchant species for a longer time until all the cavities are opened. This means that undesired damaging can occur in the underlying structures, as can be seen in Fig. 5.34.

Such architecture allows also to discuss profile and tilt control. Differently from the inertial sensors, the tilt is not related to electrical properties. Concerning the chimneys, they need to be realized in two phases, with a separate etch from each side of the wafer. In such a configuration, assuming a perfect alignment in the lithographic steps, it is important to ensure the verticality (zero tilt) of the etch so that the two stumps of the chimney will match each other (see Fig. 5.35). Also in this case, the tilting derives mainly from hardware non-ideality, but process parameters also play a role.

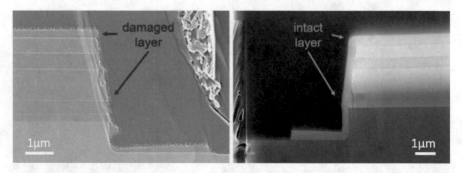

Fig. 5.34 The picture on the left is taken from a fast-etch-rate wafer region and shows the outermost layer of the stack damaged by the exposure to the etchant species. On the right, the same structure from a low-etch-rate wafer region

Fig. 5.35 On the left, good chimney etch; on the right, a bad one

5.3.2.2 Cavity Etch

The application of MEMS technology to microphones has led to the development of tiny devices with very high performances [23] (see Chap. 15 of the present book for further details), as well as to the realization of Piezoelectric Micromachined Ultrasonic Transducer (pMUT) devices allowing gestures recognition in three dimensions. In these kinds of products, challenges related to the definition of very large cavities can be encountered and will be presented in this paragraph.

Both microphones and pMUT are designed with a membrane acting as acoustic sensor, which separates the front part of the device from the so-called back cavity (see Fig. 5.36 and Fig. 5.37).

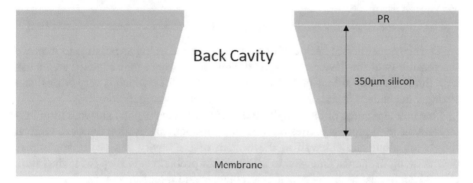

Fig. 5.36 Cross section sketch of a MEMS back cavity. Silicon etch is performed on the back side of the device, while the front side is where the membrane is placed

Fig. 5.37 SEM cross section inspection of a MEMS back cavity

Fig. 5.38 Back cavity etch performed with different chuck temperature. On the left, etch process performed at 10 °C, on the right at 17 °C (SEM inspection of the bottom of the cavity). With higher chuck temperature, roughness occurrence at the bottom of the cavity is mitigated thanks to less polymer deposition

The abovementioned back cavity should be created as big as possible to improve the sensitivity of the devices. Thus, its definition involves a high exposed area mask etch through a deep silicon etch Bosch process (Fig. 5.1), which is very critical and strongly dependent from hardware and process setup configuration.

One first challenge of back cavity definition is to keep a good sidewall morphology. As the etch process carries on, due to insufficient passivation removal, vertical roughness along the sidewall at the bottom of the cavity can occur, sometimes degenerating in shards. One way to mitigate this problem is to keep the temperature of the chuck very high to deposit less polymer and therefore remove it in a more efficient way (see Fig. 5.38).

Another challenge is to keep same process performances across different etching chambers. One critical parameter of the cavity etch is the etch enlargement with respect to mask definition, which is given by etch sidewall angle. One way to keep it under control is to measure the overlap from the cavity back side through a dedicated structure (see Fig. 5.39). This measurement gives the amount of silicon remaining after silicon etch at the bottom of the cavity, which is directly linked to the sidewall angle: low overlap value means high cavity etch enlargement, and vice versa.

Studies have shown that sidewall angle changes by modulating the bias power set point in Watt [W] in the process recipe due to a more directional plasma, causing therefore overlap values to vary. Figure 5.40 shows that this variation is ~3 μm/W and the highest the bias power, the lowest the overlap values.

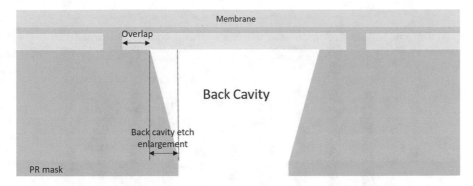

Fig. 5.39 Overlap definition at the bottom of the back cavity (sketch is upside down with respect to the direction of etching)

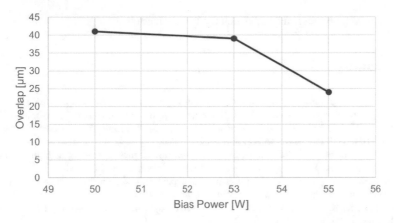

Fig. 5.40 Experimental relationship between bias power set point and overlap values

Sometimes, in a production environment, different etching chambers perform differently in terms of power delivered, because bias generators and also other hardware parts despite being the same on paper, are physically different. Thus, the trend shown in Fig. 5.40 can be used to tune plasma directionality and to align final result on wafers. One example of different reactors variability is shown in Fig. 5.41, where bias power set point is shown for different chambers. Chamber to chamber bias power differences can span from 5 W up to 30 W to align overlap values! This is a good example to understand that, once process parameters are fixed for one chamber, different hardware parts can introduce variability which has to be addressed.

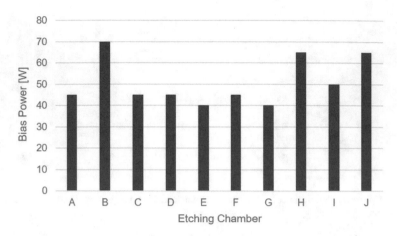

Fig. 5.41 Bias power set point modulation for different reactors

References

1. Tachi, S., Tsujimoto, K., & Okudaira, S. (1988, February). Low-temperature reactive ion etching and microwave plasma etching of silicon. *Applied Physics Letters, 52*(8), 616–618.
2. Laermer, F., & Schilp, A. (1993). *Method of anisotropically etching silicon.* German Patent No DE4241045 (US Patent No 5501893, Mar. 26, 1996).
3. Blauw, M. A., et al. (2002). Advanced time-multiplexed plasma etching of high aspect ratio silicon structures. *Journal of Vacuum Science and Technology B, 20*, 3106–3110.
4. Rangelow, I. W. (2003). Critical tasks in high aspect ratio silicon dry etching for microelectromechanical systems. *Journal of Vacuum Science and Technology A, 21*, 1550–1562.
5. Chung, C. K., & Chiang, H. N. (2004). *Inverse RIE lag of silicon deep etching.* NSTI-Nanotech 2004, www.nsti.org, ISBN 0-9728422-7-6, Vol. 1.
6. Ayón, A. A., Braff, R., Lin, C. C., Sawin, H. H., & Schmidt, M. A. (1999). Characterization of a time multiplexed inductively coupled plasma etcher. *Journal of the Electrochemical Society, 146*(1), 339–349.
7. Zlatanov, N. *Advanced plasma processing for semiconductor manufacturing.* Semicon West 2000, San Francisco.
8. Munro, S. (2009). *Notch free SOI etching on the STS ICP-RIE.* NanoFab, University of Alberta. https://www.nanofab.ualberta.ca/wp-content/uploads/2009/07/Notch-Reduction-on-the-STS-ICP-RIE.pdf
9. Owen, K. J., Van Der Elzen, B., Peterson, R. L.., & Najafi, K. *High aspect ratio deep silicon etching.* MEMS conference 2012.
10. Tang, Y., Sandoughsaz, A., Owen, K. J., & Najafi, K. (2018). Ultra deep reactive ion etching of high aspect-ratio and thick silicon using a ramped-parameter process. *Journal of Microelectromechanical Systems, 27*(4), 686–697.
11. Laermer, F., Schilp, A., Funk, K., & Offenberg, M. *Bosch deep silicon etching: Improving uniformity and etch rate for advanced MEMS applications.* MEMS conference 1999.
12. Kaajakari, V. *Micromechanical resonator.* US Patent No US 2009/0189481 A1, Jan. 23, 2009.
13. Gattere, G., Rizzini, F., Corso, L., Alessandri, A., Tripodi, F., & Gelmi, I. (2018). *Experimental investigation of MEMS DRIE etching dimensional loss.* The 5th IEEE international symposium on inertial sensors and systems.
14. Kempe, V. (2010). *Inertial MEMS – Principles and practice.* Cambridge University Press.

15. Gattere, G., Rizzini, F., Corso, L., Alessandri, A., Tripodi, F., & Paleari, S. (2019). *Geometrical and process effects on MEMS dimensional loss: A frequency based characterization*. The 6th IEEE international symposium on inertial sensors and systems.
16. Jansen, H., de Boer, M., Wiegerink, R., Tas, N., Smulders, E., Neagu, C., & Elwenspoek, M. (1997). Rie lag in high aspect ratio trench etching of silicon. *Microelectronic Engineering, 35*, 45–50.
17. Roland, J. P. (1985). Endpoint detection in plasma etching. *Journal of Vacuum Science & Technology A, 3*, 631.
18. Weinberg, M. S., & Kourepenis, A. (2006, June). Error sources in in-plane silicon tuning-fork MEMS gyroscopes. *Journal of Microelectromechanical Systems, 15*(3).
19. Merz, P., Pilz, W., Senger, F., Reimer, K., Grouchko, M., Pandhumsoporn, T., Bosch, W., Cofer, A., & Lassig, S.. *Impact of Si DRIE on vibratory MEMS gyroscope performance*. Transducers and Eurosensors conference 2007.
20. Hwang, G. S., & Giapis, K. P. (1997). On the origin of the notching effect during etching in uniform high density plasmas. *Journal of Vacuum Science & Technology B, 15*, 70–87.
21. Laermer, et al. *Plasma etching method having pulsed substrate electrode power*. US Patent No US 6,926,844 B1, Aug. 9, 2005.
22. Mayurika, J. (2018). *Study on MEMS (Micro Electro Mechanical Systems)*.
23. Widder, J. et al. (2014). *Basic principles of Mems microphones (EDN)*.

Chapter 6
Optical Lithography

Aldo Bortolotti, Nadia Galimberti, Marco Salina, Martina Scolari,
and Lorenzo Tentori

6.1 Introduction

Optical lithography is a patterning process technique for the realization of accurate structures. This technology permits to transfer a pattern to a photosensitive material using a light source. The photosensitive material, PR (PhotoResist), deposited on the substrate, is irradiated through a glass mask on which a patterned opaque chromium layer is located. Depending on the photoresist type, the exposed or unexposed areas are then dissolved using a developer solution, defining the pattern of the device structures.

In the subsequent processes, the pattern transferred to the PR during the lithography step can be reproduced on the underlying substrate (silicon, metal, oxide layers, etc.) through Dry or Wet etch processes, or can be used to define the pattern of a layer deposited onto the mask surface, as described in Fig. 6.1. In other words, MEMS manufacturing can be described as the stacking of multiple 2D structures, defined on top of each other using lithographic techniques, to fabricate 3D structures and devices.

Depending on the complexity of the devices, the ability to tightly control the critical dimensions and layer-to-layer alignment are fundamental considerations for building MEMS.

A. Bortolotti · N. Galimberti (✉) · M. Salina · M. Scolari · L. Tentori
ST Microelectronics, Analog MEMS and Sensors Group, MEMS Technology and Design R&D,
Agrate Brianza, Monza Brianza, Italy
e-mail: aldo.bortolotti@st.com; nadia.galimberti@st.com; marco.salina@st.com;
martina.scolari@st.com; lorenzo.tentori@st.com

© Springer Nature Switzerland AG 2022   169
B. Vigna et al. (eds.), *Silicon Sensors and Actuators*,
https://doi.org/10.1007/978-3-030-80135-9_6

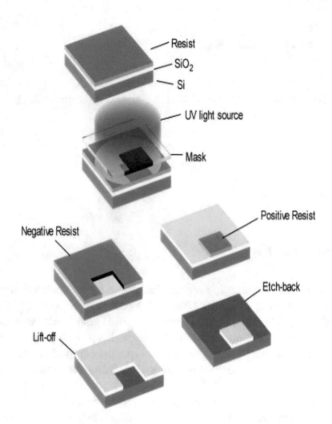

Fig. 6.1 Technology steps definition

The typical process steps involved in the lithography processes are shown in Fig. 6.2. Dehydration bake and primer dispensing are used to prepare the substrates for the resist coating which can be performed using spin/spray processes. Then a soft bake step is executed before the alignment and the following exposure. A post exposure bake for solvent evaporation is then required before the development step and an optional hard bake can be necessary at the end of the process flow to slope and strengthen the photoresist. All these lithography steps and the tools used to realize them are chosen to fulfil the process and design requirements.

Lithographic process requirements for MEMS application are often very different from those related to other technologies as IC chip and Transistor fabrication. They include the definition of thick photoresist with high aspect ratio that can be used for the realization of thick metallic structures by ECD (Electro Chemical Deposition) or for the creation of freely moving structures through deep dry etch step.

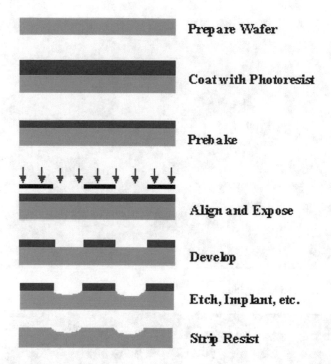

Prepare Wafer

Coat with Photoresist

Prebake

Align and Expose

Develop

Etch, Implant, etc.

Strip Resist

Fig. 6.2 Lithography process steps

Lithography lift-off applications are also quite common in MEMS fabrication. This method, used to generate a negative sidewall photoresist profile, allows the structuring of materials that cannot be etched easily, such as noble metals. Another requirement typical of MEMS manufacturing is conformal coating to cover substrate with high topography. Moreover, MEMS often require the patterning of devices' structures on both sides of the substrate (front and back sides), on standard or bonded wafers. For these reasons, the lithography exposure tools need to be able to manage wafers with high topography and warpage and to realize bottom side alignment and IR alignment.

In addition, in MEMS devices, it may be desirable to pattern polymeric materials, to create flow channel for permanent bonding application for inkjet or Micromirror devices, introducing different dedicated tool and materials into the fabrication process, as well as additional material functionality.

Examples of photolithography in MEMS are shown in Figs. 6.3 and 6.4.

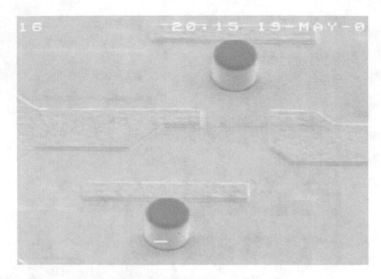

Fig. 6.3 Photoresist pillar definition

Fig. 6.4 Lithography mask definition

6.2 Thick Photoresist [1, 2]

6.2.1 Introduction

The increasing demand for high aspect ratio MEMS structures has driven the need to perform photolithography process with thick photoresist. Increasing the photoresist thickness, the aspect ratio could become a difficult matter. Taking in account a photoresist thickness of 30 μm a 3:1 aspect ratio means a photoresist pillar with 10 μm base and 30 μm height. In this situation, it is frequent to run into structures collapsing.

Thick photoresist masking is used mainly for deep silicon dry etch step [3] and ECD (Electro Chemical Deposition) step [4]. In the first case, the photoresist must endure ions etching, while, in the second case, it must be resistant to wet chemicals used during ECD step. In both cases, the process time can be much longer than standard semiconductor process, so the photoresist must be more resistant than standard ones.

Due to exposure time increasing with photoresist thickness, exposure process could be a bottleneck for standard photoresist. To avoid this situation, it is possible to find several Chemical Amplified Photoresist (CAR) in the market. Photo activation, for chemical amplified photoresist, is performed through an exposure and a curing step, and both are faster than lithography processes for standard photoresist.

6.2.2   Method and Tools

Thick photoresist coating step could be performed with different techniques such as standard spin-on coating [5, 6], dry film lamination [7] or spray coating [8]. All these techniques are used also for thin photoresist, so, the same equipment can be used.

A particular care must be dedicated to the exposure tool. To have a good photoresist profile definition, it is mandatory to have enough depth of focus of the exposure tool optics. The exposure equipment mainly used in these situations are i-line stepper with low numerical aperture or 1x mask aligner. In contrast to mask aligner, where the wafer is exposed in a single shot through the photomask pattern, steppers use photomasks in which, not the whole wafer pattern is present, but only a certain number of rows and columns of dies are present. With this technique, with a single exposure (single "step"), only a part of wafer surface is exposed meaning that to cover the entire wafer surface, the pattern on the reticle has to be exposed repeatedly across the surface of the wafer in a grid, see Sect. 6.5.2. These exposure tools could use different wavelengths: i-line stepper uses a source of Hg atoms to emit light and from the entire spectrum, with an optical filter, only the i-line is selected.

6.2.3   Case Study 1: Thick Photoresist for ECD (Electro Chemical Deposition)

Chemical Amplified Resists (CAR) [9, 10] are a family of photoresist that are activated not only by exposure step but, to be completely soluble on develop solution, they also need to receive a thermal budget. For CAR, a thermal step on

Fig. 6.5 ECD defect due to chemical amplified photoresist contamination

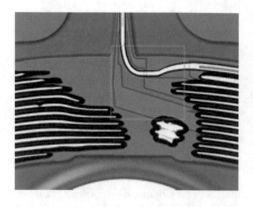

a hot plate after exposure is mandatory. Due to this double activation mechanism, Chemical Amplified Photoresists are much faster than the standard ones because a high exposure dose for activation is not needed. In general, the thicker the photoresist is, the higher is the exposure dose needed to generate a complete photoresist reaction, this is the reason why Chemical Amplified Resists are a valid choice when high film thickness is needed. Another advantage of CAR is that they are compatible with standard photoresists so they can be used on the same tracks already present in the semiconductor industries. On the other hand, chemical amplified photoresists are sensible to airborne contamination. Basis elements present in the cleanroom air could inhibit cross link activation, causing photoresist residuals during the development step. Contaminant elements could also be present on wafer surface. In this case, a surface pretreatment is necessary to avoid lithography defects.

Figure 6.5 shows a portion of die where metal structures are not present, this defect is related to an incomplete photoresist develop process caused by wafer surface contamination.

When it is necessary to reach high photoresist thickness, it could be useful to choose a photoresist with high viscosity because, due to less volatile content, it is possible to obtain high thickness with lower rotation speed during coating step. This situation increases coating stability and guarantees a wider process window. The drawback related to the presence of less volatile content is that the final photoresist thickness is not defined only by rotation speed during coating step, as happens for standard resist, but it depends also to thermal treatments performed during coating process.

In this case study, a thick CAR process setup is described. In particular, the thick photoresist mask is used to define thick metal structures through electro chemical deposition process for pico-projector fabrication.

The selected photoresist must be resistant to ECD chemistry and it has to respect the aspect/ratio requirement of 3:1. Figure 6.6 shows a cross section SEM image of the photoresist after develop step.

The photolithography process characterization has been performed through Bossung plot analysis obtained with a standard Focus/Expo Matrix methodology

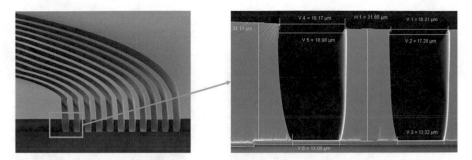

Fig. 6.6 In the left image, a SEM view of a cross section performed on a 30 μm thick photoresist is shown. In the right image, a detail of photoresist pillar is reported

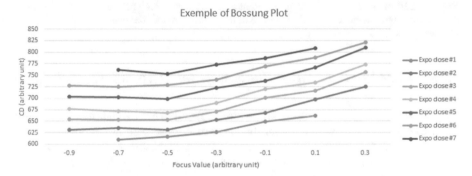

Fig. 6.7 In this graph, an example of Bossung plot is reported

(FEM). Focus/Expo Matrix methodology consists in the preparation of a wafer, already coated with the photoresist under evaluation, in which each exposure step is performed with different exposure dose and focus value. In this way, it is possible to evaluate on a single wafer different process conditions. Photoresist performances and exposure process window are evaluated through a CD (Critical Dimension) measurement on metrological SEM equipment and by the realization of two graphs: one reporting on X axis the focus values and on Y axis the CD measured for each exposure dose evaluated, while in the second, exposure doses are reported on X axis for each focus value explored. In Fig. 6.7, an example of a generic Bossung plot is reported.

After the characterization, the thick photoresist is implemented in MEMS process flow and the following electrochemical growth is performed. Figure 6.8 shows an optical image of an array of MEMS devices fabricated using thick metal structures. These structures are copper coils that permit MEMS device actuation. A detail of these structures at SEM, after the electrochemical deposition step, is also shown in Fig. 6.8.

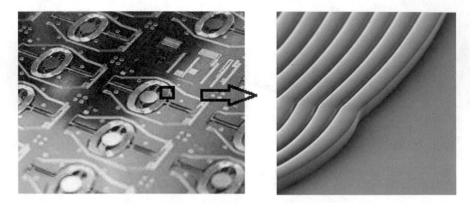

Fig. 6.8 Left image shows an optical image of a MEMS device on which thick metal structures are present. Right image shows a SEM image of metal structure (copper coils) after ECD step and photoresist removal

6.2.4 Case Study 2: Thick Photoresist for Deep Silicon Etch

As previously stated, some lithographic levels related to MEMS fabrication, require the use of photoresists with a higher thickness than those used in "standard" levels. This happens, for example, when a deep silicon etch step is involved and the photoresist must be able to withstand the following etch step. In this case study, the photoresist characterization performed to carry out a 400 μm deep silicon dry etch is described.

Different thicknesses of the selected photoresist have been evaluated and, in this paragraph, the 40 μm resist thickness is presented. The coating recipe has been optimized for 40 μm thickness and the exposure step has been performed on a mask aligner equipment. For the correct characterization of a new photoresist, there are many parameters to control and define. Starting from the coating step, the spin curve, which represents the thickness of the resist as a function of the rpm (revolutions per minute), needs to be evaluated together with the photoresist uniformity on the wafer at the desired thickness. An example of spin curve graph is shown in Fig. 6.9.

The stability of the coating process is then assessed performing marathons of repeatability which consist, typically, in the thickness measurement of the resist on 25 bare silicon wafers. In this case, the exposure step is performed on a mask aligner tool with 1X field, where mainly three parameters need to be evaluated: the dose, the exposure gap, which is the distance between the wafer and the photomask during exposure, and the alignment gap, which is the distance between the wafer and the photomask during alignment. In this case, different doses are selected to evaluate and define the best exposure dose. The development step has been characterized and optimized defining the correct times and temperatures of the post exposure bake. The number and duration of the development puddles, define how long the chemical developer stays on the wafer to get a complete photoresist development.

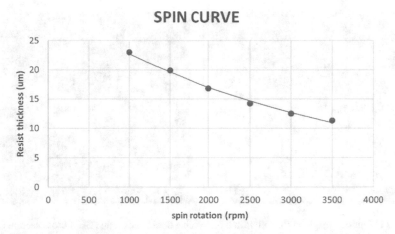

Fig. 6.9 Example of a photoresist spin curve

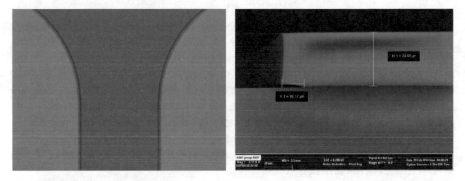

Fig. 6.10 Optical image on the left with a visible double edge, and SEM cross section on the right with a visible resist detachment

To complete the evaluation, a careful optical inspection and an evaluation with a Scanning Electron Microscope (SEM) cross section is necessary. A particular attention needs to be paid to unwanted effects during this phase. In this case, indeed, the adhesion of the resist to the substrate is a very important parameter. If the photoresist is not well adherent to the substrate, the subsequent deep silicon etch may fail, creating situations that lead, for example, to incorrect critical dimensions (CD), defects, and even the failure of the etch step.

In case of resist adhesion issue, through an optical inspection, it is possible to detect a double PR edge as shown in Fig. 6.10. In the same figure, a cross section of the same structure is reported: the photoresist, whose thickness is very close to the 40 μm target, shows a vertical profile with a partial detachment from the substrate. To improve the photoresist adhesion, different trials can be performed. In particular pre wet step, Post Exposure Bake and development step have to be deeply investigated.

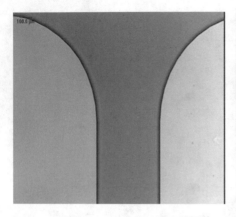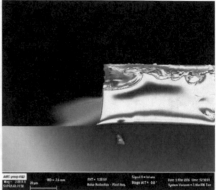

Fig. 6.11 Optical image on the left with a NO visible double edge, and SEM cross section on the right with NO resist detachment

Sometimes could happen that, to completely solve photoresist adhesion issue, a priming process with a dedicated adhesion promoter [11] must be implemented. The adhesion promoter is used to modify the substrate surface affinity guaranteeing the resist adhesion. SiO_2, in the form of quartz, glass, or silicon with (native) oxide as well as some metals, tends to form polar OH bonds on its surface after a sufficiently long exposure to atmospheric humidity. In this way, the substrates result hydrophilic ("water loving") and exhibit a poor affinity for the non-polar or low polar resin molecules of the photoresist. To make hydrophobic such a substrate surface (water-repellent and thus photoresist-attractive), the non-polar molecules of the adhesion promoter are chemically bond on it.

The implementation of the adhesion promoter improves the adhesion with the substrate as shown in Fig. 6.11.

6.3 Dry Film [12, 13]

Dry film resists are photosensitive dielectric materials used typically in technology for 3D Through-Silicon Vias and two-dimensional wafer bonding applications. Other non-3D applications for these photoresists include stress buffer and RDL (Redistribution Layer) [14, 15] dielectrics for wafer-level packaging [16, 17].

In this paragraph, the application of W2W bonding will be treated.

Adhesive wafer bonding is a technique that uses an intermediate layer for bonding (typically a polymer). The main advantages of using this approach are low process temperature (maximum temperatures well below 400 °C), surface planarization, and tolerance to particles. Evaporated glass, polymers, spin-on glasses, resists, and polyimides are some of the materials suitable for use as intermediate layers for bonding. The main properties of the dielectric materials, required for a large field of versatile applications/designs, can be summarized as: isotropic dielectric constants, good thermal stability, low CTE (Coefficient of Thermal Expansion) and Young's modulus, and a good adhesion to different substrates.

The use of this polymeric layer yields to a permanent bonding between device and cap wafers at low temperature (<200 °C) and low bow of the bonded wafer pair. Low stress on the bonded wafers is a key parameter for device functionality and subsequent process steps.

Another important advantage in the use of this dry film as adhesive layer is the simple definition of bonding areas by means of standard lithography process. Adhesive layer has a dedicated design at wafer edge to increase robustness of the bonded pair during post-processing steps.

6.3.1 Process Steps

The Dry film lithography process is divided into five different steps: lamination & delamination, reflow as optional step, exposure, post exposure bake, and development.

For each process step, the main process parameters are reported in the process scheme below, Fig. 6.12.

6.3.2 Case Study: Lamination on Cavities

In some MEMS devices process flows it is required to define the dry film as permanent bonding material on wafers with cavity already in place, typically with high topography (>100 μm) (see Fig. 6.13).

For these type of applications, the main material requirements are good adhesion on substrate and thickness uniformity.

Starting from the standard process flow, the reflow step should be removed to avoid any bonding material reflow in the cavities. All the other process parameters are the same as defined for a standard lamination on flat wafer.

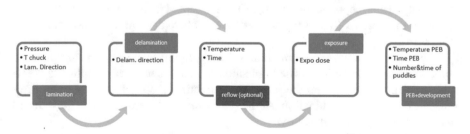

Fig. 6.12 Dry Film Process Steps

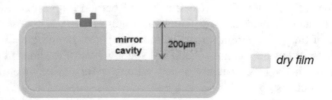

Fig. 6.13 Wafer section

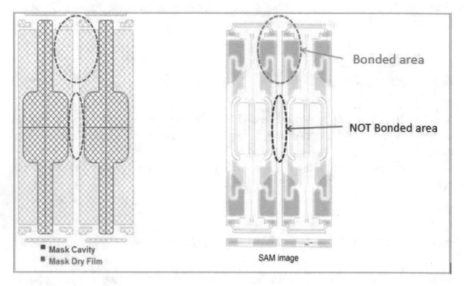

Fig. 6.14 Layout of Dry film material and silicon cavity compared with a SAM image. The dark area indicates the bonded zone, the bright area is the not bonded zone

The characterizations necessary to validate the process are mechanical profilometer measurements, for dry film thickness, as well as SAM (Scanning Acoustic Microscope) inspection on bonded wafer pairs for detection of bonding quality.

With the standard process steps (as reported in Fig. 6.12), the dry film thickness uniformity is unacceptable (about $20\mu m$) and the SAM images confirm these results showing non bonded (light grey colored) areas where dry film thickness is thinner (Fig. 6.14).

The cavity effect on the lamination and lithography process brings to introduce some process modifications. The lamination pressure is reduced to optimize "tenting" dry film on cavity, and the lamination direction is optimized to have uniform pressure during wafer lamination. At the same time, the "thermal budget" is reduced to avoid material reflow in cavity.

After retuning the process steps, the dry film uniformity is good (uniformity$\cong$ $1\mu m$), and SAM image confirms this result showing that all dry film areas are properly bonded (Figs. 6.15 and 6.16).

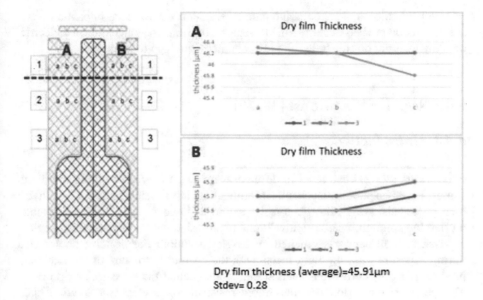

Dry film thickness (average)=45.91μm
Stdev= 0.28

Fig. 6.15 Profilometer measurement – trial

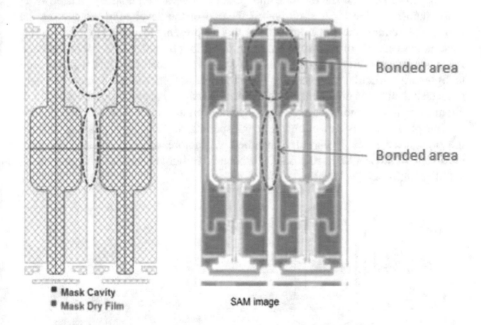

Fig. 6.16 Layout of Dry film material and silicon cavity compared with a SAM image. The Dry film area is completely and properly bonded

The lamination and lithographic parameters defined have been implemented in the production process of some MEMS sensors, like mirrors, that require a dielectric bonding material on substrates already patterned with high topography.

6.4 Negative Photoresist [18, 19]

6.4.1 Introduction

Photoresists [20, 21] can react in different ways when they are exposed to a light source. Based on photoresist chemical formulation, it is possible to have a "positive" tone photoresist, when the photoresist region exposed to light becomes soluble on develop solution, or a "negative" tone photoresist, when the photoresist region exposed to light cannot be melted on develop solution. For negative photoresist during exposure step, the light reacts with the chemical structure of the material, producing a polymerization reaction that is also called the cross-linking process. The polymerized region becomes more resistant to development solution and, due to this process, the photoresist exposed to light during the exposure step remains on substrate surface, while unexposed regions are completely washed away with photoresist developer solution. At the end of the developing process, on the wafer will remain the opposite pattern that is present on mask. In Fig. 6.17, a schematic overview of the cross-linking process is reported. Cross-linking process is a chemical reaction that happens to a negative photoresist when it is exposed to light at a particular wavelength. The molecules of photoresist exposed to light react chemically and create link with molecules nearby. After the formation of these chemical bonds, photoresist is no longer soluble on develop solution in that region.

The photoresist is not completely transparent, so, the light, passing through the photoresist layer, is partially absorbed by photoresist itself, causing the non-verticality of its profile. In Fig. 6.18, a schematic of light absorption and photoresist profile definition is reported.

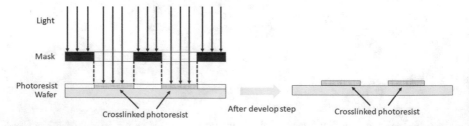

Fig. 6.17 Cross-linking process for negative photoresist. The exposed photoresist regions react to light producing polymerization, these regions become more resistant to development solution and remain on wafer surface after develop step

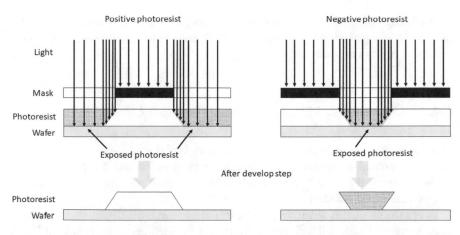

Fig. 6.18 Patterned photoresist shows different shapes in relation to their tone. The photoresist is not completely transparent, so, the surface of photoresist absorbs more energy than the photoresist in contact with the substrate. This phenomenon has a major impact close to boundaries of patterned structures. Due to the partial energy absorbed by structure boundaries, the photoresist profile is not vertical. Positive tone photoresist shows a positive profile while negative tone photoresist shows a negative profile

The partial absorption of light produces a negative profile for negative tone photoresist. This peculiar behavior is useful for some applications, such as metal definition, through lift-off technique. Indeed, negative photoresists can be used when it is necessary to define layers that, due to the process flow and/or to wafer topography, cannot be wet or dry etched. In this case, the metal is sputtered after photoresist definition, and it is removed through photoresist dissolution on dedicated solvent. The negative profile helps solvent infiltration and promotes photoresist lifting. In this way, only metal layer that has been sputtered on the substrate remains on wafer.

For negative photoresists, the material that remains on substrate after developing step is polymerized. The cross-linking process gives to the photoresist more stability and the photoresist is less susceptible to profile deformation during further thermal steps. This is the reason why it is preferable to use a negative photoresist when a higher temperature stability is needed during further process steps.

Some negative photoresists are chemically designed to become permanent after a dedicated curing step. For this type of photoresists, there is a temperature limit over which they are no longer soluble in wet chemical photoresist removal solution.

In most cases, negative photoresists are compatible with standard lithographic equipment used for positive photoresist. Some negative photoresists are soluble in TMAH [22, 23] (Tetramethylammonium hydroxide) aqueous solution as well as standard positive photoresist, but others are soluble in solvent solution; in these cases, a dedicated tool is mandatory to perform develop step.

Particular attention must be paid to mask design. In fact, as shown in Fig. 6.18, the photoresist pattern that remains on the substrate is inverted with respect to the structures present on the mask.

6.4.2 Case Study: Metal Definition Through Lift Off Process

In MEMS fabrication, it is sometimes necessary to define a metal layer that is not easy to be etched using standard techniques present in semiconductor fabs. In this situation, through a lift-off process [24], a negative tone photoresist could be used to avoid the metal etch step.

An example of this situation is represented by a MEMS device that requires a fixed pressure inside MEMS cavity during all its lifetime, such as some MEMS gyroscopes. To avoid pressure increasing inside the cavity, due to material outgassing, a metal with gettering properties must be introduced in the MEMS process flow. When the sealing between device wafer and cap wafer is performed, usually this happens during wafer-to-wafer bonding, the gettering material is activated through a thermal step. After the activation, it absorbs the gas desorbed by MEMS device during its life.

To perform a lift-off process is necessary to define a negative tone photoresist on which the metal layer will be sputtered. In order to complete the lift-off process is necessary to perform a dedicated wet resist removal step.

The negative profile of the photoresist generates a shadow effect that creates metal layer interruptions during sputtering step close to photoresist edge. During the wet resist removal step, the chemical solution penetrates through the getter layer breakages, melting down the photoresist. The metal layer sputtered on photoresist is removed with the photoresist itself. In Fig. 6.19, a schematic of lift-off process is reported.

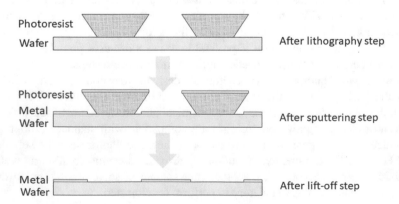

Fig. 6.19 A schematic of lift-off process is reported. Sputtering step is performed after photoresist patterning. Photoresist and metal layer on it will be removed during lift-off step

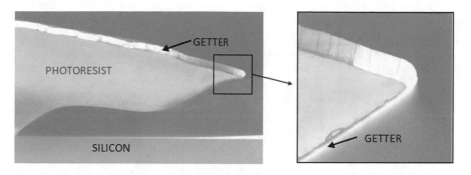

Fig. 6.20 A SEM cross section of photoresist after metal sputtering step is reported. The good conformal properties of gettering layer permit metal layer to be spattered on photoresist side wall

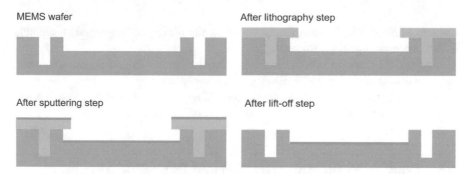

Fig. 6.21 Schematic of lift-off process for a MEMS device is shown

In most of the cases, standard metal deposition processes are characterized by poor metal step coverage that leads to discontinuity of metal layer on negative photoresist profile, allowing wet chemical removal of photoresist during the lift-off process.

It may happen, such as for some gettering material, that the metal step coverage obtained during sputtering step is very good. Due to this behavior, the metal layer will be sputtered also on photoresist side walls. In this situation, metal layer is not completely interrupted by shadow effect, but some layer breakages are present, and the lift-off process can be performed anyway. Figure 6.20 shows that the metal layer is present also on the photoresist side wall.

When the metal layer has a good step coverage, the use of silicon cavity sidewall can be useful to increase shadow effect during sputtering step, Fig. 6.21.

In this example, the presence of cavities is used to increase the efficiency of the properties of negative tone photoresist to form negative profile. Close to cavities' side walls, the photoresist thickness is higher than outside the cavity region. The partial absorption of light by negative photoresist creates a large area where only the upper part of photoresist is cross-linked. The lower part of photoresist, that is

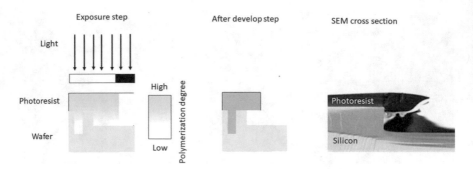

Fig. 6.22 A schematic of exposure process for negative tone photoresist in presence of cavity is illustrated. In the right image, a SEM cross section of a real MEMS gyroscope cap wafer is reported

not cross-linked, will be completely washed away during developing step, creating a large undercut under photoresist.

Figure 6.22 shows a schematic of the partial light absorption phenomenon and the results obtained on a real MEMS gyroscope cap wafer.

The large shadow area created by photoresist and ceiling trench side-wall permits a complete metal layer removal through the lift-off process.

6.5 Front to Front, Infrared (IR) and Back Side Alignment (BSA) [25, 26]

6.5.1 Introduction

Alignment, together with resolution, is one of the most critical aspects in lithography. Alignment marks placement, definition and recognition are very important steps to be taken into consideration once a new process flow is developed in all technologies and, especially, in MEMS fabrication. They determine the alignment strategy and, together with the desired resolution, the exposure tool to be used. The most common alignment method used in semiconductor industry is the Front to Front Alignment in which both, the mask to be defined and the alignment marks to be recognized, are on the same side of the wafer and they are visible from the surface, as described in Fig. 6.23.

However, the fabrication of microelectromechanical systems (MEMS) devices often requires processing and patterning both sides of the wafer. In these cases, the patterns on the backside of the wafer must be accurately aligned with respect to the features already defined on the front-side. The working principle of BSA (Back Side Alignment) is shown schematically in Fig. 6.24.

Another alignment option, quite common in MEMS fabrication, is Infrared (IR) alignment. This is used when thick epitaxial growths cover the alignment marks,

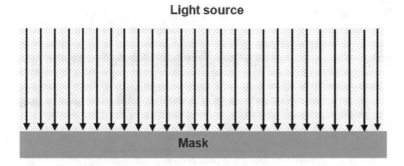

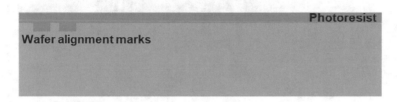

Fig. 6.23 Front to Front alignment scheme

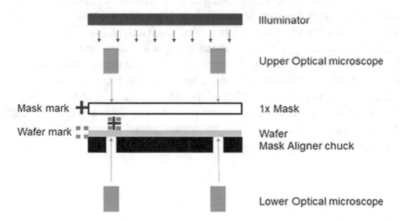

Fig. 6.24 Back Side Alignment scheme

or after wafer to wafer bonding processes to align the patterns on the two wafers. Different wafer bonding processes are used in medium and large volume production of MEMS (e.g., accelerometers, gyroscopes), Silicon-on-Insulator (SOI) substrates, inkjet products, and other consumer products. In many of these applications, after thinning and polishing the wafer pair, the backside of the thin device wafer needs to be patterned with one or more mask levels. Standard spin coating and developing

Fig. 6.25 IR Alignment scheme

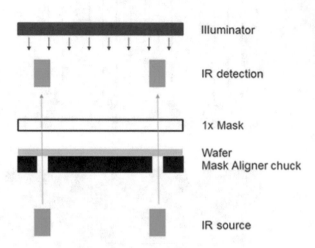

processes used for single wafer can be applied also to bonded wafers. The exposure of the resist-coated surface requires the alignment of the mask to the features on the device wafer front side, which is buried in the bond interface

Figure 6.25 shows the working principle of IR alignment: an infrared (IR) source is used to illuminate features on the back side of silicon wafer so that these can be "viewed" from the front-side of the wafer, since silicon is somewhat transparent to IR radiation. A limitation to this process is that the wafer must not have any IR-opaque layers (such as gold, aluminum metallization) in the alignment mark regions on both the wafer sides. Another limitation of the IR technique to be taken into consideration is given by the doping of the silicon which must be less than $1E^{19}$ at/cm^3 otherwise the IR radiation cannot pass through the wafer and the alignment would be compromised. In any case, the region around the alignment marks needs to be transparent to IR-radiation.

6.5.2 Method and Tools

The Front to Front Alignment strategy can be implemented working both in contact/proximity exposure (mask aligner) or in projection exposure (scanner, stepper). The basic principles of these different exposure strategies are shown in Fig. 6.26.

In contact printing, mask and wafer are in contact with each other, while in proximity printing there is a gap (usually from 10 to 40 um) between the mask and the wafer. Contact printing provides higher resolution compared to proximity printing, but the risk of mask damage is higher. With proximity printing, the damage

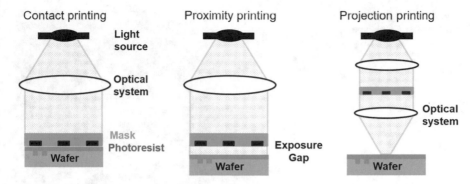

Fig. 6.26 Working principles of different exposure strategies

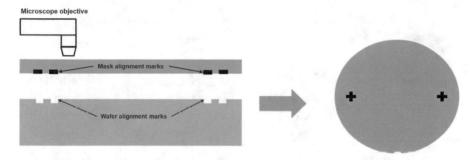

Fig. 6.27 Mask aligner Front to Front alignment scheme

to the mask is reduced and no optical element between mask and wafer is used. The main drawback is the diffraction effect caused by the exposure gap, which limits the resolution.

The exposure tool which works in contact/proximity exposure is called mask aligner. Its magnification is 1X, which means that the whole wafer is exposed in a single shot and the reticle pattern is transferred as it is to the photoresist.

In these tools, Front to Front alignment is performed by matching dedicated alignment marks on the wafer with the correspondent marks on the reticle as shown in Fig. 6.27.

The same principle is used also for BSA and IR alignment of which, mask aligners are typically provided.

After development, an alignment evaluation is performed by checking directly the marks used to align, with optical, see Fig. 6.28, or IR microscope.

On the other hand, in projection printing, once the light passes through the reticle, optical elements are used to focus and reduce the reticle image, which is projected onto the wafer. In this case, the complexity of the exposure system grows and the resolution and alignment accuracy increase. There are two major projection

Fig. 6.28 Alignment marks pattern after development

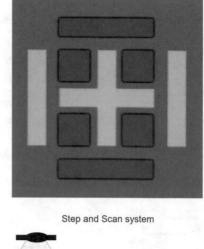

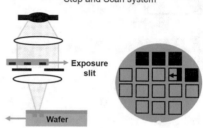

Fig. 6.29 Schematic of step and repeat (steppers) and step and scan (scanners) systems working principles

lithography printers: step and repeat printers (steppers) and step and scan printers (scanners). In Fig. 6.29, the schematic of these systems is shown.

When the wafer is processed in these tools, the pattern on the reticle is reduced by a constant factor (2X, 4X, 5X) and exposed repeatedly across the surface of the wafer in a grid. In steppers, the wafer is the only one which moves from one shot location to another under the lens, while scanners work by moving simultaneously the wafer and the reticle until a whole field is exposed, thanks to the presence of an exposure slit.

The Front to Front Alignment in steppers and scanners can follow different strategies depending on the exposure tool provider. The alignment is not performed matching directly reticle and wafer, but the reticle is aligned to the projection lens, thanks to dedicated marks, and then, the same alignment is performed with the wafer. Usually, after a pre-alignment step, the fine alignment is performed using two off-axis scopes, placed before the projection lens, for X and Y shift compensation. Thanks to the high stage motion accuracy and to the optical system complexity, the alignment accuracy is much better than in case of mask aligner.

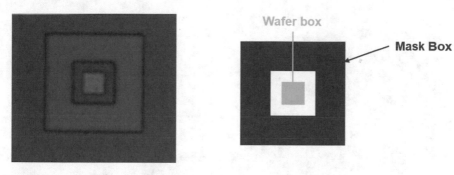

Fig. 6.30 Box in Box for alignment evaluation

Table 6.1 Lithographic performances of different exposure strategies used in MEMS fabrication

Exposure strategies	Magnification	Light source	Front-to-Front Alignment accuracy	BSA Alignment accuracy	IR Alignment accuracy	Resolution
Contact	1X	Broadband	$\leq 5\ \mu m$	$\leq 5\ \mu m$	$\leq 7.5\ \mu m$	~ 1.5 μm
Proximity	1X	Broadband	$\leq 5\ \mu m$	$\leq 5\ \mu m$	$\leq 7.5\ \mu m$	~ 5 μm
Stepper projection	2-4-5X	i-line (365 nm)	$\leq 0.3\ \mu m$	NA	NA	~ 0.35 μm

After development, the goodness of the alignment can be evaluated with dedicated box in box structures present on both the reticle and the wafer, as shown in Fig. 6.30.

Despite the most common tools used in MEMS fabrication for IR and BSA alignment are mask aligner, in the last years, exposure tool providers are also offering steppers with these features.

Depending on resist thickness, desired resolution, and alignment accuracy requirements, both i-line steppers and broadband mask aligners are used for MEMS fabrication. In Table 6.1, the typical performances of the most common exposure tools used for MEMS fabrication are listed.

Generally, Front to Front alignment guarantees a better accuracy compared to other alignment strategies, and for this reason, is used for wafers fabrication whenever possible. Moreover, there are some situations, specifically referred to MEMS, in which this alignment approach is mandatory, as discussed in the following paragraphs.

6.5.3 Case Study 1: Double Exposure

In MEMS fabrication it is often important to protect the wafer edge to prevent damages and to guarantee high wafers quality, due to the presence of critical steps, such as deep Si etch, wafer to wafer bonding, or HF release. At the same time, it

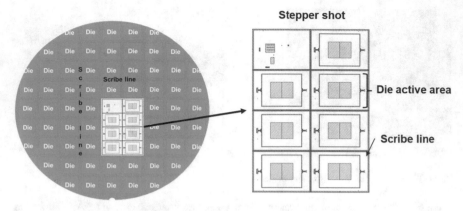

Fig. 6.31 Scheme of die scribe lines on wafer

can be necessary to pattern certain structures on the whole wafer, till the edge, to guarantee the back end processability of the wafer, as in the case of scribe lines.

Scribe lines are areas, between one die and another, used to separate individual die at the end of wafer processing, Fig. 6.31. To allow the dicing step, the scribe lines must be free of materials which prevent cutting and must be patterned on the whole wafer area. In this situation, it is necessary to protect the wafer edge, avoiding the definition of any active pattern and, at the same time, define the scribe lines on the whole wafer for good dicing.

As previously explained, when a wafer is exposed with mask aligner, the reticle pattern is transferred in one shot, as it is, to the photoresist. In this case, it is possible to change the pattern just in a specific area without affecting the others. At the same time, the resolution and the alignment accuracy are limited by the proximity printing technology.

On the other hand, in steppers and scanners, higher resolution and alignment accuracy can be achieved, but the reticle pattern is repeated in a grid of shots onto the wafer, which means that any change to the pattern will affect each shot.

Combining the flexibility of mask aligners together with the high resolution and alignment accuracy of projection printers, it is possible to obtain the desired scribe line definition: a double exposure approach is used. For the definition of active devices and related scribe lines, a stepper exposure is performed (Fig. 6.32a) and then, before the development step, an exposure in mask aligner is executed opening only the scribe lines at the wafer edge, as shown in Fig. 6.32b. For both the exposure steps, a Front to Front alignment on the same level is used in order to reduce the misalignment error. The result, coming out from the combination of the two exposure steps, is shown in Fig. 6.32c.

Another situation, in which the same double exposure approach is used, is the definition of the die tracking. This is a number printed on each single die in order to address it uniquely, also after assembling. Usually, it is patterned with a metal level to be well visible also at the end of the wafer process flow. For what was

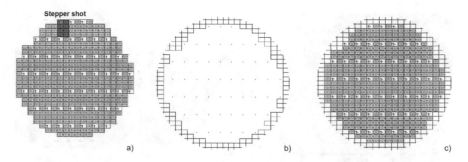

Fig. 6.32 Double exposure scheme. In (**a**), the stepper exposure is shown. In (**b**), the mask aligner exposure of the scribe line at the wafer edge is represented; while in (**c**), there is the result coming out from the combination of the two

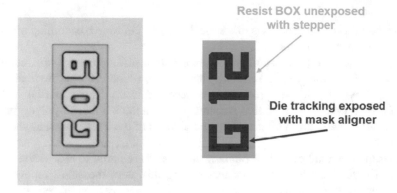

Fig. 6.33 Die-tracking double exposure

previously stated, it is not possible to print a unique number on each die with steppers, because the reticle pattern is repeated in a grid of shot. Therefore, a mask aligner is needed. At the same time, to guarantee the desired alignment accuracy and resolution, a projection exposure is essential for the definition of the metal level. For this reason, a first exposure is performed in stepper to define the active metal area and an unexposed box is left in each die in correspondence with the die-tracking number which is patterned with a second exposure with mask aligner, Fig. 6.33.

6.5.4 Case Study 2: Alignment Marks Optimization in Thick Epipoly

Thanks to its high alignment accuracy and resolution, projection printing with Front to Front alignment is the most used exposure technique in MEMS fabrication. Dedicated alignment marks are defined to guarantee the proper alignment; however,

Fig. 6.34 SEM image of a Gyroscope and schematic cross section of a THELMA MEMS mass

during the fabrication processes, these marks can be deformed, becoming unrecognizable by the exposure tools.

To guarantee the processability, an optimization of alignment marks could be necessary. This is the case of THELMA technology, which is the acronym of Thick Epitaxial Layer for Microgyroscopes & Accelerometers. These inertial sensors, used for rotation and acceleration sensing, require a thick epitaxial polycrystalline Si (Silicon) growth, up to 60 μm, for the definition of the active mass, as shown in Fig. 6.34.

After the epitaxial growth, a planarization step is required to reduce the roughness and defectiveness of the surface. During this step, the small topographies defined under thick Si and reproduced on its surface, can be erased, including the alignment marks necessary for the following lithographic steps, Fig. 6.35. In addition, the marks can be deformed due to the different crystallinity of the grown Si in the marks area. The intersection between monocrystalline Si, where the growth starts on the monocrystalline silicon of the substrate in correspondence of the alignment marks, and polycrystalline Si, where the growth starts on other substrates, can make the marks no longer recognizable by the exposure tool.

To solve these issues, before the Si growth, appropriate cavities which land directly on the Si of the substrate must be defined. The alignment marks are then defined at the bottom of these pools, and their shape must be optimized to prevent excessive deformation related to the mixing of monocrystalline and polycrystalline Si. During thick Si growth, the cavity is reproduced on the surface, with the marks at the bottom, protected from erasure during the planarization step. This expedient allows also to grow monocrystalline epitaxial Si in the alignment marks area, reducing the resulting defectiveness and deformation related to thick polycrystalline Si growth, Fig. 6.36.

To find the most stable condition in terms of defectiveness, different alignment marks, shapes, and cavity dimensions need to be evaluated and tested on steppers to verify their correct recognition. Figure 6.37 shows different alignment marks tested as they are from the layout and as they appear after the epitaxial growth.

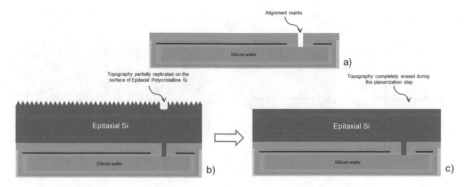

Fig. 6.35 Schematic representation of the alignment marks erasure with the planarization step. (**a**) Alignment marks before epitaxial Si growth; (**b**) Alignment marks after epitaxial Si growth; (**c**) Alignment marks after planarization

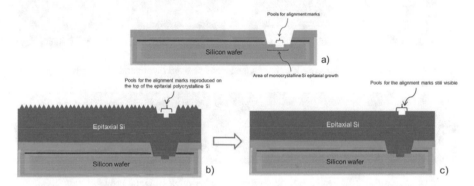

Fig. 6.36 Schematic of the alignment marks cavities in case of thick Si epitaxial growth. (**a**) Alignment marks before epitaxial Si growth; (**b**) Alignment marks after epitaxial Si growth; (**c**) Alignment marks after planarization

The combination between different marks, shapes, and cavity dimension that gives the best result in terms of alignment marks recognition must be applied, with the purpose of guaranteeing a stable alignment during stepper exposure. In this way, the necessary performances of steppers in terms of resolution and Front to Front alignment accuracy can be achieved.

6.5.5   Case Study 1: Pressure Sensors' IR Alignments

In the manufacturing process of the pressure sensor using the VENSEN technology (see Chap. 16), to allow adequate alignment of the mask after thick epi growth, as explained in Sect. 2.3.4 (Case of study: Single crystal silicon membrane for piezoresistive pressure sensors), has been chosen to proceed with IR alignment

Layout

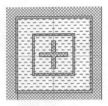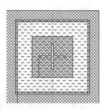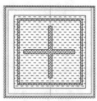

After epitaxial growth and planarization

Fig. 6.37 Different alignment marks, shapes, and cavities from layout and after epitaxial growth

Fig. 6.38 Example of the grown litho stepper marks after thick epitaxial growth

because of the presence of a thick epitaxial silicon (EPI) (about 60 µm) covering all the previously created signs and making them unusable in stepper tool for any further alignment.

At first, an evaluation of different types of thick EPI growth recipes and different depths of pillars etching were necessary to evaluate the goodness of the grown signs, see Fig. 6.38. All the trials failed the alignment in stepper tools with standard front side alignment technique.

Based on the obtained results, dedicated lithographic marks were introduced at the beginning of the process flow to avoid undesirable topographies. These marks are visible only with IR alignment, which became mandatory for this application together with the use of Double Side Polished (DSP) substrates to avoid any interference introduced by the substrate roughness.

Using a mask aligner tool with IR alignment capability it was possible to print new stepper marks after the 60 µm EPI growth (see Fig. 6.39) aligning on the buried marks previously printed (Fig. 6.40).

To understand how the buried marks were recognized with the IR technique in mask aligner, a lithographic evaluation was performed with different EPI growth

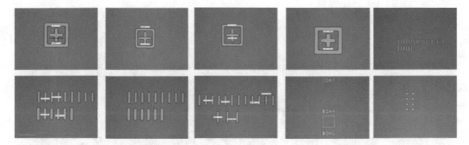

Fig. 6.39 Stepper marks mask in order to print new signs after thick epi growth using a mask aligner tool with IR technique

Fig. 6.40 Comparison between layout and IR image of Silicon wafer

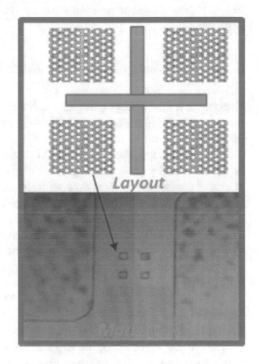

recipes. The evaluation was performed in terms of signs' recognition by the tool and stability of the signs' definition from wafer to wafer and lot to lot. Once the best condition was found on bare Si wafers, production lots were processed in that condition showing good results in terms of both definition and alignment measurement.

In addition, considering the final structure of the pressure sensor shown in Fig. 6.41, it was necessary to use the IR alignment also to define the mask on Cap 1 wafer. This mask has to be aligned on marks previously defined on sensor wafer.

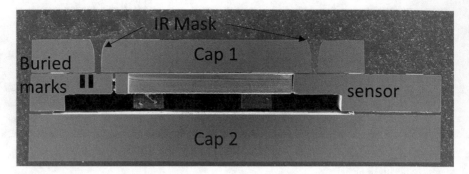

Fig. 6.41 Cross section of three wafer bonded with IR lithography and etch on top wafer Cap1 (see Chap. 16)

6.5.6 Case Study 2: BSA for Back Side Mask

Every time a new lithographic step needs to be performed, together with the alignment strategy, it is necessary to accurately evaluate the most suitable photoresist to be used depending on the application.

In this paragraph, the evaluation of the photoresist implemented for the definition of the back-cavity mask, a lithographic step performed on the back side of the wafer at end of the device process flow with the aim of releasing the moving mechanical structures typical of MEMS devices, is presented.

For this level, a Back Side Alignment technique is mandatory to perform the mask step. In addition, it is observed that the subsequent back cavity deep etching step is critical in terms of PR consumption and the thick PR used in standard lithography step is not enough. For this reason, a new resist designed specifically for dry etch step is needed and its complete characterization is necessary.

Two different photoresist thicknesses, 40 μm and 17 μm, have been evaluated on bare silicon wafers. The optimization of the coating conditions to obtain the right thickness, uniformity, and back wafer cleaning are parameters to keep under control. At the same time, the exposure doses for the two thicknesses need to be determined and the development process to be optimized in terms of number of puddles and time. Moreover, an adhesion promoter can be introduced to improve the resist adhesion to the substrate, as described in Sect. 6.2.4. The back side mask exposure is then evaluated to guarantee the correct alignment of the wafer with the mask and the evaluation of the maximum Q-time (delay time between exposure and post exposure bake and delay between post exposure bake and development) is necessary to guarantee the processability along the process flow. Finally, the alignment measurement recipe is set up. Figure 6.42 shows schematically the final process flow, with the main process parameters, for back-side mask definition.

Generally, the BSA is used in all those cases in which we have the need to flip upside down the wafer and define the mask for the subsequent deep silicon etch. In

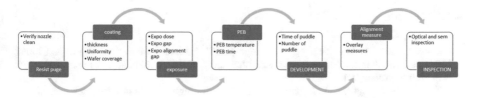

Fig. 6.42 Schematic process flow for back side mask for an autofocus project (see Chap. 21)

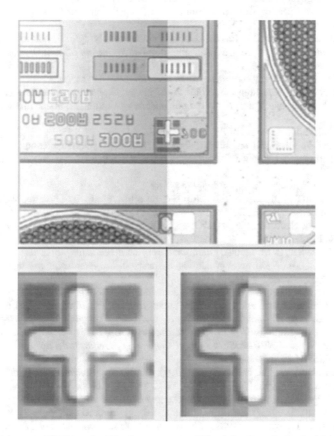

Fig. 6.43 IR image of BSA marks. The four squares will be defined on the back side of the wafer, while the cross is patterned on the front side of the wafer

these cases, the alignment marks have been previously designed on the front side of the wafer, as shown in Fig. 6.43.

Using a mask aligner tool with IR and BSA technologies, we can guarantee a good alignment front to back and preserve the performance of the device.

References

1. Chris A. Mack, Field guide to optical lithography, SPIE Field Guide Series Vol. FG06, (: 2006)
2. P. Rai-Choudhury: Handbook of microlithography, micromachining, and microfabrication, microlithography Vol 1 (1997)
3. Tang, Y., Sandoughsaz, A. & Najafi, K. (2017). Ultra high aspect-ratio and thick deep silicon etching (UDRIE). *2017 IEEE 30th International Conference on Micro Electro Mechanical Systems (MEMS), Las Vegas, NV*, pp. 700–703. https://doi.org/10.1109/MEMSYS.2017.7863504.
4. Flack, W., Nguyen, H.-A., Washio, Y., Ohgake, M., & Rosenthal, C. (2007). Characterization of a high-photospeed positive thick photoresist for lead-free solder electroplating, 65194D. https://doi.org/10.1117/12.712165
5. Koukharenko, E., Kraft, M., Ensell, G. J., & Hollinshead, N. (2005). A comparative study of different thick photoresists for MEMS applications. *Journal of Materials Science: Materials In Electronics.*
6. Kaneko, K. (2012). *Spray coating negative tone resists.* SUSS report.
7. Wangler, N., Beck, S., Ahrens, G., Voigt, A., Grützner, G., Müller, C., & Reinecke, H. (2012). Ultra thick epoxy-based dry-film resist for high aspect ratios. *Microelectronic Engineering, 97.*
8. Yu, L., et al. (2006). *Journal of Physics: Conference Series, 34,* 155.
9. Ito, H. (2003). Chemical amplification resists: Inception, implementation in device manufacture, and new developments. *Journal of Polymer Science Part A: Polymer Chemistry, 41,* 3863–3870.
10. Chemically amplified photoresist EP0388343A2
11. PHOTORESIST ADHESION PROMOTER, Inventor: Philip A. Lamarre, Waltham, Mass, Patent Number: 5,296,333, Date of Patent: Mar. 22, 1994
12. Doering, R., & Nishi, Y. (2007). *Handbook of semiconductor manufacturing technology* (2nd ed.). CRC Press.
13. May, G. S., & Sze, S. M. (2004). *Fundamentals of semiconductor fabrication.* Wiley.
14. Garrou, P. (2000). Wafer level packaging has arrived. *Semiconductor International, 23*(12), 119.
15. Liu, F., et al. (2018, May). Organic damascene process for 1.5- μ m panel-scale redistribution layer technology using 5- μ m-thick dry film photosensitive dielectrics. *IEEE Transactions on Components, Packaging and Manufacturing Technology, 8*(5), 792–801.
16. Giacomozzi, F., Mulloni, V., Resta, G. & Margesin, B. (2015). *MEMS packaging by using dry film resist, 2015 XVIII AISEM Annual Conference, Trento, Italy,* pp. 1–4. https://doi.org/10.1109/AISEM.2015.7066828.
17. Flores, G. Flack, W., Tai, E., Mack, C. (1994). *Lithographic performance in thick photoresist applications.* Hsiung, K., Tsai, C., Chen, L.C., Yen, G. (2008). *Novel high resolution epoxy based dry film material for wafer level packaging application.*
18. Madou, M. (2002). *Fundamentals of microfabrication.* Boca Raton.
19. Moss, S. J., & Ledwith, A. (1987). *The chemistry of the semiconductor industry.* Blackie & Son Limited.
20. Willson, C. G., & Stewart, M. D. (2001). Photoresists. In K. H. J. Buschow, R. W. Cahn, M. C. Flemings, B. Ilschner, E. J. Kramer, S. Mahajan, & P. Veyssière (Eds.), *Encyclopedia of materials: Science and technology.* Elsevier.
21. Reichmanis, E., & Thompson, L. F. (1989). Polymer materials for microlithography. *Chemical Reviews, 89*(6), 1273–1289.
22. Jones, S. W. *Photolithography.* ICKnolwledge LLC.
23. Silsesquioxane polymers, method of synthesis, photoresist composition, and multilayer lithographic method, US6340734B1
24. Wolf, S., & Tauber, R. N. *Silicon processing for the VLSI era Vol. 1.* Lattice Press.

25. Rai-Choudhury, P. (1997). Chapter 6: Metrology methods in photolithography. In *Handbook of microlithography, micromachining, and microfabrication, photolithography Vol 1.*
26. Franssila, S., & Tuomikoski, S. (2015). Chapter 20 – MEMS lithography. In M. Tilli, T. Motooka, V.-M. Airaksinen, S. Franssila, M. Paulasto-Kröckel, & V. Lindroos (Eds.), *Micro and nano technologies, handbook of silicon based MEMS materials and technologies* (2nd ed.). William Andrew Publishing.

Chapter 7
HF Release

Maria Carolina Turi, Ilaria Gelmi, and Raffaella Pezzuto

7.1 Introduction

Release process consists in removing "the floor" of a 3D structure that will result in free standing with some anchor points at the end of the release process itself (Figs. 7.1 and 7.2); deposited thin film layer acts as a temporarily mechanical layer (or sacrificial layer – "the floor" mentioned above) onto which the MEMS structural device will be built by patterning and etching techniques commonly used in semiconductor manufacturing (Fig. 7.3).

The way to remove the sacrificial layer has a crucial role because it needs to be mainly isotropic and properly set up to prevent any structure's collapse due to the removal technique itself. Sacrificial materials involved in the manufacturing process determine the nature of the release process itself: in some cases, polymers or organic materials, like polyimide, are employed so an isotropic oxygen plasma can be used to release the mechanical structure [1, 2]. Fluorine-based dry etch process [3] or XeF_2 Plasma are used when silicon or polysilicon layer is used as sacrificial layer [3], and metal or dielectrics are used as structural layer. However, in many cases, the sacrificial material is generally a silicon oxide (deposited either by LP/PECVD or thermally grown in furnaces): this approach suggests hydrofluoric acid (i.e., HF) as the most appropriate etchant for the oxides. This acid exists in liquid and anhydrous phases and it can be available on different commercial equipment: so, what is the best choice to carry out a proper HF release process for MEMS manufacturing?

M. C. Turi (✉) · I. Gelmi · R. Pezzuto
ST Microelectronics, Analog MEMS and Sensors Group, MEMS Technology and Design R&D, Agrate Brianza, Monza Brianza, Italy
e-mail: maria.turi@st.com; ilaria.gelmi@st.com; raffaella.pezzuto@st.com

© Springer Nature Switzerland AG 2022
B. Vigna et al. (eds.), *Silicon Sensors and Actuators*,
https://doi.org/10.1007/978-3-030-80135-9_7

Fig. 7.1 3D structure patterned on sacrificial layers

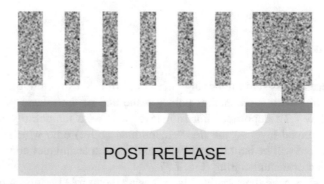

Fig. 7.2 3D structure free standing after release

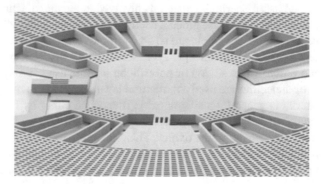

Fig. 7.3 Free-standing gyroscope

In the following paragraphs, some hints are given on how to approach this process step based on the requirement of the device, on the exposed layers, on the geometries involved, paying attention to the process parameters and all related issues that can be encountered during the process development.

7.2 MEMS RELEASE: Dry or Wet?

The choice of the best method to release a MEMS, is based on the process
architecture, the materials' compatibility and the geometries involved. Starting from
the process architecture and the materials' compatibility, the process to remove a
sacrificial oxide can be either anisotropic (plasma dry etch) or isotropic (typically
wet etch). The anisotropic etch is strictly related to the process flow used to fabricate
a MEMS and it is commonly used in micromirror devices where a metallic layer's
reflectivity needs to be highly preserved for device performance and a corrosion
protection is needed in case of acids' exposure. This case, however, is not generally
encountered in the "release process" as commonly known.

Figure 7.4 represents the concept of Isotropic and Anisotropic etch.

The Fig. 7.5 shows an example of anisotropic release while the Fig. 7.6 shows an
example of isotropic release.

Silicon dioxide is a widely used material for sacrificial layers, partly due
to its well-known etching properties in HF solutions and its integration in IC
manufacturing processing. This chapter is focused on the release process carried
out with Fluoridric Acid (HF) either in solution or vapor phase (so called vHF).

Starting from the equipment point of view, the HF release can be carried out in
wet bench with liquid HF (40%–49% conc.) or single/batch equipment equipped
with anhydrous HF. In the following sections, more details will be provided.

In general, sensing devices as accelerometers, gyroscopes, resonators, are
designed with submicron dimensions for minimal spacing: these features are not
well compatible with liquid phase etches due to the capillary forces that take place
between the liquid and the mechanical parts in minimum spacing geometries. For
such kind of devices, the preferable release process is carried out in vapor phase.
When the movable mass is released, the surface tension due to liquid immersion can

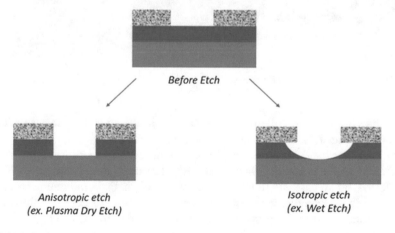

Fig. 7.4 Anisotropic and Isotropic etch

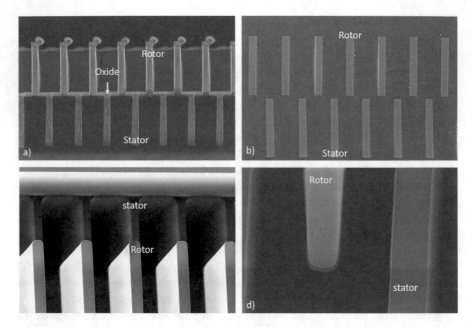

Fig. 7.5 Example of Anisotropic Release (Plasma Dry Etch): (**a**) Before Release (SEM cross section of Structural Layer1 [named rotor], Structural layer 2 [named stator] and Oxide in between); (**b**) after Release (SEM cross section); (**c**) and (**d**) after Release (SEM Top view of the bottom of the rotor)

Fig. 7.6 Example of Isotropic Release (vHF)

induce moving parts to collapse one to another in the so-called "release stiction" phenomena (Fig. 7.7). This phenomenon is irreversible.

On the other hand, acoustic membranes as microphones or ultrasound sensing membranes as Piezoelectric Micromachined Ultrasonic Transducers (PMUT), can be affected by release stiction as well, even though their minimum spacing is generally higher than the one used in accelerometers or gyroscopes. In these cases, a wet release can be set up properly with attention to the rinsing and drying steps to prevent membrane collapse, as shown Fig. 7.8.

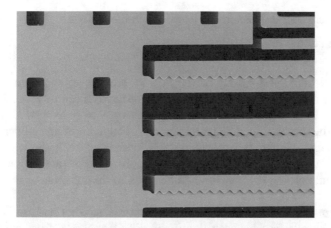

Fig. 7.7 SEM picture of two adjacent cantilevers stuck together (lateral stiction example)

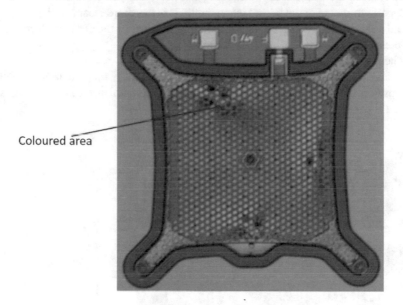

Coloured area

Fig. 7.8 Stiction on a microphone: colored parts are membrane touching

Once the released elements contact the ultra-clean, flat substrate surface below, the two surfaces permanently stick together, a condition referred to as "stiction" [8].

Stiction is a notorious case of malfunctioning of MEMS devices and it can be addressed to different root causes:

1. Capillary forces
2. Hydrogen bridging
3. Electrostatic forces
4. Van Der Waals forces

Two typical stiction phenomena described in literature are:

- In use stiction
- Release stiction

The *in-use stiction* is strictly correlated with the device functioning, because of electrostatic forces present between adjacent surfaces. It can be generally resolved by acting on the design of the mems [4], for example, by reducing the contact area, increasing the mechanical spring stiffness and/or applying, some proper coatings.

The *release stiction* is, instead, strictly correlated with the process [4, 5], especially if it is a wet process. In this case, capillary forces are responsible for the phenomenon, causing the structures to be stuck to the substrate or to the adjacent surfaces.

So, the main cause of permanent stiction between adjacent surfaces is the surface tension at the interface between liquid and air. Once the surfaces are in contact, further forces are introduced – including van der Waals, electrostatic and hydrogen bridges – which make the failure irreversible. It is therefore clear how it is necessary to find a sacrificial oxide removal method that can reduce the capillary forces due to the presence of liquids during the reaction.

The Fig. 7.9 schematizes the phenomenon of stiction.

Release Stiction can occur either vertically or out of plane, where the structures collapse on the substrate (Fig. 7.10) or laterally between surfaces facing the same plane (Fig. 7.11).

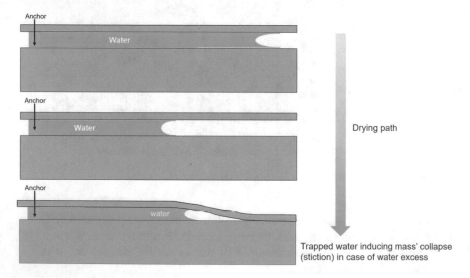

Fig. 7.9 Scheme of release stiction

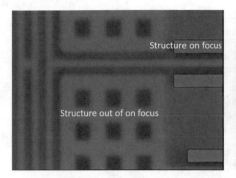 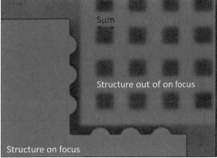

Fig. 7.10 Optical picture of z-axis out of focus structures, typical hints of release stiction (Vertical stiction)

Fig. 7.11 SEM picture showing two folded springs stuck together laterally (Lateral Stiction)

7.3 Process Control

Release process monitoring and characterization are carried out using various destructive and non-destructive techniques, with manual or automated inspection tools, some are used as in-line, others are used for the correct process set-up. The most commonly used techniques are: film thickness measurements, visible or IR range optical inspection, scanning electron microscopy (SEM), optical interferometric profilometer and tape test.

Film thickness measurements are used to find the etch rate of each material or residual layer thickness in different etchants.

Visible or IR range optical inspections are used to detect any sort of defects on wafers, residues of sacrificial layer under the free masses, stiction phenomena,

Fig. 7.12 On the left, IR inspection after release on trench not correctly open, SEM picture of residues on the right

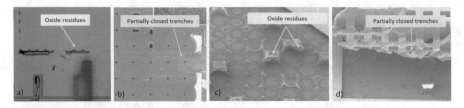

Fig. 7.13 (**a**) Optical inspection top view after tape test; (**b**) SEM inspection on backside mass after tape test; (**c**) and (**d**) SEM Inspection top view

infiltrations or detachment of the anchors. Mechanical or optical interferometric profilometers are also often used to identify stiction or structures' bending. Depending on the materials involved in the device, the undercut length can be determined by measuring etch front location from optical or IR microscopy and use it as in-line process control.

Etch profiles are analyzed by SEM cross-sections: moreover, in case of no-transparent material to IR, this technique can also be used to measure lateral etch rate under the structural layer resulting in the undercut measure.

This technique cannot be used as an in-line process monitor due to its destructive nature.

The Fig. 7.12 shows an example of an IR and SEM inspection after release when the trench is not correctly open and oxide residues are evident (lighter areas – red circled in the picture).

Tape test is a technique used to check release process in general and mainly used in characterization phase. It consists in using a tape to peel off the released structures: when a structure is partially released, it is not well removed by the tape and a residual oxide is visible below. This technique is still valid when substrates are not IR visible. Figure 7.13 shows examples of optical and SEM inspection after tape test.

7.4   Vapor HF Release (vHF Release)

As anticipated in the introduction, this section is focused on anhydrous HF for MEMS release because one of the major technology issues in surface micromachining is how to release compliant microstructures without a stiction in order to make them free standing [6]. Inertial MEMS, like Gyros and Accelerometers, are the most suitable products for vHF release because it must happen in the almost total absence of water in order to avoid the "release stiction".

The basic of the release process in vHF is always constituted by three steps:

- Humidification: surface pre-treatment with OH^- (catalyzer)
- Oxide etch: using an appropriate vHF flow
- Purge or surface drying: using N_2 flow and/or vacuum

The humidification is a surface treatment needed to adsorb the catalyst on the oxide surface to then initiate the reaction with the addition of vHF. Without a proper surface treatment, the etching process is scarcely initiated, and uniformity can be poor; an organic surface contamination can be an additional factor for a poor surface conditioning, so particular attention to surface cleaning must be considered [6].

The chemical etch reaction involved in the vapor HF release process can be summarized as:

$$SiO_2 \text{ (s)} + 4HF \text{ (v)} \xrightarrow{OH^-} SiF_4 \text{ (g)} + 2H_2O \text{ (l)}$$

The above reaction highlights that water is also a reaction by-product in liquid phase, so, to prevent stiction, it is very important to purge the reaction products very well.

The final purge step depends on the equipment used. However, since wet rinse is not considered as an option, the materials exposed to the HF can leave some residuals after etching if not properly treated. In addition, it has been found that using alcohol with short carbon chains (methanol, for example) as catalyst instead of water vapor, suppresses the formation of solid residue on the etched surface layer and results in improved etch characteristics [6].

7.4.1   Equipment on the Market

In the last decade of the '90s, ST has been one of the pioneers involved in the MEMS field, using anhydrous HF with water in a single wafer process tool, commonly used for a fast cleaning pre-deposition step, for a longer time in order to release sensing devices [7].

Table 7.1 Technolgies on the market

Equipment type	Wafers/run	Chamber process	Vacuum	HF source	Catalyst	Temperature
A	1	Teflon	No	Anhydrous	Water	RT
B	1	Stainless steel	Yes	Anhydrous	Water	Heated
C	25	Stainless steel	Yes	Anhydrous	Alcohol	Heated
D	25	Teflon	No	HF conc.	Alcohol	Heated

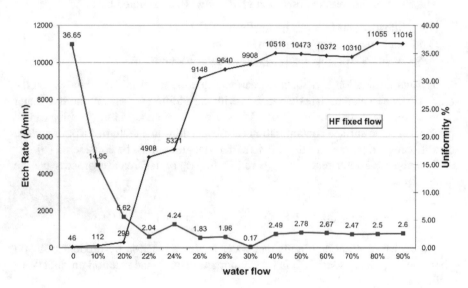

Fig. 7.14 Etch Rate trend with fixed HF content and increasing water flow

Later, different companies started to develop their own HF release tool with some differences, both in terms of hardware features, materials involved, reagents and single or batch process. In Table 7.1, a general summary of available technology in the market is shown.

The key feature for a proper vapor release process relies on the proper ratio between the HF and the catalyst. Indeed, in the absence of a catalyst, the etch reaction does not take place. In Fig. 7.14, an etch rate trend on thermal oxide is shown for fixed HF quantity and varying water content at room temperature.

From the chemical reaction point of view, more details are present in the following sections.

Fig. 7.15 Example of Polysilicon wing: Buried interconnection anchored by oxide residual

7.4.2 Process Requirements & Characterization: Inertial Sensors

Before setting up the process, the MEMS structure design needs to be known very well to determine the etch step duration. The process block consisting in humidification, etch and purge, needs to be modulated in terms of time and chemistry and the three steps can be repeated as many times to reach the desired undercut. After the surface pre-treatment, the vapor HF passing through the holes of the mechanical mass to release reaches the sacrificial oxide, starting its etching. During this etch, any structure embedded in the sacrificial oxide will be partially released: in particular, polysilicon routing, will be partially released, creating a polysilicon wing (Fig. 7.15, known as undercut).

This feature, if inserted in a dedicated vernier structure, can be used as the in-line process control of the release process (Figs. 7.16, 7.17, 7.18 and 7.20).

The undercut target will then be defined based on the device's layout (the maximum length to release).

An excessive attack of the oxide below the polysilicon interconnection can be a problem. In fact, a long polysilicon wing can become more fragile, and it can be prone to breakage during device functioning, creating a reliability problem (Fig. 7.19). Some process integration solutions can be implemented to prevent polysilicon wing formation using materials resistant to HF: this allows even more knobs for MEMS design in terms of mass, device sensitivity and area.

As mentioned above, in case of accelerometers and gyros, the vapor HF (vHF) release process set-up needs to consider two main parameters:

(a) The geometry of the epitaxial mass to release
(b) The total thickness of any oxide layer to be removed to make the epitaxial mass free to move.

About the involved geometries, the maximum distance to be released in the active area needs to be considered. With the same thickness of oxide to be released, this distance can determine the length of the process. The Fig. 7.21 shows an example of different geometries: the left image shows a maximum length to release of 7um while in the right one this is three times longer (21 μm).

Since the vHF release is a pure isotropic etch, in case of square or rectangular holes, the maximum distance to consider is the diagonal. The Fig. 7.22 shows the

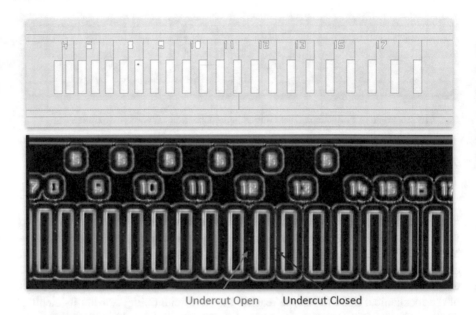

Undercut Open   Undercut Closed

Fig. 7.16 Undercut structure for in-line process control: (top) Design structure; (bottom) in-line optical dark field (DF) inspection

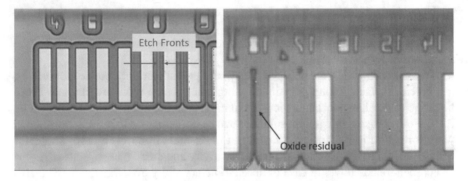

Fig. 7.17 Undercut structure: In line IR optical picture

Fig. 7.18 Undercut structure, SEM View

Fig. 7.19 (a) Example of undercut under a poly interconnection (SEM picture) – (b) Poly wing breakage top view (SEM picture)

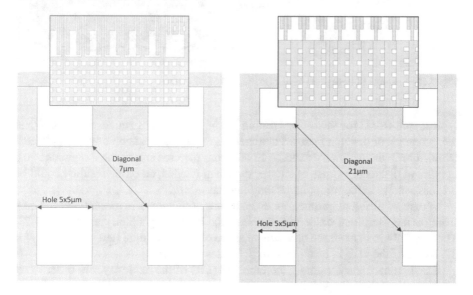

Fig. 7.20 Partial released MEMS structures example

Fig. 7.21 Different Layouts – lateral distance between trench holes: 7 μm diagonal (left); 21 μm diagonal (right)

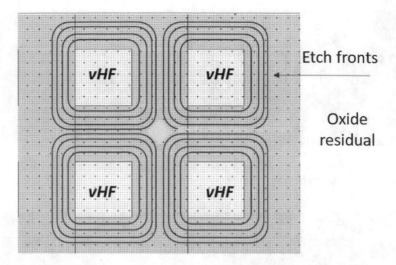

Fig. 7.22 vHF shape

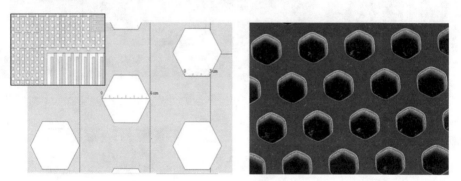

Fig. 7.23 Example of hexagonal trench (layout and SEM picture)

typical etch fronts for the vapor HF and its shape in case of square features. When the geometry of the etch features changes (as rectangle, hexagon or other), the etch fronts mimic the geometry exactly. In addition, the crossing of the diagonals is the maximum distance: in case of shorter etch than needed, an oxide residual could be present exactly in the midpoint of the diagonal.

In case of different geometries as hexagon, no diagonal is present and the etch front on the sacrificial oxide will replicate the hexagon geometry as in Figs. 7.23. The Fig. 7.24 is an example of a typical shape of oxide residual (left on purpose) for the fronts of the vHF in hexagonal trench holes geometry.

It is then necessary to know the etch rate (ER) of the oxide layer in order to set up the process time properly.

The Table 7.2 shows examples of ER on Thermal oxide using different process conditions.

Fig. 7.24 Example of a typical shape for the fronts of the vHF (Hexagonal trench geometry)

Table 7.2 ER comparison example

Equipment	Temperature	Pressure	Catalyzer	HF Flow	ER [Å/min]
A	28 °C	760 Torr	H_2O	0,200 slm	2500
C	45 °C	125 Torr	Alcohol 1	0,700 slm	2300
			Alcohol 2	0,700 slm	1800

However, ER is mainly linked to the gas mixture and ratio of reagents used: on more advanced equipment, where pressure and temperature can be set up by hardware settings, more process knobs are possible [8]. The introduction of a stopping layer below a buried interconnections in the process flow of a standard motion MEMS device represents a very advantageous innovation. This stopping layer induced a significant change in the basic of the release process: the total oxide to be removed is lower than in a standard device without stopping layer in terms of oxide volume to remove. The silicon substrate is kept protected by the initial oxidation thanks to the stopping layer's resistance to vHF etch. At this stage, the only rate determining factor for the release process is the geometry to release: in other words, even if the oxide to be removed is reduced in terms of thickness, the process length is, however, driven by the geometry. In fact, from the design point of view, the distance to release is generally much longer and the lateral etch needs to cover this distance (Fig. 7.25).

For the evaluation of the release process, a partial release is recommended to know how the oxide etch proceeds.

As mentioned in the "process control" section, for the inspection, optical microscope and SEM (Scanning Electron Microscopy) after "tape test" or IR microscope are recommended. These methods of inspection allow observing if oxide residues are present (pyramids) under the structural layer.

Figures 7.26, 7.27, 7.28 and 7.29 are typical pictures collected during process characterization.

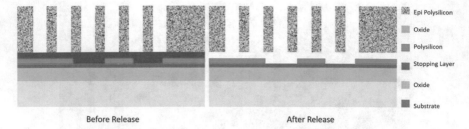

Fig. 7.25 Cross section before and after Release of Mems with HF stopping layer

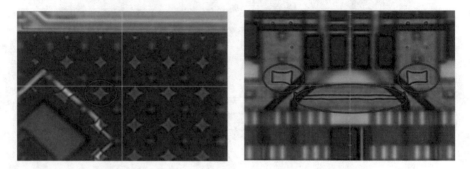

Fig. 7.26 IR inspection: Oxide residuals (red circled)

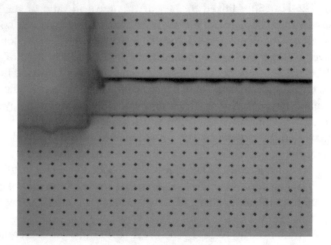

Fig. 7.27 Optical Inspection: Oxide residuals after tape test appear like dark spots

7.5 Wet HF Release

As explained in the Wet etch chapter, another way to release the MEMS structures is using a wet etch (wet release) of the sacrificial layer. Wet etching is an attractive alternative to dry etching because of its low cost.

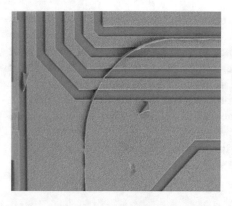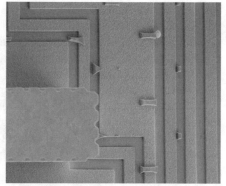

Fig. 7.28 SEM Inspection: Oxide residuals after tape test

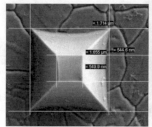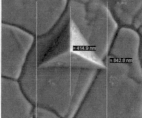

Fig. 7.29 Examples of Oxide Pyramids

Sacrificial layers are positioned underneath structural layers, and after release, the structure will result free standing with some anchor points at the end of the release process. This unlike "standard" wet etch used in IC technology requires a very wide undercut under the structural layer, therefore a lateral etch relatively long, depending on the architecture and design of the MEMS.

The choice of the wet release over the dry release requires a series of considerations that must be made a priori in the choice of the materials and the layout of the device as we will see in the following sections.

7.5.1 Hardware Requirements

Traditional wet bench uses a DIW (deionized water) rinse after concentrated HF etch and Marangoni dryer [14]. During drying, wafers submerged in water are slowly passed through an IPA (isopropyl alcohol) vaporized with N_2. The wafers are dried by the surface tension gradient between IPA and H_2O.

For the wet HF release, the traditional wet bench configuration is not as suitable as required, in fact the Marangoni drying is not "strong" enough to remove water from the cavities and created gaps by the release itself thus affecting the process by the so-called release-stiction.

For this reason, a customized wet bench for MEMS release is required and it consists in an additional tank with a liquid whose surface tension is much lower than water and it is inserted between the DI rinse and the drying station.

A basic wet bench configuration comparison is shown in the Table 7.3 below.

To ensure better control and repeatability of the etch rate, the wet release equipment is equipped with a temperature sensor (for example, a thermocouple) and a heater so as to be able to control the system temperature with a dedicated algorithm and they also use a constant known concentration.

7.5.2 Process Requirements & Controls

In the wet approach, etch rate depends on etchant bath temperature and agitation, etchant concentration, deposition process of the material to be etched, film composition and/or thermal treatments [9] film morphology, surface contamination, bath lifetime, etc.

The etch rates of different materials commonly used as sacrificial or structural layer in MEMS devices in different HF concentration bath at temperatures of 21 °C are listed in Table 7.4 as reference examples.

High etch rates of the sacrificial layers are preferable to reduce process times, increase throughput and, at the same time, decrease the exposure time to HF of the other materials present in the device structure.

Table 7.3 Wet bench configuration comparison

	Process	
Tank	Standard wet etch process	Wet HF release process
Chemical bath	HF 40%–49%	HF 40%–49%
Rinse	H2O	H2O
Custom rinse	No	IPA
Drying	N2 + IPA (Marangoni)	N2 + IPA (Marangoni or not) N2 only

Table 7.4 Etch rate in concentrated HF at 21 ° C

Materials Etch rate	Thermal Oxide μm/min	PECVD SiO2 silane based μm/min	TEOS μm/min	LPCVD SiN stochiometric μm/min	LPCVD SiN Si rich μm/min
HF 40%	0.71	2.4	2	0.56	0.09
HF 1%	0.0015	0.0085	0.003		
BOE (NH4F):HF = 7:1	0.15		0.17		0.0006

The etch selectivity and compatibility with the other materials present in a MEMS device are important parameters to choose the chemical solution to be used to release the device, and these parameters must be kept in mind during architecture definition and design phase. Indeed, a chemical solution not only etches the sacrificial layer to be removed but it could attack more or less quickly the other materials exposed. In general, the interaction between the etchant bath and materials is to be considered because it could lead to some process marginalities as materials' adhesion, materials' consumption, materials' embrittlement caused by galvanic corrosion. So, the etchant must have selectivity to the sacrificial layer as low as possible compared to the structural layer and other exposed materials. A strong synergy between process and design is necessary when there is no space for materials' improvement, but there can be space to help the process. In fact, in some cases, it has been proven helpful to add holes in the structural layer for etchant local access to reduce the maximum lengths to be released and therefore decrease the etch times, thus improving materials' consumption.

A fundamental difference between wet etch and vapor dry etch release is water phase used in the two processes. In the first, water is in its liquid phase, while in the second, it is in vapor phase: this is a significant disadvantage in the stiction prevention. After the wet release process step, a rinse in DI water is necessary to stop the reaction; this water could remain trapped in cavities or in the gaps between the thin layers and structural layer. It is difficult to remove water trapped among various layers of the structure because of the relative magnitude of surface tension and capillary forces that develop between the structural features and the substrate [4, 5]. To avoid release stiction, a wet release typically requires several rinsing steps to ensure that the solution is removed from the devices, followed by an optimized drying process. The point is, what rinsing liquid is preferable to remove water in the cavities and gaps and so to deplete the stiction phenomenon?

These rinses must have a surface tension smaller than water, to have a perfect spreading on the device's structures. In this perspective, Isopropyl Alcohol (Isopropanol IPA) [10] is a good candidate to reduce the surface tension of water because its surface tension is smaller than water; furthermore, it is commonly used in microelectronic industry and furthermore, it is very soluble in water. In addition, the so-called Marangoni effect, well-known in IC semiconductor manufacturing for wafers drying, is based on the addition of IPA to DI water.

The Table 7.5 shows the surface tension for different types of solvents.

A variety of other drying techniques have been developed over the years, some simple like spin-drying or Marangoni style dryers, or more sophisticated like supercritical drying process [11], freeze drying [12], the use of polymer support to avoid pull-down by drying liquid [13].

All the characterization techniques are the same as described in the previous sections. However, in the wet HF release process, to prevent the etchant bath degradation causing a possible rework thus inducing undesired materials' exposure to the chemical bath, a robust method consists in monitoring the etch rate test on a reference material before any release run.

Specific examples are then described in the following paragraph.

Table 7.5 Surface tension values for different solvents

Liquid	Temperature (° C)	Surface tension (Dynes/cm)
Water	20	72.75
Acetamine	85	39.3
Acetone	20	23.7
N-Butanol	20	24.6
Ethanol	20	24
Isopropanol	20	22
Glycerol	20	63.4
Ethylene glycol	20	47.7

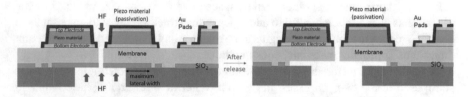

Fig. 7.30 PMUT device cross section view before (left) and after (right) release

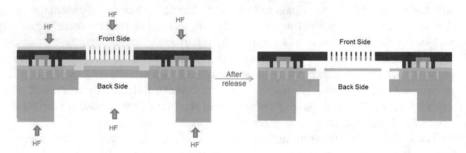

Fig. 7.31 Microphone device cross section view before and after release process

7.5.3 MEMS Applications

In this paragraph, two cases of wet release will be presented: a MEMS PMUTs (Piezoelectric Micromachined Ultrasound Transducer) for 3D gesture recognition and a more complicated MEMS microphone with two membranes to release one movable (the sensing) and one fixed (the reference). The process conditions are the same: concentrated HF (40%) at 21 °C, silicon dioxide as sacrificial layer and polysilicon as structural layer. However, these two MEMS devices have other materials that can behave differently in the HF release bath. The basic process architecture is similar for both cases that have two accesses for the liquid HF: from the front side and from the back side, as shown in schematics of Figs. 7.30 and 7.31.

These two schematics show the stratigraphy of each device before and after release.

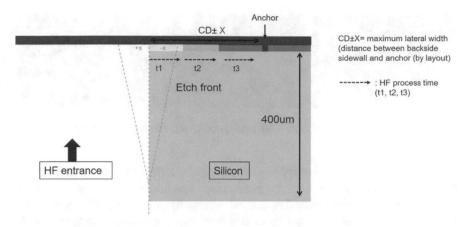

Fig. 7.32 Release process set up is correlated with the CD back side variation

The purpose of the release process is to remove the sacrificial silicon oxide layers, so to create the desired gaps: these gaps allow the free movement of the membranes.

As already mentioned in the previous vHF section, the device layout knowledge is a pre-requisite for a proper process set up.

In both devices' cases, the maximum lateral width to release corresponds to the distance between the first anchor and the back side chamber sidewall (Fig. 7.32). However, in device manufacturing, some process deviations from the ideality of the design layout must be considered. Figure 7.32 represents the possible variation from ideality of the maximum length to release due to back side chamber etch that could create a tapered profile of the silicon sidewall (either positive or negative with respect to the verticality). A second parameter impacting the ideality of the design layout is the misalignment of the back side chamber mask with respect to the anchor mask on the front (Fig. 7.32).

A proper formulation of the above consideration can be summarized as following:

$$Maximum\ lateral\ width\ (CD, um) = [a + b + c]$$

Where:

a = [layout distance between first anchor and backside chamber]
b = [mask misalignment]
c = [DRIE etch slope]

For the sake of simplicity, when approaching the process window set-up, the above formula can be simplified, including the mask misalignment in the overall "backside chamber variation":

$$Maximum\ lateral\ width\ (CD, um) = [a + d]$$

d = [back chamber variation] = $b + c$

Fig. 7.33 Cross sections for 30 (left) – 40 (center) – 50 (right) minutes release time

Starting from the preliminary layout's evaluations, the method to determine the proper HF dip time is based on the average etch rate (um/min) calculation that can be determined experimentally. The following proportional calculation, starting from the etch rate data (um/min), can be used to determine the time needed to release the maximum lateral width:

[um released per time unit]: [time unit] = [Maximum lateral width (um)]: [time needed]

This way allows a proper HF dip etch time selection, including an over etch percentage to cover any process shift and guarantee the process window robustness required.

In summary, the characterization on such devices is as following:

- Etch rate evaluation on several partial etch trials (average ER determination)
- Process control by means of optical/IR inspection or destructive SEM cross section (based on materials involved in the device itself)

In microphone, all the materials are compatible with optical and/or IR inspection, while in pMUT device, materials are not transparent to IR wavelength, so a destructive analysis is required. Figure 7.33 shows some examples of SEM cross sections for different HF dip etch times.

Figure 7.34 shows examples of microphone optical pictures after tape test during process characterization. Furthermore, in a production environment, the release process is always performed with a new fresh chemical solution tested with etch rate before any production run to overcome the in-line process monitor.

A second aspect to consider is the materials' compatibility with concentrated HF in these devices. For PMUT, all materials exposed to HF have a high compatibility with this process and an almost zero etch rate in HF, so no issue, even for long etch times, is foreseen. For microphone, on the contrary, some structural materials are doped and their reactivity to HF is not zero. Some countermeasure should be considered to minimize any detrimental effect on the structural materials due to prolonged HF exposure on the device robustness. From the layout point of view, some design modifications are preferable to reduce the maximum lateral width to release: based on its process architecture, this solution is possible by adding some holes in the poly membrane to allow HF access to the buried oxide, thus reducing HF dip time but releasing the device completely (Fig. 7.35).

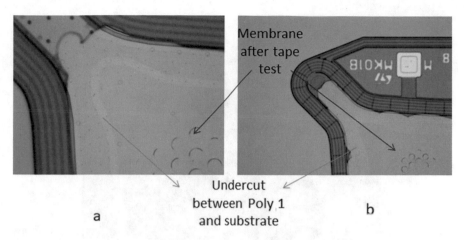

Fig. 7.34 Optical image after tape test after two dfferent HF dip times

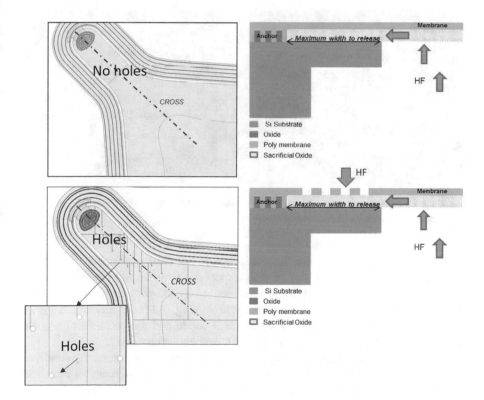

Fig. 7.35 Examples of Two Microphone layouts (without holes, with holes)

Fig. 7.36 Optical picture of Poly discoloration after tape test

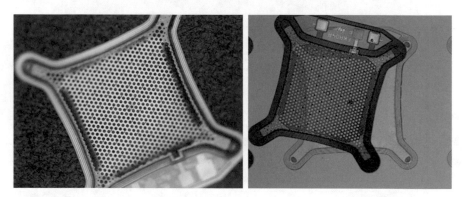

Fig. 7.37 Optical images after tape test after release for poly back plate corroded (left) and membrane and the back plate delamination (right)

Nevertheless, such exposure time reduction is not always effective in reducing other side effects that could happen with respect to the different materials involved in the process flow. A typical case is represented by the polysilicon discoloration (Fig. 7.36) or its delamination from the substrate (Fig. 7.37). As mentioned above, in the microphone process flow, polysilicon is doped and the presence of a noble metal for contacting pad induce the so-called "galvanic corrosion".

Galvanic corrosion happens when an oxidizing agent, in this case, O_2 dissolved in HF solution, is reduced at the gold surface (pads area) inducing an electrical current flow that initiates oxidation of the polysilicon (membrane and back plate). Corrosion is driven by the difference in potential between the two electrodes. Corrosion takes place on polysilicon and develops preferably along its grain boundaries.

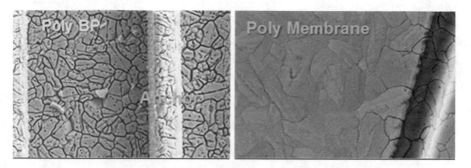

Fig. 7.38 SEM image of Poly BP & Poly Membrane in Anchor Zone

The result of the corrosion can be observed in the optical microscope like a discoloration in the back plate and membrane and the back plate delamination in the anchor zone. Figure 7.37 shows an optical picture after release for polysilicon membrane and polysilicon back plate corroded and for membrane and the back plate delamination. In Fig. 7.38, the different polysilicon morphology for back-plate and membrane due to HF corrosion is shown. The morphological difference relies on the doping level of the two polysilicon materials.

There are several anti-corrosion methods that can be used: surfactant addition in the HF bath, absence of light during release process, O_2 depletion, PS electrical polarization and sacrificial anode.

For example, surfactants act as "wettable" agents for the exposed interfaces; this may allow a uniform spread of the corrosion all over the surface, depleting the main discoloration effect at optical inspection.

However, SEM inspection confirms that the corrosion is still present, but in a more uniform way. The HF etches the grains' boundary of the polysilicon, more evident in the back plate because it is more doped as in Fig. 7.38. In the anchor areas, hints of galvanic corrosion are represented by HF infiltrations detectable between the polysilicon and substrate. In this case, the HF etches the grains' boundary in the polysilicon, reaching the oxide inside the structure and making such structure weak. This problem can be easily observed by SEM cross section (Fig. 7.39), but also with IR Microscope (Fig. 7.40), where oxide areas between adjacent anchors are hollowed out and so well visible.

The last effect that needs to be addressed in wet release process regime is stiction that could happen mainly due to liquid ambient, of course, or presence of undesired defectiveness like particles, or organic contamination like polymers.

In the presented microphone devices' architecture, as described above, there are two gaps to be created with the release process: one between the substrate and the polysilicon membrane and the second between the polysilicon membrane and the polysilicon back plate. Stiction could happen when these two membranes are touching each other, or the polysilicon membrane touches the silicon substrate.

Different types of microscopes (optical in dichroic mode, IR ones) or optical profilometers equipped with a proper handling for holed wafers can be used to

Fig. 7.39 Infiltration in anchor SEM cross section after 80 minutes release time

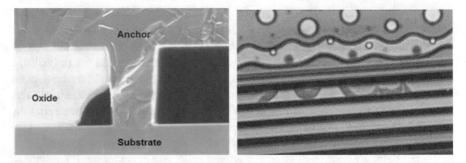

Fig. 7.40 SEM cross section (left) and IR (right) microscope for anchor infiltrations

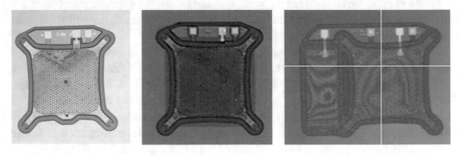

Fig. 7.41 Optical pictures in dichroic mode (left) IR (center) and optical profilometer (right) for microphone stiction affected

detect stiction. Figure 7.41 shows the appearance with different inspection tools when microphone device is stiction affected. Colored areas on the left, darker areas in the center and the right are all signs of stiction.

Fig. 7.42 Optical inspection shows an absence of stiction

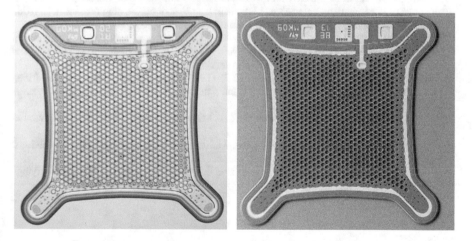

Fig. 7.43 Microphone Optical and SEM top view after release

As stated in the wet bench configuration section, the introduction of IPA rinse has been disruptive in stiction elimination (Fig. 7.42): the overall beneficial effect is minimizing the surface tension due to liquid HF and the affinity of IPA to any organic contaminant (Fig. 7.43).

References

1. Bartek, M., & Wolffenbuttel, R. (1998). Dry release of metal structures in oxygen plasma: Process characterization and optimization. *Journal of Micromechanics and Microengineering, 8*(2), 91.
2. Bagolini, A., Pakula, L., Scholtes, T., Pham, H., French, P., & Sarro, M. (2002). Polyimide sacrificial layer and novel materials for post-processing surface micromachining. *Journal of Micromechanics and Microengineering, 12*(4), 385.
3. Arana, R., de Mas, N., Schmidt, R., Franz, A. J., Schmidt, M. A., Jensen, K. F., & Micromech, J. (2007). *Microeng., 17,* 384.
4. Tasy, N., Sonnenberg, T., Jansen, H., & Legtenberg, R. (1996). Stiction in surface micromachining. *J. Micromech. Microeng, 6,* 385–397.
5. Legtenberg, R., Elders, J., & Elwenspoek, M., Stiction of surface micromachined structures after rinsing and drying: Model and investigation of adhesion mechanisms presented, In: *International Conference on Solid-Stat.*
6. *Journal of Microelectromechanical Systems,* Vol. 6, no. 3 1997, September) Dry release for surface micromachining with HF vapor-phase etching Yong-Il Lee, Kyung-Ho Park, Jonghyun Lee, Chun-Su Lee, Hyung Joun Yoo, Chang-Jin (CJ) Kim, Member, IEEE, and Yong-San Yoon.
7. Hanestad, R., Butterbaugh, J. W., Ben-Hamida, A., & Gelmi, I. (2001, September 28). Stiction-free release etch with anhydrous HF/water vapor processes. *Proceedings of SPIE 4557, Micromachining and Microfabrication Process Technology VII.*
8. Handbook of Silicon Based MEMS Materials and Technologies. Vapor Phase Etch Processes for Silicon MEMS Chapter | 25.
9. Witvrouw, A., Du Bois, B.,, De Moora, P., Verbista, A., Van Hoofa, C., Bendera, H. & Baerta, K. (2000). A comparison between wet HF etching and vapor HF etching for sacrificial oxide removal. *Proceedings of SPIE 4174-The International Society for Optical Engineering.*
10. Park, J. G., Lee, S. H., Ryu, J. S., Hong, Y. K., Kim, T. G., & Busnaina, A. A. (2006). Interfacial and electrokinetic characterization of IPA solutions related to semiconductor wafer drying and cleaning. *Journal of the Electrochemical Society, 153,* G811–G814.
11. Mulhern, G. T., Soane, D. S. & Howe, R. T. (1993). Supercritical carbon dioxide drying of microstructures. In Proceedings of 7th Int. Conf. on Solid-State Sensors and Actuators (Transducers'93) (Yokohama, 1993).
12. Guckel, H., Sniegowski, J., Christenson, T. R., & Mohney, S. (1989). Fabrication of micromechanical devices. *Sensors, 20,* 117–122.
13. Orpana, M. & Korhonen, A. O. (1991). Control of residual. In Proceedings of 6th International Conference on Solid-State Sensors and Actuators (Transducers'91).
14. Marra, J., & Huethorst, J.A.M. (1991). Langmuir 7 "Physical principles of Marangoni drying" 2748.

Chapter 8
Galvanic Growth

Giuseppe Visalli, Riccardo Gianola, and Linda Montagna

8.1 Electroplating Principles

Electrochemical deposition is a reduction of a metal ion from an aqueous or organic solution. The metal ion $M_{aq}{}^{z+}$ is deposited from the solution to the substrate which serves as cathode [1–6].

The generic reaction is reported below:

$$M_{aq}{}^{z+} + Ze^- \triangleright M_{ad} \tag{8.1}$$

Two possible electron sources can activate the reduction reaction: an external power supply or the presence of a reduction agent. In the latter case, an electroless deposition is taking place, while in the first case, electroplating occurs. Different steps are responsible for the reaction occurrence: mass transfer (diffusion), charge transfer, chemical reaction, and crystallization [3–8]. When an electrode is part of an electrochemical cell where current is flowing, its potential will differ from the equilibrium one. Being E_0, the equilibrium potential of the electrode, and E(I), the potential of the same electrode when the current is flowing, the difference between these values is defined as overpotential (η) [2, 5, 7, 9].

$$\eta = E(I) - E_0 \tag{8.2}$$

G. Visalli · R. Gianola (✉) · L. Montagna
ST Microelectronics, Analog MEMS and Sensors Group, MEMS Technology and Design R&D, Agrate Brianza, Monza Brianza, Italy
e-mail: giuseppe.visalli@st.com; riccardo.gianola@st.com; linda.montagna@st.com

© Springer Nature Switzerland AG 2022
B. Vigna et al. (eds.), *Silicon Sensors and Actuators*,
https://doi.org/10.1007/978-3-030-80135-9_8

Table 8.1 Mean current value for each different current mode where I_m is the mean current, I_p is the peak current, I_c is the cathodic current, I_a is the anodic current, t_{OFF} is the OFF time, t_{ON} is the ON time, t_c is the cathodic pulse, and t_a is the anodic pulse

Current Mode	Mean Current, Im
Direct current (DC)	$I_m = I_p$
Unipolar pulsed current (PC)	$I_m = \frac{I_p t_{ON}}{t_{ON} + t_{OFF}}$
Bipolar pulsed current (PC)	$I_m = \frac{I_c t_c - I_a t_a}{t_c + t_a + t_{OFF}}$

Since the electroplating process is controlled by mass transfer, charge transfer, chemical reaction, and crystallization, the total overpotential is defined as the sum of each individual overpotential. The slowest step is defined as the rate determining step [2–7, 9]. Electroplating processes can be performed either by using direct current or pulsed current. In Direct Current (DC) mode, current is ON during the entire deposition process. In Pulsed Current (PC) mode, the current (or the potential) is shifted between two values, generating a series of pulses. Each pulse has an ON time in which current is applied and an OFF time in which no current is applied. By controlling the ON/OFF time and the pulse amplitude, it is possible to influence mass transport condition. During the OFF time, all the positive ions can migrate towards the cathode interface, helping regenerate the double layer interface [10]. Pulsed Current can be unipolar (all pulses are in one direction), or bipolar (cathodic pulse is followed by an anodic one in which current is reversed and vice versa) [3, 4, 7, 11–16]. An important parameter that characterizes PC mode is the duty cycle, which represents the percentage of the cycle time in which the current is ON [16] and is defined as:

$$Duty\ Cycle\ (\%) = 100 \cdot \frac{t_{ON}}{t_{ON} + t_{OFF}} \tag{8.3}$$

The product of duty cycle and peak current gives the mean current value as shown in Table 8.1. When referring to duty cycle, it is important to mention frequency as well. Frequency is defined as the reciprocal of cycle time and is therefore strictly linked to the duty cycle [14, 16]:

$$\nu = \frac{1}{t_{ON} + t_{OFF}} \tag{8.4}$$

Figure 8.1 illustrates the current-time functions for direct and pulsed current.

8.2 Process Integration

Specific integration processes are needed to assure electroplating occurrence on a silicon wafer. Barrier and a seed layer deposition by PVD or CVD technique is

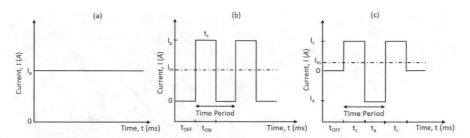

Fig. 8.1 Current vs. Time graph for (**a**) Direct Current Mode (DC), (**b**) Unipolar Pulsed Current Mode (PC) and (**c**) Bipolar Pulsed Current Mode (PC). Where I_m is the mean current, I_p is the peak current, I_c is the cathodic current, I_a is the anodic current, t_{OFF} is the OFF time, t_{ON} is the ON time, t_c is the cathodic pulse, and t_a is the anodic pulse

the very first step. Barrier layer has two main roles: to assure adhesion between the substrate and seed layer and to avoid diffusion of the plated metal towards the substrate. Seed layer is usually composed of a conductive metal to make the front of the wafer conductive and can be either unpatterned or patterned along wafer surface: plating occurs in both cases with different resistive paths. Furthermore, the seed layer can be different from the metal that will be plated depending on metals' compatibility and adhesion. On top of the seed layer, a photoresist mask is defined. Electrochemical deposition will occur only through the patterned open areas. But firstly, it is important to consider photoresist chemical compatibility with the plating baths and to assure that photoresist adhesion is not compromised by baths environment (pH, organic additives concentration, bath temperature). Once compatibility is verified, wafers are patterned by photolithography process (photoresist coating, expo, and development, ref. Chap. 6). A vertical resist profile is mandatory to respect the critical dimensions defined by mask layout both on top and at the bottom of the defined feature. Photoresist residues, called "scum", are often left during the development process especially in small open areas. Before the ECD step, a descumming process (oxygen plasma treatment) is needed to improve surface wettability and to remove resist scum at the bottom of a defined structure as shown in Fig. 8.2.

It is mandatory that descumming process does not modify photoresist profile, especially when high aspect ratio structures are involved. Excessive plasma treatment not only removes scum but etches and modifies resist walls profile as well. In these conditions, a vertical photoresist profile and therefore a straight ECD structure cannot be guaranteed (Fig. 8.3).

Post descumming, wafers are ready to perform electrochemical deposition. When electrochemical reaction occurs, metal ions are deposited from the chemical bath onto wafer surface. Photoresist removal and seed-barrier etch (to avoid short circuits) are then performed. These operations can be repeated subsequently as many times as needed. Figure 8.4 is a schematic view of the integration process flow, while Fig. 8.5 is a top view SEM image of an electrochemical plated structure.

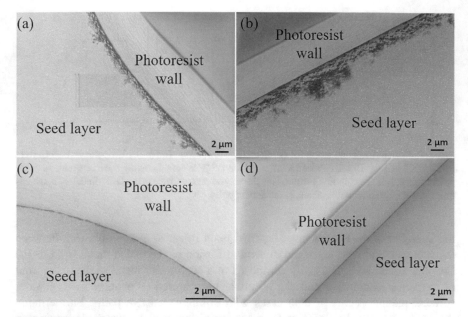

Fig. 8.2 SEM images before and after descumming process: (**a**), (**b**) scum residues at the bottom of a structure after photolithography mask definition; (**c**), (**d**) no scum present post oxygen plasma treatment

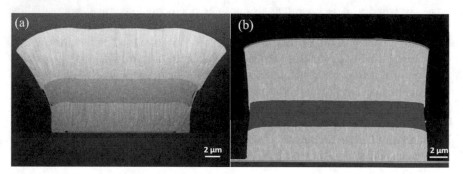

Fig. 8.3 SEM images about descumming effects on photoresist profile and ECD structure. (**a**) Excessive plasma treatment badly modifies structure profile. (**b**) Vertical profile structure successfully achieved, thanks to an appropriate plasma treatment

8.3 Electroplating Tool

Electrochemical deposition in microelectronics can be very challenging due to the small size (μm or sub-μm) of the areas in which plating will occur.

Specific tools are then required to ensure a continuous solution agitation and filtration as well as a fine control of temperature and other parameters able to influence the deposition process. Industrial wafer plating tools (most commonly

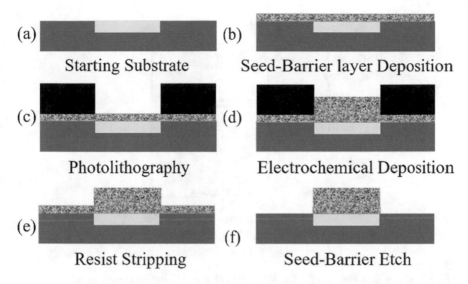

Fig. 8.4 Schematic view of a generic process flow (**a**) starting Substrate, (**b**) seed and barrier deposition, (**c**) resist coating and photolithography, (**d**) substrate preparation (descum, pre-wet) and Electrochemical Deposition, (**e**) resist stripping and (**f**) seed and barrier etch

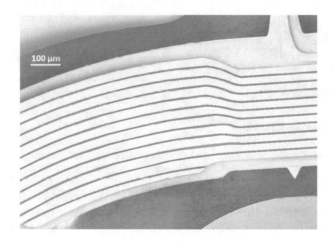

Fig. 8.5 Top view SEM image of an electrochemical plated coil

single wafer tools) are multi chamber clusters provided with pre-wet chambers, plating chambers, and rinse/dry chambers.

Pre-wet chambers allow to perform a proper wetting of the wafer surface prior to performing the ECD step: water and/or diluted acidic solutions can be used. Rinse/dry chambers are used for rinsing and drying the wafer surface once the electroplating process is completed; furthermore, when metal stack plating is performed, these chambers can be used for intermediate rinse to avoid cross

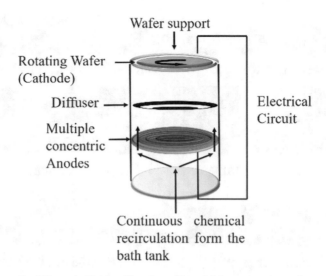

Fig. 8.6 Example of Fountain Plaiting Chamber with multiple concentric anodes

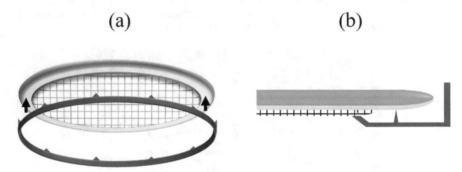

Fig. 8.7 3D view (**a**) and cross section (**b**) showing contact ring pins contacting seed layer (■) at the edge of the wafer where PR mask (#) has been removed by means of EBR

contamination between different plating baths. To optimize the space occupied by an ECD cluster, the pre-wet chambers are often configured to perform the rinse/dry function as well.

Fountain plating chambers are cells equipped with metallic anodes immersed in the plating solution. Wafer, once immersed in the solution, acts as the cathode (Fig. 8.6).

Electrical circuit between the anode and the wafer surface is assured by a wiring connection and, cathode-side, by a contact ring made of conductive fingers that electrically contact the seed layer at wafer edge. To have an exposed seed layer at wafer edge, a process by which photoresist is removed from the outer edge of the wafer is required, this process is called Edge Bead Removal (EBR) (Fig. 8.7).

Anodes can be classified according to the materials they are composed of. Inert anodes are made of an inert metal and do not dissolve in the ECD bath (e.g., platinized titanium). Soluble anodes (monolithic or pellets) are usually made of the same metal that will be plated and they dissolve in the plating solution (e.g., copper, nickel). Concentration gradient generated at cathodic surface and asymmetric distribution of the cathodic area (especially at the wafer edge if lithography blading has been performed) are the main causes of a poor Within Wafer Uniformity (WWU) in terms of deposit thickness. ECD chambers were originally equipped with a single monolithic anode which could not compensate these undesired effects. Nowadays, plating chambers can be equipped with multiple concentric anodes providing the possibility to properly modulate the local current density and therefore to improve the WWU [10, 17, 18]. Main fluid dynamic of the plating process is controlled by recirculation flow rate between a high-volume main tank and the plating chambers, where a lower volume of the plating bath gets in contact with anode and cathode. This configuration allows to work both with a high-volume batch (which grants a longer bath lifetime) and with a fine control of plating cell fluid-dynamics parameters (due to the low volume of the plating chambers). The flow at wafer surface can be changed by modifying the design of a diffuser interposed between the anode and the cathode that is able to influence local mass transportation, final metal thickness and uniformity composition. Mass transfer control is granted by wafer rotation (Revolution per Minute, RPM, set by software) and (if equipped) by the forced convection created with a proper device able to generate a turbulent flow at the wafer interface [10, 17]. Furthermore, bath chemical composition, temperature, and pH are in-line controlled by an automatic analyzer-replenisher. Plating recipes can be properly modulated thanks to *in-line parameters control set* provided by software. It is possible to set a current tolerance, the desired duty cycle, to control wafer RPM during plating process and to introduce conditioning steps by varying chamber positions, wafer rotation, and other parameters. Hardware/recipe settings and a fine bath control by automatic analyzer-replenisher concur in obtaining unique morphological and physical proprieties of the deposit. Each recipe must be customized for each plating bath depending on what results should be achieved.

8.4 Stacked Metal Plating

Stacked metal plating can be defined as the electrochemical deposition of two or more metals through the same photoresist mask obtaining a series of layers one onto the other. It is required in order to obtain complex wiring interconnections with specific mechanical and physical proprieties if no metal or alloy match the desired qualities. For example, stacked Au-Ni is needed to improve both fatigue resistance and electrical conductivity, Au-Sn is widely used to form eutectic bonding, Ni-Cu is important to avoid copper oxidation in the environment. When performing stacked metal plating, it is important to guarantee both photoresist compatibility for all the involved chemical baths and good metals layer adhesion. Layer's adhesion

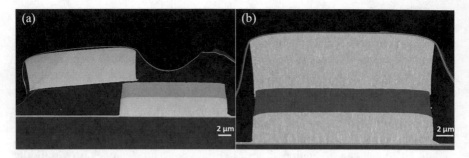

Fig. 8.8 Examples of stacked Au-Ni-Au plating. (**a**) No Gold Strike intermediate step between Ni and Au brings to Au layer lift due to bad adhesion and (**b**) successful Ni-Au adhesion, thanks to Gold Strike intermediate step

may be poor if a native oxide will form on the first metal layer surface on which the second metal will be plated. If the first metal is a noble one (e.g., Au), no native oxide will form even if rinse and dry occur between the two ECD steps. If native oxide formation kinetic is slow, a good adhesion can be obtained by subsequentially plating the second metal. Generally, a maximum time between the two plating processes is imposed. In this case, only a rinse after the first metal deposition is required between the two plating steps. Maintaining the first metal surface covered by a thin water layer between the two ECD steps is helpful as well. If native oxide formation kinetic is fast, a deoxidation intermediate step is needed to guarantee layers adhesion. This step can be performed by means of a selective acidic treatment (high etch rate on native oxide without damaging the metal surface) and/or using a specific intermediate ECD bath. To explain the latter case, let's consider the example of gold plating on top of nickel. Just after deposition, the top of Ni layer tends to oxidize and thus the gold adhesion on this surface is very poor. To overcome this issue, a treatment in chloridric acid could be used to remove the Ni oxide and regenerate pure Ni surface, anyway the following movement into the gold plating chamber could re-oxidize the Ni surface. So, specific baths with a high hydrogen evolution (due to the poor bath efficiency) are needed to allow an effective deoxidation of the underlying Ni surface and to cap it with a thin metal layer to prevent further oxidation. A low efficiency gold ECD bath with this function is usually named Gold Strike bath. Its composition is sulfite-gold bath-like (see paragraph 8.6.1), but with a much lower gold content (less than 1 g/L), allowing the hydrogen evolution needed for Ni deoxidation, followed by a thin Au cap layer deposition. Figure 8.8 shows how Gold Strike intermediate step is mandatory when Au on Ni stacked plating occurs.

8.5 Alloy Plating

To obtain specific chemical, physical, and mechanical proprieties, alloy plating is needed. Alloy is a combination of metals (or metals and other elements) and presents peculiar and/or synergistic properties with respect to single metals by which it is composed, often being a valid candidate for many applications. Alloys may be industrially prepared by reduction, powder metallurgy/compression or by melting and cooling the mixture at different rates, obtaining different composition and structures according to the relative state diagram [19]. An applicable technique in MEMS fabrication is electrochemical deposition of alloy, also called "alloy plating". The mechanisms through which co-deposition from an electrolyte solution of two (or more) metals takes place are driven by a huge number of variables. Rarely the metal ratio in the plated alloy is the same as the one in the bath. To understand the system behavior, single metals deposition potential and Tafel diagram are a good starting point even if they will not exactly predict the metal composition in the plated alloy. The presence of complexing agents, organic additives, and cross-influenced co-deposition of metals cause strong deviations from system ideality [20]. Two main groups of co-deposition mechanisms have been identified: "normal alloy plating systems" characterized by the preferential deposition of the more noble metal and the "abnormal co-deposition systems" in which the more noble metal does not preferentially deposit [21]. Furthermore, co-deposition mechanism is influenced by plating bath conditions at the extreme proximity of the cathode: electric charge transfer and redox half-reactions occur at the solid-liquid interface between solution and electrodes, forming many planes where temperature, charge, agitation, efficiency, pH, and species activities are different from those in the bulk solution, according to the Electric Double Layer theory (EDL) [22]. Consequently, deposit composition is strongly dependent on the plating reactor, especially in terms of fluid-dynamic configuration and plating bath composition control (e.g., batch, semi-batch, continuous systems). If a through-mask plating is performed, EDL theory must consider the fluid-dynamic conditions inside the micrometric pitch where the structure will grow. The set of all these complex variables makes difficult, if not impossible, to find a general model able to predict alloy composition for every plating system. Frequently it is more practical to define a specific but simplified empiric model starting from experimental results, considering it valid just for the used plating system.

8.6 Electroplating Processes

A wide selection of metals and alloys are used in MEMS fabrication. Many of them can be electrochemically deposited. In the following paragraphs, the most common ECD baths used for MEMS fabrication will be examined focusing on their main features and applications.

8.6.1 Gold Plating

Gold is widely used as metallization in several MEMS technology (RF, optoelectronics, IC packaging) thanks to its peculiar properties, such as corrosion resistance,
high conductivity, and solderability [23]. Electrodeposition of gold can be achieved
using baths containing Au cyanaurate (I) complexes that are really stable in
operating conditions: $Au(CN)_2^-$ complex has a very high stability coefficient
$(K \sim 10^{38})$. Cyanide baths have been characterized over the years in terms of deposit
properties and process conditions [24, 25]. Even if the bath does not originally
contain free cyanide (CN^-), the ion is generated during the deposition process.
Cyanide ions are known to be dangerous for most of the positive photoresist, causing
resist cracking and/or lifting by attaching the interface with the seed layer. So,
despite the bath stability, alternative cyanide-free plating bath have been developed
in the recent years, even for safety concern in managing cyanide [26]. Gold (I)-
Thiosulfate complex is quite stable with a stability constant of about 10^{26} [27]. In
principle, it could be a good candidate for a gold plating bath, but free thiosulfate
ion is unstable in aqueous environment (as shown in Eq. 8.5).

$$S_2O_3^{2-} \rightleftarrows S + SO_3^{2-} \tag{8.5}$$

The ion disproportion is minimal at high pH and low thiosulfate concentration,
while it is promoted in neutral pH environment and higher thiosulfate concentration.
These factors don't guarantee a potentially plating bath stability. Gold (I)-sulfite
complex $Au(SO_3)_2^{3-}$ has a stability constant of about 10^{10} [28] and, even if its value
is much lower than the ones of cyanide and thiosulfate complexes, is used for most
of the commercial plating baths for its good compatibility with positive photoresist
[29]. Special care must be applied in managing gold sulfite-based baths because the
complex tends to follow a disproportion reaction causing the precipitation of pure
gold and providing the formation of Au (III) as follows:

$$3\left[Au(SO_3)_2\right]^{3-} \rightarrow 2Au + \left[Au(SO_3)_4\right]^{5-} + 2\,SO_3^{2-} \tag{8.6}$$

This reaction occurs not only during plating operating condition but even during
stand-by and it is strictly correlated with bath pH due to the sulfite ions hydrolysis:

$$SO_3^{2-} + H_2O \rightarrow SO_{2(g)} + 2OH^- \tag{8.7}$$

In order to increase bath stability, a tight pH control is mandatory as well as the
presence of a stabilizing additive. In all the commercially available bath, a vendor
proprietary additive (i.e., ethylendiamine, 2,2′-dipyridine) is added to stabilize the
gold complex thus reducing the disproportion [30]. Another issue affecting bath

Table 8.2 Typical Au electroplating bath composition

Parameter	Range
Gold (g/L)	10–18
Sulfite (g/L)	30–80
Brightener (mg/L)	10–40
pH	7.5–8.5
Temperature	25–45

stability is the sulfite oxidation to sulfate. To prevent it, the oxygen evolution due to recirculation must be kept under control and nitrogen bubbling into the tank can be an option to remove oxygen from the bath. Many sulfite commercial baths have quite high pH value (>9) that could cause photoresist etching. Thanks to the presence of buffering (phosphates, citrates, and carbonates are the most used) and stabilizer agents, recent baths have a pH value closer to neutral (~8), showing very good compatibility with positive and negative photoresist. Recently, some mixed thiosulfate-sulfite baths have been deeply studied and characterized [31], but so far, no commercial bath is available for semiconductor applications. Metals or semimetals are often added to plating baths as brightener [32, 33] to reduce surface roughness. The brightener co-deposits with gold and reduces the grain size, acting as an inhibitor of the crystal growth during the deposition. According to some work [34], the grain size is strictly linked to the hardness, so smaller grain brings harder deposits. A typical Gold commercial bath composition is shown in Table 8.2.

For some packaging and MEMS applications, where thicker deposits (>10 μm) are required, high deposition rate is mandatory to increase throughput and to avoid very long process times that could impact on photoresist stability in the bath. Increasing the current density can have a very bad impact on within-wafer uniformity and mainly on within-die-uniformity when structures with different sizes are present. To overcome these limitations, a proper hardware and fine process conditions set up for several plating parameters (duty cycle, frequency, temperature, paragraph 8.3) are required [35]. One of the main applications for gold plating in MEMS technology is the wafer-to-wafer thermocompression bonding (ref Chap. 11), where two wafers with gold metallization are intimately put in contact to build up a junction. To get a uniform bonding over the entire wafer, the single structure must be flat and smooth and bump thickness must be uniform in different wafer positions. As already stated above, plating process fine-tuning is mandatory to meet all these stringent requirements. As an example, in Fig. 8.9, the bump shape change vs. frequency is shown: higher frequency allows to reach a flat bump.

Bump roughness is also temperature-dependent, as shown in Fig. 8.10.

8.6.2 Copper Plating

The central role of the copper in semiconductor technologies is mainly related to its high electrical conductivity and to its resistance to electromigration. Copper-

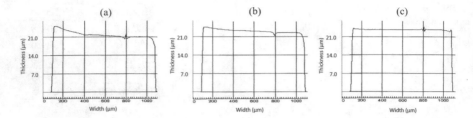

Fig. 8.9 Au plated bump shape trend at different current frequency (**a**) 8.3 Hz, (**b**) 83 Hz, and (**c**) 833 Hz

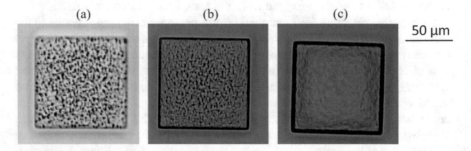

Fig. 8.10 Optical images of Au bumps showing different roughness for different plating temperature: (**a**) 50 °C, (**b**) 35 °C, and (**c**) 25 °C

Table 8.3 Typical Cu-electroplating bath composition

Parameter	Range
Copper (g/L)	16–40
Sulfuric acid (g/L)	100–200
Chloride (mg/L)	15–60
Additive A ml/L)	1–5
Additive B (ml/L)	5–25
pH	1–3
Temperature	25–40

plating baths are most often sulfate-based, while alternative baths, such as copper fluoroborate-based, are less common [36]. More details on copper plating processes can be found in literature [37, 38]. Sulfate-based baths are chemically very stable and usually work close to room temperature. The main electrolyte is copper sulfate dissolved in sulfuric acid and organic additives are usually added together with chlorine ions to help stabilize the anode dissolution. A typical copper-plating bath composition is shown in Table 8.3.

The presence of organics into the bath brings some impurities to be co-deposited with copper. Therefore, just after plating, the as-deposited Cu film is unstable by a crystallographic and electrical point of view: it undergoes, even at room temperature, to a self-annealing process that causes a drop of the resistivity and an evolution of the microstructure [39]. As alternative, an annealing in inert atmosphere

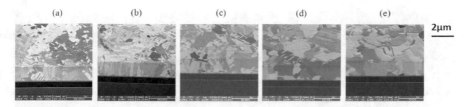

Fig. 8.11 SEM-FIB images of (**a**) self-annealed, (**b**) annealed at 100 °C, (**c**) 200 °C, (**d**) 300 °C, and (**e**) 450 °C electroplated copper

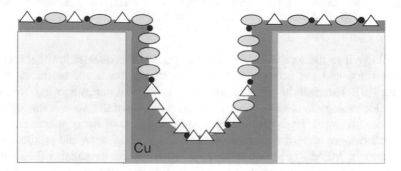

Fig. 8.12 Schematics for bottom-up filling mechanism and additives' behavior: • Brightener, ◯ Suppressor, △ Anti-suppressor. Anti-suppressor tends to accumulate at the bottom of the features

at high temperature can be done to bring the resistivity down to the bulk value and to stabilize the film structure [40] before subsequent process steps. As shown in Fig. 8.11, a big difference in grain structure is evident between a self-annealed and a thermally treated copper film despite the same resistivity. The higher the annealing temperature, the higher the grain size.

The organic additives play a very important role in the copper ECD. They can be divided into three categories: suppressor (high molecule size with big steric effect and a weak physical surface adsorption), anti-suppressor or leveler (medium molecular size with polarization effect and a stronger physical surface adsorption), and brightener (small molecules which tend to co-deposit with copper). The huge influence that the organic additives have on the shape, the profile, the roughness, and the uniformity of copper deposit can be seen in some papers [41] where it is clearly shown how an accurate management of additive concentrations can bring about the desired bump shape and film roughness.

The additive's role is determined by their effect during plating process (Fig. 8.12), especially in the damascene- and vias-filling processes.

Copper damascene plating is widely employed to fill vias used as metal interconnection in advanced CMOS technologies and it is always followed by CMP (Chemical Mechanical Polishing) to remove the excess copper plated on wafer surface, ensuring no shorts between structures [42]. In this technology, the main target is to completely fill the trenches by speeding up the deposition in inner regions

Fig. 8.13 SEM images of copper grain and roughness as a function of deposition rate. (**a**) 0.13 μm min^{-1} (**b**) 0.25 μm min^{-1}, and (**c**) 0.8 μm min^{-1}

and limiting it to the top structures. Different additive concentrations between the wafer surface and the bottom of the via are crucial for a successful via-filling process [43]. Through-the-mask copper electrochemical deposition has become a suitable technology to achieve copper bumps on the top of the wafer having specific geometrical features. Even if the increasing exploitation of the copper is due to its high-performance electrical properties, metal shape, roughness, and profile are key parameters in MEMS field technologies as well. Plating parameters (i.e., current density) can have a big impact on the deposit grain and roughness. As an example, in Fig. 8.13, the grain evolution vs. deposition rate is shown.

Copper ECD is widely used in high aspect ratio structures definition. In MEMS field, one of the most innovative and challenging applications is the definition of copper coils used for electromagnetic mirror actuation, whose spire has an aspect ratio of about 4:1. First, the pre-wet step (mandatory for surface wettability, see Sect. 8.3) must assure the adhesion of the high aspect ratio photoresist. By performing a standard pre-wet with high pressure DI water/chemical, photoresist adhesion cannot be guaranteed and short-circuiting between copper spires can occur, as shown in Fig. 8.14a; a low DI water/chemical pressure pre-wet step is then necessary. To achieve such pre-wet, a proper control of DI water/chemical pressure needs to be accomplished by implementing a proper spray-bar in the pre-wet chambers (Fig. 8.14b).

Once the proper pre-wet has been set, electrodeposition fine-tuning process must occur. The deposition rate plays a very important role, especially if both high and low aspect ratio structures must be plated through the same mask. The best compromise between high and low deposition rates needs to grant both acceptable throughput and good control of the deposit in terms of within wafer, die, and feature uniformity. A SEM cross section of the coil spires helps to understand whether the fine-tuning process has been successfully achieved and the spires' shape matches the requirements (Fig. 8.15).

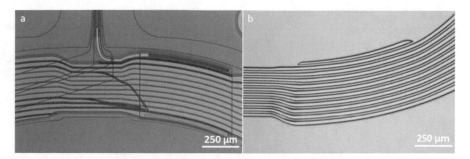

Fig. 8.14 Effect of pre-wet step: (**a**) high DI water/chemical pressure pre-wet causes a bad photoresist adhesion, photoresist lift, and shorts between copper spires, while in (**b**) a controlled pre-wet step can guarantee a good adhesion and no short-circuiting

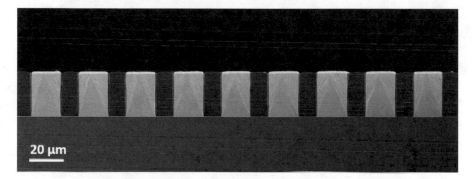

Fig. 8.15 Cross section SEM image of coil spires used for electromagnetic mirror actuation

8.6.3 Nickel Plating

Nickel plating commercial baths are generally acidic, Nickel sulfate-based or Nickel sulfamate-based. Usually, bromide or chloride Nickel salts are added if a Nickel soluble anode is used. These species increase and stabilize the anode dissolution by removing the native oxide which forms on its surface. Ni deposition is in competition with hydrogen ions conversion in hydrogen gas. This side reaction reduces the bath efficiency and could damage the photoresist and/or the Ni film. Moreover, it could modify the local pH, causing nickel oxide formation and co-deposition. For these reasons, a buffering system is needed to keep a stable pH value close to the cathodic area. Historically, the most common buffer is boric acid even if its poor solubility limits the concentration to a maximum of 30–40 g/L. Often a surfactant (wetting agent) is present to reduce the surface tension at wafer surface. Finally, a stress reducer is sometimes present (Saccharin or its compounds are the most used) [44]. Typical bath composition is shown in Table 8.4.

Temperature is usually higher than 40 ° C to increase boric acid solubility, to allow higher current density and, consequently, higher deposition rate. Usually,

Table 8.4 Typical Ni
electroplating bath
composition

Parameter	Range
Nickel bromide (g/l)	50–70
Nickel Sulfamate (g/l)	240–400
Boric acid (g/l)	25–40
Stress reducer (g/l)	0.1–1.2
Surfactant (ml/l)	0.2–1
pH	3–4.5
Temperature	40–65

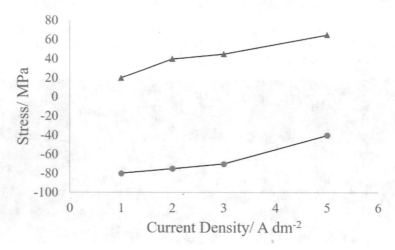

Fig. 8.16 Stress evolution (MPa) vs. Current Density (ASD) for • Ni sulfamate-based bath with stress reducer additives and ▲ Ni sulfate-based bath

an increase in bath temperature is linked to a film stress decrease. In MEMS technology, mainly for thick deposits, film mechanical properties are of paramount importance. One of the main requirements is a stress-free and stress-gradient-free film. Modifying bath parameters (i.e., nickel concentration, anode activator, and organic additives) and plating conditions (i.e., temperature, current density, waveform) can have a direct effect on several mechanical properties, such as stress, hardness, Young's module, and crystallography [45, 46]. Stress evolution with temperature can be very critical for MEMS structures as well. Furthermore, both Ni salts (sulfate vs. sulfamate and chloride vs. bromide) and organic additive choices are responsible for Ni deposit stress. As an example, in Fig. 8.16, the film stress behavior vs. current density for two different baths is plotted.

Table 8.5 Typical Sn electroplating acid bath composition

Parameter	Range
TiN (g/L)	50–70
MSA acid (g/L)	100–200
Antioxidant (mL/L)	5–40
Brightener (g/L)	2–15
Wetter (mL/L)	5–20
pH	1–2
Temperature	20–40

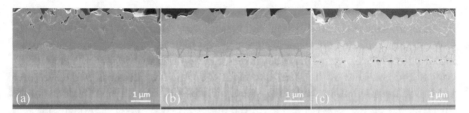

Fig. 8.17 SEM cross section images of Au-Sn bumps. Intermetallic compounds formation vs. time. (**a**) post Au-Sn plating, (**b**) after 1 week from Au-Sn plating and (**c**) after 4 weeks from Au-Sn plating

8.6.4   Tin Plating

Tin is widely used in electronics, mainly for wafer level packaging and bonding processes because several solders (whose purposes are electrical conduction, mechanical support, and heat dissipation) are tin-based and can be chosen, according to the melting temperature, for eutectic bonding.

Tin can be electrodeposited from alkaline stannate baths or acidic stannous Sn(II) baths [47]. The latter are most commonly sulfate or methanesulfonate-based and both suffer from oxidation of Sn(II) to Sn(IV), causing the formation of insoluble colloidal salts. The choice between alkaline and acidic baths can be done according to the required deposit characteristics (throwing power, roughness). As for other plating baths, organic additives as brightener, wetting agent, and antioxidant are usually added to the main bath component (see Table 8.5).

Gold-tin soldering (80%Au: 20%Sn by weight percent) has historically been employed in the microelectronics industry for several applications (i.e., die attach) due to relatively low melting eutectic temperature (about 278 ° C). Plating tin on top of gold bump requires a stable photoresist in both plating baths and good thickness uniformity to keep the proper ratio between Au and Sn. Just after plating, both gold and tin start a reciprocal interdiffusion, bringing to the formation of several intermetallic compounds at the Au-Sn junction [48]. Figure 8.17 shows the intermetallic compounds formation vs. time: as the intermetallic compounds' growth increase, cracks due to Kirkendall voids [49] between Au-Sn original interface increase as well.

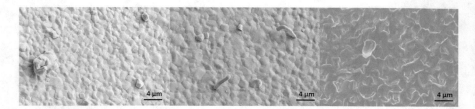

Fig. 8.18 SEM images of whiskers formation due to Cu diffusion towards Sn layer

As an alternative to gold-tin stacked plating, few attempts have been done to co-deposit gold and tin from the same electrolytic bath as an alloy having the proper metal ratio [50]. Copper-Tin solder layers can be obtained by plating sequentially copper and tin. One of the main issues for the Cu-Sn metal stack is the whiskers formation that is mainly due to the stress induced by the copper diffusion into the tin layer. When the stress is too high, the upper tin oxide layer is broken and very thin Sn extrusions appear on top of the structure, as shown in Fig. 8.18 [51–53].

8.6.5 Ni-Fe Alloy Plating

Ni-Fe alloys are widely used in many technical fields due to their peculiar properties. Permalloy (Ni:Fe 80:20 w/w) is known for its very high magnetic permeability, being of industrial interest as transformer magnetic core and for magnetic shielding applications. The very low thermal expansion coefficient of Invar (Ni: Fe 36:64 w/w) makes it an eligible material for mechanical parts of high precision instruments such as clocks and seismic creep gauges [54–56]. Ni-Fe alloy plating is classified as anomalous co-deposition [57–59]. A common mechanism proposed is reported in Eqs. (8.8, 8.9 and 8.10).

$$M^{2+} + OH^- = MOH^+ \tag{8.8}$$

$$MOH^+ + e^- = MOH_{ad} \tag{8.9}$$

$$MOH_{ad} + e^- = M + OH^- \tag{8.10}$$

Adsorbed intermediates formation in sub-reaction (8.9) is the rate determining step. Dissociation constant of $FeOH^+$ is lower than the $NiOH^+$ one [58]. For this reason, $FeOH^+$ concentration at the cathode is higher than $NiOH^+$ concentration. $FeOH^+$ adsorption is therefore promoted and $FeOH_{ad}$ sterically prevents $NiOH^+$ adsorption on the cathode surface, acting as inhibitor for Ni^{2+} reduction. This

Fig. 8.19 SEM view of ECD permalloy structures

mechanism indicates pH at the cathode interface as one of the most important variables which define the deposit composition.

8.6.5.1 Case Study: Ni:Fe 80:20 Through Mask Alloy Plating

Permalloy is the commercial name of a Ni 80%-Fe 20% (w/w%) alloy whose magnetic properties were discovered in 1923 [60]. Commercial permalloy alloys typically have relative permeability μ_r of around 100,000, compared to several thousand for ordinary steel. Electroplating of permalloy with different ECD baths and plating conditions has been widely studied [61–64]. This case study describes the development of a process to electroplate 10 μm thick Ni-Fe structures (Fig. 8.19).

The first step in alloy plating development is to identify an ECD bath whose chemical composition permits to reach both the desired deposit metals ratio and the stability of its own performances over lifetime (idle time and process time). A vendor proprietary sulfate-based chemistry has been used (Table 8.6).

Chemical composition, temperature, and pH of the bath are in-line controlled by an automatic analyzer-replenisher. A fountain plating chamber with multiple nickel soluble anodes has been used. In Ni-Fe plating temperature, fluid-dynamic conditions, current density, pH, and bulk ions concentration contribute together to the final deposit composition. DOE (Design of Experiment) considering all the cited variables at the same time would be too expensive in terms of number of experiments

Table 8.6 ECD bath composition for a sulfate-based chemistry

Parameter	Range
Density (g/mL)	1.0–1.4
Nickel (g/L)	45–60
Iron (II) (g/L)	1.0–3.0
Chloride (g/L)	20–35
Sulfate (g/L)	40–65
$NiCl_2*6H_2O$ (g/L)	75–105
$NiSO_4*6H_2O$ (g/L)	110–170
pH	1.0–3.2
Boric acid (g/L)	15–45

to be performed. To reduce dimensions number through which we need to explore our process windows, some parameters have been fixed to a specific value. These values have been chosen starting from macro-considerations on bath stability and plating tool hardware features. First, nickel and iron concentration ratio in the bath must be kept as constant as possible. Ni salts concentration has shown good stability along the bath life time. Slight deviations from target value may occur only if long idle periods occur causing an excessive Ni-anode dissolution in the bath. Fe (II)/Fe (III) ratio in the bath must be kept as high as possible in order to avoid low efficiency and deviations from standard alloy composition (Fe^{3+} ions have different deposition potential compared to Fe^{2+} ions). Fe (II) oxidation to Fe (III) is promoted by oxygen development at the anode during ECD process. Oxidation rate depends on operating condition: the higher the current density, the higher the oxidation rate. Fe (II) oxidation is possible in idle time as well: turbulent bubbling recirculation and temperatures above 40 ° C should be avoided due to faster bath oxygenation and consequent higher Fe (II) oxidation rate. For these reasons, temperature has been set between 30 and 40 °C and the plating chamber flow kept within 7.0 and 11.0 L/min. Bath pH value at the cathode interface influences co-deposition process as well, but it is important to consider that pH value is strongly dependent on the bulk starting value. Considering the stability limit given by the bath supplier (pH 3.2) and the drop of the efficiency under pH 2.0, the final pH has been fixed between 2.5 and 3.0. At this pH value, an acceptable efficiency of 75–80% and a good Fe (II) stability has been verified up to 4 months bath lifetime. Fluid-dynamic parameters (tunable through recipe and plating tool hardware options, see paragraph 8.3) strongly influence the double layer condition at the cathode and, therefore, the co-deposition mechanism. In through-the-mask plating, resist thickness and open area define the volume of the micro-pitches through which electroplating takes place. Fluid dynamics conditions inside these openings are very difficult to be modeled. In this case study, plating occurs through a 3 to 12 μm thick photoresist, 100 to 200 μm^2 open structures with an exposed area from 1% to 3%.

Fluid-dynamics parameters have been explored starting from plating tool hardware features. There are three main parameters which can be tuned to influence fluid-dynamics at wafer interface:

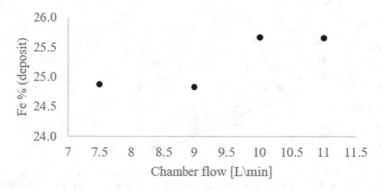

Fig. 8.20 Chamber flow influence on deposit composition (current density DC $= 24.5$ A·dm^{-2})

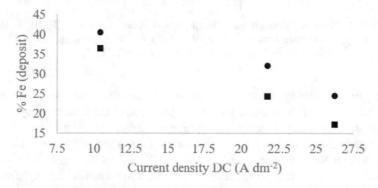

Fig. 8.21 Turbulence device activation influence on deposit composition. • turbulence device on; ■ turbulence device off

1. *Wafer rotation*. At higher Revolution Per Minute (RPM) values, differences between fluid-dynamic conditions at wafer edge (where tangential fluid speed is higher) and wafer center increase. Considering that chamber flow may vary from 7.0 to 11 L/min (pump hardware limits), it is possible to assume that RPM values from 0 to 10 do not influence fluid dynamics. Over 10 RPM thickness within wafer uniformity is not acceptable. RPM values choice is therefore mainly a matter of within wafer uniformity, especially if the device generating turbulent flow at wafer interface is off (see point 3). RPM value was set to 2–10 RPM, which permits to obtain a within-wafer uniformity <6% .

2. *Plating chamber flow*. It slightly influences deposit composition (Fig. 8.20). As mentioned, it is set to 7.0–10.0 L/min for turbulence reduction reasons (more bubbling drives to higher Fe(II) oxidation rate).

3. *Turbulent flow at wafer interface generation*. It strongly influences the Fe content in the deposit, drastically changing the flow conditions at the interface: the flow turns from laminar to turbulent. If turbulence is high, Fe deposition is promoted. For our purpose, the device is set to "off". (Fig. 8.21).

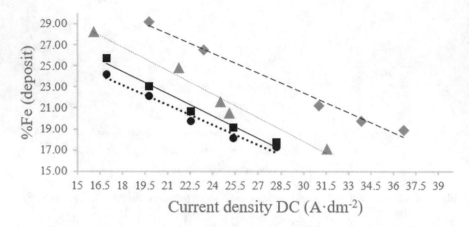

Fig. 8.22 %Fe deposited through mask in function of total current and [Fe(II)] in the bath with T = 30 °C, chamber flow = 7.5 L/min and pH = 2.5. • [Fe (II)]$_{bath}$ = 1.75 g/L; ■ [Fe(II)]$_{bath}$ = 2.0 g/L; ▲ [Fe(II)]$_{bath}$ = 2.1 g/L; ♦ [Fe(II)]$_{bath}$ = 2.25 g/L

The process window in terms of deposit composition has been explored in function of [Fe (II)] concentration in the bath and current density maintaining constant the previously described parameters and the nickel salts concentration. Many experiments have been performed adding Fe (II) to a fresh new bath, leading to the results reported in Fig. 8.22.

Linear proportionality correlation exists between current density and %Fe deposited. Furthermore, the higher [Fe (II)] concentration in the bath, the higher %Fe deposited at the same current density. Different [Fe (II)] bath content drives to trend lines with almost the same slope but with different intercepts with Y axis. The distance between different trend lines is negligible when [Fe (II)] concentration in the bath is around 1.8–1.9 g/L, while it is quite wide when [Fe (II)] concentration in the bath is higher than 2.0 g/L. In case of complex ECD mask layout, alloy composition dependence by local current density must be considered due to significantly different open areas in different wafer or die zones. Higher local open areas will lead to lower current densities and a higher Fe content, while lower local open area will generate structures with a lower Fe content (Fig. 8.23).

8.6.6   Gold Deplating-Electrochemical Wet Etch

Gold is one of the most noble metals, therefore gold seed layer removal by means of chemical etching can be done only with very aggressive etching solutions (ref Chap. 9). If gold seed etch is performed without a patterned mask (blanket etch), etchant might have poor selectivity on other materials exposed on the wafer surface. Moreover, if high topographies (e.g., several μm thick electroplated bumps) are

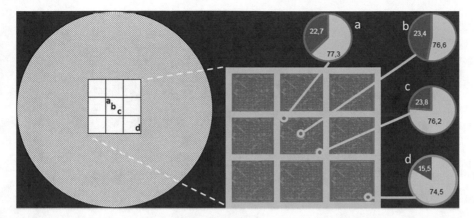

Fig. 8.23 Deposit composition in function of local open area on a patterned wafer; ▪: photoresist, □: patterned die; a, b, c, d: points measured by X-Ray Fluorescence, ▪: Fe weight %, ▪: Ni weight %

Table 8.7 Typical deplating bath composition

Parameter	Range
Thiocyanate (g/l)	20–50
pH	10–12
Temperature	25–45

present, efficient seed removal at the bottom of them can be done only with very long etching time that could further damage these structures (i.e., increase in roughness and high width variation). A valid alternative is to electrochemically etch gold with a technique called deplating. In this application, wafers act as anode and gold is dissolved in the solution with proper complexing agents able to stabilize the Au^+ ions. In the past, commercially available solutions were based on cyanide or thiourea due to the good stability of gold complexes [65, 66]. Due to safety concern (solutions with free cyanide or thiourea) thiocyanate-based or iodide-based solutions [67] are considered as a valid alternative (see Table 8.7 for typical bath composition).

When deplating process starts, gold seed layer is continuous and the efficiency is high due to its low resistivity. Going on with the process, some wafer areas are preferentially etched due to the different current density and/or fluid-dynamic local conditions (Fig. 8.24). Once the seed layer is completely etched in these zones, the preferential resistive path to etch the residual gold seed is through the barrier layer, generally much more resistive than gold, causing a drop of the efficiency. Therefore, the overall efficiency of the deplating process strictly depends on the landing barrier layer.

Moreover, barrier layers can be damaged after deplating process due to chemical and electrochemical compatibility issues (Fig. 8.25); sometimes specific landing layers require a specific deplating bath in terms of organic additives and/or bath composition.

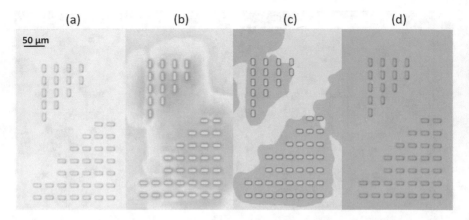

Fig. 8.24 Progress of Au deplating over process time (t). (**a**) continuous seed layer (t = 0), different etch progress in different areas (**b**) t = t' and (**c**) t = t", (**d**) seed deplating completion (t = t$_{end}$)

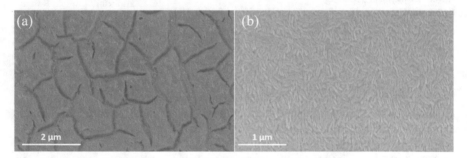

Fig. 8.25 TiW barrier landing layer after deplating process. (**a**) damaged TiW surface due to electrochemical incompatibility with Au deplating bath and (**b**) typical TiW grains surface after deplating process thanks to a different Au deplating bath composition

One of the main applications for gold electrochemical etch is the gold seed layer removal after gold plating: the seed layer and part of the plated gold are removed at the same time. In this case, no undercut will occur because of the contemporary lateral etch of the ECD structures. One of the preferred embodiments thereof is to have the same etch rate for the seed layer and ECD structures obtaining, after deplating, the same gold bump thickness as the one achieved after plating (Fig. 8.26).

Fine tuning of process parameters (current density, flow rate, temperature) is important to get a complete seed removal without damaging the plated structures and the barrier layer exposed to the solution once the seed has been removed. Different etch rates may occur on PVD (seed) and ECD (bumps) gold. This phenomenon is due to the different surface roughness generated by the two different deposition techniques: ECD gold is generally rougher than PVD gold by at least one order of magnitude in terms of Total Index Reading (TIR, mechanical profilometer

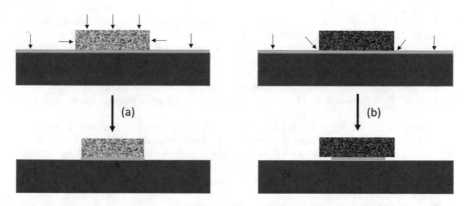

Fig. 8.26 Scheme of (**a**) Au seed layer deplating in presence of Au ECD bumps (no undercut) and (**b**) generic chemical wet etch of Au seed layer in presence of inert bumps (e.g., Ni; undercut occurs)

linear scan). The higher roughness of ECD bumps leads to a higher local surface area and therefore to a lower current density in respect of the seed layer. Different electrochemical regimes will occur on the two areas depending on what average current is set by recipe, leading to different process efficiency and therefore etch rates. A suggested solution is to set an average current which permits to have the same regime in both PVD and ECD surface regardless the roughness values.

References

1. Landolt, D. (2002). Electrodeposition science and technologies in the last quarter of twentieth century. *Journal of the Electrochemical Society, 149*(3), S9–S20.
2. Bard, A. J., & Stratmann, M. (2007). *Encylopedia of Electrochemistry, 5*, 3.
3. O'M Bockris, J., & Reddy, A. K. N. (2002). Modern electrochemistry, Vol. 1, 2nd ed.
4. O'M Bockris, J., Reddy, A. K. N, Gamboa-Adelco, M. (2002). Modern electrochemistry, Vol. 2A, 2nd ed.
5. Paunovic, M., & Schlesinger, M. (1998). *Fundamentals of electrochemical deposition*. Wiley.
6. Schlesinger, M., & Paunovic, M. (2000). *Modern electroplating*. Wiley.
7. Ghodssi, R., & Lin, P. (2011). *Ch 3: MEMS materials and process handbook* (pp. 147–186). Springer.
8. Datta, M., & Landolt, D. (2000). Fundamental aspects and applications of electrochemical microfabrication. *Electochimica Acta, 45*, 2535–2558.
9. Crow, D. R. (1998). *Principles and applications of electrochemistry*. Stanley Thornes (Publishers) Ltd..
10. Kim, B., & Ritzdorf, T. (2004). Electrical waveform mediated through-mask deposition of solder bumps for wafer level packaging. *Journal of the Electrochemical Society, 151*, C342–C347.
11. Datta, M., & Landolt, D. (1985). Experimental investigation of mass transport in pulse plating. *Surface Technology, 25*, 97–110.
12. IBL, N. (1980). Some theoretical aspects of pulse electrolysis. *Surface Technology, 10*, 81–104.

13. Landolt, D., & Marlot, A. (2003). Microstructure and composition of pulse plated metals and alloys. *Surface and Coatings Technology, 169–170*, 8–13.
14. Chandrasekar, M. S., & Pushpavanam, M. (2008). Pulse and pulse reverse plating- conceptual, advantages and applications. *Electrochimica Acta, 53*, 3313–3322.
15. Thiemig, D., Lange, R., & Bund, A. (2007). Influence of pulse plating parameters on the electrocodeposition of matrix metal nanocomposites. *Electrochimica Acta, 52*, 7362–7371.
16. Pearson, T., & Dennis, J. K. (1991). Facts and fiction about pulse plating. *Transactions of the IMF, 69*(3), 75–79.
17. Young-joo, O., Chung, S.-h., & Lee, M.-s. (2004). Optimization of thickness uniformity in electrodeposition onto a patterned substrate. *Materials Transactions, 45*(10), 3005–3010.
18. Woodruff, D. J., & Hanson, K. M. Electroplating Apparatus with Segmented Anode Array, United States patent US 006497801B1, Dec 24,2002.
19. Sharma, B. K. (1991). *Industrial chemistry (including chemical engineering)* (pp. 1666–1691). Goel Publishing House.
20. Faust, C. L. (1941). Alloy plating. *Transactions of The Electrochemical Society, 80*(1), 301–327.
21. Brenner, A. (1963). *Electrodeposition of alloys: Principles and practice*. Academic.
22. Paunovic, M., & Schlesinger, M. (1998). *Fundamentals of electrochemical deposition* (1st ed., pp. 42–48). Wiley.
23. Reid, F. M., & Goldie, W. (1974). *Gold plating technology*. Electrochemical Publications Limited.
24. Rapson, W. S., & Groenewald, T. (1978). *Gold usage*. Academic Press, Inc.
25. Wilkinson, P. (1986). *Gold Bulletin, 19*, 75.
26. Green, T. A. (2007). *Gold Bulletin, 40*(2), 105.
27. Dimitrijevic, S., Rajcic-Vujasinovic, M., & Trujic, V. (2013). *International Journal of Electrochemical Science, 8*, 6620.
28. Kato, M., & Okinaka, Y. (2004). *Gold Bulletin, 37*, 37–44.
29. Traut, J., Wright, J., & Williams, J. (1990). *Plating Surface Finishing, 77*, 49.
30. Honma, H., & Kagaya, Y. (1993). *Journal Electrochemical Society, 140*, L135.
31. Osaka, T., Kodera, A., Misato, T., Homma, T., Okinaka, Y., & Yoshioka, O. (1997). *Journal of Electrochemical Society, 144*, 3462.
32. D.G. Foulke in F.M. Reid and W. Goldie. (1974). *Gold plating technology* (p. 52). Electrochemical Publications Limited.
33. Watanabe, H., Hayashi, S., & Honma, H. (1999). *Journal of the Electrochemical Society, 146*, 574.
34. Lo, C. C., Augis, J. A., & Pinnel, M. R. (1979). *Journal of Applied Physics, 50*, 6887.
35. D. Erikson, G. Visalli, A. Lodi, L. Castoldi, N. Thorp, P. McHugh, Wilson, (2007) Wilson, IEEE/SEMI Advanced Semiconductor Manufacturing Conference (ASMC), 222–226. Stresa, Italy
36. Weisenberger, L. M., & Durkin, B. J. (1994). Copper plating. *Surface Engineering, 5*, 167–176.
37. Mikkola, R. D., & Chen, L. (2000). *Proceedings of the IEEE2000 international interconnect technology conference* (pp. 117–119).
38. Liu, Y., Wang, J., Yin, L., et al. (2008). *Journal of Applied Electrochemistry, 38*, 1695–1705.
39. Lagrange, S., Brongersma, S. H., Judelewicz, M., Saerens, A., Vervoort, I., Richard, E., Palmans, R., & Maex, K. (2000). *Microelectronic Engineering, 50*, 449.
40. Castoldi, L., Morin, S., Visalli, G., Fukada, T., Ouaknine, M., Roh, E. H., & Yoo, W. S. (2003). ESC Meeting in Paris.
41. Castoldi, L., Lodi, A., & Visalli, G. (2008). *ECS Transactions, 16*(14), 23–28.
42. Nguyen, V. H., Hof, A. J., van Kranenburg, H., Woerlee, P. H., & Weimar, F. (2001). *Microelectronic Engineering, 55*, 305.
43. Vereecken, P. M., Blinstead, R. A., Deligianni, H., & Andricasos, P. C. (2005). *IBM Journal of Research and Development, 49*(1), 3–18.

44. Li, Y., Yao, J., & Huang, X. (2016). *International Journal Metallurgy Material Engineering, 2*, 123.
45. Luo, J. K., Pritschow, M., Flewitt, A. J., Spearing, S. M., Fleck, N. A., & Milne, W. I. (2006). *Journal of the Electrochemical Society, 153*(10), D155–D161.
46. Luo, J. K., Flewitt, A. J., Spearing, S. M., Fleck, N. A., & Milne, W. I. (2004). *Materials Letters, 58*, 2306.
47. Tan, A. C. (1993). *Tin and Solder plating in the semiconductor industry (chs 8–10)* (p. 197). Chapman and Hall.
48. Xu, H., Vuorinen, V., Dong, H., & Paulasto-Krockel, M. (2015). *Journal of Alloys and Compounds, 619*, 325–331.
49. Mita, M., Miura, K., Takenaka, T., Kajihara, M., Kurokawa, N., & Sakamoto, K. (2006). *Materials Science and Engineering B, 126*, 37–46.
50. Sun, W., & Ivey Materials, D. G. (1999). *Science and Engineering, B65*, 111–122.
51. Baated, A., Kim, K.-S., & Suganuma, K. (2011). *Journal of Materials Science: Materials in Electronics, 22*, 1685.
52. Oberndorff, P., Dittes, M., & Crema, P., & Chopin, S. 8th IPC/JEDEC Int. Conf. on Lead-Free Electronic Components and Assemblies, Dec. 2004, Boston MS, USA.
53. Dittes, M., Oberndorff, P., & Petit, L. (2003) Tin Whisker formation - results, test methods and countermeasures, 53rd Electronic Components and Technology Conference, 2003. Proceedings., 2003, pp. 822–826.
54. Hirano, T., & Fan, L.-S. (1996). Invar electrodeposition for MEMS application. *Micromachining and Microfabrication Process Technology II, 2879*, 252–259.
55. Nagayama, T., Yamamoto, T., & Nakamura, T. (2019). Electroplating and characterization of invar Fe–Ni alloy films from plating baths containing organic acids. *ECS Transactions, 89*(7), 65–80.
56. Dubin, V. M., Lisunova, M. O., Kovalenko, I. O., Walton, B. L., Downey, G., Su, J., & Witt, K. (2016). Invar electroplating for controlled expansion interconnects. *ECS Transactions, 75*(7), 33–40.
57. Hessami, S., & Tobias, C. W. (1989). A mathematical model for anomalous Codeposition of nickel-Iron on a rotating disk electrode. *Journal of the Electrochemical Society, 136*, 3611–3616.
58. Nakano, H., Matsuno, M., Oue, S., Yano, M., Kobayashi, S., & Fukushima, H. (2004). Mechanism of anomalous type electrodeposition of Fe-Ni alloys from sulfate solutions. *Materials Transactions, 45*(11), 3130–3135.
59. Harris, T. M., & Clair, J. S. (1996). Testing the role of metal hydrolysis in the anomalous electrodeposition of Ni-Fe alloys. *Journal of the Electrochemical Society, 143*(12), 3918–3922.
60. Permalloy, A. (1923). New Magnetic Material of Very High Permeability H. D. Arnold, G. W. Elmen First published: July 1923 *Bell Labs Technical Journal 2*(3): 101–111.
61. Venkatasetty, H. V. (1970). Electrodepostion of thin magnetic Permalloy films. *Journal of the Electrochemical Society, 117*(3), 403–407.
62. Kovac, Z. (1971). The Effect of Superimposed A.C. on D.C. in Electrodeposition of Ni-Fe Alloys. *Journal of The Electrochemical Society, 118*(1), 51–57.
63. Grimmett, D. L., Schwartz, M., & Nobe, K. (1993). A comparison of DC and pulsed Fe-Ni alloy deposits. *Journal of the Electrochemical Society, 140*(4), 1136–1340.
64. Leith, S. D., Ramli, S., & Schwartz, D. T. (1999). Characterization of NixFe1-x (0.10 < x < 0.95) electrodeposition from a family of Sulfamate-chloride electrolytes. *Journal of the Electrochemical Society, 146*(4), 1431–1435.
65. Jeffrey, M. I., & Ritchie, I. M. (2000). *Journal of Electrochemical society, 147*(9), 3257–3262.
66. Wei, D., Chai, L., Ichino, R., & Okido, M. (1999). *Journal of Electrochemical Society, 146*(2), 559–563.
67. Hu, Z., & Ritzdorf, T. (2007). *Journal of Electrochemical Society, 154*(10), D543–D549.

Chapter 9
Wet Etching and Cleaning

Gianluca Longoni, Davide Assanelli, and Cinzia De Marco

9.1 Introduction

MEMS microfabrication keeps raising compelling challenges concerning wet processes, being either wet etching or wet photoresist stripping and cleaning strategies. Primarily, large material variety involved in MEMS fabrication, and sometimes unconventional for standard microelectronics industry, imposes an in-depth study of the compatibility among chemical moieties and an evaluation of compounds long-time reliability. Chemicals involved in wet processes, indeed, need to be evaluated for what concern not only their efficacy in fulfilling their technical aim, but also in merit to chemical or physical interference they may be troubled with in presence of unprotected metals or particularly stressed or compromised surfaces. In most cases, even without exotic materials utilization, uncommon wet strategies must be implemented due to silicon patterning treatments peculiar of MEMS microfabrication. That is the case of deep recesses and high aspect-ratio structures achieved by Deep Reactive Ion etching (RIE), whose cleaning requires dedicated chemicals and care. Materials and chemicals integration in complex process flows must also involve a careful evaluation of the hardware of choice for wet processes. Micromechanical systems, by definition, are composed of freed and moving portions (proofing and seismic masses, cantilevers, clamped membranes and rotor) which must not be subjected, to avoid the increase of mechanical failure and yield loss, to harsh handling at the front-end microfabrication level. In this frame, wet cleaning and photoresist stripping techniques can hardly rely on a physical, rather than a chemical, contribution. The use of scrubbing and jet-spray, as well as spinning tools must be thus carefully evaluated, if not precautionarily excluded.

G. Longoni (✉) · D. Assanelli · C. De Marco
ST Microelectronics, Analog MEMS and Sensors Group, MEMS Technology and Design R&D,
Agrate Brianza, Monza Brianza, Italy
e-mail: gianluca.longoni@st.com; davide.assanelli-ams@st.com; cinzia.de-marco@st.com

© Springer Nature Switzerland AG 2022   259
B. Vigna et al. (eds.), *Silicon Sensors and Actuators*,
https://doi.org/10.1007/978-3-030-80135-9_9

This heterogeneous set of prerequisites is pushing MEMS industry toward process hybridization which involves and combines dry and wet techniques particularly for what concern photoresist removal and polymer cleaning. In the following paragraphs, for this reason, the reader might bump into technological process blocks description in which dry and wet processes are intertwined, and these will be the cases where a fully wet approach can't simply suffice.

Wet cleaning hardware types are commonly shared with wet etching, since a part from minor due differences related to the adopted chemistries accountable for photoresist and polymer removal rather than chemical etching, both deal with solutions and liquid chemicals. Hardware types can be categorized into three main groups

- Wet benches
- Spray tools
- Single wafers spinning tools

The choice of the correct tool should be done in function of silicon substrate characteristics, device layout, and process flow. The wafer thickness range in MEMS technology can be very wide, starting from 120 μm up to standard thickness, which is 725 μm, but in case of bonded wafers, the total thickness can reach 1 mm or higher. For each wet etching, wet photoresist stripping, and cleaning strategy, the utilized tool will be quoted and its choice motivated. Briefly, spray tools are commonly used for solvent cleanings and photoresist removal, albeit metals' wet etching can also be performed. Their tumbling barrel-like chamber, however, whose rotation can reach 2000 rpm during final drying procedure, is not advisable for thin wafers or substrate densely patterned with movable parts. Wet benches, in these circumstances, would be more appropriate. Wet benches can be used for a wide range of wafers, from thin to bonded wafers, and can be used even for holed wafers thanks to the vertical handling, that involves only the bevel of the wafer, avoiding any kind of vacuum or chucking. Wafers are not moreover subjected to mechanical stress since no spinning is employed in the drying process, on the contrary, only hot nitrogen purging is commonly used. A possible drawback for thin wafers could be their sticking during lifting from tanks. Wafer support bars and carriers, if specifically designed for thin wafers housing, can in these cases avoid stiction. Single wafer spinning tools are versatile and vastly used for both cleaning and wet etch procedures. In single wafer spinning procedure, high RPM wafer rotation is exploited, during chemical dispensing, to achieve a uniform wafer coverage and a reproducible residence time of the chemically active moieties on wafer surface. Lastly, equipment that combine wet tanks and single wafer chambers with pressurized spray chemistries can be very useful in MEMS cleaning steps where the stripping of "modified" photoresist and polymer removal is required.

9.2 Wet Etching

9.2.1 Overview

Wet etching procedures encompass all the micromachining processes aimed at selectively removing functional layers employing conveniently formulated aqueous solutions. Wet etching, compared to dry etching, offers, among the advantages, the benefit of considerably lower processing costs, parceled into chemicals cost, hardware expenditure, and complexity. Many chemistries are at disposal for wet etching, and similarly, a high number of etching parameters can be modulated such as bias (undercut), etch rate, selectivity, overetching, feature size control, and etch anisotropy. The latter has its maximum relevance when dealing with preferentially oriented crystalline layers or substrates [1]. In such cases, etchant molecules' interaction with crystalline facets' chemical terminations rules the etch rate magnitude. More in general, being driven by almost exclusively chemical surfaces interactions, a foremost importance is occupied by layers surface conditions in wet etching. This aspect includes surface chemical terminations, surface roughness and exposed area, contaminants surface adsorption (such as water, halides, metallic contaminants, or volatile organic compound VOC [2]), and surface passivation operated by oxides and hydroxides of the underlying layers and their constituents. Conversely, isotropic wet etching, which is peculiar of the majority of etchant species, allows the control of etching variables functional to other technological steps, such as metallization deposition or stress relief. That is also the case of sloping profiles angle tuning, which however carries with itself modification of the initial feature dimensions.

9.2.2 Silicon Wet Etch

9.2.2.1 Isotropic Wet Etching of Silicon

Silicon wet etch can be performed following two approaches: isotropic and anisotropic etching. The first one is the method of choice when silicon wafer uniform stress relief is sought, especially after physical wafer thinning through chemical mechanical grinding (CMG) and Chemical-Mechanical polishing (CMP) process [3]. Silicon micro defects, crystalline defects, dislocations, and microcracks [4], induced by the above-mentioned thinning techniques, are effectively removed employing silicon etching chemistry involving a strong silicon oxidant, namely nitric acid (HNO_3) and a silicon oxide etching agent, i.e., hydrofluoric acid (HF). This process is effectively used in MEMS fabrication as a surface finishing and polishing step, after wafer thinning to target functional thickness, commonly performed on the whole surface of wafer backside. Isotropic etch is however poorly suggested for silicon wafer etch through photoresist mask or with hard-masking approach. The scarce control on silicon undercut makes this technique unsuitable for silicon patterning. In brief, silicon etching is perpetrated by silicon oxidation first and contextual silicon oxide removal. Etch rates in the range of 10–40 μm per minute are typical. To achieve a high uniformity of silicon etching, hardware is critical. Considerable high nonuniformity is achieved in etching baths: uneven and uncontrolled bath agitation and gas evolution (nitric oxide, NO) are strongly deleterious for silicon etch uniformity and consequently for stress homogeneous relief. For this reason, single wafer spinning tools are strongly preferred over immersion ones. Products of the overall Eq. 9.4 reported below, (H_2SiF_6) are moreover rapidly spun off from the spinning wafer, avoiding side-products precipitation. Equations 9.2, 9.3 and 9.4 are intended to clarify oxidant formation NO_2, moiety responsible, together with molecular oxygen, and HF for silicon oxidation [5], and silicon oxide etch, respectively.

$$4\,HNO_3 \rightarrow 4\,NO_2 + 2\,H_2O + O_2 \tag{9.1}$$

$$4\,NO_2 + 2\,Si \rightarrow 2\,SiO_2 + 4\,NO \tag{9.2}$$

$$2\,SiO_2 + 12\,HF \rightarrow 2H_2SiF_6 + 4\,H_2O \tag{9.3}$$

$$2\,Si + 4\,HNO_3 + 12\,HF \rightarrow 4\,NO + 6\,H_2O + 2\,H_2SiF_6 + O_2 \tag{9.4}$$

In $HF : HNO_3$ mixture, if HF concentration approaches zero, silicon etch rate rapidly drops. On the other hand, since pure concentrated HF is inert on un-passivated silicon surface and uncapable of silicon oxidation, no etch occurs if HNO_3 is solely present [6]. The etch rate values previously reported (10-40 um per minute) relates to HF(48%):HNO3(70%):H2O = 3:2:5 mixture composition.

Inertness of single-crystalline silicon surface in HF solution is however undermined by pH condition of the etching medium: as it has been indeed revealed by Jakob and co-workers, buffered solutions (HF:NH4F or HF:NaOH mixtures) at different alkaline pH values expose (111) silicon surface to increasing etch rates in proportion with alkaline pH values. This trend unveiled the silicon oxidation dependence with hydroxyl (OH^-) ion concentration in the etchant [7]. Together with the etchant composition, silicon surface wettability has a relevance in limiting silicon chemical etching. Wetting agents are often employed to increment silicon wettability with a positive outcome on etching uniformity. Among wetting agents, acetic acid (CH3COOH) and phosphoric acid (H3PO4) are often taken into consideration. In image (b) of Fig. 9.1, etch rates dependence on the ternary HF:HNO3:CH3OOH mixture composition is schematized. It must be noted that, as HF concentration increases, higher dependence of the etch rate with temperature condition has to be faced. In general, an Arrhenius dependence, of the type expressed in Eq. 9.5, between reaction rate and temperature, is verified for isotropic etching in acidic mixtures [8].

$$dlnk\Big/dT = E_a\Big/RT^2 \tag{9.5}$$

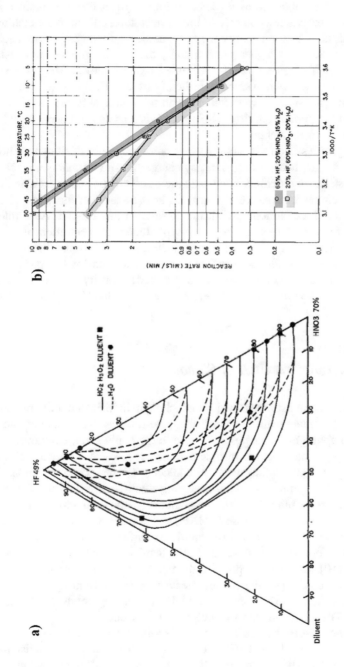

Fig. 9.1 (a) Etch rate dependence with etchant composition, diluent, either water or acetic acid, is included in the solution; (b) Arrhenius plot for H_2O diluted isotropic etching solution with different acids composition: light-blue highlighted data stand for 20% HF:60% HNO_3:20% H_2O, while red-highlighted stand for 65% HF:20% HNO_3:15% H_2O. Reprinted with permission of IOP Publishing

For mixture composed by 20% HF:60% HNO$_3$:20% H$_2$O, a two-segment Arrhenius plot can be obtained as depicted in image (b) of Fig. 9.1, indicating a higher activation energy below 30 °C. At low temperature, silicon oxide etching is preferential, and the activation energy is mostly associated with oxidation reaction induced by nitric acid. At higher temperatures, the activation energy is instead associated with the diffusion-limited reaction of oxide removal. This is the last condition which allows to obtain smoother silicon surfaces. Having verified a strong dependence of silicon isotropic wet etch rate with solution temperature, accuracy in keeping solution temperature control within ± 1 °C is thus deemed mandatory for a reproducible and uniform etching.

When dealing with silicon wet etching, electronics properties of silicon are not to be neglected since they heavily influence any charge-transfer mechanism involving silicon surface states, including silicon oxidation and silicon oxide etching. This feature is of foremost importance not only in acid media wet etch, but also in silicon etch in alkaline solutions, as illustrated later in this chapter. As a rule of thumb, an increment in n-type as well as p-type dopant concentration of silicon substrate always leads to a raise in isotropic etch rate in HF-HNO$_3$-based acidic media, as consequence of the incremented charge carrier density. For particularly heavily doped substrates, with a nominal surface carrier density of 3×10^{22} cm^{-3} (n^+), an etch rate of 0.8 nm min^{-1} is registered in even diluted HF solution at 40 °C [9].

9.2.3 Anisotropic Wet Etch of Silicon

Anisotropic etch is demonstrated as valuable technique in MEMS microfabrication, since it enabled the manufacturing of intricate three-dimensional structures such as proof masses, diaphragms, trenches for microfluidic application, cantilevers, etc. Contrary to isotropic etching, anisotropic etching proceeds at different etch rates according to exposed crystalline facets in contact with the etching solution. The latter, in particular, is common in alkaline etching solutions based on KOH, NaOH, CsOH, RbOH, but most often NH_4OH and quaternary ammonium hydroxides such as tetra-methyl ammonium hydroxide (TMAH) and ethylene di-amine pyrocatechol (EDP), with the possible addition of alcohols as surfactant to modulate etchant interaction with silicon surface. A valid electrochemical model has been proposed by Seidel et al. [10] for the description of the interaction of alkaline solution with silicon surface. According to the model, electrons transfer from hydroxyl group toward silicon conduction band happens via silicon dangling bonds and backbond surface states. The different energy levels of the backbond surface states are responsible for the anisotropic effect, as they have a direct consequence on activation energy of backbonds breaking. Different etch rates, in the context of anisotropic etching, thus directly derive from peculiar interaction of etching moieties (e.g., hydroxyl anions, OH^-) with crystalline facets terminations, that is, Si-exposed atoms on silicon surface and silicon dangling bonds, in direct contact with etching solution. Seeking clarity, two competitive mechanisms have always to be taken into account when silicon anisotropic etching is considered: OH^- adsorption and electron transfer, accountable for etching action as illustrated before, and counter-ions adsorption (such as NH_4^+, tetramethyl ammonium cations, and other organic compounds), responsible for etching blocking condition [11]. It has been suggested that tetramethyl ammonium cations behave similarly to IPA-isopropyl alcohol molecules in guiding silicon surface structuring, despite they do not intervene on surface tension of the solution [12]. For the further considerations about anisotropic etching velocities and geometries, except for where otherwise specified, <100>-oriented Si substrate will be always considered, as it is the most common substrate employed in MEMS manufacturing. As a common rule, planes with the lowest etching rate are prone to progressively develop as the etching proceeds [13]. That is the case for {111} crystallographic planes family on a < 100>-oriented Si substrate, as reported in image (a) in Fig. 9.2. (111) Plane, toward which the lowest etch rate is registered in alkaline solutions, typically intersects bottom (100) ones at the peculiar angle of 125.3° (54.7° if groove (100)-top is considered) [14]. Added to this according to how the etch masking is operated, significant over etching (also referred to as undercutting) must be considered. As the misalignment of a square test-mask increase in respect with <110> direction normal to (100) planes, the undercutting also gains significance. This aspect, whose relevance can be often underestimated, is crucial in containing microstructures oversizing. In MEMS fabrication, TMAH based anisotropic etching has been suggested as smoothening procedure aimed at reducing surface roughness [15]. This approach can be also employed for curtailment of surface defects and asperities induced by Deep silicon Reactive Ion dry Etch (DRIE), a widespread process involved in the definition of several hundreds of micrometers-wide circular cavities on the backside of, for instance, MEMS acoustic transducers (Images (b) and (c) in Fig. 9.2).

9.2.4 Dielectrics Wet Etch

Etch rates, in aqueous solutions, for dielectrics layers such as SiO_2, Si_xN_y, $Si_xO_yN_z$, and Al_2O_3 are strictly bound to dielectrics layer chemical and physical properties, conveyed by their deposition methods. It has been widely verified, for example, that thermal oxide, due to its higher density, etches more slowly than LPCVD (Low-Pressure Chemical Vapor Deposition) silicon oxide. Other factors can contribute in modifying etch rates, such as chemical composition [16] (hydrogen trapped in dielectrics deposited by plasma-enhanced CVD) and annealing post processes [17]. HF aqueous solution is the reference etching moiety for silicon oxides and nitrides as well. Etch rates are linear with HF concentration for dilutions in the $H_2O : HF$ 10:1 and 100:1 range, and they progressively get weaker with use. A more stable etchant, characterized by a higher compatibility with photoresist masking,

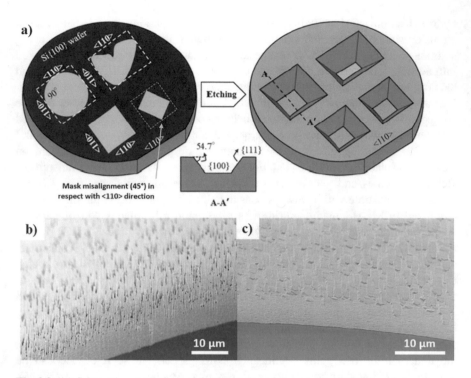

Fig. 9.2 (a) Schematic representation of anisotropic etch by means of alkaline solution through different mask shapes. Reprinted with permission of Springer; (b) and (c) SEM micrographs refer to silicon walls of 500 μm-wide cavity, etched via DRIE, before and after TMAH smoothing procedure, respectively

is guaranteed by Buffered Hydrofluoric acid solution (BHF) or Buffered Oxide Etchant (BOE), which is a mixture composed by 40% NH_4F and 48% HF, in the ratio 5:1 or 10:1, having the latter, half of the etch rate of the former. A mixture commonly employed is the 7:1 at the temperature of 21 °C. In these conditions, the etch rate of 880 Å min^{-1} can be registered on thermally grown oxide, while a roughly three-fold etch rate (3000 Å min^{-1}) is obtained on LPCVD silicon oxide. A ratio of 20:1 is also frequently used, if milder etch rates are desired [17], and some exemplifying etch rates are reported in Table 9.1. NH_4F concentration is precociously kept below 10–15% since at higher concentration not only the etch rate starts decreasing, due to complexation of HF_2^- by NH_4^+ cation, but also the risk of $(NH_4)_2SiF_6$ precipitation starts to rise significantly. BOE etching profile and taper of silicon oxide can be conveniently modified by intervening on photoresist masking characteristics. BOE etching is fundamentally isotropic on SiO_2, and when a good photoresist adhesion is guaranteed, semicircular etching walls are formed. On the contrary, the partial delamination of photoresist, from dielectric surface, favors the withdrawal of the silicon oxide upper front, originating a more reentrant profile, and,

Table 9.1 Etch rate in Buffered Oxide Etchant (20:1) of widespread SiO_2 layers employed in integrated circuit manufacturing, being APCVD and LPCVD atmospheric and low-pressure chemical vapor deposition techniques, respectively

Layer	Etch rate (BOE 20:1) [17]
Thermal oxide	168 Å min^{-1}
LPCVD (Teos + O_2)	318 Å min^{-1}
APCVD (SiH_4 + O_2)	570 Å min^{-1}
APCVD + anneal 500 °C	444 Å min^{-1}
APCVD + anneal 600 °C	426 Å min^{-1}
PECVD (SiH_4 + N_2O)	246 Å min^{-1}

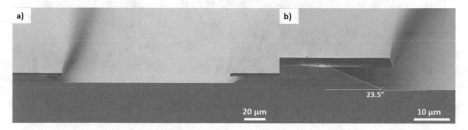

Fig. 9.3 (**a**) SEM cross section of groove etched in silicon oxide by means of BOE. Overhanging layer is formed by the photoresist. (**b**) Higher magnification of oxide tapered wall with a pitch angle of 23.5°

as a consequence, flattens the incident angle (pitch angle) between silicon oxide wall and substrate (Fig. 9.3).

As previously mentioned, HF can be a valid compound for silicon nitride wet etching; nevertheless, it hardly succeeds in being selective toward silicon oxide. When wet etching has to be selectively targeted to nitride etching, H_3PO_4 is highly preferred [18], and etching is commonly operated heating a 85% H_3PO_4 solution to 165 °C and reaching an etch rate of 100 Å min^{-1}[19]. SiO_2 selectivity, expressed as the ratio between Si_3N_4 and SiO_2 etch rate, can be further improved by hydrofluorosilicic acid (H_2SiF_6) namely a species which inhibits SiO_2 dissolution reaction.

9.2.5 Metal Wet Etch

9.2.5.1 Gold Wet Etch

Owing its high electrical conductivity, resistance to galvanic corrosion, good bond ability, gold layers deposition, and patterning find wide application in MEMS manufacturing. Continuous scaling down of MEMS dimension however is casting new challenges on gold layer thickness and critical dimension uniformity. Despite lower critical dimension loss, gold dry etch, by means of plasma and ion-

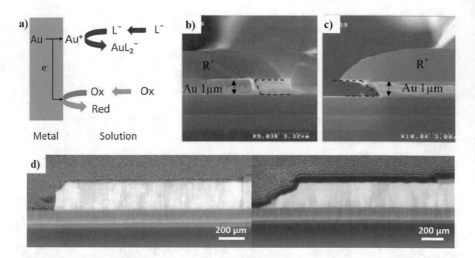

Fig. 9.4 (**a**) Illustrated mechanism of gold etching in a complexing-oxidizing medium and reentrant profile with a positive (**b**) and negative (**c**) pitch depending on the underlying layer. If a metallic layer is employed, galvanic coupling is promoted and gold negative slope is obtained. (**d**) SEM micrograph cross sections of gold etch achieved with (left) and without (right) photoresist stabilization

milling, carries with itself the limitation of low volatility of the etching products, recently overcome by chlorine-based and hydrogen-based plasma. Gold wet etching chemistries encompass cyanide-, iodide-, bromide-, and chloride-containing solutions. Chemically speaking, gold etching is achieved through synergistic action between complexation and gold surface potential modulation. The former action is fulfilled by the presence, in the gold etchant formulation, of a ligand, or Au ion complexing agents, as Au(I) and Au(III) ions' solvation by water molecules is thermodynamically impeded [20]. In the above-mentioned solution, this action is guaranteed by cyanide and halide anions which stabilize Au cations by complexing them as they are formed at anodic sites. Gold surface potential modulation is instead operated by an oxidizing agent, by accepting electrons' donation from gold surface cathodic sites, as illustrated in image (a) in Fig. 9.4: gold etch is indeed electrochemical in nature, a controlled corrosion process [21]. Among the cited etchant solution, the iodine/iodide system [22, 23] tends to be preferred over cyanide-based gold etching, due to lower toxicity and environmental impact. In the iodide system, the oxidant is the triiodide moiety $\left(I_3^-\right)$ formed by the reaction between iodine and iodide ion $(I_2 + I^-)$, which typically ensures etch rates between 0.5 and 1 μm min^{-1}, at room temperature, and with an iodide concentration of 0.2 M, suitable for thick layers etching. Gold wet etch is commonly isotropic in nature. However, a reentrant profile can be observed in particular conditions or with peculiar underlying layers such as Cr, Ti, or TiW [24]. The explanation of the phenomenon must be referred to as galvanic coupling between gold layer and underlying metal, and cathodic and anodic areas redistribution which happens once the underlying layer emerges during etching. As it happens, oxidant reduction can be sustained on the same cathodic area, in which the underlying metal replaces gold, while gold etching anodic areas are sensibly reduced, since only the structures' sidewalls remained exposed. In this context, anodic current density increases, and so does the gold etch rate, especially in the close vicinity to gold-underlying layer interface, where the metals' coupling resistive path is shorter. The reentrant profile which generates at this point translates in a gold slanted profile, with a negative pitch, especially for thick gold layers (see image (b) and (c) in Fig. 9.4). It has to be kept in mind that while anodic reaction, namely gold oxidation, can be under kinetic as well as mass diffusion control, cathodic reaction, i.e., oxidant reduction, is mostly under oxidant mass diffusion control, and agitation in the etching system must be thus considered as a variable not to be neglected in the optimization of etching parameters [25]. The effect of different underlying metal layers and different gold etching tools on gold etching rate has been widely investigated and quantified too.

Worth mentioning is the relevant role of gold etchant additives, aimed at pH regulation, in modifying structures profile pitch. Weakly acidic pH (5–6) favors a steeper gold profile rather than gold etchant kept at acidic pH. Ultimately, photoresist stabilization also promotes steep gold profiles rather than nonstabilized photoresist, as can be spotted in image (d) in Fig. 9.4.

9.2.5.2 Aluminum and Aluminum Alloys Wet Etch

Aluminum, Aluminum–Copper (Al-Cu), and Aluminum–Silicon–Copper (Al-Si-Cu) wet etch is commonly operated in H_3PO_4–HNO_3 acidic aqueous solutions. In this ambient, HNO_3 replicates what has been previously discussed for isotropic silicon etching: Aluminum surface oxidation to Al_2O_3 is perpetrated by nitric acid, while phosphoric acid is accountable for Al_2O_3 dissolution. The etching profile, in orthophosphoric acid-based solution, demonstrated to be highly sensitive to surfactant presence. In particular, a slanted or tapered aluminum profile can be obtained if acetic acid, with the ratio 1:8 in respect with H_3PO_4, is added to the solution in the so-called PAN mixture (Phosphoric-Acetic-Nitric acid solution) [26]. Aluminum profile taper can be moreover varied acting on etching bath temperature and acetic acid concentration, obtaining customized profile angles spanning between 30° and 90°. This aspect is of foremost importance when it comes to passivate the metallization with dielectrics or insulating layers: their step coverage, that is the thickness of dielectric at the border between aluminum and the substrate, on a tapered aluminum profile will be thicker compared to that on a merely vertical aluminum profile.

Aluminum and aluminum alloys wet etch can be performed through dipping in wet bench equipment (batch tool) as well as in single-wafer spin process tools. The former is the method of choice when tapered aluminum profile together with an improved etch uniformity is sought. Because acetic acid addition to obtain $H_3PO_4 : HNO_3 : CH_3COOH : H_2O$ 16:7:2:1 composition, combined with single-

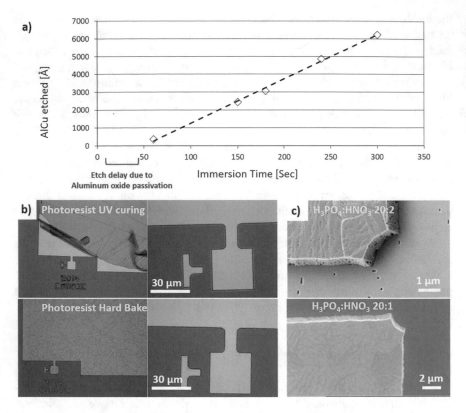

Fig. 9.5 (**a**) etch rate curve on Al-Cu (0.5%) alloy in H_3PO_4:HNO_3:CH_3COOH: H_2O solution, calculated etch rate value is 1220 Å min^{-1}; (**b**) Photoresist UV curing (upper optical images) and photoresist hard bake effect (lower images) on photoresist lifting occurrence during rinse and drying procedure at the end of Al metal etch in a wet bench equipment. (**c**) HNO_3 concentration effect on aluminum vertical wall after wet etch in single-wafer tool

wafer spin process tool hardware, promotes massive lifting of the photoresist mask, leading to an unbearable dimensional loss.

Regardless of the equipment employed, in evaluating etching time, it must be taken into consideration that a certain delay (few tens of seconds) is commonly observed before aluminum surface is attacked by etchant, as can be observed in image (a) in Fig. 9.5. This feature is due to aluminum Al_2O_3 passivation, which can be tens of Angstrom thick and surely retard etching of bulk aluminum. As previously described, acetic acid addition to etching mixture is mandatory as profile tapering mediator. This addition, however, might negatively impact photoresist adhesion on aluminum substrate, and photoresist lifting can happen during rinse and drying, at the end of the etching procedure in a wet bench equipment (image (b) in Fig. 9.5). Proper photoresist stabilization needs to be implemented in these circumstances. Aluminum etch in single-wafer spin process tool, based on $H_3PO_4 - HNO_3$ solution, gave results comparable to those obtained via aluminum dry etching, in terms of profile cleanliness, absence of aluminum walls pitting, and walls verticality, as illustrated in the lower image of inset (c) in Fig. 9.5. In respect with dry etching, aluminum wet etching performed in spin process tools guarantees better performances also in terms of mask selectivity, a feature that severely limits RIE application to thick aluminum layers definition [27]. Etching performed with a 20:1 $H_3PO_4 : HNO_3O$ mixture, kept at 50 °C, was effectively employed for obtaining steep aluminum profile (equivalent to a 90° angle), together with the attack uniformity of 5%, referred to 200 mm wafers. The higher content of nitric acid in 20:2 $H_3PO_4 : HNO_3$ solution is deleterious for aluminum surface and walls integrity, as it promotes intergranular and generalized corrosion (see top optical image of inset (c) in Fig. 9.5).

9.2.5.3   Ti and TiW Wet Etch

Effective solutions for Titanium etching are diluted hydrofluoric solution, orthophosphoric acid solutions, sulfuric acid solution, and ammonia-hydrogen peroxide (also known as APM or SC1 solutions). As mentioned for other metallization, acids concentration and etching baths temperature strongly affect measured etch rates. Concerning H_3PO_4 solutions, the optimum condition has been achieved with mildly concentrated solution (61% by weight) at 80 °C [28], which ensures a combination between the acceptable etch rate of 185 Å min^{-1} with a minimum photoresist erosion (under-etching condition limited to less than 2 μm). Sulfuric acid, blessed with noticeably higher etch rates on Ti layers (632 Å min^{-1} can be achieved in 46% H_2SO_4 solution at 80 °C), is often discarded for photoresist incompatibility. For the quality of the etching, ammonia-hydrogen peroxide mixture is preferred in microfabrication to both sulfuric and orthophosphoric acids. It has been demonstrated by Verhaverbeke et al. that in presence of ammonia, under the form of ammonium hydroxide (NH$_4$OH), the oxidizing species is not only H_2O_2 but also OH_2^- [29]. Since the OH_2^- is a strong oxidant also for copper, in contact with whom Cu^{2+} ions are easily complexed by ammonium cations, SC1 solution is not suitable for Ti etching in presence of unprotected copper layers. In this context, diluted HF (1% by weight) shows higher selectivity and is widely employed when copper must be the titanium etch landing layer. Diluted HF solutions suits well the purpose of titanium etching. Titanium layer with a thickness of 200 Å, deposited via physical vapor deposition, is effectively etched by employing HF (47%): H_2O mixture in the weight ratio of 1:100 at the temperature of 25 °C, dispensed for less than a minute in a single-wafer spin process tool. Ti residues are sometimes left after the etching process, as reported in image (b) in Fig. 9.6. Formation mechanism of the mentioned residues has not been fully understood yet. They occur both in recesses of the device structures and in relatively unshielded areas, where vertical structuring of the device is not an obstacle to fluid dynamics. No Titanium surface contamination has been detected in such areas prior to Titanium wet etch. However, a plasma pretreatment in a mildly reducing atmosphere containing forming gas (H$_2$/N$_2$ 5% mixture) prevented residues accumulation during etching (image (c)

Fig. 9.6 (**a**) SEM image of device portion after Ti etching achieved with diluted HF solution. Ti etching is operated having Cu as landing layer. Higher magnification images of the red-framed area are provided in image (**b**) and (**c**). (**b**) refers to the case in which no plasma surface pretreatment is employed so that the landing copper surface appears covered in Ti residues. (**c**) Device surface after Ti etch in this case a reducing plasma pretreatment was employed prior to Ti etching

in Fig. 9.6). This presumably suggests an excessively oxidized titanium surface (higher thickness of titanium oxide passivation layer) which delays bulk titanium etch by HF.

TiW layers, with Ti:W 1:9 weight ratio (0.3:0.7 atomic) on average, are easily etched in hydrogen peroxide solutions (30% by weight). Marginalities related to this process relate to inhomogeneous composition of common TiW layers deposited by physical methods (PVD), along vertical direction. It is widely accepted that top and bottom layers richer in Ti are formed during sputtering, and the bulk phase composition, instead, approaches the nominal Ti:W 1:9 ratio of the TiW alloy [30, 31]. Exposed layer richer in Ti readily passivates forming thick TiO_x which in turn triggers an induction time, ranging from few seconds to few minutes, before bulk TiW is etched, similarly to pure Ti and Al, as previously discussed. Another significant drawback of H_2O_2 TiW etching is the dusty etching residue which is commonly detected with electron microscopy inspection of processed surface (image (a) in Fig. 9.7). This defect often plagues TiW layer patterning for contacts and signal routing realization on MEMS devices. Recently, residue composition has been confirmed to be composed of tungsten oxide since only W and O were detected via EDX analysis (image (b) in Fig. 9.7). Recent studies highlighted that hydrated tungsten oxides ($WO_x \cdot nH_2O$) can be partially reduced in weak acidic media [32]. H_2SO_4 solution (0.01 M, pH = 2), whose compatibility with exposed materials has been thoroughly assessed, has been thus sprayed on the surface, followed by a rinse and dry step in a single-wafer spray process tool and demonstrated effective in removing TiW etch residues.

9.2.5.4 Cu Wet Etch

Copper is commonly etched employing acidic aqueous solutions based on $H_2O_2 : H_3PO_4 : CH_3COOH : H_2O$ mixture in the ratio 1:1:1:20 [33]. Isotropic copper wet etching, for its poor control on under-etching and critical dimensions loss, is not often employed when high resolution and structure high definition are strictly required when submicron level is sought [34]. As reported elsewhere in this

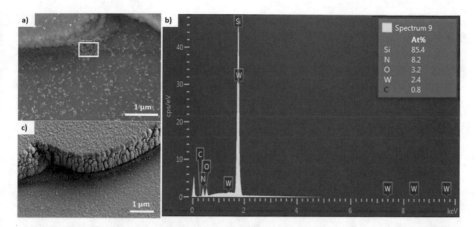

Fig. 9.7 (**a**) SEM micrograph of wet etch residues left after TiW wet etch in hydrogen peroxide solution (30% in weight) performed at 55 °C in a batch spray tool. Scale 1 μm. (**b**) EDX spectrum of the framed area in image (**a**). (**c**) SEM image of device surface after 120-second deoxidation performed in 0.1 M H$_2$SO$_4$ solution (pH = 2)

volume, copper is however a desirable layer in MEMS microfabrication, for copper and gold electrochemical deposition (ECD) process. In these circumstances, it is useful to have copper layer protected by a top cap layer, which can be selectively and conveniently removed by wet etching, when uncovered copper layer must be exposed for electrochemical growth. Copper layer, in this case, is commonly referred to as ECD seed layer. Typically, thin copper layers can be etched via also alkaline media. Ammonia-based solutions are among the most effective wet etching media for copper due to high ammonium ion affinity to copper as a ligand, forming stable $Cu(NH_3)_4^{2+}$ species [35]. Copper ion complexation must follow an effective copper oxidation triggered by strong oxidant moiety such as hydrogen peroxide. Addition of other complexing agents (acetic acid, pyrophosphoric acid, ethylenediaminetetraacetic acid), effectively segregating $Cu(NH_3)_4^{2+}$ ions, can further facilitate copper etch by tipping reaction equilibria of the following reactions toward the complexed copper products, limiting disproportionation reaction consuming peroxide reactant [36].

$$Cu + 2H_2O_2 \rightarrow Cu^{2+} + H_2O + O_2 Cu + 2H_2O_2 \rightarrow Cu^{2+} + H_2O + O_2$$

$$Cu^{2+} + 4(NH_4)^+ 4OH^- \rightarrow Cu(NH_3)_4^{2+} + 4H_2O.$$

Alkaline (around pH 10) ambient for copper etching is beneficial for layer selectivity, particularly in presence of Ni, NiFe alloys, and Au metals: the etching rate, in respect to these three metals, is three order or magnitude higher on copper. As mentioned at the beginning of this paragraph, hydrogen peroxide is a common oxidant also in acidic copper etchant. Other than phosphoric acid-based etchant, $H_2O_2 : HCL : H_2O$ with a mixture components ratio of 1:3:2 at 30 °C, and $H_2SO_4 : H_2O_2$ with a composition ratio 50:1 at 120 °C, are commonly employed solutions [37]. As a final quote, $FeCl_3 : HCl$ has also been suggested as copper etchant in front-end microelectronics fabrication, allowing the definition of structures with dimensionalities as low as 600 nm. In such mixture, Cl^- exerts the role of copper ligand, while Fe^{3+} species is accountable for copper layer oxidation.

9.3 Wet Cleanings

The cleanings and resist removal steps involved in the different MEMS technologies platforms will be treated in this paragraph. The cleanings are commonly thought as steps aimed to remove unwanted materials such as resist or dry etch byproducts and polymers, but their effectiveness is closely related to device functionality and possible failure modes. Moreover, they may influence following process steps, such as lithography whose quality is sensitive to surface conditions, and can play a central role in surface preparation, and conditioning before dielectrics or metal layers deposition. Cleaning in MEMS technologies consists in a wide family of processes, starting from standard cleaning and surface preparation which involve highly oxidant chemistries up to dedicated cleaning sequences with proper solvents applied to critical process steps of MEMS fabrication. If standard chemistries are very effective for Silicon and dielectrics treatments, wider families of solvents are needed for other applications, when metallization or specific dielectric layers are combined in a MEMS process flow. In the following paragraphs, this wide field of processes will be presented, starting from consolidated techniques in microelectronics and then explaining specific solutions tuned for specific MEMS applications, enlightening the constrains and the criticalities of each cleaning step.

9.3.1 Standard Wet Surface Preconditioning and Wet Resist Removal

Contamination removal has a fundamental role in MEMS technology, as it is already for Front-end of line Integrated Circuit (FEOL IC) manufacturing, since impurities and process particulates on patterned wafers are one of the primary reasons of device failure [38]. MEMS failure mechanism can be due to stiction, mechanical fracture, cracks, delamination, stresses, and electrical shorts as shown in Fig. 9.8.

The purpose of cleaning is to eliminate the defectiveness deposited on the wafer during the whole micromachining process flow that can compromise active device area functionality. In front-end operations, there is nearly a cleaning step before

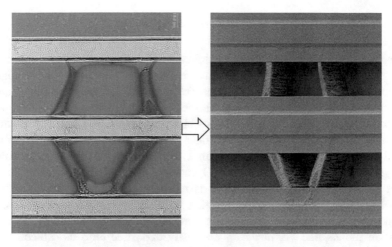

Fig. 9.8 Residues left on wafer surface during an improper rinse causing silicon strings between trenches after trench etch step and compromising electromechanical functionality of MEMS device

every thermal process and after every process step like plasma etching, high-current implantation, etc. Therefore, there is at least one cleaning per mask, covering 30–40% of a typical MEMS process flow. Main contamination sources in a Clean Room can be organics coming from resist residues or volatile products, particles and inorganics coming from environment, wafer handling and chemical/metals contamination mainly from previous process steps [39]. Most of contaminants are removed in a wet environment by chemical action helped by mechanical forces [40]. The wafers are immersed in aqueous chemicals or solutions dispensed on their top, followed by an effective rinsing and dry steps to avoid a new source of contamination (e.g., watermarks [41]) and to allow a safe wafer handling. Particle deposition in liquids depends on the exposed substrates, on particle materials and surface electrostatic repulsion, and attraction forces [42]. In general, the golden rule is that at high pH (in alkaline media) the particle deposition is avoided, while it is favored in acidic media [43]. Standard wet cleanings used in the first part of the process flow, i.e., when metallization layers are not yet present, can be classified as:

- Presurface conditioning (i.e., before thermal or Silicon Epitaxial growth, dielectrics or metal deposition, and implantation steps)
- Postsurface treatments (i.e., photoresist and polymer removal post Implantation step)

9.3.2 Wet Surface Preconditioning Methods

The terms "surface preparation" or "surface conditioning" are often used in place of "cleaning," since the goal is in fact to chemically prepare a contamination free surface for the subsequent process. The most efficient wet cleaning sequence developed almost 60 years ago by Kern et al. and still in use nowadays is the RCA cleaning sequence [43–45]. This wet sequence consists in a hot mixture of ammonium hydroxide (NH_4OH), hydrogen peroxide (H_2O_2), and deionized water (H_2O) so called SC-1 (Standard Clean 1), followed by a hot aqueous solution of Hydrochloric acid (HCl) and H_2O_2 named SC-2 (that stands for Standard Clean 2). SC-1 chemistry sometimes is referred as ammonia peroxide mixture (APM). It is a high pH (~9–10) oxidizing solution able to remove particles, light organic contamination, and complexing some metals such as Cu and Zn [46]. Other transition metals like Fe, however, are easily deposited onto the Silicon surface from SC-1 solutions [47]. SC-1 removes contaminants by a combination of slightly etching the SiO_2 or Si substrate that makes an undercut below particles and an electrostatic repulsion of the particle and the surface due to the high pH of the solution [48]. The etching of silicon (Si) in SC-1 is linked to two reactions: The first reaction is with species (HO^{2-}) that oxidizes the Si surface forming a thin chemical oxide (in a range of 20–50 Å), while the second reaction between silicon oxide and hydroxyl ions (OH^-) removes the oxide layer. As reported in Eqs. 9.6 and 9.7, the removal rate of Si is increasing with the concentration of hydroxyl groups in the APM solution and can cause Si pitting or microroughness [49].

$$Si + 2\,OH^- + H_2O \rightarrow Si\,O_3^{2-} + 2\,H_2 \qquad (9.6)$$

$$Si\,O_2 + 4\,OH^- \rightarrow \left(Si\,O_4^{4-}\right)_{aq.} + 2\,H_2O \qquad (9.7)$$

Silicon microroughness is one of the reasons why SC-1 solution needs to be diluted as much as possible, reducing the ammonia content, compared to the original RCA formulation. The SC-2 or HPM (hydrochloric acid and hydrogen peroxide mixture) aqueous solution is needed after APM step to have an effective removal of alkaline and a wide range of metal contaminants [50]. In presurface conditioning cleanings, there is not a heavy metal contamination (i.e., Pd, Au) on the wafers; therefore, even HPM solution can be consistently diluted. A way to improve particle removal in RCA cleaning in wet benches is adding a physical force like Megasonics agitation in SC-1 tank [38]. Moreover, a very dilute RCA in a single pass usage without chemistry recirculation has the following advantages:

– It is possible to drain the chemicals after each cleaning.
– No peroxide chemistry decomposition issues [51].
– Higher bath cleanliness.
– No cross-contamination issues.

– Availability of flexible equipment which use a single tank for the whole RCA cleaning.
– Less chemical consumption.

For his oxidizing nature, RCA cleaning is not compatible with metal layers and photoresists [52]. In some cases, such as deposition of epitaxial silicon, it may also be necessary to have a hydrogen-terminated surface that is free from native or chemical oxide, avoiding Silicon substrate roughness or damage. Diluted hydrofluoric acid (dHF) is used for this purpose as the last step before final rinsing and drying of a wafer with Si surface exposed. Such "HF-last" processing results in a hydrophobic surface compared to the hydrophilic surface left by RCA or SPM cleaning [53] and needs queue time monitor between the cleaning step and the subsequent process to avoid native oxide regrowth

9.3.3 Postprocessing Surface Treatments

Postprocess cleanings are done by dry and/or wet methods and involve removal of all the contaminants and organic material coming from "dirty" steps like reactive ion etching (RIE) that leaves sputtered residues on the resist and on the sidewalls. Cleaning treatments must be done even after an implant processing that leaves a "crust" of hardened resist full of Carbon-rich products, very hard to remove only by wet chemistry. Therefore, a typical postprocess cleaning is done by using a plasma step for photoresist bulk and crust ashing, followed by a wet cleaning process that removes any remaining organic residues. A hot solution of sulfuric acid (H_2SO_4) and hydrogen peroxide H_2O_2 mixture (so called SPM or Piranha solution), is normally used as wet chemistry for organic matter stripping. It is a strong exothermic reaction (>120 °C), showed in Eq. 9.8, that allows a fast-chemical decomposition of all organic materials thanks to atomic oxygen formation, a strong oxidizing species.

$$H_2SO_4 + H_2O_2 \rightarrow H_3O^+ + HSO_4^- + O \qquad (9.8)$$

SPM cleaning efficiency is time dependent, due to sulfur salt formation related to the reaction with organic contaminants. To prevent salt formation, a hot rinsing or an SC1 step is added straight after SPM process. Cleaning processes can be done in different types of equipment like immersion wet benches, centrifugal spin-spray batch, and single-wafer systems or scrubbers. Tables 9.2 and 9.3 show typical chemicals concentration, temperature ranges of usage, cleaning sequences, and type of contamination removals for the most used pre- and postcleaning processes in the first part of MEMS manufacturing.

Rinsing between chemical steps is basically removing the bulk chemical, but the most critical parts of the sequence are the final rinsing and drying steps. For instance, if the wafers are hydrophobic (i.e., after HF step), the highly active H-

Table 9.2 Standard chemistry mixtures used for presurface conditioning and posttreatments cleanings in the first part of MEMS process flow

Chemistry mixture common nomenclature	Chemical concentration	Temperature range [°C]
$NH_4OH:H_2O_2:H_2O$ (SC-1/APM)	1:1:100 ÷1:1:200	70 ÷ 80
$HCl:H_2O_2:H_2O$ (SC-2/HPM)	1:1:100 ÷1:1:200	70 ÷ 80
$H_2SO_4: H_2O_2$ (SPM)	2:1÷4:1	90÷130
$HF:H_2O$ dHF	1:100	30

Table 9.3 Typical cleaning sequence used for presurface conditioning and posttreatments cleaning in the first part of MEMS process flow

Cleaning step level	Typical sequence	Contamination removal type
Prethermal oxidation or dielectrics deposition	SC-1/rinse/SC-2-rinse/dry (RCA)	Organics, Inorganics, some metals
Preannealing	SC-1/rinse/SC-2/rinse/dry (RCA)	Organics, Inorganics, some metals
Pre-silicon Epitaxial growth or poly-Si deposition	dHF/rinse/dry	Native oxide, chemical oxide, trace of metals
Premetal deposition	dHF/rinse/dry	Native oxide, chemical oxide, trace of metals
Post RIE and dry PR removal	dHF/rinse/SPM/rinse/SC-1 (or RCA)/rinse/dry	Organics, particles, metals
Post implantation	SPM/rinse/SC-1(or RCA)/rinse/dry	Organics, particles, metals
Postchemical mechanical planarization (CMP)	SC-1/rinse/SC-2/rinse/dry (RCA)	Organics, Inorganics, some metals, slurry residues
Post dielectric or metal deposition	Rinse /dry (scrubbing)	Inorganic, particles

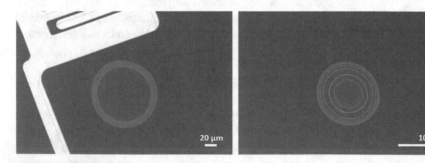

Fig. 9.9 SEM image of a water spot left on top of a hydrophobic Si surface after a not efficient sequence of final rinse and dry

terminated surface can easily attract contaminant particles. Moreover, microdroplets on wafers surface can evaporate quickly leaving watermarks mainly composed of silicates [39], while macrodroplets (mm large) evaporate in seconds leaving residues like the one in Fig. 9.9.

High aspect ratio features are common in MEMS products, hence a bad drying quality of active areas together with hydrophobic surfaces can induce particle generation and compromising the device performances. The key challenge is to perform the drying process, which typically involves moving parts, by avoiding particle adhesion on wafers. For this reason, different types of dryers have been developed based on surface sensitivity, from basic spin dryer with and without hot N_2 flow to very sophisticated systems like the well-known Marangoni drying procedure that use Isopropyl alcohol (IPA) which, due to its low surface tension, displaces water when it condenses on the wafer surface [54]. Some types of dryers permit to have chemicals in the drying tank, and it is used to perform HF last processes. In this way, it is possible to avoid water–air crossings and watermarks residues thanks to an *in situ* nitrogen purge.

9.3.4 MEMS Solvent-Based Cleanings

In this paragraph, all those processes that cannot be done with traditional chemistries (such as SPM, SC1, SC2, etc.) will be described. Some solutions for polymer removal after dry etch will be presented, describing the tools and the solvents used to obtain layers cleaning for two main fields: dielectrics and metal dry etch. Some specific cases will be treated in detail to understand which are the challenges raised by MEMS fabrication, explaining how these solutions have been put in place, and showing final results by SEM pictures and defectivity data analysis.

Three cases of study will be described enlightening all the aspects and criticalities of MEMS cleaning:

- Deep silicon etch
- Thin Piezoelectric film dry etch
- Lift-off after metal deposition

These solutions are useful to understand how resist removal and polymer cleaning are fundamental to guarantee device quality and reliability.

Dry etching can exert a strong physical action on photoresist, which can, in most of the cases, lead to encrusting and surface hardening of the organic photoresist. Passivation phase of the anisotropic deep reactive ion etching (DRIE) is moreover source of a conspicuous amount of polymeric matter [55].

As mentioned in the previous paragraphs, all the organic materials left by etch byproducts and resist residues must be removed, because buried organic residues can cause reliability issues such as corrosions on metals layer, stiction of free structures, or device performance degradation.

In case of dielectrics (Fig. 9.10) and metal (Fig. 9.11) etch, polymer removal and DRIE by-products removal processes are done by employing amine-based formulations (such as Hydroxylamine or Aminoethoxy ethanol), organic compounds, and specific inhibitors to prevent metal corrosions [56]. These solvents are used at

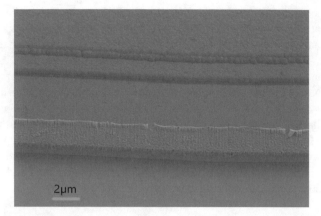

Fig. 9.10 SEM picture of a dielectric layer after dry etch definition and photoresist removal. It is possible to see the polymer deposition on patterned layer profile. This accumulation is typically at the layer profile, where the substrate passivation occurs during plasma etch, and where the Photoresist mask facilitates polymer deposition

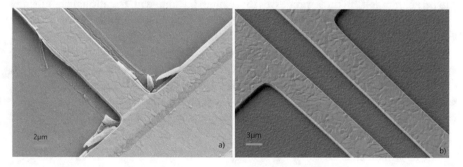

Fig. 9.11 SEM picture of an AlCu Metal after dry etch and photoresist removal. In panel (**a**), it is shown the large amount of polymers that can redeposit on the etched substrate, and should be removed by use of solvent. In panel (**b**) the same metal layer after solvent cleaning

relatively low temperatures (60–80 °C) to preserve composition and increase bath life.

Metal dry etch produces chloride-rich polymers that form large residues and veils that deposit on the patterned metal strips or pads (Fig. 9.10). Chloride migration from polymer to metallization walls is highly likely, especially in presence of condensed water, leading to localized corrosion. Dry etch polymer removal has to be thus thorough and must happen within a sufficiently short time span, in order to avoid corrosion triggered by environmental humidity. These solvents are typically used in spray solvent tools in which a bowl of 25–100 wafers is rotating during chemical dispense.

Rinse and dry steps are critical too. A first rinse is generally done in IPA, followed by DI rinse (CO_2 flow is added in case of basic solvents to balance pH and avoid its drift toward alkaline environment potentially risky for metal corrosions insurgence [57]). Wafer drying is done with high rotation (up to 2000 rpm) and hot N_2 flow.

In some cases, standard plasma ashing (Resist removal by Oxigen based plasma) followed by solvent polymer removal is not enough to guarantee structure cleaning. This is the case of thick metal etch or multiple stacked layers (auto aligned etches, for multiple stacks) which need thick photoresist masks that may require photo stabilization processes. After such lithography treatments and long dry etch, plasma photoresist is really hardened and large amount of polymer is attached to the structures. Plasma ashing induces polymer deposition on metal strips, and long solvent cleaning is not always enough.

High pressure dispenses of solvent combined with wafer immersion at high temperature can be a good solution for both resist and polymer removal. This kind of process requires specific tools which are composed of immersion tanks and single-wafer stations, with chemical dispense at high pressure.

In MEMS technologies, every cleaning must be thought in the contest of the specific case under study. Compatibility between exposed materials (dielectrics or metals) and cleaning chemicals must be thoroughly assessed.

DMSO-based solvents are a big family of chemistries, and they are used for all the mentioned cleanings, but they are often very different from one another, in relation to selectivity and cleaning potential.

There's a subfamily of DMSO-based solvents which are used for Bosch process cleaning, because they contain specific amines particularly effective in removing fluoride-based polymers on silicon substrate (TMAH, Dimethyl propyl ammonium hydroxide). Silicon pitting corrosion is however a risk using such alkaline mixtures, with potential consequent CD loss on particularly sensitive structures (e.g., doped polysilicon, which can have higher etch rates in aggressive solutions) [58, 59].

Wet resist removal by use of DMSO solvents is often done in presence of metals (metals stripes already patterned, or metallization used for specific applications such as mirrors or actuators, which are Au, TiW, AlCu, Pd, and others), and for this reason, they contain etchants inhibitors: every solvent has its own characteristics and needs to be chosen in relation to the specific application.

The cleaning efficiency of such process steps can be done only by SEM inspection because optical inspection is not able to detect eventual change in layer morphology or insurgence of micro corrosion events (Fig. 9.12).

Negative photoresist is a family of materials that are used for specific processes:

- Liftoff—because they can have specific profile useful for mask removal after metal deposition.
- Deep silicon etch—because they can reach higher thickness
- Electro Chemical Deposition—they are generally stable and can sustain long processes in aggressive chemical baths

These photoresists are highly cross-linked (the cross-linking can be amplified by thermal treatments at high temperatures during lithography steps) and can be removed with alkaline strippers with high pH values (>9/10).

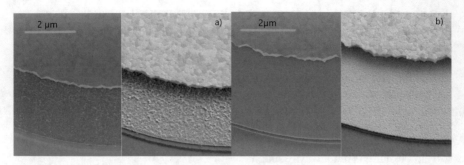

Fig. 9.12 SEM picture of a typical multilayer metallization used for interconnections in MEMS technologies. The metal strip stack is composed by Au as top layer, TiW, and Ti autoaligned under Gold. In panel A, it can be appreciated the TiW and Ti corrosion due to nonselective solvent action, while in panel B for comparison a standard TiW/Ti layer morphology

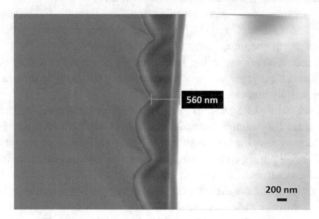

Fig. 9.13 SEM picture of a trench cross section after Bosch Etch process. It can be appreciated the passivation polymer deposited over the scallops

9.3.5 Case of Study: Cleaning Procedure After Deep Silicon Etch

Silicon dry etch in MEMS technologies can be a very wide field, from the point of view of etch depth (ranging from 1 to 500 μm), aspect ratios (up to 1:100), or exposed areas (from 1% to 90%). This paragraph will present a specific case, interesting from the point of view of cleaning: the definition of the movable part of the device, which in the case of micromirror technology is the Rotor. This is the core of the application, because it consists in the silicon structure where the mirror metallization will be realized, and for this reason, its cleaning needs to be perfect, not to cause metal mirror corrosions or defectivity that could cause failure during device lifetime (Fig. 9.13).

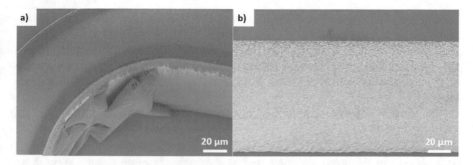

Fig. 9.14 SEM pictures of trench silicon sidewalls. In panel (**a**), residues of Bosch process detaching from silicon structures. In panel (**b**), a clean surface

The case under study is the dry etch of a monocrystalline silicon wafer with target depth of 400 μm, with exposed area on the wafer about 30%. The critical part of this cleaning is related to the large amount of fluorinated polymer that is present on the silicon sidewall as a consequence of a long etch recipe, in fact its thickness at the top of the structures can reach 500 nm or more depending on dry etch process conditions (see Fig. 9.14).

A typical photoresist mask for deep silicon etch can range from 15 up to 30 μm, depending on device topography, and during long Bosch etch, the hardening can be critical for resist removal (for this application, the mask is done with a positive photoresist without high temperature processes such as Hard Bake or oven, to guarantee vertical resist profile and consequent vertical Silicon sidewall without significant CD loss).

For these reasons, a simple dry resist removal followed by a cleaning in a DMSO-based solvent could be insufficient.

To guarantee complete cleaning, it has been necessary to set up a sequence aimed to remove the photoresist crust before removing the photoresist bulk; this can be done by combining a Descum process followed by a solvent cleaning. Descum process is an O_2-based plasma at low temperature (<80 °C), and it can be done in an RF chamber (if it is needed a physical effect to remove polymer crust) or in a microwave chamber (for a soft etch).

The polymer removal can be done using Hydrofluoroethers (HFE), which are resist selective and acts only on polymer, detaching it from the silicon substrate.

This step can be done in a wet bench, at low temperature (<23 °C), because this family of solvents is very volatile and needs to be handled in dedicated tools with vapor recovery systems to reduce chemical waste.

Once the polymer has been completely removed from silicon sidewalls and photoresist top, it is possible to perform the photoresist mask removal. This step can be done by using high temperature $O_2/N_2H_2/N_2$ plasma (typical PR stripping) in microwave chambers. Polymer residues could persist after dry photoresist stripping, on silicon surface. A further wet cleaning might be in these cases desirable and effective in dissolving organic compounds leftovers.

This wet removal can be done by use of DMSO-based solvents, with specific additives for Silicon cleaning.

At this level, it is often possible to have metals exposed on the wafers (metal routing or metallization used for device actuation), and for this reason, the choice of the chemical is critical from the point of view of selectivity versus the specific layers.

The cleaning sequence described has been implemented in different process flows for MEMS fabrication and guarantees proper cleaning after deep silicon etch: in Fig. 9.14 are reported SEM pictures that show a silicon sidewall with residues detaching from the substrate compared to a completely cleaned surface, with no residues. On MEMS devices (as actuators, micromirrors...), it is important to obtain deep cleaning, because polymer can detach during the device lifetime and redeposit elsewhere generating malfunctioning or mirror corrosions.

9.3.6   Case of Study: Cleaning Post PZT Etch

PZT actuators are composed of top electrode and PZT multiple layers which are patterned by two subsequent dry etch steps performed with a single photoresist mask, and a bottom electrode which is patterned by use of a dedicated mask. The process step that will be described in this paragraph is the top electrode and PZT etch, which is a very physical etch with consequent resputtering of a large amount of material, combined with a thick polymer deposition on the etched structure. The Photoresist needs to sustain a physical double layer etch, with consequent polymer hardening and formation of a thick photoresist crust. In Fig. 9.15 is reported a SEM picture of a PZT structure before resist removal and in Fig. 9.16 is reported the elemental composition and TEM analysis of the crust that shows a massive presence of resputtered metals (Ti, W, Pt, Pb, and Zr).

Photoresist mask removal done with O_2-based plasma in a microwave chamber at high temperature, followed by a cleaning in a spray solvent tool with an amine-based solution, is not successful since:

– After Plasma Ashing, a large amount of polymer deposits on top layer surface, as it can be seen in Fig. 9.17 where TiW top electrode layer is fully covered by white residues.
– Amine-based chemical clean needs to be very long, with consequent visible effect on top electrode, which starts to degrade after such long recipe.

In general, these flaws have been found for large etch areas and for thick PZT layers (up to 2 μm), which are very common for MEMS actuators devices.

Cleaning by use of DMSO-based chemistries in a dedicated tool which combines high-pressure chemical dispense and immersion in hot bath showed good results, even though some expedients are needed. Typical process requires immersion in a soaking station with DMSO at T ~ 90 °C, plus high-pressure chemical dispense

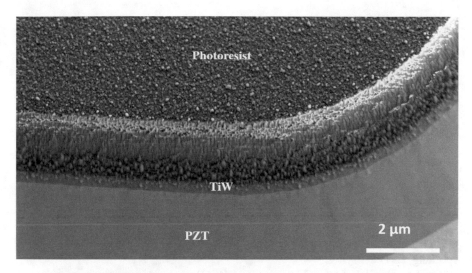

Fig. 9.15 TiW and PZT layers after dry etch and before resist removal. Photoresist mask after plasma etch, and the polymer on the etched structure profile

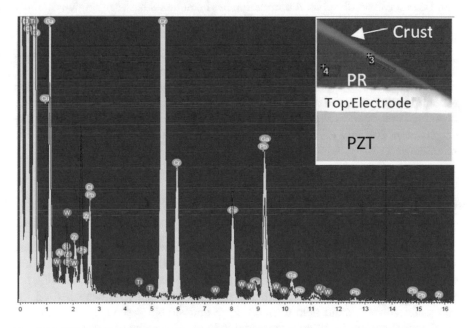

Fig. 9.16 TEM analysis (top-right inset) and EDX elemental analysis (main window) of the crust on top of etched photoresist shows a massive presence of resputtered metals. EDX mapping area number 3 refers to yellow spectrum while EDX mapping area number 4 refers to red-lined spectrum

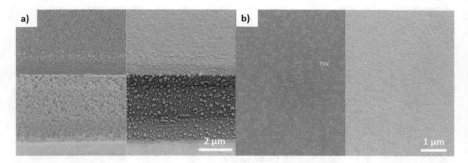

Fig. 9.17 Ashing residues on (**a**) vertical and (**b**) horizontal top electrode surfaces after O_2-based plasma photoresist stripping

(with a nozzle pressure ranging between 1800 and 2000 psi), followed by IPA rinse, spinning, and N_2 drying.

For large PZT devices, which means big PZT areas, a large quantity of veils and residues have been found on top layer surface after wet removal, and surface analysis confirmed that this was part of the resist crust that can be burned with plasma ashing, and DMSO cannot dissolve. A correlation has been found between residues number (as an example, see below the results collected on 25 wafers by automatic optical inspection) and soaking time in DMSO[1] (Fig. 9.18).

Because of this evidence, the process has been modified, removing the immersion step, increasing process time and dispense pressure of DMSO in the single-wafer chamber. The result is a good resist removal without residues redeposition (confirmed by AOI), and a good cleaning of PZT layer's profile; SEM picture after cleaning is reported in Fig. 9.19. It is important to underline that DMSO is highly selective on TiW and PZT, and it can be considered a good candidate to preserve metal layer integrity.

9.3.7   Case of Study: Lift-off Process

The technology platform of MEMS Gyroscopes relies on controlled atmosphere inside the device in order to let it work properly [60]. Volatile by-products left on wafers could degas modifying atmosphere within cavity over time, causing not reliable and stable performances, hence a Getter material [61] deposited on the cap wafer surface could help to trap all the undesirable gaseous species. Due to its chemical composition, getter material is hard to be etched by standard etching

---

[1]For this kind of tools which combine different process stations (wet station for soaking, single wafer high-pressure module, and rinse module) soaking time is not fixed, but depends on wafer handling and process time in all the modules. Hence, for a full cassette (from wafer 1 to wafer 25), lot soaking time is typically a curve (see Fig. 9.18).

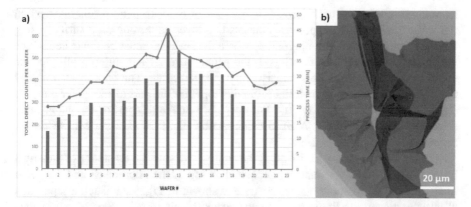

Fig. 9.18 The graph reported in (**a**) helps to highlight the dependence of the defectivity on top electrode (resist crust residues after resist removal) on the soaking time in DMSO. In fact, on the *x* axis are reported 23 wafers of a process cassette, and for every single wafer the blue histogram reports the total number of defects, while the overlapped orange line represents the soaking process time for every single wafer. The overlapping shows clearly the correlation between the number of defects and the immersion process time. On panel (**b**) is reported a SEM picture of a resist crust residue on the Top electrode layer

Fig. 9.19 SEM picture (in-lens and secondary electrons micrographs are reported on the left and right of the image respectively) of a TiW/PZT layers stack after dry etch and proper cleaning. No residues or polymers are visible

techniques, so the lift-off approach is required in order to pattern this material on the wafer surface. Liftoff process consists in a layer patterning on a surface using a specific photoresist with open areas where the material must be deposited. Once the mask is exposed and developed, the getter material is sputtered all over the wafer surface both on the photoresist and on the substrate. The photoresist removal dissolves the lithography material removing together the part of the getter that was deposited on resist top ("lift-off") but does not remove the part of the getter layer deposited on the substrate. A SEM cross-section Fig. 9.20 shows a MEMS device before and after Getter Lift-off.

The key for a successful lift-off is the ability to ensure the existence of a distinct break between the layer material deposited on top of the photoresist and the getter material deposited on the wafer substrate as shown in Fig. 9.21. Such a separation allows the dissolving liquid to reach and attack the photoresist. Several parameters must be considered to get good lift-off results:

– Create an undercut profile by using a mask with a negative slope.
– During getter film deposition, the temperatures cannot be high enough to burn the photoresist.

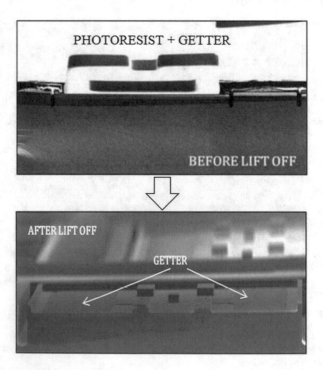

Fig. 9.20 SEM cross-sections before and after Lift-off process. The getter layer is left only inside the cavities. Scale 200 μm

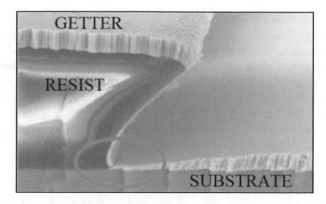

Fig. 9.21 This SEM image indicates a break between the getter material deposited on top of the photoresist and on the wafer substrate necessary to have an effective lift-off process. Scale 50 μm

Fig. 9.22 SEM images of PR and Getter residues left on trenches after a lift off process in solvent spray tool with a long recipe of DMSO solvent dispensed at 70–80 degrees and high rpm. Scale 100 μm

- Adhesion of the deposited getter film on the substrate must be very good.
- High compatibility of all the materials exposed with the solvent used for the lift-off process.

The solvents, widely used in microelectronics and suitable to dissolve, are NMP (N-Methyl Pyrrolidone), Acetone, or DMSO solvents. Low boiling point and safety risk for acetone and high toxicity for NMP products make DMSO a good candidate for the lift-off process [62]. The equipment used is also a very important choice for an efficient getter lifting. Batch spray tools have a poor capability in photoresist lifting (even for very long process times and high process temperatures for the solvent), leaving residues mainly within deep trenches as shown in Fig. 9.22.

This hardware configuration is not suitable for lift-off because the solvent has a limited "residence time" which does not allow a complete resist removal on patterned wafers. Moreover, the nozzles can spray the solvent on the wafer with a limited max pressure, therefore getter/resist residues are not removed efficiently from the wafer surface. Another possibility is to run the lift-off in wet benches with proper agitation or sonics to allow a long soaking time, but in presence of deep cavities, the photoresist could be difficult to be completely removed. A tool having a soaking station with hot solvent at 70–80 degrees followed by a single wafer high pressure chamber (HPC) in a range of 2000–3000 psi and a rinsing station with Isopropyl Alcohol (IPA) and Nitrogen, could achieve a good cleaning capability and an efficient getter lift-off process. Physical cleaning is fundamental for a complete photoresist plus Getter removal without leaving residues on wafer surface and inside cavities and trenches, together with the chemical cleaning done by a good photoresist stripper like DMSO.

References

1. Dutta, S., Imran, M., Kumar, P., Pal, R., Datta, P., & Chatterjee, R. (2011). *Microsystem Technologies, 17*, 1621.
2. Ruzyllo, J. (1998). Proceedings of the fifth international symposium on cleaning technology in semiconductor device manufacturing, proceeding volume 97- electrochemical society 35.
3. Landesberger, C., Paschke, C., Spöhrle, H-P., & Bock, K. (2014). Handbook of 3D Integraion: 3D process technology, chapter 16: Backside thinning and stress-relief techniques for silicon wafers, pp. 207–226.
4. Pei, Z. J., Billingsley, S. R., & Miura, S. (1999). *International Journal of Machine Tools and Manufacture, 39*, 1103–1116.
5. Kobayashi, H., Maida, A. O., Takahashi, M., & Iwasa, H. (2003). *Journal of Applied Physics, 94*(11), 7328–7335.
6. Williams, K. R., Gupta, K., & Wasilik, M. (2003). *The Journal of Microelectromechanical Systems, 12*(6), 761–778.
7. Jakob, P., & Chabal, Y. J. (1991). *The Journal of Chemical Physics, 95*, 2897.
8. Schwartz, B., & Robbins, H. (1961). *Journal of the Electrochemical Society, 108*, 365–372.
9. Liu, L., Lin, F., Heinrich, M., Aberle, A. G., & Hoex, B. (2013). *ECS Journal of Solid State Science and Technology, 2*(9), 380–383.
10. Seidel, H., Csepregi, L., Hueberger, A., & Baumgärtel, H. (1990). *Journal of the Electrochemical Society, 137*, 3612.
11. Zubel, I., Barycka, I., Kotowska, K., & Kramkowska, M. (2001). *Sensors Actuators, A, 87*, 163–171.
12. Kramkowska, M., & Zubel, I. (2009). *Procedia Chemistry, 1*, 774–777.
13. Zubel, I., & Barycka, I. (1998). *Sensors Actuators, A, 70*, 250–259.
14. Pal, P., & Sato, K. (2015). *Micro and Nano Systems Letters, 3*, 6.
15. Sundaram, K. B., Vijayakumar, A., & Subramanian, G. (2005). *Microelectronic Engineering, 77*, 230–241.
16. Adams, A. C. (1986). Silicon nitride and other insulator films. In J. Mort & F. Jansen (Eds.), *Plasma Deposited Thin Films* (Vol. 5). CRC.
17. Spierings, G. A. C. M. (1993). *Journal of Materials Science, 28*, 6261–6273.
18. Sundaram, K. B., Sah, R. E., Baumann, H., Balachandran, K., & Todi, R. M. (2003). *Microelectronic Engineering, 70*, 109–114.
19. Seo, D., Bae, J. S., Oh, E., Kim, S., & Lim, S. (2014). *Microelectronic Engineering, 118*, 66–71.
20. Nichol, M. J. (1980). *Gold Bulletin, 13*(2), 46–55.
21. Green, T. A. (2014). *Gold Bulletin, 47*, 205–216.
22. Eidelloth, W., & Sandstrom, R. L. (1991). *Applied Physics Letters, 59*, 1632.

23. Muraa, G., Vanzi, M., Stangoni, M., Ciappa, M., & Fichtner, W. (2003). *Microelectronics Reliability, 43,* 1771–1776.
24. Gabette, L., Segaud, R., Fadloun, S., Avale, X., & Besson, P. (2009). *ECS Transactions, 25*(5), 337–344.
25. Köhler, M. (1999). *Etching in microsystem technology.* Wiley-VCH.
26. Tsujimuraa, T., & Makita, A. (2002). *Journal of Vacuum Science and Technology B, 20*(5), 1907–1913.
27. Scotti, G., Kanninen, P., Kallio, T., & Franssila, S. (2014). *The Journal of Microelectromechanicalsystems, 23*(2), 372–379.
28. Matylitskaya, V., Partel, S., & Kasemann, S. (2015). International conference on applied physics of condensed matter and of the scientific conference advanced fast reactors. *Proceedings 21, International Nuclear Information System,* 49(22).
29. Verhaverbeke, S., & Parker, J. W. (1998). *Proceedings of the fifth international symposium on cleaning technology in semiconductor device manufacturing* (Ruzyllo J & Novak RE (Eds.), p. 184) Pennington, USA.
30. Hill, M. (1980). *Solid State Technology, 53.*
31. van den Meerakker, J. E. A. M., Scholten, M., & van Oekel, J. J. (1992). *Thin Solid Films, 208,* 237–242.
32. Nave, M. I., & Kornev, K. G. (2017). *Metallurgical and Materials Transactions A, 48,* 1414–1424.
33. Castoldi, L., Visalli, G., Morin, S., Ferrari, P., Alberici, S., Ottaviani, G., Corni, F., Tonini, R., Nobili, C., & Bersani, M. (2004). *Microelectronic Engineering, 76,* 153–159.
34. Hampden-Smith, M. J., & Kodas, T. T. (1993). *MRS Bulletin, 18*(6), 39–45.
35. Huang, Q., & Podlaha, E. J. (2005). Selective etching of CoFeNiCU/Cu multilayers. *Journal of Applied Electrochemistry, 35*(127–132).
36. Wu, Y. B., Ding, G. F., Wang, H., & Zhang, C. C. (2011). *Efficient solution to selective wet etching of ultra-thick copper sacrificial layer with high selective etching ratio,* 16th International Solid State Sensors, Actuators and Microsystems Conference.
37. Choi, T.-S., & Hess, D. W. (2015). Chemical etching and patterning of copper, silver, and gold films at low temperatures. *ECS Journal of Solid State Science and Technology, 4*(1), N3084–N3093.
38. Merlijn van Spengen, W. (2003). MEMS reliability from a failure mechanisms perspective. *Microelectronics Reliability, 43*(7), 1049–1060.
39. Singh, Contamination in MEMS processes and removal methodology, IJME Vol 3 Iss 1 Paper 8, 375–378.
40. Schwartzman, S., Mayer, A., & Kern, W. (1985). *Megasonic Particle Removal from Solid-State Wafers, RCA Rev, 46,* 81.
41. Itoh, H. (1998). *Ultraclean surface processing of silicon wafer: Secrets of VLSI manufacturing* (T. Hattori, Ed., pp 503-507). Springer.
42. Riley, D. J., & Carbonell, R. G. (1993). *Journal of Colloid and Interface Science, 158,* 259.
43. Kern, W., & Puotinen, D. A. (1970). *RCA Review, 31*(6), 187.
44. Kern, W. (1990). In J. Ruzyllo & R. E. Novak (Eds.), *First international symposium on cleaning technology in semiconductor device manufacturing, 90-9* (p. 3). The Electrochemical Society.
45. Kern, W. (1990). In J. Ruzyllo & R. E. Novak (Eds.), *First international symposium on cleaning Technology in Semiconductor Device Manufacturing, 90-9* (p. 3). The Electrochemical Society.
46. Gale, G. W., Cui, H., & Reinhardt, K. A. (2018). Aqueous cleaning and surface conditioning processes. *Handbook of silicon wafer cleaning technology,* pp. 201–202.
47. Mori, Y., Uemura, K., Shimanoe, K., & Sakon, T. (1995). *Journal of the Electrochemical Society, 142,* 3104.
48. Hiemenz, P. C., & Rajagopalan, R. (1997). *Principles of colloids surface chemistry* (3rd ed.). Marcel Dekker.
49. Celler, G. K. (2000). Etching of silicon by the RCA standard clean 1. *Electrochemical and Solid-State Letters, 3*(1), 47–49.

50. Kern, F. W., & Gale, G. W. (2000). In Y. Nishi & R. Doering (Eds.), *Handbook of semiconductor manufacturing technology* (p. 87). Marcel Dekker.
51. Takahashi, I., Kobayashi, H., Ryuta, J., Kishimoto, M., & Shingyouji, T. (1993). *Japanese Journal of Applied Physics, 32*, L1183.
52. Chiarello, R. P., Parker, R., Helms, C. R., Chen, W., Tang, S., & Cook, L. J. (1997). In G. Higashi, M. Hirose, S. Raghavan, & S. Verhaverbeke (Eds.), *Symposium proceedings, science technology for semiconductor surface preparations* (Vol. 477, p. 533). Materials Research Society.
53. Kasi, S., & Liehr, M. (1990). *Applied Physics Letters, 57*(20), 2095.
54. Marra, J., & Huethorst, J. A. M. (1991). *Langmuir, 7*, 2748.
55. Pilloux, Y. (2017). *China semiconductor technology international conference (CSTIC)*, 12–13 March 2017, Shangai, China.
56. Kvakovszky, G., McKim, A., & Moore, J. C. (2007). *ECS Transactions, 2*, 11.
57. Foley, R. T., & Nguyen, T. H. (1980). *Journal of the Electrochemical Society, 127*, 2563.
58. Bertagna, V., Erre, R., Rouelle, F., & Chemla, M. (1999). *Journal of the Electrochemical Society, 146*(1), 83–90.
59. Seidel, H., Csepregi, L., Heuberger, A., & Baumgartel, H. (1990). *Journal of the Electrochemical Society, 137*, 11.
60. Kobayashi, S., Hara, T., Oguchi, T., Asaji, Y., Yaji, K., & Ohwada, K. (1999). *Double-frame silicon gyroscope packaged under low pressure by wafer bonding*. In *Transducers'99* (pp. 910–913).
61. Kullberg, R. C. (2009). Review of vacuum packaging and maintenance of MEMS and the use of getters therein. *Journal of Micro/Nanolithography, MEMS, and MOEMS, 8*(3), 031307.
62. Kvakovszky, G., McKim, A. S., & Moore, J. (2007). A review of microelectronic manufacturing applications using DMSO-based chemistries. *ECS Transactions, 11*(2), 227–234.

Chapter 10
Piezoelectric Materials for MEMS

**Andrea Picco, Paolo Ferrarini, Claudia Pedrini, Angela Cimmino,
Lorenzo Vinciguerra, Michele Vimercati, Alberto Barulli,
and Carla Maria Lazzari**

10.1 Introduction

The piezoelectric effect links directly mechanical and electrical quantities, resulting in a particularly interesting effect for sensor and actuator applications. The transduction can be done in both directions, i.e., by obtaining charging from the application of mechanical force (direct piezoelectric effect) and vice versa (converse effect). The discovery of the piezoelectric effect dates to 1880, when it was detected and systematically studied by the Curie brothers in several minerals found in nature. Together with the development of new applications starting from the First World War, new synthetized materials were found with greatly enhanced piezoelectric and ferroelectric properties. The most important example is lead zirconate titanate (PZT), for its outstanding properties that make it the first choice for applications, including MEMS. Since the second war post period, the use of piezoelectric materials propagated in a wide variety of applicative fields, from very simple technologies of daily use like gas lighters to very sophisticated electronic devices like high-frequency filters for telecommunications and a wide variety of sensors and actuators.

The original version of this chapter was revised: The author name "Paolo Ferrari" has been changed to "Paolo Ferrarini". The correction to this chapter is available at https://doi.org/10.1007/978-3-030-80135-9_28

A. Picco (✉) · P. Ferrarini · C. Pedrini · A. Cimmino · L. Vinciguerra · M. Vimercati · A. Barulli · C. M. Lazzari
ST Microelectronics, Analog MEMS and Sensors Group, MEMS Technology and Design R&D, Agrate Brianza, Monza Brianza, Italy
e-mail: andrea.picco@st.com; paolo.ferrarini@st.com; claudia.pedrini@st.com; angela.cimmino@st.com; lorenzo.vinciguerra@st.com; michele.vimercati@st.com; alberto.barulli@st.com; carla.lazzari@st.com

© Springer Nature Switzerland AG 2022, corrected publication 2022   293
B. Vigna et al. (eds.), *Silicon Sensors and Actuators*,
https://doi.org/10.1007/978-3-030-80135-9_10

The piezoelectric effect is possible only in crystals belonging to the noncentrosymmetric point groups, except for crystal class 432. This is because in these materials the mechanical deformation of the unit cell, that is composed of positive and negative charges, is geometrically associated to the formation of an electric dipole. The sum of all the dipoles in the material generates the macroscopic piezoelectric response of the material. Of the 20 piezoelectric crystal classes, only 10 can maintain an electric polarization in the absence of an electric field. The preferred direction of polarization is referred to as polar axis. Since this remanent polarization depends on the temperature, because of the thermally induced deformation of the unit cell, the materials in these 10 classes are called pyroelectric. Finally, among the 10 pyroelectric crystal classes, materials can be found in which the direction of the remanent polarization can be reversed under the application of a sufficiently strong electric field: these materials are called ferroelectric.

It is beyond the scope of this text to give a thorough dissertation about the foundations of piezoelectricity and the large number of different piezoelectric materials that can be used in industrial applications. Here, the attention will be focused on the peculiar features given by the integration of piezoelectric materials on silicon wafer for the fabrication of piezoelectric MEMS, and two important materials will be discussed: lead zirconate titanate (PZT) and aluminum nitride (AlN).

The first fact to be pointed out is that piezoelectric materials integrated on wafer by means of silicon planar technology come in the form of thin films, i.e., with thickness in the order of few microns. Piezoelectric actuators for MEMS are extended areas of thin piezoelectric films covering large portions of the mechanical structures they have to deflect. The electrodes needed to apply the electrical field to the piezoelectric material are thin conductive films themselves, and they are found below and above the thin piezoelectric layer. For this reason, they are referred to as bottom electrode and top electrode, respectively; they can be made of the same material or not, depending on the technology. The resulting system made of bottom electrode-piezoelectric film-top electrode is often named "piezo stack" in this chapter. In the well-established manufacturing technologies, each of the three layers constituting the piezo stack are deposited on the entire wafer surface, and successively patterned by means of photolithography and etching processes. The fundamental aspects of these processes will be illustrated in this chapter, in the cases of PZT- and AlN-based stacks.

After several decades of development, a well-established know-how exists about bulk piezoelectric materials. When piezoelectric thin films are considered, however, the situation changes quite sensibly, and some precautions must be taken in transferring the concepts valid in the bulk domain to the thin film systems. In the case of thin piezoelectric films integrated on silicon for MEMS application, for instance, the boundary conditions are very different in that the piezoelectric film is tightly clamped to the substrate. This firstly means that the piezoelectric coefficients linking electrical and mechanical quantities as they are found in the basic piezoelectric formalism must be modified by effective coefficients including the clamping effect of the substrate. The interaction between the substrate and the piezoelectric thin film is fully acting also during the film deposition, thus being able to deeply affect the film properties. It is of paramount importance to know and master these effects to obtain a film with good quality and performance. As a third item, it must be reminded that in thin films also the interaction of the piezoelectric material with the electrodes gains importance as the thickness is reduced: special care must be devoted to this aspect to guarantee the proper performance and reliability of the piezoelectric film.

10.2 PZT

10.2.1 Material Overview

Among ferroelectric and nonferroelectric materials that exhibit piezoelectric behavior, Lead Zirconate Titanate (PZT) shows the best piezoelectric properties suitable for thin membrane or cantilever actuation, and for this reason, it is widely used for MEMS application [1, 2]. Its high piezoelectric coefficients, Curie temperature T_C, breakdown voltage, and dielectric constant [3] along with the possibility to modulate the piezoelectric performances by custom doping [4, 5] make PZT a better material compared to others such as BaTiO3, ZnO, or AlN. However, the toxic nature of lead compounds implies a series of complications in terms of equipment safety requirements to satisfy local laws (as ROHS directive for EU [6]) as well as integration problems in semiconductor facility's line due to cross contamination.

PZT is a solid solution of Lead Zirconate ($PbZrO_3$) and Lead Titanate ($PbTiO_3$) both having a Perovskite ABO_3 crystal structure in which the Ti^{4+} or Zr^{4+} cation occupies the center of a parallelepiped (B site) whose corner accommodates the larger Pb^{2+} cations (A site), while the face-centered O^{2-} anions form an octahedron surrounding the central atom (see Fig. 10.1) Above T_C, the crystal symmetry is cubic and the material is paraelectric. During cooling, below T_C, the oxygen octahedron and the central cation undergo a distortion, breaking the cubic crystal symmetry and giving rise to a dipole moment centered on the central cation and oriented in the opposite direction with respect to the oxygen displacement direction. Depending on the Zr/Ti ratio and temperature, the crystal cell changes its symmetry: the tetragonal Ti-rich phase is separated by a Morphotropic Phase Boundary (MPB) from the rhombohedral Zr-rich phase as showed in the phase diagram reported in Fig. 10.2.

PZT crystallographic orientation is driven by substrate/bottom electrode (BE) characteristics [7], and film usually grows in an epitaxial way controlled by nucleation. Final film texture also depends on the deposition technique and process parameters, generally giving rise to films in which (100), (110), and (111) are the main orientations. Preferential orientation can be achieved using buffer layers (i.e., LaNiO, MgO, PbTiO2, TiO2) acting as seed [8].

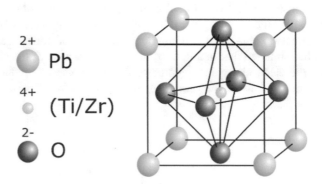

Fig. 10.1 Scheme of a tetragonal perovskite PZT cell

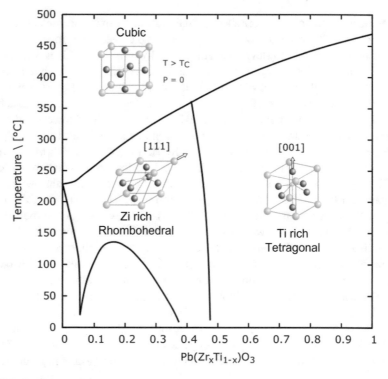

Fig. 10.2 PZT phase diagram

10.2.2 *PZT Deposition*

The main techniques used for piezoelectric film deposition are of chemical type such as Metal Organic Chemical Vapor Deposition (MOCVD) and Chemical Solution Deposition (CSD) or physical such as Pulsed-Laser Ablation (PLA) and Radio

Table 10.1 advantages and drawbacks of different PZT deposition techniques. In the table, the symbols indicate the level of strength or weakness of each item, related to each technique, according to the following notation: very good/strong plus (++), good/plus (+), average (0), poor/minus (−)

	MOCVD	PLA	Sputtering	CSD
Deposition technique	Chemical reaction of organic gaseous precursor at substrate interface	Ejection of the depositing material by target ablation using a short impulse of high energetic focused laser	Physical vapor deposition: atoms are deposited on the substrate after being removed from the target by accelerated ions	Deposition of chemical precursors followed by decomposition of organics and crystallization at high temperature
Film uniformity	0	0	+	++
Film conformity	+	+	+	−
Stoichiometry control	−	0	0	+
Composition change	0	0	0	++
Throughput	−	0	+	0
Industrialization	−	−	++	+
In-situ crystallization	Y	Y	Y	N

Frequency (RF) magnetron sputtering. CSD and Sputtering are so far the most widely used in industry due to their reliability, material performance, and ease of fabrication and integration. Each technique presents its own advantages and drawbacks summarized in Table 10.1. The two techniques of industrial interest are Sputtering and Chemical Solution Deposition. The latter allows better control of uniformity and quality, but it suffers from quite low throughput. Sputtering, though being much faster, is more challenging from the point of view of process tuning and control.

Physical Vapor Deposition (PVD) or Sputtering of PZT film requires a specific hardware configuration with an RF magnetron sputtering chamber and a strict control on the chuck temperature. The sputtering process is performed at high temperature allowing the crystallographic orientation and avoiding subsequent high temperature annealing steps. The temperature chuck uniformity is a key factor that can be challenging especially for larger wafer size, such as 200 mm.

With respect to physical deposition, CSD methods have the great advantage of a simpler and cheaper process and equipment configuration as well as a very accurate control in terms of stoichiometry. This can be easily achieved changing the relative concentration of chemical precursors in the solution, or doping elements. The most widely used CSD process is the sol-gel route in which the solution containing the precursors is deposited by spin coating followed by a Rapid Thermal Annealing (RTA) to induce the material crystallization. Very uniform film without voids or cracks and with high piezoelectric performance can be obtained with this technique, making it the most suitable for PZT mass production. The main drawbacks of sol-gel spin-coating process are the low throughput also due to the need of an *ex situ* crystallization, the inability to form conformal and uniform film on high topography, and the presence of chemical gradients through the film thickness. This last aspect is the most important in influencing the piezoelectric properties and can mainly be ascribed to the loss of lead as PbO [9, 10] and the formation of Ti/Zr gradient inside the film during the high temperature crystallization steps [11].

If the former gradient can be easily balanced incrementing the lead precursor concentration in a certain percentage with respect to the right stoichiometry [12], the latter can be reduced only using two or more different chemistry with a different Zr/Ti concentration ratio, increasing the process complexity and hardware requirements in case of automatic dispense [13].

10.2.3 Sol-Gel Chemistry

Chemical precursors, solution solvent, and composition play an important role on the formation of piezoelectric films. Typical precursors are metal alkoxides and metal salts, which undergo hydrolysis and polycondensation reactions to form a colloid (sol). The use of alkoxides precursors over Metal-Organic Decomposition (MOD) compounds has the advantage to obtain films with a very well-controlled final stoichiometry due to the formation of trimetallic complex. Moreover, because alcohol derivatives of metals can be ultrapurified via distillation and recrystallization, materials with high purity can be obtained. The solution stoichiometry can be varied to obtain films with different Zr/Ti ratio or even adding metals dopants in order to modulate the piezoelectric performances of the material. Usually, an excess of Pb (in the order of 10% to 15%) is used to compensate for lead loss during pyrolysis and mainly during RTA. The lead loss mechanism involves the metal migration over porous interphase boundaries during crystallization of perovskite phase. More is the organic part to be pyrolyzed or lower the crystallization rate, more is the film porosity resulting in a higher lead loss. Unbalanced composition can give rise to unwanted phases during the annealing step, like fluorite or pyrochlore, that can affect the overall piezoelectric performance. For instance, a Zr excess slows down the crystallization process promoting fluorite phase, while a high Ti content forces the onset of perovskite crystallization at lower temperatures and enhances (111) texture.

10.2.4 Lead-Free Sol–Gel Chemistry

PZT is composed of about 60 wt.% of lead, which rises ecological concerns: the pollution due to lead oxide (PbO) evaporation during the sintering process

Table 10.2 Composition and properties of lead-free materials. From ref. [18]

System	Tc (°C)	d_{33} (pC/N)	K
BaTiO$_3$	130	140	1400
BNT	310	64	302.6
BNT-BT	280	125	625
BNT-BT-Nb$_2$O$_5$	250	149	1230
BNT-BT-CeO$_2$ + La$_2$O$_3$	–	162	831
BNT-BT-MnCO$_3$	243	160	–
BKT-BT-MnCO$_3$	174	117	2300
BNT-BK-BT	125	150	–
BNT-BZT	244	147	8814
BNT-BKT-SrTiO$_3$	292	185	868
BNKLi-BT	210	205	1040
(K,Na)NbO$_3$	400	95	500

has greatly hindered the applications of Pb-based ferroelectric materials. The 25 European Union (EU) included PZT as a hazardous substance in its legislature to be substituted by safe materials ["Waste Electrical and Electronic Equipment" (WEEE) and "**R**estriction **o**f the use of certain **H**azardous **S**ubstances in electrical and electronic equipment" (RoHS)] [14, 15, 16].

Therefore, developing a lead-free and environmental-friendly material with a piezoelectric coefficient comparable with that of PZT (200–710 pC/N) is highly desirable. In recent years, diverse lead-free systems, both bulk ceramics [17] and thin films, are being investigated. Basically, the lead-free systems are (i) perovskites, i.e., BNT, BaTiO3 (BT), KNbO3, NaTaO3, etc., and (ii) non-perovskite, i.e., bismuth layer structured ferroelectrics (BLSF) and tungsten-bronze-type ferro-electrics. While the perovskites are suitable for actuator and high-power application, BLSF seems to be a candidate for ceramic filter and resonator applications. The perovskite-type ferroelectrics are hopeful candidates for lead-free piezoelectric ceramics because its anisotropy in piezoelectric properties is large compared to other ferroelectrics. A list of lead free piezo materials and their properties is presented in Table 10.2, including Barium Titanate, Bismuth-alkaline metal Titanates, and Niobates [18].

Among these, (K, Li, Na)(Nb, Ta)O3 and (Ba, Ca)(Zr, Ti)O2 are the most investigated lead-free piezoelectrics families, and (1-x) Ba(Zr0.2Ti0.8)O3-x (Ba0.7Ca0.3)TiO3, abbreviated as BZT–BCT, and (K,Na)NbO3, abbreviated as KNN, are considered as the most promising candidates.

In 2009, Wenfeng Liu and Xiaobing Ren have researched BZT–BCT ceramics fabricated by a conventional solid-state reaction method. They reported that the existence of a cubic-rhombohedral-tetragonal (C-R-T) triple point in the phase diagram locating at x 32% and T = 57 °C can cause high piezoelectricity, with values comparable with that of PZT transducer. They showed a high piezoelectric coefficient d_{33} up to 620 pC/N at a relatively low Curie temperature Tc $\sim$ 93 in bulk ceramics with a perovskite structure and a complex composition of 0.5Ba(Ti0.8Zr0.2)O3–0.5(Ba0.7Ca0.3)TiO3 (BZT–0.5BCT) [19]. Single crystal form of BZT–BCT composites at the morphotropic phase boundary (MPB) has shown giant piezoelectric coefficient $d_{33} \sim$ 1500–2000 pC/N [20]. By optimizing the poling conditions for Ba(Zr0.2Ti0.8) (Ba0.7Ca0.3)O3 ceramic composites, a huge piezoelectric coefficient $d_{33} \sim$ 630 pC/N with 56% planar electromechanical factor was observed [21]. Using the sol–gel technique, lead-free BZT–BCT ceramics were prepared by Yang et al. [22], whose dielectric maximum was above 9000 with a maximum of piezoelectric coefficient $d_{33} \sim$ 400 pmV$^{(-1)}$.

In 2004, Saito et al. revealed the tremendous potential of textured polycrystals as a substitute for PZT ceramics, which in turn stimulated widespread scientific interest in developing lead-free alternatives, exhibiting a piezoelectric constant d_{33} of above 300 pC/N, and a peak d_{33} of 416 pC/N [23]. Strategies in terms of enhancing electromechanical properties within the KNN system mainly focus on phase-boundary engineering, domain engineering, and optimization of processing. Wu et al. observed a large d_{33} of 570 pC/N in polycrystalline KNN ceramics by simultaneously shifting dual-phase transition temperatures and constructing a distinct phase structure [24]. Li et al. achieved ultrahigh electromechanical properties with d_{33} of 700 pC/N and a planar electromechanical coupling factor of 76% (measure of the conversion efficiency between electrical and mechanical energy) in highly textured KNN ceramics [25]. In terms of d_{33}, KNN materials are indeed comparable to soft commercial PZT (350–700 pC/N). However, a life-cycle analysis conducted according to ISO standards showed that due to the methods used for extraction of niobium from its ore, the environmental impact of KNN is several times greater than that of PZT. In addition, niobium ores will soon fall under the EU Conflict Minerals Regulation (EU2017/821), which limits the import of minerals to Europe from conflict-affected areas.

Unfortunately, all these goals refer to ceramic materials, not useful for microelectronics applications, while sol–gel alternatives do not reach the properties' level met by lead containing materials, which is the absolute necessary level for actual applications. Electrical (piezoelectric) properties are inferior when compared with lead-containing ceramic and cannot be stably achieved throughout a wide temperature range. Moreover, the properties obtained in the laboratory cannot generally be stably achieved at a mass production scale. There are still many remaining issues needing to be solved to achieve mass production of practical products. Adding to that, even in the case that mass production technology is achieved, the required properties for substituting almost all the applications cannot be obtained.

10.2.5   Sol–Gel Process

Sol–gel process is a well-established technique used to produce both bulk and ceramic films. The process is based on the conversion of metal-organic monomers in solution into colloidal particles (sol) that acts as precursors for the formation of a gel, an aggregate with a composition varying from discrete particles to a polymer network. The gel formed by the progressive solvent loss and colloidal particles aggregation is a diphasic system containing both liquid and solid phases. Gel evolves toward a dense ceramic (film or bulk) by means of different annealing treatments. Low temperature step promotes solvent loss and film densification (accompanied by a significant material shrinkage), while further steps at higher temperature cause first the pyrolysis of organics and then the evolution from an amorphous material into a crystalline one through nucleation and grain growth [26]. Sol–gel process is cheaper and requires in general lower temperature compared to other process like powder syntherization. Furthermore, this technique is more versatile and allows the material to be deposited in different ways: dip coating and spin coating to obtain films or casting to obtain bulk ceramics, fibers, aerogels, or even powders.

10.2.6   Sol–Gel Deposition Process

PZT deposition by CDS is a rather complex process compared to other process used in MEMS fabrication. Due to the formation of Pb-Si compound and to the high diffusivity of lead on silicon, it is not possible to deposit PZT directly on silicon. Conductive oxides like IrO, $SrRuO_3$, and $LaNiO_3$ can be used as bottom electrode even if platinum metallization is the most employed due to the high conductivity, resistance at high temperature, and low mismatch with PZT lattice. To avoid lead diffusion through Pt grains, SiO_2/TiO_2 barrier is usually employed to isolate the silicon substrate [27]. To obtain thick films (in the order of microns) suitable for MEMS devices actuation, the deposition-annealing steps must be reiterated until the desired thickness target is reached. In the coating step, M layers (usually 3 or 4) are piled up (each dried and pyrolyzed) before being annealed at high temperature to get the complete crystallization. The process is then repeated N times to reach a film with a total of M x N layers (see process scheme in Fig. 10.3). Generally, this deposition scheme leads to untextured PZT on Pt (111), but a preferential orientation can be easily achieved using the right seed layer: $PbTiO_3$ leads to (100) orientation, while TiO_2 has the effect to induce a final (111) texture [28].

Another common and versatile way to drive the wanted orientation exploits the use of a single PZT layer over a range of 60–100 nm acting as seed layer [29]. Sol–gel composition (both precursors [30] and elements stoichiometry [31]), pyrolysis [32], and RTA temperatures as well as film thickness are the major parameters influencing the final film texture (see Fig. 10.4). The mechanism underlying the formation of a specific orientation is rather complex and involves the formation of intermediate species at the electrode/seed interface. Generally, the formation of Pb–Pt intermediates promotes (111) orientation, while the PbO formation induces (100) texture. All the processes reducing Pb oxidation at Pt interface cause the formation of Pb_x–Pt_y intermetallic phase [33]; among these: the slow oxygen diffusion due to high film thickness (or a too rapid annealing) and the oxygen consumption due to the oxidation of organic species contained in the sol–gel [32]. The piezoelectric $e_{31,f}$ coefficient reaches a maximum when the Zr/Ti ratio is close to the MPB composition

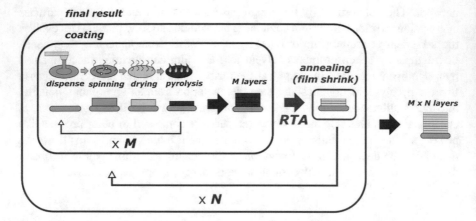

Fig. 10.3 Sol–gel process scheme. The process is repeated several times in order to obtain a thin PZT film with thickness in the order of few microns

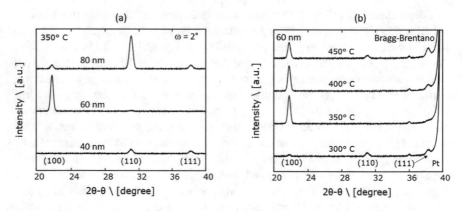

Fig. 10.4 Effect of process parameters on the XRD spectra of PZT seed layer. (**a**) Influence of seed layer thickness and (**b**) influence of hard bake (HB) temperature

and the crystallographic orientation is (100) enhancing the material performance [34, 35, 36].

Available commercial PZT solutions allows for the growth of single layer ranging from 40 nm to a maximum of 200 nm. After crystallization, the film undergoes a shrink of about 24% almost independently to the thickness reached after spin coating. The maximum thickness before each crystallization step (as well as the maximum final film thickness) reachable without observing cracks depends in general on the used chemistry and the stress evolution of the layer constituting the substrate stack. An important contribution to the overall stress of the system, formed by the film and the bottom electrode, comes from the PZT film orientation and thickness. Stress becomes less tensile at the increasing of film thickness and more tensile at the increasing of (111) orientation. Sol–gel solution is spin-cast on a silicon wafer substrate directly in contact with the bottom electrode. Just after deposition, a significant amount of solvent is present into the film and is removed by evaporation both during spinning (during this phase begins the gelation of the sol) and drying into the hot plate. The less volatile organics components are pyrolyzed at higher temperatures in an oxygen environment (usually 300–400 °C) leading to an amorphous material [37]. All the major parts of the organic matrix must be removed before crystallization to avoid cracks and voids. As the previous parameters, also rapid thermal annealing (RTA) plays a paramount role in the final film morphology. Depending on the nature of the substrate (i.e., the bottom electrode or the buffer layer) and film composition after pyrolysis, the transition from the amorphous phase to a perovskite structure occurs starting from 500 °C up to 750 °C, leading to a polycrystalline-textured film. During RTA, an oxygen gas flow is required in order to guarantee the right oxygen content in the final perovskite structure. A deficit of oxygen produces poor quality film, while an oxygen excess can cause film delamination or high residual stress.

10.2.7 Sol–Gel Deposition Tools

PZT coating requires dedicated spin-coater tools equipped with high temperature hot plates for pyrolysis (able to reach at least 400 °C) and customized safety devices to avoid volatile lead compounds leakage into the environment as well as dedicated drain to collect the wasted chemical. Additional hot plates at lower temperature (100–250 °C) can be used as drying station between the coating step and the pyrolysis step to enhance film uniformity and texture. In fact, the early stage of film formation is determinant for the following crystal growth and grain orientation. At least two dispense line should be present suitable to manage dedicated chemicals for the seed layer and the film bulk and even more to obtain a gradient-free film. Both pump and pressurized canister can be used as chemical delivery system maintaining the chemical feeder under an inert and dry atmosphere in order to avoid unwanted hydrolysis reactions compromising the viscosity of the chemical. Coating cups should be equipped with ancillary cleaning system to avoid chemical build up and dying along the cup walls causing particles. Edge bead Removal (EBR) must be integrated at the end of the coating step using a proper solvent (usually the main solvent used in the sol–gel) because the film becomes almost insoluble after any thermal treatment. Material pile-up near the removed zone is potentially dangerous during the RTA step as it can cause film delamination and the consequent release of particles (see Fig. 10.5 panel (a)). To overcome this issue, the material deposited must be dried, evaporating the major part of the solvent trapped in the xerogel matrix and increasing the spinning time (at fixed rpm) until a steady thickness is reached.

The rapid thermal annealing (RTA) is the necessary step to crystallize the PZT layer after the sol–gel deposition. The RTP tool is required to have a lamp-based system to heat the wafers to high temperatures in a very short time. After reaching the designated temperature, the tool must keep the wafer at this temperature without any temperature oscillation and then a controlled ramp-down temperature

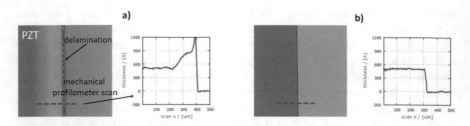

Fig. 10.5 (**a**) PZT pile-up giving rise to delamination after RTP due to insufficient spinning time. (**b**) Flat and neat profile of the deposited film obtained increasing the spinning time

is performed. The tool must then be equipped with a pyrometer able to measure and control the wafer temperature, which results to be the key parameter of all RTA processes. The RTA tool has also to support an oxygen flow line needed to reach the correct perovskite structure during the RTA process and a pump for clearing away the volatile products of lead compounds created during the process. Moreover, two quartz liners are required, one above and one below the wafer stage, to keep the lamps cleaned during the tool life. They are changed and cleaned after some cycles of RTA process.

10.2.8 Process Control

Crystallographic orientation and film thickness are the main parameters to be controlled during the fabrication process. The main techniques available to measure in line the thickness of unpatterned PZT films are X-Ray Reflectometry (XRR), polarimetry, and optical reflectometry, with the last one being most effective in providing in a very fast way (and nondestructive manner) accurate and reproducible thickness values of both amorphous and crystalline PZT layers. Dedicated recipes with different refraction and absorption indexes (n & k) values must be employed to measure as deposited or crystallized seed layer films as well as a fine tuning is needed to build up recipes able to measure in a linear way PZT film ranging from 200 nm to 2000 nm.

Sol–gel deposition process, by its own multistep nature, does not allow an accurate control of the final stack thickness even if the reproducibility and film uniformity are very good with standard deviations usually less than 1%. Nevertheless, single layers can be deposited with high accuracy, essential requirement to get a precise control of the preferential crystallographic orientation of the seed layer. In order to control this parameter, influencing the mechanical (stress), electrical (leakage, dielectric constant, and breakdown voltage), and piezoelectrical (piezoelectric coefficients) film properties, diffractograms have to be collected in Bragg–Brentano geometry in a 2theta angular range spanning from 20 to 40 degree to take in account for PZT (100), (110), and (111) lattice planes. To establish the preferential orientation of the textured film, both absolute and relative intensities of those peaks must be considered even if the former strongly depends on the substrate, barrier/bottom electrode, and film thickness as well. Relative intensity with respect to the other PZT peaks (110 and 111) is defined as in Eq. 10.1:

$$I_{rel\ PZT\ (100)} = \frac{I_{PZT(100)}/0.82}{I_{PZT(100)}/0.82 + I_{PZT\ (110)} \times 1 + I_{PZT(111)}/0.62} \tag{10.1}$$

where the absolute intensities I are multiplied by the relative normalization coefficients taken from powder spectra.

10.2.9   PZT Film Physical and Chemical Characterization Techniques

Besides the techniques used as process control to assess the film quality after deposition, other kinds of measurements are usually employed to characterize the film in terms of chemical–physical or piezoelectric properties. This chapter reviews SEM/TEM, SPM, nanoindentation, XRD, EDX, and TOF/SIMS techniques used for the morphological, mechanical, and chemical–physical characterization of both PZT seed layer and thick film and Double Beam Laser Interferometer (DBLI) together with Parametric Testing (PT) techniques utilized for the measurement of piezoelectric properties of patterned PZT. On Pt bottom electrode, the morphology and orientation of a thin PZT seed layer depend on the deposition process parameters as well as on the characteristics of the metallization as the grain size and deposition temperature. Perovskite structure growth mechanism is rather complex and comprises the formation of an intermediate pyrochlore phase at the initial stage of seed formation. The loss of Pb due to the evaporation in form on PbO and the simultaneous diffusion through Pt grain boundaries leads to the formation of pyrochlore nanocrystalline zones interspersed among PZT perovskite grains. SEM and TEM images (see Fig. 10.6) reveal nonhomogeneous morphology in which perovskite (A) and nanocrystals (B) coexist to give rise to a polycrystalline film with a preferential (100) texture.

The chemical and mechanical properties of the nanocrystalline phase have been studied by means of EDX and ultra-nano indentation techniques respectively, the former confirming the lead deficiency in the pyrochlore nanocrystalline zone and the latter pointing out a lower hardness with respect to the perovskite grains. SPM shows that seed layer is nonplanar with depressed zone corresponding to the pyrochlore phase (see Fig. 10.7).

A large amount of pyrochlore phase causes higher leakage currents and reduces the percentage of piezoelectric active area lowering the piezoelectric coefficients of the material. Furthermore, pyrochlore nanocrystalline zones have a lower etch rate in chlorine-based chemistries, causing possible residuals both in wet and dry etch processes (see Fig. 10.8).

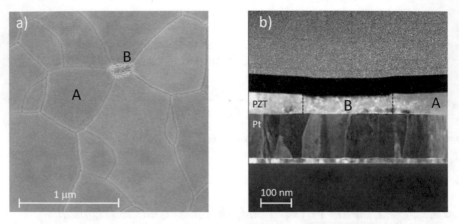

Fig. 10.6 (a) SEM planar view and (b) TEM cross-section images of PZT seed layer

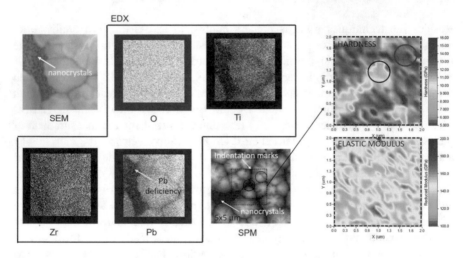

Fig. 10.7 EDX and nanoindentation characterization of pyrochlore zones in the PZT seed layer

For these reasons, this unwanted phase must be reduced for using the seed layer sol–gel chemistries with an excess of lead with respect to the final stoichiometry, usually in the order of 10% to 20%. The use of unbalanced chemistries is also useful to reduce the chemical gradients that are developed along the film height during each RTA step. Due to the lower crystallization temperature of $PbTiO_3$ with respect to $PbZrO_3$, the bottom part of the film has a higher Ti/Zr concentration ratio, while the interface is lead depleted due to PbO evaporation. EDX and TOF-SIMS analysis can be used to measure in a qualitative and semiquantitative way respectively the chemical gradients inside the film, showing how a sawtooth pattern is formed in correspondence of each interface formed by the annealing (see Fig. 10.9, panel (a)).

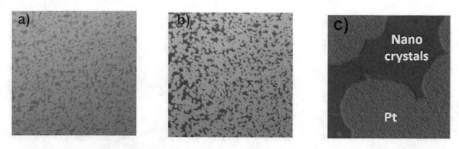

Fig. 10.8 (**a**) PZT seed layer before and (**b**) after wet etch. (**c**) SEM details of residuals due to presence of pyrochlore nanocrystalline areas

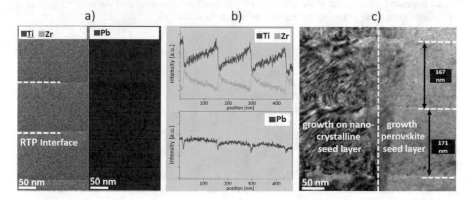

Fig. 10.9 Analysis of Zr/Ti gradients inside the PZT Sol–Gel film. Panel (**a**): TEM/EDX images evidencing the concentration of Ti (blue), Zr (yellow) and Pb (red) inside the PZT film. Panel (**b**): same data of panel (a), as a function of position in the film. Panel (**c**): detail of TEM images of the PZT film, evidencing two grains with nanocrystalline and perovskite PZT

These interfaces are also clearly visible in TEM images as well as the columnar crystal growth and the defectivity deriving from the nanocrystalline zone into the seed layer (see Fig. 10.9, panel (b)). However, it is worth noting how, even in the case of a nonideal seed layer, the perovskite grains tend to expand even in the in-plain direction other than the out-of-plane. The growing process can be easily followed using an optical microscope at 100X magnification. As show in Fig. 10.10, the nanocrystalline zones tend to reduce as the film became thicker and completely disappear if their extension is smaller with respect to the perovskite grains dimension.

On the other hand, the evolution of PZT film during growth can be easily monitored measuring film stress and crystallographic properties such as PZT(100) peak intensity and position. As pointed out in graphs in Fig. 10.11, PZT film stress is correlated to PZT(100) peak position, decreasing toward less tensile values as the thickness increases.

The main (100) peak shift toward lower values of 2theta angle suggests a stress relaxation during film growth confirmed by wafer bow measurements. The film becomes less tensile as the film increases in thickness until a steady constant value

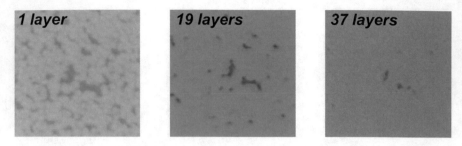

Fig. 10.10 Disappearance of nanocrystalline zones with film growth. The nanocrystalline phase is the dark area in the images. It can be seen that its surface diminishes as the number of deposited layers increases. This is due to the fact that PZT crystalline areas promote the ordered growth of following layers, thus reducing the amount of nanocrystallites

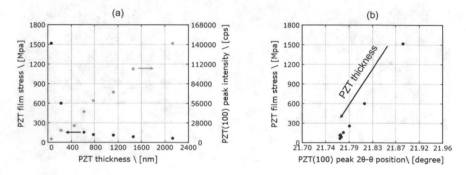

Fig. 10.11 (a) PZT film stress and XRD intensity as a function of film thickness and (b) correlation of film stress with PZT(100) peak position

is reached. The abrupt change in stress noticed during the seed layer formation can be ascribed to the bottom electrode metallization lattice reorganization due to the PZT annealing temperature much greater than the metal deposition temperature. This fact is confirmed by the Pt (111) peak position shift and intensity increase after seed annealing. The overall orientation of the seed layer is maintained at the end of the deposition process even if relative percentage of film oriented along the (100) direction increases with the film thickness (see Fig. 10.12).

10.3 Study Case: First Wafer Effect

10.3.1 Introduction

As for all the MEMS devices, the in-line testing procedures are important to give a feedback on the process flow. For this kind of devices, beyond the Parametric Test

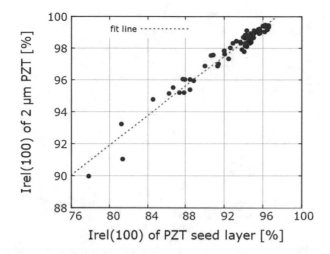

Fig. 10.12 Correlation between PZT (100) relative intensity of the seed layer and 2-micron-thick film

(PT), a mechanical testing is performed on moving structures (i.e., membranes or cantilevers).

The case study hereby reported deals with a recurrence anomaly found on different products, generally for the first wafer of the lot and sometimes also for the second one. This anomaly has been detected either at the stress or at the crystallographic orientation measurement of the PZT 2 μm layer showing a higher stress and a lower percentage of the PZT peak along the (100) direction, respectively. This difference in stress and crystallographic orientation has direct consequences on the final behavior of the device.

10.3.2   Study of PZT 2 μm—Stress and XRD

The study has been done on an autofocus (AF) device, where the stress of the layers is very important for its functionality. The core of this product is a membrane constituting the AF lens and the actuation driving that exploits the piezoelectric properties of a 2 μm PZT layer. For a correct functionality of the device, the lens bow must be negative to guarantee focus to infinite while, when voltage is applied to the PZT stack, the membrane deflects in a positive way to change the lens bow for close focusing (see Chap. 21 of this book).

The PZT stack is one of the key layers driving the lens bow in absence of applied field. Due to its importance, the PZT stress has been monitored on many wafers. By plotting the PZT stress value with respect to the lens deflection, it has been found that some wafers deviate from the average population (see Fig. 10.13). By taking a

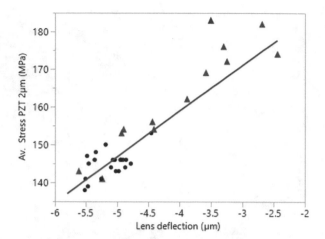

Fig. 10.13 Correlation between average stress of PZT2μm [MPa] and lens deflection [μm] measured at the end of the process flow

better look to the results, it is possible to see that this anomaly arises for the first wafers of almost every lot.

To confirm and better understand this anomaly, the measurement of crystallographic orientation has been performed on the same wafers. By considering the relative intensity of the peak (100) of PZT 2 μm, it is possible to find a very good correlation between stress and crystallographic orientation of PZT 2 μm, showed in Fig. 10.14, as already reported in literature [38]. The PZT film mostly [100] oriented had a less tensile stress, while, if the PZT texture is more complex, a higher stress was measured. The red triangles represent the wafers processed as first wafer before the PZT deposition.

This suggested that other steps preceding the PZT deposition may influence the PZT growth and finally the device behavior.

10.3.3 Titanium Stuffing

In a few lots, it has been noticed that not only one but two "first wafers" showed this stress anomaly. From the commonality analysis of the process chambers present in the process tools, it was found that the only equipment which could give this effect was the RTP used for the oxidation for the titanium barrier. It is important to point out that in this process the TiO_2 formation is achieved in two steps (pure titanium deposition by PVD, and *ex situ* TiO_2 formation by RTP in oxygen atmosphere in a separate equipment). This approach is convenient because it exploits standard equipment for the process, and it's able to achieve the formation of a TiO_2 layer with

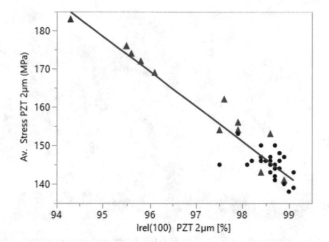

Fig. 10.14 Correlation between average stress and Irel (100) of PZT 2 μm of all the measured lots. The red triangles represent the wafers processed as first wafer before the PZT deposition

good quality and low roughness with respect to reactive ion sputtering. On the other hand, special attention must be paid to the oxidating conditions and atmosphere.

Actually, by studying and better understanding the RTP tool and the process, it has been understood that the anomaly on the first wafers processed at the titanium oxidation occurred in two cases:

1. If the tool switched from nitrogen to oxygen processes.
2. If the waiting time between one lot and the other one was more than 2 min. Due to a hardware configuration, after this idle time, a purge flow in nitrogen was activated, and it caused the anomaly. By substituting this nitrogen flow with one in oxygen, the anomaly disappeared.

The cause is probably due to a different crystallization phase of the TiO_2 which is not preferential for the correct crystallization of the bottom electrode and consequently the PZT [39, 40]. In Fig. 10.15, a reference wafer (blue line) is compared to the two anomalous wafers. The reference sample is crystalline TiO_2 (rutile) oriented along (200) in the growth direction, while the other cases show a less oriented crystalline TiO_2. Furthermore, in the anomalous wafers is also present a peak along (110), not seen in the reference. The cause of such differences in TiO_2 orientation may be on the first stages of oxidation which may enable nucleation and growth plane directions other than (002) when the chamber environment is out of complete conditioning. In conclusion, a different orientation of TiO_2 induces different orientation of BE and PZT along [111] and [100], respectively.

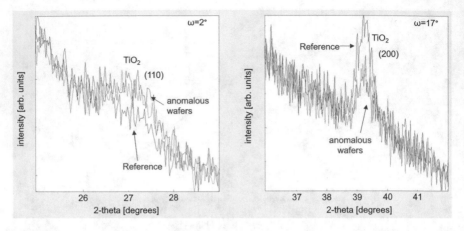

Fig. 10.15 XRD measurements of TiO$_2$. The blue line refers to the reference wafer. It has a higher intensity of the peak (200) and a lower intensity of the peak (110) with respect to the two first wafers

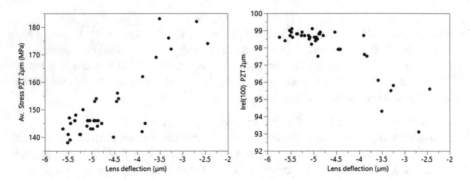

Fig. 10.16 Correlation of the lens deflection and the stress and Irel (100) of PZT 2 μm

10.3.4 Actuator Deflection vs. XRD and Stress

As previously shown, the correlation between stress and Irel (100) with the lens deflection is present and almost linear: if the PZT does not mainly grow along [100], also the device functionality is affected and the lens bow is higher. Moreover, due to this linearity, it was possible to define a threshold value on the stress and the relative intensity of the peak (100) of PZT 2 μm to screen the in-spec and the out-of-spec wafers, according to the proper specification limits (see Fig. 10.16). So, by using the XRD and stress measurement tools, it is possible to screen the devices at the PZT deposition, well before the lens deflection measurements.

10.4   PZT-Based Piezoelectric Stack Etch

10.4.1   Introduction to Piezoelectric Stack Etching

The Piezoelectric stack for MEMS applications is composed of multiple layers: a conductive layer as top electrode, a piezoelectric material, and a second conductive layer as bottom electrode. The electrode layers typically fall under the categorization of conductive materials and metals and can usually be etched by conventional techniques. Piezoelectric layers, instead, while being dielectric materials, often contain elements that are extraneous to conventional etching techniques. Specifically, PZT contains extremely nonvolatile elements such as lead and zirconium which can be challenging for conventional dry etching techniques.

The common approach for plasma etching is based on the volatilization of the component elements of the film to be etched by using compatible chemistries. The desired results are byproducts with a very low boiling point that can be easily extracted from the chamber in gaseous form and expelled through the pumping system of the tool. This approach is not applicable for materials containing lead and zirconium because their byproducts in fluorine and chlorine-based chemistries have extremely high boiling points; for comparison, the boiling points of common semiconductor processes byproduct are reported in Table 10.3.

The alternatives can utilize a wet etch approach, nullifying the volatility requirement, or employ a plasma etching chamber with specific characteristics that can allow the correct processing of nonvolatile materials.

10.4.2   Wet Etch

Wet etch approach is based on the removal of material by chemical reactions in a solution and subsequent dilution of reacted species. The wet approach has peculiarities that can render it superior or inferior to dry etching depending on the application. See Table 10.4 for a brief sum-up of advantages and disadvantages.

Table 10.3 Etching byproducts boiling point

Element to be etched	Etchant	Byproduct	Byproduct Boiling Point
Al	Chlorine-based	$AlCl_3$	180 °C
Si	Chlorine-based	$SiCl_4$	57 °C
	Fluorine-based	SiF_4	−86 °C
Zr	Chlorine-based	$ZrCl_4$	331 °C
	Fluorine-based	ZrF_4	905 °C
Pb	Chlorine-based	$PbCl_2$	950 °C
	Fluorine-based	$PbF2$	1293 °C

Table 10.4 Wet etch characteristics

Advantages	Disadvantages
Higher material selectivity	Low anisotropic etch potential
Nearly no mask consumption	Lower critical dimension control
Low dependence on exposed area	Irregular etch profile for polycrystalline materials

Table 10.5 Dry etch characteristics

Advantages	Disadvantages
High critical dimension control	Nonvolatile species generation
Controllable iso/anisotropic etch	Low throughput
Tunable selectivity versus multiple materials	Low material selectivity
Endpoint capability	Mask consumption

When tackling materials such as PZT, the main limitation of wet etching is the presence of various crystalline phases and orientations having different etch rates. In particular, the layer-by-layer deposition of the sol–gel processes creates concentration gradients that can result in the formation of irregular etch profiles. This, in turn, can result in irregular side-wall aspects and lower than desired dimensional control. The characteristics mentioned and the significant presence of undercut can be deleterious to piezoelectric actuators application for which devices are subjected to fatigue and cyclic deformations.

10.4.3   Dry Etch

The etching phenomenon in plasma chambers is based on the progressive removal of elements and clusters of materials by a combination of physical and chemical interaction of the surface with the plasma. See Table 10.5 for a brief sum-up of advantages and disadvantages of dry etching over wet etching. Generally, dry etch offers greater accuracy in dimensional control, and this requirement being paramount for MEMS fabrication, dry etching techniques are often chosen for the definition of PZT-based piezoelectric devices.

As anticipated, dry etching strongly relies on the evaporation or sublimation of the byproducts and, if the species obtained are not sufficiently volatile, they can redeposit on the wafer surface leading to undesired effects. The main issue can be the reduction of etching speed or even an etch-stop condition for which the rate of redeposition is equal or surpasses the rate of material removal.

To avoid this effect, a physical contribution is needed, allowing the removal of material from the wafer surface. The physical interaction is achieved by the formation of positively ionized species that are accelerated toward the wafer surface by an electrical potential. Commonly used ions are obtained from noble gases and other species that can be easily ionized. Since in the specific case of PZT etching ion bombardment is essential, it must be possible to achieve low working pressures (below 50 mTorr) to reduce the anelastic collisions between species while also applying large acceleration potentials (over 100 V). Only the combined use of ionic bombardment and chemical interaction generates conditions that allow material to be removed from the wafer surface. To summarize, the tool requirements for these materials are:

- High density plasma generation at RF frequency
- Low pressure operation (<50 mTorr)
- Bias voltage for ion acceleration with secondary low frequency generator
- Reactive species (fluoride or chloride-based)

While these characteristics allow for dry etching and material removal, another consequence of the generation of nonvolatile byproducts is the formation of a film of material on the inside components of the etching chambers. This effect is not an issue for sample wafers preparation and low volume processing but can readily become critical for production ramp up. The materials deposited on the chamber insides are mostly composed of the heavy metal lead and zirconium and can lead to issues in:

- Chamber performance
- Wafer quality
- Chamber reliability

The result of the issues illustrated is that the chamber might need frequent cleaning and removal of material from its inside components. In a production environment, these maintenance events are not desirable and decrease productivity while increasing tool running costs. The next paragraph is dedicated to the subject, exploring in further detail how the cleanings can be reduced, and the approaches that can guarantee stable chamber performance.

10.4.4   Process Chamber Stability and Mean Time Between Cleanings

Plasma chambers require periodic maintenance to avoid drifts in their processing capabilities. One of the most important types of preventive maintenance regularly performed on these tools is the so-called wet cleaning. During said procedure, the chamber is brought back to atmospheric pressure, and at room temperature, the internal components are either cleaned *in situ* or substituted by clean ones, sending the used ones to a washing station, and an overall check of the process kit conditions is performed. The average time between two of these preventive maintenance actions is defined as Mean Time Between Cleanings (MTBC).

One of the main challenges connected to the etching of PZT is achieving an acceptable target for the MTBC of the chambers employed. Although the impact on productivity of frequent cleanings is clear, it is important to clarify why these chambers require more care with respect to related metal dry etch tools. It has been stated before that PZT contains heavy metal elements that cannot be pumped away in gaseous form and will tend to accumulate on the chamber internal components, progressively forming a film. This condition is commonly found in reactive ion etching equipment and is not specific to piezoelectric materials; in this case, however the velocity at which the film is formed is much higher than for conventional dry etching processes like silicon or silicon oxide etching. The process conditions can be influenced and modified by the rapid accumulation of material, and degradations might be visible even after few hours of process time.

The main consequences of uncontrolled material buildup can be grouped by their effects: impact on process performance (etch rate, uniformity, etc.) and impact on wafer defect count. The first occurrence is caused by the interaction of the accumulated substances with the electromagnetic fields necessary for plasma generation. Altering the transmission of power causes shifts in the system inductance that are either reflected in process parameters shifts or in process performance. This effect can be avoided by specifically designed components that limit material deposition on parts that are critical for power transfer. The wafer defect count can also dramatically increase if particles precipitate from the plasma chamber components on the wafer during processes; this occurrence is commonly referred to as flaking and can be avoided by either geometrically shielding the wafer or by carefully monitoring the material adhesion on the chamber components. However, once the material adhesion is impaired, processing must be stopped and a chamber cleaning procedure is required.

Considering the previous discussion, the management of material deposition on chamber components is paramount to achieve high MTBC results. Two methods are commonly employed when regulating film formation and accretion:

- No wafer-less autoclean (NO WAC) approach
- Wafer-less autoclean (WAC) approach

The first approach allows all the nonvolatile materials to accumulate on the process kit elements that are positioned around the wafer inside the chamber. This method aims at keeping the redeposited byproducts stable for the entire chamber lifetime. The second approach is instead based on the removal and pumping away of the byproducts and materials that have been etched from the wafer. Defined as a wafer-less auto clean (WAC) step, this strategy relies on the use of a plasma step that runs without any wafer inside the process chamber. The aim of the WAC is resetting the status of the etch chamber after each processed wafer, exposing both the chamber internal parts and the wafer chucking surface to the plasma. While this can be a valid approach to prolong the MTBC, WAC is only available in specifically designed chambers. The chucking surface, for example, must be composed of adequate materials (such as thick ceramic layers instead of anodized aluminum), so as not to be damaged by the exposure to the plasma without the protection of a wafer. The use of incorrect chamber components and materials when combined with the necessary aggressive cleaning chemistries can lead to their progressive and irreversible deterioration.

Due to all the limitations illustrated, the MTBC of chambers dedicated to these materials cannot currently achieve the levels of other more consolidated and commonly available technologies. Nonetheless, it is possible to reduce the frequency of preventive maintenance and allow production to run consistently by carefully tailoring WAC/NO WAC approaches to each specific process and by pairing them with purposely developed chamber components.

Having defined the etching properties of the piezoelectric materials and the tool requirements for their processing, the following paragraphs are dedicated to a detailed overview of etching processes exploring their characteristics and requirements for the definition of the piezoelectric stack in example.

10.4.5   Dry Etch Process Requirements for PZT-Based Piezoelectric Stack

The piezoelectric stack under examination has been described in the previous chapters, and it suffices to remember that 3 different materials need to be etched for the definition of the piezoelectric actuator: TiW layer as top electrode, PZT layer as active piezoelectric material, and Pt as bottom electrode. Each of these steps has different requirements and challenges, the main to be considered are:

1. Masking approach: the materials to be etched require aggressive chemistries with high delivered power, and the masking material must be able to endure the process without excessive modification. Furthermore, the nonvolatility of the materials to be etched plays an important role, with the risk of having material redeposit on the mask deforming the geometries and hindering subsequent removal of the masking medium.
2. Etch selectivity: when one material in a stack is etched, terminating the process on a different film, this last one is known as landing layer. If a stack is composed of multiple materials to be etched in series, it can be important to preserve the integrity of the landing layers, particularly if they act as electrical conductors. The ability of an etching process to consume a layer while preserving another one is known as selectivity.
3. Endpoint capability: for the same reasons explained in the previous point, it is important to have the ability to stop once the landing layer is reached. This can be achieved by employing a plasma emission spectrometer that can detect a variation in the species present in the plasma and subsequently stop the etching step when complete material removal is achieved and before excessive consumption of the landing layer takes place.

If the above conditions are met, etching of the complete stack can be attained. The masking approach exemplified in the following paragraphs employs 2 different photoresist masking steps:

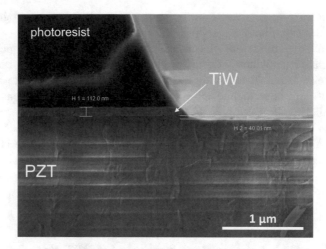

Fig. 10.17 SEM cross-section after TiW etching

- Mask1 is used for the definition of the top electrode and PZT active layer.

  The use of the same mask for the first two layers allows automatic alignment of the first electrode and the piezoelectric material.
- Mask2 is used for the definition of the bottom electrode.

  The use of a different mask for this layer allows the definition of contact pads to allow interconnections to reach the bottom electrode.

The etching steps for the 3 different materials are detailed separately in the following paragraphs. The tool of choice for all 3 layers in this example is an RF plasma etch vacuum chamber with 1250 W 13.56 MHz planar TCP and a secondary generator at lower frequency to introduce bias voltage and allow physical sputtering.

10.4.6   TiW Etch

TiW is a metal alloy commonly used and etched for conventional semiconductor applications. In the specific case of a Piezoelectric stack, two main constraints are present: allowing a complete etching of the material to avoid micromasking of the PZT layer and having enough control on the photoresist mask consumption. TiW can be readily etched in metal etching tools with a plasma composed of Cl_2 and BCl_3. Cl_2 acts as the main etching reactant, while BCl_3 is used to remove the oxidation layer from the material. This allows etch rates well over 1000 Å/min while maintaining adequate photoresist selectivity. A typical SEM cross-section after etching is shown in Fig. 10.17. One can observe the consumption of the underlying PZT layer and deduce that the process chamber will be contaminated by the presence of lead and zirconium.

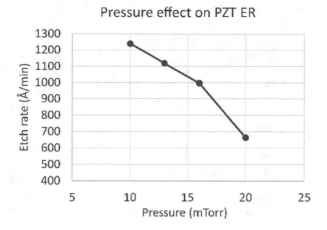

Fig. 10.18 Effect of varying pressure on PZT etch rate

10.4.7   PZT Etch

PZT is a very challenging material for plasma etching applications, and the correct process conditions must be identified to obtain significant etch rates while maintaining the other etch requirements. PZT etch rate can be substantially increased by lowering the process pressure if a stable and dense plasma can be generated. This is due to the contribution of ion bombardment to the etch chemistry: the lower the pressure, the higher the final velocity that ions can reach before impacting the surface of the material. The trend of PZT etch rate increase with lowering of pressure is reported in Fig. 10.18.

To achieve reasonable selectivity to the photoresist mask, the chemistry must also be tuned. Varying the ratio and nature of the chemically-active etching species can influence greatly the photoresist etch rate. The effect of using BCl_3 as oxidant instead of Cl_2 is clearly visible on photoresist etch rate in Fig. 10.19.

Another fundamental challenge related to PZT etching is, once again, connected to the low volatility of the reaction byproducts. The risk connected to having strong ion bombardment is the redeposition of material on the wafer leading to structure geometry degradation and formation of so-called fencing on the structures. This can be avoided by changing the photoresist profile, achieving a more sloped angle that reduces material accumulation on the sidewall of structures. Two SEM images are provided to visualize the issue: Both results are obtained by using the same etching process, but one employs a nonsloped photoresist with initial angle above 85° while the second employs a photoresist that has been reflowed to achieve an angle below 70°. It is apparent (see Fig. 10.20) that the non-reflowed photoresist causes the build-up of significant resputtered material and cannot be considered a valid approach. The takeaway is that the photoresist profile should be carefully tuned for PZT etching applications.

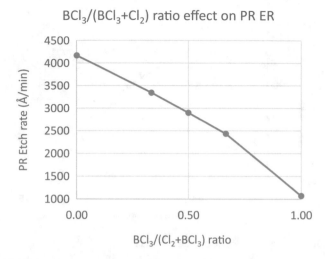

Fig. 10.19 BCl3/Cl2 ratio effect on photoresist etch rate

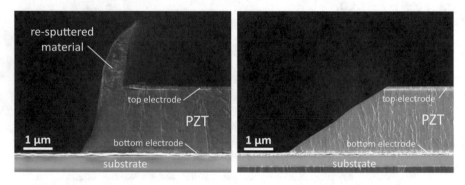

Fig. 10.20 Comparison of etch profile after photoresist removal. On the left a case of nonsloped photoresist leading to sever resputtering, on the right the photoresist is reflowed to achieve sloping and avoid re-sputtering

One final requirement is avoiding excessive consumption of the landing layer, the platinum bottom electrode. The ER on platinum is mainly driven by the presence of bombarding species in the plasma and by the acceleration potential (bias voltage). Argon is known to have high sputtering yields [41] when used to dislodge platinum atoms, and thus, its presence in the plasma must be diluted by other species to increase PZT to Pt selectivity. Studies show that an increase of BCl_3 gas flow significantly reduces overall ion current density in a $BCl_3/Cl_2/Ar$ plasma [42]. Therefore, the overall ability to preserve platinum can be increased by reducing imposed bias and increasing the ratio of BCl_3. Both these actions come at the expense of etch rate on PZT as visible in Fig. 10.21.

The chemistry selected for this example is based on BCl_3, CHF_3, and Ar. CHF_3 is added to the final recipe after observing that its presence both increases etch rate

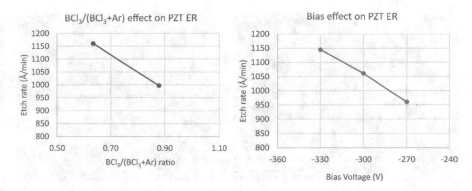

Fig. 10.21 Left panel: effect of BCl3 and Ar ratio on PZT etch rate. Right panel: effect of Bias Voltage on PZT etch rate

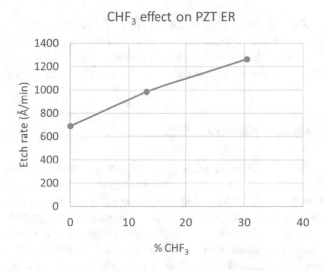

Fig. 10.22 Effect of varying the percentage of CHF3 gas flow on PZT etch rate

and improves plasma uniformity on the wafer, and its effect on etch rate is shown in Fig. 10.22.

The proposed etch chemistry achieves etch rates over 1000 Å/min. In Fig. 10.23, an SEM cross-section and a TEM close-up on the bottom electrode layer show how both requirements of conserving photoresist and not damaging the Pt bottom electrode can be achieved by etch chemistry fine tuning. The ability to control the length of the etch step is guaranteed by an endpoint system as mentioned previously. The suggested emission lines for PZT etching monitoring are those connected to lead compounds, the one considered in this example is the 406 nm, the wavelength corresponding to one of Pb(I) emission lines.

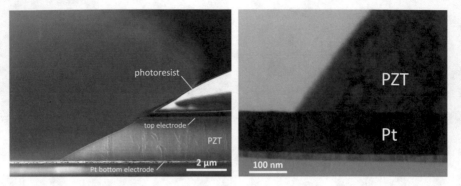

Fig. 10.23 Left panel: SEM cross-section after PZT etching step; right panel: TEM close-up on the Pt bottom electrode after PZT etch

Once the PZT material is consumed, the signal intensity drops since lead is no longer present in the etch byproducts and platinum is relatively impervious to the stated etch chemistry. Once the PZT etching step is completed, the wafer is cleaned and masked again for the subsequent platinum bottom electrode etching process.

10.4.8   Pt Bottom Electrode Etch

The bottom electrode layer is composed by platinum and titanium oxide in a thickness ratio of 5:1, and the main contributor to etching time is the platinum layer and is therefore the focus of this paragraph. Although platinum is a noble metal and reacts only weakly with conventional etching chemistries, its sputtering is readily performed by Ar ions. Literature reports various examples for how the addition of a reactive gas can improve platinum etching performances. Both fluorine and chlorine chemistries can be used resulting in different etch profiles and characteristics. In this example, a combination of Ar and Cl_2 is explored in further detail: the presence of chlorine is believed to give rise to compounds such as platinum (II) chloride that can be more easily removed by the physical action of argon ions with respect to pure platinum. The studied plasma chemistries have a Cl_2 to Ar ratio between 2:1 and 1:2 depending on patterned geometries and exposed areas, and the chamber pressure is kept in the range between 5 and 15 mTorr to facilitate ion bombardment effects. The effect of varying the Cl_2 and Ar volumetric ratio is illustrated in Fig. 10.24, showing that higher chlorine contents reduce both Pt etch rate and selectivity to photoresist. This is also confirmed in literature examples [43].

Nonetheless, a certain percentage of Cl_2 is required to both improve plasma uniformity and to avoid the accumulation of Pt or its nonvolatile byproducts on the etch sidewall, such as those visible in Fig. 10.25. It is also reported that the platinum chloride byproducts are water soluble [44, 45] and, if present in moderate quantities,

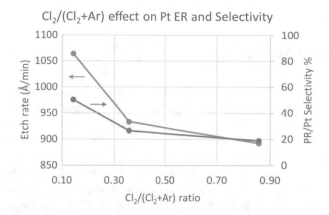

Fig. 10.24 Effect of varying Cl_2 and Ar ratio on Pt etch rate (blue line) and on photoresist/platinum selectivity (orange line)

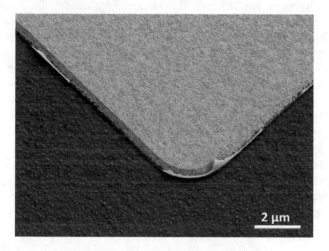

Fig. 10.25 SEM top view of material accumulated at etch sidewall

can be removed by subsequent cleaning processes, thus reducing the impact of their accumulation at etch sidewalls.

Given these conditions, etch rates around 1000 Å/min can be achieved while maintaining adequate photoresist selectivity. A final consideration goes to the underlying silicon oxide landing layer, and its consumption usually needs to be limited to avoid variations in final device performance. This can be achieved both by increasing etch chemistry selectivity and by employing plasma emission spectroscopy to determine the end of the etching step as illustrated in the previous steps.

10.5 AlN

10.5.1 Material Overview

AlN is a group III–V semiconductor of great technological importance due to several peculiar properties, such as high thermal conductivity, high hardness (11–15 GPa), high resistance to temperature, high electrical resistivity (1011–1013 Ω cm), wide band gap (6–6.2 eV), and high velocity of acoustic waves [46]. All these features make AlN thin films suitable for a broad range of applications: Bulk Acoustic Resonators (Fbar), Bulk Acoustic Wave filters (BAW), Surface Acoustic Wave filters (SAW), protective coatings, insulating and surface passivation layers in microelectronics, functional layers in high power, high frequency, and optical devices.

Piezoelectric AlN films are also being currently investigated as potential actuator/sensing layers in MEMS. For this kind of application, the most important characteristics that the layer should have are a good piezoelectric response and a controlled residual stress. The piezoelectric behavior of AlN thin films depends on their crystal properties (preferential c-axis orientation), presence of impurities, type of substrate, crystal morphology (grain size and rocking curve), film stoichiometry (depending on Ar:N_2 flow ratio), and surface roughness. Since MEMS devices like PMUT are very sensitive to the residual film stress, a lot of work has been done in terms of both process parameters optimization and tool hardware configuration in order to control the stress for all desired characteristics: stress average, stress uniformity across wafer and across entire film thickness.

In microelectronics, AlN thin films can be deposited by CVD, pulsed laser deposition, DC/RF reactive sputtering (PVD), and molecular beam epitaxy. Due to its low cost, versatility, and low temperature deposition, the reactive sputtering is the most used technique for AlN deposition.

10.5.2 AlN Deposition

AlN thin film deposition has become one of the most promising fields in MEMS because of its good compatibility with design on silicon substrates and a large electro-mechanical coupling constant.

Deposition using DC magnetron reactive sputtering has the advantage over other deposition methods due to its better process parameters control and low deposition temperature. AlN DC reactive sputtering process is based on the ion bombardment of a pure Al target (cathode) using a high DC voltage source. Through momentum transfer, atoms close to the Al target surface became volatile, reacting with Nitrogen and producing compounds that are transported as a vapor to the substrate (anode) on which film growth happens.

Table 10.6 DC magnetron reactive sputtering typical process parameters

Process parameters	Numerical value
Deposition temperature	300 °C
DC pulsed power	7500 W
Cathode voltage	$(250 \div 260)$ V
Process pressure	5.5×10^{-3} mbar
Ar:N$_2$ flow ratio	1:3 (for stoichiometric AlN)

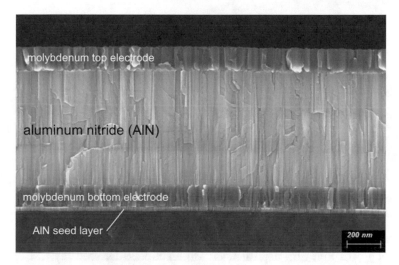

Fig. 10.26 SEM (Scanning Electron Microscope) picture of a piezo stack cross-section, showing AlN film (002) preferred orientation. From bottom to top: AlN seed layer, Mo bottom electrode, AlN piezo film, Mo top electrode

Mechanical, electrical, thermal, chemical, and optical properties of deposited AlN films can be defined during the deposition by controlling different process parameters such as sputtering power, distance between sample and target, gas flow ratio (Ar:N$_2$), process pressure, and temperature. In Table 10.6, the typical process parameters are listed.

AlN thin film characteristics are strictly related to its crystal structure. To grow films with the c-axis oriented perpendicularly to the substrate surface, i.e., with (002) preferred orientation, it is necessary to provide the formation of a well-textured Mo bottom electrode which is usually deposited on AlN seed layer of 30 nm (see Fig. 10.26).

The study of the thin film orientation and structure can be characterized by X-Ray Diffraction (XRD). AlN piezo crystallographic orientation (002) quality can be evaluated through the Rocking Curve (RC) FWHM (Full Width at Half Maximum), that is, the lower is the value, the higher is the film quality.

The substrate contribution for depositing a well-textured AlN thin film is very important; the contribution of different substrates has been studied in the following cases of study:

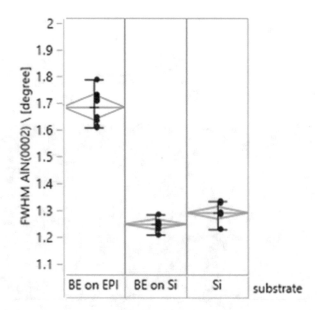

Fig. 10.27 RC FWHM plot based on 9-point map measurements

(a) AlN 1 μm thin film deposited on Mo electrode/AlN seed layer/Si epitaxial grown (EPI)

(b) AlN 1 μm thin film deposited on Mo electrode/AlN seed layer/Si substrate (100)

(c) AlN 1 μm thin film deposited directly on Si substrate (100)

The results of this analysis are plotted in Fig. 10.27: the impact of both the substrate and the presence of AlN seed layer on film quality can be appreciated. The RC value difference between case (a) and case (b) is mostly related to the higher roughness of EPI substrate with respect to Si one. Furthermore, the difference between case (b) and case (c), even if small, highlights even more the importance of having a seed layer to give the AlN film the (002) preferred orientation.

Ionic bombardment during deposition causes several types of defects in the crystal structure also inducing a residual stress which should be controlled by tuning the process parameters. Since deposition conditions also influence other film properties (such as crystal orientation, stoichiometry, and thickness uniformity), AlN PVD modules are usually equipped of an RF bias module and a tunable matching network in order to control both the average and distribution of the film stress on wafer (see Fig. 10.28).

AlN thin film stress can be quantified by considering both the average and its distribution. Tuning of the AlN average stress is feasible to compensate the stress variation due to the aging of target and shields (lifetime effect). The stress tuning can be accomplished by exploiting a particular relationship between RF bias applied power and the AlN film average stress, without affecting the stress uniformity.

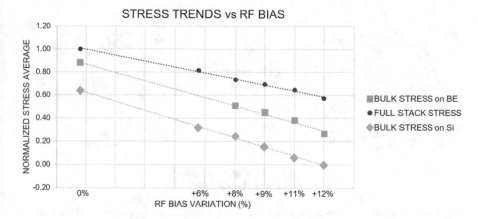

Fig. 10.28 Stress trends as a function of RF bias for: AlN 1 μm film deposited on AlN seed/bottom electrode (blue square); full piezo stack including AlN seed/bottom electrode/piezo AlN/top electrode (purple circle); AlN 1 μm film deposited on Si (100) substrate (gray rhombus)

Table 10.7 AlN film main properties

Material characteristics	Typical values
Deposition rate	$(50 \div 70)$ nm/min
Film thickness	$(200 \div 1000)$ nm
Film thickness non-uniformity	< 1% (on 8" Si wafers)
Film conformality	Good
Stress mean on wafer	Adjustable $(-250 \div 250)$ MPa
Stress range on wafer	<200 MPa (on 8" Si wafers)
Rocking curve FWHM (002)	<1.5 degrees (for 1000 nm films on Mo electrode/AlN seed/Si)
d_{33} parameter	5 pm/V

On the other hand, AlN stress profile can be controlled by varying the magnet position on the target backside, the target–substrate distance, or the process pressure keeping fixed the $Ar:N_2$ ratio to guarantee a stoichiometric film.

In Table 10.7, AlN film main properties are illustrated.

10.5.3 Scandium Doping

AlN piezoelectric performances could be enhanced by Sc doping. Aluminum Scandium nitride (AlScN) with wurtzite crystallographic structure (see Fig. 10.29) has been demonstrated to be an excellent material due to high piezoelectric performance and high maximum operating temperature. For a strong piezoelectric response, highly textured film with c-axis orientation is desired.

According to recent studies [47], the piezoelectric response of the thin film depends on the Sc concentration as shown in Fig. 10.30. The piezoelectric coeffi-

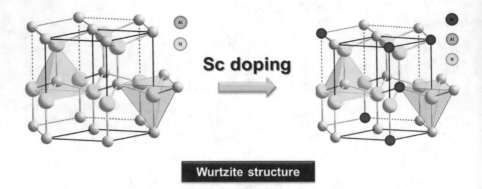

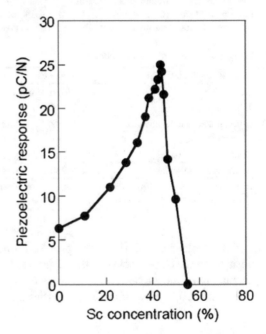

Fig. 10.29 Wurtzite structure for AlN and AlScN: Sc doping consist in the substitution within the lattice of Al atoms with Sc atoms

Fig. 10.30 $Al_{1-x}Sc_xN$ piezoelectric response as a function of Sc concentration

cient d_{33} gradually increases as the Sc concentration increases from 0% to 43%: in this range, columnar grains like those found in undoped AlN thin films can be observed and the crystal structures are only present in the hexagonal phase. When the Sc concentration is 43%, the alloy exhibits a peak in d_{33} of about 24.6 pC/N, and the $Sc_{0.43}Al_{0.57}N$ alloy shows the largest achievable piezoelectric coefficient among the tetrahedrally bonded semiconductors. Between 43% and 55%, the piezoelectricity is lowered due to both hexagonal and cubic phases' coexistence, while above 55%, the crystal structure changes totally to the cubic phase, which is a stable phase of ScN and does not have piezoelectricity.

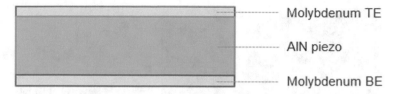

Fig. 10.31 Schematic view of Mo/AlN/Mo piezoelectric stack

AlScN films can be deposited by the cosputtering method or by pulsed DC reactive sputtering from an AlSc alloy target (PVD mono-source). The cosputtering method is the most versatile for material characterization but affected by low deposition rates. The monosource configuration, instead, is better in terms of deposition rate, but the greatest limitation lies in the difficulty of manufacturing alloy targets with gradually increasing Sc concentration that could be susceptible to cracking and particles' defectivity.

10.6   AlN Etch

Due to the 11.53 eV per atom high bond energies existing between nitrogen and aluminum, AlN material is very hard to etch using conventional dry etching techniques (i.e., Fluorine based). For this reason, the use of a dry etching tool able to generate a high-density plasma is very important.

The stack to realize an AlN piezoelectric thick film-based MEMS must be composed by a metallic layer, for example, Molybdenum, which acts as TOP ELECTRODE (TE), the BULK piezo layer (AlN), and a second metallic layer (Molybdenum) which acts as BOTTOM ELECTRODE (BE). A schematic view of this stack is illustrated in Fig. 10.31.

In order to perform the Molybdenum/AlN/Molybdenum dry etch steps, a standard TCP (Transformer Coupled Plasma) metal etch tool has been identified as the best choice. Chlorine-based chemistry has been identified as the most appropriate to etch these layers.

10.6.1   TE Etch

Molybdenum is a transition metal and can be etched using dry etch tool and a proper chemistry. To define the TE and the piezoelectric active layer, a dedicated photoresist mask can be used. Through this mask, it is possible to etch both the Moly and the AlN layers into the same chamber. Due to their very different chemical properties, two different chemistries (recipes) are needed. The selected chemistry for Moly etch is a mixture of Cl_2, Ar, and O_2 gases which is able to do an anisotropic

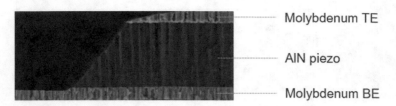

Fig. 10.32 SEM image after TE Mo/AlN etch

etch of the material, and it allows to reach good Etch Rate (~3000 Å/min, similar to the Etch Rate on other metal etches). The selectivity with respect to photoresist is extremely important.

10.6.2 AlN Bulk Etch

AlN has piezoelectric properties, but it is an insulator, so a different chemistry is necessary to obtain a good anisotropic etch of the material. The most appropriate chemistry found is based on a BCl_3/Cl_2 ionized plasma. Through a correct tuning of all the main recipe parameters, it is possible to obtain also in this case a good Etch Rate (~3000 Å/min), good photoresist selectivity, and good Molybdenum selectivity. The last aspect is mandatory in case of piezoelectric device because the Molybdenum under the AlN layer acts as BOTTOM ELECTRODE and its electrical properties are directly proportional and strongly dependent on its thickness (after etching). As mentioned above, the TE etch and the AlN BULK etch are realized with the same photoresist mask. In Fig. 10.32, a SEM image after these etches is reported.

10.6.3 BE Etch

The Molybdenum BOTTOM ELECTRODE etch is the same as the TOP one. In this case, however, a "soft landing" overetch step is necessary to preserve the landing layer thickness. This step needs to be opportunely set up to minimize the landing layer consumption.

10.6.4 MTBC

The MTBC (Mean Time Between Clean) of a metal etching chamber is normally shorter than MTBC of the other dry etching tools (Silicon, dielectric, etc.), due to the low-volatile or nonvolatile nature of the byproduct species and their accumulation on the chamber walls. In the case of Moly and AlN etch, an optimization work related to the chamber conditioning before each of the two etches has been done. This guarantees to maintain the chamber in the right condition from the etching point of view (ER, uniformity, selectivity, etc.) and helps preventing an early failure of the chamber. Thanks to these measures, it is possible to obtain a good MTBC, which is comparable with the other metal etch chambers (hundreds of RF hours).

10.6.5   Endpoint Detection

Since many years the endpoint detection system has been integrated within all the dry etching chambers. Through the light intensity variation of the plasma at a certain wavelength (depending on the etching materials and the chemistry used), it is possible to detect the interfaces between the etching and the landing layer.

The ENDPOINT detection ensures the correct etching time for both Mo and AlN layer, heavily limiting the landing layer consumption.

10.7   Piezoelectric Materials: Figure of Merit

The piezoelectric effect can be well explained by means of coupled equations, linking four fundamental quantities: the electrical displacement D, the electrical field E, the strain S, and the stress T. The first two quantities are rank-1 tensors, while the latter two are rank-2 tensors. Two most widely used formalisms for the coupled equations are the strain-charge formalism (Eqs. 10.2 and 10.3), and the stress-charge formalism (Eqs. 10.4 and 10.5). In these equations, s^E and c^E are the mechanical compliance and stiffness at constant E; ε^T and ε^S are the dielectric tensors under fixed stress or strain, respectively; and finally, e and d are the piezoelectric tensors representing the direct piezoelectric effect. The converse piezoelectric effect is expressed by means of their transpose e^t and d^t. In the International System of Units (SI), the coefficients in the e tensor are expressed in units of C/m^2: if multiplied by an electric field (in N/C, Eq. 10.3), they give a stress (N/m^2), whereas, when multiplied by a strain (adimensional, Eq. 10.4), they give a surface charge density (C/m^2), corresponding to the electrical displacement. Coefficients in the d tensor are in m/V. When multiplied by an electrical field (in V/m, Eq. 10.1), they give an adimensional parameter (strain), whereas, multiplied by a stress (N/m^2, Eq. 10.2), they give again a surface charge density (electrical displacement).

$$\{S\} = \left[s^E\right]\{T\} + \left[d^t\right]\{E\} \tag{10.2}$$

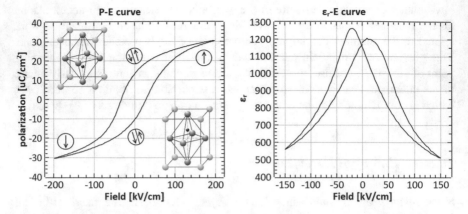

Fig. 10.33 Left panel: polarization-field curve of a PZT thin film, measured by means of a current integrator (Aixacct TF Analyzer). The values of maximum positive and negative polarization ($P_{max}+$ and $P_{max}-$) are set equal by the instrument. Right panel: dielectric constant versus applied field. The different heights of the two peaks are linked to the different nature of the electrodes (in this case, TiW as top electrode and Pt as bottom electrode)

$$\{D\} = [d]\{T\} + \left[\varepsilon^{T}\right]\{E\} \tag{10.3}$$

$$\{T\} = \left[c^{E}\right]\{S\} - \left[e^{t}\right]\{E\} \tag{10.4}$$

$$\{D\} = [e]\{S\} + \left[\varepsilon^{S}\right]\{E\} \tag{10.5}$$

The electrical displacement is the first quantity of interest when characterizing a new piezoelectric film. In practice, it is the charge density accumulated on the two electrodes of the thin film capacitor, and it can be easily measured. With reference to Fig. 10.33 and Eq. 10.3, the electrical field applied to a capacitor can be varied, and the corresponding charge can be plotted in a polarization loop curve (or P–E loop).

Two methods can be used for obtaining the P–E loop (see Fig. 10.34). The first consists in measuring the voltage V_{ref} across a reference capacitor, put in series with the device under test (DUT), in a so-called Sawyer–Tower configuration [48]. From the knowledge of the reference capacitance, it is immediate to know the charge accumulated on the DUT electrodes. The other method consists in measuring the current injected on the capacitor electrodes by means of an operational amplifier, with the input maintained at virtual ground. The integration with respect to the time of the electrical current gives the charge accumulated on the electrodes. This configuration has the advantage of being very flexible, but on the other hand, it introduces a systematic uncertainty in the absolute value of the charge, since this

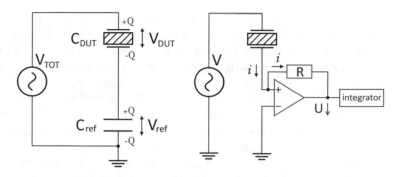

Fig. 10.34 Left scheme: Sawyer–Tower circuit. The voltage across the reference capacitor is measured in order to obtain the charge accumulated in the DUT. If C_{ref} is much higher than C_{DUT}, the voltage drop on the reference capacitor is negligible with respect to $V_{DUT} \approx V_{TOT}$. Right scheme: virtual ground method. The charge on the DUT is obtained by the integration of the current flowing in the circuit, given by U/R

is obtained by integration: an arbitrary offset can be set in the P–E loop, without having any effect on the measured current.

For a ferroelectric material as depicted in Fig. 10.33, the P–E loop shows a noticeable hysteresis due to the existence of two stable polarization states, distributed inside the material in zones of coherent polarizations (domains) [49]. At the maximum positive polarization (P_{max}^+), all the domains are aligned in the same direction. When the field is decreased to zero, the domains partially lose their orientation and a residual polarization P_r is measured (P_r^+). If the field is then reversed to negative values, the domains start to switch, until the total polarization reaches zero in correspondence of the negative coercive field (E_c^-). If the field is increased along the negative semi-axis, the polarization becomes negative, until reaching the maximum negative value (P_{max}^-). From this point, by reversing the variation of the electrical field toward positive values, the lower branch of the P-E curve is drawn, reaching the point of negative residual polarization (P_r^-) and the positive coercive field (V_c^+) in complete analogy to the higher branch.

Differently from above, in which the field varies over a wide range of values (large signal mode), there is another interesting way to measure the variation of charge versus applied field, in which a small ($\ll V_c$) oscillating voltage is superimposed to the voltage ramp ranging from V_{max} to V_{min} (small signal measurement) [50, 51]. The dielectric capacitance ε_r is obtained from the measured capacitance C as $\varepsilon_r = C*d/S/\varepsilon_0$, where d is the film thickness, S is the capacitor area, and ε_0 is the dielectric permittivity of the vacuum. Also in this case a hysteresis is observed, with its maxima corresponding to the coercive fields in the P–E curve.

Among all the piezoelectric coefficients, two of them are of particular interest in practical cases. The first is e_{31}. This coefficient expresses the coupling between a field directed along the vertical axis and the stress in the horizontal plane (see Eq. 10.4). This is very important for MEMS, for two reasons. First, the applied electrical field is always perpendicular to the wafer surface due to the thin film configuration; second, the MEMS structures are actuated by exploiting the in-plane stress generated by the piezoelectric thin film. For example, the in-plane stress is used for deflecting membranes and cantilevers along the vertical direction. The e_{31} coefficient thus possesses an immediate meaning and application in MEMS devices. The other interesting coefficient is d_{33}, i.e., the deformation of the piezoelectric film in the vertical direction, under application of a vertical electric field. The meaning of this parameter is quite intuitive; though not being involved directly in the actuation of MEMS structures, it is important from the point of view of the film characterization and measurement, as will be explained later.

Up to now, general considerations about piezoelectric coefficients for a free-standing piezoelectric film have been done. However, in real devices, the piezoelectric thin film is always clamped to a substrate (the MEMS structure to be actuated), and this boundary condition leads inevitably to a mixing of out-of-plane and in-plane coefficients when one tries to find out the ratio between the applied field and the in-plane stress or the out-of-plane strain. It can be demonstrated that the pure e_{31} and d_{33} coefficients cannot be measured singularly, but two alternative, "effective" quantities must be considered [52]. For a piezoelectric film clamped to a perfectly rigid substrate (meaning that it cannot be deformed), the effective e_{31f} coefficient and the effective d_{33f} coefficient are expressed by Eqs. 10.6 and 10.7. Regarding Eq. 10.6, it is possible to clearly see the contribution of the in-plane clamping through the d_{31} coefficient, that multiplies a quantity depending entirely on mechanical parameters of the film (s_{13p}, s_{11p}, and s_{12p}). A similar expression holds for e_{31f}: in Eq. 10.7, $Y_{1,p}$ and ν_{12p} are the Young's modulus and the Poisson's ratio of the piezoelectric film.

$$\frac{S_3}{E_3} = d_{33,f} = d_{33} - 2d_{31} \frac{s_{13,p}^E}{s_{11,p}^E + s_{12,p}^E} \tag{10.6}$$

$$\frac{T_1}{E_3} = e_{31,f} = \frac{d_{31}}{s_{11,p}^E + s_{12,p}^E} = d_{31} \frac{Y_{1,p}}{\left(1 - \nu_{12,p}\right)} \tag{10.7}$$

In order to measure the e_{31f} coefficient of a film, several methods can be used. The first is based on the direct piezoelectric effect: it consists in measuring the electrical displacement D_3 induced by the application of an in-plane mechanical strain S_1, by means of the four-point bending technique (4PB) [53]. In practice, this is achieved by cutting the substrate, on which the piezo film is clamped, into an appropriate cantilever shape. The substrate is supposed to have thickness h. Then, the cantilever is placed in a tool that applies a force F by means of four rods, as depicted in the left panel of Fig. 10.35; the distance between the internal rods, L_1, is known. Under the action of the load force, the cantilever bends, and the displacement u_{cant} of its midpoint with respect to the undeflected plane can be measured by means of an interferometer. The in-plane strain S_1 is readily obtained from geometrical considerations, as illustrated in Eq. 10.8. The top and bottom

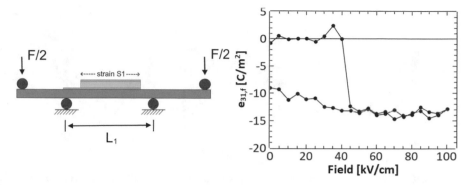

Fig. 10.35 Left: explanatory scheme of four point bending (4PB) technique for the measurement of the e_{31f} coefficient exploiting the direct piezoelectric effect. Right: results of a 4 PB measurement on unpoled PZT in presence of different bias field (courtesy Emmanuel Defay, Luxembourg Institute of Science and Technology)

electrodes are connected to a galvanometer, in order to measure the charge generated on the capacitor. The electrical displacement D_3 induced by the deformation is obtained by measuring the induced charge Q, divided by the electrode area A. Then, e_{31f} is directly derived as shown in Eq. 10.9, in which a useful approximation is displayed under the reasonable hypothesis that $u_{cant} \ll L_1$. A constant electric field can be imposed to the piezoelectric stack during the measurement: thus, the same measurement can be repeated under different bias electric fields, obtaining the results depicted in Fig. 10.35 (right panel). With reference to the same figure, it is possible to see that, as far as the bias field is below the coercive field of 40 kV/cm, the domains are randomly oriented and no net piezoelectric effect is observed. Above the coercive field, a strain-induced charge appears and an e_{31f} value of 14 C/m^2 is measured. Returning to zero bias, the e_{31f} remains high due to the residual polarization, though reduced for the partial loss of domain orientation.

$$S_1 = \frac{h \cdot u_{cant}}{(l_1/2)^2 + u_{cant}^2} \tag{10.8}$$

$$e_{31f} = \frac{Q}{A} \cdot \frac{(l_1/2)^2 + u_{cant}^2}{h \cdot u_{cant} \cdot (1 - v_s)} \sim \frac{Q l_1^2}{4Ah \cdot u_{cant} \cdot (1 - v_s)} \tag{10.9}$$

Another method for obtaining the e_{31f} coefficient of a piezoelectric film is based on the converse effect: this means that the in-plane stress T_1 induced by the application of an electric field E_3 is measured. In principle, the same experimental setup used for 4PB can be exploited, with the difference that u_{cant} must be monitored as a function of the applied bias field E_1, without the application of any mechanical load. As an alternative, the deflection of a clamped-free cantilever can be used,

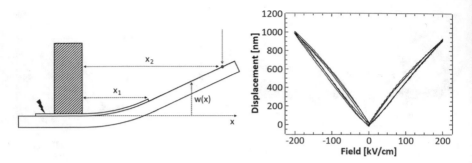

Fig. 10.36 Left: explanative scheme for the measurement of e_{31f} coefficient based on cantilever deflection (converse piezoelectric effect). Right: displacement of a cantilever with respect to applied field. Two unipolar measurements were taken separately, for positive and negative polarization, respectively. Note the small asymmetry in the actuation efficiency with respect to the two polarizations

as shown in Fig. 10.36. The substrate holding the piezo film is again shaped as a cantilever, which then is clamped at one end by means of a proper sample holder. The piezo film extends on the cantilever surface for a length x_1 from the clamping point. The electrical field is applied to the film, and the cantilever bends upward. The profile of the cantilever can be described by a function $w(x)$, where w is the vertical displacement at position x. Usually, the displacement is probed by means of an interferometer or a Doppler vibrometer, at a fixed position x_2. Depending on the geometrical/mechanical parameters of the cantilever and the piezo film, the expression of the in-plane stress can be obtained as shown in Eq. 10.10 [54]:

$$T_1 = -\frac{1}{3} \cdot \frac{Y_s \cdot h_s^2}{(1 - v_s) c_f} \cdot \frac{u(x_2)}{x_1 [2x_2 - x_1]} \cdot \frac{1}{h_p} \tag{10.10}$$

Where h_s and h_p are the thickness of the substrate and the film, Y_s and v_s are the Young's modulus and Poisson's ratio of the substrate, and c_f is the ratio between the actuator width (i.e., the piezoelectric capacitor) and the cantilever width. An example of tip displacement versus applied field is shown in Fig. 10.36 (right panel). The e_{31f} coefficient is readily obtained by dividing the stress T_1 by the applied field E_1.

Both the methods described above are simple in principle, but they have the disadvantage of being destructive for the substrate (in the MEMS framework, the substrate is the silicon wafer). A third, nondestructive approach for the measurement of the e_{31f} coefficient exploits the removal of the hypothesis of the substrate being perfectly rigid, i.e., nondeformable. If the substrate can be locally deformed by the action of the piezoelectric film, the expression for d_{33f} must be modified as shown in Eq. 10.12, where v_s and Y_s are the Poisson's ratio and Young's modulus of the substrate. The key point is that the resulting $d_{33f,meas}$ includes an additive contribution describing the substrate deformation, which is proportional to the film

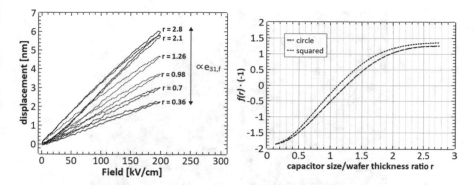

Fig. 10.37 Left: displacement-field characteristics of clamped squared PZT thin film capacitors with different size, measured by Double Beam Laser Interferometry (DBLI). The thickness of the wafer is 715 μm; the side length of the capacitors ranges from 250 μm to 2000 μm. The value of r (size to wafer thickness ratio) is indicated for each case. Right: plot of the $f(r)$ function used in Equations 10.11 and 10.12, for squared and circular capacitors

e_{31f}, through a peculiar function f that depends uniquely on the ratio r between the piezoelectric capacitor width and the substrate thickness [55, 56]. This is demonstrated in Fig. 10.37, left panel, in which the measured z-axis deformation of a clamped lead zirconate titanate (PZT) film is shown, for different size of the measured capacitor. The plot puts into evidence how the measured displacement increases for the same applied field as the capacitor size increases from a minimum size of 250 μm to a maximum size of 2000 μm on a 715-μm thick wafer substrate, until it tends to saturate for a size/substrate thickness ratio r between 2.1 and 2.8. The difference in the measured displacement on the different capacitors is due to the deformation of the silicon substrate, operated by the PZT thin film through the e_{31f} (for reference, see eq. 10.11) coefficient, as expressed by the second term in Eq. 10.12. The $f(r)$ function contains only geometrical information; it was obtained by Sivaramakrishnan et al. by means of numerical simulation for square and rounded capacitors on silicon substrates, and it is plotted in the right panel of Fig. 10.37.

$$d_{33f,meas} = d_{33,f} + f(r) \cdot s_{13,s} \cdot e_{31,f} \tag{10.11}$$

$$d_{33f,meas} = d_{33,f} - f(r) \cdot \frac{v_s}{Y_s} \cdot e_{31,f} \tag{10.12}$$

From the framework explained above, a simple method to extract the e_{31f} parameter consists in measuring $d_{33f,meas}$ for several capacitors with different r ratio. Then, if $d_{33f,meas}$ is plotted versus the value of $f(r)$ for each capacitor, a linear trend

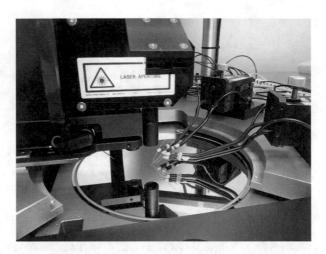

Fig. 10.38 Image of a Double Beam Laser Interferometer for an 8-inch wafer. The optical system for the delivery of the laser beam to the front surface of the wafer is well visible, together with the tips to apply the voltage on the devices

is expected, whose intercept at $f(r) = 0$ is the true d_{33f} of the piezoelectric film and whose slope is directly proportional to e_{31f}. Since the two constants in Equation 10.12 (the substrate Young's modulus Y_s and the substrate Poisson's ratio ν_s) are known, the value of e_{31f} is readily obtained.

As evident from Fig. 10.37, the displacement of a clamped piezoelectric field is in the order to few nanometers. To perform such fine measurements, a Double Beam Laser Interferometer (DBLI) can be used (see Fig. 10.38) [57]. In this equipment, a laser beam is split in two parts (one for reference and one for measurement). An interference pattern is obtained by the superposition of the reference and the measuring laser beams, which has been reflected in sequence from the top surface of the piezo capacitor and the bottom surface of the wafer. This allows eliminating the spurious contribution of undesired out-of-plane wafer displacements, thus being sensitive only to the thickness variations of the film-wafer system. The variation of thickness due to the film actuation changes the length of the optical path travelled by the measuring beam, thus causing a change in the interference pattern that is detected by an optical sensor (usually a photodiode). On the other hand, in case of displacement of the substrate, the phase shift at the reflection on the front side is exactly compensated by an opposite phase shift at the reflection on the back side. As a conclusion, the use of DBLI measurements together with Eq. 10.12 represents a powerful and nondestructive tool for the quantitative evaluation of e_{31f} in piezoelectric films.

Another important piece of information that must be gained for a piezoelectric thin film regards the reliability of the material [58]. This means how robust is the film with respect to device operation, and how long it can be put at work

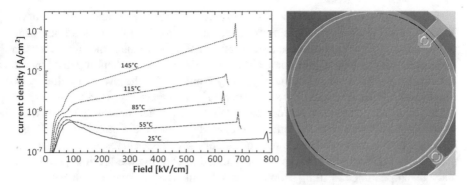

Fig. 10.39 Left: current-field curves of passivated, undoped PZT thin film capacitors for different temperatures, measured with a B1500A device analyzer. The voltage ramp used is of the staircase type, having steps of 1 V with duration of 0.5 seconds (2 V/s resulting speed). Note the increasing trend of the current with temperature. The small current peaks at the end of each curve are due to the mechanical breakdown (tiling) of the PZT thin film. Right: optical image of a tiled PZT capacitor. Cracks are visible on the entire capacitor surface

without degradation of its performance [59]. For piezoelectric films, two main failure mechanisms are the electrical and mechanical breakdown. In the electrical breakdown, a discharge between the top and bottom electrode is observed while the film is actuated, leading to short circuit between the electrodes or even loss integrity in large portions of the film; in the mechanical breakdown, the physical continuity of the film is broken by means of cracks propagating into the material, even in absence of catastrophic electrical failure. To investigate these aspects, a useful characterization is the measurement of current versus voltage curves, for different temperatures. Figure 10.39 shows the results of this measurement taken on an unpoled, 2-μm-thick undoped PZT film, topped with a TiW electrode. The current peak at low voltage is related to the polarization of the unpoled film. The current diminishes versus the applied field at room temperature, reaching a plateau. By raising the temperature, the current increases rapidly, reaching a factor of more than 100x at 145° C with respect to the conduction at room temperature. A higher current flow is experimentally correlated with a shorter lifetime, so a fast degradation of the material with respect to the temperature can be expected in this case. There is not a universal law linking the current conduction to the film lifetime, and the relationship between the two must be verified from case to case. Nevertheless, a comparative prediction can be done by looking at these characteristics for different films. As a general operative rule, these measurements should be done on passivated capacitors, i.e., actuators covered with a protective insulating thin layer (alumina, CVD silicon oxides) after opening of vias to the electrodes and fabrication of metal leads for contacting. This prevents spurious effects affecting the film robustness, greatly reduces the variability of results, and allows obtaining clear trends with respect to the film properties.

In Fig. 10.39, a current peak is observed at the end of each current-voltage curve. Though being not electrically catastrophic, this is the signature of the interesting phenomenon of mechanical breakdown [60], or tiling, in which the film cracks in many little pieces (tiles), still connected to the electrodes and kept together by the electrodes and the passivation, but physically separated for the entire film thickness. The aspect of a tiled capacitor is shown in the right panel of Fig. 10.39. Usually, the tiling phenomenon appears at a critical value of the applied field. Interestingly, the onset of tiling depends on the temperature, as clear from the plot in Fig. 10.39. For this reason, the characterization of current versus voltage curves is very useful in order to design properly the experiments aimed at deriving the lifetime prediction, since the appearance of tiling should be avoided in order to have reliable results.

The extraction of device lifetime is a tricky task, especially for MEMS, because not only the behavior of a clamped film should be investigated, but rather the real case of the functioning device under realistic operation, to consider all the possible factors affecting the actuating film including mechanical stresses and fatigue. Nevertheless, Time-Dependent Dielectric Breakdown measurements on clamped or released structures can give some hints about the intrinsic robustness of a piezoelectric film. In these experiments, a constant voltage is applied for a prolonged time, at a given temperature (Constant Voltage Stress, CVS). A matrix of voltage/temperature conditions is usually performed, with several DUTs, enough to obtain a good statistic of the time to failure in each stressing condition. The time to failure detection is based on the application of a threshold to the conduction current registered during the experiment. In most cases, a Weibull analysis [61] (see Fig. 10.40) is applied to extract the characteristic lifetime and shape parameter for each condition; then, a model for the temperature and the voltage acceleration is assumed, to find the activation energies for each process and the related acceleration factors. With these acceleration factors available, a prediction of the lifetime in operating conditions (temperature and voltage) can be done.

The reliability of piezoelectric thin films is a very complex subject, in which many interactions can be present. Here, an example is given about the use of two different sol–gel solutions for the deposition of PZT. One is undoped, while the other is doped with niobium. As can be seen from Fig. 10.41, the current conduction at high temperature is greatly reduced in the case of doped PZT, and the tiling occurrence is avoided for applied fields up to 1000 kV/cm. These considerations suggest that the lifetime should be longer: this was indeed confirmed by TDDB experiments.

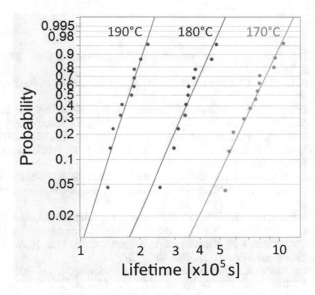

Fig. 10.40 Example of Weibull probability analysis of CVS experiments carried out at different temperatures. Each dot is a capacitor failed at a given time, indicated on the x-axis. All capacitors were maintained at the same bias voltage during the stress test. The characteristic lifetime decreases with increasing temperature. A thermal activation energy can be extracted from this analysis. Courtesy of I. Pedaci (STMicroelectronics - Arzano)

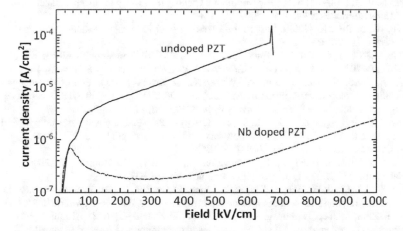

Fig. 10.41 Comparison of current-voltage curves, measured at 145° C, of undoped and Nb-doped PZT. The voltage is applied to the bottom electrode, with the top electrode grounded. In the doped material, the current is lower for about two orders of magnitude. Note the occurrence of tiling at 700 kV/cm in undoped PZT, whereas Nb-doped PZT is much more robust with respect to mechanical breakdown

References

1. Defay, E. (2011). *Integration of ferroelectric and piezoelectric thin films*. ISTE Ltd and John Wiley & Sons, Inc.
2. Setter, N. (2005). *ELECTROCERAMIC-BASED MEMS fabrication-technology and applications*. Springer.
3. Cain, M. G. (2014). *Characterisation of ferroelectric bulk materials and thin films*. Springer.
4. Ko, S. W., Zhu, W., Fragkiadakis, C., Borman, T., Wang, K., Mardilovich, P., & Trolier-McKinstry, S. (2019). *Journal of the American Ceramic Society, 102*, 1211–1217.
5. Kwok, K. W., Tsang, R. C. W., Chan, H. L. W., & Choy, L. C. (2004). *Journal of Applied Physics, 95*, 1372–1376.
6. EU DIRECTIVE 65/2011 as amended by EU DELEGATED DIRECTIVE 863/201.
7. Okamura, S., ABE, N., Otani, Y., & Shiosaki, T. (2003). *Integrated Ferroelectrics, 52*, 127–136.
8. Millon, C., Malhaire, C., & Barbier, D. (2004). *Sensors and Actuators A, 113*, 376–381.
9. Guo, Q., Cao, G. Z., & Shen, I. Y. (2013). *Journal of Vibration and Acoustics, 135*, 1–9.
10. Senkevich, S. V., Pronin, I. P., Kaptelov, E. Y., Sergeeva, O. N., Il'in, N. A., & Pronin, V. P. (2013). *Technical Physics Letters, 39*, 400–403.
11. Es-Souni, M., & Piorra, A. (2001). *Materials Research Bulletin, 36*, 2563–2575.
12. Zhu, W., et al. (2019). *Journal of the American Chemical Society, 102*, 1734–1740.
13. Bassiri-Gharb, N., Bastani, Y., & Bernal, A. (2014). *Chemical Society Reviews, 43*, 2125–2140.
14. EU Dir. 2011/65/EU, https://ec.europa.eu/environment/waste/rohs_eee/adaptation_en.htm
15. ELV Dir. 2000/53/EC, https://ec.europa.eu/environment/waste/elv/legislation_en.htm
16. Bell, A. J., & Deubzer, O. (2018). Lead-free piezoelectrics—The environmental and regulatory issues. *MRS Bulletin, 43*, 581–586.
17. Lead-free piezoceramics – Ehere to move on? *Journal of Materiomics, 2* (2016), 1–24.
18. Panda, P. K. (2009). Review: Environmental friendly lead-free piezoelectric materials. *Journal of Materials Science, 44*, 5049–5062.
19. Liu, W., & Ren, X. (2009). Large Piezoelectric Effect in Pb-free Ceramics. *Physical Review Letters, 103*, 257602.
20. Bao, H., Zhou, C., Xue, D., Gao, J., & Ren, X. (2010). A modified lead-free piezoelectric BZT–xBCT system with higher TC. *Journal of Physics D: Applied Physics, 43*, 465401.
21. Su, S., Zuo, R., Lu, S., Xu, Z., Wang, X., & Li, L. (2011). Poling dependence and stability of piezoelectric properties of Ba(Zr0.2Ti0.8)O3-(Ba0.7Ca0.3)TiO3 ceramics with huge piezoelectric coefficients. *Current Applied Physics, 11*, S120–S123.
22. Liu, X., Jiang, Z., & Han, J. (2010). Preparation and characterization of (Ba0.88Ca0.12)(Zr0.12Ti0.88)O3 powders and ceramics produced by Sol-Gel process. *Advanced Materials Research, 148–149*, 1062–1066.
23. Saito, Y., Takao, H., Tani, T., Nonoyama, T., Takatori, K., Homma, T., Nagaya, T., & Nakamura, M. (2004). Lead-free piezoceramics. *Nature*, 432.
24. Xu, K., Li, J., Lv, X., Wu, J., Zhang, X., Xiao, D., & Zhu, J. (2016). Superior piezoelectric properties in potassium-sodium niobate lead-free ceramics. *Advanced Materials, 28*, 8519.
25. Li, P., Zhai, J., Shen, B., Zhang, S., Li, X., Zhu, F., & Zhang, X. (2018). Ultrahigh piezoelectric properties in textured (K,Na)NbO3 -based lead-free ceramics. *Advanced Materials, 30*.
26. Coffman, P. R., Barlingay, C. K., Gupta, A., & Dey, S. K. (1996). *Journal of Sol-Gel Science and Technology, 6*, 83–106.
27. Cao, J., et al. (2006). *Journal of Applied Physics, 99*, 1–7.
28. Muralt, P., et al. (1998). *Journal of Applied Physics, 83*, 3835–3841.
29. Es-Souni, M., & Piorra, A. 36, 2563–2575, (2001).

30. Kotova, N. M., Vorotilov, K. A., Seregin, D. S.,& Sigov, A. S. 50, 661–666, (2014).
31. Gong, W., Li, J., Chu, X., Gui, Z., & Li, L. (2004). *Journal of Applied Physics, 96*, 590–595.
32. San-Yuan, C., & I-Wei, C. (1998). *Journal of the American Ceramic Society, 81*, 97–105.
33. Zhao, J. S., et al. (2006). *Journal of the Electrochemical Society, 153*, F81–F86.
34. S. Trolier-McKinstry, Journal of Electroceramics, P. Muralt, 12, 7–17, (2004).
35. Kalpat, S., et al. (2001). *Japanese Journal of Applied Physics, 40*, 713–771.
36. Du, X., Belegundu, U., & Uchino, K. (1997). *Japanese Journal of Applied Physics, 36*, 5580–5587.
37. Wang, L., Yu, J., Wang, Y., & Gao, J., 19, 1191–1196, (2008).
38. *Journal of Applied Physics 96*, 590 (2004).
39. Cao, J.-L., et al. (2006). Structural investigations of Pt/ TiOx electrode stacks for ferroelectric thin film devices. *Journal of Applied Physics, 99*(11), 114107.
40. Okamura, S., et al. (2003). Influence of Pt/TiO2 bottom electrodes on the properties of ferroelectric Pb (Zr, Ti) O3 thin films. *Integrated Ferroelectrics, 52*(1), 127–136.
41. Wasa, K. (2012). *Handbook of sputtering technology* (2nd ed), pp. 41–75.
42. An, T.-H., Park, J.-Y., Yeom, G.-Y., Chang, E.-G., & Kim, C.-I. (2000). Effects of BCl3 addition on Ar/Cl2 gas in inductively coupled plasmas for lead zirconate titanate etching. *Journal of Vacuum Science and Technology A, 18*(1373–1376).
43. Shibano, T., Nakamura, K., Takenaga, T., & Ono, K. (1999). Platinum etching in Ar/Cl2 plasmas with a photoresist mask. *Journal of Vacuum Science & Technology A, 17*(799).
44. Lide, D. R. (Ed.). (1991). *Handbook of chemistry and physics* (72nd ed.). Chemical Rubber.
45. Wuua, D.-S., Kuoa, N.-H., Liaoa, F.-C., Horngb, R.-H., & Leec, M.-K. (2001). Etching of platinum thin films in an inductively coupled plasma. *Applied Surface Science, 169–170*(15), 638–643.
46. Hariharan, R., & Raja, R. (2017). Investigation on micro structural, mechanical and tribological properties of aluminum nitride (AlN) coating deposited by RF magnetron sputtering. *International Journal of Latest Trends in Engineering and Technology.*
47. Teshigahara, A., Hashimoto, K., & Akiyama, M.. (2012). Scandium aluminum nitride: Highly piezoelectric thin film for RF SAW devices in mutli GHz range. *IEEE International Ultrasonics Symposium.*
48. Sawyer, C. B., & Tower, C. H. (Feb. 1930). Rochelle salt as a dielectric. *Physics Review, 35*(3), 269–273.
49. Damjanovic, D. (2005). Hysteresis in piezoelectric and ferroelectric materials. In I. Mayergoyz & G. Bertotti (Eds.), *The science of hysteresis* (Vol. 3). Elsevier.
50. Bolten, D., Boettger, U., & Waser, R. (2003). *Journal of Applied Physics, 93*(3), 1.
51. Bolten, D., Lohse, O., Grossmann, M., & Waser, R. (1999). Reversible and irreversible domain wall contributions to the polarization in ferroelectric thin films. *Ferroelectrics, 221*(1), 251–257.
52. Defay, E. (2011). *Integration of ferroelectric and piezoelectric thin films*, ISBN 978-1-84821-239-8.
53. Prume, K., Muralt, P., Calame, F., Schmitz-Kempen, T., & Tiedke, S. (2007). *IEEE Transactions on Ultrasonics, Ferroelectrics, and Frequency Control, 54*(1).
54. Mazzalai, A. et al. (2013). Simultaneous piezoelectric and ferroelectric characterization of thin films for MEMS actuators, Joint UFFC, EFTF and PFM Symposium.
55. Sivaramakrishnan, S., Mardilovich, P., Mason, A., Roelofs, A., Schmitz-Kempen, T., & Tiedke, S. (2013). *Applied Physics Letters, 103*, 132904.
56. Sivaramakrishnan, S., Mardilovich, P., Schmitz-Kempen, T., & Tiedke, S. (2018). *Journal of Applied Physics, 123*, 014103.
57. Gerber, P., Roelofs, A., Lohse, O., Kügeler, C., Tiedke, S., Böttger, U., & Waser, R. (2003). Short-time piezoelectric measurements in ferroelectric thin films using a double-beam laser interferometer. *The Review of Scientific Instruments, 74*, 2613–2615.

58. Polcawich, R. G., Feng, C. N., Kurtz, S., Perini, S., Moses, P. J., & Trolier-McKinstry, S. (2000). AC and DC electrical stress reliability of piezoelectric lead zirconate titanate (PZT) thin films. *IMAPS, 23*, 85–91.
59. Daniel Chen, H., Udayakumar, K. R., Li, K. K., Gaskey, C. J., & Cross, L. E. (1997). Dielectric breakdown strength in sol-gel derived PZT thick films. *Integrated Ferroelectrics, 15*, 89–98.
60. Zhu, W., Borman, T., DeCesaris, K., Truong, B., Lieu, M. M., Ko, S. W., Mardilovich, P., & Trolier-McKinstry, S. (2019). *Journal of the American Ceramic Society, 102*, 1734–1740.
61. Jung In Yang, Dissertation, Pennsylvania State University 2016.

Chapter 11
Wafer-to-Wafer Bonding

Laura Oggioni, Matteo Garavaglia, and Luca Seghizzi

11.1 Introduction

Wafer-to-wafer bonding is a unique and peculiar process step in front-end manufacturing. It is a key step in MEMS process flow enabling the manufacturing of complex devices. Without this technique which allows the realization of sensors and actuators that are present at consumer and industrial level, many features in our smart devices would not exist, and the IoT (Internet of Things) world would not be possible.

It is widely known that any two surfaces can be joined together by a suitable adhesive material. The same idea applies to wafer to wafer (hereafter abbreviated "W2W") bonding, and this is exactly what makes it such an interesting and dynamic process: as device design and manufacturing needs evolve, bonding materials, techniques, and equipment features change and improve. This makes process engineering work very interesting and challenging, since new characterizations, setups, and issues arise at a rapid pace.

In the semiconductor industry, W2W bonding has been first used by SOI wafer manufacturers. Several proprietary methods have been developed and brought to the substrate market since the end of the last century by IBM, SOITEC, SiGen, and many others.

Starting from the 1990s, a new wave of MEMS devices developed from the academic field toward high-volume manufacturing, and the consequence is the explosion of new bonding methods such as glassfrit, adhesive, metal-to-metal, eutectic, and several tool makers which came to the scene enabling an industrial approach [1]. Those have been challenging years which brought the invention

L. Oggioni (✉) · M. Garavaglia · L. Seghizzi
ST Microelectronics, Analog MEMS and Sensors Group, MEMS Technology and Design R&D,
Agrate Brianza, Monza Brianza, Italy
e-mail: laura.oggioni@st.com; matteo.garavaglia@st.com; luca.seghizzi@st.com

© Springer Nature Switzerland AG 2022
B. Vigna et al. (eds.), *Silicon Sensors and Actuators*,
https://doi.org/10.1007/978-3-030-80135-9_11

and development of accelerometer, gyroscopes, bolometers, pressure sensors, and medical devices which are nowadays in many applications.

The further development of bonding techniques as well as improved tool capabilities eventually opened the field to a superior level of highly integrated 3D devices (RAM, BSI) [2].

In the last fifteen years, the need of more complex process solution for 3D integration and of thinner wafers was the driving force for temporary bonding and debonding development. A major revolution considered that for more than a century only permanent bonding existed. These new techniques have allowed the development of micromirrors, microspeakers, and microlenses which are now getting on the market for intelligent and autonomous driving, for the augmented and virtual reality, and for the smart glasses' markets.

11.2   Bonding Classification

All the permanent bonding methods can be classified in two major categories:

1. Bonding without intermediate material
2. Bonding with intermediate material

The direct bonding without an intermediate layer includes any type of bonding where the surfaces to be joined are put directly into contact. If any material is used in between the wafers, the bonding method falls into the second category. Starting from this general classification, it is common practice to define the so-called bonding tree (Fig. 11.1).

11.2.1   Direct Wafer Bonding

The upper branch comprises the most common direct wafer bonding technologies. Fusion bonding is based on the intimate contact between two very smooth surfaces (~0.2 nm RMS). At this level of roughness, a bonding wave is initiated by a tip, and it propagates through the whole wafer surface. Van der Waals or weak hydrogen bonds ensure that the wafer pair is held in place. To enforce the atomic interaction, usually a high temperature (~1000–1100°C), thermal treatment is done to complete the dominating chemical reaction [3]. Because surface properties strongly affect the bonding strength, a recent approach involves the use of a plasma for the chemical activation of the surfaces to be bonded. By this method, the anneal temperature may be greatly reduced in the range of 300–400°C, making this technique very useful for 3D integration of CMOS devices [4].

Anodic bonding is a legacy W2W method that has been first described in 1969 by Pomeranz and Wallis [5]. Anodic bonding uses Na^+ mobility in sodium-rich glasses under an electric field to create a secure interface between a glass and a

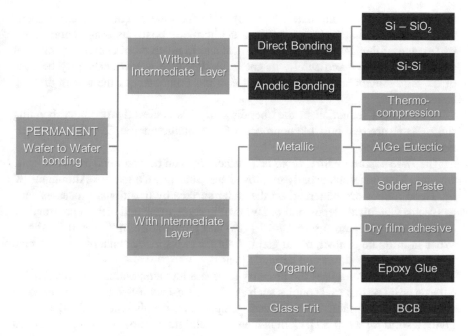

Fig. 11.1 W2W bonding tree

silicon substrate. The wafer pair is placed under an electric field and heated: Na+ ions move from the surface toward the negative side of the field, and the oxygen reacts with silicon forming a strong covalent bond. Because of the presence of Na, this technique is not CMOS compatible, and it is mainly used to seal passive devices such as bulk micromachined pressure sensors or fluidic devices.

For all the direct fusion bonding methods, the surface cleanliness is paramount, and every small defect or previous process residual greatly jeopardizes the overall yield, inducing voids or bubbles at the bonding interface.

11.2.2 Intermediate Layer Wafer Bonding

As per definition, the intermediate layer wafer bonding requires a gluing agent between the surfaces. The most widely used material in MEMS applications is glassfrit.

Glassfrit is a suspension of glass particles in an organic binder [6], which enables glassfrit to be deposited and defined on wafers by screen printing technique. After the dispense, the paste is annealed in an air recirculated oven where solvents are first outgassed, then the organic binder is burnt, and eventually the paste is glazed to reach a glass-like status. The printed wafer, normally the cap, is coupled to a device wafer, heated to 430–440 °C for a few minutes applying force to ensure a good thermal coupling and intimate contact between the wafers. The glass partly melts and reflows forming a strong, reliable, and hermetic bond. Its usage comprises a wide range of devices in the consumer field and in much more demanding markets such as automotive and safety. Its main disadvantage is low scalability because of the high silicon area consumption due to the limitations of the screen-printing technique.

A consistent reduction in the "bonding area" is achieved with metal bonding rings. There are two possible approaches for "metallic bonding": thermocompression and eutectic.

The *Thermocompression Bond* is a method that can be described as a solid-state welding. The materials usually employed are Gold [7], Copper, or Aluminum [8, 9]. These metals are patterned on the wafer surface by traditional processes such as Electro Chemical Deposition or PVD with proper mask and etch steps. Bonding requires high pressure (up to 10–15 MPa) and temperature in 300–450 °C range. When in intimate contact, metal grains from the two surfaces undergo a solid-state reaction forming a stable and hermetic joint.

The *Eutectic Bond* uses the property to have a low temperature (below 450 °C) eutectic point of material couples such as Al-Ge, Au-In, Cu-SN, and Au-Si to name only the most common; at the eutectic temperature, metals react to form an alloy giving a solid joint [10, 11]. Complementary metal rings whose thickness is defined by the atomic % at the eutectic point are brought into contact and heated just above the eutectic temperature. A solid-state reaction occurs with the formation of a near-liquid phase that solidifies upon controlled cooling leaving a solid joint.

Quite similar is the *Solder Bond* with Au-Sn as the most common pair [9]. The bonding proceeds through the formation of a liquid phase during the heating of the wafer pair, the bump reflows, and upon cooling, as previously described, a stronger intermetallic phase ensures the joint between the surfaces. Both eutectic and solder bonds are quite sensitive to the presence of native oxide, and some attention should be paid at surface preparation procedures.

The term *Adhesive Bonding* includes in its perimeter a wide range of materials: thermosetting polymers such as benzocyclobutene (BCB), a thermosetting polymer, Epoxy Glue, SU-8, a commonly used epoxy-based negative photoresist, and positive or negative tone polyimides [12]. A standard sequence for adhesive bonding includes surface preparation by cleaning and drying, coating with an adhesion promoter, if required, coupling and bonding by means of heating and pressure, a final cure, if necessary, to reach the required polymerization stage of the adhesive material (see the following figure where polymerization details are given for BCB material); curing step can be either a thermal anneal or an irradiation with UV or Visible light. A more comprehensive study of adhesive bonding materials can be found in [12]. The advantage of Adhesive bonding is the low temperature and the forgiveness toward slight surface defectivity, since the materials are a few microns thick and can compensate for any surface unevenness. The major disadvantage is that adhesive bonding is not hermetic, and a long-term leak cannot be tolerated in some applications.

11.3   Temporary Bonding and Debonding Overview

Recently, the needs for much more complex 3D process architecture paved the way to the development of multiple methods for temporary bonding and debonding. The first needs arose when it was required to reduce CMOS wafer thickness for thinner and thinner die assembly. The first widely used industrial approach was the Wafer Support System proposed by 3M around 2005 followed in later years by a wide range of options.

Typically, these methods join a carrier or support wafer (Silicon or Glass) to a "device" wafer, which is then processed with traditional IC steps such as grinding, polishing, mask etch, and deposition. The postprocess temperatures, the nature of the chemicals to be used, and the workability under high vacuum conditions are the driving forces in selecting among the possible solutions. In analogy with the Bonding Tree, a Temporary Bonding tree can be defined as well (Fig. 11.2).

Two major classes of temporary bonding materials can be identified: tape based, and polymer based.

Tape-based options use an adhesive tape to join the carrier wafer to the device wafer. The postprocessing is not usually performed in harsh condition because of the poor vacuum performances of the most common tapes and the low chemical compatibility. The release is done by heating the pair or by UV irradiation to lower the adhesion force of the tape, and the carrier is then easily removed by lifting. Nitto and Sekisui [13] offer a wide range of solutions.

Polymer-based options comprise a wide variety of materials and debonding techniques; every method is based on an organic polymer as adhesive material to

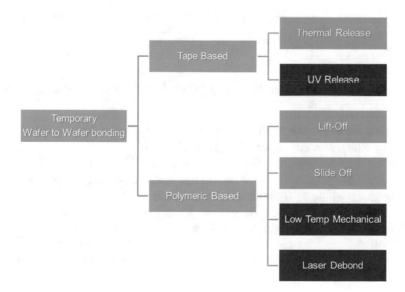

Fig. 11.2 Temporary W2W bonding tree

• Slide off

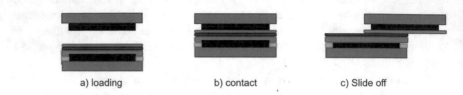

a) loading   b) contact   c) Slide off

Fig. 11.3 Schematic sequence of slide off debonding

• Lift off

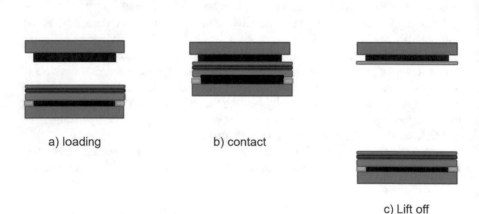

a) loading   b) contact

c) Lift off

Fig. 11.4 Schematic sequence of lift off debonding

join carrier and device wafers. What really makes a difference is the release method. There are several options:

1. Heating the wafer pair to soften the polymer and proceed with a slide off of the carrier with respect to the device (Fig. 11.3).
2. Heating the wafer pair as before and proceed with a lift off (Fig. 11.4).
3. Break the bonding layer at the wafer edge with a blade and promote a fracture through the whole surface (mechanical debond).
4. Irradiate with a UV laser the glue to destroy a thin release layer previously deposited on the glass carrier and proceed with lift off (UV debond).
5. Lift off debond after laser ablation through the carrier wafer. This is the latest developed debonding technique, and it offers the advantage of being independent

on device design and features since no or very low debonding force is applied; it is carried out at room temperature.

Many commercial solutions are available on the market, companies like BSI, Nissan Chemical, and Fujifilm offer their own proprietary material, and it is up to the end user to select the most suitable method and integrate it in the device flow.

11.4 Bonding Characterization

A challenge of wafer-to-wafer bonding is to properly characterize bonding joint. Currently, there are numerous methods available, both destructive and nondestructive. The former includes all the inspections that look at the bonding interface, such as Scanning Acoustic Microscopy (SAM), Infrared Microscopy, or Visible Light Microscopy; these techniques are usually used for in-line process control. The latter includes lab methods such as cross-sections of the bonding interface using scanning electron microscopy (SEM), transmission electron microscopy (TEM), plus other methods such as Shear, Pull and Chevron tests, which measure the strength of the bonding joint; and Maszara's test, specifically dedicated to bonding interface energy evaluation [9, 14–16].

11.4.1 SAM and IR Inspection

Infra-Red (IR) microscopy and scanning acoustic microscopy (SAM) are in-line inspections to screen bonding quality at wafer level. Which method is applied depends on the bonding technique used: IR inspection is the most suitable for AlGe eutectic bonding and certain types of adhesive bonding, while SAM inspection is widely employed for direct, glassfrit, metallic and polymer bonding.

SAM is undoubtedly the most common and recognized inspection method to assess bonding quality [15].

Scanning acoustic microscopy works by directing focused ultra-sound from a transducer to a sample, coupled with the sound source via water. Sound hitting the object is either scattered, absorbed, reflected (scattered at 180°) or transmitted (scattered at 0°). Reflected sound is measured by the emitting transducer. As the transducer scans the sample surface in the xy plane, an image is thus formed by plotting the reflected beam intensity (Figs. 11.5 and 11.6). The "time of flight" of the pulse is defined as the time taken for it to be emitted by an acoustic source, reflected by the sample and received by the detector. The time of flight can be used to determine the distance of the inhomogeneity from the source given knowledge of the speed through the medium.

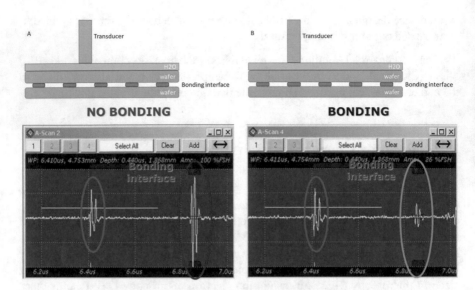

Fig. 11.5 Examples of SAM inspections. In figure **a**, the transducer is positioned over nonbonded area. In figure **b**, the transducer is positioned over well-bonded area

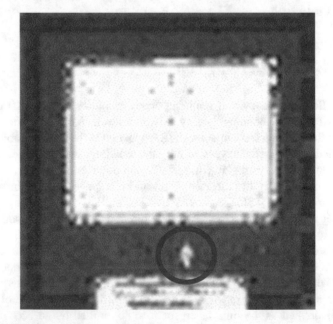

Fig. 11.6 Typical SAM image of a MEMS device

The resolution of the image is limited either by the physical scanning resolution or the width of the sound beam (which in turn is determined by the frequency of the sound): reasonable image resolution for bonding in-line SAM inspection

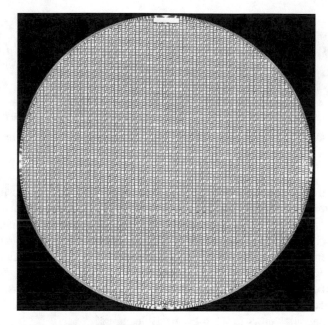

Fig. 11.7 Typical SAM image of a full wafer

ranges between 20 and 50 μm/pixel. Another limitation of this technique is the long inspection time: equipment suppliers are trying to improve wafer throughput by either having multiple acquisition channels or multiple wafers inspected at the same time. Also, a combination of the two is possible (Fig. 11.7).

SAM inspection images are processed to identify defects according to defined criteria. In-house image analysis software has been developed to automatically find bad pixels, i.e., bonding defects. Dice identified as "no good" are automatically rejected at electrical wafer sort (EWS).

IR inspection is the recommended method to check bonding quality of AlGe-bonded devices. In this case, short wavelength infrared imaging systems are used to perform quality inspection of semiconductor wafers. Silicon is transparent to IR light. Front-side or back-side illumination of the silicon-based devices with infrared light enables imaging of wafer alignment marks, microcracks, and bonding joints.

11.4.2 SEM and TEM Inspections

Scanning electron microscope (SEM) is a type of electron microscope that produces images of a sample by scanning the surface with a focused beam of electrons. The electrons interact with atoms in the sample, producing various signals that contain information about the surface topography and composition of the sample.

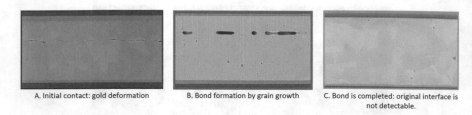

A. Initial contact: gold deformation | B. Bond formation by grain growth | C. Bond is completed: original interface is not detectable.

Fig. 11.8 Gold-gold thermocompression bonding: characterization by means of SEM cross-section

The electron beam is scanned in a raster scan pattern, and the position of the beam is combined with the intensity of the detected signal to produce an image. The signals used by an SEM to produce an image result from interactions of the electron beam with atoms at various depths within the sample. Various types of signals are produced including secondary electrons (SE), reflected or back-scattered electrons (BSE), characteristic X-rays and light (cathodoluminescence) (CL), absorbed current (specimen current), and transmitted electrons. Secondary electron detectors are standard equipment in all SEMs, but it is rare for a single machine to have detectors for all other possible signals.

Secondary electrons have very low energies on the order of 50 eV, which limit their mean free path in solid matter. Consequently, SEs can only escape from the top few nanometers of the surface of a sample. The signal from secondary electrons tends to be highly localized at the point of impact of the primary electron beam, making it possible to collect images of the sample surface with a resolution of below 1 nm. Back-scattered electrons (BSE) are beam electrons that are reflected from the sample by elastic scattering. Since they have much higher energy than SEs, they emerge from deeper locations within the specimen and, consequently, the resolution of BSE images is less than SE images (Fig. 11.8).

Transmission electron microscope (TEM) is an instrument where a beam of electrons is transmitted through the sample to form an image. Samples are usually less than 100 nm thick. An image is formed from the interaction of the electrons with the specimen as the beam is transmitted through it. The image is then magnified and focused onto an imaging device.

Contrast is produced, called "image contrast mechanisms". Contrast can arise from position-to-position differences in the thickness or density ("mass-thickness contrast"), atomic number ("Z contrast", referring to the common abbreviation Z for atomic number), crystal structure, or orientation ("crystallographic contrast" or "diffraction contrast"). Each mechanism gives a different kind of information, depending not only on the contrast mechanism but on how the microscope is used—the settings of lenses, apertures, and detectors. Therefore, TEM analysis returns an extraordinary variety of nanometer- and atomic-resolution information, in ideal cases revealing not only where all the atoms are but what kinds of atoms they are and how they are bonded to each other. For this reason, TEM is regarded as an essential tool for nanoscience in both biological and materials fields.

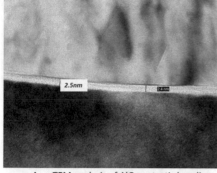

A.   TEM analysis of AlGe eutectic bonding

B.   TEM oxygen map to highlight oxide layer at bonding interface

C.   TEM analysis of AlGe eutectic bonding

D.   TEM oxygen map to highlight oxide layer at bonding interface

Fig. 11.9 TEM analysis of: (**a**) AlGe eutectic bonding, (**b**) Oxide layer at bonding interface, (**c**) AlGe eutectic bonding, (**d**) Oxide layer at bonding interface

A good example of TEM analysis applied to wafer-to-wafer bonding characterization is aluminum germanium eutectic bonding: an oxide layer at the bonding interface is detected, whose thickness depends on surface preparation of both germanium and aluminum strips prior to bonding. If oxide thickness is greater than about 5 nm, then no metal intermixing occurs (Fig. 11.9).

Enlarged TEM analysis of AlGe bonding with about 2.5 nm of oxide at the interface (see oxygen map to the left): there is aluminum and germanium intermixing as shown in the Ge and Al map (center and right) (Fig. 11.10).

Shear test method, pull, chevron tests, and Mazdara method give a good understanding of the bonding material and the bonding strength achieved. They are usually done as part of development and characterization phase to determine the integrity of materials and manufacturing procedures, and to evaluate the overall performance of the bonding, as well as to compare various bonding technologies with each other.

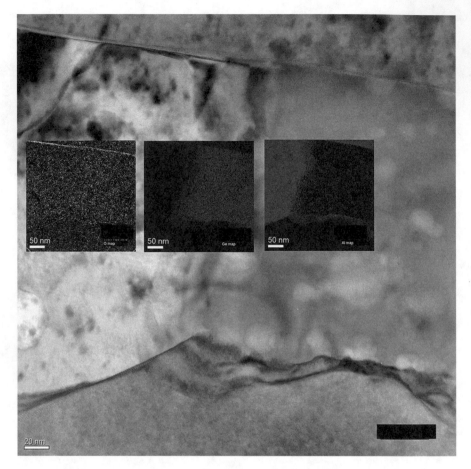

Fig. 11.10 Oxide at the interface is 2.5 nm thick, therefore good Germanium-Aluminum inter-mixing occurs

Measuring the applied force at which failure occurs and inspecting the visual appearance of the broken joint gives insight on the failure mechanism and best process conditions.

11.4.3 Pull Test

Bond strength is measured by pulling apart the bonded sample (usually a die) and recording the maximum force applied at which the sample breaks. As the bond separates, fracture surfaces are separated enabling failure mode analysis.

A major drawback of pull test is sample preparation, since each sample to be measured is to be joined to the two pulling heads and properly fixed (Fig. 11.11).

Fig. 11.11 Schematic drawing of pull test apparatus used to measure the fracture force by pull test

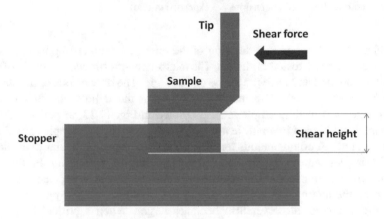

Fig. 11.12 Schematic drawing of shear test apparatus

11.4.4 Shear Test

Shear test is like pull test, but the applied force is parallel to the bonding surface, and there is no need of sample preparation. It is widely used for adhesive as well as metallic bonding, either thermocompression or eutectic, while it is not suitable for brittle materials such as glassfrit (Fig. 11.12).

11.4.5 Micro-Chevron Test

It is commonly used to characterize the wafer bonding strength as well as the damage behavior of bonded interfaces. The evaluation is determined by analyzing the fracture toughness K_{IC}, as is common in fracture mechanics, that is, expressing the material resistance against crack propagation [16].

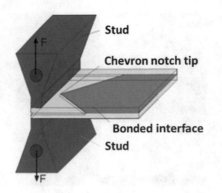
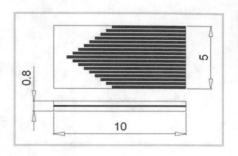

Fig. 11.13 Schematic drawing of micro-chevron test apparatus (left) and (right) typical micro-chevron structure. (Reported dimensions are expressed in mm)

In Fig. 11.13, a schematic drawing of the micro-chevron test apparatus as well as the structure to be tested is shown. The micro-chevron structure is made of strips that will be bonded and have the shape of a triangle. The dimensions of the structure are in the range of several millimeters, and alpha is the angle of the chevron notch tip (70 degrees is usually employed). As shown in Fig. 11.13, to perform the test, the samples are loaded in a pull tester with the free end of the samples glued to the pull tester studs. A homogeneous tensile load is applied perpendicular to the bonded area: During testing, a crack initiates at the chevron tip. Increasing the force, two opposite effects are observed. First, the resistance to the crack expansion increases because of the augmented bonded area of the triangular shaped pattern. Second, as the crack grows, the lever arm becomes larger. When a critical length ac is reached, there is an uncontrollable crack expansion the destruction of the specimen is initiated.

The fracture toughness K_{IC} can be calculated by

$$K_{IC} = \frac{F_{max}}{B\sqrt{W}} Y_{min}$$

Where:

- F_{max} corresponds to the maximum force applied during the test at which the critical crack length occurs.
- B is the length and W is the width of the sample.
- Y_{min} is the minimum of a geometric function that is generally determined by FEM simulations.

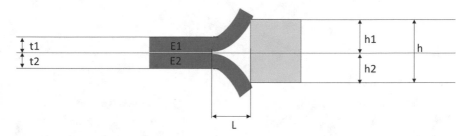

Fig. 11.14 Crack opening showing blade thickness, top and bottom wafer thickness, Young's modulus, and crack length.

11.4.6 Mazdara's Method

It is a destructive technique because it requires blade insertion at the interface between the bonded wafers to initiate a crack (see Fig. 11.14 for more details).

As the crack propagates, its length can be measured using IR inspection. By the following formula, bonding energy can be calculated [17].

$$\gamma = \frac{3E_1t_1^3 E_2t_2^3 h^2}{16\left(E_1t_1^3 + E_2t_2^3\right)L^4}$$

11.5 Hardware Description

Bonding equipment used in semiconductor front-end manufacturing are usually made up of two main modules: aligner for wafer-to-wafer alignment, and bonder to process (bond) the aligned wafer pair. In recent years, *in situ* alignment has emerged to improve alignment accuracy (* "Chapter 29 – Wafer-bonding equipment", Viorel Dragoi, Paul F. Lindner, EV Group E. Thallner GmbH, Sankt Florian am Inn, Austria).

Basic hardware configuration can be summarized as in the following scheme (Figs. 11.15 and 11.16).

Equipment suppliers make hardware configuration modular to meet specific customer process/application requirements in a flexible way. Standard modules available include (please refer to Fig. 11.17):

Alignment can either be achieved with or without alignment keys: In the latter case, it is often referred to as "mechanical" or "notch to notch" alignment. The equipment centers the wafers with respect to each other using the notch (for 8" wafers) and the wafer edge; this method is not the most accurate, having a 50 to 100 μm alignment accuracy at best, but it could be done without the aligner module (provided chambers are appropriately configured), thus saving footprint, initial equipment cost, and throughput.

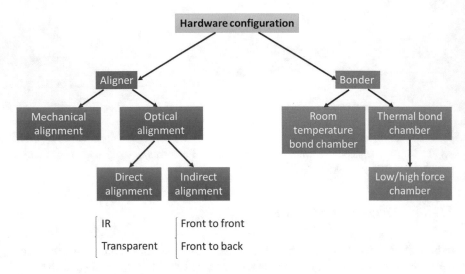

Fig. 11.15 Typical bonder hardware configuration

Fig. 11.16 Image of a Gemini-automated production wafer bonding system. (Courtesy by EVG)

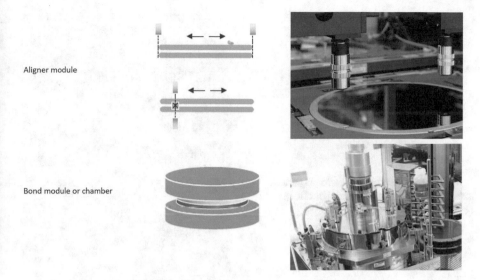

Fig. 11.17 Sequences of steps to obtain a perfect wafer to wafer direct bonding process at low temperature with EVG Gemini tool. (Courtesy by EVG)

In case alignment keys are used to align, depending on the illumination method, it can either be "direct", IR, or transparent, or "indirect", and alignment keys are on the front of both wafers or on the front of one wafer and on the back side of the other. Optics to search for the alignment keys use external references for calibration purposes (Fig. 11.18).

Once the wafer pair is properly aligned, it is moved into the chamber for bonding. Wafers are positioned onto a moving support (typically referred to as "bond tool" or "bond chuck" or "bond frame") which mechanically holds the wafer pair in place and keeps the wafers separated by inserting spacers in between. These are removed during bonding after the proper gas and pressure set point inside the chamber are reached (Figs. 11.19 and 11.20).

Nowadays aligner modules are capable of alignment accuracy of 0.5 μm or even less. All moving parts such as clamps or spacers could add misalignment to the wafer stack. Periodic maintenance should be performed on such parts to ensure that no friction, no deformation, or no loosening of bolts or screws occurs which could lead to misalignment. Moreover, nonparallelism between top and bottom tables of the aligner module and a faulty clamp insertion or mechanism are other misalignment contributors.

Bond chucks or frames are often responsible for misalignment issues: Clamp force might not be evenly set causing a shift in one direction of one wafer with respect to the other. Clamp and flag designs have been improved over the years to reduce W2W misalignment (Fig. 11.21).

Bond tools deform over time due to continuous stress and strain caused by repetitive heating cycles. In Fig. 11.22, a deformed frame is schematically shown.

Typical front side alignment sequence:

1. Top optics look for alignment keys
    on bottom wafer (front side),

2. Bottom optics look for alignment
    keys on top wafer (front side),

3. Top and bottom optics calibrate
    before each alignment.

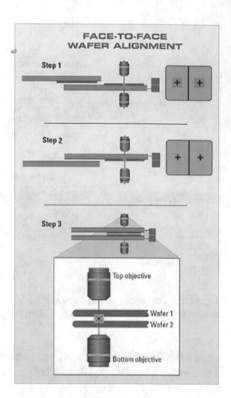

Fig. 11.18 Alignment procedure normally used in aligner modules. (Courtesy by EVG)

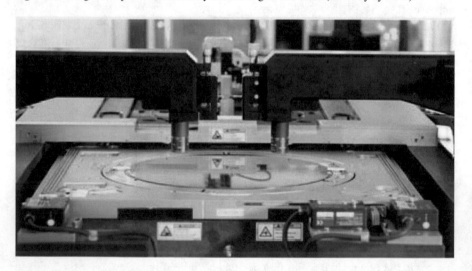

Fig. 11.19 Smart view alignment stage of Gemini. (Courtesy of EVG)

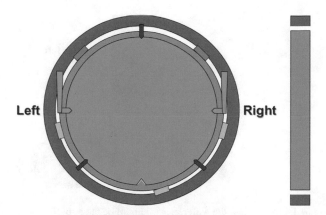

Fig. 11.20 Schematic of a bond chuck well settled not inducing wafer to wafer misalignment. Clamps are shown in green at the left and right, spacers or flags are 3 and shown in violet. Gray region inside the frame (shown brown) is the pressure insert, where wafers sit during bonding

Old design   New design

Fig. 11.21 Schematic view of flag designs. Old design (left) could easily lead to misalignment issues since spacer shape and tip could deteriorate. New design (right) is more reliable and stable over continuous usage. User can decide thickness and length of the spacers according to specific process needs

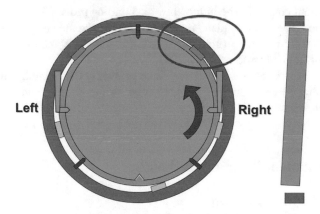

Fig. 11.22 Example of a deteriorated bond chuck inducing misalignment

Pressure insert is no longer centered with respect to the frame, and it is even sitting on a different plane (right hand side of the image), causing a rotation of the top wafer with respect to the bottom one and resulting in misalignment.

The other major component of the equipment is the bonding chamber/s. Depending upon specific needs and application, one aligner module feeds more chambers,

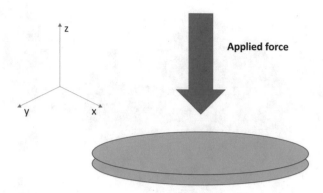

Fig. 11.23 Applied force to bond the wafer pair: it should be the only force acting on the wafers (z direction) to minimize misalignment occurring in the bonding chamber

typically up to four, to improve wafer throughput in front-end manufacturing, since bonding process takes much longer than alignment procedure. Bonding chambers could be used for room temperature bonding (RT) as in the case of fusion/direct bonding or for thermal bonding. The latter type could be rated as low force or high force chamber. The former can apply a force of up to 10 KN and it is used for glassfrit and organic bonding (BCB, epoxy glue), while the latter can apply a force of up to 100 KN and it is suitable for metallic thermocompression as well as eutectic bonding.

Misalignment could also be induced by the bonding process inside the chamber: if the resultant force in the *xy* plane is other than zero, and the layer making the joint goes through a liquid phase (glassfrit, BCB, and eutectic), then there is a strong chance for misaligned wafers (Fig. 11.23).

Resultant force is zero when clamp force is evenly distributed, bond frame is planar, bonding force is applied perpendicular to the wafers, and no rotation is induced by the piston.

11.6 Permanent Bonding Techniques

11.6.1 Glassfrit Bonding

Probably the most widely used bonding process in MEMS industry is glassfrit bonding: a very forgiving process with respect to surface nonuniformities (roughness, particles, topography), easy to integrate in the MEMS process flow allowing hermetic sealing and very high and stable process yield. However, it suffers of W2W misalignment, and it does not allow aggressive device shrinkage since glassfrit strips are defined by printing technique through a stencil and not via a standard lithographic process [6].

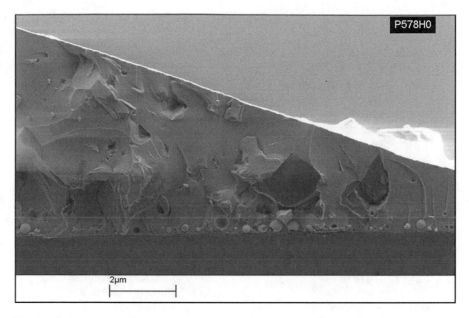

Fig. 11.24 SEM image of the fracture surface of a glassfrit layer after screen printing and firing

11.6.1.1 Screen Printing Technique

Glassfrit is a PbO-Zn-B_2O-SiO_2 paste (lead-zinc-borosilicate paste with ceramic fillers) with a low melting point, deposited onto wafers by screen printing using a stencil as a mask (Fig. 11.24).

Stencil is made by a frame holding a tensed mesh with an embedded photosensitive polymer. Apertures in the photosensitive polymer are defined by the device layout (Fig. 11.25).

As the squeegee moves back and forth on the stencil spreading the paste, the paste is deposited onto the underlying wafer where apertures are defined in the mesh (Fig. 11.26).

A thermal treatment (normally referred to as "firing") is required after screen printing step to evaporate solvents, burn the binder, and glaze the deposited paste. Detailed information on the thermal profile is given in Fig. 11.27.

Considering for example an inertial sensor device made of a sensor and a cap wafer, the glassfrit is generally applied on the cap wafer by screen printing, which is bonded to the sensor wafer afterward. During the bonding process, the glass material softens and wets the surface of the sensor wafer, which is held in close contact to the cap wafer by an applied force. During the final cooling step, a glassfrit layer of approx. 5 to 8 μm thickness is formed because of the bonding process.

Figure 11.28 shows a scanning electron microscope image of the glassfrit layer bonded to a silicon wafer. It can be observed, that at the glass–silicon interface lead precipitations are formed according to the following reaction:

a)

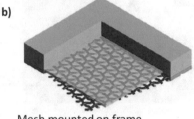

stainless steel mesh
d: wire diameter
w: opening

c)

Exposure of photosensitive polymer

b)

Mesh mounted on frame,
Coated with photosensitive polymer

d)

Defined aperture after exposed
polymer developing.

Fig. 11.25 Stencil details: For an 8 inch wafer, stencils are 10 × 10 inches

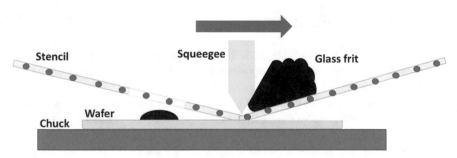

Fig. 11.26 Principle of screen printing

$$2PbO + Si \rightarrow 2Pb + SiO_2$$

The bonding interface must withstand the additional process steps in F.E. and in B.E., such as wafer thickness reduction and the pressure applied during molding. Furthermore, the functionality and the performance of the devices must be guaranteed over the entire product lifetime (e.g., the vacuum inside the bonded cavity).

To evaluate the reliability and lifetime of glassfrit joints, the microstructural and mechanical properties were characterized. For this purpose, the influence of the formation of lead precipitations at the glassfrit layer–silicon interface and the

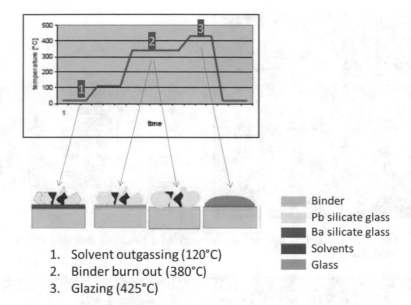

1. Solvent outgassing (120°C)
2. Binder burn out (380°C)
3. Glazing (425°C)

Binder
Pb silicate glass
Ba silicate glass
Solvents
Glass

Fig. 11.27 Details of the firing step

$$2PbO + Si \rightarrow 2Pb + SiO_2$$

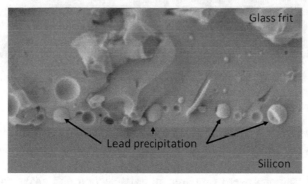

Glass frit

Lead precipitation

Silicon

Fig. 11.28 SEM cross-section of glassfrit–silicon interface

influence of the bonding temperature on the strength properties of the glassfrit joints were investigated using pull and micro-chevron tests.

Those tests were carried out at the Fraunhofer Institute for Microstructure of Materials and Systems IMWS in Halle, to investigate the fracture strength and fracture toughness of the glassfrit-bonded interfaces. The design of the mask layout and the preparation of the test wafers were done at STMicroelectronics (Fig. 11.29).

To study the formation and the effect of lead precipitations, silicon substrates were printed with glassfrit paste according to the mask layout and bonded to wafers

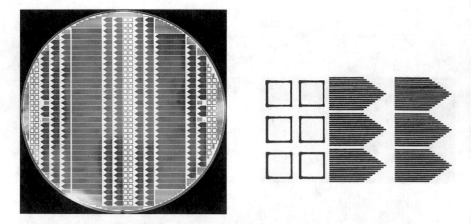

Fig. 11.29 SAM image of the bonded test mask (left) and detailed view showing pull and micro-chevron test structures (right)

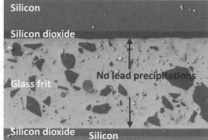

Fig. 11.30 (**a**) SEM cross-sections of a silicon–glassfrit–silicon sample lead precipitations have been formed. (**b**) SEM cross-section of a SiO_2–glassfrit–SiO_2 sample: no lead precipitations have been formed. (Images: Fraunhofer IMWS)

with different surface layers: native oxide, silicon oxide (1 μm thick), and Al (100 nm).

Afterward, samples were prepared in cross-section and analyzed by SEM at Fraunhofer IMWS to investigate the microstructure of the resulting interfaces. It was found that lead precipitations are formed at silicon–glassfrit interfaces. At interfaces with thin surface layers, either a SiO_2 or a metallic layer, between glassfrit and silicon substrate, no lead precipitations are formed (Fig. 11.30).

Figure 11.31 shows the results of pull tests at glassfrit-bonded frame samples with different interfaces. The results point out that the formation of lead precipitations at the interface does not influence the resulting tensile strength.

The influence of the temperature of the bonding process on the interface strength properties was also determined by pull and micro-chevron tests. Figure 11.32 shows a Weibull plot of the fracture stress obtained by pull tests at glassfrit samples that were bonded at different process temperatures using a confidence range of 90 %.

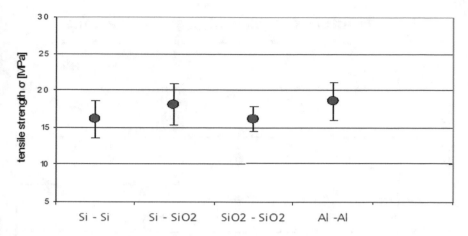

Fig. 11.31 Tensile strength of glassfrit-bonded samples with different bonding interfaces

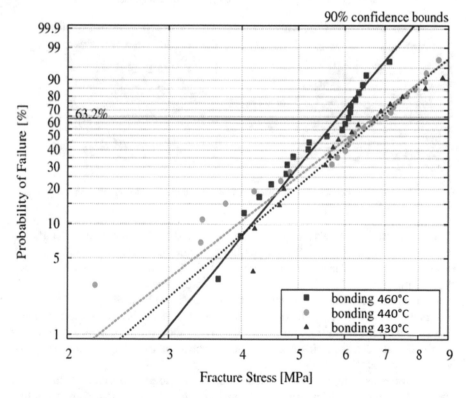

Fig. 11.32 Weibull plot of the failure probability as a function of the fracture stress of glassfrit samples that were bonded at different process temperatures

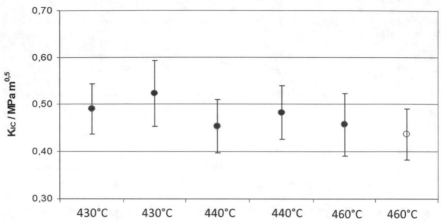

Fig. 11.33 Fracture toughness K_{IC} for samples bonded at different temperatures

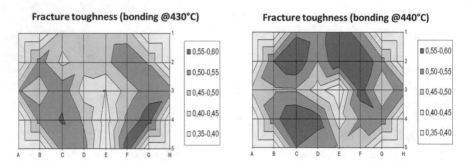

Fig. 11.34 Fracture toughness K_{IC} across the wafer for samples bonded at different temperatures

The Weibull plot shows that bonding at the highest process temperature of 460 °C results in the highest Weibull modulus but a smaller characteristic fracture stress. Bonding at lower process temperatures of 430 °C and 440 °C leads to a lower Weibull modulus and a higher characteristic fracture stress.

Beside pull tests, the micro-chevron test was used to characterize the interface of wafers that were bonded at different temperatures. The resulting fracture toughness K_{IC} ($K_{IC} \approx 0.5$ MPam$^{1/2}$), shown in Fig. 11.33, shows no significant differences for the investigated samples and thus no influence of the temperature during the bonding process (Fig. 11.33).

As numerous samples were taken from each wafer at different locations, a fracture toughness distribution across the wafer has been plotted, shown for two examples in Fig. 11.34.

The results show that microstructure diagnostics and mechanical characterization give valuable feedback to process engineers about the working conditions of the equipment and help to properly address maintenance procedures. Furthermore, it enables a high robustness of the process steps during manufacturing and thus helps to increase manufacturing yield and device reliability.

11.6.2 Au–Au Thermocompression Bonding

Thermocompression metallic bonding uses both temperature and compression to join two layers without the formation of a liquid phase. Gold is a suitable material for this bonding method; besides, gold does not grow a native oxide like Ge, Al, or most metals, making it easier to use in a device process flow.

The effect of compression is to plastically deform the metal to put gold present on both wafers to be joined in contact at the microscopic level, thus reducing local thickness nonuniformities and smoothing the surface; the effect of temperature is to enable gold new grain formation to erase the bonding interface.

Compression can be achieved not only by applying force to the wafer pair (latest chamber configurations can apply up to 100 kN and market request is to increase this figure), but also by improving the microscopic gold surface, a smooth surface ensures a good contact. ECD (Electro Chemical Deposition) process must provide a uniform gold deposit across the wafer because thickness nonuniformity could prevent a good gold contact. Besides, gold strips profile must be flat, no peaks at the edges, or a dome-shaped profile. Device design plays a major role, since the smaller the bonding area is, the higher the pressure applied at a given force.

Gold structure impacts bonding quality and depends on deposition method, which could be sputtering (PVD), evaporation, or electroplating.

Electro-plated films are usually composed of small grains with a preferred crystallographic orientation: misoriented grains in a very oriented bulk are very high, which is suitable to achieve bonding since misoriented grains are responsible for abnormal grain growth. Also, electro-plated metals contain a small amount of "grain refiner"; for electro-plated gold, thallium, arsenic, copper, bismuth, and lead are the most widely used. They reduce grain size and tend to segregate (Fig. 11.35).

Thallium (refiner) concentration in electro-plated gold is a critical parameter to achieve good bonding (migration of the grain boundaries across the interface): in the graph below leakage yield loss percentage is plotted as a function of Tl concentration (ppm) in the plating bath. Leakage yield loss is measured at electrical wafer sorting (EWS) and it is a parameter closely related to hermeticity achieved with gold–gold bonding: the higher the yield loss, the worse is the bonding quality of the wafer (Fig. 11.36).

The reason for this behavior is found in the Gold–Thallium phase diagram (Fig. 11.37).

There is a mixed liquid phase (Au+Tl) in equilibrium with Au when the Tl/Au ratio in weight is higher than 0.2% (Au and Tl have about the same molecular weight) and at a temperature higher than 217 °C. This mixed liquid phase at the grain boundaries where the thallium tends to segregate is responsible for the physical

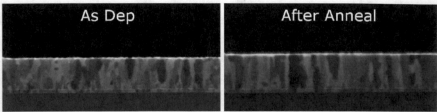

SEM cross section of PVD gold as deposited (left) and after annealing (380°C, 30min.). No crystallization is detectable.

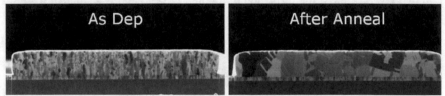

SEM cross section of ECD gold with TI refiner as deposited (left) and after anneal (380°C, 30min.). Crystallization is clearly detectable.

Fig. 11.35 SEM cross-sections of PVD Gold (top images) pre and post annealing and SEM cross-section of ECD (Bottom Images) pre and post annealing

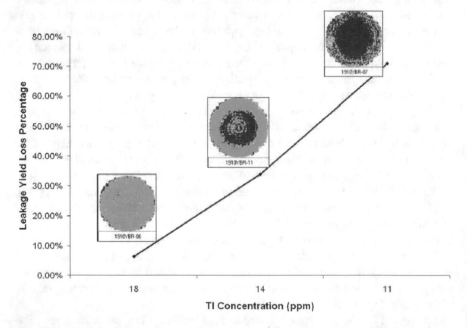

Fig. 11.36 Leakage yield loss measured at EWS as a function of TI refiner concentration is the gold plating both

grain migration through the bonding interface. This liquid phase works as a lubricant for the grain boundary motion.

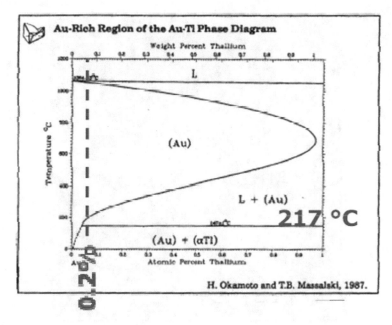

Fig. 11.37   Gold–Thallium phase diagram

11.6.3   AlGe Eutectic Bonding

Eutectic bonding Aluminum-Germanium offers many advantages over Au–Au thermocompression bonding: for sure it is cheaper and easier to integrate in the device and cap wafer process flows, at the same time it offers the same shrinkage capability while it suffers of misalignment due to liquid phase; fortunately, process induced misalignment can be managed via design [10, 11].

A widely used method to both set-up and monitor AlGe bonding process is via IR inspection: automatic microscopes available on the market inspect user defined areas on the wafer to detect bonding defects. As a rule of thumb, good bonding means almost black ring, bad bonding means a spotted or completely white ring. Hereafter, an example with both conditions within the same die (Fig. 11.38).

SEM cross-sections have been made in regions where bonding ring is either black or white at IR inspection to validate this screening and in-line control method (Fig. 11.39).

To prevent defective bonding, a process window with respect to bonding temperature is recommended as well as a measure with thermocouple wafer of the real temperature experienced by the wafer during processing.

Using this procedure, it was found that the real temperature experienced by the wafer during the process is not uniform across the wafer, but there is a difference of about 10 °C between center and edge. Therefore, the only way to set-up a more stable and reliable process was to increase set-point temperature. Later, a process

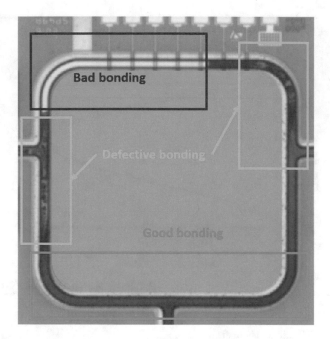

Fig. 11.38 IR inspection of a bonding ring (AlGe eutectic bonding)

window was run to check for any inconvenience eventually caused by a too high temperature. Trials have been evaluated using IR inspection, warpage/bow, and shear test measurement (Figs. 11.40, 11.41 and 11.42).

Bonding temperature in the range 440–450 °C is the most suitable considering hardware available, IR inspection yield maps, bow, and shear test data.

11.6.4 Adhesive Bonding: BCB

BCB is an interesting material for MEMS application: it is a thermosetting polymer, suitable for permanent adhesive wafer bonding, which is very robust and resistant to various acid and solvents. It is commercially available both as non and photosensitive material, it can either be spin or spray coated, but it can also be deposited on the wafer by layer transfer technique as it will be shortly discussed. After W2W bonding, an annealing step is necessary for material polymerization. More details are given in Fig. 11.43 (courtesy of Dow chemicals).

BCB layer has been successfully integrated as glue material in piezoelectric actuated inkjet heads, which require at least 3 wafers to be bonded together (Chap. 19). The resistance of BCB to various types of ink has been extensively proven by dedicated reliability trials. The bonding strength of the two joints (such is this case involving three wafers) has been evaluated and positively assessed by means of micro-chevron tests. The process flow for BCB integration (layer transfer) is hereinafter reported.

SEM cross section of the AlGe bonding ring where it looks white at IR inspection (see right image). Anomalous metal layer is detected, confirming poor bonding.

IR image detail of the bonding ring.

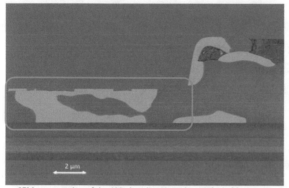

SEM cross section of the AlGe bonding ring where it looks black at IR inspection (see right image). Good Al and Ge intermixing is highlighted, confirming good bonding.

IR image detail of the bonding ring.

Fig. 11.39 SEM cross-section and IR image highlighting a good and a poor bonding condition

11.6.4.1 Layer Transfer

This coating procedure is useful when only some specific parts of a device are to be coated with the bonding layer, and topography is such to prevent usage of more standard methods such as spin or spray coating. Please refer to the following scheme (Fig. 11.44).

For a cap wafer as shown above, BCB spinning or spray coating cannot be used due to deep trench presence and a step of at least 100 microns deep, together with the requirements of quite small dimensions and tight alignment. Besides, BCB is to be used for bonding and must be only on top surfaces, not inside trench or cavity bottom for best device performance. The only technique capable to overcome these constraints is layer transfer. Hereafter, you can find a schematic description (Figs. 11.45 and 11.46).

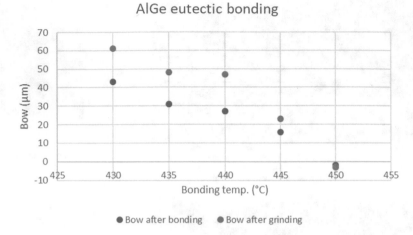

Fig. 11.40 Warpage after bonding and after grinding as a function of bonding temperature

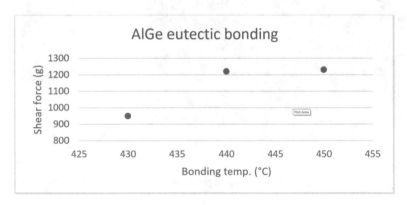

Fig. 11.41 Shear force of AlGe-bonded samples as a function of bonding temperature

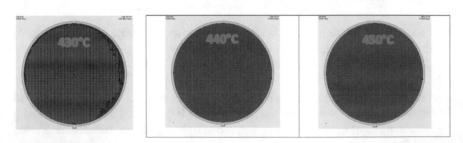

Fig. 11.42 IR microscopy maps as a function of bonding temperature. Red dots are dice classified as defective due to marginal bonding

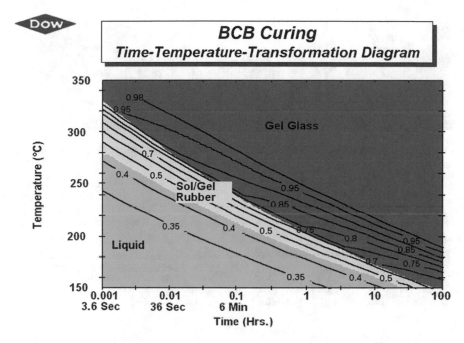

Fig. 11.43 Time-temperature transformation diagram for BCB material. Typical polymerization conditions after bonding step are temperature of 250 °C for at least 30 min. (Courtesy of DOW Chemicals)

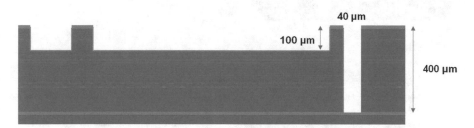

Fig. 11.44 Cap wafer cross-section: topography is really challenging and prevents spray or spin coating techniques from being usefu l

Fig. 11.45 Cap wafer cross-section: BCB bonding material must be only on top structure, not inside trenches or at cavity bottom for best device performance

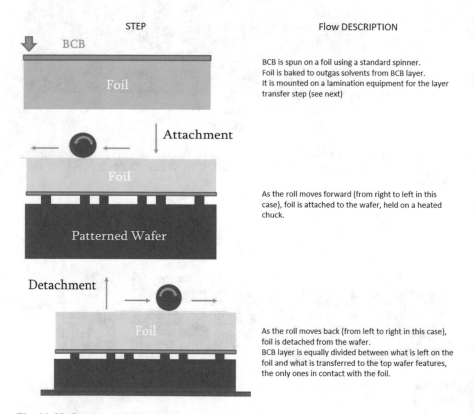

STEP   Flow DESCRIPTION

BCB is spun on a foil using a standard spinner.
Foil is baked to outgas solvents from BCB layer.
It is mounted on a lamination equipment for the layer transfer step (see next)

As the roll moves forward (from right to left in this case), foil is attached to the wafer, held on a heated chuck.

As the roll moves back (from left to right in this case), foil is detached from the wafer.
BCB layer is equally divided between what is left on the foil and what is transferred to the top wafer features, the only ones in contact with the foil.

Fig. 11.46 Screen printing sequence

11.7 Temporary Bonding

11.7.1 Introduction to Temporary Bonding Approach

As described at the beginning of this chapter, new MEMS technologies require more complex design and architecture. In some applications, it is required to work both front and wafer backside as well as to process very thin substrates (thickness less than 100 μm). A temporary bonding approach is needed to enable standard processing (handling) of thin wafers and to protect the front while working on the back side.

A typical temporary flow involving temporary bonding and debonding is reported in Fig. 11.47.

As previously mentioned, there are different types of debonding techniques. In Fig. 11.48, a schematic process flow for the most widely used methods is given.

Depending on the temporary material and debonding technique employed, cleaning procedure may vary: after slide off and laser debonding solvent cleaning is effective in removing the temporary material residues; while detaping is used after mechanical debonding. The latter method could induce moving structures breakage depending on device design (Fig. 11.49).

As for bonding, also debonding equipment is modular, letting the customer choose which and how many modules of each type to insert in the main frame. Available modules include: debonding module, whose configuration depends on technique employed (laser, slideoff/lift off, mechanical), cleaning module, film frame mounter in case device wafer is supported by a frame after debonding (Fig. 11.50).

11.7.2 Temporary Bonding Study Case—Microcracks

Depending on device wafer features, temporary bonding material could lead to detrimental effects. For example, when trenches are present on device front side and their depth is such that only a thin silicon membrane is left on the backside as shown in Fig. 11.51, cracks could be detected in the back-side membrane after temporary bonding or during post processing, as shown in Fig. 11.51.

To explain the phenomenon, experimental and simulation tests were carried out to characterize the temporary bonding material considering its rheology, which plays a major role in affecting the workability of the wafers.

11.7.2.1 Simulation

Temporary bonding material supplier provided Young's modulus and CTE data as a function of temperature. As seen in Fig. 11.52, when temperature rises from room

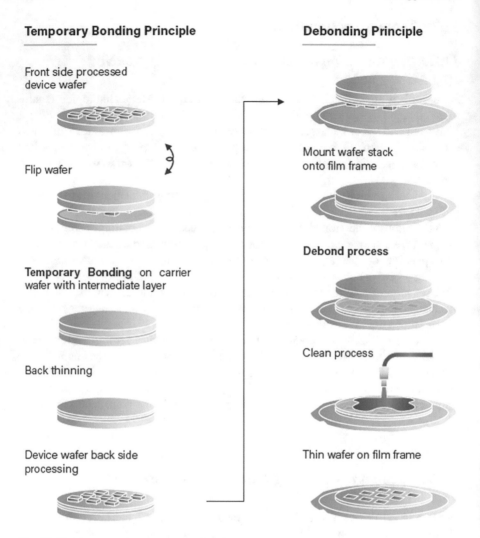

Fig. 11.47 Temporary bonding approach and post processing

temperature to 100 °C, CTE sharply increases (red line), while Young's modulus slowly decreases (green line).

A FEM analysis with COMSOL was used to predict the stress on the back-side membrane given the geometrical parameters and the material properties. As a result of the simulation, the blue line in Fig. 11.52 was computed showing a stress curve with a peak value at a temperature of 85 °C. The graph shows that if the temperature remains below 70 °C, the material does not expand and maintains its stiffness. On the other hand, when the temperature increases over 95 °C, the material expands and becomes less stiff.

Thermal Slide-off Debond

Laser Debond

Mechanical Debond

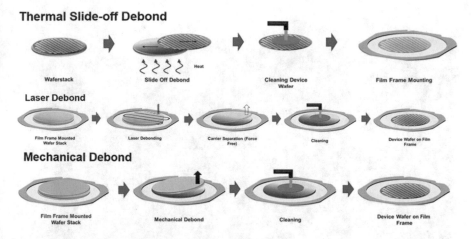

Fig. 11.48 Debonding methods schematic flow comparison

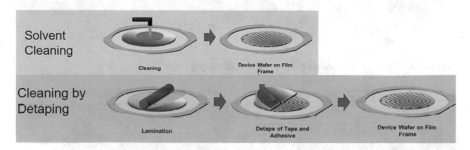

Fig. 11.49 Cleaning method comparison to remove temporary bonding material after debonding

The above results agree with the material properties provided by the material supplier, which state that TG occurs at 85 °C for the temporary material employed. Therefore, the transition from glassy to plastic phase leads to a change of material properties, and the maximum stress value occurs at the material's glass transition temperature.

Given the studied geometry, it was possible to predict the membrane deflection as pictured in Fig. 11.53. Membrane displacement was computed as a function of temperature: From 84 °C to 90 °C, a maximum of 180 nm displacement is predicted.

Simulation shows that the maximum stress is 1.5 GPa in temperature range 84 °C–90 °C. This value is well above the silicon breakage limit (1 GPa). It means that the membrane has a high probability to break, while the thermoplastic material goes through TG. Moreover, a membrane bow of 180 nm was predicted at the TG temperature.

Fig. 11.50 Automated debonding equipment. (Courtesy by EVG)

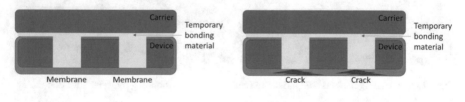

Temporary bonding step   *Cracks detected in the backside membrane*

Fig. 11.51 During temporary bonding post processing, cracks were detected in the back-side membrane

11.7.2.2 Thermal Cycle Measurements

To verify the predicted values computed with FEM analysis, a thermal cycle has been done while measuring the deflection of the back-side membrane with an interferometer.

- After temporary bonding to a carrier wafer, system shows a membrane inflection of 100 nm at room temperature (Fig. 11.54).

- Heating up the system on a hot plate from room temperature (RT) to 100 °C, material expansion caused no membrane deformation as measured with the interferometer.

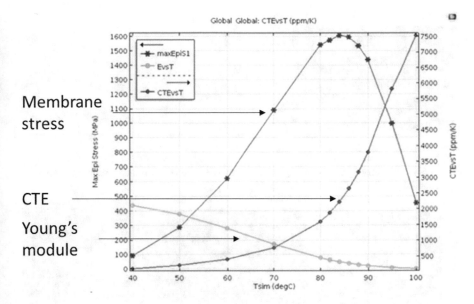

Membrane
stress

CTE

Young's
module

Fig. 11.52 CTE (red line), Young's modulus (green line), and the resulting stress (blue line)

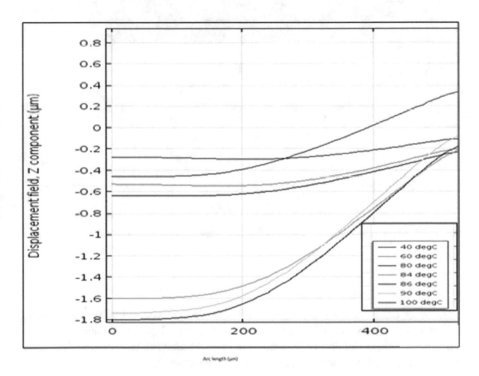

Fig. 11.53 Membrane displacement as function of the temperature for the given geometry $0.5 \times 1 \, mm^2$

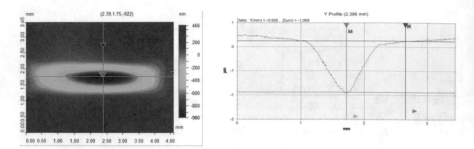

Fig. 11.54 Geometry 1 mm × 400 μm measured with the interferometer showing membrane deflection inward due to the material shrinking

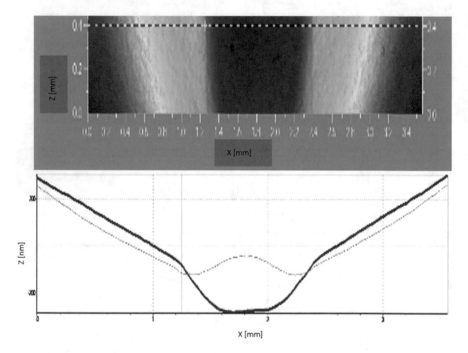

Fig. 11.55 Membrane deflection while cooling down the system. A max deflection of 200 nm is measured

Cooling the system from 100 °C to RT, membrane deflects by 200 nm reaching a final inflection of 100 nm (same as to begin with) (Fig. 11.55) due to material shrinkage. This deflection is reached at the glass transition temperature, TG equal to 85 °C. At this point, there are two possible scenarios:

Table 11.1 Summary of four of the most widely used bonding techniques

	Glass frit	Au-Au	AlGe	BCB
Temperature (°C)	430-440	380	435-445	150-175*
Force (kN)	< 10	> 40	> 40	< 3
Time (min.)	≈ 10	≈ 20	≈ 8	≈ 30*
Pros.	Cost, reliable, hermetic seal, forgiving, process control	Strip width/device shrinkage, no native oxide, hermetic seal, alignment, process control	Strip width/device shrinkage, hermetic seal	Versatile material, forgiving, process control, resistant to harsh environment
Cons.	Strip width/device shrinkage, misalignment	Process integration, cost, process control	Misalignment, native oxide (surface preparation)	Non hermetic seal, process time, misalignment
Notes	Temperature driven	Force driven	Temperature driven	* + polimerization

- A membrane mechanical failure happens because the stress is too high.
- A whip effect due to the stress release suddenly happens and the membrane reaches a final bow value of 100 nm (same as measured after bonding).

11.8 Final Remarks

Before closing this chapter, a summary of the four permanent bonding processes discussed is given in the Table 11.1, highlighting process conditions, pros, and cons for each one.

References

1. Gosele, U., Stenzel, H., Reiche, M., Martini, T., Steinkirchner, H., & Tong, Q.-Y. (1996). History and future of semiconductor wafer bonding. *Solid State Phenomena, 47–48*, 33–44.
2. Tong, Q.-Y., & Gösele, U. (1998). *SemiConductor wafer bonding: Science and technology.* Wiley-Interscience. ISBN:978-0-471-57481-1
3. PloÈûl, A., & KraÈuter, G. (1998). *Wafer direct bonding: Tailoring adhesion between brittle materials.* Max-Planck-Institut fuÈr Mikrostrukturphysik. Accepted 12 October 1998.
4. Wallis, G., & Pomerantz, D. I. (1969). Field assisted glass-metal sealing. *Journal of Applied Physics, 40*(10), 3946–3949. https://doi.org/10.1063/1.1657121. Bibcode:1969JAP....40.3946W
5. Plach, T., Hingerl, K., Tollabimazraehno, S., Hesser, G., Dragoi, V., et al. (2013). Mechanisms for room temperature direct wafer bonding. *Journal of Applied Physics, 113*, 094905. https://doi.org/10.1063/1.4794319
6. Knechtel, R. (2005). Glass frit bonding: An universal technology for wafer level encapsulation and packaging. *Microsystem Technologies, 12*, 63–68. https://doi.org/10.1007/s00542-005-0022-x

7. Tsau, C. H.-H. (2003). *Fabrication and characterization of wafer-level gold thermocompression bonding*. Massachusetts Institute of Technology, Department of Materials Science and Engineering.
8. Malika, N., Schjølberg-Henriksen, K., Poppe, E., Takloc, M. M. V., & Finstada, T. G. (2014). Al Al thermocompression bonding for wafer-level MEMS sealing. *Sensors & Actuators: A. Physical, 211*, 115–120.
9. Ramm, P., Lu, J. J.-Q., & Taklo, M. M. V. (2012). *Handbook of wafer bonding*. Wiley-VCH.
10. Baum, M., Jia, C., Haubold, M., Wiemer, M., Schneider, A., Rank, H., Trautmann, A., & Gessner, T. (2010). *Eutectic wafer bonding for 3-D integration*.
11. Wolffenbuttel, R. F., & Wise, K. D. (1994). Low-temperature silicon wafer-to-wafer bonding using gold at eutectic temperature. *Sensors and Actuators A, 43*, 223–229.
12. Niklaus, F., Enoksson, P., Griss, P., Kälvesten, E., & Stemme, G. (2001). Low-temperature wafer-level transfer bonding. *Journal of Microelectromechanical Systems, 10*, 525–531.
13. Niklaus, F., Enoksson, P., Griss, P., Kälvesten, E., & Stemme, G. (2001). Low-temperature full wafer adhesive bonding. *Journal of Micromechanics and Microengineering, 11*, 100–107.
14. Wandelt, K. (2019). *Surface and interface science: Volume 9: Applications I/Volume 10: Applications II*. Wiley-VCH.
15. *Proceedings of the fourth international symposium on semiconductor wafer bonding: Science, technology, and applications*, 1998.
16. Munz, D., Bubsey, R. T., & Srawley, J. E. (1980). *International Journal of Fracture, 16*(4), 359.
17. Vallin, Ö., Jonsson, K., & Lindberg, U. (2005). *Adhesion quantification methods for wafer bonding*. Uppsala University, The Angstrom Laboratory.

Part III
MEMS Sensors

Chapter 12
Linear and Nonlinear Mechanics in MEMS

Claudia Comi, Alberto Corigliano, Attilio Frangi, and Valentina Zega

12.1 Introduction

Mechanical issues, and particularly mechanical reliability, are extremely important in all the phases of MEMS design and in the development of relevant production technologies [1]. Design for reliability is the right approach to obtain successful commercial devices.

Proper design and fabrication of micro-devices must ensure the correct functioning both under the forecast design conditions and under unpredictable, possibly dangerous situations like accidental drop, mechanical and electrical shock, harsh environment. Other phenomena that can negatively influence the MEMS response are a vast set of non-ideal behaviours that bring the microsystem response far from the design range.

Mechanics and its interaction with other physics plays a major role in a proper design for reliability approach and in controlling possible non-ideal behaviours that bring the microsystem response far from the design range, which is usually kept in the linear regime.

In most cases, linear responses are required to simplify the read-out control circuit and to obtain predictable devices with low dispersion in the final properties. All nonlinear phenomena related to the mechanics of microsystems must be well understood to obtain effective designs that can be transformed into real products.

Typical mechanical components of microsystems are mechanical oscillators; their dynamic response can become complex and highly nonlinear, e.g., due to the interaction between membrane and bending regimes in slender beams or plates, due

C. Comi · A. Corigliano (✉) · A. Frangi · V. Zega
Department of Civil and Environmental, Politecnico di Milano, Milano, Italy
e-mail: claudia.comi@polimi.it; alberto.corigliano@polimi.it; attilio.frangi@polimi.it; valentina.zega@polimi.it

© Springer Nature Switzerland AG 2022
B. Vigna et al. (eds.), *Silicon Sensors and Actuators*,
https://doi.org/10.1007/978-3-030-80135-9_12

to temperature variations, to the internal contacts between surfaces at low distance, to the electro-mechanical coupling induced by electrostatic actuation.

The purpose of this chapter is to describe some aspects of the complexity of microsystems from a mechanical perspective with particular reference to nonlinear mechanical problems in MEMS inertial sensors focusing on real devices designed and studied by the authors.

The chapter is organized as follows.

Section 12.2 contains a brief introduction on mechanical and multi-physics problems in MEMS, with the focus on the behaviour of mechanical oscillators.

In Sect. 12.3, the Hamilton's principle is first introduced as a tool to formulate approximate semi-analytical solutions for mechanical oscillators in the linear and nonlinear regime, then various examples of application to real cases are discussed.

Section 12.4 contains a presentation of an efficient numerical method for the modelling and simulation of complex nonlinear responses, with a particular focus on resonating devices.

Other interesting nonlinear and complex problems in mechanical microsystems are discussed in Sect. 12.5, again with reference to real cases.

In Sect. 12.6 some closing remarks are presented.

12.2   Mechanical and coupled problems in MEMS

We start recalling in Sect. 12.2.1 the equation governing the dynamic behaviour of a one degree-of-freedom (dof) mechanical oscillator and the main properties of its dynamic response. Notwithstanding its simplicity, this reference mechanical oscillator in the linear and nonlinear regimes shows many of the typical responses that can be found in MEMS. In many practical examples a 1 dof model is built for the preliminary design phases, for the device control, and for the coupling with read-out circuits.

The subsequent Sects. 12.2.1–12.2.5 are devoted to the most important and most frequently found multi-physics and nonlinear problems in microsystems.

12.2.1   One degree-of-freedom oscillator

Let us consider the device shown in Fig. 12.1: a material point of mass m, constrained by an ideal mass-less extensional spring of stiffness $k \geq 0$ and subject to a viscous damping force $b\dot{u}$ and to an additional external force F.

The equation of motion reads

$$m\ddot{u} + b\dot{u} + ku = F \qquad (12.1)$$

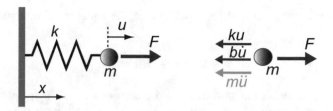

Fig. 12.1 One dof oscillator

and must be supplemented by initial conditions on displacement and velocity, i.e.:

$$u(0) = u_0, \quad \dot{u}(0) = \dot{u}_0. \tag{12.2}$$

In Eq. (12.1), first and second derivatives with respect to the time variable t have been denoted with a superposed dot and two superposed dots, respectively.

Equation (12.1) is linear and with constant coefficients, its general solution can be obtained as the sum of the general integral of the homogeneous equation $u_h(t)$, for $F = 0$, and of a particular solution of the whole equation $u_p(t)$:

$$u(t) = u_h(t) + u_p(t). \tag{12.3}$$

$u_h(t)$ represents the *free vibration* of the system for assigned initial conditions. Considering the undamped case corresponding to $b = 0$, Eq. (12.1) for $F = 0$ can be rewritten as

$$\ddot{u} + \omega^2 u = 0 \tag{12.4}$$

with

$$\omega \equiv 2\pi f_0 \equiv \sqrt{\frac{k}{m}}, \tag{12.5}$$

where ω is the undamped circular natural *frequency*, its units are radians per second [rad/s], f_0 is the frequency and is expressed in cycles per second or Hertz.

The solution of the Eq. (12.4) is given in terms of trigonometric functions as

$$u(t) = A_1 \cos \omega t + A_2 \sin \omega t \tag{12.6}$$

where A_1 and A_2 are constants that can be obtained from the initial conditions:

$$u(0) = u_0 = A_1, \quad \dot{u}(0) = \dot{u}_0 = A_2 \omega \tag{12.7}$$

When considering damped free oscillations, the equation of motion is usually written as

$$\ddot{u} + 2\xi\,\omega\dot{u} + \omega^2 u = 0, \tag{12.8}$$

where $\xi = b/(2\omega m)$ is the un-dimensional damping factor. The magnitude of the damping factor ξ compared to unity can be used to distinguish three cases: for $\xi < 1$, under-damped case, the motion is oscillatory with decreasing amplitude; for $\xi \geq 1$, critically damped and over-damped cases, the motion is non-oscillatory, its amplitude decays monotonically except possibly for one zero crossing.

The particular integral $u_p(t)$ to the solution of the Eq. (12.1) is considered here in the case of harmonic forces due to their importance in MEMS applications:

$$F(t) = F\sin(\omega_F t), \tag{12.9}$$

where F is the amplitude of the assigned force and ω_F its angular frequency.

The particular integral in this case is an oscillating function of amplitude u_F which has the same angular frequency of the assigned force and a phase shift φ:

$$u(t) = u_F \sin\left(\omega_F t + \varphi\right), \tag{12.10}$$

$$u_F = \frac{F}{k\sqrt{\left(1 - \left(\frac{\omega_F}{\omega}\right)^2\right)^2 + 4\xi^2\left(\frac{\omega_F}{\omega}\right)^2}}, \tag{12.11}$$

$$\tan\varphi = \frac{2\xi\left(\frac{\omega_F}{\omega}\right)}{\left(\left(\frac{\omega_F}{\omega}\right)^2 - 1\right)}. \tag{12.12}$$

The complete solution of Eq. (12.1) to a given external force and assigned initial conditions is then obtained as

$$u(t) = u_h(t) + u_p(t) = e^{-\xi\omega t}\left(B_1\cos\omega_d t + B_2\sin\omega_d t\right) + u_F\sin\left(\omega_F t + \varphi\right), \tag{12.13}$$

being B_1 and B_2 two constants to be determined enforcing the initial conditions (12.2), obtaining:

$$B_1 = u_0 - u_F\sin\varphi, \quad B_2 = \frac{1}{\omega_d}(\dot{u}_0 + \xi\omega u_0) - \frac{u_F}{\omega_d}(\omega_F\cos\varphi + \xi\omega\sin\varphi). \tag{12.14}$$

For every damping level below the critical one the amplitude of the forced oscillation, represented in Fig. 12.2, has a maximum reached when the frequency of the external force is near to the frequency of the oscillator. When the damping is zero, the maximum becomes infinite. The phenomenon which brings to a high

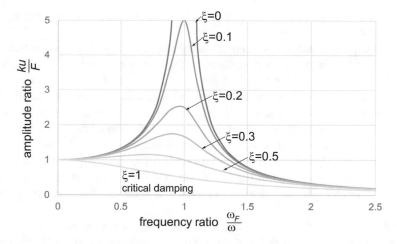

Fig. 12.2 One dof oscillator: amplitude as a function of the ratio between the frequency of the external force and the frequency of the oscillator at varying damping factor

increase in the amplitude for a given frequency of the external input is called *resonance* and the corresponding frequency is called *resonance frequency*. Forcing the system at resonance is common in microsystems in which a device must be kept oscillating at the highest possible amplitude with a minimum amount of power consumption. We define these devices as *resonators* or *resonant devices*.

Another interesting feature of the one dof forced response is the fact that when the frequency of the external input goes to zero or it is near to zero, the amplitude of the forced response is equal to F/k which corresponds to the static response of the elastic spring subject to a force with intensity F, i.e., no dynamic effects are felt by the oscillator. This is also the reason why the term multiplying the value F/k in Eq. (12.11) is simply called *amplification factor*. This particular situation is exploited in many practical examples concerning microsystems, typically in accelerometers, where the variation in time of the external signal to be measured, i.e., the acceleration, is usually very slow with respect to the resonant frequency of the device and can therefore be considered as *quasi-static*.

Considering an external force with a frequency much higher than the frequency of the oscillator, a de-amplification with respect to the purely static response F/k, is obtained (see Fig. 12.2) until when the amplitude can be considered negligible, also this circumstance is exploited in the design of microsystems.

Considering now the complete response given by Eq. (12.13), it can be observed that the first part goes to zero at increasing time whenever the damping factor is non-zero; this means that, after a sufficient amount of time, the prevailing term will always be the one given by the forced oscillation.

As already remarked, many MEMS resonant structures can be studied through the simple Eq. (12.1). This means being able to determine the coefficients m, b, k and the external force F corresponding to the specific case studied.

The mass m must be determined looking for the mass participating at the dynamic mode that one wants to describe.

The damping coefficient b needs in many MEMS a complex separate computation of the major damping sources, again for the specific movement described by the single dof considered. Damping is usually mainly caused by the fluid surrounding the moving mechanical structure, by anchor losses and thermoelastic dissipations. Other sources of solid damping are present due to, e.g., internal rearrangements of crystals in polycrystalline materials or to the scattering of the acoustic phonons associated with the resonant mode with thermal phonons both in the Landau-Rumer regime and the Akhiezer regime [1, 2]. Complexity of damping phenomena can in some case oblige to abandon the simplified and linearized version $b\dot{u}$ appearing in Eq. (12.1) and considering more complex and nonlinear functions of kinematic quantities.

The stiffness k should be computed applying structural mechanics; this is standard in the linear regime but it can become more complex in nonlinear regimes due to, e.g., coupling of bending and membrane effects in beams.

Coming to the external forcing term, F, one must be aware that external forces in MEMS are in many cases generated by multi-physics interactions, e.g., through electrostatic, electro-thermal, or piezoelectric actuators. Therefore relevant, although simplified, governing equations must be considered for each of these phenomena, as briefly discussed in the subsequent subsections.

Furthermore, because of the small dimensions of Microsystems and due to the increasing request of improved performances, it is frequent to activate the nonlinear dynamic behaviour during the regular functioning of the MEMS resonant device. MEMS designers must then be able to simulate such nonlinear behaviour and to identify the possible sources of nonlinearities (e.g., geometric nonlinearities, fluid-structure interaction, material nonlinearities, and so on). The one dof description can in many cases be still valid but Eq. (12.1) will be supplemented by nonlinear terms, e.g., quadratic, cubic, or higher order functions of the displacement, as will be discussed through various examples in Sects. 12.3, 12.4 and 12.5.

In the following subsections, we briefly discuss the five major sources of loading and multi-physics coupling in MEMS which can also be the origin of nonlinear responses. In Sect. 12.5 other sources of nonlinearities will be discussed.

12.2.2 Mechanical forces and nonlinearities

The internal reactions of deformable bodies in MEMS depend on displacements, deformations, and stresses to which deformable parts are subject. During the regular working conditions stresses are usually low with respect to the elastic limit of the material and the mechanical behaviour can then be considered as linear elastic [3]. This situation changes when the device is subject to non-common working conditions like in the cases of accidental impacts, when the stress levels can cause fracture. Fracture phenomena always imply a nonlinear response due to the variation of instantaneous material and structural stiffness related to the propagation of cracks. Accidental drop events and fracture in microsystems have been extensively studied, e.g., in [4–8], they represent a highly nonlinear mechanical response in MEMS, that must be studied and governed with the approaches of Fracture Mechanics.

Inside MEMS there are often oscillating parts which are kept in motion by means of on-board actuators. The frequency of oscillations can be high, in the order of tens of thousands of Hertz like in micro-gyroscopes or even very high, in the order of millions of Hertz, like in resonators; this means that during the expected lifetime the oscillating part will undergo billions of cycles. This situation can imply problems related to *mechanical fatigue* or *subcritical crack propagation*, a nonlinear phenomenon well studied in metallurgy which appears for stress levels which are much lower than those able to cause instantaneous fracture. Experimental evidences show that the phenomenon can appear also in silicon MEMS, usually for stress levels that are much higher than those used in regular working conditions for oscillating parts and for very high number of cycles (see, e.g., [9–12]).

The *internal reactions* of deformable bodies can depend nonlinearly on the level of displacement and deformation also in the linear elastic regime, due to the so-called nonlinear geometric effects. As a meaningful example, when the flexural oscillations of a beam are small, the bending behaviour is completely decoupled from the axial one and the study of the deformable body can be carried out in the geometrically linear regime. In some meaningful cases, as those discussed in the Sects. 12.3.2.1 and 12.3.2.2, the axial response affects the bending one, then a *geometrically nonlinear response* is activated with the bending stiffness dependent on the level of axial force in the beam. Depending on the structural configuration, this effect can induce the so-called *mechanical hardening or softening* with interesting implications on the nonlinear dynamic response (see, e.g., [13, 14]). Geometric nonlinearities in MEMS usually end-up in a hardening behaviour of the mechanical structure, however, in the literature there are examples of micro-structures that exhibit softening mechanical behaviours [15–19].

12.2.3 Electrostatic forces and nonlinearities

Electrostatics and its many implications play a major role in the study and design of MEMS also in a mechanically oriented vision since it requires the study of a fully coupled electro-mechanical problem and is the source of several nonlinear phenomena.

Consider, for instance, the case where the driving force is generated by a parallel plate configuration through a potential difference between the movable part of the MEMS (*shuttle*) kept at voltage $V = V_p$ and fixed electrodes kept at voltage $V = 0$ as depicted in Fig. 12.3.

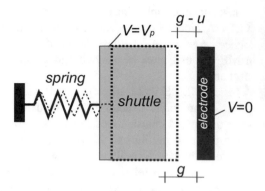

Fig. 12.3 Scheme of the parallel plate driving

In the situation shown in Fig. 12.3 the movable plate is connected through a linear elastic spring with stiffness k to a fixed point and is kept parallel to a second fixed plate.

Considering the movable plate with a mass m, and neglecting damping sources, the equation of motion governing its displacement u, is similar to the one studied in Sect. 12.2.1:

$$m\ddot{u} + ku = F. \tag{12.15}$$

In the present case the external force F is the resultant of electrostatic interactions acting on the surface of the movable plate and can be computed as

$$F = \frac{\epsilon_r \epsilon_0 S}{2} \left(\frac{\bar{V}}{g - u} \right)^2, \tag{12.16}$$

being $\epsilon_r \epsilon_0$ the product of relative and absolute permittivity, S the surface of the parallel plate, g the initial gap between the parallel surfaces.

At difference with the example considered in Sect. 12.2.1, the obtained equation is nonlinear in the displacement; hence we must now deal with a nonlinear oscillator. Its study can be done with the tools of nonlinear dynamics.

In practical applications, the Eq. (12.15) is studied with the voltage difference considered as a given function of time, e.g., periodically variable in time, and a large variety of possible responses can be obtained.

It is useful to develop the electrostatic force as a function of the displacement; the reference value for the displacement is chosen equal to the solution of the static equilibrium equation, i.e., Eq. (12.15) without the inertial term:

$$u_0 = u_{eq} : \quad k u_{eq} - \frac{\epsilon_r \epsilon_0 S}{2} \left(\frac{\bar{V}}{g - u_{eq}} \right)^2 = 0. \tag{12.17}$$

The displacement u can then be considered as the sum of u_{eq} and an increment Δu:

$$\Delta u \equiv (u - u_0) = (u - u_{eq}), \quad u = u_{eq} + \Delta u. \tag{12.18}$$

The equation of motion can be rewritten as follows:

$$m\ddot{u} + ku - \frac{\epsilon_r \epsilon_0 S}{2} \left(\frac{\bar{V}}{g - u} \right)^2 = 0 \approx \tag{12.19}$$

$$m\Delta \ddot{u} + \left(k - \frac{\epsilon_r \epsilon_0 S \bar{V}^2}{(g - u_{eq})^3} \right) \Delta u - \frac{\epsilon_r \epsilon_0 S \bar{V}^2}{2} \left[\frac{3\Delta u^2}{(g - u_{eq})^4} + \frac{4\Delta u^3}{(g - u_{eq})^5} + \dots \right] = 0.$$

The obtained equation is now expressed in the unknown Δu, which is the variation of the displacement with respect to the static equilibrium position.

Depending on the chosen order for the development of the electrostatic force, we obtain a linear or a nonlinear oscillator of second, third, and higher orders.

As a general remark, from Eq. (12.19) it can be seen that the nonlinear effects are amplified by the voltage, considered as a given quantity. Depending on the electromechanical properties of the device, it will be possible to detect a maximum value of given voltage below which the nonlinear effects become negligible.

As it will be discussed in the following sections, usually the optimal response of microsystems in terms of the whole device controlled by an electronic circuit is obtained in the linear regime, hence the study of nonlinearities as the ones contained in Eq. (12.19) is of paramount importance to understand the working regime of a device and to forecast possible misfunctioning. In some peculiar cases a nonlinear response can be also advantageous to obtain better performances.

If we consider only the terms up to the linear one in Eq. (12.19), we obtain the linear oscillator equation:

$$m\Delta \ddot{u} + (k - k_{elec}) \Delta u = m\Delta \ddot{u} + \left(k - \frac{\epsilon_r \epsilon_0 S \bar{V}^2}{(g - u_{eq})^3} \right) \Delta u = 0, \tag{12.20}$$

where the so-called electrostatic stiffness k_{elec} has been defined. The linear oscillator described by Eq. (12.20) has the frequency depending on the assigned voltage difference between the capacitor plates, i.e.:

$$\omega = \sqrt{\frac{(k - k_{elec})}{m}} = \sqrt{\frac{\left(k - \frac{\epsilon_r \epsilon_0 S \bar{V}^2}{(g - u_{eq})^3} \right)}{m}}. \tag{12.21}$$

From the above expression we understand that the electro-mechanical interaction has the effect of reducing the effective stiffness. This in turn reduces the natural frequency of the device and has important implications in the design of resonant microsystems or inertial microsystems in which resonant devices are inserted.

Let us consider again the static equilibrium Eq. (12.17), it is interesting to consider it as a condition giving the voltage as a nonlinear function of the static displacement. The plot has a maximum characterized by the following values of displacement and electric potential:

$$u_{eq} \equiv u_{pi} = \frac{d_0}{3},$$

$$\bar{V} \equiv \bar{V}_{pi} = \sqrt{\frac{8}{27} \frac{k}{\epsilon_r \epsilon_0 S} (g)^3}. \qquad (12.22)$$

u_{pi} and V_{pi} are called *pull-in* displacement and Voltage; when these levels are reached, the static equilibrium can no more be guaranteed and the movable parallel plate suddenly goes toward the fixed one.

In the dynamic response, when the static equilibrium position u_{eq} is the one at pull-in, the natural frequency of vibration around the static equilibrium position goes to zero; this in turn implies that the oscillating motion transforms into a divergent one and a dynamic pull-in phenomenon shows up. More generally, dynamic pull-in should be discussed for the electro-mechanical coupled problem applying the notion of stable or unstable motion and looking for the situations that transform a stable vibrating motion into an unstable one.

12.2.4 Fluid-structure interaction

An important source of dissipative effects and possible nonlinearities is the fluid-structure interaction of solid portions moving at high frequency inside MEMS boxes in which gas at various pressures is contained. This interaction, called *fluid damping* [20, 21], happens, e.g., in resonant accelerometers and in the majority of micro-gyroscopes. The quantitative evaluation of fluid damping must be based on the accurate representation of the fluid-structure interaction at varying internal pressures. Different regimes must be distinguished for the fluid; they go from the standard fluid dynamics represented by Navier-Stokes equations, which hold at atmospheric pressure, to rarefied gas dynamics with non-deterministic, statistical descriptions like Boltzmann equation, which hold at low pressure. In peculiar situations, e.g., for the movement of micromirrors plates, it is not possible to simplify the fluid damping with linear terms and nonlinear effects must be considered [22, 23].

In microsystems, another typical fluid-structure interaction problem is the one governing the behaviour of a micro-pump where a deformable membrane interacts with a fluid with the purpose to move the fluid. A precise evaluation of the whole system behaviour needs the solution of the fully coupled fluid-structure interaction problem.

A situation somewhat similar to the one of micro-pumps concerns the Micro Ultrasonic Transducers (MUT) or Piezo Micro Ultrasonic Transducers (PMUT) which send and/or receive acoustic waves in the surrounding fluid (air or liquid) domain.

In general terms, a fluid-structure interaction problem is met whenever a fluid in motion interacts with a deformable solid in such a way that the mechanical response of the solid depends on the behaviour of the fluid and the fluid motion in the region surrounding the solid is influenced by the mechanical response of the solid.

Solving a fluid-structure interaction problem thus means solving simultaneously the equations governing the solid and those governing the fluid. In many cases this is a very difficult task and a lot of numerical procedures have been ad-hoc formulated. In a more general view, the fluid-structure interaction problem can be considered as a multi-physics or coupled problem.

12.2.5 Thermal effects

Thermal phenomena can have an important influence on the mechanical response of microsystems. In general, temperature always influences the mechanical response through the presence of thermal strains and through the mismatch of the coefficient of thermal expansion (CTE) of the various materials composing the devices. Moreover, material parameters like the Young's modulus and CTE also change with temperature [24, 25] thus deeply influencing the frequency response of the MEMS resonant devices. A typical situation in which thermal stresses arise in a deformable elastic body is when the body is non homogeneous and made with different parts having different CTEs; in this case also uniform temperature distributions cause noncompatible thermal strains due to the CTE mismatch in neighbouring materials.

The temperature dependence of the resonant frequency of MEMS devices has been extensively studied in the literature especially in resonators for real time clocking applications [26–28]. In particular, the temperature dependence of the natural frequency has been investigated in combination with nonlinearities, as in the example of Sect. 12.3.2.2.

CTEs mismatch and variable temperature conditions during fabrication can be the source of *residual stresses* at the end of the fabrication process, which in turn can produce nonlinear effects in the dynamic response of oscillating beams and plates. This source of nonlinearity is strictly process-related and in many cases very difficult to dominate with simple modelling. Residual thermal stresses originate due to a combination of temperature gradient, CTE mismatch, kinematic constraint, and the way in which different parts of the body are warmed and/or cooled during the fabrication process. Residual stresses are of paramount importance in thin film production and they must be estimated and taken into consideration during design of microsystems.

Heat conduction is present in MEMS also in the form of coupled effects, in particular the study of the thermo-mechanical coupling is very important to understand the so-called *solid damping* in which vibration kinetic energy is partially transformed into thermal energy. When the pressure inside the MEMS box reduces to very low values, near to vacuum conditions, fluid damping becomes negligible, while an important source of damping comes from the interaction of thermal and mechanical fields in the thermoelastic responses. This form of solid damping, called *thermoelastic damping* or *TED* [29–32], cannot be easily eliminated; it can be only reduced by careful design of the deformable portions: reducing the volumetric part of the deformation makes them less prone to thermoelastic damping [26, 33].

If one considers the material behaviour out of the elastic regime, thermo-mechanical coupling involves more complex phenomena than in standard thermoe-lasticity. As an example, it must be recalled that anelastic phenomena as plasticity or viscosity are accompanied by dissipative effects which induce local temperature increments, thermal strains, and mechanical property variations which should be described by suitable constitutive laws as in the case of thermo-visco-plasticity. At the limit at very high temperatures, an elasto-plastic solid can behave like fluids, in this case there is a very strong thermo-mechanical coupling as, e.g., in metal forming processes.

Before concluding this section, it is important to mention the Joule effect, which is another important coupling phenomenon due to the transformation of electrical energy into thermal energy whenever a current flows in a conductor. The power dissipated in heat is given by the product of the electric current and the voltage variation across the conductive element; taking into account the Ohm's law, it can also be expressed as voltage drop to the second power divided by the electric resistance.

Electro-thermal coupling coming from Joule effect can be taken into consideration by inserting in the thermal power balance an internal heat source given by Joule effect. In the case in which the electric parameters, like resistivity, can be considered as independent from the temperature, the electric and the thermal problem can be solved in sequence. More generally, the electric and the thermal problem must be solved together as fully coupled.

The Joule effect can be exploited in MEMS to create electro-thermo mechanical actuators that are capable of exerting a high force with respect to the electrostatic ones.

12.3 Analytical methods for linear and nonlinear problems in MEMS

The purpose of the present section is to show how the equations of motion for a large class of MEMS can be obtained through the Hamilton's principle and a proper discretization procedure.

Four meaningful examples are also reported. They illustrate the effectiveness of the reduction of complex systems to one degree-of-freedom (dof) models for the description of MEMS resonant devices. The simple models are endowed with analytical or semi-analytical solutions, which give predictions in good agreement with experimental results.

12.3.1 Hamilton's principle for a discrete formulation

The mechanical evolution of a body subject to conservative forces can be characterized by the *Lagrangian* function $\mathcal{L}$, defined as the difference between the kinetic energy $\mathcal{T}$ and the potential energy, Π: $\mathcal{L} = \mathcal{T} - \Pi$. The integral of the *Lagrangian* between two time instants t_1 and t_2 is called the *action functional*.

Hamilton's principle states that the true evolution of a system between two specified states $\mathbf{u}_1 = \mathbf{u}(t_1)$ and $\mathbf{u}_2 = \mathbf{u}(t_2)$ is a stationary point of the *action functional*, i.e., it is a solution of the functional equation

$$\int_{t_1}^{t_2} [\delta \mathcal{T}(\dot{\mathbf{u}}) - \delta \Pi(\mathbf{u})] dt = 0. \tag{12.23}$$

If non-conservative forces are present, this principle states that the sum of the time variation of the difference between kinetic and potential energies and the work done by the non-conservative forces over any time interval t_1 and t_2 equals zero, for any varied path $\delta \mathbf{u}$ from time t_1 and t_2 with $\delta \mathbf{u}(t_1) = \delta \mathbf{u}(t_2) = 0$:

$$\int_{t_1}^{t_2} \delta[\mathcal{T}(\dot{\mathbf{u}}) - \Pi(\mathbf{u})] dt + \int_{t_1}^{t_2} \delta W_{nc}(t) dt = 0, \tag{12.24}$$

δW_{nc} being the virtual work of non-conservative forces. If the Lagrangian of a system is known, then the equations of motion of the system may be obtained by a direct substitution of the expression for the Lagrangian into the Euler–Lagrange equation:

$$\frac{\partial \mathcal{L}}{\partial \mathbf{u}} - \frac{d}{dt} \frac{\partial \mathcal{L}}{\partial \dot{\mathbf{u}}} = 0. \tag{12.25}$$

A discrete formulation is obtained by approximating the displacement field $\mathbf{u} = \mathbf{u}(\mathbf{x}, t)$ as

$$\mathbf{u}(\mathbf{x}, t) = \sum_{i=1}^{n} \boldsymbol{\Psi}_i(\mathbf{x}) U_i(t), \tag{12.26}$$

where $\boldsymbol{\Psi}_i(\mathbf{x})$ are suitably chosen spatial shape functions and $U_i(t)$ are the dof of the problem. Substituting Eq. (12.26) into Eq. (12.25) one reduces the dynamic behaviour of the deformable body to the oscillation of a n-dof system, possibly nonlinear.

Often in MEMS devices just one dof can be retained ($U_1 = U$). For instance, the oscillations of resonators constituted by slender beams can be described by the transverse displacement

$$u(x, t) = \Psi(x)U(t), \tag{12.27}$$

where Ψ is the eigenfunction, normalized with the maximum value, and $U(t)$ represents the displacement of the corresponding point (often the mid-point of the beam).

The nonlinear dependence on U of the Euler–Lagrange equation is often transformed in polynomial expressions after development in Taylor series expansion of the nonlinear functions, [34]. The equation of motion can thus be expressed as follows:

$$m\ddot{U} + b\dot{U} + \sum_{j=1}^{p} k_j U^j = F(t). \tag{12.28}$$

The linear case corresponds to $p = 1$. The nonlinear terms arise from geometric and/or electrostatic nonlinearities. The expressions of the equivalent mass m, damping coefficient b, stiffnesses k_j, and force F depend on the specific problem considered and will be given in the following subsections for some meaningful cases. In general, $F(t)$ contains the external forces, if any, and the actuation forces. The electrostatic actuation, besides introducing linear and nonlinear stiffness terms, can also introduce nonlinearity in $F(t)$ for high AC voltage, [35]. However, as in the real cases considered in the following, the AC voltage is small with respect to the DC voltage, the simple harmonic form $F(t) = \overline{F} \cos(\omega t)$ will be considered.

When two or more dof are retained in Eq. 12.26, a system of nonlinear equations is obtained.

12.3.1.1 Duffing oscillator

The dynamic of several MEMS devices, after reduction to a one dof system, can be well described by a Duffing oscillator. Its equation of motion is a special case of Eq. (12.28) with $p = 3$ and $k_2 = 0$, namely

$$m\ddot{U} + b\dot{U} + k_1 U + k_3 U^3 = \overline{F} \cos(\omega t). \tag{12.29}$$

To solve Eq. (12.29) several methods can be used. The successive approximations method, [13], the harmonic balance method, [36, 37], the perturbation methods [38] such as the method of multiple scales [39] have been developed and applied to this problem.

Here we adopt the successive approximations method. The solution is hence expressed as, see, e.g., [13] for details:

$$U = \bar{U} \cos(\omega t) - \frac{\bar{U}^3}{32} \frac{k_3}{k_1} \cos(3\omega t). \tag{12.30}$$

In Eq. (12.30) ω is the actual value of the angular frequency and $\bar{U}$ is the oscillation amplitude; they are related to the amplitude $\overline{F}$ of the driving force by

$$\left(\frac{\overline{F}}{k_1}\right)^2 = \left(2\left(1 - \frac{\omega}{\omega_0}\right)\bar{U} + \frac{3}{4}\frac{k_3}{k_1}\bar{U}^3\right)^2 + \left(\frac{b}{m\omega_0}\bar{U}\right)^2 \tag{12.31}$$

with $\omega_0 = \sqrt{\frac{k_1}{m}}$ angular eigenfrequency of the linear problem.

For given mass, stiffness, damping parameters and forces, Eq. 12.31 allows to obtain the forced frequency response of the nonlinear resonator.

Figure 12.4 shows the oscillation amplitude versus the normalized frequency for different combinations of the parameters. The linear response is represented by the central black curves ($k_3 = 0$). When $k_3 < 0$, the resonance curves are bent on the left, exhibiting the so-called *soft spring* effect. When $k_3 > 0$, the resonance curves are bent on the right, *hard spring* effect; thick and thin lines correspond to high and low quality factor Q, respectively. In the nonlinear case there is a region of bi-stability, i.e., a region where, for a given frequency, there are two stable solutions (marked by P_{st} in Fig. 12.4) and a third unstable solution (marked by P_{inst} in Fig. 12.4). For very low values of Q, this region disappears, even in the nonlinear case, as shown by the dashed line in Fig. 12.4.

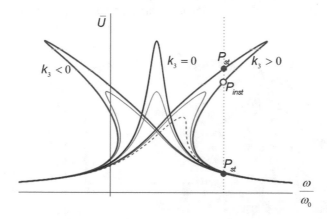

Fig. 12.4 Oscillation amplitude versus normalized frequency. Softening ($k_3 < 0$), linear ($k_3 = 0$) and hardening behaviour ($k_3 > 0$); thick and thin lines correspond to high and low Q, respectively.

12.3.2 Application to MEMS Devices

The present subsection is intended to give an overview of possible nonlinear and coupled responses in microsystems, with particular reference to the dynamic response of electrostatically actuated mechanical oscillators, important components of inertial MEMS.

12.3.2.1 Oscillator for a resonant accelerometer showing hardening and softening behaviour

As a first example of the key role of resonators in MEMS and of their analytic modelling, let us analyse the silicon uniaxial resonant accelerometer proposed in [40] and shown in Fig. 12.5a.

The accelerometer is composed of a square inertial mass, two suspension beams (or springs, horizontal in Fig. 12.5) and two resonators (thin vertical beams in Fig. 12.5). The resonators are attached to the substrate at one end and to the springs at the other end, at small distance d_1 from the anchor point. They are actuated at resonance and sensed by two parallel electrodes attached to the substrate.

During functioning, the external acceleration a causes the translation of the inertial mass m, which, in turn, induces tension and compression of the same magnitude in the two resonators (Fig. 12.5b). As a result, the resonance frequency of the two oscillators changes, providing a differential sensing of the external acceleration.

To increase the sensitivity of the accelerometer, the frequency change should be increased. This leads to design very slender beams that can easily enter the geometrically nonlinear regime and to design small gaps between the electrodes with consequent strong electrostatic nonlinearity in parallel plates.

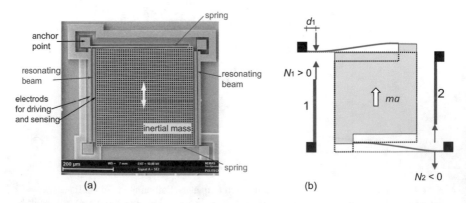

Fig. 12.5 (a) SEM image of the resonant accelerometer. (b) Effect of external acceleration a. Rielaborated from: [40], Fig. 1 and 6. Reproduced with permission of IEEE

Fig. 12.6 Schematic view of the L-shaped resonator with the actuation scheme

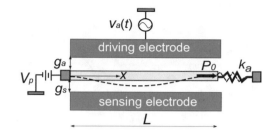

The resonators can be analysed as slender beams clamped at one end and constrained through an equivalent axial spring k_a (accounting for the little arm d_1) at the other end, see Fig. 12.6. Expressing the transversal displacement as in Eq. (12.27), one obtains the one dof Eq. (12.28) with $p = 3$ and with the equivalent mass, stiffnesses, and damping coefficient expressed as, [41]

$$m = \int_0^L \rho A \Psi^2 dx, \tag{12.32}$$

$$k_1 = k_{m1} + k_G + k_{e1}, \tag{12.33}$$

$$k_{m1} = \int_0^L EJ(\Psi'')^2 dx, \tag{12.34}$$

$$k_G = P_0 \int_0^L (\Psi')^2 dx, \tag{12.35}$$

$$k_{e1} = -\epsilon_0 w V_p^2 \left(\frac{1}{g_a^3} + \frac{1}{g_s^3} \right) \int_0^L \Psi^2 dx \tag{12.36}$$

$$k_2 = k_{e2} = -\frac{3}{2} \epsilon_0 w V_p^2 \left(\frac{1}{g_a^4} - \frac{1}{g_s^4} \right) \int_0^L \Psi^3 dx, \tag{12.37}$$

$$k_3 = k_{m3} + k_{e3} = \frac{1}{2} \frac{k_a EA}{k_a L + EA} \left[\int_0^L (\Psi')^2 dx \right]^2 - 2\epsilon_0 w V_p^2 \left(\frac{1}{g_a^5} + \frac{1}{g_s^5} \right) \int_0^L \Psi^4 dx, \tag{12.38}$$

$$b = \int_0^L b^* \Psi^2 dx, \tag{12.39}$$

where A is the cross section, J is the inertia moment, E is the Young's modulus, P_0 is the axial force induced by external acceleration, V_p is the bias voltage, w is the out-of-plane thickness of the electrodes, and b^* is the viscous coefficient.

The linear mechanical stiffness accounts for the bending stiffness of the beam k_{m1} and for the geometrical stiffness k_G which depends on the external acceleration through the axial force. The effect of the first-order electrical stiffness is a downward shift of the resonator natural frequency. For ideal structures, with symmetric gaps $g_a = g_s$, the nonlinear second order term disappears (Eq. 12.37) thus recovering the case of a Duffing oscillator discussed in Sect. 12.3.1.1. The third order stiffness (Eq. 12.38) has a mechanical and an electrostatic contribution. The first one is always positive (*hard spring* effect) and accounts for the membrane effect: it disappears if $k_a \to 0$, i.e., if there is no axial constraint and it is maximum for the doubly clamped beam, without the transversal short arm ($k_a \to \infty$). The second term, due to the electrostatic actuation, is negative (*soft spring* effect) and grows

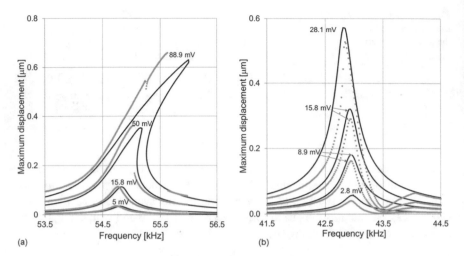

Fig. 12.7 Spectral response of the L-shaped resonator for different actuation voltages: comparison experiments (orange points)—theoretical prediction (black lines). (**a**) $V_p = 4V$; (**b**) $V_p = 9V$

with the bias potential V_p, thus it can mitigate the hardening effect associated with the third order term of the elastic stiffness. As pointed out in [42], by properly choosing the bias voltage one can compensate the mechanical nonlinearity with the electrical one in order to extend the linear behaviour to a wider range of actuation voltages. Figure 12.7 shows this effect: with a bias voltage $V_p = 4V$ a strong *hard spring* effect is visible (Fig. 12.7a), while an almost linear behaviour is obtained with $V_p = 9V$ (Fig. 12.7b). In both cases a good agreement was found between experimental results (orange) and numerical predictions obtained with the one dof model (black) [41].

12.3.2.2 Double Ended Tuning Fork for a resonant accelerometer showing hardening behaviour under varying temperature conditions

This second example highlights the effects of the temperature variation in the nonlinear dynamic response of a MEMS oscillator. Reference is made to the Double Ended Tuning Fork resonator (DETF) of the resonant accelerometer studied in [43, 44] and shown in Fig. 12.8. The functioning of this accelerometer is similar to the one described in the previous section, a DETF is employed in this case and comb fingers, instead of parallel plates, are used for actuation and sensing.

Figure 12.8a and b show a SEM image of the DEFT resonator and its resonant mode. The effect of the external acceleration, acting along the *x*-axis on the inertial mass (not shown in the figure), is an inertial force that, through a leverage mechanism, is transmitted to the DEFT as an axial force P_0.

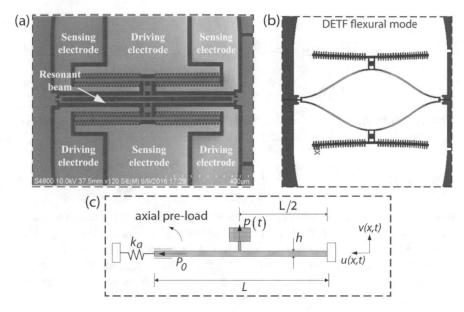

Fig. 12.8 (a) SEM Close-up view of the DETF resonator discussed in [44]. (b) Schematic view of the DETF resonator. (c) Schematic view of a single resonant beam

A single resonant beam of the DEFT can be schematized as shown in Fig. 12.8c, where the axial spring k_a is introduced to take into account the leverage system and the force exerted by the driving comb fingers on the resonator is modelled through the transversal dynamic load $p(t)$.

When the device is subject to significant temperature variations ΔT with respect to a reference temperature $T(0) = T_0$, the silicon properties change and modify the dynamic response of the system. To model this effect, in [44] the temperature-dependent expressions of the Young's modulus and of the thermal expansion coefficient proposed in [45] and [46] were considered, namely

$$E(T) = E(298.16\,\text{K})\left(1 + TCE_1 \cdot \Delta T + TCE_2 \cdot \Delta T^2\right)$$
$$\alpha(T) = -4 \times 10^{-12}T^2 + 8 \times 10^{-9}T + 4.7 \times 10^{-7}, \tag{12.40}$$

with $TCE_1 = -63.82\text{ppm/K}$, $TCE_2 = -51.99\text{ppb/K}^2$, and $E(298.16\text{K}) = 168.9\text{GPa}$ for $<110>$ monocrystalline silicon.

As detailed in [44], the nonlinear dynamics of the resonator under varying temperature conditions is again governed by Eq. (12.28) with $p = 3$. However, the equivalent mass and stiffnesses differ from those given by Eq. (12.32) and Eqs. (12.35)–(12.38). In particular, the presence of comb fingers attached to the resonator entails an additional term in the equivalent mass m_{comb}, furthermore, the linear behaviour of comb fingers actuation and sensing results in zero electrostatic stiffnesses $k_{e1} = k_{e2} = k_{e3} = 0$. The mechanical stiffnesses are temperature dependent and read

$$k_{m1} = \int_0^L E(T)J\,(\Psi'')^2 dx, \tag{12.41}$$

$$k_G = (P_0 - \frac{k_a E(T)A \cdot \alpha(T)\Delta T}{k_a L + E(T)A} L) \int_0^L (\Psi')^2 dx, \tag{12.42}$$

$$k_3 = k_{m3} = \frac{1}{2} \frac{k_a E(T)A}{k_a L + E(T)A} \left[\int_0^L (\Psi')^2 dx \right]^2. \tag{12.43}$$

Note that, as in this case $k_3 = k_{m3} > 0$, only hard spring behaviour is expected. The equivalent load depends on the temperature variation and reads

$$F = p(t) + \frac{2\alpha(T)\Delta T E(T)J}{h} \int_0^L \Psi'' dx, \quad \text{with } p(t) = \frac{\partial C}{\partial x} V_P v_a, \tag{12.44}$$

where $v_a = |v_a| \cos(\omega_F t)$ is the actuation voltage, C is the capacitance measured by the comb fingers, V_p is the bias voltage, and h is the beam height (see Fig. 12.8c).

Figure 12.9 shows the frequency responses of the resonator measured at different temperatures (curves marked by stars). Both the natural frequency of the resonator and the maximum vibration amplitude at resonance decrease with increasing

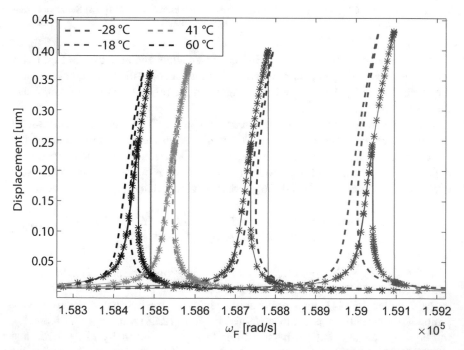

Fig. 12.9 Frequency response of the DETF resonator discussed in [44] at different temperatures in the range [−30°C; 60°C]. A driving voltage of 10mV of amplitude and a bias voltage of 10V is applied to the device

temperatures. This effect is well reproduced by the one dof model: the solution of Eq. (12.28) with the parameters defined by (12.41)–(12.44), shown by dashed lines in Fig. 12.9 is in good agreement with the experimental results.

12.3.2.3   Torsional resonator for a resonant accelerometer showing softening behaviour

The one dof nonlinear oscillator model can also be applied to resonant structures having angular oscillations. This is the case of the micromirror that will be discussed in the next section and of the so-called torsional resonators.

The torsional resonators considered in this section are the sensing components of the out-of-plane resonant accelerometer proposed in [47] and shown in Fig. 12.10.

The working principle of this device is the following. The torsional resonators, marked by A and B in Fig. 12.10a, are kept at resonance during the functioning by electrostatic actuation (Fig. 12.11). They are attached to the inertial mass which, due to an out-of-plane acceleration, tilts around the axis $a - a$ and makes the gap g_0 between the resonators and the electrodes change. Since the electrostatic stiffness depends on the gap, it changes and makes the frequency vary in opposite directions in resonators A and B, thus allowing for a differential acceleration sensing.

The rotation of the resonator is expressed as $\theta(x, t) = \Theta(t)\Psi(x)$, with Ψ a proper shape function and $\Theta(t)$ is assumed as the single dof of the oscillator. The dynamic equilibrium equation is again expressed by Eq. (12.28) with $U(t) \equiv \Theta(t)$. To be able to obtain good predictions also for relatively high values of actuation voltage, and to explore the limits of the Duffing model, we consider here $p = 5$. Since in this case the dof is an angle, the equivalent mass, stiffnesses, and forcing term change, accordingly, their dimensions and read

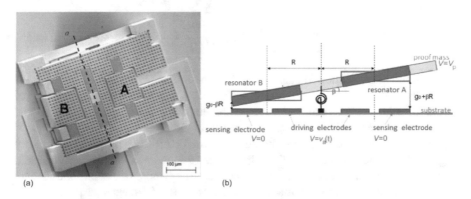

(a)   (b)

Fig. 12.10   (a) SEM image of the out-of-plane accelerometer with two torsional resonators A and B; (b) schematic side view of the accelerometer inclined by an external acceleration. Source: [48], Fig. 1. Reproduced with permission of Elsevier

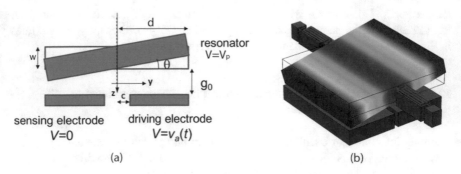

Fig. 12.11 (**a**) Electrostatic actuation scheme of the torsional resonator, (**b**) first torsional mode of the torsional resonator. The contour plot of the displacement field is shown in colour

$$m = 2 \int_0^L \rho J_p \Psi^2 \mathrm{d}x + \rho J_{mass}, \tag{12.45}$$

$$k_1 = k_m + k_{e1}, \tag{12.46}$$

$$k_m = 2 \int_0^L G J_t (\Psi')^2 \mathrm{d}x, \tag{12.47}$$

$$k_{e1} = -\frac{2\epsilon_0 B}{g_0^3} V_p^2 \left(\frac{d^3}{3} - \frac{c^3}{3} \right), \tag{12.48}$$

$$k_3 = k_{e3} = -\frac{4\epsilon_0 B}{g_0^5} V_p^2 \left(\frac{d^5}{5} - \frac{c^5}{5} \right), \tag{12.49}$$

$$k_5 = k_{e5} = -\frac{6\epsilon_0 B}{g_0^7} V_p^2 \left(\frac{d^7}{7} - \frac{c^7}{7} \right), \tag{12.50}$$

$$F(t) = \frac{\epsilon_0 B}{g_0^2} V_p v_a(t) \left(\frac{d^2}{2} - \frac{c^2}{2} \right), \tag{12.51}$$

where d and c are defined in Fig. 12.11, J_t is the torsional moment of inertia, G is the shear elastic modulus, J_p is the polar moment of inertia of a single spring of length L, ρJ_{mass} is the centroidal mass moment of inertia of the rigid mass, and $F(t)$ is now the external torque applied to sustain the oscillation.

It is worth noting that the mechanical torsional stiffness is linear also for very large angles, hence the only nonlinearity comes from the parallel plate electrostatic actuation. The even order stiffness terms are zero for symmetry reasons and k_{e3} and k_{e5} are negative: this implies a softening effect in the frequency response of the torsional resonator.

In Fig. 12.12a the frequency response curve for a torsional resonator (the geometric dimensions used are the ones of the torsional resonator of the z-axis

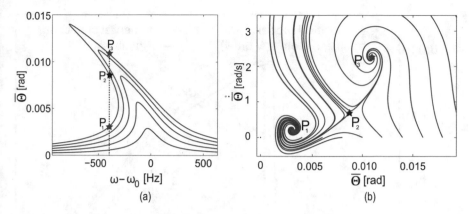

Fig. 12.12 Dynamic response of the torsional resonator discussed in [50]. (**a**) Frequency response curve with $V_p = 6$V and $|v_a| = 10, 20, 30, 40, 50$ mV; (**b**) trajectories of the system in the state plane: the two stable steady state solutions (P_1, P_3) and the unstable one (P_2) are marked with the red and black stars, respectively

resonant accelerometer studied in [49]) is shown for different values of the actuation voltage $|v_a|$ at the fixed polarization voltage $V_p = 6$V.

Similarly to what evidenced in Fig. 12.4 for the hard spring case, also for the present soft spring behaviour there is a range of ω where Θ becomes a multivalued function and for a given ω three solutions are available (see Fig. 12.12a, points P_1, P_2, and P_3). The critical value of the actuation voltage $|v_a|$ after which a bi-stable behaviour occurs is

$$|v_{a_c}| = \sqrt{\frac{8b^3\omega_0^3}{3|k_3|}\frac{2g_0^2}{\epsilon_0 L V_p(d^2 - c^2)}}. \tag{12.52}$$

Figure 12.13, taken from [51], shows the comparison between the analytical prediction obtained with the third order and fifth order approximations of the electrostatic moment and the experimental results on a torsional resonator fabricated through the Thelma[c] surface micromachining process by ST-Microelectronics. From Fig. 12.13 it is evident that for high values of the actuation voltage when the highly nonlinear regime is entered, the analytic prediction becomes less reliable and the inclusion of other terms in the Taylor approximation would not improve the results significantly. On the other hand, the numerical solution obtained by time integration (without Taylor expansions approximation), also shown in Fig. 12.13, well reproduces the experimental response of the resonator thus proving the effectiveness of the proposed one degree-of-freedom model.

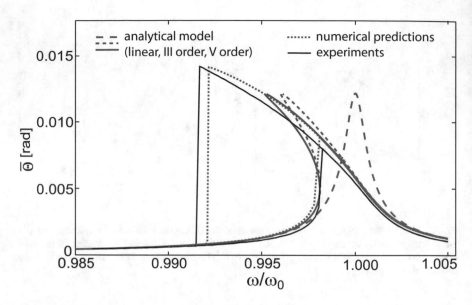

Fig. 12.13 Frequency response of the torsional resonator discussed in [51], for $V_p=5$V and $|v_a|=$ 100 mV: linear, III and V order analytical approximations (orange curves), numerical prediction (dotted blue) and experimental curves (continuous black); both forward and backward frequency sweeps are shown

12.3.2.4   Micromirror showing nonlinearity

Micromirrors are often composed by a plate, attached to the substrate by torsional springs, that during functioning rotates according to its first mode. Even though, as already remarked, torsional springs exhibit a linear behaviour also for very large rotation angles, some nonlinearity, of mechanical nature, can be observed due to particular geometric configurations. Also in that cases the one dof nonlinear oscillation can effectively be employed.

We consider here the micromirror shown in Fig. 12.14 constituted by a circular plate, clamped to the substrate through two torsional beams that allow the rotation of the mirror around the axis a-a. The piezoelectric actuation is realized through four actuating beams clamped at one end to the substrate and attached to the mirror plate by folded springs. The eigenmode of the mirror, characterized by the rotation of the plate, is shown in Fig. 12.15a for half of the device.

The numerical analysis of the dynamic response of the system in the geometrically nonlinear regime is quite demanding due to the complex geometry. One can then search for a simplified approach. Actually, the dynamic of the system is well described by neglecting the deformation of the circular plate (rigid body assumption) and by substituting the deformable elements by equivalent springs. The resulting one dof system is shown in Fig. 12.15b.

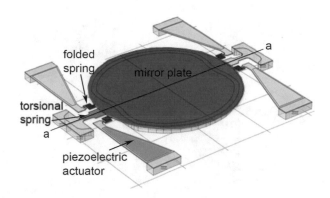

Fig. 12.14 MEMS micromirror with piezoelectric actuation

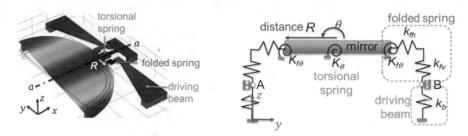

Fig. 12.15 One dof discrete parameters scheme of the mirror

The mirror is reduced to a 2-D system, the mass is lumped into a single rigid body, constrained in both spatial directions, which can only rotate around the axis (a-a). The effect of the torsional beams is lumped into the torsional stiffness K_θ, the effect of the actuating beam is lumped into the stiffness k_b and the effect of the folded spring is lumped into the vertical, k_{fv}, torsional, $K_{f\theta}$, and horizontal, k_{fh}, springs. The rollers at points A and B represent the horizontal constraints of the driving beams. The degrees of freedom of the proposed model is the tilting angle, since the vertical displacement of the driving beams is not independent. The linear stiffnesses introduced in the discrete model can be evaluated by analytic formulas on simplified geometries or numerically computed on the actual geometries of the single deformable elements as the ratio between an applied force (or torque) and the corresponding displacement (or angle), see [52].

Even though all springs are considered as linear, large rotations result in a nonlinear relation between the restoring elastic torque $M(\theta)$ and the angle θ. The nonlinear relation can be obtained imposing the equilibrium in the deformed configuration, see Fig. 12.16 and reads

$$M(\theta) = (K_\theta + 2K_{f\theta})\theta + 2k_{fh}R^2(1-\cos\theta)\sin\theta + \frac{k_{fv}k_b}{k_{fv}+k_b}R^2\sin 2\theta. \quad (12.53)$$

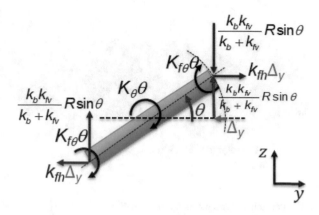

Fig. 12.16 Equilibrium in deformed configuration of the one dof scheme of the mirror

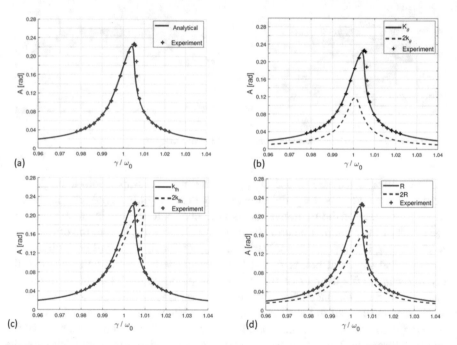

Fig. 12.17 (a)Frequency response curves of analytical model and experimental points at 15.2 V, (b) influence of torsional stiffness K_θ, (c) influence of folded springs stiffness k_{fh}, and (d) influence of distance R

Equation (12.53) can be approximated by a third order polynomial expression, thus reducing the nonlinear behaviour of the mirror to that of a Duffing oscillator with a hard spring effect. Figure 12.17a shows the comparison between experimental data and model prediction: a good agreement is observed. The discrete model allows for an easy parametric study which evidences the role of different parameters on the global response of the micromirror. In particular, the effects of the stiffness of the torsional spring K_θ, of the stiffness of the folded springs k_{fh} and of the distance R from the folded springs and the rotation axis are considered and the results obtained by doubling each one of these parameters are displayed in Figs. 12.17b,c, and d, respectively. The torsional spring only governs the linear behaviour: doubling its value one reduces the maximum amplitude by a factor slightly lower than two (see dashed line in Fig. 12.17b). Conversely, the folded spring is responsible of the nonlinear behaviour: the hard spring effect is higher when doubling k_{fh} (see dashed line in Fig. 12.17c). The distance R has a significant effect on both the linear and nonlinear behaviour, see Fig. 12.17d where the continuous blue line represents the response obtained with the real value of R , while the dashed red curve is the frequency response obtained doubling its value.

12.4   Numerical methods for linear and nonlinear problems in MEMS

Analytical approaches are indeed useful in particular at the design level, but often MEMS are highly complex structures that cannot be reduced to simple models. This has motivated the progressive development of numerical techniques for the different multi-physics phenomena involved with the ultimate goal of enabling an accurate prediction of MEMS response. While in the linear regime such capabilities are available for moderate-size models in commercial codes for mechanics, electrostatics, and thermoelasticity, on the contrary large models and, in particular, nonlinear effects still pose severe challenges.

A comprehensive introduction to modern numerical techniques is out of the scope of this book, and we will limit ourselves to the discussion of their application to the specific MEMS family of resonators. Resonators lend themselves to the creation of a reduced model in the nonlinear regime in dynamics without resting on specific structural theories or simplified formulas. We will briefly focus on three main topics: geometric nonlinearities, nonlinear electrostatic forces, and dissipation, encompassing fluid damping, thermoelasticity and anchor losses.

We will validate our procedure on the Double Ended Tuning Fork (DETF) resonator shown in Fig. 12.18 fabricated in polysilicon by ST-Microelectronics [53] through the Thelma[c] surface process in near-vacuum conditions (pressure $p <$ 50 μbar). The material properties of polysilicon utilized in these simulations are: Young's modulus $E = 167\,\text{GPa}$, density $\rho = 2330\,\text{Kg/m}^3$, and Poisson's ratio $\nu = 0.22$.

The two resonating beams are 375 μm long and have an out-of-plane thickness t of 24 μm. Actuation and detection of the motion of the structure are achieved by means of four electrodes placed on the two sides of the beams. The DETF oscillates in its first flexural mode with $f_0 = 550\,\text{kHz}$ in accordance with a standard FEM modal analysis. A prototype has been fabricated in polysilicon

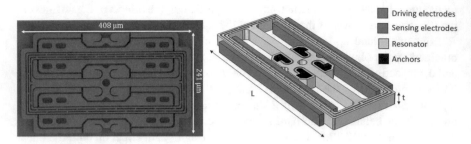

Fig. 12.18 SEM image and CAD model of the DETF1

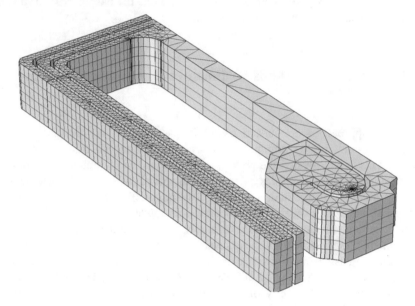

Fig. 12.19 FEM model of a quarter of the resonator

12.4.1 Methodology

The resonators analysed undergo non-infinitesimal, but still moderate transformations. It is worth stressing that the large transformations (rotations) occurring in some MEMS like micromirrors require dedicated approaches that will not be discussed herein. Also, we will only outline a possible procedure to trace the nonlinear frequency response function (FRF) of the actuated mode.

Even for the simple MEMS considered, FEM models are rather large. For instance, the mesh for one-quarter of the structure depicted in Fig. 12.19 is made of approximately 10^4 quadratic wedge elements. Moreover, the low packaging pressure induces quality factors often larger than 10^4. As a consequence, classical *brute force* time-marching schemes have a prohibitive cost and suitable reduced order techniques are advisable.

When the response becomes nonlinear, the resonator oscillates with a form that is still very similar to the linear main mode (MM). Other high frequency modes can impact on the stress distribution and the global stiffness but have limited influence on the electrostatic forces and on fluid dissipation. It is thus still feasible to formulate a one dof model governed by a nonlinear modal coordinate that, for the tuning fork MEMS investigated herein, is the maximum in-plane displacement q of the resonating beams.

In the following Sects. 12.4.1.1–12.4.1.3 we will detail the numerical treatment of geometric, electrostatic, and damping nonlinearities as a function of the selected governing parameter [54].

12.4.1.1   Geometric nonlinearities

A standard approach to enforce mechanical equilibrium conditions for a structure in non-infinitesimal transformations is to apply the Principle of Virtual Power in the reference configuration

$$\int_{\Omega_0} \rho \ddot{\mathbf{u}} \cdot \mathbf{w}\, d\Omega + \int_{\Omega_0} \mathbf{S} : \mathrm{sym}(\nabla^T \mathbf{w} \cdot \mathbf{F}) d\Omega = \int_{S_0} \mathbf{f} \cdot \mathbf{w}\, dS, \quad \forall \mathbf{w} \in \mathcal{C}(0),$$

$$(12.54)$$

where $\mathbf{w}$ is a test function, $\mathbf{u}$ are unknown displacements, $\mathbf{S}$ is the second Piola Kirchhoff stress tensor, $\mathbf{F}$ is the transformation gradient. If small strains are considered, which is typical of MEMS structures, then the following constitutive law holds (Saint-Venant Kirchhoff law)

$$\mathbf{S} = \mathcal{A} : \mathbf{e} \qquad \mathbf{e} = \frac{1}{2}\left(\nabla \mathbf{u} + \nabla^T \mathbf{u} + \nabla^T \mathbf{u} \cdot \nabla \mathbf{u}\right),$$

$\mathcal{A}$ being the constant stiffness tensor and $\mathbf{e}$ the Green-Lagrange strain tensor. As well known, the $\mathbf{S} : \mathrm{sym}(\nabla^T \mathbf{w} \cdot \mathbf{F})$ term generates at most cubic nonlinearities in terms of the unknown displacement field $\mathbf{u}$. Finally, $\mathbf{f}$ are external forces exerted on the surface of the structure (accounting, e.g., for electrostatic pressure and fluid forces).

Several reviews discuss the reduced order modelling of structures at the macroscale governed by Eq. (12.54) (e.g., [55]). Classical techniques resting on the definition of a reduced modal basis where the selection of modes can be based either on physical insight or on Proper Orthogonal Decomposition techniques [56]. Alternative approaches are formulated within the framework of nonlinear Normal Modes (NNM) [57, 58] which are the extension of the classical modal basis of linear problems. Of particular interest for the present investigation is interpretation of NNM as periodic solutions. If a FE model is available, these can be obtained resorting, e.g., to the Harmonic Balance Method [59], but the numerical cost of such approaches rapidly grows with the model size since they require solving several large nonlinear systems of equations.

Recently, a technique tailored for MEMS applications has been proposed and validated in [60]. This procedure can be classified as an implicit static condensation approach [55, 57, 61] in which only one master mode (MM) is retained in the reduced basis.

If $\psi(\mathbf{x})$ denotes the displacement field of the linear MM, mass normalized, the nonlinear stiffness is evaluated by statically forcing the structure with suitable body forces $\mathbf{F}$ which are proportional to $\psi(\mathbf{x})$: $\mathbf{F} = \beta\psi(\mathbf{x})$, similarly to what proposed in [61]. The strong motivation for this choice, besides simplicity, is that these loads are a very good approximation of inertia forces occurring during the vibration and the associated stiffness is particularly accurate.

The range of the load multiplier β is prescribed so as to cover the expected displacements in the structure. For example, for the resonator of Fig. 12.18, a fraction of the gap between the resonator beams and the drive and sense electrodes will be covered. Next, a series of static nonlinear analyses are run spanning the β space. Let $\mathbf{u}(\beta; \mathbf{x})$ denote the solution for a given β and $q(\beta)$ the midspan deflection. We assume that the map $q(\beta)$ is invertible to give $\beta(q)$; invertibility is guaranteed in the small perturbation limit and it can be reasonably guessed that it holds also for mild nonlinearities. Practically, the inversion is performed numerically through fitting procedures. As anticipated, on the contrary we approximate the acceleration as

$$\ddot{\mathbf{u}}(\mathbf{x}, t) = \psi(\mathbf{x})\ddot{q}(t). \tag{12.55}$$

This means that we admit that the linear MM is a good approximation for the inertia force distribution. In [60] it is shown that these assumptions lead to a very simple final form of the ROM:

$$\frac{1}{\psi_\mathrm{q}}\ddot{q} + \beta(q) = F(t), \tag{12.56}$$

where ψ_q is the nondimensional midspan deflection of the modal shape $\psi(\mathbf{x})$, $F(t)$ is the load participation factor, i.e., the projection on ψ of the surface loads (e.g., of electrostatic or fluidic nature, as comment later):

$$F = \int_{S_0} \mathbf{f} \cdot \psi \, dS. \tag{12.57}$$

Equation (12.56) shows that the nonlinear stiffness term actually coincides with the load multiplier and provides not only a clear physical interpretation but also an indication of how to perform the inversion of the nonlinear manifold.

This procedure is rather simple and can be implemented in any commercial code.

The computed stiffness $\beta(q)$ for the resonator of Fig. 12.18 is plotted in Fig. 12.20 which presents only the nonlinear contribution. As a first approximation,

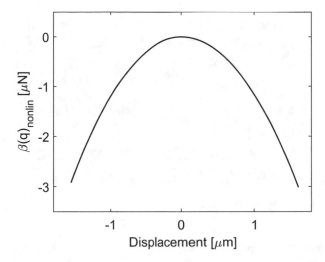

Fig. 12.20 Numerical nonlinear stifness

the resonating beams can be represented as clamped-clamped beams of length L for which classical analytical formulas of the nonlinear stiffness are available. However, ignoring the flexibility of the anchoring regions is only a rough approximation. Moreover, the presence of slots in the beams, inserted to reduce thermoelastic dissipation, imposes to define an equivalent section which is highly arbitrary. As an example, we consider two different values for the in-plane thickness: the real one of 15 μm and an equivalent value of 9.8 μm, respectively. In the latter case the thickness has been decreased by an amount corresponding to the aperture of the slots. The analytical predictions of the natural frequencies are 960 kHz and 531 kHz in the two cases, respectively, the natural frequency of the DETF being 550 kHz. Anyway, even if in the latter case the linear stiffness is almost exact, its nonlinear behaviour remains too hardening.

12.4.1.2   Electrostatic Nonlinearities

Electrostatic analyses are routinely performed on MEMS devices, but the prediction of nonlinear electrostatic forces and nonlinear read-out current requires specific provisions. Progress in computing capabilities is already making fast FEM analyses of complex 3D structures viable. For instance, in [62] concepts of material derivatives have been applied to compute the electrostatic torque and torque derivative for the comb fingers of MEMS micromirrors. However, at the present the most efficient numerical tool is still represented by fast multipole integral equations that have reached a high level of maturity in the last decades (see, e.g., [63]).

We will focus, for simplicity, on the layout of Fig. 12.21 which represents schematically the resonator under consideration. Let Ω_1 denote the drive electrodes,

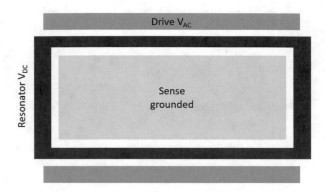

Fig. 12.21 Typical layout of a resonator

Ω_2 the resonator body, and Ω_3 the sense electrodes. A constant voltage $V_2 = V_{DC}$ is applied to the resonator, while a time-dependent signal is imposed on the drive $V_1 = V_{AC} \sin \omega t$, with $V_{DC} \gg V_{AC}$. The sense electrodes are kept to virtual ground.

In general the conductor potentials V_i are assigned, and the unknown field is the distribution of charge surface density σ on the conductors, which is computed solving the following integral equation:

$$V_i(\mathbf{x}) = \int_S \frac{1}{4\pi} \frac{1}{r} \frac{\sigma(\mathbf{y})}{\varepsilon_0} dS, \qquad \forall i, \forall \mathbf{x} \in S_i, \tag{12.58}$$

where S is the collection of the $S_i = \partial \Omega_i$. It is worth recalling that the electrostatic pressure (force per unit surface) $\mathbf{f}$ exerted on the surface of a conductor at point $\mathbf{x}$ of unit normal vector $\mathbf{n}$ is directly associated with the main unknown of the problem according to the formula $\mathbf{f} = \sigma^2/(2\varepsilon_0)\mathbf{n}$. Moreover, since only the conductor surfaces are discretized, and the dynamics of electromagnetic forces is much faster than the oscillation of the resonators, it is straightforward to repeat the analysis following the movement of the resonators. Indeed, the displacement is approximated as $\mathbf{u} = \boldsymbol{\psi} q$ and the coordinates of the surface elements are updated according to $\mathbf{x} + \boldsymbol{\psi} q$ and $\sigma(\mathbf{x}, q)$ can be computed as a function of q, adopting a quasi-static approach.

For a given q, let $\tilde{\sigma}_i(\mathbf{x}, q)$ denote the charge distribution corresponding to a fictitious problem where a unit potential is imposed only on Ω_i: $V_i = 1$ and $V_j = 0$, $j \neq i$. Each field $\tilde{\sigma}_i(\mathbf{x}, q)$ is the solution of (12.58) with the specified potentials. Thanks to the linearity of (12.58) at fixed q, the total charge on every conductor can be expressed as

$$\sigma(\mathbf{x}, q) = \sum_i \tilde{\sigma}_i(\mathbf{x}, q) V_i. \tag{12.59}$$

The nonlinear load participation factor (12.57) thus becomes

$$F_e(q) = \int_{S_2} \frac{\sigma^2}{2\varepsilon_0} \psi_n dS = \sum_{i,j} \left(\int_{S_2} \frac{\tilde{\sigma}_i \tilde{\sigma}_j}{2\varepsilon_0} \psi_n dS \right) V_i V_j, \tag{12.60}$$

where $\psi_n = \boldsymbol{\psi} \cdot \mathbf{n}$ and the integral is limited to the resonator surface S_2 since the sense and drive electrodes have zero velocity. In the specific case of Fig. 12.21, from (12.60):

$$F_e(q) \simeq \tilde{F}_{e1}(q) V_{DC}^2 + \tilde{F}_{e2}(q) V_{DC} V_{AC} \sin \omega t, \tag{12.61}$$

where we have neglected terms in V_{AC}^2 and:

$$\tilde{F}_{e1}(q) = \int_{S_2} \frac{\tilde{\sigma}_2^2}{2\varepsilon_0} \psi_n dS, \quad \tilde{F}_{e2}(q) = \int_{S_2} \frac{\tilde{\sigma}_1 \tilde{\sigma}_2}{\varepsilon_0} \psi_n dS.$$

The former term acts as a nonlinear spring, while the latter represents the nonlinear forcing term of the system. It is worth stressing that the nonlinear spring effect induces a softening behaviour in the dynamic response of the resonator and must be combined with the $\beta(q)$ coming from the geometric nonlinearities. As a consequence, depending on the entity of the displacements and on the voltages applied to the electrodes, the frequency response of the resonator can show a hardening or softening behaviour. If the total charges Q_i on the conductors and the capacitance coefficients C_{ij} coefficients are

$$Q_i(q) = \int_{S_i} \sigma \, dS = C_{ij}(q) V_j, \quad C_{ij}(q) = \int_{S_i} \sigma_j \, dS \tag{12.62}$$

the current measured on the sense is

$$I_3(q, \dot{q}) \sim \partial_\psi [C_{32}] \dot{q} \, V_{DC} = \left(\int_{S_2} \frac{\tilde{\sigma}_3 \tilde{\sigma}_2}{\varepsilon_0} \psi_n dS \right) V_{DC} \dot{q}. \tag{12.63}$$

It is worth stressing that $\tilde{F}_{e1}(q)$, $\tilde{F}_{e2}(q)$, and $I_3(q, \dot{q})$ are nonlinear functions of the midspan displacement and can be precomputed in a selected range $[q_{min} : q_{max}]$ and then interpolated whenever required.

With reference to the resonator under consideration, in Fig. 12.22 we compare the electrostatic terms $\tilde{F}_{e1}$ and $\tilde{F}_{e2}$ computed numerically with the classical analytical formulas for parallel plates capacitors. In this latter case we consider a Taylor series of seventh order to obtain a polynomial approximation of the electrostatic pressure and project each term on the analytical modal shape function. Fringe field effects are taken into account according to Palmer's formula [64]. From Fig. 12.22, it is evident that the analytical model underestimates the electrostatic terms, thus underlining the need of a more complete and accurate numerical model. The current measured

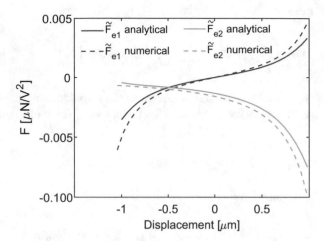

Fig. 12.22 Numerical and analytical estimations for electrostatic spring and forcing terms for unitary voltages

by the sensing electrodes is computed in-line from Eq. (12.63) while integrating Eq. (12.66) in terms of the displacement q of the flexible arm of the DETF resonator.

12.4.1.3 Thermoelastic and fluid dissipation

The dissipation level in a MEMS is typically measured by means of the so-called quality factor Q defined as $Q = 2\pi W/\Delta W$ where ΔW and W are the energy lost per cycle and the maximum value of energy stored in the device, respectively. Dissipation is determined by many causes [65, 66], each one connected to an amount of energy loss and the total ΔW is computed as the sum of the different contributions, assuming perfect decoupling. We will briefly discuss three potential dissipation sources: anchor, thermoelastic, and fluid dissipation.

Anchor losses are due to the scattering of elastic waves from the resonator into the substrate. Since the latter is typically much larger than the resonator itself, it is assumed that all the elastic energy entering the substrate through the anchors are eventually dissipated. In order to reproduce this dissipation numerically specific absorbing boundary conditions, and the Perfectly Matched Layer (PML) approach, in particular, has gained attention in the dedicated literature. The idea of employing the PML technique for microsystems has been proposed by [67]. A careful validation of the procedure for cantilever beams has been presented in [68] and experimental benchmarks with 3D MEMS resonators have been later described in [69–71] where it has been shown that anchor losses can be estimated with very good accuracy. It is important to remark that these analyses do not require any calibration of constitutive parameters and can be considered exact to within the limits of the assumption that the energy scattered through the anchors is dissipated.

However, the device analysed herein has a tuning fork configuration which minimizes the associated anchor losses which will be hence neglected in what follows.

Thermoelastic dissipation is associated with the bending deformation of the resonator, although it can be mitigated by the presence of slots [72] (see Fig. 12.18). The quality factor Q_{TED} is computed using the standard numerical approach [65] implemented in several commercial codes and will not be discussed herein. It is evaluated once for all in the reference configuration and is considered constant during the analysis. We estimate a thermoelastic quality factor $Q_{TED} = 68000$.

Some comments are worth stressing for fluid dissipation. Despite the pressure inside the package is very low, the effect of fluid damping on the overall quality factor cannot be considered negligible even at very low displacement ranges and increases with the displacement magnitude. At the pressure levels typical of the resonators addressed, the gas flow develops in the so-called free-molecule regime and its effects of the structure can be computed through the integral equation model proposed in [73] and validated in [74]. In [75] a simplified approach has been formulated based on precomputed tables for recurrent structures encompassing parallel plates, comb fingers, and masses with etched holes. The approach belongs to the family of integral equation techniques based on the collisionless Boltzmann equation [76, 77] and is deterministic. It is intrinsically free of statistical noise issues at the low speeds characteristic of MEMS which limit the applicability of Monte-Carlo techniques [78]; moreover it is essentially an ab-initio fully 3D simulation which can in principle account for any type of structure. The attention is limited to working frequencies f generating velocities which are small with respect to thermal velocity, i.e., $d \times f/\sqrt{2\mathcal{R}T_0} \ll 1$ (d is a typical displacement, $\mathcal{R}$ is the universal gas constant divided by the molar mass, and T_0 is the package temperature). We also assume that the displacement $\mathbf{u}$ of the deformable MEMS admits the classical decomposition $\mathbf{u}(\mathbf{x}, t) = \boldsymbol{\psi}(\mathbf{x})q(t)$. As a consequence, also the force exerted by the gas molecules on the MEMS surfaces admits the decomposition $\mathbf{t}(\mathbf{x}, t) = \mathbf{f}(\mathbf{x})\dot{q}(t)$ and yields, when integrated on the MEMS surface, the dissipative term of the 1D model:

$$\int_S \mathbf{t}(\mathbf{x}, t) \cdot \boldsymbol{\psi}(\mathbf{x})\mathrm{d}S = \left(\int_S \mathbf{f}(\mathbf{x}) \cdot \boldsymbol{\psi}(\mathbf{x})\mathrm{d}S\right)\dot{q} = B\dot{q}. \qquad (12.64)$$

The constant B can be more conveniently expressed as

$$B = \tilde{B}\rho_0\sqrt{2\mathcal{R}T_0} = \tilde{B}p_0\sqrt{2/(\mathcal{R}T_0)}, \qquad (12.65)$$

where $\tilde{B}$ has the dimensions of a surface, depends only on the problem geometry and, given the complexity of MEMS devices, must be estimated numerically.

The dissipation is proportional to the package pressure and depends nonlinearly on the displacement amplitudes of the flexible arms of the resonator (see, e.g., Fig.12.23). A quasi-static calculation is performed placing the structure in its actual configuration $\mathbf{x} + \mathbf{u} = \mathbf{x} + \boldsymbol{\psi}(\mathbf{x})q(t)$ and obtaining an estimate of $\tilde{B}(q)$

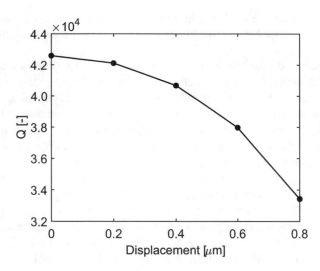

Fig. 12.23 Total quality factor Q as function of the DETF displacement. A pressure of 30 μbar inside the MEMS package is considered

for a given range of midspan displacement q. Indeed this quasi-static approach is reasonable if the shuttle velocity remains small with respect to thermal velocity of molecules.

It is worth stressing that the nominal pressure inside the package is in principle fixed by the fabrication process at the bonding level, but technological spread affects its value. As a consequence, a first set of experimental data in the linear regime are generally utilized to identify the real pressure inside the package where the resonator is encapsulated.

The overall quality factor $Q(q)$ is finally obtained as a function of the displacement q for the given pressure by combining the two contributions, i.e., $1/Q = (1/Q_f(q) + 1/Q_{TED})$.

Gas damping simulations are performed for a unit nominal working pressure since fluid damping is proportional to p and can be easily scaled. The pressure inside the package has been identified as $p = 30\ \mu$bar by fitting the linear experimental response of the resonator. This value is compatible with the specifications given by the fabrication process (i.e., $p < 50\ \mu$bar) and is adopted for the other numerical estimations in Fig. 12.24. The trends of the numerically estimated total Q as a function of the midspan displacement of the DETF are reported in Fig. 12.23 for $p = 30\ \mu$bar.

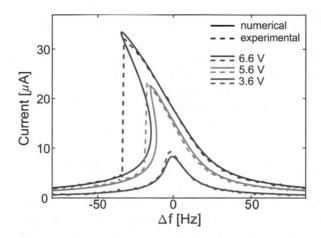

Fig. 12.24 Frequency response of the DETF resonator. Numerical predictions and experimental measurements are reported in solid and dotted lines, respectively. $V_{AC} = 100\,\text{mV}$, $V_{DC} = 3.6\text{V}$, 5.6V, and 6.6V

12.4.2   Reduced order model and validation

Once all the different contributions have been precomputed separately, it is possible to write the equation of motion governing the dynamics of the resonators:

$$\frac{1}{\psi_q}\ddot{q} + \frac{1}{\psi_q}\frac{\omega_0}{Q(q)}\dot{q} + \beta(q) - \tilde{F}_{e1}(q)V_{DC}^2 = \tilde{F}_{e2}(q)V_{DC}V_{AC}\sin(\omega t), \qquad (12.66)$$

where ω_0 is the linear natural frequency of the flexural mode.

Equation (12.66) can be integrated with any method of choice. An option is the so-called brute force approach in which, for any fixed forcing frequency ω, a nonlinear time-integrator (e.g., Newmark or Runge-Kutta schemes) is applied to the differential equation until a steady state is reached. The maximum amplitude is recorded before moving to the next value of frequency in an upward or downward sweep. As a consequence, unstable branches of the response cannot be simulated with this approach. Alternatively, a continuation method with arc length control can be applied to obtain both stable and unstable branches. The latter option is adopted for the analyses of this chapter.

Figure 12.24 presents the comparison between experimental and numerical responses for different amplitudes V_{DC} when $V_{AC} = 100\,\text{mV}$. A very good agreement is found both qualitatively and quantitatively. Electrostatic nonlinearities, that induce softening in the dynamic response of the resonator, are dominating with respect to the mechanical hardening in this specific device.

12.5   Other mechanical and coupled problems in MEMS

Many other devices in which complex multi-physics and nonlinear behaviours arise would deserve attention. In this section a brief overview of the main nonlinear and multi-physics problems emerging in MEMS and not discussed in the previous sections is reported without the ambition of being exhaustive.

12.5.1   PiezoMEMS and material nonlinearities

Piezoelectric materials are widely employed in MEMS and their usage is becoming fundamental in particular in resonators, Piezoelectric Micromachined Ultrasonic Transducers (PMUT), and micromirrors. For instance, devices like PMUT are attracting a lot of interests in recent years for their intriguing applications in medical acoustic imaging, rangefinders and fingerprint recognition, non-destructive testing, velocity sensing, and three-dimensional object recognition. A PMUT can be schematized as a suspended layered membrane, in which one of the layers is made of piezoelectric material. When a voltage difference is applied on the two sides of the piezoelectric layer, the membrane starts oscillating at its resonant frequency, thus emitting acoustic waves in the surrounding fluid. The multi-physics nature of such devices is then evident: interactions between electrical, mechanical, and acoustic fields are indeed unavoidable. Moreover, to give a reasonable estimation of the damping, thermo-viscous losses in the fluid domain must be also considered. Finally, nonlinear dynamic regimes can be entered when the vibrating membranes exploit the membrane-bending coupling and when residual stresses and pre-deformed configurations induced by the fabrication process are not negligible [79].

An important aspect is the manifestation of nonlinearities in ferroelectric materials like PZT employed as thin films with typical thickness of few microns. The drive voltage applied across the thin film induces electric fields in the order of 10^7 V/m and generates a strongly nonlinear material response. The classical linearized theory of piezoelectricity does not apply and must be replaced with the general Landau-Devonshire theory of ferroelectrics [80]. An isolated idealized ferroelectric crystal below the Curie temperature T_C has several spontaneous polarization states (SPS) in which atoms find a stable equilibrium with microscopic displacements relative to the lattice. Sol-gel deposited thin films often come in the form of polycrystals, i.e., regions of homogeneous crystallographic orientation separated by amorphous fixed grain boundaries. Within a grain, the material organizes itself in domains with uniform polarization separated by sharp interfaces (domain walls) that can move according to the enforced boundary conditions. The evolution and reorientation of domains is the microscopic origin of the macroscopic nonlinear hysteretic behaviour of ferroelectric materials. In [81, 82], where a FEM implementation of the Phase Field Method is applied, efforts have been directed towards the direct numerical simulation of these phenomena and of their influence on the macroscopic polarization hysteresis loop.

Once the polarization history is available, through simulation or direct experimental measurements, it must be accounted for when imposing the equilibrium of the device. Equation (12.54) is enriched with a new term at the RHS:

$$\int_{\Omega_0} \rho \ddot{\mathbf{u}} \cdot \mathbf{w} \, d\Omega + \int_{\Omega_0} \mathbf{S} : \text{sym}(\nabla^T \mathbf{w} \cdot \mathbf{F}) d\Omega \qquad (12.67)$$

$$= P^2 \int_{\Omega_P} \Big[c_{13}(\varepsilon_{11}[\mathbf{w}] + \varepsilon_{22}[\mathbf{w}]) + c_{33}\varepsilon_{33}[\mathbf{w}] \Big] d\Omega \qquad \forall \mathbf{w} \in \mathcal{C}(0)$$

which represents the forcing due to the piezo actuation. Ω_p denotes the piezo-films domain, c_{ij} are constants derived from the mechanical and electrostrictive properties of piezoelectric materials and P denotes the magnitude of the polarization vector. This formulation holds under the assumption that the piezo-patches undergo infinitesimal transformations only. It is worth mentioning that piezoelectric materials exploited in MEMS are deposited as thin films on a plane that we take here orthogonal to the $\mathbf{e}_3$ direction and it can be safely assumed that the polarization vector has the form $\mathbf{P} = P\mathbf{e}_3$. From (12.67) it is clear that the polarization intervenes in a nonlinear manner and strongly impacts the overall response.

12.5.2   Material nonlinearities induced by high doping levels

Despite material nonlinearities can be usually neglected in single-crystal silicon or in polysilicon with a low level of doping, it has been recently shown that a completely different treatment must be deserved in case of highly doped silicon. In [83], three types of bulk acoustic mode resonators are fabricated in highly doped single-crystal silicon and their frequency responses are experimentally tested. It has been proved that material nonlinearities and crystal orientation strongly influence the dynamic responses of such devices and that the standard sources of nonlinearities (i.e., geometric and electrostatic) can become even negligible.

Crystallographic orientation and high doping levels of single-crystal silicon are widely exploited in MEMS resonators to improve their frequency stability against temperature variations. In [24–26, 84], it is proved that there is an intrinsic minimum in terms of frequency drift in temperature for each n-doping (i.e., Phosphorous) level of silicon and that it decreases by increasing the doping level. With a n-doping level higher than 10^{20} cm^{-3}, a MEMS resonator can indeed achieve the temperature stability of its quartz counterpart, thus representing a valid alternative for clocks applications. Moreover, it is shown that such minimum can be achieved by properly orienting the structure with respect to the crystallographic axes on the wafer. As an example, for a Double Ended Tuning Fork fabricated in single-crystal silicon with doping level of $7.26 \, 10^{19}$cm^{-3}, the frequency drift moves from 2100 ppm down to

160 ppm when the orientation angle between the [100] wafer axis and the resonator's structure increases from $0°$ to $45°$.

To optimize the design of single-crystal silicon MEMS resonators in terms of frequency stability, the temperature and doping dependence of the elastic constants (i.e., c_{11}, c_{12}, and c_{44} being a material with cubic symmetry) and material parameters (i.e., thermal expansion coefficient, specific heat and thermal conductivity) must be then considered together with the orientation of the structure on the silicon wafer as shown in [26, 84].

12.5.3 Low distance and contact forces

Miniaturization is one of the main goal of a MEMS designer, the reduction of the gaps between moving components and electrodes is then desirable if the overall footprint of MEMS devices wants to be reduced without deteriorating the performances. As already observed in this chapter, gaps are usually in the order of some micrometres and can be reduced to less than 1 μm in the case of resonators or other high-performance devices. For this reason, interaction forces between approaching surfaces must be considered together with the contact dynamics arising in case of shocks or other unwanted external sources.

Stiction is a highly nonlinear phenomenon that arises in MEMS devices when different mechanical components significantly reduce their distance. It is usually considered as a serious reliability issue in MEMS since it can lead to an irreversible contact between components, thus compromising the regular functioning of the MEMS device. Stiction is governed by forces arising at very narrow distances between surfaces. In presence of humidity, for example, stiction phenomena are induced by capillary attraction, while in dry conditions the always present Casimir/Polder and van der Waals forces are the main sources [66, 85, 86].

Other sources of nonlinearities come from repeated contacts of surfaces of oscillating parts against fixed or different movable components. These undesired phenomena are often accidental and for this reason very difficult to predict and control. Moreover, they result in complex dynamic responses that have been studied in the literature through ad-hoc mechanical structures. In [87], for example, a frequency comb induced by contact dynamics in MEMS is put in evidence both numerically and experimentally, while in [88] an assessment of the Newmark method is proposed to compute chaotic vibrations of impacting oscillators.

12.5.4 Internal resonance in MEMS

Internal resonance appears in MEMS resonant devices when the natural frequencies of distinct modes of the MEMS oscillator satisfy a commensurate relationship. It consists in a strong coupling between the modes and in an energy transfer among them. It has been widely studied from the modelling point of view [89] and only recently it has been exploited as an innovative sensing principle for MEMS gyroscopes [90] and to stabilize the oscillation frequency of nonlinear self-sustaining micromechanical resonators [91].

12.5.5 Parametric resonance in MEMS

Parametric resonance occurs in a mechanical structure when it is parametrically excited and oscillates at one of its resonant frequencies. Parametric excitation differs from forcing since the action appears as a time varying modification on a system parameter. In MEMS resonant devices, usually, parametric resonance arises when the modal stiffness (i.e., k in the one degree-of-freedom model) is modulated at two times the natural frequency of the oscillator. Different sources can lead to such complex dynamic behaviour and for this reason it is quite popular in MEMS. For the sake of brevity, in the following, we will discuss only two meaningful examples of parametric amplification in MEMS studied by the authors: self-induced parametric amplification in a Disk Resonator Gyroscope (DRG) and electrostatically parametrically actuated micromirror.

The dynamic behaviour of an oscillator undergoing parametric resonance is described by the Mathieu equation. It is a linear differential equation with variable coefficients. By considering U as the one degree-of-freedom of the system, it reads

$$\ddot{U} + \frac{\omega}{Q}\dot{U} + \left[1 + \frac{\Delta k}{k}\cos(2\omega t)\right]\omega^2 U = \frac{F}{m}\cos(\omega t + \phi), \tag{12.68}$$

where $\lambda \equiv \frac{\Delta k}{k}$ is the fractional stiffness change, F the amplitude of the time-dependent forcing term, Q the quality factor and ω the natural frequency. When $\lambda = 0$, the device is a linear resonator and the sensitivity to force at the resonance frequency ω is Q/k. When $\lambda \neq 0$, the device is a parametric resonator and the excess parametric gain depends on the phase ϕ of the 1ω signal F relative to the 2ω *pump*.

When other sources of nonlinearities (e.g., electrostatic actuation or high quality factor) are present, the nonlinear Mathieu equation must be taken into account:

$$\ddot{U} + \frac{\omega}{Q}\dot{U} + \left[1 + \frac{\Delta k}{k}\cos(2\omega t) + \frac{k_3}{k}U^2\right]\omega^2 U = \frac{F}{m}\cos(\omega t + \phi), \tag{12.69}$$

where k_3 is the new cubic term modelling other nonlinear sources.

A lot of work has been done in particular to study the influence of cubic nonlinear terms in the stability of such devices [92]. In [93], for example, Rhoads and Shaw use the method of averaging to compute theoretical steady-state responses for a parametrically amplified Duffing oscillator operating in open loop and conclude that stable open-loop operation is possible at the cost of decreased performance or bi-stable behaviour. In [94] it is shown that stable, parametrically amplified operation in the presence of cubic nonlinearities is possible without suffering from jump instabilities, hysteresis, or degraded performance if the close-loop operation is considered.

12.5.5.1   Disk Resonating Gyroscope (DRG)

Disk or ring gyroscopes operate using two orthogonal flexural radial vibration modes coupled through the Coriolis force. In general, their mode shapes can be described through $\cos(n\vartheta)$ and $\sin(n\vartheta)$, where n is the mode number. In the following, the 2ϑ modes, which are separated from each other by $45°$, are considered.

During the regular functioning, the first 2ϑ mode, i.e., drive mode, is electrostatically driven and when an external angular rate acts on the device, the Coriolis force activates the motion of the second 2ϑ mode, i.e., sense mode, that allows the electrostatic read-out through properly designed electrodes. When the large displacement regime is entered, nonlinear mechanical coupling between the two degenerate modes of the DRG (i.e., drive and sense modes) leads to self-induced parametric amplification.

Figure 12.25a shows the lumped element model describing the phenomenon: the nonlinear elastic effects induced by the mechanical coupling of the drive and sense modes induce a modulation Δk of the stiffness of the sense mode that depends on the drive axis displacement. Being the 2ϑ mode shape two-fold symmetric, the

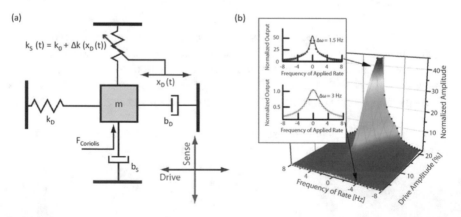

Fig. 12.25 (**a**) Lumped element model of gyroscope: the displacement of the drive axis modulates the stiffness of the sense axis at twice the resonant frequency, thus parametrically amplifying Coriolis force and electrostatic inputs to the sense axis. (**b**) Experimental response of the DRG to external rate. Inset shows the measured frequency response at small and large amplitudes of the driving displacement, indicating the reduced bandwidth observed at large amplitude due to the artificial increase in Q. Source: [95], Fig. 3 and Fig. 4. Licensed under a Creative Commons Attribution 4.0 International License http://creativecommons.org/licenses/by/4.0/

stiffness variation does not depend on the sign of the displacement and Δk can be described through a rectified sine wave whose phase shift relative to the drive axis displacement is $\phi = 45°$, i.e., phase relationship between the 2ω pump and 1ω signal waveforms in Eq. (12.68).

Self-induced parametric amplification was first observed by measuring the gyroscope's sensitivity to rotation rate S_Ω as a function of the amplitude of the driven mode $\overline{x}_D$ [95]. When the amplitude of the drive mode is small $\overline{x}_D < 2.5\%$, the frequency response exhibits the expected Lorentzian shape with $\Delta\omega/2\pi = 3$ Hz. As $\overline{x}_D$ is increased, the scale-factor increases at a rate much greater than $\overline{x}_D$: an 8-fold increase in $\overline{x}_D$ results in a 67-fold increase in S_Ω and a two-fold reduction in $\Delta\omega$ as shown in Fig. 12.25b.

12.5.5.2 Electrostatic Micromirrors

Micromirrors are MEMS merged with microoptics and involve sensing or manipulating optical signals on a very small size scale, using integrated mechanical, optical, and electrical systems. The micromirror studied in [96] consists in a central circular

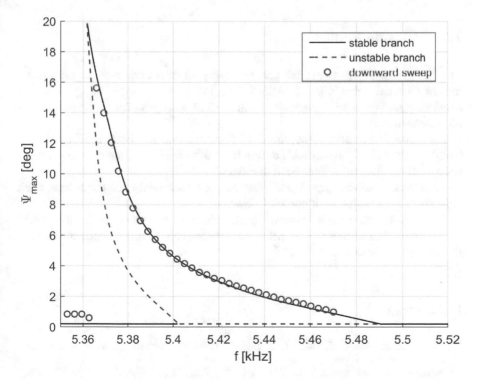

Fig. 12.26 Mirror max opening angle for a sinusoidal excitation at 5 kHz, $V_0 = 80$ V; experimental downward sweep (discrete symbols) and numerical continuation (continuous line)

reflecting surface suspended through two coaxial torsional springs. Electrostatic actuation and read-out of the mirror's torsional mode is achieved through four sets of 33 fingers anchored to trapezoidal regions directly attached to the central structure. It is worth noting that, due to symmetry, the electrostatic torque around the torsional axis of the mirror generated by these sets of comb fingers vanishes in the rest configuration. However, for some combinations of the input voltage amplitude and frequency, it is possible to activate the mirror rotation through such electrostatic scheme by exploiting the parametric amplification phenomenon.

In the experiments of Fig. 12.26 the mirror is excited with a sinusoidal wave at frequency near ω_0, the torsional eigenfrequency of the mirror and the phase between the driving signal and the mirror oscillation is kept constant by a closed loop. Being the expression of the electrostatic force proportional to the voltage at the power two, it follows that the forcing term is at $2\omega_0$ and the mirror rotation is activated by parametric resonance.

In Fig. 12.26, a comparison between experimental data and numerical prediction is reported when the micromirror is actuated with a sinusoidal voltage bias of $V_0 = 80$ V. Considering that only stable branches can be reproduced in the experiments, the agreement with numerical data is in general impressive.

12.6 Closing Remarks

The main purpose of the present chapter is to give an overview of the most often met mechanical behaviours in MEMS. The focus has been on the dynamics of oscillating parts and on their various interactions with physical phenomena other than mechanics.

As a simple prototype, the one dof schematization has been considered as a paradigm for MEMS dynamics. Starting from this simple mechanical model, basic sources of nonlinearities have been discussed.

Analytical, semi-analytical, and wide purpose numerical approaches have been discussed for the solution of complex devices.

From the examples briefly shown in this chapter, the reader can have a glance on the interesting complexity of mechanically dominated multi-physics interactions that always occur in MEMS.

References

1. Corigliano, A., Ardito, R., Comi, C., Frangi, A., Ghisi, A., & Mariani, S. (2018). *Mechanics of microsystems*. Wiley.
2. Rodriguez, J., Chandorkar, S. A., Watson, C. A., Glaze, G. M., Ahn, C. H., Ng, E. J., Yang, Y., & Kenny, T. W. (2019). Direct detection of akhiezer damping in a silicon mems resonator. *Scientific Reports, 9*, 2045–2322.

3. Hopcroft, M. A., Nix, W. D., & Kenny, T. W. (2010). What is the young's modulus of silicon? *Journal of Microelectromechanical Systems, 19*(2), 229–238.

4. Ghisi, A., Fachin, F., Mariani, S., & Zerbini, S. (2009). Multi-scale analysis of polysilicon mems sensors subject to accidental drops: Effect of packaging. *Microelectronics Reliability, 49*(3), 340–349.

5. Mariani, S., Martini, R., Ghisi, A., Corigliano, A., & Simoni, B. (2011). Monte Carlo simulation of micro-cracking in polysilicon mems exposed to shocks. *International Journal of Fracture, 167*(1), 83–101.

6. Son, D. I., Kim, J. J., & Kwon, D. I. (2005). Fracture behavior of single- and polycrystalline silicon films for mems applications. *Key Engineering Materials, 297-300*, 551–556.

7. Reedy, E. D., Boyce, B. L., Foulk, J. W., Field, R. V., de Boer, M. P., & Hazra, S. S. (2011). Predicting fracture in micrometer-scale polycrystalline silicon mems structures. *Journal of Microelectromechanical Systems, 20*(4), 922–932.

8. Kahn, H., Tayebi, N., Ballarini, R., Mullen, R., & Heuer, A. (2000). Fracture toughness of polysilicon mems devices. *Sensors and Actuators A: Physical, 82*(1), 274–280.

9. Xiong, X., Wu, Y., & Jone, W. (2008). Material fatigue and reliability of mems accelerometers. In *2008 IEEE International Symposium on Defect and Fault Tolerance of VLSI Systems* (pp. 314–322).

10. Merlijn van Spengen, W. (2012). Static crack growth and fatigue modeling for silicon mems. *Sensors and Actuators A: Physical, 183*, 57–68.

11. Huang, Q. W., Li, X. G., & Wang, Y. H. (2014). Analysis of mechanical fatigue behavior for mems structures. In *Sensors, Mechatronics and Automation*, vol. 511 of *Applied Mechanics and Materials* (pp. 565–568). Trans Tech Publications Ltd.

12. Langfelder, G., Longoni, A., Zaraga, F., Corigliano, A., Ghisi, A., & Merassi, A. (2008). A polysilicon test structure for fatigue and fracture testing in micro electro mechanical devices. In *SENSORS, 2008 IEEE* (pp. 94–97).

13. Lifshitz, R., & Cross, M. C. (2009). *Nonlinear Dynamics of Nanomechanical and Micromechanical Resonators*.

14. Strogatz, S. H. (2007). *Nonlinear Dynamics And Chaos. Studies in Nonlinearity*.

15. Cho, H., Jeong, B., Yu, M.-F., Vakakis, A. F., McFarland, D. M., & Bergman, L. A. (2012). Nonlinear hardening and softening resonances in micromechanical cantilever-nanotube systems originated from nanoscale geometric nonlinearities. *International Journal of Solids and Structures, 49*, 2059–2065.

16. Ganapathia, M., & Politb, O. (2017). Dynamic characteristics of curved nanobeams using nonlocal higher-order curved beam theory. *Physica E, 91*, 190–202.

17. Krylov, S., & Dick, N. (2010). Dynamic stability of electrostatically actuated initially curved shallow micro beams. *Continuum Mechanics and Thermodynamics, 22*, 445–468.

18. Pan, K.-Q., & Liu, J.-Y. (2011). Geometric nonlinear dynamic analysis of curved beams using curved beam element. *Acta Mechanica Sinica, 27*(6), 1023–1033.

19. Hajjaj, A. Z., Alcheikh, N., & Younis, M. I. (2017). The static and dynamic behavior of mems arch resonators near veering and the impact of initial shapes. *International Journal of Non-Linear Mechanics, 95*, 277–286.

20. Fedeli, P., Frangi, A., Laghi, G., Langfelder, G., & Gattere, G. (2017). Near vacuum gas damping in mems: Numerical modeling and experimental validation. *Journal of Microelectromechanical Systems, 25*(5), 890–899.

21. Bao, M., & Yang, H. (2007). Squeeze film air damping in mems. *Sensors and Actuators A: Physical, 136*(1), 3–27.

22. Nabholz, U., Heinzelmann, W., Mehner, J. E., & Degenfeld-Schonburg, P. (2018). Amplitude- and gas pressure-dependent nonlinear damping of high-q oscillatory mems micro mirrors. *Journal of Microelectromechanical Systems, 27*(3), 383–391.

23. Zaitsev, S., Shtempluck, O., Buks, E., & Gottlieb, O. (2012). Nonlinear damping in a micromechanical oscillator. *Nonlinear Dynamics, 67*(1), 859–883.

24. Ng, E. J., Hong, V. A., Yang, Y., Ahn, C. H., Everhart, C. L. M., & Kenny, T. W. (2015). Temperature dependence of the elastic constants of doped silicon. *Journal of Microelectromechanical Systems, 24*(3), 730–741.
25. Jaakkola, A., Prunnila, M., Pensala, T., Dekker, J., & Pekko, P. (2014). Determination of doping and temperature-dependent elastic constants of degenerately doped silicon from mems resonators. *IEEE Transactions on Ultrasonics, Ferroelectrics, and Frequency Control, 61*(7), 1063–1074.
26. Zega, V., Frangi, A., Guercilena, A., & Gattere, G. (2018). Analysis of frequency stability and thermoelastic effects for slotted tuning fork mems resonators. *Sensors, 18*(7), 2157.
27. Jaakkola, A., Prunnila, M., Pensala, T., Dekker, J., & Pekko, P. (2015). Design rules for temperature compensated degenerately n-type doped silicon mems resonators. *Journal of Microelectromechanical Systems, 24*, 1832–1839.
28. Shin, D., Heinz, D., Kwon, H.-K., Chen, Y., & Kenny, W. (2018). Lateral diffusion doping of silicon for temperature compensation of mems resonators. In *2018 IEEE Intern. Symp. on Inertial Sensors and Syst.* (pp. 1–4).
29. Lifshitz, L., & Roukes, M. L. (2000). Thermoelastic damping in micro- and nanomechanical systems. *Physical Review B, 61*, 5600–5609.
30. Prabhakar, S., & Vengallatore, S. (2009). Thermoelastic damping in hollow and slotted microresonators. *Journal of Microelectromechanical Systems, 18*(3), 725–735.
31. Asadi, S., & Sheikholeslami, T. F. (2016). Effects of slots on thermoelastic quality factor of a vertical beam mems resonator. *Microsystem Technologies, 22*(11), 2723–2730.
32. Abdolvand, R., Johari, H., Ho, G. K., Erbil, A., & Ayazi, F. (2006). Quality factor in trench-refilled polysilicon beam resonators. *Journal of Microsystem Technologies, 15*, 471–478.
33. Candler, R. N., Duwel, A., Varghese, M., Chandorkar, S. A., Hopcroft, M. A., Woo-Tae Park, Bongsang Kim, Yama, G., Partridge, A., Lutz, M., & Kenny, T. W. (2006). Impact of geometry on thermoelastic dissipation in micromechanical resonant beams. *Journal of Microelectromechanical Systems, 15*(4), 927–934.
34. Tiwari, S., & Candler, R. N. (2019). Using flexural MEMS to study and exploit nonlinearities: a review. *Journal of Micromechanics and Microengineering, 29*(8), 083002.
35. Rhoads, J. F., Shaw, S. W., & Turner, K. L. (2010). Nonlinear dynamics and its applications in micro-and nanoresonators. *Journal of Dynamic Systems, Measurement and Control, Transactions of the ASME, 132*(3), 1–14.
36. Veijola, T., Mattila, T., Jakkola, O., Kiihamaki, J., Lamminmaki, T., Oja, A., Ruokonen, K., Sepa, H., Seppala, P., & Tittonen, I. (2000). Large-displacement modeling and simulation of micromechanical electrostatically driven resonators using the harmonic balance method. *IEEE MTT-S International Microwave Symposium Digest, 1*, 99–102.
37. Hosen, M. A., Chowdhury, M. S. H., Ali, M. Y., & Ismail, A. F. (2017). An analytical approximation technique for the duffing oscillator based on the energy balance method. *Italian Journal of Pure and Applied Mathematics, 37*, 455–466.
38. Nayfeh, A. H. (1981). *Introduction of perturbation techniques.* New York.
39. Nayfeh, A. H., & Mook, D. T. (1995). *Nonlinear oscillations.* Wiley Classic Library Edition.
40. Comi, C., Corigliano, A., Langfelder, G., Longoni, A., Tocchio, A., & Simoni, B. (2010). A resonant microaccelerometer with high sensitivity operating in an oscillating circuit. *Journal of Microelectromechanical Systems, 19*(5), 1140–1152.
41. Tocchio, A., Comi, C., Langfelder, G., Corigliano, A., & Longoni, A. (2011). Enhancing the linear range of mems resonators for sensing applications. *IEEE Sensors Journal, 11*(12), 3202–3210.
42. Rhoads, J. F., Shaw, S. W., Turner, K. L., Moehlis, J., DeMartini, B. E., & Zhang, W. (2006). Generalized parametric resonance in electrostatically actuated microelectromechanical oscillators. *Journal of Sound and Vibration, 296*(4), 797–829.
43. Zhang, J., Su, Y., Shi, Q., & Qiu, A. P. (2015). Microelectromechanical resonant accelerometer designed with a high sensitivity. *Sensors (Switzerland), 15*(12), 30293–30310.

44. Zhang, J., Wang, Y., Zega, V., Su, Y., & Corigliano, A. (2018).   Nonlinear dynamics under varying temperature conditions of the resonating beams of a differential resonant accelerometer. *Journal of Micromechanics and Microengineering, 28*(7), 075004.
45. Melamud, R., Hopcroft, M., Jha, C., Kim, B., Chandorkar, S., Candler, R., & Kenny, T. W. (2005).   Effects of stress on the temperature coefficient of frequency in double clamped resonators. In *Digest of Technical Papers - International Conference on Solid State Sensors and Actuators and Microsystems, TRANSDUCERS '05* (vol. 1, pp. 392–395). IEEE.
46. Zhu, Y., Corigliano, A., & Espinosa, H. D. (2006).   A thermal actuator for nanoscale in situ microscopy testing: design and characterization. *Journal of Micromechanics and Microengineering, 16*(2), 242.
47. Comi, C., Corigliano, A., Ghisi, A., & Zerbini, S. (2013).   A resonant micro accelerometer based on electrostatic stiffness variation. *Meccanica, 48*(8), 1893–1900.
48. Caspani, A., Comi, C., Corigliano, A., Langfelder, G., Zega, V., & Zerbini, S. (2014b).   Dynamic nonlinear behavior of torsional resonators in mems. *Journal of Micromechanics and Microengineering, 24*, 095025 (9pp).
49. Caspani, A., Comi, C., Corigliano, A., Langfelder, G., Zega, V., & Zerbini, S. (2014a). A differential resonant micro accelerometer for out-of-plane measurements. *Procedia Engineering, 87*, 640–643.
50. Comi, C., Corigliano, A., Zega, V., & Zerbini, S. (2015). Optimal design and nonlinearities in a z-axis resonant accelerometer. In *16th International Conference on Thermal, Mechanical and Multi-Physics Simulation and Experiments in Microelectronics and Microsystems (EuroSimE)* (pp. 1–6), Budapest, Hungary.
51. Comi, C., Corigliano, A., Doti, M., Garatti, A., Langfelder, G., & Zega, V. (2016). Torsional microresonator in the nonlinear regime: experimental, numerical and analytical characterization. In *Proceedings Eurosensors 2016* (pp. 1–6), Budapest, Hungary.
52. Manzotti, M. (2020).   *On the non-linear behaviour of piezoelectrically actuated resonant micro-mirrors.* Master Thesis.
53. Corigliano, A., De Masi, B., Frangi, A., Comi, C., Villa, A., & Marchi, M. (2004). Mechanical characterization of polysilicon through on-chip tensile tests. *Journal of Microelectromechanical Systems, 13*(2), 200–219.
54. Zega, V., Gattere, G., Koppaka, S., Alter, A., Vukasin, G. D., Frangi, A., & Kenny, T. W. (2020).   Numerical modelling of non-linearities in mems resonators. *Journal of Microelectromechanical Systems, 29*(6), 1443–1454.
55. Mignolet, M. P., Przekop, A., Rizzi, S. A., & Spottswood, S. M. (2013).   A review of indirect/non-intrusive reduced order modeling of nonlinear geometric structures. *Journal of Sound and Vibration, 332*(10), 2437–2460.
56. Kerschen, G., Golinval, J. C., Vakakis, A., & Bergman, L. (2015).   The method of proper orthogonal decomposition for dynamical characterization and order reduction of mechanical systems: An overview. *Nonlinear Dynamics, 41*, 147–169.
57. Renson, L., Kerschen, G., & Cochelin, B. (2016). Numerical computation of nonlinear normal modes in mechanical engineering. *Journal of Sound and Vibration, 364*, 177–206.
58. Touzé, C., & Amabili, M. (2006).   Non-linear normal modes for damped geometrically non-linear systems: application to reduced-order modeling of harmonically forced structures. *Journal of Sound and Vibration, 298*, 958–981.
59. Krack, M., Panning-von Scheidt, L., & Wallaschek, J. (2013).   A method for non-linear modal analysis and synthesis: Application to harmonically forced and self-excited mechanical systems. *Journal of Sound and Vibration, 332*, 6798–6814.
60. Frangi, A., & Gobat, G. (2019). Reduced order modelling of the non-linear stiffness in mems resonators. *International Journal of Non-Linear Mechanics, 116*, 211–218.
61. McEwan, M., Wright, J., Cooper, J., & Leung, A. (2001). *A finite element/modal technique for nonlinear plate and stiffened panel response prediction.*
62. Frangi, A., Guerrieri, A., & Boni, N. (2017).   Accurate simulation of parametrically excited micromirrors via direct computation of the electrostatic stiffness. *Sensors, 17*(4), 779.

63. Frangi, A., & Di Gioia, A. (2005). Multipole BEM for the evaluation of damping forces on mems. *Computational Mechanics, 37*, 24–31.
64. Palmer, H. B. (1937). The capacitance of a parallel-plate capacitor by the Schwarz-Christoffel transformation. *Transactions of the American Institute of Electrical Engineers, 56*(3), 363–366.
65. Ardito, R., Comi, C., Corigliano, A., & Frangi, A. (2008). Solid damping in micro-electro-mechanical systems. *Meccanica, 43*, 419–428.
66. Ardito, R., Corigliano, A., & Frangi, A. (2013). Modelling of spontaneous adhesion phenomena in micro-electro-mechanical systems. *European Journal of Mechanics-A/Solids, 39*, 144–152.
67. Bindel, D. S., & Govindjee, S. (2005). Elastic PMLs for resonator anchor loss simulation. *International Journal for Numerical Methods in Engineering, 64*, 789–818.
68. Frangi, A., Bugada, A., Martello, M., & Savadkoohi, P. (2013). Validation of PML-based models for the evaluation of anchor dissipation in mems resonators. *European Journal of Mechanics - A/Solids, 37*, 256–265.
69. Frangi, A., Cremonesi, M., Jaakkola, A., & Pensala, T. (2013). Analysis of anchor and interface losses in piezoelectric mems resonators. *Sensors and Actuators A: Physical, 190*, 127–135.
70. Segovia-Fernandez, J., Cremonesi, M., Cassella, C., Frangi, A., & Piazza, G. (2015). Anchor losses in AIN contour mode resonators. *Journal of Microelectromechanical Systems, 24*(2), 265–275.
71. Frangi, A., & Cremonesi, M. (2016). Semi-analytical and numerical estimates of anchor losses in bistable mems. *International Journal of Solids and Structures, 92-93*, 141–148.
72. Ghaffari, S., Ng, E. J., Ahn, C. H., Yang, Y., Wang, S., Hong, V. A., & Kenny, T. W. (2015). Accurate modeling of quality factor behavior of complex silicon mems resonators. *Journal of Microelectromechanical Systems, 24*(2), 276–288.
73. Cercignani, C., Frangi, A., Lorenzani, S., & Vigna, B. (2007). BEM approaches and simplified kinetic models for the analysis of damping in deformable mems. *Engineering Analysis with Boundary Elements, 31*(5), 451–457. Innovative Numerical Methods for Micro and Nano Mechanics - Part I.
74. Frangi, A., Fedeli, P., Laghi, G., Langfelder, G., & Gattere, G. (2016). Near vacuum gas damping in mems: Numerical modeling and experimental validation. *Journal of Microelectromechanical Systems, 25*(5), 890–899.
75. Fedeli, P., Frangi, A., Laghi, G., Langfelder, G., & Gattere, G. (2017). Near vacuum gas damping in mems: Simplified modeling. *Journal of Microelectromechanical Systems, 26*(3), 632–642.
76. Frangi, A. (2009). A BEM technique for free-molecule flows in high frequency mems resonators. *Engineering Analysis with Boundary Elements, 33*, 493–498.
77. Frangi, A., Ghisi, A., & Coronato, L. (2009). On a deterministic approach for the evaluation of gas damping in inertial mems in the free-molecule regime. *Sensors and Actuators A: Physical, 149*, 21–28.
78. Bird, G. A. (1994). *Molecular gas dynamics and the direct simulation of gas flows.*
79. Massimino, G., Colombo, A., D'Alessandro, L., Procopio, F., Ardito, R., Ferrera, M., & Corigliano, A. (2018). Multiphysics modelling and experimental validation of an air-coupled array of PMUTs with residual stresses. *Journal of Micromechanics and Microengineering, 28*(5), 054005.
80. Devonshire, A. F. (1954). Theory of ferroelectrics. *Advances in Physics, 3*, 85–130.
81. Fedeli, P., Kamlah, M., & Frangi, A. (2019). Phase field modeling of domain evolution in ferroelectric materials in the presence of defects. *Smart Materials and Structures, 28*, 035021–9.
82. Fedeli, P., Cuneo, F., Magagnin, L., Nobili, L., Kamlah, M., & Frangi, A. (2020). On the simulation of the hysteresis loop of polycrystalline PZT thin films. *Smart Materials and Structures.*

83. Yang, Y., Ng, E. J., Polunin, P. M., Chen, Y., Flader, I. B., Shaw, S. W., Dykman, M. I., & Kenny, T. W. (2016). Nonlinearity of degenerately doped bulk-mode silicon mems resonators. *Journal of Microelectromechanical Systems, 25*(5), 859–869.
84. Zega, V., Opreni, A., Mussi, G., Kwon, H.-K., Vukasin, G., Gattere, G., Langfelder, G., Frangi, A., & Kenny, T. W. (2020). Thermal stability of DETF mems resonators: numerical modelling and experimental validation. In *2020 IEEE 33rd International Conference on Micro Electro Mechanical Systems (MEMS)* (pp. 1207–1210).
85. Hariri, A., Zu, J., & Ben Mrad, R. (2007). Modeling of wet stiction in microelectromechanical systems (mems). *Journal of Microelectromechanical Systems, 16*(5), 1276–1285.
86. Heinz, D. B., Hong, V. A., Ahn, C. H., Ng, E. J., Yang, Y., & Kenny, T. W. (2016). Experimental investigation into stiction forces and dynamic mechanical anti-stiction solutions in ultra-clean encapsulated mems devices. *Journal of Microelectromechanical Systems, 25*(3), 469–478.
87. Guerrieri, A., Frangi, A., & Falorni, L. (2018). An investigation on the effects of contact in mems oscillators. *Journal of Microelectromechanical Systems, 27*(6), 963–972.
88. Moorthy, R., Kakodkar, A., Srirangarajan, H., & Suryanarayan, S. (1993). An assessment of the newmark method for solving chaotic vibrations of impacting oscillators. *Computers and Structures, 49*(4), 597–603.
89. Hajjaj, A., Alfosail, F., & Younis, M. (2018). Two-to-one internal resonance of mems arch resonators. *International Journal of Non-Linear Mechanics, 107*, 64–72.
90. Sarrafan, A., Azimi, S., Golnaraghi, F., & Bahreyni, B. (2019). A nonlinear rate microsensor utilising internal resonance. *Scientific Reports, 9*, 8648.
91. Antonio, D., Zanette, D. H., & Lopez, D. (2017). Frequency stabilization in nonlinear micromechanical oscillators. *Nature Communications, 8*, 15523.
92. Mond, M., Cederbaum, G., Khan, P. B., & Zarmi, Y. (1993). Stability analysis of the non-linear Mathieu equation. *Journal of Sound and Vibration, 167*(1), 77–89.
93. Rhoads, J. F., & Shaw, S. W. (2010). The impact of nonlinearity on degenerate parametric amplifiers. *Applied Physics Letters, 96*(23), 234101.
94. Zega, V., Nitzan, S., Li, M., Ahn, C. H., Ng, E., Hong, V., Yang, Y., Kenny, T., Corigliano, A., & Horsley, D. A. (2015). Predicting the closed-loop stability and oscillation amplitude of nonlinear parametrically amplified oscillators. *Applied Physics Letters, 106*(23), 233111.
95. Nitzan, S. H., Zega, V., Li, M., Ahn, C. H., Corigliano, A., Kenny, T. W., & Horsley, D. A. (2015). Self-induced parametric amplification arising from nonlinear elastic coupling in a micromechanical resonating disk gyroscope. *Scientific Reports, 5*, 9036.
96. Frangi, A., Guerrieri, A., Carminati, R., & Mendicino, G. (2017). Parametric resonance in electrostatically actuated micromirrors. *IEEE Transactions on Industrial Electronics, 64*(2), 1544–1551.

Chapter 13
Inertial Sensors

Giorgio Allegato, Lorenzo Corso, and Carlo Valzasina

13.1 Inertial Sensors: An Historical Background

Inertial Sensors were one of the first product family developed using MEMS technology. The first paper about a MEMS accelerometer dates to 1979 when a bulk micromachined accelerometer was designed and first prototypes were manufactured at Stanford University [1]. Device working principle was based on piezoresistive sensing, and the MEMS structure was based on a glass-silicon-glass bonded wafer stack. Following this publication, it took about 20 years to have first MEMS accelerometers available on the market; the market driver for development was the replacement of switches for airbag deployment on automotive applications. In 1991, Analog Device commercialized the first industrial and high-volume MEMS accelerometer [2], realized by surface micromachining and based on capacitive sensing principle. About 10 years later, MEMS accelerometers from different suppliers entered the consumer market finding high-volume application in game consoles and expanding in the following years to smartphones, laptops, and IOT/wearable devices for more intuitive user interface and activity tracking.

As for the accelerometers, also MEMS gyroscopes required some decades from the first prototypes before achieving technology maturity for high-volume manufacturing. Driver application was still automotive for advanced safety systems for vehicle stability control.

G. Allegato · L. Corso
ST Microelectronics, Analog MEMS and Sensors Group, MEMS Technology and Design R&D, Agrate Brianza, Monza Brianza, Italy
e-mail: giorgio.allegato@st.com; lorenzo.corso@st.com

C. Valzasina (✉)
ST Microelectronics, Analog MEMS and Sensors Group, MEMS Technology and Design R&D, Cornaredo, Italy
e-mail: carlo.valzasina@st.com

© Springer Nature Switzerland AG 2022
B. Vigna et al. (eds.), *Silicon Sensors and Actuators*,
https://doi.org/10.1007/978-3-030-80135-9_13

First MEMS gyroscopes for automotive were commercialized in the late 1990s and expanded to consumer market in the late 2000s when they found application in smartphones for better user interface and optical image stabilization, gaming, virtual, and augmented reality applications.

During the last years, inertial sensors evolved versus multiaxes integrated systems based on accelerometers and gyroscopes combination (6-axis) or with additional 3-axis magnetometers (9-axis) for fully integrated Inertial Measurement Units (IMUs) [3].

The availability of low-cost sensors and growing market for low-accuracy devices put rapidly MEMS inertial sensors at the top of MEMS device diffusion ranking, a process now called the "MEMS Second Wave" [4].

The development of consumer inertial sensors partially leveraged on the available know-how coming from "First MEMS Wave" of the 1990s but required huge developments in several fields of MEMS design and manufacturing. Inertial MEMS processes were streamlined, and foundries increased wafer size from 4 or 6-inch to 8-inch wafers. Design switched from complex architectures to simpler solutions, exchanging cutting-edge performances with higher yield and shorter time to market. Wafer-to-wafer bonding with hermetic wafer-level encapsulation allowed the transition from ceramic hermetic packages to Systems in Package (SiP) encapsulated with inexpensive, overmolded techniques (Land Grid Array (LGA) or Quad-flat No Leads (QFN) packages being the most common). Testing and calibration process was greatly developed, in close cooperation with test equipment manufacturers, increasing calibration throughput up to the range of thousands of units per hour (UPH).

These technological advances transitioned Inertial MEMS Sensors industry from jewel-like manufacturing to low-cost, high-volume, high-yield processes.

In the context of MEMS inertial sensors expansion, STMicroelectronics was the leader in the MEMS consumerization wave introducing on the market the first high-volume accelerometer for gaming application in the Nintendo WII User Interface in 2006 [5] and introducing the first worldwide integrated 3-axial gyroscope for smartphones in 2010. Today, STMicroelectronics holds a leadership position in the inertial sensor field also in automotive and industrial areas.

In the following paragraphs, accelerometers and gyroscope principles of working will be introduced, and the impact of process parameters on device performances will be presented. The STMicroelectronics "THELMA" technology for the manufacturing of inertial sensors will be presented, and some examples of technology solution to address specific product and application requirements will be described.

13.2 Capacitive MEMS Accelerometers

13.2.1 Accelerometer Working Principles

Accelerometers are sensors used to transduce acceleration (input signal) into electrical output. These sensors could take different naming, depending on the intended acceleration input: Accelerometers are intended to sense the dynamic acceleration

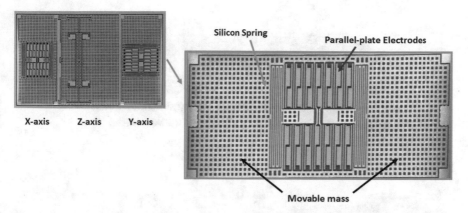

Fig. 13.1 SEM image of MEMS Accelerometer, with highlights on the layout of the movable mass, spring, and electrodes

resulting from motion in an inertial frame of reference, while inclinometers are accelerometers intended to sense the projection of Earth's gravitational field on sense axis. The common measurement unit used to describe accelerometer characteristics is g, corresponding to the freefall acceleration due to Earth's gravitational field at sea level, i.e., (1 g = ~9.81 $\frac{m}{s^2}$).

The most common and widespread MEMS accelerometers are based on capacitive transduction. The structure used to realize a MEMS single-axis accelerometer is a silicon mass suspended via deformable silicon cantilevers [6]. In the presence of external acceleration, the mass is subjected to a force and the thin cantilevers deforms, reaching an equilibrium position. The displacement of the movable mass with respect to the null-force position is transduced into capacitive change with dedicated electrodes.

Detailed description of accelerometer working principles and underlying mathematical models can be found in [7, 8].

An SEM image of a single-axis, surface micromachined MEMS accelerometer realized with STMicroelectronics THELMA process is reported in Fig. 13.1. The perforated silicon plate acts as the inertial mass, while thin polysilicon suspension is used as deformable springs. A set of polysilicon fingers are attached to the movable inertial mass and constitute one plate of the variable capacitor. The other plates are polysilicon-fixed counterelectrodes attached to the substrate with anchoring points (Fig. 13.2).

In Fig. 13.3 is shown the schematic representation of such accelerometer, as well as the transduction chain of the system. We can schematically divide a MEMS capacitive accelerometer into a mechanical part and an electrical transduction part.

The Mechanical model of an accelerometer is usually simplified as one Degree-of-Freedom (DOF) mass-spring-damper system. While this lumped model can be used to describe the main behavior of the microsystem, a more refined model may be used to consider second-order effects such as spurious vibration modes and other nonidealities.

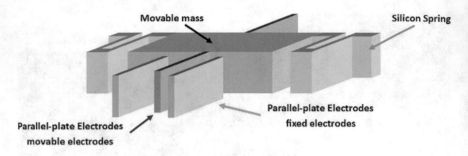

Fig. 13.2 Capacitive MEMS accelerometer schematic

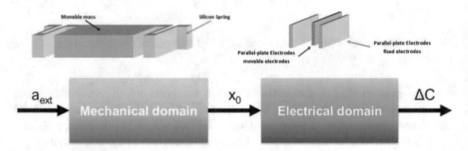

Fig. 13.3 Capacitive MEMS transduction chain

Spring-mass-damper system with one Degree of Freedom (DOF) is shown in Fig. 13.4 and is described by three main parameters: m is the movable mass of the system, b_x is the damping term, and k_x is the elastic stiffness. The general equation of motion is therefore:

$$m\frac{d^2x}{dt^2} + b_x\frac{dx}{dt} + k_x x = ma_{ext} \tag{13.1}$$

where a_{ext} is the external acceleration acting on the system. Solving for the easiest condition of constant acceleration, we find that the equilibrium position of the system is

$$x_0 = \frac{m}{k_x}a_{ext} = \frac{1}{\omega_0^2}a_{ext} \tag{13.2}$$

$$\omega_0 = \sqrt{\frac{k_x}{m}} \tag{13.3}$$

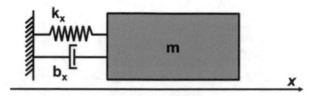

Fig. 13.4 Lumped parameters model of mechanical part of a MEMS accelerometer: the spring-mass-damper system

Where ω_0 is the resonant (angular) frequency of the dynamic system. Equation (13.2) is therefore a measure of low-frequency sensitivity of the mechanical portion of the system in Fig. 13.3.

If input acceleration is not constant, we can treat Eq. (13.1) as a harmonic oscillator, whose solution is of the form

$$x(t) = A_1 \cos(\omega_0 t) + A_2 \sin(\omega_0 t) \tag{13.4}$$

where A_1 and A_2 are constants determined by the initial conditions of the system. The behavior of the system near resonance is dependent on the quantity

$$\xi = \frac{b_x}{2\sqrt{mk_x}} \tag{13.5}$$

known as damping ratio; when ξ is higher than 1, the system is overdamped and input near resonance is quenched, while for $0 < \xi < 1$, the system is underdamped, and resonance peak appears at the resonance frequency. This is of paramount importance when dealing with accelerometer response to external vibration disturbances. The frequency response for different values of the Quality Factor Q ($Q = \frac{1}{2\xi}$) is reported in Fig. 13.5.

Electrical transduction portion of a capacitive MEMS accelerometer is realized with fingers acting as fixed and movable plates of a capacitor. The most common configurations are comb-finger and parallel plate sensing, shown in Fig. 13.6.

Comb-finger sensing is based on the sliding of capacitor plates at fixed gap, and the change in capacitance is achieved with the change of the facing surface area. This method offers the advantage of good linearity for large displacements but suffers from lower sensitivity per unit of rest capacitance C_0. Basic expression for rest capacitance C_0 and linear sensitivity $\frac{dC}{dx}$ of comb-finger is reported in Eq. (13.6)

$$C_0 = \varepsilon_0 \frac{W\,L_0}{g}; \quad \frac{dC}{dx} = \frac{d}{dx}\left(\varepsilon_0 \frac{W\,L(x)}{g}\right) = \varepsilon_0 \frac{W}{g} \tag{13.6}$$

Parallel plate sensing is instead based on changes of the gap between electrodes, at fixed facing area. This sensing configuration has the advantage of increased sensitivity per unit of rest capacitance C_0 with respect to comb-finger, but has the

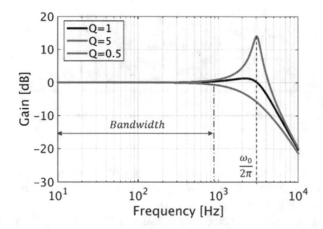

Fig. 13.5 Frequency response of spring-mass-damper system for different Quality factors – ξ. With low Q factor, the frequency response is overdamped and no amplitude amplification is present; with high Q factor, the frequency response is underdamped and resonance peak is visible at resonance frequency

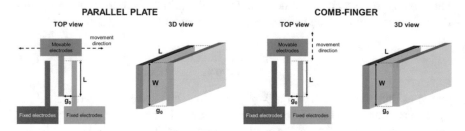

Fig. 13.6 Parallel plates vs. comb finger sensing electrodes for capacitive MEMS

drawback of nonlinear capacitance-displacement response, as can be seen from Eq. (13.7)

$$C_0 = \varepsilon_0 \frac{W\,L}{g_0}; \frac{dC}{dx} = \frac{d}{dx}\left(\varepsilon_0 \frac{W\,L}{(g-x)}\right) = -\varepsilon_0 \frac{W\,L}{(g-x)^2} \sim -\varepsilon_0 \frac{W\,L}{g_0^2} \qquad (13.7)$$

We can derive the complete MEMS accelerometer sensitivity to external acceleration in case of parallel plate sensing, for the sensing chain described in Fig. 13.2, we can combine Eqs. (13.7) and (13.2) to get

$$\frac{\Delta C}{\Delta a_{ext}} = C_1 - C_2 = 2\frac{1}{\omega_0^2}\varepsilon_0 \frac{W\,L}{g_0^2} \qquad (13.8)$$

where the equation considers a typical differential configuration of sense electrodes but neglects the readout biasing effect, known as electrostatic spring softening.

In parallel plates' structures, the gap between plates is varied, and therefore the electrostatic force acting on opposite plates will be dependent on position. The electrostatic force directed along the x axis has a magnitude:

$$F = \frac{1}{2}\frac{dC}{dx}V^2 = \frac{1}{2}\frac{\varepsilon A}{(g_0 - x)^2}V^2 \qquad (13.9)$$

The dependency of the force to position is crucial as it acts as elastic negative stiffness and changes the actual stiffness of the mechanical spring adding a term:

$$k_{el} = -\varepsilon_0\frac{WL}{g^3}V^2 \qquad (13.10)$$

This negative electrostatic stiffness is opposed to the spring elastic stiffness. When the voltage is increased over the value $V_{pull-in}$:

$$V_{pull-in} = \sqrt{\frac{8kg_0^3}{27\varepsilon_0 A}} \qquad (13.11)$$

The electrostatic stiffness overcomes the elastic stiffness k, and the plates collapse in the so-called pull-in effect.

13.2.2   Accelerometer Specifications and Requirements

When describing inertial sensor performances, the correct option is to refer to the IEEE Standard for Inertial Sensor Terminology 528–2019 [9], in which all useful terms are specified.

When comparing accelerometer datasheets, however, terms may vary from different vendors, requiring some effort to compare different products.

Depending on the application, requirements for MEMS accelerometer may vary significantly; as an example, Zero-Offset value and its drift over external conditions (i.e., temperature, lifetime, etc.) are of critical importance for inclinometers, while its value is almost negligible for automotive-safety high-g accelerometers for airbag firing.

It is therefore of little sense to compare accelerometer intended for different application in a single table to looking for the best performer; in Table 13.1 is nevertheless given a broad picture of how requirement for MEMS accelerometer varies over different applications.

Table 13.1 MEMS accelerometers requirements for different applications

	Consumer user interface	Automotive ESP	Structural health monitoring inclinometer	Automotive airbag firing
Input-range [g]	2–16	6–20	0.25–1	120–400
Zero-offset [mg]	60–120	50	2–5	1000
Zero-offset vs. T [mg/K]	0.5–2.0	0.1–0.3	0.05	50
Noise [μg/$\sqrt{Hz}$]	100	–	20	–
Bandwidth [Hz]	100	60	5	400
Power consumption [mW]	0.5	5	2	2

13.2.3 MEMS Accelerometers Design Principles

During the design phase of a MEMS accelerometer, several aspects should be considered: MEMS design always turns out to be a co-design of mechanical transducer, signal conditioning ASIC, and Package.

The basic input parameters for accelerometer design are related to sensor requirements, as well as ASIC parameters and general requirements. Typical examples of such requirements are:

- *Application Requirements*: Full Scale, Resolution or Noise, Input Signal Bandwidth, Zero-Offset stability, Scale Factor stability, linearity, etc.
- *Mission Profile*: operating environmental conditions (temperature, humidity, etc.), vibration and shock levels (operative and maximum survival), expected lifetime, etc.
- *ASIC characteristics*: Readout voltages, capacitance change per unit acceleration.
- *General Requirements*: Sensor Size, Target Cost, etc.

Starting from these requirements, a feasibility study is performed to sketch the basic sensor structure and provide a first-round estimation of MEMS electromechanical performance parameters.

One important tradeoff during the design phase of capacitive accelerometers is represented by the conflicting requirements of increasing as much as possible the mechanical sensitivity of the transducer vs. the full scale and robustness of the sensor.

By maximizing the mechanical sensitivity, i.e., increasing the displacement of the movable mass at reference input acceleration, the transducer reaches lower noise levels and better stability.

In fact, most of the external disturbances such as packaging and soldering stresses, temperature-related stresses, etc. could be modeled as fixed amount of imposed displacement on the transducer movable mass. A high displacement per unit input acceleration will result in a lower input-related effect of the external disturbance.

The high mechanical sensitivity is usually achieved with low frequency of the spring-mass-damper equivalent system, as can be seen from Eq. (13.2). Typical frequencies for consumer MEMS capacitive accelerometers are in the range of few kHz.

With an open-loop readout architecture, movable mass displacement per unit input acceleration is closely related to full scale, i.e., maximum acceleration input allowed before touching stoppers or having pull-in effects. Lower the frequency, lower the full-scale achievable.

A lower resonant frequency for the transducer is also related to lower stiffness of the spring or higher mass, as can be seen from Eq. (13.3). If the system has lower spring stiffness, the mechanical restoring force will be consequently lower, resulting in lower safety margin against stiction. The resonant frequency can be lowered by increasing movable mass and keeping high spring stiffness, at the expenses of transducer dimensions.

Some of the tradeoffs described above can be bypassed using more complex architectures, such as closed-loop force-to-rebalance systems [10] or Frequency Modulated readouts [11]. This comes with the drawback of increased system complexity and increased power consumption.

The design phase aims at predicting sensor performances in all possible process combinations; therefore, not only the typical sensor parameters should be considered, but also the outcome of different process variations.

To design and evaluate Robust sensor Performances [12], different modeling techniques could be employed. Finite Element Modeling (FEM) and analytical or semianalytical modeling can be used to extract typical values of MEMS parameters (e.g., Resonant frequency, rest capacitance, etc.), and the sensitivity of MEMS performances to process parameters.

As an example, process parameters such as structural Epi-poly Silicon thickness and dimensional loss (CD-loss) due to lithography and silicon etching could impact the resonant frequency of an accelerometer.

FEM modal analysis can be applied to the typical MEMS accelerometer geometry to obtain the frequency of first vibration modes.

In Fig. 13.7 are presented the first two vibration modes of the MEMS accelerometer of Fig.13.1, as obtained from an FEM modal analysis in the simulation software ANSYS.

Figure 13.8 shows how changes in process can affect the cross-section of a spring, and specifically, how CD loss variations can result in different sections of a beam while MEMS structural layer thickness variations can alter the thickness of the spring. Sensor mass is also changed by process variations.

Selected variations to the geometry can be applied in the FEM simulation, considering the variability of the process, and "sensitivity" function of resonance frequency vs. process parameters could be extracted. By performing several FEM simulations in different process corners, it is possible to evaluate how MEMS parameters are affected by process variations.

First Vibration Mode   **Second Vibration Mode**

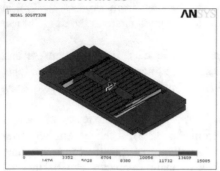 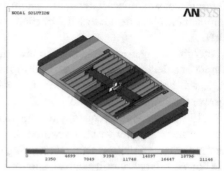

Fig. 13.7 First two vibration modes of the MEMS accelerometer of Fig. 13.1, as obtained from an FEM modal analysis. Color scale represents total displacement in the vibration mode

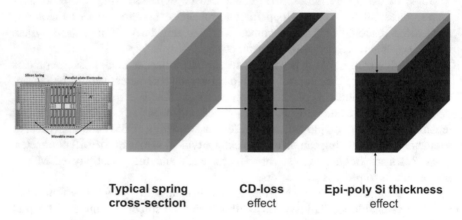

Typical spring cross-section   **CD-loss** effect   **Epi-poly Si thickness** effect

Fig. 13.8 Effects of process on the section of a silicon beam used as a spring: CD loss is the dimensional loss due to Silicon etching process, resulting in reduced beam cross-section; thickness of the beam is influenced by polysilicon growth/removal

Such method is known as *corner analysis* and offers the possibility to investigate MEMS parameter dependency on process variations as well as MEMS performances at the very edges of process specifications (i.e., process corners), while keeping the number of simulations reasonably low. It can therefore be applied to time-consuming simulations such as FEM.

13.3   Capacitive MEMS Gyroscopes

13.3.1   Gyroscope Working Principles

Gyroscopes are sensors capable of converting angular rate (input signal) into electrical output. Angular rate is the amount of angle covered in unit time during a rotation and is measured in degrees per hour [°/h] or degrees per second [dps].

Gyroscopes can be classified as rate gyroscopes, if the measured quantity is angular rate as described before, or Rate-Integrating (angle) gyroscopes, if the physical quantity they measure is the time integral of Angular Rate, i.e., they directly provide the angular change from a specific reference time [13].

Rate-integrating gyroscopes are especially suitable for navigation applications and usually exploit precession of vibration via energy transfer in two vibration modes as the result of external motion [14]. The high symmetry required for these MEMS to operate has prevented them to achieve mass production and commercial success. Rate gyroscopes are instead the most widespread type of MEMS gyroscope in the market.

The principle of operation of rate MEMS gyroscope is the excitation of a vibration mode in response to Coriolis Force acting on a proof mass. Coriolis Force is the apparent force arising in noninertial frame of references as a combination of the linear velocity of the object and the rotation rate of the noninertial frame

$$F_c = -2mv_x \wedge \Omega_{ext} \tag{13.12}$$

Where m is the mass of the moving system, v_x is the linear velocity in one direction, and Ω_{ext} is the external rotation rate.

The basic implementation of a single-axis rate gyroscope is a 2 Degree-of-Freedom (DOF) mechanical system in which one vibration mode (drive axis) is externally excited in resonance and the second vibration mode is excited because of Coriolis Force [15].

The equation of motion for the drive axis x and the sense axis y may be written as:

$$m_x \ddot{x} + c_x \dot{x} + k_x x = F_x \sin(\omega t) \tag{13.13}$$

$$m_y \ddot{y} + c_y \dot{y} + k_y y = F_c = -2m\dot{x}\Omega_z \tag{13.14}$$

where m_x is the mass, c_x is the dissipation coefficient, k_d is the stiffness, and $F_x \sin(\omega t)$ is the force applied to drive axis x, while m_y is the mass, c_y is the dissipation coefficient, k_y is the stiffness of the sense axis y, and F_c is the Coriolis force acting on the sense axis as a result of a rotation Ω_z around z axis.

When such a configuration is achieved with discrete springs and masses, the system is called Coriolis Vibratory Gyroscope (CVG). More complex implementations

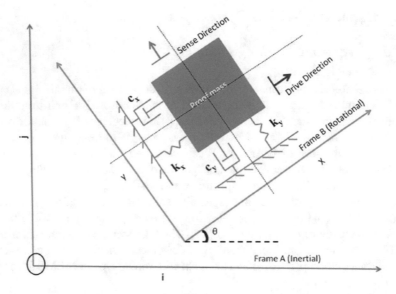

Fig. 13.9 Schematic representation of MEMS Coriolis Vibrating Gyroscope. The mechanical element can be described by 2-Degree-of-Freedom oscillator: a spring-mass-damper system models the drive direction x, and another spring-mass-damper system models the sense direction y. When the frame of reference of the movable mass (x,y) is rotated with respect to an inertial reference (i,j), Coriolis Force couples the drive and sense dynamics

of this 2-DOF mechanical system may rely on solid disks [16] or rings of material [17]. A schematic representation of a Vibrating Coriolis Gyroscope is given in Fig. 13.9.

As for accelerometers, the most widespread principle used for actuation and sensing in MEMS CVG gyroscopes is electrostatic actuation and capacitive sensing.

Electrostatic actuation is the generation of a force in a capacitor because of applied potential on the two armors.

The force acting on a capacitor plate is

$$F_d = \frac{1}{2} \frac{\partial C_d}{\partial x} \Delta V^2 \qquad (13.15)$$

where $\frac{\partial C_d}{\partial x}$ is the derivative of the capacitance value along motion direction, and ΔV is the voltage difference between capacitor armors.

In the sense direction, the system can be treated as a capacitive accelerometer where input signal is Coriolis acceleration. Therefore, all the equations described in the previous chapter can be applied.

When dealing with drive and sense axis frequencies, two different configurations are possible:

1. Drive frequency and sense frequency are placed at a fixed distance. The gyroscope is said to operate at mode-mismatched condition.
2. Drive frequency and sense frequency overlap (to a certain amount of accuracy). The gyroscope is said to operate in mode-matched condition.

Mode-matched operation greatly enhances the mechanical sensitivity of the sense axis, due to amplification of sense movement at resonance, but this operation mode requires active frequency steering loops and electromechanical feedback loops to cancel gyroscope nonidealities and drift of mechanical parameters [18].

Practically, most of the MEMS CVG gyroscopes in the market fall in the mode-mismatched operation mode. Sensitivity is not maximized, but a low-power, open-loop sense architecture can be used.

Ideal 2-DOF resonators have completely separated drive and sense mechanical dynamics, but in reality, a coupling may occur. As can be seen in Eqs. (13.15) and (13.16), when a coupling term k_{xy} is introduced in the equations of motion (13.12) and (13.13), the drive x direction and the sense y direction equations are coupled; therefore, a displacement in drive direction will result as a parasitic displacement in y direction and will be displaced as an output signal.

$$m_x \ddot{x} + c_x \dot{x} + k_x x = F_x \sin(\omega t) \tag{13.16}$$

$$m_y \ddot{y} + c_y \dot{y} + k_y y + k_{xy} x = F_c = -2m\dot{x}\Omega_z \tag{13.17}$$

The spurious signal arising from such coupling had a 90° phase difference from Coriolis signal and is therefore called quadrature signal. Coupling coefficient is a consequence of fabrication imperfections, specifically anisotropic silicon etching resulting in nonvertical definition of the flexures, that in turn results in nonideal beam's cross-section [19, 20].

When a drive beam has not vertical etching, the drive electrostatic force applied along the X axis will result in parasitic displacement in the Y direction, as described in Fig. 13.10.

In commercial consumer products, complex MEMS designs are used to achieve three-axis sensitivity. The mechanical element can be described as a 4-DOF oscillator with one vibration mode dedicated to drive and the other three vibration modes dedicated to Coriolis Force sensing along the three orthogonal directions. An example of this structure can be found in Fig. 13.11, showing STMicroelectronics gyroscope used in LSM6DS3 6-axis Inertial Measurement Unit (IMU). The main vibration modes corresponding to drive and sense dynamics are presented in Fig. 13.13.

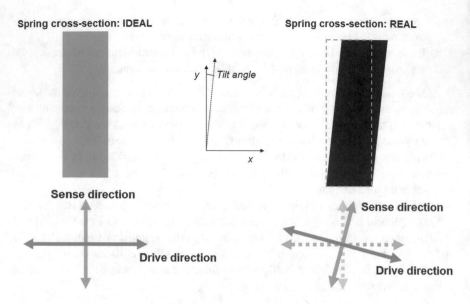

Fig. 13.10 Description of the spring tilt impact on Gyroscope sense/drive displacement

13.3.2 Gyroscope Specifications and Requirements

Performance indicators of MEMS CVG gyroscopes can be found in the IEEE Standard Specification Format Guide and Test Procedure for Coriolis Vibratory Gyros [21], on top of the already mentioned IEEE Standard for Inertial sensors terminology [9].

The main parameters defining MEMS gyroscope performances are:

Bias (Offset or Zero-Rate Level ZRL) is the averaged output of the sensor measured over a specified time in absence of input rotation. Bias is measured in degrees per second [dps] or degrees per hour $[\frac{\circ}{h}]$. ZRL behavior may be described as a function of external conditions, such as acceleration sensitivity, temperature sensitivity, temperature gradient sensitivity, temperature hysteresis, and vibration sensitivity. Bias characteristics are usually described in terms of Allan Variance [22], whose components Angle Random Walk (ARW), Bias Instability (BI), and Rate Random Walk (RRW) are a measure of random bias drift. Allan Variance is a powerful method to identify noise sources in inertial sensors and is computed from gyroscope output by clustering the data at given averaging time, computing the two-sample variance of these clusters and plotting the resulting variance as a function of the averaging time.

Scale factor (Sensitivity) is the ratio of a change in output to a change in the input intended to be measured. It can be measured in [V/dps] for analog-output sensors or in [LSB (Least Significant Bit)/dps] for digital sensors. Different error classes are used to describe deviation from the ideal output characteristic: scale factor error,

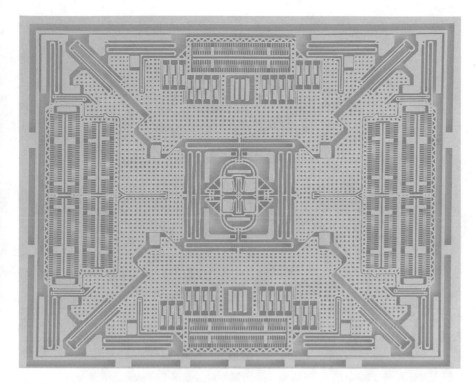

Fig. 13.11 SEM Top View of the LSM6DS3 3-axis gyroscope. Vibration modes of this structure used for drive and sense motion are reported in Fig. 13.13

nonlinearity, asymmetry, and stability against environmental disturbances such as temperature. These deviations from ideal curve are usually measured in [%] or [ppm].

Rate noise density is the spectral density of noise components expressed as [dps/$\sqrt{\text{Hz}}$] and refers to the high-frequency noise terms characterized by a white-noise spectrum on the gyro rate output due to correlation time much shorter than the sample time. Noise density has a connection with ARW, and it can be derived from on Allan Variance plot by reading the slope line at T = 1.

Power consumption is the power dissipated by the gyroscope system (MEMS and the control electronics) and is measured in [W]. For low-power devices, the common way to report power consumption is to provide the current drained from power supply during operation in [μA] (Fig. 13.12).

In Table 13.2 are reported typical values of gyroscope performance parameters as a function of the intended application; as can be seen, the typical values of bias stability and scale factor accuracy span several orders of magnitudes. Commercial MEMS gyroscopes are widely diffused in consumer and industrial applications stretching to low-end tactical, while other gyroscope technologies are used for higher accuracy applications (Table 13.3).

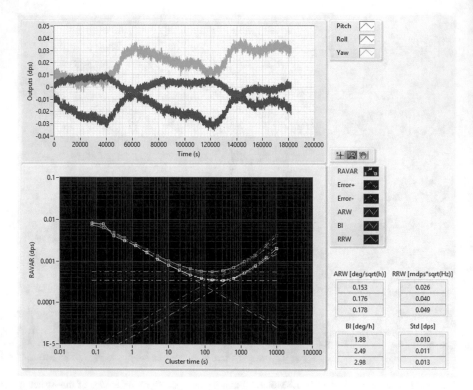

Fig. 13.12 Allan Variance plot. Allan Variance is a measure of stability due to random errors and is used to highlight different noise components in sensors. On the top a series of samples of gyroscope output (with no rate input provided) is reported. The same dataset is analyzed with Allan Variance method in the plot below. This plot is divided in three main regions: the $-1/2$ slope region is a measure of Angular Random Walk AWR (White Noise), the flat region is a measure of Bias Instability (Flicker Noise, or $\frac{1}{f}$), and the $+1/2$ slope region is a measure of Rate Random Walk RRW (Red Noise, or $\frac{1}{f^2}$)

13.3.3 MEMS Gyroscopes Design Principles

The design principles and tradeoffs of MEMS gyroscopes are like the ones of MEMS accelerometers described in the previous chapter. For gyroscopes also, several aspects should be considered and MEMS design is in practice a co-design of technological process, mechanical transducer, signal conditioning ASIC, and package.

The design always starts from sensor requirements and ASIC parameters. The choice of system architecture (how to cancel nonidealities, how to drive the transducer and to read out Coriolis signal, where to set drive and sense resonances, etc.) is fundamental and directly derives from the stability, power consumption, and size requirements.

Table 13.2 Classification of gyroscopes with respect to application

Grade	Application	Bias stability $\left[\frac{^\circ}{h}\right]$	Scale factor accuracy [ppm]
Consumer	User interface	30–1000	5000–10,000
Industrial and low-end tactical	Guided ammunitions	1–30	100–5000
Tactical	Platform stabilization	0.1–30	10–100
High-end tactical	Missile navigation	0.1–1	1–10
Navigation	Aerospace inertial navigation	0.01–0.1	<1
Strategic	Submarine navigation	0.0001–0.01	<1

Table 13.3 Reports the typical parameters for consumer and automotive MEMS gyroscopes

	Consumer user interface	Automotive ESP
Input-range [dps]	2000–4000	125–300
Zero-rate level ZRL [dps]	10	2
ZRL vs. T [dps/K]	0.02	0.005
Noise [dps/$\sqrt{Hz}$]	0.007	0.005
Bandwidth [Hz]	400	60
Power consumption [mW]	1	30

With a preliminary feasibility study, it is possible to translate basic sensor parameters (mass, frequencies, capacitance, etc.) into preliminary estimation of MEMS electromechanical performance parameters.

Usually, the architecture of choice for consumer MEMS gyroscopes is closed loop sensing, open loop driving, and electronic open-loop quadrature compensations, a very simple architecture that can achieve reasonably good noise and stability performances with superior power consumption figures.

For automotive Electronic Stability Program (ESP) gyroscopes, the stability requirements are tighter than for consumer counterparts; therefore, usually a closed-loop electromechanical quadrature compensation architecture is used.

Most of the existing MEMS CVG gyroscopes are operated in mode-mismatched, with drive-to sense frequency separation of approximately 5%. The choice of frequency separation, or frequency mismatch, is a critical choice because it defines the tradeoff between gyroscope stability vs. electromechanical sensitivity.

Reducing the frequency mismatch, the mechanical sensitivity is maximized, because Coriolis Force, which is modulated at drive frequency, excites the sense vibration mode in a region closer to resonance. The drawback is that sense transfer function has a steeper phase variation close to sense resonance; therefore, the

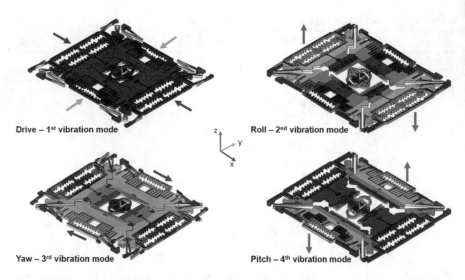

Fig. 13.13 First four vibration modes of a MEMS three-axial gyroscope of Fig. 13.11. These vibration modes are obtained with FEM modal analysis and show Drive Mode, actuated in closed loop with electrostatic force, and Roll, Yaw and Pitch modes, forced by Coriolis force when the gyroscope rotates around y, z, and x axes

coherent demodulation used to separate Coriolis and quadrature signal is affected by larger errors because of environmental disturbances. This in turns translates into larger quadrature signal leakage to Coriolis signal and therefore larger bias drifts, as explained in [23].

This tradeoff could be overcome with usage of closed-loop electromechanical quadrature compensation, that drastically reduces the quadrature signal at MEMS output and therefore reduces the requirement of perfectly coherent demodulation. Electromechanical steering of sense motion to reduce quadrature with dedicated electrodes is well known and widespread used method to improve gyroscope stability, at the price of higher power consumption, larger mechanical structures, and increased system complexity.

As well as for accelerometer, Frequency Modulated readouts have been proposed for gyroscopes [24] and realized with commercial MEMS technologies [25].

Also, for MEMS gyroscopes, the prediction of sensor performances in all possible process combinations enables the achievement of robust sensor performances.

In Fig. 13.13 are presented the first four vibration modes of a MEMS three-axial gyroscope, as obtained from an FEM modal analysis in ANSYS.

In Table 13.4 is shown how structural Epi-poly Silicon thickness CD-loss process parameters impact the resonant frequency of a gyroscope, and therefore the critical parameter of frequency mismatch. This table is obtained by corner analysis with FEM simulations.

Corner Analysis anyhow does not offer any reliable information on the effective distribution of MEMS parameters and tends to overestimate the importance of

Table 13.4 Impact of CD-loss & Epi-poly Thickness on the resonance frequency of Gyroscope vibration modes

CD_loss [μm]	typ	−3σ	+3σ	typ	Typ
Epi-poly Si thickness [μm]	typ	typ	typ	-3σ	+3σ
1st mode frequency [Hz]	20,000	19,000	21,000	20,000	20,000
2nd mode frequency [Hz]	20,850	19,800	21,890	20,690	21,010
3rd mode frequency [Hz]	20,850	19,980	21,720	20,850	20,850
4th mode frequency [Hz]	20,850	19,830	21,870	20,700	21,000

highly unprobable process configurations. A combination of three independent process parameters at 3-sigma values occurs in approximately one-over ten million devices, and designing a system coping with such unlikely configurations could result in poor efficiency and reduced competitiveness, since we can easily screen this part out from production and reduce the overall burden on system design.

Monte Carlo Analysis is a more efficient way to evaluate the distribution of MEMS parameters but requires knowledge of the process parameter statistical distribution and a huge number of simulations to be useful. A faster method to compute spreads of MEMS parameters is a semianalytical model: Analytical formulas are used to compute values of MEMS parameters (sensitivity, frequency, etc.); process spreads are considered by introducing process dependencies in analytical formulations derived from selected FEM and corner analysis.

Figure 13.14 shows the Monte Carlo distribution of the frequency mismatch parameter for the gyroscope described before. As can be seen comparing Fig. 13.14 with Table 13.4, the frequency values obtained with corner analysis represents extreme cases if compared with Monte Carlo simulation. Monte Carlo analysis in fact predicts the full distribution of frequency difference between first Drive mode and second Roll mode, while corner analysis in Table 13.4 only gives point predictions for the different values of frequency in extreme process conditions.

It should be now clear the critical role that process variations have on the design of robust MEMS sensors. It is therefore of paramount importance to adjust MEMS process and MEMS design to achieve target parameters in all process conditions, in a Process-Product Co-Design.

13.4   THELMA Technology Introduction

THELMA is the ST-proprietary technology for the manufacturing of inertial sensors, i.e., accelerometers and gyroscopes. THELMA is an acronym for "**TH**ick **E**pitaxial **L**ayer for **M**icro-gyroscopes and **A**ccelerometers". THELMA technology architecture is based on a surface micromachining process in which a thick epitaxial polysilicon film is used as the structural layer and an oxide film as the sacrificial layer. THELMA technology allows to manufacture micromechanical components which detect movements like acceleration or rotation and translate them into an

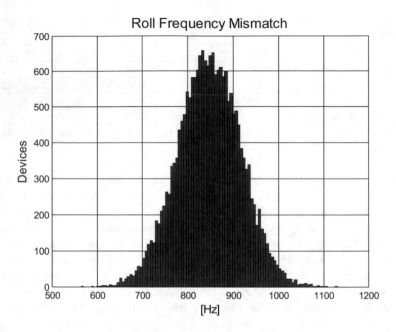

Fig. 13.14 Results of Monte Carlo simulation of frequency difference between first Drive mode and second Roll mode of the gyroscope reported in Fig. 13.11 and whose vibration modes are reported in Fig. 13.13. First, a semianalytical model of the two vibration frequencies is built, highlighting the impact of process parameters in the two frequencies. Then a random set of process parameters is given as input for the model, and the resulting frequency distribution for the two modes is computed. The distribution shown here is the sample-by-sample difference of the two frequency distributions

electrical signal which is read and translated into an analog or digital signal, by an external application-specific IC (ASIC).

As discussed in the previous paragraphs, THELMA sensing principle is capacitive, and it's based on the gap change between moveable and fixed electrodes induced by both in-plane and out-of-plane movements. The key device layers which enable this sensing mechanism are:

- Thick epitaxial polysilicon layer: The film thickness can range from 15 to 60 μm. This layer is patterned by a dry etch process to design the device components like the seismic mass, springs, and in-plane electrodes. Displacement between fixed and moveable electrodes translates into a capacitance change which allows the detection of in-plane movements (Fig. 13.2).
- Buried polysilicon layer: This film is used to design electrical interconnects and buried electrodes below the seismic mass. Out-of-plane displacement of the seismic mass can be detected by a capacitive change between the seismic mass itself and the buried electrodes. Figure 13.15 shows a schematic picture of an out-of-plane capacitor and the principle of out-of-plane capacitive signal reading: The rotation of the inertial mass along an in-plane axis, under an external

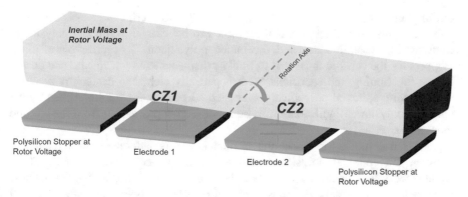

Fig. 13.15 Out-of-plane sensing principle: Inertial mass rotation around an in-plane axis under z-acceleration induces a change on the capacitance of the out-of-plane electrodes (CZ1 and CZ2)

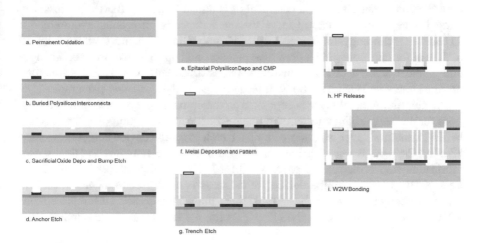

Fig. 13.16 THELMA process flow

acceleration along the z direction, induces a gap change between epitaxial polysilicon and buried polysilicon electrodes.

– A sacrificial oxide layer: This film defines the vertical gap between the buried polysilicon film and the epitaxial polysilicon film of the seismic mass. The oxide layer is removed by a vapor HF process in the areas where the seismic mass must be released and can move under external acceleration or rotation.

A more detailed description of the THELMA manufacturing flow is reported in Fig. 13.16 and described hereafter.

THELMA sensors are manufactured on silicon substrates. The first process step is the realization of a permanent oxide layer (13.16a) to insulate the buried polysilicon layer from the substrate. After permanent oxidation, the buried polysilicon is deposited by LPCVD and patterned to design interconnects and electrodes for out-of-plane sensing structures (step 13.16b). A TEOS sacrificial layer is deposited afterward; this layer thickness defines the z gap between the epitaxial polysilicon used for the seismic mass and the buried polysilicon. To realize out-of-plane stoppers for the seismic mass and to limit its displacement after release and during device functioning, the sacrificial layer is partially etched to design "molds" for the following epitaxial polysilicon layer (step 13.16c); these "molds", filled by the epitaxial polysilicon, will result into bump structures protruding from the seismic mass backside.

After bump molds formation, the sacrificial layer is patterned to design anchors between the epitaxial polysilicon and buried polysilicon (step 13.16d): These anchors work both as electrical contacts between the two layers and as mechanical anchors to keep the seismic mass anchored on specific poly areas after the final MEMS release process.

Following sacrificial layer patterning, structural epitaxial polysilicon film is deposited and planarized by a chemical mechanical polishing (CMP) process step (step 13.16e). Final structural layer thickness depends on product requirements; THELMA technology implements epitaxial polysilicon thickness values from 15 to 30 μm for consumer and automotive products and up to 60 μm for specific high-end applications where high-sensitivity values are required. This last technology option, known as THELMA-60, will be discussed in the following paragraphs.

An aluminum metal layer is deposited on the epitaxial polysilicon film to design pads for device probing and wire bonding. After metal definition (step 13.16f), the MEMS structural layer is patterned by DRY etch to design device components (step 13.16g): the seismic mass, springs, and in-plane sensing electrodes. The final sensor fabrication step is the release process (13.16h): The sacrificial layer is removed by a vapor HF process, and the device inertial mass is made free to move. To achieve a more effective release process, the seismic mass is patterned with release holes which provide the access to the HF vapors for the sacrificial oxide removal (Fig. 13.17).

To protect the moving components of the inertial sensors during assembly steps and operation and to better control the environment in which the sensors work, THELMA products are packaged at wafer level by a wafer-bonding process. The sensor wafer is bonded to a cap wafer by a glass-frit glue layer, which consists of a low melting temperature glass. Wafer-bonding process is performed under a controlled atmosphere and at a fixed pressure setpoint, which are defined at front-end manufacturing on the wafer bonder equipment.

The cap wafers include through-silicon holes on top of the pads areas which allow the access to metal pads for device testing and wire bonding. Cavities can be designed also on top of the device area with different functions; specific design allows to realize stopper areas to limit the out-of-plane displacements of the seismic mass and to reduce mechanical shocks' impact on device functionality. Containment trenches for glue layer overflow are designed on top cap to prevent glass-frit excess in the device cavity.

Figure 13.18 shows an SEM cross-section of the bonded die.

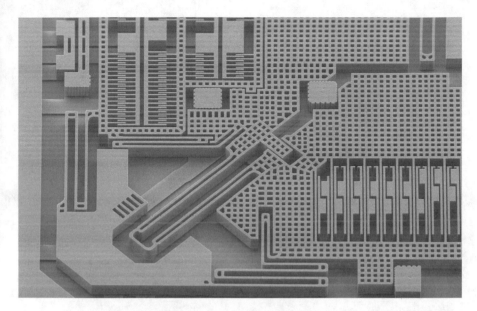

Fig. 13.17 Detail of a THELMA gyroscope with the holed device inertial mass, springs, and electrodes

Fig. 13.18 Cross-section View of a MEMS singulated die with pad sensor and cap dice, bonding glue layer, and pad window on the top cap

13.4.1 Process Specificities: Wafer-Level Package (WLP): Vacuum Level through Getter Technology

In gyroscope products, the driving principle is based on the resonating structures electrostatic actuation through parallel plates electrodes. To reduce the mass damping during the driving movement and to keep low actuation voltages and power consumption, wafer bonding at low pressure is required, typically lower than 1 mbar.

To achieve low vacuum levels inside the gyroscope wafer level package, a getter layer [26] is deposited on the cap wafer. The getter layer is a solid material that, if chemically active, can adsorb the reactive gas species from the surrounding environment without desorbing the species themselves during its lifetime. Getter

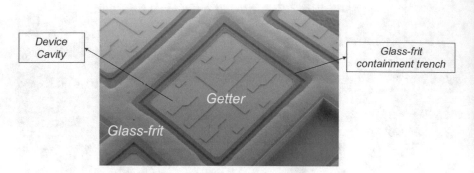

Fig. 13.19 SEM top view of a gyro cap wafer with getter layer inside the device cavity

layers used for MEMS are typically Zr- or Ti-based alloys which are thermally activated, i.e., start to adsorb the environment noninert gas species at temperature higher than an activation threshold. Adsorbed gas species inside the MEMS cavity typically include H2, H_2O, O_2, CO, CO_2, N2, CH4, and more complex hydrocarbons.

Figure 13.19 shows an SEM top view of a gyro cap wafer with glass-frit layer, glass-frit containment trenches, and getter inside the device cavity area. The getter layer is thermally activated during the temperature ramp-up of the wafer-bonding process and absorbs not inert gases which are present in the bonding atmosphere and that can result from surface degassing effects.

By the proper selection of the bonding gas atmosphere with inert (e.g., Ar, Ne, Kr) and not inert gases (e.g., N2, N2H2) in the bond chamber, it's possible to achieve and control low pressure values in the final device cavity. For example, a final device cavity pressure value of 0.1 mbar can be achieved by performing the wafer-bonding process at 10 mbar with a N2 99%/Ar 1% gas mixture. Getter films integration allows also to improve the gyroscope cavity pressure stability. In fact, getter absorbs gas species desorbed from the mems surface ensuring stable pressure inside the bonding cavity during device lifetime and better pressure uniformity from part to part and from run to run.

THELMA platform allows the integration on the same die of an accelerometer and a gyroscope device by a custom bonding layer design and getter integration. These devices are typically named COMBO, since they result from the combination of accelerometers and gyroscopes. On these products, the bonding layer is designed with a double frame to separate gyroscope and accelerometer cavities. Getter deposition is performed in the gyro cavity only to achieve a lower cavity pressure versus the accelerometer one. An SEM top picture of the cap of a COMBO product is shown in Fig. 13.20.

The mechanism which allows to achieve different pressure values in the two device cavities is illustrated in Fig. 13.21. During wafer-to-wafer-bonding process, sensor and cap wafers are put into contact between them; by increasing process temperature, getter is thermally activated, resulting in the absorption of noninert

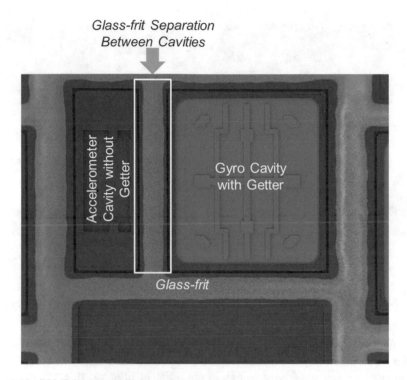

Fig. 13.20 COMBO cap die SEM top view. Accelerometer and gyro cavities are separated by glass-frit; getter is deposited on gyro cavity only

gases present inside the gyro cavity. Due to the glass-frit layer layout which separates the accelerometer and the gyroscope in two different cavities, getter does not absorb noninert gases in the accelerometer cavity and final cavity pressure is defined by the bonding process setpoint. By this approach, it is possible to achieve different pressure values in the two device cavities. For example, by using a bonding gas mix with 1% Ar content, it is possible to keep the accelerometer cavity at a pressure of several tens of mbar versus the gyroscope cavity with a pressure lower than 1 mbar.

This solution allows to keep overdamping on accelerometer side by keeping the benefit of low damping, like low power consumption and low noise, on gyroscope side.

13.4.2 THELMA-60 Technology Platform

THELMA-60 is a process option for the THELMA technology which allows to manufacture inertial sensors with a 60-μm-thick epitaxial polysilicon. A thicker mass has several benefits in terms of performances.

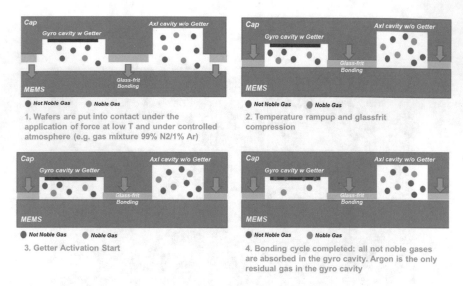

1. Wafers are put into contact under the application of force at low T and under controlled atmosphere (e.g. gas mixture 99% N2/1% Ar)

2. Temperature rampup and glassfrit compression

3. Getter Activation Start

4. Bonding cycle completed: all not noble gases are absorbed in the gyro cavity. Argon is the only residual gas in the gyro cavity

Fig. 13.21 Getter integration description on COMBO product: double pressure cavity approach

A thicker mass allows to realize in-plane capacitors with larger electrodes area which result into a larger capacitance and a higher in-plane sensitivity. THELMA-60 allows to reach sensitivity values in the order of pF/g vs. fF/g values for the traditional THELMA sensors. As an alternative to the increase of sensitivity, a thicker mass allows to put into a smaller area, a device with a specific capacitance value resulting in the shrinkage of the device itself (Fig. 13.22).

The use of a thicker mass allows to also reduce the sensitivity of the device to the Brownian noise, since it is inversely proportional to the sensor mass.

A thicker mass allows the design of out-of-plane sensing devices with larger mass and higher sensitivity. In fact, sensitivity is inversely proportional to the sensor frequency and proportional to the mass of the sensor (Fig. 13.23).

Based on the above considerations, THELMA-60 allows to manufacture high-sensitivity sensors by surface micromachining for applications where complex and expensive bulk micromachining processes were typically used. THELMA-60 bridges the gap between surface and bulk micromachining by enabling the manufacturing of high-end sensors with the benefits of cost effectiveness, flexibility, smaller size, and high yields.

THELMA-60 technology is currently implemented on high-performance applications for medical and industrial market. One example of THELMA-60 application is the manufacturing of high-sensitivity and ultralow-power accelerometers for patient activity monitoring in healthcare applications, like peacemakers and other implantable cardiac devices.

In the industrial field, THELMA-60 technology is used to manufacture sensors for seismic oil and gas exploration. In fact, traditional bulky geophones can be replaced by THELMA-60 products, resulting in better performances, cost effective-

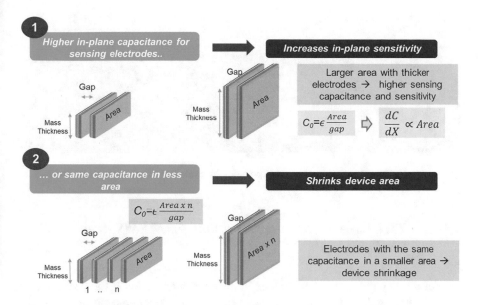

Fig. 13.22 Thicker inertial mass on THELMA-60 allows an increased in-plane sensitivity or the device area shrinkage

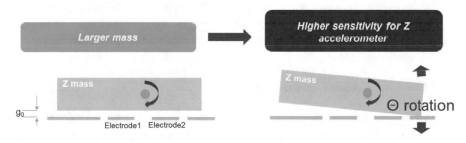

Fig. 13.23 Higher sensitivity is allowed on THELMA-60 for z accelerometers

ness, and space saving. These sensors' working principle is based on the "echo" effect (Fig. 13.24). They are used to detect the vibrations emitted by a source and reflected by the soil and the underlying geological layer to investigate the soil composition and explore gas and soil reservoirs. THELMA geophones exploit THELMA-60 technology benefit and high-vacuum THELMA capability, resulting in high-sensitivity and ultralow noise.

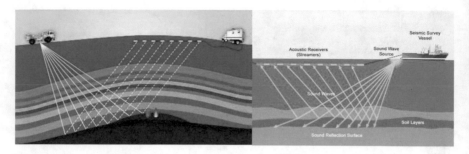

Fig. 13.24 Geophones allow geological exploration by exploiting the echo principle, through the detection of the acoustic reflected waves

13.4.3 THELMA Technology Solutions for MEMS Area Shrinkage

One of the key challenges for sensors in the consumer market is the package area and thickness reduction, which allows integration of these components on smaller portable devices or the integration of more components on the same area. Package area reduction can be achieved by the reduction of the MEMS die area which results also in the increase of the numbers of functional dice present on wafer and a reduction of the single die cost. In this paragraph, some examples of THELMA technology options which allow the die area reduction will be presented.

13.4.3.1 SMERALDO Technology

"SMERALDO" is the Italian translation of the English word "EMERALD", but also a technology option of the THELMA platform which allows the manufacturing of the final MEMS die without "pad windows" openings, thanks to the integration of through silicon vias. The final aspect of the MEMS die is a full silicon cubic stone but with a valuable device inside, here the name of "SMERALDO" like the precious stone.

In "SMERALDO" technology, the MEMS die area shrinkage is allowed by the removal of a dedicated area for metal pads in the sensor layout: In fact, through silicon vias take the signal from buried polysilicon interconnects to the pads which are placed on the back of sensor die, in the area which is already filled by the glass-frit bond layer on MEMS die area. Figure 13.25 shows a 3D schematic of the SMERALDO MEMS architecture. By this solution, it is possible to achieve a die area shrinkage of 20% by removing the dedicated area device pads, without a significant impact on wafer cost. Figure 13.26 shows a schematics of device layout change and area shrinkage.

"SMERALDO" technology exploits a low-cost process to realize through silicon vias. A schematic TSV process flow description is depicted in Fig. 13.27.

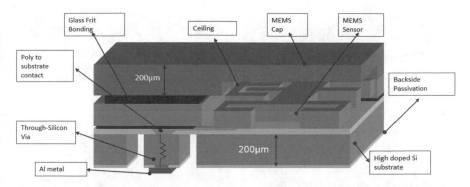

Fig. 13.25 3D schematics of the Smeraldo architecture with thorough substrate silicon via below the glass-frit bond layer

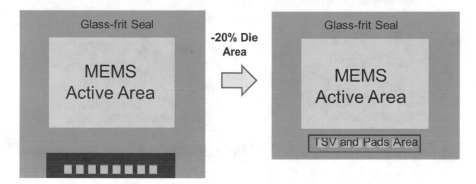

Fig. 13.26 Die area shrinkage by integration of TSV and pads on glass-frit bond seal frame area

The MEMS sensor substrate is a high doped substrate with a resistivity in the order of 1 mohm*cm; signal from MEMS sensor is brought to the substrate by vias opened in the THELMA permanent oxide (13.27a). After glass-frit wafer bonding, the sensor silicon substrate is grinded to a final thickness value of about 200 μm (13.27e), metal for pads is deposited and patterned on sensor substrate backside (13.27f), and silicon "pillars" are patterned by a deep dry etch on areas where poly to substrate vias were designed (13.27g). By this approach, conductive silicon vias that are "pillar-shaped" are fabricated which extract electrical signal from the hermetically sealed device area to the pads placed on the sensor backside.

For MEMS products in LGA package, silicon pillars can be insulated among them by the resin layer which is used for molding during the final LGA package fabrication (Fig. 13.28).

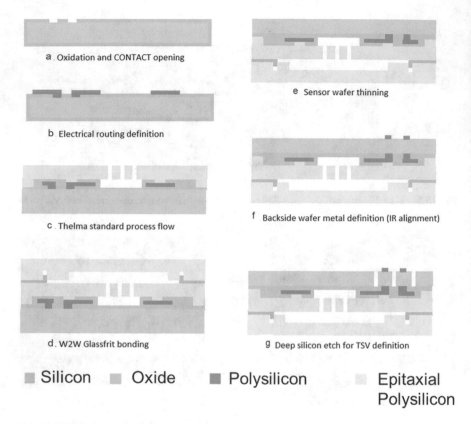

a. Oxidation and CONTACT opening

b Electrical routing definition

c. Thelma standard process flow

d. W2W Glassfrit bonding

e Sensor wafer thinning

f Backside wafer metal definition (IR alignment)

g Deep silicon etch for TSV definition

■ Silicon   ■ Oxide   ■ Polysilicon   ░ Epitaxial Polysilicon

Fig. 13.27 Schematic Smeraldo process flow

13.4.3.2 VIA FIRST Option

THELMA-VIA FIRST is an option of the THELMA process that allows the integration of Via-First Through Silicon Vias (TSV) in the sensor wafer. Unlike the SMERALDO TSVs, whose silicon pillars are surrounded by the molding resin in the device package, THELMA-VIA FIRST TSVs use silicon oxide as dielectric isolation between the vias and the sensor substrate. This approach improves the vias electrical insulation strength and allows the integration of multilevel routing on the back side of the sensor wafer. Figure 13.29 shows the schematic architecture of a MEMS device with TSV realized with this approach.

In THELMA Via First, the MEMS sensor substrate is highly doped with a resistivity in the order of 1 mohm*cm like in the SMERALDO technology option; the electrical signals from the MEMS sensor are brought to the substrate by the contact opened in the THELMA permanent oxide. Via first sensor process flow starts with the vias trenches definition. Electrical isolation between vias and substrate is obtained by vias filling with thermal oxide and undoped polysilicon. A CMP step

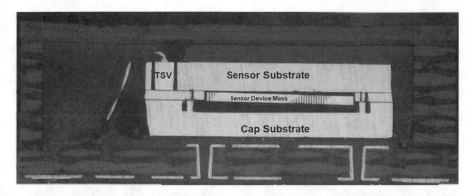

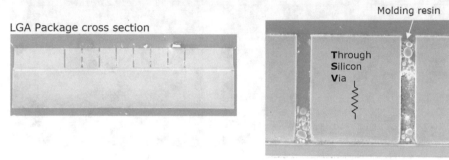

Fig. 13.28 Cross-section of a Smeraldo device in package

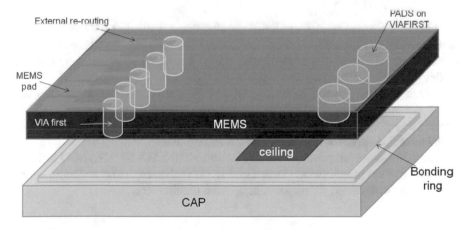

Fig. 13.29 3D schematics of a MEMS device realized with VIA FIRST approach

is then performed to remove polysilicon filling on top of the permanent oxide and provide the right planarity to allow the next steps of sensor flow. Figure 13.30 shows an SEM cross-section of the vias filled with polysilicon and oxide and a detail of the top of the vias with oxide and polysilicon filling.

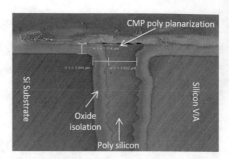

Fig. 13.30 SEM cross-section of VIA FIRST definition on a sensor wafer

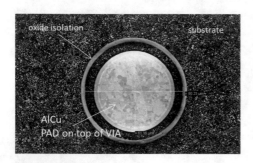
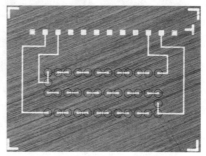

Fig. 13.31 SEM and optical top view of the backside of sensor wafer at the end of the process: a top view of a single via and a structure with multiple interconnected vias are shown

After W2W bonding, the VIAS are revealed through silicon back lapping of the sensor wafer leaving a flat surface that allows further deposition of dielectric and electrical contacts with the external part of the vias. Metal routing can be realized in this phase.

Figure 13.31 shows the finished surface of the sensor wafer after grinding and metal rerouting layer deposition and patterning. A detail of a single via is shown and a daisy chain structured for vias testing where multiple vias are present and metal rerouting layer for vias interconnects among them.

THELMA Via First provides an alternative to THELMA SMERALDO technology for products where device area shrinkage is requested, and a rerouting of the metal layer is requested for package integration requirements.

13.4.3.3 THELMA-PRO: THELMA with PROtective Permanent Oxide Coating

One of the key process steps in THELMA inertial sensor manufacturing is the MEMS release process, in which the sacrificial oxide layer is removed by a vapor HF etch process. During this process, permanent oxide is partially etched below the buried polysilicon interconnects and cause the presence of polysilicon "wings"

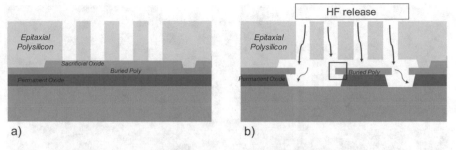

Fig. 13.32 Schematics of a THELMA device before (**a**) and after (**b**) HF MEMS release. Polysilicon wings are red squared in the (**b**) picture

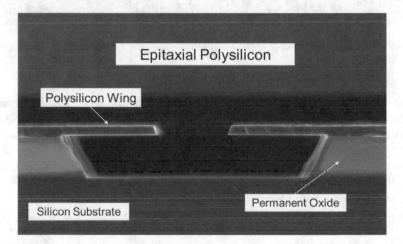

Fig. 13.33 SEM cross-section of polysilicon wings

(Figs. 13.32 and 13.33). These structures represent a potential area of mechanical weakness of the devices, since "wings" breakage can be induced during mechanical shocks in case of sensing mass collision. For this reason, HF vapor etch time must be tightly controlled both to remove completely the sacrificial oxide layer below the inertial mass and to reduce polysilicon "wings" overhang for a robust mechanical structure.

During the product design, the HF release effects on MEMS layers must be considered. In fact, buried polysilicon must be large enough to prevent the interconnected structures to be fragile, typically at least 30 μm. In addition, the release holes density in the epitaxial polysilicon layer must ensure the full removal of the sacrificial oxide layer without inducing a large overhang on polysilicon "wings"; typically, 5 micron distance between neighboring release hole is applied.

The constrains described above translate into a limitation in the device shrinkage both for the space needed to design the buried interconnects and for the lower mass density on the inertial sensor device induced by the high release holes density.

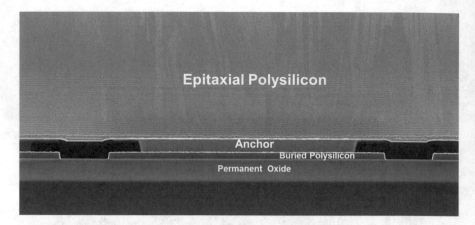

Fig. 13.34 SEM cross-section of a THELMA-PRO device after HF release: no underetch below polysilicon is present

To overcome these limitations and to boost the shrinkage of THELMA MEMS products, a process option of the THELMA technology platform was developed under the name of THELMA-PRO. In this technology, an HF-resistant protective coating layer is deposited on top of the permanent oxide before the polysilicon interconnects deposition. The first result of this layer introduction is the removal of the polysilicon "wings". Figure 13.34 shows an SEM cross-section of a THELMA-PRO sensor product after HF release.

Due to "wings" removal, polysilicon interconnects width is no more affected by considerations related to the HF release process. In addition, longer HF release steps can be implemented and release holes distance on the inertial mass can be increased, resulting in a higher sensor mass density versus standard THELMA technology. Typically, in THELMA-PRO polysilicon interconnects, width can be reduced to few micron and release holes distance is increased from 2 to 3 times the values used for standard THELMA. Figure 13.35 shows a comparison between the release holes densities of two THELMA and a THELMA-PRO product and the SEM top pictures of two gyroscopes manufactured by these technologies.

The combined effect is a 30% area shrinkage on product die thanks to the reduction of the space required from interconnects and from the inertial mass.

THELMA-PRO technology is currently used in high-volume production on both accelerometers and gyroscopes for consumer, industrial, and automotive markets. Die area shrinkage has been the key driver for THELMA-PRO introduction on consumer products. Polysilicon "wings" removal and the consequent improvement of the device mechanical robustness have been the driver for THELMA-PRO adoption also for the automotive market and in general to those applications where robustness to high-g mechanical shocks is required.

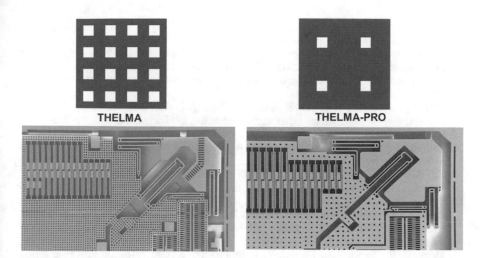

Fig. 13.35 THELMA and THELMA-PRO release hole densities

13.5 Conclusions

Thanks to dedicated MEMS technologies and evolving design principles, MEMS inertial sensors have dramatically improved in the last decade in terms of accuracy, quality and reliability, and cost.

At the beginning of MEMS accelerometer era, MEMS sensors were available only with very limited full-scale, bulky packages and power consumption in the mW region. Nowadays low-power accelerometers for consumer and wearable electronics applications offer state-of-art power consumption in appealing small packages, with good noise and accuracy.

The evolution of MEMS gyroscopes has followed a similar path, increasing the number of available sense axis from 1 to 3 and further decreasing size and power consumption. The further improvements of MEMS gyroscopes will proceed in strong reduction of power consumption to enable low-power applications and further improvement of gyroscope performances to enable high-value applications.

The key recipe for such roadmap is the evolution of technology capability in terms of reduced process variability and improved process architectures toward complete 3D structures, and the parallel evolution of better modeling of physical phenomena at a smaller accuracy scale.

References

1. Roylance, L. M., & Angeli, J. B. (1979, December). A batch-fabricated silicon accelerometer. *IEE Transactions on Electron Devices, ED-26*(12).
2. https://www.analog.com/media/en/technical-documentation/obsolete-data-sheets/2044696ADXL50.pdf

3. Vigna, B. (2011) *Tri-axial MEMS gyroscopes and six degree-of-freedom motion sensors*. In International electronic device meeting.
4. Edwards, C. (2009). *MEMS: The second wave*. E&T – Engineering and Technology. https://eandt.theiet.org/content/articles/2009/07/mems-the-second-wave/, Monday, July 6.
5. https://investors.st.com/news-releases/news-release-details/stmicroelectronics-drives-gaming-revolution-nintendos-wiitm
6. Lemkin, M., & Boser, B. E. (1996). *A micromachined fully differential lateral accelerometer*. In Proceedings of custom integrated circuits conference, San Diego, CA, USA, pp. 315–318.
7. Kempe, V. (2011). *Inertial MEMS: Principles and practice*. Cambridge University Press.
8. Corigliano, A., Ardito, R., Comi, C., Frangi, A., Ghisi, A., & Mariani, S. (2018). *Mechanics of microsystems*. Wiley.
9. 528–2019 – IEEE Standard for Inertial Sensor Terminology. https://standards.ieee.org/content/ieee-standards/en/standard/528-2019.html
10. Lemkin, M. A., Boser, B. E., Auslander, D., & Smith, J. H. (1997). *A 3-axis force balanced accelerometer using a single proof-mass*. In Proceedings of international solid state sensors and actuators conference (Transducers '97), vol. 2, Chicago, IL, USA, pp. 1185–1188. https://doi.org/10.1109/SENSOR.1997.635417.
11. Marra, C. R., Tocchio, A., Rizzini, F., & Langfelder, G. (Oct. 2018). Solving FSR versus offset-drift trade-offs with three-Axis time-switched FM MEMS accelerometer. *Journal of Microelectromechanical Systems, 27*(5), 790–799.
12. Robust sensor Performances.
13. Shkel, A. M. (2006). *Type I and type II micromachined vibratory gyroscopes*. In Proceedings of the IEEE/ION position location navigation symposium, April 2006, pp. 586–593.
14. Meyer, A. D., Rozelle, D. M., Trusov, A. A., & Sakaida, D. K. (2018). *milli-HRG inertial sensor assembly – A reality*. In Proceedings of the IEEE/ION position location navigation symposium, April 2018.
15. Bernstein, J., Cho, S., King, A. T., Kourepenis, A., Maciel, P., & Weinberg, M. (1993). *A micromachined comb-drive tuning fork rate gyroscope*. In Proceedings of the MEMS '93, pp. 143–148.
16. Johari, H., & Ayazi, F. (2006). *Capacitive bulk acoustic wave silicon disk gyroscopes*. In Proceedings of the IEEE electron devices meeting, December 2006, pp. 1–4.
17. Ayazi, F., & Najafi, K. (1998). *Design and fabrication of high-performance polysilicon vibrating ring gyroscope*. In: IEEE MEMS, Heidelberg, Germany, pp. 621–626.
18. Prikhodko, I. P., Gregory, J. A., Clark, W. A., Geen, J. A., Judy, M. W., Ahn, C. H., & Kenny, T. W., *Mode-matched MEMS Coriolis vibratory gyroscopes: Myth or reality?*, In 2016 IEEE/ION Position Location and Navigation Symposium (PLANS), pp. 1–4.
19. Weinberg, M. S., & Kourepenis, A. (2006). *Journal of Microelectromechanical Systems, 15*(3), 479–491.
20. Izadi, M., Braghin, F., Giannini, D., Milani, D., Resta, F., Brunetto, M. F., et al. (2018). *A comprehensive model of beams' anisoelasticity in MEMS gyroscopes, with focus on the effect of axial nonvertical etching*. In 2018 IEEE International Symposium on Inertial Sensors and Systems (INERTIAL).
21. IEEE Standard 1431–2004 (2004). *IEEE standard specification format guide and test procedure for Coriolis Vibratory Gyros*, IEEE Standard 1431–2004.
22. El-Sheimy, N., Hou, H., & Niu, X. (2008). Analysis and modeling of inertial sensors using Allan variance. *IEEE Transactions on Instrumentation and Measurement, 57*(1).
23. Facchinetti, S., Guerinoni, L., Falorni, L. G., Donadel, A., & Valzasina, C. (2017). *Development of a complete model to evaluate the Zero Rate Level drift over temperature in MEMS Coriolis vibrating gyroscopes*. In 2017 IEEE International Symposium on Inertial Sensors and Systems (INERTIAL), March 2017, pp. 125–128.
24. Kline, M., Yeh, Y., Eminoglu, B., Najar, H., Daneman, M., Horsley, D., & Boser, B. (2013, January). *Quadrature FM gyroscope*. In 2013 IEEE 26th international conference on Micro Electro Mechanical Systems (MEMS), pp. 604–608.

25. Minotti, P., Dellea, S., Mussi, G., Bonfanti, A. G., Facchinetti, S., Tocchio, A., et al. High scale-factor stability frequency-modulated MEMS gyroscope: 3-Axis sensor and integrated electronics design. *IEEE Transactions on Industrial Electronics, 65*(6), 5040–5050.
26. Moraja, M., Amiotti, M., & Longoni, G., MST. (2003). *Patterned getter film wafers for wafer level packaging of MEMS*, Munich, October 2003.

Chapter 14
Magnetometers

Dario Paci, Anna Cantoni, and Giorgio Allegato

14.1 Magnetic Sensors Overview

14.1.1 Hall Sensors

Talking about magnetic sensors, it is worth to start with Hall sensors. Hall effect was discovered in 1879 by Edwin Herbert Hall and has been fundamental for a huge number of sensors developed by semiconductor industry. Basically, Hall effect is a consequence of Lorentz force acting on charge carriers [1]. Considering a plate of conductive material as the one shown in Fig. 14.1, when a current I flows from terminal A to terminal B, a Lorentz force is applied to carriers due to an external magnetic field B. This force F_L deflects carriers in orthogonal direction, and, therefore, an electric field E is generated between C and D terminals. Thus, a voltage between C and D is generated. This voltage or Hall voltage V_H can be found when the electrical and Lorentz forces are in equilibrium:

$$F_T = F_E + F_L = qE + qvxB = 0 \qquad (14.1)$$

where q is the carrier charge and v its velocity, which can be assumed to be equal to the drift velocity v_d for all the carriers flowing in the plate of Fig. 14.1, according to the smooth-drift approximation [1], so that Lorentz force can be correlated to current

D. Paci (✉)
ST Microelectronics, Analog MEMS and Sensors Group, MEMS Technology and Design R&D, Cornaredo, Italy
e-mail: dario.paci@st.com

A. Cantoni · G. Allegato
ST Microelectronics, Analog MEMS and Sensors Group, MEMS Technology and Design R&D, Agrate Brianza, Monza Brianza, Italy
e-mail: anna.cantoni@st.com; giorgio.allegato@st.com

© Springer Nature Switzerland AG 2022
B. Vigna et al. (eds.), *Silicon Sensors and Actuators*,
https://doi.org/10.1007/978-3-030-80135-9_14

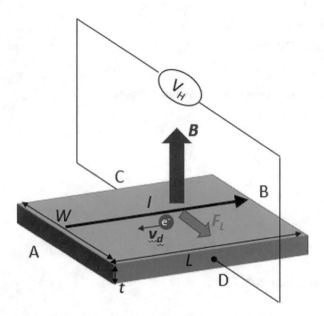

Fig. 14.1 Working principle of a Hall plate. Charge carriers are electrons

I flowing in the plate, being $I = Wdnqv_d$, where W and d are width and thickness of the plate, while n is the carrier density. Assuming now the carriers to be electrons and vectors directions as in Fig. 14.1 and defining $V_H = EW$, after straightforward calculation, the Hall voltage can be written as:

$$V_H = \frac{1}{qnd} I B_z = \mu \frac{W}{L} V B_z \qquad (14.2)$$

where B_z is the component of magnetic field perpendicular to the plate, L the plate length and V the voltage between A and B driving the bias current I.

The simplified structure in Fig. 14.1 can be integrated quite easily in a technological platform typical of semiconductor industry by exploiting a diffused or implanted well in silicon substrate (Fig. 14.2). Generally, the well is low doped (i.e., small carrier density and large mobility) to guarantee high sensitivity to the magnetic field. Furthermore, n-type well is preferable due to the higher mobility of electrons vs. holes in silicon. Due to the key role of mobility in Hall Effect, some sensors have even been developed in compound semiconductors such as GaAs, InAs, or InSb, which show mobility from 6 to 60 times larger than n-type silicon sensors [2]. On the other hand, one important advantage of Hall sensors is the integrability of the sensitive element on the same chip with analog front end and digital part. Using compound semiconductors, this advantage is lost. Referring to Fig. 14.2, an oxide layer isolates different active areas as well as the silicon from metals above the Hall sensor. A p-well layer is used to insulate the Hall cell from the n-type substrate. The cell is completed by contacts to silicon and metals to

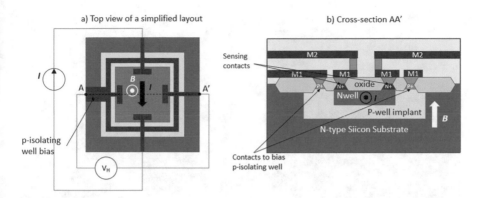

Fig. 14.2 (**a**) Simplified Hall cell layout in a process with several metal levels as CMOS or BCD. (**b**) Cross-section taken along a plane including contacts to sense Hall voltage. I is the bias current and B is the magnetic field. Contact and via plugs are in green

bias the sensor and collect the signal. Looking at the structure in Fig. 14.2, one of the most important point of strength of the Hall sensor is the simplicity of the structure, very close to a conventional implanted resistance, already included in a typical CMOS or BCD (Bipolar, CMOS, DMOS) process; at most implant dose of the well could be optimized for carrier mobility maximization. Furthermore, Hall sensors are well known and reliable devices, have good bandwidth, are practically not limited with respect to full scale, since they are quite linear vs. magnetic field, and they are not based on ferromagnetic elements that can saturate. Thus, they are quite suited for current sensing and as position sensors or switches combined with strong magnets, especially for industrial and/or automotive applications where there are high magnetic fields and/or currents. An aspect which can be limiting for Hall sensor accuracy in its typical structure (Fig. 14.2) is the so-called piezo-hall effect according to which sensitivity of a Hall sensor is affected by mechanical stresses applied to silicon [3–5]. The working mechanism is associated to piezoresistivity.

Indeed, defining the current-related sensitivity S_I from Eq. (14.2) so that $V_H = S_I I B_z$, the carrier density is affected by mechanical stress, and, thus, it is possible to write it as a function of mechanical stress tensor σ, as [4]:

$$S_I = S_I^0 \left(1 + \frac{\Delta S_I (\sigma)}{S_I^0} \right) \tag{14.3}$$

where S_I^0 is the current-related sensitivity of the unstressed Hall sensor. The relative sensitivity change $\Delta S_I (\sigma) / S_I^0$ is a function of the principal stress, and it is not affected by shear stress due to crystal structure of silicon [5]:

$$\frac{\Delta S_I (\sigma)}{S_I^0} = P_{12} \left(\sigma_{xx} + \sigma_{yy} \right) + P_{11} \sigma_{zz} \tag{14.4}$$

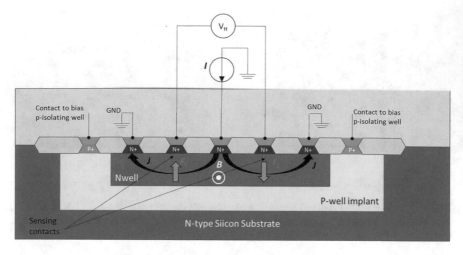

Fig. 14.3 Schematic cross-section of a VHS (Vertical Hall Sensor)

where P_{11} and P_{12} are the only components of piezo-hall coefficient tensor which are not zero in case of silicon, while σ_{xx} and σ_{yy} are principal stresses parallel to chip surface and σ_{zz} the principal stress component perpendicular to the chip.

This leads to a drift of sensitivity due to stress introduced by package after molding injection, as well as a drift of sensitivity after thermal cycle or generally after any other event leading to plastic deformation of the package. Of course, this results in sensitivity drift over time and loss of accuracy of the magnetic field reading. The usual proposed solution to mitigate this effect is to introduce close to the Hall sensor a piezoresistive element, usually a Wheatstone bridge of diffused resistors in silicon, which provides a signal proportional to the stress, not affected by magnetic field, which can be used to compensate the Hall sensitivity drift vs. mechanical stress. The method is complicated by the fact that both piezoresistive coefficients and piezo-hall coefficients depend on temperature, but good results have been nonetheless reported in literature [3, 4].

To close this summary of Hall sensors' main features, it must be considered that the basic structure of Hall sensor, i.e., the Hall plate (Figs. 14.1 and 14.2), is sensitive only to the component of magnetic field perpendicular to the chip. This could be limiting for many applications, such as position, angular, and speed rotation sensors, as well as compass. Thus, the Vertical Hall Sensor (VHS) has been proposed [1, 6]. The Vertical Hall Sensor exploits a "vertical movement" of current flow in the chip to obtain sensitivity to magnetic field component parallel to die. Referring to Fig. 14.3, a current is driven in the central contact, so that when planar component of magnetic field parallel to the chip B is zero, the voltage V_H is zero since the two sides of the device are symmetric. On the other hand, when external magnetic field is directed as in Fig. 14.3, Lorentz Force brings electrons far from the sensing contact on the right, while it brings electrons close to the left sensing contact. In this way, a voltage drop is generated between sensing contacts. Unfortunately, this

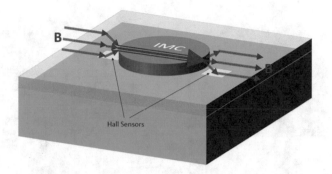

Fig. 14.4 Sketch of a Hall sensor + Integrated Magnetic Concentrator (IMC) and its effect on an external magnetic field B parallel to the chip

approach has the limit of leading to low sensitivity. Furthermore, offset management is more complex than in standard Hall plates, since chopping technique used to cancel offset [7] requires a symmetry between sensing and bias port so that it is possible to exchange them during the measurement.

A second approach (Fig. 14.4) is to introduce an "Integrated Magnetic Concentrator" (IMC), i.e., a ferromagnetic layer with high magnetic permeability, which is able to deflect the magnetic field lines, so that a magnetic field parallel to the chip can be bent close to the edge of the concentrator: In this way, a component perpendicular to the chip is locally generated, allowing a Hall plate positioned in such a position to be sensitive to a magnetic field parallel to the die [8]. This allows even to amplify the field close to the sensor, leading to devices with sensitivity and resolution higher than standard Hall sensors [9]. On the other hand, the introduction of a ferromagnetic element can reduce the range of linearity of the sensor due to the typical saturation mechanism of ferromagnetic materials. The required ferromagnetic layer can be integrated on top of standard CMOS or BCD process by sputtering or ECD and must be a soft ferromagnetic layer like Nickel Iron.

14.1.2 Lorentz Force Magnetometer

A Lorentz Force Magnetometer is a Micro Electro-Mechanical System (MEMS), whose movements are driven by Lorentz force. Sensing these movements allows to measure the magnetic field who drives them [10]. The working mode of this device could be explained looking to the simplified structure in Fig. 14.5a. A current is driven from an anchor to another of a suspended mechanical mass supported by flexural springs to bias the structure. If no magnetic field is applied, the structure does not move, and the two-sensing capacitance C_{S1} and C_{S2} are equal. If there is a magnetic field in z direction, the Lorentz force

$$F_L = lI \times B \qquad (14.5)$$

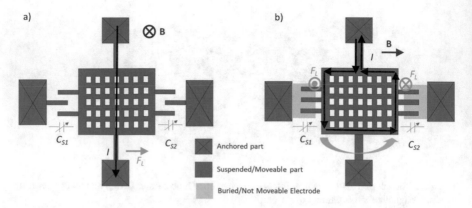

Fig. 14.5 Working principle of a Lorentz force magnetometer: (**a**) sensing of magnetic field perpendicular to the chip; (**b**) sensing of magnetic field parallel to the chip

deflects the structure in plane, and there is a difference between capacitance C_{S1} and C_{S2}, which is the electrical quantity giving the measure of magnetic field. The structure can be stimulated also with a current close to the resonance to amplify movements. In-plane components of magnetic field can be detected if the bias current flows in a loop on a suspended structure as in Fig. 14.5b, so that a torque is applied to the structure. In this case, torsional springs are required. Such a kind of devices can be fabricated with a surface micromachined process quite similar to the one used in general for accelerometer and gyroscope modules, making them interesting for the integration on the same die with inertial sensors and consequently allowing a compact solution for 6-axis and 9-axis modules.

They have the same advantage as standard Hall sensors in terms of full scale, and they do not suffer of saturation issues that sensors including ferromagnetic layers have, as Hall including Integrated Magnetic Concentrator (IMC). On the other hand, being mechanical structures, they should be designed to sustain mechanical shock and can be affected by drift induced by package stress. Nonetheless, some recent papers show very interesting performance in terms of high resolution and bandwidth [11].

14.1.3 Fluxgate and Magnetoinductive

A magnetic field can be detected also by monitoring the effect of the magnetization on a ferromagnetic element which is used as the core of a transformer or as the core of an inductor. In the first case, the device is a Fluxgate, in the second a magnetoinductance. As discrete devices, they are quite interesting with respect to the resolution: A Fluxgate can even reach the resolution of hundreds fT [12]. Integrated versions of Fluxgate have been proposed, as well [13]: the ferromagnetic

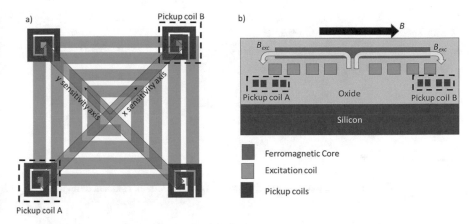

Fig. 14.6 Sketch of a 2-axis Fluxgate based on planar inductors with field B_{exc} generated by excitation coil during a half period (**a**) Top view; (**b**) cross-section along one of the two sensitivity axes

element can be integrated on top of CMOS/BCD process like in the case of IMC used for Hall sensors. The easiest way to use such a core is to combine it with planar inductors (Fig. 14.6), not wrapped around the coil, and this limits the performance of the sensor, since the primary (excitation coil) magnetic flux is less effectively coupled with core, as well as the secondary (pickup coils). Better should be to wrap coil around the core, but it makes integration process more complex.

Referring to Fig. 14.6, when an alternate current is driven through the excitation coil, a magnetic field $\boldsymbol{B_{exc}}$ is generated on the core. Current is fixed so that $\boldsymbol{B_{exc}}$ is high enough to saturate the core, in a direction during the first semiperiod, in the opposite direction in the second semiperiod. Changes in core magnetization are sensed by pickup coils since lead to voltage pulses across pickup coils, following the magnetic flux variation. If magnetic field B to be sensed is zero, the voltage pulses are equal in both the pickup coils and in both the excitation semiperiod. If B is in the direction shown in the Fig. 14.6b, the right side of the core is closer to the saturation than the left side when $\boldsymbol{B_{exc}}$ is in the direction shown in Fig. 14.6b (first excitation semiperiod), and this leads to a shorter and smaller pulse at pickup coil B than at the pickup coil A. The opposite happens on the second semiperiod, when $\boldsymbol{B_{exc}}$ is in the opposite direction. Collecting the difference between pickup coils A and B, the average of the signal is proportional to magnetic field B.

In case of magneto-inductive sensors, the working principle is easier, but still can be based on nonlinear function $B(H)$ of a ferromagnetic material (Fig. 14.7) [12]. Indeed, if a ferromagnetic material is used as the core of an inductor, inductance is strongly dependent on core permeability which changes in dependence on the magnetization of the core, i.e., the working point on $B(H)$ curve. Referring to the hysteresis loop in Fig. 14.7 and assuming to move on the upper branch, if the magnetic field to be sensed H is zero, the slope of curve $B(H)$ and consequently incremental magnetic permeability μ is quite high, while when H increases to

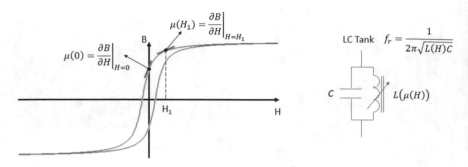

Fig. 14.7 Fundamentals of magneto-inductive sensor; f_r is the LC tank resonance frequency

H_1, the slope of the curve and consequently the magnetic permeability decreases. Thus, a change of the external magnetic field H results in a change of inductance, following the change of magnetic permeability. This change can be easily read if we put a capacitance C in parallel to the inductor, and we use this LC tank in the feedback circuit of an oscillator to fix the circuit oscillating frequency. Both fluxgate and magneto-inductive sensors provide good level of resolution, making them interesting for compass application, but their integrated version on silicon shows some limits. First, in the case of Fluxgate, the need to saturate the ferromagnetic layer leads to significant power consumption. Second, both fluxgate and magneto-inductive need very long ferromagnetic core (several hundred of microns or better millimeters of length) to allow sensitivity to the single component of magnetic field, since it minimizes the demagnetization in sensitivity direction and maximizes it in cross-axis direction [14].

Obtaining it is quite easy for the two in-plane components, i.e., it costs only in terms of chip size, but it becomes extremely tough for out-of-plane components. Thus, a three-axis sensor based on integrated magneto-inductive or fluxgate sensors should be obtained at package level by means of a die vertically assembled to sense z axis or by sensing z axis with another transduction mechanism (e.g., Hall sensor).

14.1.4 Magnetoresistive

The magnetoresistive effect is the property of a material which changes its resistance according to the external magnetic field. It is a significant effect only in some ferromagnetic material, and it was observed on samples of iron the first time in 1857 by William Thomson (Lord Kelvin). In these samples, the resistance was a function of the direction of the external magnetic field, showing higher resistance in case of current flowing in direction parallel to the magnetic field and lower resistance in case of current flowing perpendicularly to the magnetic field. It was the first observed case of Anisotropic Magnetoresistance (AMR) [15].

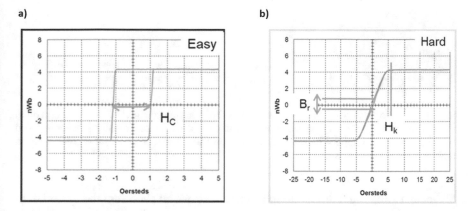

Fig. 14.8 Typical hysteresis loops for an anisotropic ferromagnetic element suited for AMR sensors (e.g., NiFe): (**a**) Easy axis; (**b**) hard axis. Source: characterization measurements of NiFe film from STMicroelectronics AMR process

The Anisotropic Magnetoresistance in ferromagnetic material, i.e., the dependence of resistance on angle between magnetic field and current density vector, found an explanation at microscopic scale by considering the effect on electron scattering of the magnetic field, leading to the conclusion of higher scattering probability for charge carriers traveling parallel to magnetic field.

Magnetoresistive materials show usually also an anisotropic behavior with respect to magnetization vs. magnetic field curve. Thus, it is possible to define an easy axis along which the hysteresis loop $M(H)$ has small coercive field H_c and is almost square, so that ideally it allows the material to be magnetized or at about $+M_{sat}$ or at about $-M_{sat}$, where M_{sat} is the saturation magnetization. Perpendicularly to the easy axis is the hard axis, which is quite linear, with very small hysteresis between $-H_k$ and $+H_k$, where H_k is the so-called anisotropy field (Fig. 14.8). The quite linear dependence of magnetization on magnetic field along the hard axis (Fig. 14.8b) can be exploited to obtain a sensor. In fact, an external field H along the hard axis rotates the magnetization vector so that the angle ψ between current density vector J and magnetization M changes (Fig. 14.9). The dependence of resistance R on the angle between current density and magnetization ψ can be written as [15, 16]:

$$R = R_0 - \Delta R \sin^2\psi = R_0 \left(1 - \frac{\Delta\rho}{\rho}\sin^2\psi\right) \qquad (14.6)$$

Where R_0 is the resistance when magnetic field is parallel to the current, and $\Delta\rho/\rho$ is the magnetoresistivity coefficient. Magnetoresistivity coefficient is significant (in the range of 2–4%) for Nickel-based alloys as NiFe, NiCo, NiFeCo. NiFeMo and CoFeB have been proposed too, but, in these cases, magnetoresistive coefficient has been found to be smaller than 1% [15].

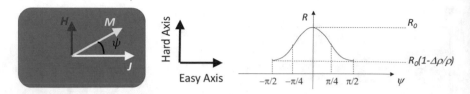

Fig. 14.9 The magnetoresistive effect: the resistance of a ferromagnetic material portion (in blue) changes according to the angle ψ between magnetization M and current density J vector, induced by a magnetic field H applied in material hard axis direction. A plot of resistance R dependence on angle ψ is represented as well

Another important effect to be considered when choosing the material to fabricate magnetoresistors is the magnetostriction, i.e., the fact that a strain is induced in a ferromagnetic film when it is magnetized, and when a film is stressed, its characteristic M(H) loop changes (inverse magnetostriction effect) [16]. Of course, this effect must be minimized, and if we consider also that H_k and, H_c must be small enough to guarantee good level of sensitivity, this limits a lot the possibility of choice in the material and practically leads to an optimum which is NiFe with composition close to Ni (80%) and Fe (20%).

AMR based on change in resistivity of NiFe resistors will be better described in Sect. 14.2. However, they have some significant advantages when it is important to resolve small fields as in compass application: They have sensitivity much better than Hall sensors, and their noise is "white", so that it can be easily reduced by average operations. Indeed, corner frequency of 1/f noise in AMR sensors is generally smaller than 10 Hz [17]. Furthermore, the almost zero magnetostriction of NiFe (80:20) [18] and consequently insensitivity to mechanical stresses ensures long-term stability of magnetoresistors performance. Finally, some architecture features shown in 14.2.2 leads to a very high stability of sensor offset vs. long-term variation and temperature, thanks to a dynamic offset cancellation.

On the other hand, integrated AMR, being based on ferromagnetic thin layers of some tens of nm, shows limited full scale (few tens of Gauss), due to the saturation of magnetic material. Finally, AMR are naturally sensitive to in-plane magnetic field but not to field component perpendicular to the chip (z-axis): being based on thin film, demagnetization factor along thickness is huge and makes the film insensitive to components perpendicular to the chip. The 3-axis sensitivity can be achieved at package level, by combining a biaxial and a monoaxial sensor as shown in Sect. 14.3.

Other magnetoresistive elements have been developed, essentially to increase sensitivity to magnetic field, i.e., increasing $\Delta\rho/\rho$. Different multilayer structures have been proposed, and among them, two show very competitive performance: the Giant Magnetoresistance (GMR) and the Tunnel Magnetoresistance (TMR).

In both cases, the basic structure includes three layers, typically few nanometers thick (Fig. 14.10): A first layer, or *pinned layer*, is a ferromagnetic hard layer, whose magnetization is fixed by the choice of material and by the fabrication process and

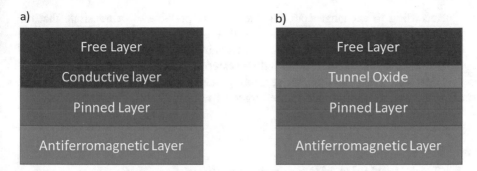

Fig. 14.10 Schematic structure of a: (**a**) GMR; (**b**) TMR

not affected by external magnetic fields, at least unless the magnetic field does not exceed a maximum, generally large, limit. Pinned layer can be Cobalt, NiO, or NiFe. Second layer is a nonferromagnetic layer which is a conductor layer in case of GMR, copper for instance (Fig. 14.10a), while is a thin insulating layer in case of TMR, enabling current conduction via Tunnel Effect (Fig. 14.10b). Finally, there is a soft ferromagnetic layer, i.e., the *free layer*, which can follow the direction of an external magnetic field rotating in the plane parallel to the chip. Free layer too can be Cobalt, NiO, or NiFe.

Both the devices work thanks to electron spin interaction with the magnetization: For example, in case of GMR, the scattering at each interface depends on direction of magnetization: One direction maximizes the scattering occurrence of spin $-1/2$ electrons, the opposite maximizes the scattering occurrence of spin $+1/2$ electrons. Thus, when magnetization of free layer is antiparallel to the pinned layer, one group of electrons is scattered at one structure interface, and the second group is scattered at the other interface, leading to maximum electrical resistance, while in case of parallel orientation, one of the two groups of electron is practically not scattered at the interfaces, leading to minimum resistance [16].

On the other hand, in TMR the electrons jump across a thin oxide thanks to tunnel effect: Also tunnel probability is affected by the spin and magnetization, so that, if magnetizations of the two layers are parallel, the probability of electron jump is maximum because one electron group probability is maximum, leading to minimum junction resistance. Instead, if magnetizations are antiparallel, both groups of electrons have small tunnel probability, leading to maximum resistance. Thus, if the free-layer magnetization is rotated by an external magnetic field, it is like to open or close a valve for electron flow.

GMR and TMR stability of pinned layer is generally improved by growing pinned layer on an antiferromagnetic layer (Fig. 14.10), i.e., a material with domain structure like ferromagnetic, but which reaches minimum of energy when domains are arranged in opposite directions so that their resultant magnetization vector is zero. Typical antiferromagnetic materials are FeMn, IrMn, NiO, and NiMn.

According to exchange coupling theory, it is possible to demonstrate that the effect of antiferromagnetic layer is to shift the ferromagnetic loop of the pinned layer, which results in a much higher field required to flip its magnetization and in a well-defined stable magnetization with no external field [16].

GMR and TMR show higher sensitivity than AMR, but are worse in terms of accuracy and resolution, due to the larger 1/f noise [17, 19], especially in case of TMR [20].

14.2 A Typical AMR Sensor Structure

The fundamental AMR structure is a simple NiFe strip of length L> > W (Fig. 14.11). This helps to better define the selectivity to direction by adding the geometry anisotropy to material anisotropy. Geometry anisotropy is defined by demagnetization field H_D [15, 16, 21]. When an external magnetic field H_{ext} magnetizes a ferromagnetic material, the magnetic dipoles at the ends of the ferromagnetic material are oriented, but not balanced due to the discontinuity ferromagnetic material/diamagnetic material (e.g., air, silicon oxide). This leads to the creation of a demagnetization field H_D, which opposes to H_{ext}. H_D is a vector, depending on the material geometry, and is proportional to material magnetization vector M_s, so that the resulting field H_{int} inside the material is:

$$H_{int} = H_{ext} - H_D = H_{ext} - NM \tag{14.7}$$

where N is a tensor defined as demagnetization factor, and it is in general not constant inside the material (it decreases going far from material edges). Formulas to calculate N for some specific geometries can be found in [14], specifically in the case of a generic ellipsoid, where the magnetization and N are constant inside the volume. N in an ellipsoid is a diagonal tensor if the reference system is parallel to ellipsoid principal axis, and it leads to larger H_D along shorter axis of the ellipsoid, while smaller H_D along the longer axis. When the three axes are equal, i.e., in the case of a sphere, H_D is 1/3 of magnetization in any direction [14].

In a geometry different from an ellipsoid like the one in Fig. 14.11, we can consider H_D as the average demagnetizing field, which is however much larger along thickness and width, making easier the magnetization along length. In this case, no analytical formulae are available, but H_D can be nonetheless extracted by FEM simulations.

If anisotropy field H_k is small enough, it is quite easy to make the length of the resistor the easier direction of magnetization for the resistor. Consequently, when no external magnetic field is applied to the AMR, the internal magnetization is aligned with resistor length, i.e., x-direction following Fig. 14.11. Indeed, considering the energy balance on magnetoresistor in Fig. 14.11, it is possible to write energy W_s as:

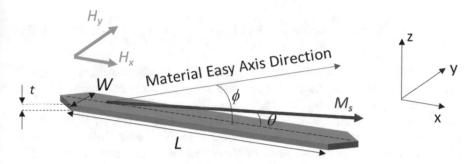

Fig. 14.11 Schematic structure of a single AMR resistor

$$W_s = \frac{1}{2}\mu_0 H_k M_{sat} \sin^2(\varphi - \theta) + \frac{1}{2}\mu_0 H_D M_{sat} \sin^2\theta \tag{14.8}$$

where M_s is the saturation magnetization of the film, H_D is the demagnetization field along y (i.e., strip width), while θ is the angle between strip magnetization and magnetoresistor length direction; ϕ is the angle between film easy axis and magnetoresistor length direction, while μ_0 is the vacuum magnetic permeability. When $H_D >> H_k$, the minimum energy occurs for $\theta = 0$.

On the other hand, if an external magnetic field H oriented along strip width (i.e., along y, $H_x = 0$) rotates the strip magnetization of an angle θ Eq. (14.8) becomes:

$$W_s = -\mu_0 H_y M_{sat} \sin\theta + \frac{1}{2}\mu_0 H_k M_{sat} \sin^2(\varphi - \theta) + \frac{1}{2}\mu_0 H_D M_{sat} \sin^2\theta \tag{14.9}$$

Assuming, to simplify the calculations, that material easy axis is aligned to magnetoresistor length ($\phi = 0$), the condition of minimum energy W_s leads to:

$$\sin\theta = \frac{H_y}{H_D + H_k} \tag{14.10}$$

Which holds for $H_y < H_D + H_k$, while for $H_y > H_D + H_k$, minimization of W_s leads again to $\theta = 0$. In this way, both sensitivity and range of measurements of the magnetoresistor are fixed.

14.2.1 Barber Poles to Linearize R(B)

If current flows parallel to resistor length in Fig. 14.11, it holds that $\psi = \theta$ in Eq. (14.6). If the value for $\sin\theta$ from Eq. (14.10) is substituted in Eq. (14.6), the dependence of resistance on an external field H_y is found to be not linear:

$$R = R_0 \left(1 - \frac{\Delta\rho}{\rho}\sin^2\theta \right) = R_0 \left(1 - \frac{\Delta\rho}{\rho}\left(\frac{H_y}{H_D + H_k} \right)^2 \right) \qquad (14.11)$$

To obtain a linear dependence of resistance on magnetic field, a layer with higher conductivity than the magnetoresistive one is generally introduced [22]. This layer can be put just above the magnetoresistive layer, in electrical contact with it. For instance, an aluminum layer can be used if magnetoresistive layer is NiFe. Since aluminum conductivity is much higher than the one of NiFe, it is easy to obtain a negligible sheet resistance of aluminum layer with respect to the NiFe one, by depositing an aluminum thickness of some hundreds of nanometers. In this condition, patterning aluminum at 45° with respect to NiFe magnetoresistor length forces the current to flow at 45° with respect to magnetization M_s when magnetic field along y is 0 (Fig. 14.12).

These conductive elements at 45° with respect to magnetoresistor are called "barber poles", since they recall the typical appearance of some old barber shop signboard.

Starting from this condition, as shown in Fig. 14.12b, if the changes of $\theta(H_y)$ are small enough, the dependence of resistivity (and thus resistance) on θ (and thus on magnetic field H_y) is linear. Again from Eqs. (14.6) and (14.10), if the angle between current and magnetization is $\psi = \theta + \pi/4$, after introduction of barber poles, dependence of R on H_y becomes:

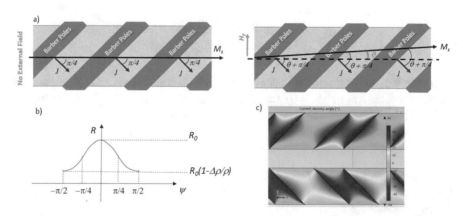

Fig. 14.12 (a) Typical barber poles structure with magnetoresistor magnetization without any external field and with a field H_y in sensitivity direction. (b) Typical dependence of resistance on the angle between magnetization and current density. (c) FEM simulation results of the effective current density direction in a real structure

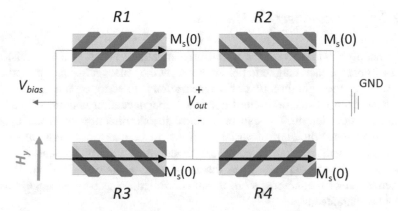

Fig. 14.13 Wheatstone bridge arrangement for an AMR sensor. $M_s(0)$ is the bar magnetization when $H_y = 0$

$$R = R_0 \left(1 - \frac{\Delta\rho}{\rho}\sin^2(\theta + \pi/4)\right) = R_0 \left(1 - \frac{\Delta\rho}{2\rho} - \frac{\Delta\rho}{2\rho}\sin\theta\,\cos\theta\right) =$$

$$= R_0 \left(1 - \frac{\Delta\rho}{2\rho} - \frac{\Delta\rho}{2\rho}\left(\frac{H_y}{H_D+H_k}\right)\sqrt{1 - \left(\frac{H_y}{H_D+H_k}\right)^2}\right)$$

$$(14.12)$$

Considering small θ and H_y, it can be approximated as:

$$\frac{R - R_0^*}{R_0^*} = \frac{\Delta R}{R_0^*} \approx -\frac{\Delta\rho}{2\rho}\frac{H_y}{H_D + H_k} \tag{14.13}$$

Where $R_0^* = R_0(1 - \Delta\rho/2\rho) \approx R_0$. An increase of Hy leads to a decrease or an increase of resistance, depending on the orientation of barber poles (see also Fig. 14.12b). If barber poles, direction of magnetization, and positive Hy are as in Fig. 14.12a, an increase of Hy leads to a decrease of the resistance, following signs in Eqs. (14.12) and (14.13). If barber poles would be 90° rotated with respect to ones in Fig. 14.12a, an increase of H_y would lead to a decrease of resistance, i.e., the angle between current and magnetization becomes $\psi = \theta - \pi/4$, so that the sign in Eq. (14.13) turns to be positive.

This degree of freedom can be exploited by connecting resistors with different barber poles orientation in a Wheatstone bridge configuration, as in Fig. 14.13, where R1 and R4 are resistances with barber poles parallel to the one in Fig. 14.12a, while R2 and R3 include orthogonal barber poles. Output voltage V_{out} in this configuration is, where V_{bias} is the bridge voltage bias:

$$V_{out} = V_{bias}\frac{\Delta R}{R_0^*} = V_{bias}\frac{\Delta\rho}{2\rho}\frac{H_y}{H_D + H_k} \tag{14.14}$$

14.2.2   Set and Reset Coils

It is generally advised to set the initial magnetization of each of the 4 resistors in Fig. 14.13 before each magnetic field measure, at the sensor start-up or periodically, to be sure that the right direction of magnetization is fixed for each magnetoresistor [23]. If no external magnetic field is applied, magnetization is quite aligned with magnetoresistors length, but can be oriented in one direction or in the opposite one. This changes the sign of variation of resistance due to a change of external field H_y, since the operating point on the curve in Fig. 14.12b can be $\psi = 45$ or $\psi = -45$, depending on the initial magnetization condition. If initial magnetization of magnetoresistors is not controlled, Wheatstone bridge working cannot be guaranteed.

To set initial magneto resistor magnetization, i.e., sensor working point, an integrated coil (i.e., the Set/Reset Coil) can be designed to generate a pulsed magnetic field, to force the magnetization of magnetoresistors in zero-field condition. In Fig. 14.14, a sketch of the set/reset coil working is represented. If a pulse of current is driven through the coil, a pulse of magnetic field is driven on the magnetoresistors, setting magnetization parallel to H_{set}. To guarantee a value of H_{set} large enough to saturate magnetoresistors and fix magnetization direction all over the resistor length, generally high amplitude pulses of currents (i.e., at least several hundreds of mA) are required. Thus, low resistance coils are required. Since in the two sides of the coil in Fig. 14.14, the current flows in opposite direction, set magnetic field is opposite at the two sides of the coil, as well as magnetization of magnetoresistors laying on coil opposite sides. Thus, barber poles orientation of one couple of resistors (R1, R3) has been rotated by 90° with respect to Fig. 14.13 to retain Wheatstone bridge functionality.

The coil can be used not only to set or refresh the magnetoresistors magnetization but also to allow offset cancellation. If a first measure is done after set pulse as in Fig. 14.14a, and $H_y > 0$, R2 and R3 are greater than in the condition of $H_y = 0$, let us say they are both equal to $R^* + \Delta R(H_y)$, while R1 and R4 both decrease and are equal to $R^* - \Delta R(H_y)$. This leads to an output of the bridge $V_{out,set}(H_y)$ given by the sum of a part dependent on H_y and an electrical offset due to the mismatch among the 4 resistance:

$$V_{out,set} = V_s + V_o = V_{bias}\frac{\Delta R\left(H_y\right)}{R^*} + V_o \qquad (14.15)$$

After this measurement, if another current pulse is driven along the coil in the opposite direction with respect to the first one (Fig. 14.14b), a reset field H_{reset} is driven on the magnetoresistors. H_{reset} leads to a 180° rotation of the magnetization of all the magnetoresistors, resulting in a change of sign of the variation of resistance with respect to the zero-field condition $\Delta R(H_y)$, for all the magnetoresistors. So that the output measured after reset $V_{out,reset}(H_y)$ is:

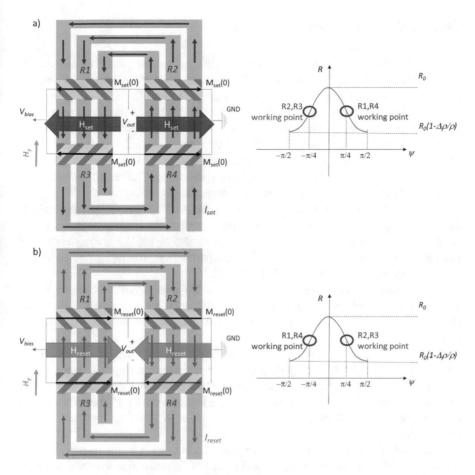

Fig. 14.14 Set/Reset coil working principle: (**a**) Set pulse effect; (**b**) Reset pulse effect

$$V_{out,reset} = V_s + V_o = -\frac{V_{bias}\,\Delta R\,(H_y)}{R^*} + V_o \qquad (14.16)$$

Since the sign of the part of output voltage dependent on H_y is changed, while the offset sign is not, it is enough to make the difference between the two measurements to obtain a result independent on the offset. Of course, this limits the bandwidth of the system because variation of field faster than the set and reset procedure cannot be detected, but it is a very powerful method to improve the accuracy of the measurement, especially because it removes offset dependence on temperature, as well as time-dependent drifts.

14.3 ST AMR Technology Overview

ST AMR technology allows the manufacturing of magnetoresistive sensors for consumer and industrial applications. These sensors allow the sensing of external magnetic fields along the three space directions (X, Y, and Z).

ST AMR technology sensing is based on the use of a NiFe film (80/20 permalloy) as the magnetoresistive layer. NiFe stripes are designed according to the sensing direction of the device; magnetization anisotropy (easy axis) is induced along the stripes' longitudinal direction by process conditions. This procedure induces the film anisotropy, together with the "shape anisotropy" given by the structure geometry. In Fig. 14.15, a typical array of magnetoresistors is shown.

When a small external magnetic field is applied on a NiFe layer perpendicular to the layer magnetization, it produces a slight change in the magnetization direction of the film, resulting in a resistivity change of the layer. For an angle of ~45° between the sense current and the layer magnetization, the resistance change is linear according to what reported in Sect. 14.2.1.

Since aluminum has a resistivity ten times lower than NiFe, NiFe stripes are covered with "barber pole" aluminum metals placed at 45° versus the NiFe stripes direction, to induce a 45° oriented electrical current versus the magnetization direction.

When an external magnetic field H_y is applied, the layer magnetization direction changes of an angle θ; for small angles, the resistance change is linear versus θ and proportional to the applied external field (Fig. 14.12b). To measure the external magnetic field, NiFe resistance change is measured by a Wheatstone bridge configuration. Opposite resistance changes are induced under the application of an external field.

A second thick aluminum layer is integrated on the sensor technology platform to manufacture the set/reset metal coils and induce NiFe magnetization status set/reset during device working, as reported in Sect. 14.2.2.

A schematic process flow of the ST AMR technology is shown in Fig. 14.16. The NiFe layer, whose thickness value is typically included between 20 and 60 nm, is deposited on an insulated silicon substrate. Magnetization anisotropy (easy axis) is induced on the NiFe layer during Physical Vapor Deposition (PVD) by a magnetic chuck. During this process, an electromagnet placed below the PVD chuck allows the control of the orientation of the magnetic domains in the permalloy film, by adjusting the current supplied to the electromagnet itself. Since NiFe layer oxidation

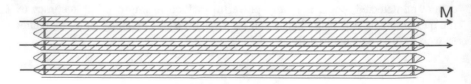

Fig. 14.15 A typical NiFe array layout in ST AMR sensors

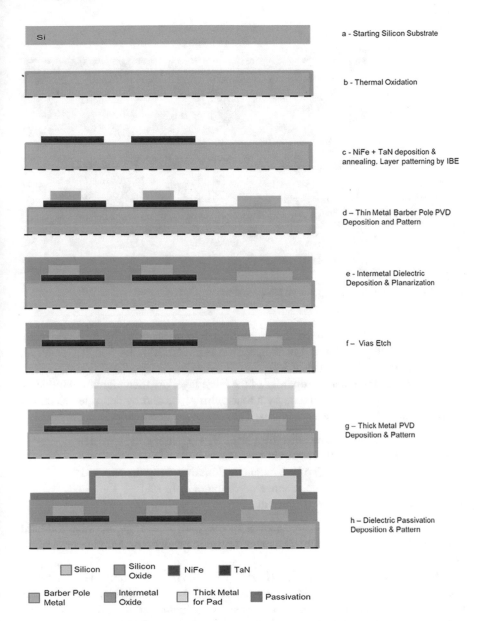

a - Starting Silicon Substrate

b - Thermal Oxidation

c - NiFe + TaN deposition & annealing. Layer patterning by IBE

d – Thin Metal Barber Pole PVD Deposition and Pattern

e - Intermetal Dielectric Deposition & Planarization

f – Vias Etch

g – Thick Metal PVD Deposition & Pattern

h – Dielectric Passivation Deposition & Pattern

| Silicon | Silicon Oxide | NiFe | TaN |
| Barber Pole Metal | Intermetal Oxide | Thick Metal for Pad | Passivation |

Fig. 14.16 Schematic process flow description of ST process to fabricate AMR sensors

can induce a degradation in the film magnetic performances, a few-nanometers-thick protective layer, e.g., TaN, is deposited after NiFe in the PVD tool, without vacuum break.

NiFe layer is then annealed in an oven tool, under a magnetic field induced by a permanent magnet, to better align the magnetic domains along the NiFe stripes'

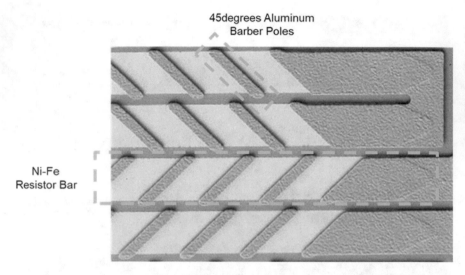

Fig. 14.17 Aluminum barber poles on NiFe resistors

direction and further enhance the magnetic performances of the film. Magnetic anneal is typically performed at temperature between 200 and 300 °C, under an external magnetic field of 0.1–0.5 T.

Electrical and magnetic properties of the NiFe layer are then monitored inline by a magnetic hysteresis loop tracer, by a four-point probe and by dedicated electrical process control monitor test structures. The monitored parameters include:

- Retentivity B_r (nWb) and anisotropy field H_k (Oe) along the hard axis
- Coercivity H_c along the easy axis
- Angular dispersion (deg) of the magnetic domains along the easy axis
- Magnetoresistive resistance change percentage: $d\rho/\rho$ (%)
- Layer sheet resistance: R_s (Ohm/Sq)

In Fig. 14.8, two experimental curves for magnetization along the easy and hard axis have been already shown. The main magnetic parameters monitored inline are shown as well in Fig. 14.8.

After film deposition and anneal (Fig. 14.16c), the layer is patterned by Ion Beam Etch (IBE) to design the stripes reported in Fig. 14.15. Following NiFe resistor patterning, a thin aluminum metal is deposited and patterned to design the "barber poles" metallization (Fig. 14.16d). Figure 14.17 shows a SEM top picture of NiFe bars covered by aluminum barber poles.

An oxide layer is then deposited and planarized to passivate the NiFe resistors and the aluminum barber poles and is etched to create vias (Fig. 14.16e–f). Since the intermetal TEOS layer defines the gap between the NiFe resistors and the set/reset metal coil, thickness control is very important to ensure the correct set/reset functionality and repeatability among different devices.

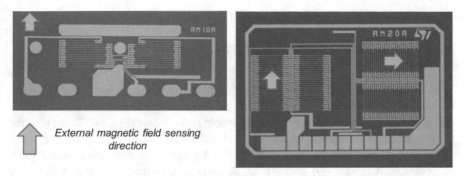

Fig. 14.18 Optical top view of a monoaxial (left) and a biaxial (right) AMR magnetic sensor

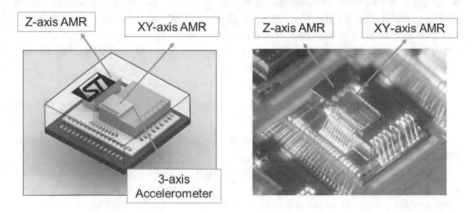

Fig. 14.19 6-axis DOF sensor with 3-axis Accelerometer and 3-axis AMR magnetic sensors

A few-micron aluminum metal is then deposited and patterned (Fig. 14.16g) to realize both the set/reset metal coils and the device pads. A CMOS-like passivation (TEOS/SiN) is deposited on top of the metal layers and patterned on the pad area (Fig. 14.16h).

ST AMR Technology allows the realization of sensors which detect the external magnetic field in directions along the die plane. Figure 14.18 shows an optical top view of a monoaxial and a biaxial AMR magnetic sensor after barber poles patterning, with the sensing directions shown by the orange arrows.

Manufacturing of a 3-axial magnetic sensor is possible by the package-level integration of a biaxial sensor die and a monoaxial sensor die placed vertically inside the package itself; $X–Y$ dice is connected to the ASIC by wire bonding; bumps are deposited on Z-axis sensor for direct interconnection on the ASIC. Figure 14.19 shows a schematic drawing and the optical picture of a 6 Degree of Freedom (DoF) sensor (3-axis accelerometer +3-axis magnetic) where ST AMR technology is used to manufacture the magnetic sensor components. AMR dice placement in package is shown.

14.4   Device Performances Versus Device Parameters

14.4.1   Sensitivity and Linearity Range

Results in Eqs. (14.10), (14.11), and (14.12) hold in a simplified condition, since magnetization is not constant all over the magnetoresistor and current is not exactly 45° all over the space between one barber pole and another (Fig. 14.12c). Thus, for a more accurate estimation of sensitivity, FEM simulations of the magnetoelectric coupling between external magnetic field and magnetoresistors are required [21]. Nonetheless, keeping H_D and H_k small ensures according to Eq. (14.12) high sensitivity of V_{out} with respect to H_y, but it reduces the linearity range. Generally, it is a good practice to keep as small as possible H_k so that linearity range and sensitivity can be ensured by a good control of the geometry and are quite independent on NiFe material properties variations. H_D can be extracted as an average value across the magnetoresistor from a magnetostatics FEM analysis, but it is possible to assume roughly that it is [16]:

$$H_D \propto \frac{t}{W} \tag{14.17}$$

Thus, it is possible to play on ratio between NiFe resistors width W and thickness t to fix the desired full scale/sensitivity.

On the other hand, another way to obtain high sensitivity is to have barber poles working efficiently in deflecting current at 45° with respect to the magnetoresistor length. According to what was described in Sect. 14.2.1, not only current to magnetization angle at 45° guarantees linearity but also maximum sensitivity, considering it as $\partial R/\partial H_y$.

Barber poles are more efficient if the electric contact between barber poles and NiFe is good and if barber poles distance is lower than magnetoresistor width.

Finally, of course, magnetoresistivity coefficient $d\rho/\rho$ has to be kept as high as possible to obtain high sensitivity.

14.4.2   Set and Reset Efficiency

Set and reset is a crucial feature of AMR sensors, and it is fundamental to make it work properly. Indeed, it allows offset cancellation but also a well-defined initial magnetization of AMR stripes, allowing low noise level. Furthermore, it recovers correct functionality of the sensor after saturation of the sensor when exposed to high magnetic fields.

Since high current is generally required to set and reset magnetoresistors, set and reset are driven by means of very short current pulses provided by the discharge of a capacitance, to limit consumptions and to avoid disturbance on power supply line.

Consequently, coil resistance must be as small as possible and, for this reason, thick metal layers are required. On the other side, a tradeoff between coil resistance and coil efficiency must be found in coil design. If we define as factor of merit of the coil the ratio between average magnetic field generated on the sensor and current driven on the coil, this coil factor increases with the number of turns and/or decreases with the turn width, i.e., in an opposite way than resistance does.

14.4.3 Cross-Axis (2-Axis Magnetometers)

When a 2-axis magnetometer is designed, the sensor includes a first Wheatstone bridge with magnetoresistors whose length is parallel to x axis, for y-axis sensing and a second Wheatstone bridge with magnetoresistors whose length is parallel to y-axis, for x-axis sensing, as shown in Fig. 14.18. Film anisotropy is fixed at $45°$ with respect to both X and Y axes, during film deposition, to obtain a condition as much symmetric as possible on both the axes.

As shown in Sects. 14.2 and 14.4.1, it is generally advised to fix $H_D >> H_k$ so that geometry dominates the resistor anisotropy, but in this case, the influence of material anisotropy rotates to a small angle the effective easy axis direction toward the $45°$ direction. Thus, the two sensing axes are not exactly orthogonal. If external field H has a component both along y (H_y) and along x (H_x), Eq. (14.9) must be modified as follows:

$$W_s = -\mu_0 H_y M_s \sin\theta - \mu_0 H_x M_s \cos\theta + \frac{1}{2}\mu_0 H_k M_s \sin^2(\varphi - \theta) + \frac{1}{2}\mu_0 H_D M_s \sin^2\theta$$

(14.18)

In a general case, Eq. (14.18) can be solved by means of numerical models [21]. If small variations of θ are assumed, it is reasonable to define cross-axis as the ratio between magnetization rotation due to H_x change $\partial\theta/\partial H_x$ and magnetization rotation due to H_y change $\partial\theta/\partial H_y$. Considering y axis (i.e., $\theta \approx 0$), the two variations can be calculated by forcing minimum condition on Eq. (14.18) (i.e., $\partial W_s/\partial\theta = 0$) and, approximating the solution for small θ, the following expression for cross-axis c_{xy} can be found:

$$c_{xy} = \left.\frac{\frac{\partial\theta}{\partial H_x}}{\frac{\partial\theta}{\partial H_y}}\right|_{H_x, H_y \to 0} = \frac{H_k}{2H_D}$$

(14.19)

This is a further evidence of the fact that it could be useful to work with $H_D >> H_k$, but also that there is a tradeoff between sensitivity ($\propto 1/H_D$) and cross-axis.

14.5 Conclusions

An overview of the most common technologies for the manufacturing of magnetic sensors and their principle of working has been presented. Special focus has been dedicated to the AMR technology, and a detailed description of the STMicroelectronics AMR technology platform for the manufacturing of single- or multiple-axis magnetic sensors has been presented. This technology is currently implemented on different ST products from standalone magnetic sensors to e-compasses (6-axis), finding applications both in the consumer and in the industrial market.

References

1. Popovic, R. S. (2003). *Hall effect devices (IoP)*. CRC Press.
2. Osamu, O. (2007). *Compound semiconductor bulk materials and characterizations*. World Scientific Publishing.
3. Ausserlechner, U., Motz, M., & Holliber, M. (November 2007). Compensation of the piezo-hall effect in integrated hall sensors on (100)-Si. *IEEE Sensors Journal, 7*(11).
4. Huber, S., Schott, C., & Paul, O. (2013, August). Package stress monitor to compensate for the piezo-hall effect in CMOS hall sensors. *IEEE Sensors Journal, 13*(8).
5. Ausserlechner, U. (2004, October). *The piezo-Hall effect in n-silicon for arbitrary crystal orientation*. In Proceedings of the IEEE sensors, Vienna, Austria, vol. 3, pp. 1149–1152.
6. Schurig, E., Demierre, M., Schott, C., & Popovic, R. S. (2002). A vertical hall device in CMOS high-voltage technology. *Sensors and Actuators A, 97–98*, 47–53.
7. Bilotti, A., Monreal, G., & Vig, R. (June 1997). Monolithic magnetic hall sensor using dynamic quadrature offset cancellation. *IEEE Journal of Solid-State Circuits, 32*(6), 829–836.
8. Schott, C., Racz, R., & Huber, S. (2005). Smart CMOS sensors with integrated magnetic concentrators, IEEE Sensors, Irvine, CA.
9. Popovic, R. S., Randjelovic, Z., & Manic, D. (2001). Integrated hall-effect magnetic sensors. *Sensors and Actuators A, 91*, 46–50.
10. Langfelder, G., Buffa, C., Frangi, A., Tocchio, A., Lasalandra, E., & Longoni, A. (Sept. 2013). Z-Axis magnetometers for MEMS inertial measurement units using an industrial process. *IEEE Transactions on Industrial Electronics, 60*(9), 3983–3990.
11. Marra, C. R., Laghi, G., Gadola, M., Gattere, G., Paci, D., Tocchio, A., & Langfelder, G. *100 nT/√Hz, 0.5 mm2 monolithic, multi-loop low-power 3-axis MEMS magnetometer*. In Proceedings of the 2018 IEEE Micro Electro Mechanical Systems (MEMS), 2018 (IEEE Journals & Magazines), pp. 748–758.
12. Tumanski, S. (2007). Induction coil sensors – A review. *Measurement Science and Technology, 18*, R31–R46.
13. Drljaca, P. M., Kejik, P., Vincent, F., Piguet, D., & Popovic, R. S. (October 2005). Low-power 2-D fully integrated CMOS fluxgate magnetometer. *IEEE Sensors Journal, 5*(5).
14. Osborn, J. A. (June 1945). Demagnetizing factors of the general ellipsoid. *Physical Review, 67*(11–12).
15. Tumanski, S. (2001). *Thin film magnetoresistive sensors (IoP)*. CRC Press.
16. Tumanski, S. (2011). *Handbook of magnetic measurements (IoP)*. CRC Press.
17. Stutzke, N. A., Russek, S. E., & David, P. (2005). Pappas: Low-frequency noise measurements on commercial magnetoresistive magnetic field sensors. *Journal of Applied Physics, 97*.
18. Bonin, R., Schneidera, M. L., Silva, T. J., & Nibarger, J. P. (2005). Dependence of magnetization dynamics on magnetostriction in NiFe alloys. *Journal of Applied Physics, 98*.

19. Jury, J. C., Klaassen, K. B., van Peppen, J. C. L., & Wang, S. X. (Sept. 2002). Measurement and analysis of noise sources in giant magnetoresistive sensors up to 6 GHz. *IEEE Transactions on Magnetics, 38*(5), 3545–3555.
20. Klaassen, K. B. (2007, February). Electrical low-frequency noise in tunneling Magnetoresistive heads: Phenomena and origins. *IEEE Transactions on Magnetics, 43*(2).
21. Spinelli, A. S., Minotti, P., Laghi, G., Langfelder, G., Lacaita, A. L., Paci, D. *Simple model for the performance of realistic AMR magnetic field sensors*. In Proceedings of the 2015 transducers – 2015 18th international conference on solid-state sensors, actuators and microsystems (TRANSDUCERS), 2015 (IEEE Journals & Magazines), pp. 2204–2207.
22. Kuijk, K., van Gestel, W., & Gorter, F. (September 1975). The barber pole, a linear magnetoresistive head. *IEEE Transactions on Magnetics, 11*(5), 1215–1217.
23. Hauser, H., Fulmek, P. L., Haumer, P., Vopalensky, M., & Ripka, P. (2003). Flipping field and stability in anisotropic magnetoresistive sensors. *Sensors and Actuators A, 106*, 121–125.

Chapter 15
Microphones

Silvia Adorno, Fabrizio Cerini, and Federico Vercesi

15.1 Microphone Overview

A microphone is an acoustic transducer aimed to convert sound into an electric signal. Sound is a pressure wave with frequency in the audio band range 20 Hz–20 kHz and pressure that ranges from 20 μPa (audible threshold) up to several tens of Pascal. To detect this physical quantity, microphones need a moving element, usually a membrane, vibrating when reached by sound, and some transduction principle to convert this vibration in an electrical signal, i.e., a voltage or charge variation.

Microphone applications are extremely wide: concert halls and public events, radio and television, phones, PCs, consumer electronics, etc. Nowadays, microphones are everywhere around us, and the increasing diffusion of personal and home assistants is raising the importance of human–machine interaction through voice.

First experiments on microphones date back to the second half of the nineteenth century, and the first commercial product, the carbon microphone, was invented in 1876 and widely adopted by the end of the century. Several types of microphones have been conceived and developed since then. In 1916, the first condenser microphone, the grandfather of most of current MEMS microphone, was invented. The ribbon microphone and the dynamic microphone appeared in the following decades.

S. Adorno (✉)
ST Microelectronics, Analog MEMS and Sensors Group, MEMS Technology and Design R&D,
Cornaredo, Italy
e-mail: silvia.adorno@st.com

F. Cerini · F. Vercesi
ST Microelectronics, Analog MEMS and Sensors Group, MEMS Technology and Design R&D,
Agrate Brianza, Monza Brianza, Italy
e-mail: fabrizio.cerini@st.com; federico.vercesi@st.com

© Springer Nature Switzerland AG 2022
B. Vigna et al. (eds.), *Silicon Sensors and Actuators*,
https://doi.org/10.1007/978-3-030-80135-9_15

In second half of twentieth century, condenser and dynamic microphones dominated professional applications (especially studio recording and live performances),
whereas the Electret Condenser Microphone (ECM) was the standard solution
for telephones and personal electronics. ECMs, proposed in 1962, are condenser
microphones with a built-in charge due to an electret layer (a thin plastic foil
able to retain electrical charge). ECMs are low-cost, high-sensitivity, wide-band
microphones, and they represented a breakthrough in the field.

The first MEMS microphone was introduced in 1983 [1]. MEMS microphone
exploits silicon micromachining techniques to offer several advantages with respect
to ECMs: smaller size, compatibility with surface mounting process (whereas ECMs
require a dedicated and more expensive solder process), higher robustness against
mechanical shocks. Moreover, MEMS microphones have an Application-Specific
Integrated Circuit (ASIC) integrated in the same package: This reduces area and
cost and allows for better performance: Microphone parameters can be adjusted via
trimming so to have high part-to-part repeatability. It took other 20 years before
MEMS microphones were ready for mass production, mainly driven by mobile
phones: In 2014, MEMS microphone volumes overtook ECM, and they are still
growing.

15.2 MEMS Microphone Description and Specification

15.2.1 MEMS Microphone Architecture

As shown in Fig. 15.1a, a typical microphone system [2, 3] is made of three
elements: a MEMS sensor that is the electromechanical part of the system, an
ASIC (Application-Specific Integrated Circuit) for readout and conditioning, and
a package enclosing the two silicon elements. Figure 15.1b reports the top view of
a real microphone system without package cap, together with the main dimensions.

MEMS sensor converts sound pressure into an electrical signal. Two transduction
principles are commonly adopted in MEMS microphones: capacitance variation or
direct piezoelectric effect. Fabrication adopts semiconductor technology blocks and
dedicated micromachining process steps, as described in Sects. 15.4 (for capacitive
microphones) and 15.6 (for piezoelectric microphones).

The ASIC has different functions: provide proper polarization to the sensor if
needed, perform signal readout, conditioning, and elaboration, i.e., buffering and
analog to digital conversion. Both analog and digital outputs are widespread. In
analog output, a sinusoidal sound pressure variation produces as output signal a
sinusoidal voltage with amplitude proportional to pressure amplitude. In the case
of digital output, output signal is a digital sequence of 0 s and 1 s representing
frequency and amplitude of the input pressure. In MEMS microphones, digital
signal is usually in Pulse Density Modulation (PDM) format: a single bit signal

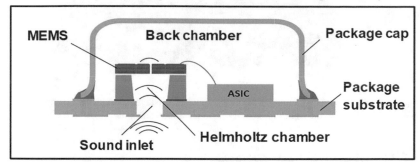

Fig. 15.1 (a) Diagram of a typical microphone system. (b) Top view of a real microphone system without package cap. Wire bonding in Au material connects MEMS to ASIC and ASIC to substrate

where the density of 1 s is proportional to amplitude of the input signal. I2S (Integrated Interchip Sound) is also another available standard.

Package is a key element in microphones [4, 5]. It protects sensing element and integrated circuit, as in all MEMS devices, but it also deeply affects performances of the microphone system. Typical microphone package is a cavity, made of a land grid array (LGA) substrate and a metal or organic cap, as shown in Fig. 15.2. Sensor and ASIC are attached to the substrate and electrically connected, usually via wire bonding. The sound port, i.e., a hole in the substrate or in the cap, allows sound entering the cavity. Dimensions of cavity and sound port have a significant effect on microphone performances, as described in Sect. 15.2.2.

Fig. 15.2 X-ray image of a bottom port microphone. A 4-layer LGA substrate and a metal can are visible. MEMS is on the right: note cavity through silicon substrate. ASIC is on the left

15.2.2 Acoustic Specification

A brief review of most significant parameters [6] for a microphone can be helpful to understand the most important constraints in their design [7] and fabrication.

Sensitivity is defined as the ratio between output signal and input pressure, where input pressure is a standard acoustic signal with amplitude 1 Pa RMS (94 dBSPL) and frequency 1 kHz.

For analog microphones, sensitivity becomes therefore the amplitude of the output voltage produced by the standard acoustic pressure input: this is measured in dBV (RMS voltage signal at 1 Pa).

In digital microphones, analog-to-digital converter (ADC) full scale must be considered since it sets the maximum output signal. Therefore, sensitivity is measured in dBFS, defined as the ratio between the output signal produced by the standard acoustic pressure input and the ADC full scale.

Commercial products commonly adopt some standard levels (i.e., −38 dBV for analog microphones and − 26 or − 41 dBFS for digital microphones) to be fully compatible with the readout chain. Part to part spread is key: State of the art is sensitivity within +/− 1 dB from part to part.

Sensitivity response is the output behavior across the audio frequency range, as shown in Fig. 15.3 [3].

All MEMS microphones have a low frequency roll-off: Sensitivity decreases toward lower frequencies. Roll-off point is defined as the frequency where signal is 3 dB lower than its 1 kHz value, as shown in Fig. 15.3. This phenomenon is due to the need of a ventilation hole in the membrane to equalize static ambient pressure: Without this hole, microphone would act as a pressure sensor, since ambient pressure is several magnitude orders higher than sound pressure. This ventilation hole acts as an alternative path for the sound, subtracting part of the signal from the membrane. Roll-off frequency can be set by ventilation hole design and is usually below 100 Hz. A typical choice of roll-off frequency is 35 Hz.

At higher frequencies, package acts as a Helmholtz resonator: A resonance peak is therefore present in the range 15–20 kHz or slightly above 20 kHz, as shown in Fig. 15.3. A wide and short sound port helps to move this peak toward higher frequencies.

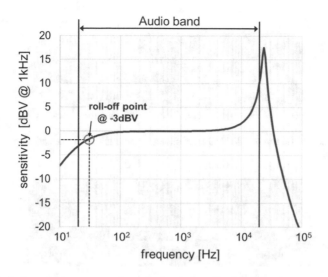

Fig. 15.3 Typical sensitivity response in audio band normalized at 1 kHz

Total Harmonic Distortion (THD) measures the alteration of the signal frequency content [6]. It is given by the formula reported in Eq. (15.1):

$$THD = \frac{\sqrt{A_2{}^2 + A_3{}^2 + A_4{}^2 + A_5{}^2 + \cdots}}{A_F},$$

(15.1)

where A_f is the amplitude of the fundamental component, and A_2, A_3, ..., A_n are the amplitudes of the 2nd, 3rd, ..., Nth components, respectively. Usually, first five harmonics are included. For a signal at 1 kHz, state of art for THD is below 0.1% at 94 dBSPL and below 4% at 130 dBSPL. Main contributors to THD are sensors intrinsic transduction nonlinearity and amplifier nonlinearity.

Signal-to-noise ratio (SNR) is the ratio between sensitivity at 1 kHz and integrated noise floor in audio band, and A-weight is usually applied. Noise contributions come from both sensor and ASIC. Sensor noise is generated by different dissipation mechanisms, both acoustic-mechanical and electrical: The most important are fluidic squeeze-film damping in capacitive microphones and material losses in piezoelectric microphones.

Microphone packages can adopt a top port or bottom port configuration, as shown in Figs. 15.4 and 15.5, respectively. The top port has sound inlet in the cap, front chamber is package inner volume, and back chamber is MEMS internal cavity. Vice versa, the bottom port has sound inlet in the substrate (below MEMS), front chamber is MEMS internal cavity, and back chamber is package inner volume.

Acoustic behavior of a MEMS microphone is dominated by two compliances in series: membrane compliance and back chamber compliance [8]. Back chamber is an additional stiffness that opposes to membrane movement, therefore affecting

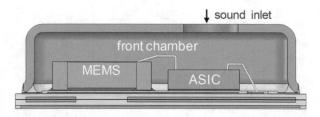

Fig. 15.4 Scheme of a top port MEMS microphone. Sound inlet is in package cap. Back chamber is MEMS cavity

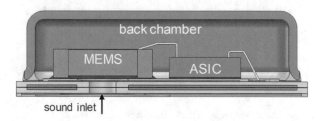

Fig. 15.5 Scheme of a bottom port MEMS microphone. Sound inlet is located in the package substrate below MEMS. Back chamber is the whole package volume

sensitivity. Back chamber compliance C_{back}, in acoustic domain, is defined as in Eq. (15.2) [9]:

$$C_{back} = \frac{V_{back}}{\rho\, c^2}, \tag{15.2}$$

where V_{back} is back chamber volume, ρ is air density, and c is speed of sound in air.

Therefore, the larger its volume, the lower the stiffening effect. To increase sensitivity, back chamber volume should be as large as possible: There is an evident tradeoff with package miniaturization.

Moreover, the sound inlet and the front chamber form a Helmholtz resonator. This is responsible for the high-frequency resonance peak described before. To shift the resonance peak toward higher frequencies, front chamber volume must be as small as possible.

Under these assumptions, bottom port configuration gives better performances, both in terms of higher sensitivity, because of larger back chamber, and flatter frequency response, because of smaller front chamber corresponding to high-frequency resonance peak.

MEMS microphones have also to face challenging reliability requests: They should survive drops, shocks, aggressive temperature, and humidity cycles. These will be discussed in detail in Sect. 15.2.3.

15.2.3   Reliability Requirements

MEMS microphone specification requires challenging reliability aspects too. Qualification strategy involves both standard ICs tests and dedicated tests based on customer requests. In tests based on standard specifications, common to ICs, test conditions are defined by standards and parts are measured in socket using standard final testing equipment. Some examples of standard test are HTOL (high-temperature operating life), HTSL (high-temperature storage life), TC (thermal cycles), ESD & LU (electrostatic discharge and latch up), and so on.

Tests based on customer specification are usually carried on with parts soldered on quality boards and tested with a dedicated bench setup. Frequently, these tests are aimed to mechanical robustness. Some examples are Mechanical Shocks (MS), Variable Frequency Vibrations (VFV), free fall, and Compressive Air Test (CAT).

Main failure mechanism for MEMS microphones is contamination. Whereas inertial sensors are hermetic, microphones have a cavity package and a hole in the substrate or in the cap (sound port) and therefore are more exposed to external contamination. This contamination is killer for microphone functionality in terms of reduced sensitivity and SNR or, in very severe cases, of complete loss of functionality. Proper choice of assembly line class is therefore very important, together with dedicated handling procedures in all manufacturing steps of the final device.

Another possible failure mode is cap detachment in the cavity package. Cap detachment is typically caused by mechanical overstresses but can also be caused by thermal stresses. Some examples of possible mechanisms causing cavity detachment are incomplete dispense of the solder paste or glue, wrong reflow profile (if the reflow profile differs from the standard one, the glue on the package can be weakened, losing its fixing capability), free fall over a rigid surface from more than 1 m, wrong cut line in the separation of single components from the strip, and force exceeding a threshold (typically 2.5 g) for tape removal. In the case of cavity detachment, a failure in any audio parameter—sensitivity, SNR, frequency response, and THD—can be observed.

Another key aspect for quality and reliability of microphones is mechanical robustness. Smartphones are the main application of MEMS microphone, and mechanical overstress related to drop event is common in the field: For this reason, phone makers have defined specific tests, such as: guided free fall test and tumbler test. Such tests are performed with components mounted on a dedicated board with cover to simulate the phone dimensions and weight, to reproduce the real mechanical stress on MEMS, and to investigate possible failure mechanisms related to package and solder joint reliability. Due to its very thin thickness, broken membrane is one of the main experienced failure modes. Backplate breakage is also possible.

15.3   Capacitive Microphones

In first approximation, as shown in the basic schematic of Fig. 15.6, capacitive MEMS microphone sensor can be modeled as a parallel plate capacitor. A movable membrane acts as the first plate of the capacitor, and the second plate known as backplate is the fixed part.

Regarding the membranes, many geometries are possible. They can be fully clamped on their entire perimeter or suspended via springs or mechanical bumps. Membrane design [10], material, and thickness have a strong impact on microphone sensitivity.

On the other hand, backplate must be much stiffer than the membrane since it is the reference side of the capacitor. Therefore, backplate is usually thicker and comprises many acoustic holes for air to pass through. Both membrane and backplate must be conductive or comprise at least one conductive layer forming the electrode.

Under acoustic pressure, membrane movement changes plates' distance and the capacitance varies accordingly with the relation reported in Eq. (15.3):

$$C = \varepsilon_0 \frac{A_{ele}}{d_0 - x} \quad \text{with} \quad x = \frac{P \cdot A_{mem}}{k}, \tag{15.3}$$

where ε_0 is vacuum dielectric constant, A_{ele} is the common area of the capacitor electrode, d_0 is the initial air gap thickness defining microphone sensitivity, k is the equivalent mechanical spring constant of the membrane together with the membrane area A_{mem}, and P is the applied acoustic pressure.

To equalize ambient pressure below and above the membrane, a ventilation hole must be included in the MEMS. In case of suspended membranes, as shown in Fig. 15.7, any small gap between membrane edge and anchored area acts as a ventilation

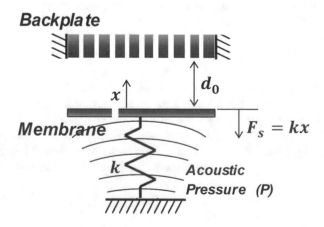

Fig. 15.6 Basic schematic of capacitive MEMS sensor

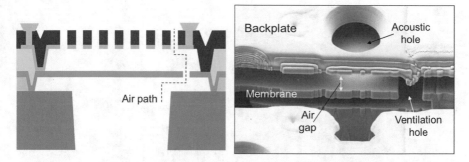

Fig. 15.7 On the left, a schematic cross-section of a capacitive MEMS microphone with suspended membrane. On the right, an FIB (focused ion beam) cut of membrane edge. The channel between membrane and substrate is the ventilation hole. Air can follow this path and escape laterally

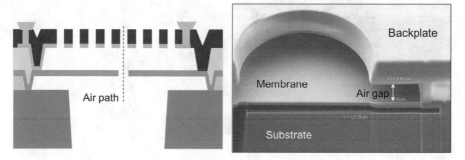

Fig. 15.8 On the left, a schematic cross-section of a capacitive MEMS microphone with fully clamped membrane. On the right, FIB cut of membrane clamped at its edge. No ventilation hole is present in this area. A dedicated hole must be placed somewhere else on the membrane

hole. Generally, an overlap between membrane and substrate is present, and the roll-off point can be controlled with a proper choice of overlap length and distance between membrane and MEMS substrate. Alternately, in case of fully clamped membrane, only membrane holes can be adopted, as reported in Fig. 15.8.

If a constant charge Q is stored on the plates of the MEMS capacitor, a variation of capacitance C induces a variation of voltage V that is read by the ASIC and processed if needed. The ASIC is not only able to read the electrical signal V as a function of the acoustic pressure P applied on the membrane, but it is also able to apply to the electrodes, in different phases, both a constant charge through the charge pump, and a bias voltage for air gap variation resulting in sensitivity trimming.

Main factors affecting MEMS sensitivity are:

- Capacitance value related to membrane area and air gap thickness
- Membrane mechanical compliance, set by its thickness, residual stress, and spring geometry (if springs are present)

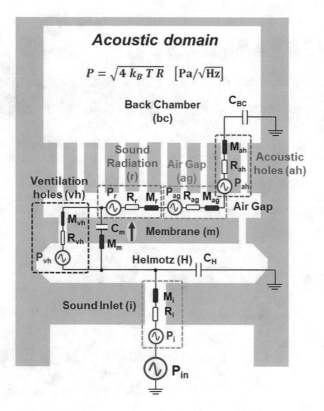

Fig. 15.9 Example of electro-mechanical-acoustic lumped-element of a capacitive MEMS microphone

- Back chamber volume
- Bias applied on the two plates

To deeply model the behavior of the microphone and to evaluate sensitivity, noise, and frequency response, a model is needed. We will describe a lumped equivalent model built exploiting the electro-acoustic analogy.

According to the direct electroacoustic analogy [11], i.e., Voltage (V) – Pressure (P), Current (I) – Volumetric flow rate (Q), and Charge (Q) – Volume (U), the fluidic-mechanical behavior of a capacitive MEMS microphone (with single backplate and membrane) can be modeled [12, 13] by the equivalent electro-mechanical-acoustic lumped-element model reported in Fig. 15.9. A similar model can be built also for a piezoelectric microphone too [14], as will be discussed in Sect. 18.5 (with changes for the different MEMS sensor elements).

Referring to Fig. 15.9, the model includes all the main acoustic and mechanical elements, in acoustic domain. We can now describe the different elements.

Membrane is described by two parameters: an equivalent compliance (C_m) and a mass, equivalent to an inductive (M_m) behavior.

Back chamber is described as a compliance (C_{bc}). Membrane and back chamber compliances are key in determining sensitivity because they set the overall compliance of the system.

Sound inlet is mainly described by the equivalent flow resistance (R_i) due to the airflow in the duct. Moreover, the combination of sound inlet flow resistance and front chamber compliance (C_H) defines a Helmholtz resonator, i.e., a resonant cavity that shapes the transfer function of the device.

Additional flow resistance elements associated to acoustic dissipation are ventilation holes (R_{vh}), air gap (R_{ag}), acoustic holes (R_{ah}), and sound radiation (R_r), which represent the friction due to radiation of sound from a vibrating plate (the membrane) in the surrounding air. All these fluidic resistances are therefore associated with the noise pressure source P [13]. Main noise sources in MEMS are due to squeeze film damping [15]. Proper design of acoustic holes in the backplate can minimize damping, but there is a tradeoff between noise reduction and sensitivity, i.e., the larger the holed area, the less the capacitance area. Furthermore, an additional squeeze film damper is the ventilation hole.

External sound pressure is an equivalent acoustic source P_{in}, whereas each noise source is an equivalent acoustic noise source P. When P_{in} is active and noise sources P are short-circuited, the model calculates displacement of the membrane under the pressure signal. With a suitable electric model (not shown in figure), variable capacitance can be calculated from this displacement and final electrical signal can be evaluated: This is the sensitivity. Both sensitivity and frequency response of the microphone system in audio band (20 Hz–20 kHz) are therefore calculated (see Fig. 15.3 as an example).

When P_{in} is short-circuited and noise sources P are active, membrane displacement due to noise sources and, from this, microphone noise in audio band can be evaluated. An example of typical Power Spectral Density (PSD) of the different noise sources of a capacitive MEMS microphone is reported in Fig. 15.10.

Comparing Figs. 15.3 and 15.10, it can be observed that noise transfer function has a resonance peak in the same position of sensitivity transfer function. So, resonance peak position plays an important role, because if it is in the center of the audio band, integrated noise value will be higher than if the peak is close to band edge (20 kHz) or outside audio band.

Microphone SNR is defined as the ratio between sensitivity at 1 kHz and A-weighted noise in audio band. The SNR value will depend on both MEMS and ASIC noise.

In general, several MEMS architectures have been proposed [10]. Some of them showed in Fig. 15.11 include:

- Membrane suspended to springs
- Fully clamped membranes, anchored on all the perimeter
- Fully suspended membranes: no spring are present, but membrane lays on pillars that space it from backplate and prevents electrical contact

All these options have some pros and cons. Suspended membranes can efficiently reject stresses, whereas fully clamped membranes require tight control of residual

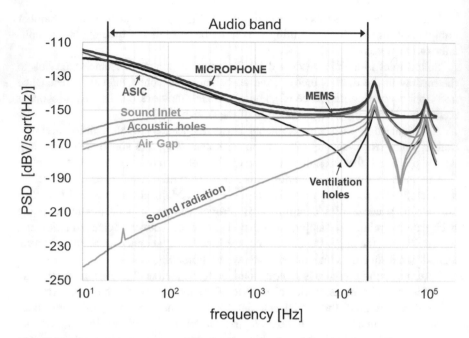

Fig. 15.10 Example of MEMS noise power spectral density. Each line is the contribution of a MEMS element; red line is total MEMS noise Power Spectral Density (PSD)

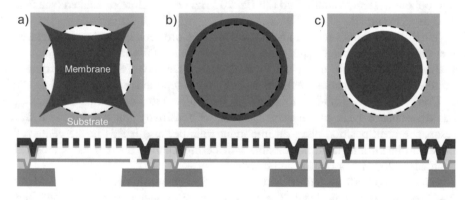

Fig. 15.11 Schematics of possible architecture for capacitive MEMS microphone. (**a**) Suspended membrane on four springs that are also anchor points; (**b**) fully clamped membrane on its perimeter; (**c**) suspended membrane laying on spacing pillars

stresses during fabrication and a careful selection of package design and material to avoid performance drifts due to external stresses.

On the other hand, suspended membranes have a ventilation hole on all membrane perimeters. This ventilation hole acts as a noise source because of squeeze film damping between membrane and silicon substrate. In a fully clamped architecture, instead, no additional noise source is present, and SNR is dramatically improved.

Moreover, perimetral ventilation hole in suspended membrane architecture sets roll-off point too. Tight overlap control is needed to avoid huge roll-off spread: back chamber etch optimization and control is mandatory.

15.4 Capacitive Microphone Technology

Capacitive microphone fabrications, as other MEMS sensors and actuators, rely on traditional electronics microfabrication techniques [16, 17]. The typical fabrication flow for such devices can be divided in four main blocks:

- Membrane and anchors fabrication
- Air gap and backplate fabrication
- Contacts and metal pads fabrication
- Cavity patterning and final device release

The microphone process flow usually starts from silicon substrate with the deposition or growth of a sacrificial layer, normally silicon dioxide, that has the function of release layer for the microphone membrane that sits on top. The first oxide layer can be patterned to fabricate anchorage points for the sensing membrane, as shown in Fig. 15.12a.

Polysilicon deposited by Low Pressure Chemical Vapor Deposition (LPCVD) is a common choice material for the fabrication of the sensing membrane, as shown in Fig. 15.12b. As mentioned, membrane thickness and residual stress are critical design parameters for the final microphone sensitivity; therefore, it is common to perform a high-temperature annealing to change polysilicon crystallography and residual stress. In case of clamped microphone membrane designs, it is important to achieve good adhesion between membrane and substrate to avoid possible delamination during the final release step.

Air gap between membrane and backplate is usually obtained through multiple depositions of sacrificial silicon oxide layers, as shown in Fig. 15.13a. The patterning of these layers is performed to fabricate backplate anchors and bumps, as shown in Fig. 15.13b. Backplate stiffness is usually achieved using thick polysilicon and silicon nitride LPCVD depositions. The ratio between silicon and nitrogen used during the silicon nitride deposition plays an important role in material compatibility

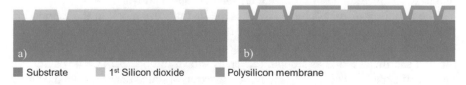

■ Substrate　　■ 1st Silicon dioxide　　■ Polysilicon membrane

Fig. 15.12 Capacitive MEMS microphone process flow: (**a**) Anchor patterning on first sacrificial layer. (**b**) Sensing membrane deposition and patterning

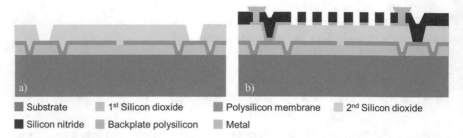

Fig. 15.13 Capacitive MEMS microphone process flow: (**a**) Second sacrificial layer for air gap definition. (**b**) Backplate stack deposition and patterning followed by metal pad definition

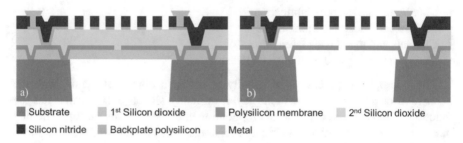

Fig. 15.14 Capacitive MEMS microphone process flow: (**a**) Cavity etching. (**b**) Final release step to etch away first and second sacrificial layers

with the final hydrofluoric release of the MEMS microphone as well as in the electrical performance of the backplate element, i.e., resistivity affecting leakage current and dielectric constant affecting capacitance.

As for the sensing membrane anchor points, proper cleaning prior to every LPCVD deposition of backplate elements is required to avoid weak interface that can cause delamination of backplate itself. Backplate is usually patterned using dry etching processing to obtain the characteristic holed design needed to reduce air squeeze damping and secondly to achieve electrical contacts to polysilicon membrane and backplate counter electrode.

The last part of the wafer front side process flow is usually the metallization layer deposition via sputtering or evaporation techniques and its patterning to obtain electrical interconnection and device bond pads, as shown in Fig. 15.13b. Typical metals used are gold, aluminum, and aluminum-copper alloys.

The pattering of microphone cavity is the last process block in the fabrication flow and consists in the dry etching of the silicon substrate beneath the sensing membrane using deep reactive ion etching or wet silicon etchants (like TMAH) and the first sacrificial oxide as stopping layer, as shown in Fig. 15.14a. The cavity has the role of exposing the sensing membrane to external sound coming from the holed package. Microphone cavities options include single and multiple volume cavities to better match back-end process flow.

The final and key front-end process step for capacitive microphone fabrication is the release process using hydrofluoric acid either in liquid or vapor phase, as shown in Fig. 15.14b. All the layers and materials mentioned above must sustain the release process for as long as required for the complete etching process of the sacrificial oxides.

Capacitive MEMS microphones may have strict constrains on device dimensions in order to meet consumer electronics specification. In addition, since cavity volume is strictly related to the microphone sensitivity performance, demanding MEMS die thickness specification must be satisfied at front-end level usually implying a critical handling during front-end process flow.

As previously mentioned from a reliability and robustness standpoint, capacitive MEMS microphones operate constantly exposed to external environment; therefore, they can be subjected to particle-induced contaminations and mechanical failure. However, during front-end process flow, one of the major failure mechanisms is associated with stiction. During the final release process, devices can be subjected to mechanical stiction of the movable parts due to capillary forces arising during sacrificial layers etching. To prevent such phenomenon, mechanical stoppers or bumps are usually fabricated to avoid membrane stiction onto the backplate counter electrode and onto the silicon substrate. Mechanical bumps reduce the contact areas between the two surfaces, thus reducing the likelihood of stiction phenomena, even though their fabrication increases device complexity [19, 20].

Another possible source of anomalous membrane displacement is the sound pressure burst and mechanical shocks that can lead to membrane stiction or mechanical failure of MEMS device. In such cases, mechanical bumps can also greatly improve the reliability in-operation performance of the device.

15.5 Piezoelectric Microphones

Piezoelectricity is a property of several materials with noncentrosymmetric crystal structure. In these materials, a deformation of the structure can change the distance between atoms, thus creating a polarization and a nonzero electric field (and vice versa).

In a piezoelectric material, a strain across the material induces a charge piling upon its surfaces (direct piezoelectric effect). On the other hand, when an electric field is applied, this generates a variation of strain in the material and, generally, a variation in its geometrical dimensions (converse piezoelectric effect).

Piezoelectric material can be integrated in silicon technologies as thin films: PZT (Lead Zirconate Titanate) and AlN (Aluminum Nitride) are the most common materials.

Piezoelectric fundamental equations can be written in different forms. We just recall the linear Eqs. (15.4) and (15.5) in stress-charge form:

$$D_k = e_{kij} \cdot S_{ij} + \varepsilon_{ki} \cdot E_i \qquad (15.4)$$

$$T_{ij} = c_{ijkl} \cdot S_{kl} - e_{kij} \cdot E_k \qquad (15.5)$$

where D is the electric displacement vector, e is the piezoelectric constants matrix, S is the strain tensor, T is the stress tensor, ε is the dielectric constants matrix, c is the elastic constants matrix, E is the electric field vector in the material, and i, j, k are matrix indices. From these equations, there is a full electromechanical coupling, i.e., both stress and electrical displacement depend simultaneously on mechanical properties (strain) and electrical properties (electric field). Particularly, the piezoelectric coefficients e couple strain and electrical displacement: the larger these coefficients, the higher the charge piled up at a given strain.

Piezoelectric MEMS microphones exploit direct piezoelectric effect. A piezoelectric material is present somewhere in the sensor movable area. Membrane movement induces a strain inside the piezoelectric material, and this generates charge piling up on the electrodes. Usually, microphones are made of movable cantilevers or membranes. Piezoelectric material is on top of these mechanical elements, sandwiched between a top and a bottom electrode. Therefore, the most important piezoelectric coefficient is e_{31}, coupling in plane strain (due to thin structure deflection) and electrical displacement along Z axis.

Typical embodiments of piezoelectric microphones are cantilever-like structure (not necessarily of rectangular shape) or membranes where piezoelectric material is deposited in the regions where strain is higher. Proper placement of the piezoelectric material is needed to achieve acceptable sensitivity and good linearity. Multilayers of piezoelectric material can be used to increase conversion efficiency. Sometimes the whole structure is made only of the piezoelectric stack, without any sustaining layer.

Most of the acoustic considerations done for capacitive microphones are valid for piezoelectric microphones too. Particularly, the interactions with the package are the same, and the need for a ventilation hole is still present. This can be a hole in the membrane or can be the gap between different cantilevers.

SNR definition is the same as described for capacitive microphones, and total noise power spectral density is given by contribution from different noise sources.

MEMS sensitivity in piezoelectric microphones is determined by both piezoelectric coupling efficiency and acoustic and mechanical effects. Therefore, sensitivity depends on:

- Capacitance value related to piezoelectric material area and thickness
- Piezoelectric coefficients
- Structure mechanical compliance, set by thicknesses, residual stresses, geometry of springs, and piezo stack
- Back chamber volume

No bias is applied by ASIC, and sensitivity trimming is achieved only by changing amplifier gain. Proper control of residual stresses, affecting mechanical sensitivity, and piezoelectric coefficients is mandatory.

Main noise source in these microphones is piezoelectric material losses: These are unwanted charge flow across the piezoelectric material due to deviation of the piezo layer from the ideal dielectric, that result in noise and/or thermal dissipation. Losses are a sort of internal friction; they vary in frequency, and they are related to the dielectric and mechanical response of the piezoelectric material to external applied field or stress. For this reason, AlN is the preferred solution for microphones due to its very low material losses. This noise source can be modeled as a current generator in parallel with piezo capacitance and resistance [21]. Resistance value is given in Eq. (15.6):

$$R_p = \frac{1}{2\pi f C_p \tan(\delta)}, \tag{15.6}$$

where f is frequency, C_p is capacitance of the piezoelectric element, and $\tan(\delta)$ is a parameter related to material losses. Noise therefore depends on both material properties and capacitance value.

The example of equivalent electro-mechanical-acoustic lumped-element model of a piezoelectric MEMS microphone [22, 23] is reported in Fig. 15.15. This allows to study the MEMS–ASIC–package interactions.

The model includes all the main acoustic and electrical elements, in acoustic and electrical domains respectively, characterizing the piezoelectric MEMS microphone system. FEM results and analytical formulas define lumped parameter values. Lumped model can be evaluated via tools such as Matlab® – Simulink.

We can now describe the different elements. In analogy with what discussed for the capacitive microphone model, membrane is represented by a compliance and a mass and back chamber by a compliance. Sound inlet is a resistance because of flow dissipation, and front chamber is modeled as a compliance.

Additional flow resistance elements associated with acoustic dissipation are ventilation holes (R_{vh}) and sound radiation (R_r).

External sound pressure is an equivalent acoustic source P_{in} as well as each noise source is an equivalent acoustic noise source P. When P_{in} is active and noise sources P are short-circuited, the model calculates sensitivity and frequency response of the microphone system in audio band (20 Hz–20 kHz). When P_{in} is short-circuited and noise sources P are active, microphone noise in audio band can be evaluated. SNR is defined as the ratio between sensitivity at 1 kHz and A-weighted noise in audio band.

Main difference with the acoustic model for a capacitive microphone is due to the lack of the backplate: Therefore, neither acoustic holes nor air gaps are present, and all the elements related to these two physical entities are missing.

On the other hand, an additional electric domain is present. In this case, we cannot simply evaluate membrane displacement and calculate capacitance variation, but we must convert membrane displacement into piezoelectric stress and stress into

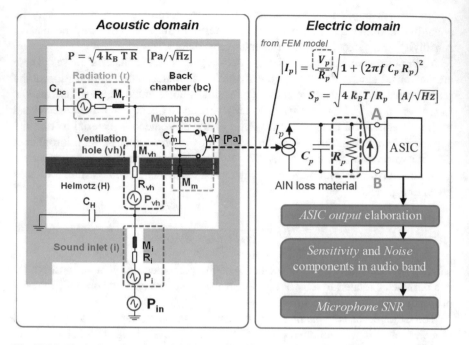

Fig. 15.15 Equivalent electro-mechanical-acoustic lumped-element model of an AlN microphone system for SNR evaluation [22]

voltage signal. The conversion from displacement into stress and charge generation can be modeled via FEM simulation. The final signal generated from the piezo material can be evaluated with the help of an electrical model of the piezo, seen as a current generator in parallel with piezo capacitance and equivalent resistance.

15.6 Piezoelectric Microphone Technology

Piezoelectric microphone fabrication has many points in common with the capacitive microphone, and the typical architectures comprise cantilever beams designs and clamped membranes with the piezoelectric elements sitting on top [18, 24].

Front-end fabrications can be divided in four main blocks:

- Anchors and mechanical structure fabrication
- Piezoelectric stack deposition and patterning
- Contacts, interconnections, and metal pads fabrication
- Cavity patterning and final device release

The fabrication begins with the first sacrificial layer deposition, usually silicon oxide, and it is followed by anchorage points patterning if present. As showed in

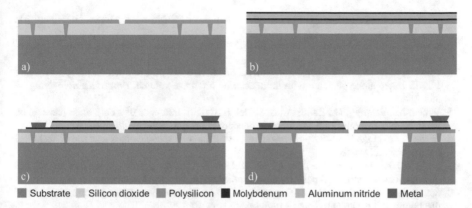

Fig. 15.16 Typical piezoelectric MEMS microphone process flow. (**a**) Deposition and patterning of the purely mechanical structure. (**b**) Deposition of the piezoelectric stack composed by bottom electrode, piezoelectric layer, and top electrode. (**c**) Piezoelectric stack patterning and metal pad definition. (**d**) Cavity etch and final release

Fig. 15.16a and b, the active structure can be realized using a mechanical layer with the piezoelectric stack deposited on top or the whole structure can be made by the piezoelectric stack itself. The so-called piezoelectric stack is composed by the bottom electrode, the active piezoelectric layer, and the top electrode layer. A common material choice for the electrode layers is Molybdenum (Mo) and Titanium (Ti), while Aluminum Nitride (AlN) is usually chosen for the active piezoelectric layer due to its low losses compared to PZT material. Multiple stacks are also a technological possibility: however, unimorph and bimorph stacks are usually the only ones adopted due to technological complexity. A passivation layer, usually AlN, can be integrated in the process flow to protect electrodes and improve device reliability versus humidity. Metal interconnection and microphone cavity patterning follow the same process steps as for capacitive MEMS microphones. A schematic cross-section of finished device is shown in Fig. 15.16d.

With respect to capacitive MEMS microphones, piezoelectric microphone architecture is usually more robust against external particles and water contaminations because there is no air gap between the two electrodes. Moreover, piezoelectric microphones are usually less susceptible to mechanical stiction of movable parts, and some architecture can be in principle dust and waterproof. On the other hand, performances are usually worse, especially SNR and roll-off control, i.e., in cantilever architectures [24].

References

1. Hohm, D., & Gerhard-Multhaupt, R. (1984). Silicon dioxide electret transducer. *The Journal of the Acoustical Society of America, 75*, 1297.

2. Widder, J., & Morcelli, A. (2014). Basic principles of MEMS microphones, EDN network, May 14.
3. AN4426 Application note – Tutorial for MEMS microphones. Technical Report, ST Microelectronics (2017).
4. Feiertag, G., Pahl, W., Winter, M., Leidl, A., Seitz, S., Siegel, C., & Beer, A. (2010). Flip chip MEMS microphone package with large acoustic reference volume. *Procedia Engineering, 5*, 355–358.
5. Winter, M., Feiertag, G., Leidl, A., & Seidel, H. (2010). Influence of a chip scale package on the frequency response of a MEMS microphone. *Microsystem Technology, 16*, 809–815.
6. Lewis, J. (2011). *AN-1112 Application note – Microphone specifications explained*. Technical report, Analog Devices.
7. Shah, M. A., Shah, I. A., Lee, D.-G., & Hur, S. (2019). Design approaches of MEMS microphones for enhanced performance. *Journal of Sensors*, Review Article, 26 pages.
8. Scheeper, P. R., Olthuis, W., & Bergveld, P. (1994). Improvement of the performance nitride diaphragm and backplate. *Sensors and Actuators A, 40*, 179–186.
9. Bay, J. (1997). *Silicon microphones for hearing aid applications* Mikroelektronik Centret, Technical University of Denmark, Copenaghen.
10. Füldner, M., Dehé, A., & Lerch, R. (2005). Analytical analysis and finite element simulation of advanced membranes for silicon microphones. *IEEE Sensors Journal, 5*, 857–863.
11. Kim, B. H., & Lee, H. S. (2015). Acoustical-thermal noise in a capacitive MEMS microphone. *IEEE Sensors Journal, 12*, 6853–6860.
12. Lee, J., Je, C. H., Yang, W. S., & Kim, J.(2012). *Structure-based equivalent circuit modeling of a capacitive-type MEMS microphone*. In International Symposium on Communications, and Information Technologies (ISCIT).
13. Kim, B.-H., & Lee, H.-S. (2015). Acoustical-thermal noise in a capacitive MEMS microphone. *IEEE Sensors Journal, 15*, 6853–6860.
14. Williams, M. D., Griffin, B. A., Reagan, T. N., Underbrink, J. R., & Sheplak, M. (2012). An AlN MEMS piezoelectric microphone for Aeroacoustic applications. *IEEE Journal of Microelectromechanical Systems, 21*, 270–283.
15. Dehé, A., Wurzer, M., Füldner, M., & Krumbein, U. (2013). *The Infineon silicon MEMS microphone*, AMA conferences, Sensors.
16. Conti, S., & Perletti, M. (2008). *Integrated acoustic transducer obtained using mems technology, and corresponding manufacturing process*. US8942394.
17. Perletti, M., Losa, S., Tentori, L., & Turi, M. C. (2015). *Integrated electroacoustic MEMS transducer with improved sensitivity and manufacturing process thereof*. US10057684.
18. Cerini, F., Adorno, S., & Vercesi, F. (2019). *Piezoelectric acoustic MEMS transducer and fabrication method*. EP3557881A1.
19. Alley, R. L., & Cuan, G. J.(1992). *R: The effect of release-etch processing on surface microstructure stiction*. In Technical Digest IEEE solid-state sensor and actuator workshop, pp. 202–207, 862.
20. Lee, Y., Park, K., Lee, J., Lee, C., Yoo, J., Kim, C., & Yoon, Y. (1997). Dry release for surface micromachining with HF vapor-phase etching. *Journal of Microelectromechanical Systems, 6*(3).
21. Levinzon, F. A. (2004). Fundamental noise limit of piezoelectric accelerometer. *IEEE Sensors Journal, 4*, 108–111.
22. Cerini, F. F., & Adorno, S. (2018). Flexible simulation platform for multilayer piezoelectric MEMS microphones with signal-to-noise ratio (SNR) evaluation. *MDPI Proceedings, 2*, 862.
23. Horowitz, S., Nishida, T., Cattafesta, L., & Sheplak, M. (2007). Development of a micromachined piezoelectric microphone for aeroacoustics applications. *The Journal of the Acoustical Society of America, 122*(6), 3428–3436.
24. Grosh, K., & Arbor, A.(2010). *Piezoelectric MEMS microphone*. US2010O2.54547A1.

Chapter 16
Pressure Sensors

Enri Duqi, Giorgio Allegato, and Mikel Azpeitia

16.1 History of MEMS Pressure Sensors

Considerable progress has been made in silicon pressure sensors from the discovery of the large piezoresistive effect in Silicon and Germanium in 1954 [1]. Together with strain gauges, piezoresistive pressure sensors were the first successful commercialized MEMS devices nearly 30 years before the term MEMS was coined by professor R. Howe [2]. The Implementation simplicity of these sensing devices with better sensitivity of 2 orders of magnitude with respect to metal strips led to the first commercialized devices [3]. A big leap forward in terms of performance came in 1962 when Tufte [4] reported the first silicon pressure sensors with piezo resistors diffused in a monocrystalline silicon membrane. Developments in integrated circuits manufacturing process technology like the anodic bonding or the anisotropic etching of silicon had a dramatic importance for all MEMS devices. Most importantly, the anodic bonding and the electrochemical etch stop played a role in the continuous improvement of the integrated MEMS pressure sensors and paved the way of capacitive pressure sensors [5]. In 1982, Kanda [6] gave some typical piezo resistance coefficients of silicon for the more common crystallographic directions, impurity concentrations, and temperatures useful for the designing piezoresistive sensors and integrated circuits. The membrane thickness, which is a key parameter for mass production also in terms of yield and calibration time,

E. Duqi (✉)
ST Microelectronics, Analog MEMS and Sensors Group, MEMS Technology and Design R&D,
Cornaredo, Italy
e-mail: enri.duqi@st.com

G. Allegato · M. Azpeitia
ST Microelectronics, Analog MEMS and Sensors Group, MEMS Technology and Design R&D,
Agrate Brianza, Monza Brianza, Italy
e-mail: giorgio.allegato@st.com; mikel.azpeitia@st.com

© Springer Nature Switzerland AG 2022   523
B. Vigna et al. (eds.), *Silicon Sensors and Actuators*,
https://doi.org/10.1007/978-3-030-80135-9_16

was initially controlled with degenerate P++ doping [7]. NovaSensor combined the use of Fusion bonding and electrochemical etch stop, which permits precise thickness membrane control without the use of large doping concentration [8]. As per the capacitive solution, the silicon bulk chemical etches and the fusion bonding permitted to seal under vacuum one of the membrane sides to create the reference vacuum cavity. Piezo resistivity was the technology of choice mainly because of the simplicity of its implementation, and it only required to make a couple of localized resistors without the need to accurately control a thin gap during the wafer-to-wafer bonding step, as for capacitive technology. Bulk micromachining technology was for long the preferred technique. The first miniaturized devices had relatively low performance if compared with today's application requirements. Conventional silicon/glass pressure sensors were fabricated by anodic bonding technique, which translates in output instability in terms of offset drift, hysteresis over temperature and humidity due to thermal mismatch or bonding interfaces' built-in stress. More recently, in the last 15 years, further improvements have been introduced in terms of technology. The bonding step for the membrane formation was eliminated by STMicroelectronics [9] and Bosch [10] with the big advantage to have thinner, smaller, and mechanically more robust chips (Fig. 16.1) .

Moreover, the sealing of the cavity does not require any wafer-to-wafer bonding, and thus, the reliability of the scaling joint is higher. Advances have been made also in the packaging of the miniaturized pressure sensors, and ST Microelectronics introduced the smallest 2 x 2 x 0.7 mm package in 2015 implemented in a full molded plastic package that does not make use of serial dispensing of soft glue for mechanical decoupling. As described in [11], batch processing has been made possible using standard molding compounds by transferring the mechanical decoupling spring inside the MEMS using a suspended membrane enabling full molded packaging for a stress-sensitive device like the pressure sensor. The above-mentioned process flows will be covered more in detail in the following sections of this chapter.

STMicroelectronics Patented technology for pressure sensor membrane formation

Monolithic monocrystalline silicon with hermetic cavity

State of the art technology

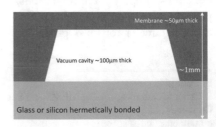

Silicon membrane bonding with glass or silicon wafer to form the cavity

Fig. 16.1 The two drawings are not in scale. On the left, a membrane manufactured with VENSENS™ process, and on the right, the same membrane realized with standard bulk micromachining process. VENSENS™ process does not require any wafer-to-wafer bonding step to seal the cavity

16.2   Conventional Piezoresistive and Capacitive Pressure Sensors: Description and Specification

Several types of pressure sensors can be built using MEMS techniques. According to the working principle, pressure sensors can be divided into piezoresistive, capacitive, optical fiber, resonant, and piezoelectric types. Here, we will discuss two of the most common: piezoresistive and capacitive. In both, a flexible layer (membrane) is created which acts as a diaphragm that deflects under pressure. In a capacitive sensor (Fig. 16.2), conducting layers are deposited on the diaphragm and the bottom of a cavity to create a capacitor. The capacitance is typically a few picofarads. Deformation of the diaphragm changes the spacing between the conductors and hence changes the capacitance. In a piezoresistive sensor (Fig. 16.3), conductive sensing elements (implanted silicon resistances) are fabricated directly on to the diaphragm. Changes in the resistance of these conductors provide a measure of the applied pressure. The change in resistance is proportional to the strain, which affects the silicon bandgap. The resistors are connected in a Wheatstone bridge network, which allows very accurate measurement of changes in resistance. The piezoresistive elements can be arranged so that they experience opposite strain (half are stretched and the other half are compressed) to maximize the output signal for a given pressure.

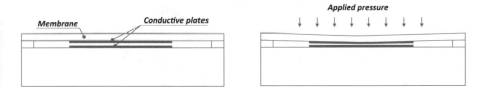

Fig. 16.2   Structure of conventional capacitive sensor

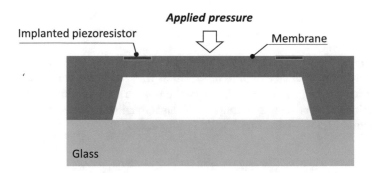

Fig. 16.3   Conventional piezoresistive pressure sensor

16.2.1 Piezoresistive Effect and Conventional Piezoresistive Pressure Sensors

Piezoresistive pressure sensors exploit the variation of resistance of a semiconductor layer under the application of an external stress. Piezo resistivity of silicon arises from a change in the electronic structure, a deformation of the energy bands because of applied stress leading to a modification of the charge carriers' configuration. The piezoresistive coefficient is the relative change of the resistance value of the resistor $\Delta R/R$ for an applied stress. This term is modulated [6] by many variables such as the doping level and type of the semiconductor and the orientation of the piezo resistor with respect to the crystal lattice (Fig. 16.4). In Figs. 1.1 and 1.2 of Chap. 1 of this book, the crystal structure of silicon and the most used crystal planes are represented.

Piezoresistive Stress on a solid can be expressed in the form of a second rank tensor with nine components (3 normal and 6 shear components) which can be further simplified to six independent components using the conditions of force balance [1, 13]. For crystalline symmetry in silicon, the piezoresistive coefficient matrix can be reduced to have only three nonzero independent components of piezoresistive coefficient tensor:

$$
\pi =
\begin{bmatrix}
\pi_{11} & \pi_{12} & \pi_{12} & 0 & 0 & 0 \\
\pi_{12} & \pi_{11} & \pi_{12} & 0 & 0 & 0 \\
\pi_{12} & \pi_{12} & \pi_{11} & 0 & 0 & 0 \\
0 & 0 & 0 & \pi_{44} & 0 & 0 \\
0 & 0 & 0 & 0 & \pi_{44} & 0 \\
0 & 0 & 0 & 0 & 0 & \pi_{44}
\end{bmatrix}
$$

The piezoresistive effect to transduce mechanical strain into resistivity on silicon can be expressed as the sum of the transversal and longitudinal components on the resistor:

$$
\frac{\Delta \rho}{\rho} = \pi_l \epsilon_l + \pi_t \epsilon_t
$$

where π_l and π_t are the longitudinal and transverse piezoresistive coefficients, and ϵ_l, ϵ_t denote longitudinal and transversal strain with respect to the resistor's current flow direction. Depending on the crystallographic orientation of the resistors, the relationship between the transversal and longitudinal stress and the three fundamental piezoresistive coefficients in silicon are:

$$
\pi_l = \frac{\pi_{11} + \pi_{12} + \pi_{44}}{2} \approx \frac{\pi_{44}}{2}, \quad \pi_t = \frac{\pi_{11} + \pi_{12} - \pi_{44}}{2} \approx -\frac{\pi_{44}}{2}.
$$

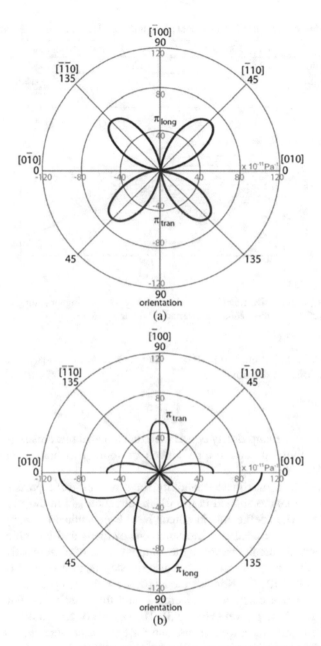

Fig. 16.4 Room temperature piezoresistive coefficients in the (100) plane of (**a**) p-type silicon (**b**) n-type silicon. These graphics predict piezoresistive coefficients very well for low doses [6]

Table 16.1 Piezoresistive tensor components in lightly doped silicon, room temperature [1]

	$\pi_{11}(10^{-11}Pa^{-1})$	$\pi_{12}(10^{-11}Pa^{-1})$	$\pi_{44}(10^{-11}Pa^{-1})$
p-Si (7.8 Ω-cm)	6.6	−1.1	138.1
n-Si (11.7 Ω-cm)	−102.2	53.4	−13.6

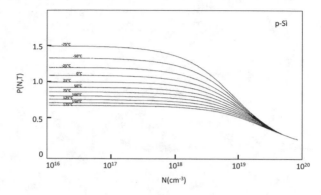

Fig. 16.5 The piezoresistive effect is dependent on temperature; a higher doping level reduces the piezoresistive effect, but also reduces the variation with temperature

for resistors aligned along the <110> crystal direction. For resistors aligned in the <100> direction, the relations are:

$$\pi_l = \pi_{11}, \pi_t = \pi_{12}$$

The values of the empirically determined fundamental piezoresistive coefficients for high resistivity p- and n-type silicon at room temperature are shown in Table 16.1.

Consequently, to maximize sensitivity for a piezoresistive pressure sensor, n-type piezo resistors are aligned in the <100> direction and in the <110> direction when p-doped. The coefficients in silicon have been studied by many researchers since Smith [1] reported them. Measurement methods and test chips have been refined to correctly predict not only the nominal value of piezoresistive coefficient values along the crystal directions but also the dependence from temperature and doping levels [4, 6, 12, 13]. Kanda's model [6] theoretically predicts the observation of the piezoresistive coefficients' decrease, and thus sensitivity, with increasing temperature and doping level (Fig. 16.5). The control of temperature, therefore, is a key factor for piezoresistive sensors, and for this reason, a temperature sensor is typically coupled with the piezoresistive pressure sensor.

Piezoresistors resistors are sensitive to temperature variation, which changes the mobility and number of carriers, resulting in a change in conductivity (or resistivity)

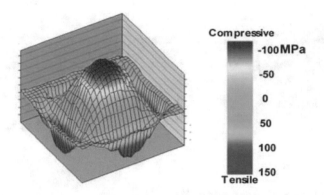

Fig. 16.6 Rectangular-shaped membrane stress distribution: The maximum values are at the edge of the membrane

and piezoresistive coefficients (sensitivity). In Fig. 16.5 [6], the piezoresistive coefficient of a p-type resistor is plotted as a function of doping level and temperature.

Piezoresistive Pressure Sensors (Conventional) Piezoresistors are usually fabricated on a thin membrane to amplify the external stress effect on them and thus increase the sensitivity of the system (Fig. 16.3). In fact, when a pressure load is applied, the membrane undergoes a deformation with a specific geometrical stress distribution depending on the membrane thickness and layout. For example, thinner membranes allow to achieve higher sensitivity values. A rectangular-shaped membrane evidences a stress distribution with maximum values at the edge of the membrane as reported in Fig. 16.6; for this reason, in this case, piezoresistors are typically placed at the membrane edges. In literature, different silicon membranes fabrication processes are reported. These include silicon substrates' anisotropic etching by TMAH or KOH with an etch stop layer [7]; membrane fabrication by the release of a buried sacrificial oxide on an SOI substrate [8]; and membrane fabrication by porous silicon [10].

The piezoresistors are connected in a Wheatstone bride configuration so that, under the application of an input voltage V_{in}, the change of resistance is translated into an output voltage V_{out}. Two resistor designs are implemented in the bridge configuration: One type will increase its resistance by means of applied pressure, while the other type will decrease it (Fig. 16.7).

The output voltage V_{out} across the Wheatstone bridge circuit is given by:

$$V_{out} = (\Delta R/R) \cdot V_{in}$$

$$\Delta R/R = \pi_l \sigma_l + \pi_t \sigma_t$$

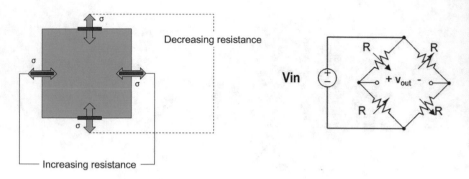

Fig. 16.7 Membrane and Wheatstone bride

where σ_l is the longitudinal stress parallel to the current direction in the resistance, σ_t is the transverse stress perpendicular to the current direction, π_l is the longitudinal piezoresistive coefficient of silicon, and π_t is the transverse piezoresistive coefficient of silicon.

16.2.2 Capacitive Pressure Sensors

Capacitive sensors (Fig. 16.2) measure the deflection of a membrane under a pressure load due to the variation of the gap thickness t_{gap}. In a similar way to static capacitors, the relation is:

$$C(P) = \varepsilon_0 \varepsilon_r \frac{A}{t_{gap}(P)} \sim \frac{1}{t_{gap}(P)}$$

where C is the measured capacitance value, ε_0 the free space dielectric constant, and ε_r is the cavity reference permittivity. A is the electrode area of the capacitor, and typically one of the electrodes is movable on top of the reference cavity. For a parallel plate capacitor, the capacitance is:

$$C(P) = \int \int \frac{\varepsilon_0 \varepsilon_r}{t_{gap} - w(r, \theta, P)} r\, dr\, d\theta$$

with $w(r, \theta, P)$ being the deflection of the membrane in circular coordinates. For circular membranes with uniform thickness, the analytical expression gives:

$$C(P) = \int_0^{2\pi} \int_0^R \frac{\varepsilon_0 \varepsilon_r}{t_{gap} - \frac{P(R^2 - r^2)}{128R}} r\, dr\, d\theta$$

Solving the integral gives:

Table 16.2 Piezoresistive & capacitive transducer comparison

Piezoresistive approach: Advantages	Challenges
Wide dynamic range Good linearity Large pressure sensitivity of monocrystalline silicon membranes Low voltage DC power supply	Implicit second-order dependence on the temperature of the piezoresistive coefficient must be compensated
Capacitive approach: Advantages	**Challenges**
Large pressure sensitivity Simpler analog front end compared to the piezoresistive sensors Lower spurious dependence on temperature than piezoresistive sensors	Vacuum sealing of the reference cavity Natural nonlinearity over pressure to be compensated Dynamic range Influenced by parasitic capacitance

$$C(P) = 8\pi\varepsilon_0\varepsilon_r\sqrt{\frac{2R}{Pt_{gap}}}\tanh^{-1}\left(\frac{R^2}{8}\sqrt{\frac{P}{2Rt_{gap}}}\right)$$

The analytical form can be expanded in Taylor series [15] to predict the membrane deflection versus pressure analytically. From the analytical expression and their natural inverse dependence on t_{gap}, the nonlinearity of the relation between the output capacitance and the applied pressure is expected. In some solutions, it has been proposed to reduce the nonlinearity at the expense of decreased sensitivity by using membranes in contact mode [16] or bossed diaphragms that present a less distorted moving plate under an applied load [17]. Similarly to the microphone membranes, the pull-in phenomenon that occurs when the electrostatic force becomes larger than the elastic force of the membrane reduces the dynamic range and must be considered in the dimensioning of the reference gap thickness. The increased pressure sensitivity and decreased temperature sensitivity are among the advantages of capacitive pressure sensors over piezoresistive pressure sensors. However, before choosing the transducer's physical principle, an accurate comparison on the linearity of the device on full scale must be done to not transfer the burden on calibration if the high sensitivity is of primary importance in the application. Also, to take full advantage of the better behavior over temperature, the capacitive solution must be sustained by an excellent mechanical decoupling in the back-end packaging method. Among the disadvantages of capacitive solutions is that large membranes bring large signal loss due to parasitic capacitance which should be taken into consideration. A summary of the advantages and disadvantages of both piezoresistive and capacitive transducers is compared in Table 16.2 [13, 14].

16.3   VENSEN Technology: Architecture and Schematic Process Flow

STMicroelectronics developed a proprietary technology named VENSEN for the fabrication of piezoresistive pressure sensors built on thin monocrystalline silicon membranes. In the following paragraphs, a schematic description of the VENSEN technology architecture and process flow will be presented. An innovating technology architecture named Bastille based on VENSEN fabrication process will be presented, which allows to integrate high-performance piezoresistive pressure sensors in small and low-cost LGA packages. The VENSEN process enables the fabrication of monocrystalline silicon membranes above an air cavity with controlled dimensions. The name is inspired by Venezia, a city whose basement is built on wooden pillars. In a similar way, the surface of the sensing membrane is built on silicon pillars, which is one of the key steps for the construction of the core of the pressure sensing element. Starting from monocrystalline silicon substrates, three main process steps contribute to the realization of the silicon membrane. The first step is a Deep Reactive Ion Etching where the silicon pillars are fabricated; the pillars array area defines the in-plane extension of the membrane. The second step is an epitaxial silicon growth in order to seal the top area of the cavity and define the membrane final thickness. The third step is an annealing at high temperature (>1100 °C) in hydrogen atmosphere, which enables the migration of the buried silicon pillars and creates the final cavity and the membrane. Membrane fabrication is reported in Fig. 16.8 by a sequence of SEM cross pictures at the three main fabrication steps.

To transduce the external pressure applied to the membrane into an electrical signal, four piezoresistors are manufactured at the membrane edges by ion implantation. Figure 16.9 shows the typical layout of a VENSEN sensor with an SEM cross-section of the single crystal membrane.

To compensate for temperature induced drift on sensor output and to provide a more accurate and stabile output control, a temperature sensor can be integrated

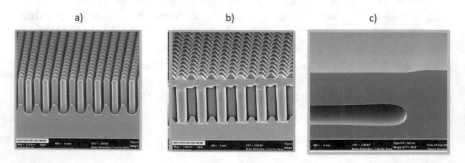

Fig. 16.8 Key steps of the membrane creation process: (**a**) pillars etch, (**b**) epitaxial growth, and (**c**) annealing

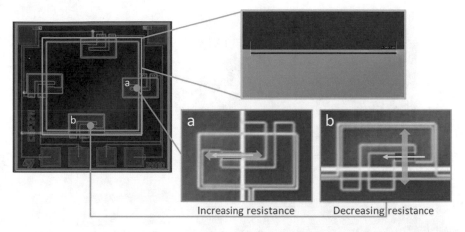

Fig. 16.9 Layout of a VENSEN sensor with an SEM cross-section of the single crystal membrane

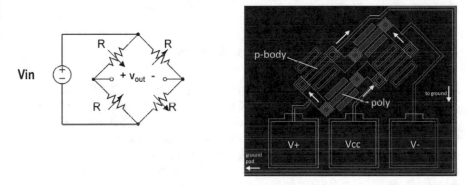

Fig. 16.10 Temperature sensor integrated on the VENSEN Platform

on the VENSEN Platform. The temperature sensor structure is a Wheatstone bridge consisting of a couple of p–Si diffused resistors (with positive temperature coefficient of resistance—TCR) and a couple of polysilicon resistors (with negative TCR). With this configuration, the resistance variation of the pressure sensor due to temperature change will be adequately compensated (Fig. 16.10).

The pressure and temperature sensors are based on the same architecture, but the resistance variation is due to different effects: piezoresistance in the first, and thermal variation of resistance in the latter. Figure 16.11 shows the layout of a VENSEN sensor where the different device components are reported: the sensing membrane, the implanted piezoresistors the temperature sensor components, and the pad area.

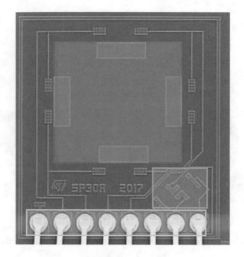

Sensing Membrane

PZR area

Temperature sensor

Pad area

Fig. 16.11 Layout of a VENSEN sensor

16.3.1   VENSEN Schematic Process Flow

In the following paragraph, a schematic process flow of VENSEN technology is described: At the very beginning of the process flow, the silicon membrane is fabricated; afterward, all the diffused areas will be defined with subsequent implantation steps for piezo resistor fabrication, contact areas in silicon substrate, and the resistors for temperature sensing element. The temperature sensor will be completed through poly silicon deposition and definition. Once all components are fabricated, an oxide layer will be deposited, and contacts will be opened before the routing definition with a metal deposition and definition. As last step, the passivation will be deposited on top of the metal and it will be etched away on the pad opening area for latter connection with ASIC device.

16.3.2   An Industrialization Challenge: Piezoresistive Pressure Sensor in a Full Molded Package, "the Bastille Architecture"

The integration of MEMS device on package is a critical point for pressure sensors since stress contribution from package must be accurately managed to avoid an undesired contribution to the overall pressure sensing. A typical package architecture used for pressure sensors is a holed cavity package (Fig. 16.12), in which the sensing element is protected from the external world by the package but connected though a sensing hole on it, to sense the external pressure.

Sensing Membrane fabrication through VENSEN process	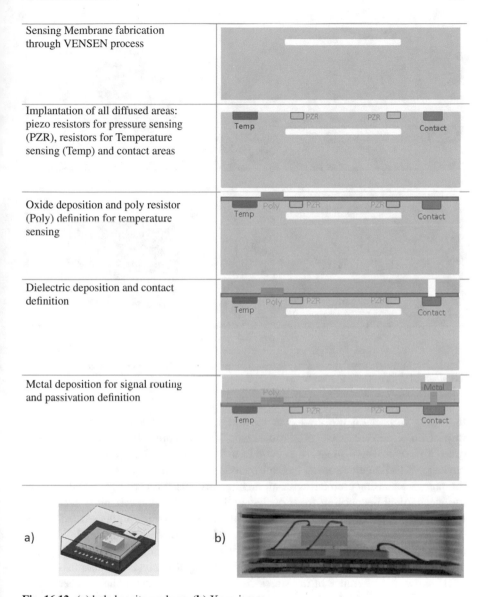
Implantation of all diffused areas: piezo resistors for pressure sensing (PZR), resistors for Temperature sensing (Temp) and contact areas	
Oxide deposition and poly resistor (Poly) definition for temperature sensing	
Dielectric deposition and contact definition	
Metal deposition for signal routing and passivation definition	

Fig. 16.12 (**a**) holed cavity package; (**b**) X ray image

In this package solution, the mechanical properties of the adhesive used are extremely relevant because the stress affects the device performances. From the other side, the dimension scalability especially required in consumer market is difficult with this kind of package due to intrinsic requirements, as a dedicated space on the top area for wire bonding. A full molded package has great advantage in terms of scalability in comparison to cavity packages, but the drawback is that

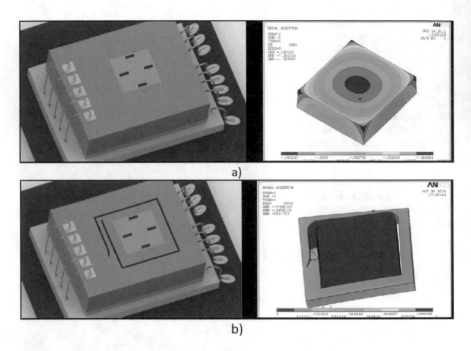

Fig. 16.13 Difference of the stress distribution in the membrane area: (**a**) in conventional membrane, (**b**) in suspended membrane

the stress impact on the device is very high, which has an extremely negative impact on a piezoresistive pressure sensor. To overcome this obstacle and integrate a piezoresistive pressure sensor in a full molded LGA (Land Grid Array) package, STMicroelectronics has introduced a revolutionary method to suspend the pressure sensing element with an architecture based on springs, which permits the mechanical decoupling from package-related stress. A mechanical decoupling system is fabricated all around the sensing element with springs etched in the substrate which relax elastically the stress induced by the package and reject the undesired contribution of the intrinsic stress and thus enhance the sensing capacity and stability of external pressure. The finite element analysis simulations in Fig. 16.13 show the difference of the stress distribution in the membrane area between a soft glue (Young modulus E = 100 kPa) die-attached sensor on a substrate and the suspended membrane mounted using die attach film (E = 2GPa).

The mechanical decoupling system is integrated on the same chip area together with the sensing element. The resultant architecture is extremely insensitive to the stress applied by package and enables the integration of the sensing element in a highly stressful full molded LGA package. The above architecture provides enormous flexibility in terms of packaging. In Fig. 16.14, a schematic cross-section is shown with the final package structure with reduced 2 x 2 x 0.7 mm dimensions [18]. Die attach film bonds the MEMS to the ASIC in a glue-less step. The MEMS

Fig. 16.14 Schematic cross-section is shown with the final full molding package

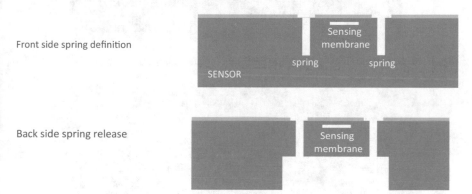

Fig. 16.15 Bastille modified (first version): dual etching process from both sides, front and back

is wire bonded to the ASIC, on top of the substrate. It consists of a wafer-to-wafer bonded cap to protect the sensing area and to enable full molded packaging.

Mechanical decoupling architecture integration on MEMS Pressure Sensors by STM has been implemented following a continuous technological development path to achieve a higher level of device compactness.

Bastille Modified (First Version)—Front and Rear Architecture The first release of this system consists of a dual etching process from both sides, that is, front and back; during the front side etch, the springs are predefined by means of a deep silicon dry etching, and afterward, they will be released in a following step by means of a backside etching step (Fig. 16.15).

The final device configuration is based on a sensor wafer sandwiched between a top cap and a bottom handle wafer in a triple wafer architecture as per Fig. 16.16. The bottom cap is a handling substrate which enables the subsequent attach of the MEMS on the ASIC. The top cap implements pressure ports with micrometric diameter to communicate with the external environment and acts as a mechanical stopper for the sensor in the Z axis at the same time (Figs. 16.16 and 16.17).

Double Membrane "Bastille" Architecture From this initial successful configuration, the aim has been to simplify the architecture and integrate the fabrication of the springs into a single wafer by means of a **Double Membrane** concept (Fig. 16.18). The key point is to exploit the ST Proprietary VENSEN process to fabricate

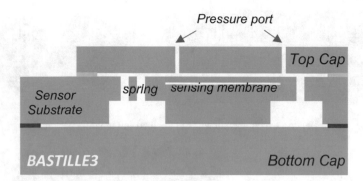

Fig. 16.16 Schematic cross-section of MEMS with 3-wafer configuration

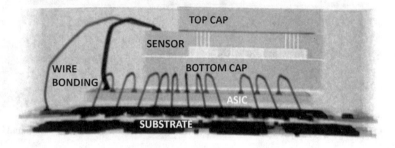

Fig. 16.17 Cross-section with X-rays of the final package configuration

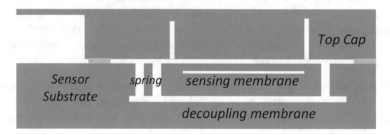

Fig. 16.18 Schematic cross-section of MEMS with two-wafer "Bastille" configuration

not only the sensing membrane, but also an additional membrane for mechanical decoupling [11, 19].

The result is a package-decoupled sensor in a single-wafer approach for both the pressure-sensing element and the decoupling spring. Advantages of the Double Membrane configuration are as follows: the stack configuration is reduced from three wafer to two wafer with consequent process simplification; the stack simplification opens the possibility for future device scalability by means of thickness reduction; and furthermore, since the mechanical decoupling structure is fabricated at the very beginning of the process flow, as it will be shown below, the possibility to

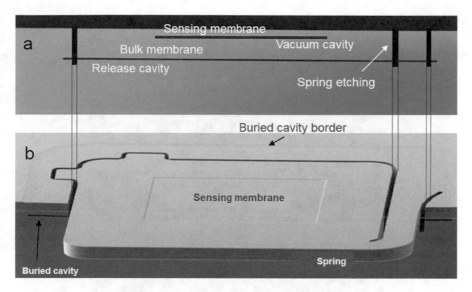

Fig. 16.19 (a) SEM view of the dual cavity forming process. (b) Spring etches on top of membrane with 3D SEM view of the released structure

integrate it on other devices different from pressure sensors can open an incredible space for innovative frontend—backend integration scenarios. In Fig. 16.19a and b, a cross-section and a tilted view of the double membrane structure is presented; the membrane for mechanical decoupling, also named bulk membrane here, lies on a buried level in the silicon substrate with the silicon etched around it down to its cavity to define the springs, which suspend and decouple the sensing membrane on top of it.

Bastille Double Membrane—Schematic Process Flow The process flow of the dual membrane process is schematically shown in Fig. 16.20. The starting step is a 50-μm-thick bulk VENSEN membrane (a) which is followed by a 10-μm-thin membrane (b). Afterward, the piezo resistors are implanted, and the temperature sensor is fabricated by means of diffused resistor implantation and polysilicon definition, followed by the passivation, and contact with metal paths (c). Next, by etching the bulk membrane in (d), the springs are defined, and the structure is released, resulting in a suspended architecture without the use of a release oxide. Finally, a silicon cap is bonded on top of the sensor wafer, and the pressure ports are defined by means of deep silicon etching together with the pad window.

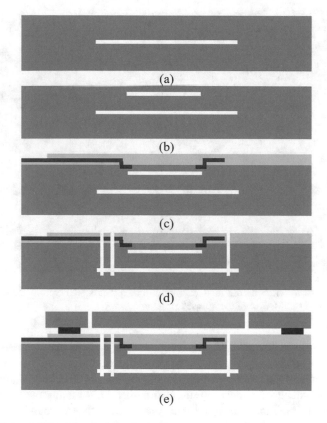

Fig. 16.20 Process flow of the dual membrane sensor

References

1. Smith, C. S. (1954). Piezo resistance Effect in Germanium and Silicon. *Physics Review, 94*(1), 42–49.
2. Chollet, F., & Liu, HB, "A (not so) short Introduction to Micro Electromechanical Systems," version 5.4, 2018., http://memscyclopedia.org/introMEMS.html
3. Kulite, "Celebrating Fifty Years", http://www.kulitesensors.com.cn/reference/KuliteHistory.pdf
4. Tufte, O. N., Chapman, P. W., & Long, D. (1962). Silicon diffused-element piezoresistive diaphragms. *Journal of Applied Physics, 33*, 3322.
5. Hatanaka, K., Sim, D. Y., Minami, K., & Esashi, M., "Silicon diaphragm capacitive vacuum sensor", in Tech. Dig. 13th Sensor symp., 1995, pp. 37–40.
6. Kanda, Y., "A graphical representation of the piezoresistance coefficients in silicon", IEEE Trans Electron Devices. 1982; 29:64–70. [Google Scholar].
7. Price, J., "Anisotropic etching of silicon with KOH-H2O isopropyl alcohol", ECS semiconductor silicon, pp. 339–353 (1973).
8. Jackson, T., Tischler, M. A., & Wise, K. D. (1981). An Electrochemical P-N Junction Etch-Stop for the Formation of Silicon Microstructures. *IEEE Electron Device Letters, EDL-2*(2), 44–45.

9. Barlocchi, G., Corona, P., Faralli, D., & Villa, F. F. (2005). Method for forming buried cavities within a semiconductor body, and semiconductor body thus made. *US Patent, 7*, 811–848.

10. Armbruster, S., Schäfer, F., Lammel, G., Artmann, H., Schelling, C., Benzel, H., Finkbeiner, S., Lärmer, F., Ruther, P., Paul, O. (2003). A novel micromachining process for the fabrication of monocrystalline Si-membranes using porous silicon. In: Proceedings of 12th international conference on solid-state sensors, actuators, and microsystems, Boston, pp 246–249.

11. Duqi, E., Baldo, L., Urquia, M. A., & Allegato, G. (2019). A Piezoresistive Mems Barometer with Thermomechanical Stress Rejection. In *2019 20th International Conference on Solid-State Sensors, Actuators and Microsystems & Eurosensors XXXIII (TRANSDUCERS & EUROSENSORS XXXIII)* (pp. 659–662). Berlin, Germany. https://doi.org/10.1109/TRANSDUCERS.2019.8808357.

12. Richter, J., Pedersen, J., Brandbyge, M., Thomsen, E. V., & Hansen, O. (2008). Piezoresistance in p-type silicon revisited. *Journal of Applied Physics, 104*(2), 023715.

13. Doll, J. C., & Pruitt, B. L. (2013). *Piezoresistor Design and Applications* (1st ed.). New York, NY, USA: Springer.

14. Kumar, S. S., & Pant, B. D. (2014). Design principles and considerations for the 'ideal' silicon piezoresistive pressure sensor: a focused review. *Microsystem Technologies,20*, 1213–1247. https://doi.org/10.1007/s00542-014-2215-7.

15. Eaton, W. P., & Smith, J. H. (1997). Micromachined pressure sensors: review and recent developments. Smart Mater. *Structure, 6*(5), 530.

16. Cho, S. T., Najafi, K., & Wise, K. D. (1990). Secondary sensitivities and stability of ultrasensitive silicon pressure sensors. In *Technical Digest, IEEE Solid-State Sensor and Actuator Workshop* (pp. 184–187). Hilton Head '90.

17. Zhang, Y., & Wise, K. D. (1994). An ultra-sensitive capacitive pressure sensor with a bossed dielectric diaphragm. In *Technical Digest, Solid-State Sensor and Actuator Workshop* (pp. 205–208). Hilton Head '94.

18. https://www.st.com/en/mems-and-sensors/lps22hh.html

19. Baldo, L., Duqi, E., & Villa, F. (2015). *Micro-electro-mechanical device and manufacturing process thereof*. US Patent, US10150666B2.

Chapter 17
Environmental Sensors

Giuseppe Bruno and Michele Vaiana

17.1 Humidity Sensors

17.1.1 Introduction

Humidity is the measure of the amount of water present in a gas like air. Humidity sensors are the devices aimed to measure this physical parameter.

Humidity sensor is one of the most important devices widely used in consumer, industrial, biomedical, environmental and other applications for measuring and monitoring humidity.

Humidity sensors can be divided into two groups, as each category uses a different method to calculate humidity: relative humidity (RH) sensors and absolute humidity (AH) sensors. Relative humidity is calculated by comparing the live humidity reading, at a given temperature, to the maximum amount of humidity for air, at the same temperature. RH sensors must therefore measure temperature in order to determine relative humidity. In contrast, absolute humidity is measured without reference to temperature.

In this chapter will be first introduced the definition of relative humidity (%RH) and its relationship with the temperature.

The two most common RH sensors are: capacitive and resistive sensors.

In the next section, we introduce the working principle of a humidity sensor, focusing mainly on capacitive sensor describing two different approaches. The first is a planar electrodes sensor that uses two different dice inside a package, the sensor die and ASIC die, the second is an interdigitated capacitive sensor integrated with

G. Bruno (✉) · M. Vaiana
ST Microelectronics, Analog MEMS and Sensors Group, MEMS Technology and Design R&D, Catania, Italy
e-mail: giuseppe.bruno@st.com; michele.vaiana@st.com

© Springer Nature Switzerland AG 2022
B. Vigna et al. (eds.), *Silicon Sensors and Actuators*,
https://doi.org/10.1007/978-3-030-80135-9_17

the electronic part in the same die. The last approach is, of course, preferable in terms of manufacturing process cost, enabling the sensor to consumer market.

17.1.2   Relative Humidity Definition

In a closed chamber containing liquid water, the kinetic energy of the water molecules depends on the Temperature: the higher the temperature, the higher the energy. At a given temperature, the water molecules on the liquid surface with a kinetic energy high enough to overcome intermolecular forces can escape from the liquid and remain trapped in the space above the liquid. As the gaseous molecules bounce around, some of them hit the surface of the water and may get trapped again. After some time, the number of water molecules leaving the surface of the liquid water will be equal to the number rejoining it. At the end, there will be an equilibrium where the number of gaseous water molecules remains statistically constant.

Since the molecules bounce around the closed chamber, they will hit the wall exerting a pressure P.

At equilibrium, this pressure is called saturated vapour pressure (P_s).

The vapour pressure P_w can be expressed by the Clapeyron relation:

$$\frac{dP_w}{dT} = \frac{L_v}{T \Delta V}$$

(17.1)

where L_v is the latent heat, T the absolute temperature in Kelvin and ΔV the volume change of the phase transition.

Assuming the specific volume of the liquid much smaller than that of the gas and assuming the vapour gas ideal ($PV = nRT$), Eq. 2.1 can be approximated by the Clausius-Clapeyron equation

$$\frac{dP_v}{dT} = \frac{(L_v P_w)}{(R_v T^2)}$$

(17.2)

where R_v is the water vapour gas constant.

Relative humidity (% RH) is defined as the ratio of the partial vapour pressure in air P to the saturated vapour pressure at a given temperature t, $Ps(t)$:

$$\%RH = \frac{P}{P_s(T)} \cdot 100$$

(17.3)

Therefore, the relative humidity in the closed chamber is exactly 100% RH when equilibrium is achieved.

The value of relative humidity change varying the temperature. The higher is the relative humidity, the higher is the relative humidity change varying the temperature.

Note that at a relative humidity of about 90% RH at ambient temperature, a temperature deviation of $\Delta T = 1^{\circ}$ C results in a change of up to 5% RH.

17.1.3 Principle of Operation

Adsorption of a gas on a solid surface is a consequence of surface energy, like surface tension. Most atoms that constitute a solid are bound on all sides by other atoms in the bulk of the solid. The atoms on the surface of the solid are incompletely bound. Due to van der Waals forces of interaction, these surface atoms are more reactive, and they attract vapour (but also gas and liquids) to satisfy the imbalance of atomic forces. (Fig. 17.1)

The adsorption phenomena is well described by the work of Dubinin-Astakhov considering the following:

$$A = R * T * \ln\left(\frac{P_s}{P}\right) \tag{17.4}$$

Where A is the differential molar work needed to transport one mole of the adsorbate (gas) to the surface of an infinitely large adsorbent based on Polanyi theory, P is the equilibrium pressure at temperature T, P_S is the saturated vapour pressure of the adsorbate, so the ratio P/P_s represents the relative humidity (% RH), and R is the ideal gas law constant (0.008314 Kj/mole) and T is the temperature in Kelvin.

Dubinin-Astakhov [1, 2, 3] states the following relation between v_0 (limiting pore fraction volume) and v (fraction volume of adsorbate):

$$v = v_0 * \exp\left[-\left(\frac{RT\ln\left(\frac{P_s}{P}\right)}{E_a}\right)\right] \tag{17.5}$$

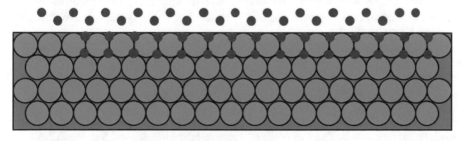

Fig. 17.1 The surface molecules (in green) are experiencing a bond deficiency; thus, it is energetically favourable for them to adsorb molecules (in red)

Where n is an empirical factor function of the mixture of adsorbate and adsorbent and E_a is the activation energy, normally below 40Kj/mol for physical adsorption.

Different materials have been used for manufacturing humidity sensors.

Ceramic materials are known to change their electrical properties, such as resistance, in the presence of humidity. The first initial adsorbed layer of water molecules is stable and ordered. The conductivity mechanism starts upon the second physiosorbed layer. Consequently, at low humidity, the conductivity is almost unchanged. For this reason, most ceramic materials are not sensitive to RH below 20%.

Semiconductors are promising candidates for sensing humidity with a high precision. At higher temperatures, oxygen molecules react with the surface molecules and are reduced to oxygen ions that are chemisorbed to the semiconductor surface. Thus, electrons are accumulated at the surface, which results in an increase of the number of holes, leading to band bending at semiconductor surfaces. Water molecules, in the nearby area, are physiosorbed to the semiconductor surface and replace the oxygen ion sites. The surface electrons depletion is neutralized. Hence, conductivity is increased. Semiconductor based humidity sensors are usually driven at temperatures around 200 °C because the signal change is higher and linear. This high temperature limits their applications.

The third type of humidity sensing material is an organic material, for example, a polymer. Polymers are the best candidate in many sensing applications due to low cost, commercial availability and easy deposition on different transducers. Photosensitive polymer adds advantages related to rapid polymerization, elimination of volatile organic solvents usage and reduction of technological process steps. A polymeric layer in contact with water vapour shows a change in its conductivity or its dielectric constant ε and this property allow using a capacitive sensor to sense the relative humidity in the ambient. (Fig. 17.2)

17.1.4 Capacitive Humidity Sensor with Planar Electrodes

The simple capacitive sensor structure is realized using two planar metal electrodes with a dielectric in between. Capacity is function of the polymer relative dielectric constant ε_r (Fig. 17.3):

$$C = \varepsilon_0 \varepsilon_r \frac{A}{d} \tag{17.6}$$

Since this structure is sensitive to electro-mechanical stress and aging, a differential structure can be used to realize a relative humidity sensor. See Fig. 17.4.

A capacitive sensor has the top plate holed in order to expose the polymeric material to the environment and permit the H_2O molecules to be adsorbed. On the other hand, the polymeric material in the reference sensor is protected from

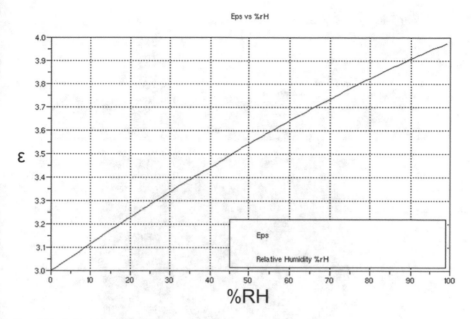

Fig. 17.2 Polymer dielectric constant versus relative humidity

$$C = \varepsilon_0 \varepsilon_r \frac{A}{d}$$

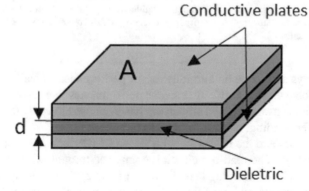

Fig. 17.3 Conventional capacitor with planar metal electrodes and dielectric in the middle

the environment by a passivated top metal plate. Passivation acts as a barrier for humidity.

For human comfort, both relative humidity and temperature are essential as temperature is needed to convert relative humidity value in absolute humidity value. Therefore, a very precise temperature sensor is integrated as close as possible to the humidity sensor element.

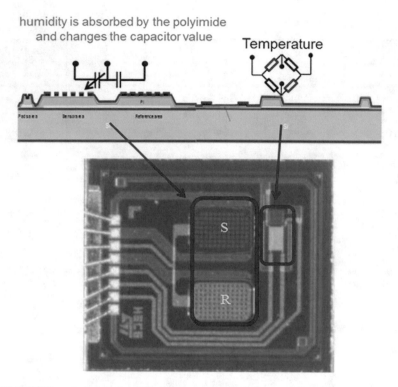

Fig. 17.4 STMicroelectronics humidity sensor top view. A parallel plate humidity sensor consists of a holed parallel plate capacitor S that change its value varying the environment relative humidity and of a fixed reference capacity R. A temperature sensor is also embedded on the silicon as close as possible to the sensor

Package plays a key role on humidity sensor performances. Ideally, package has to guarantee robustness not affecting sensor performances like responsivity. In the case of humidity sensors, a holed silicon cap has been used to cover the sensor with a wafer to wafer bonding, using a dry film as interlayer.

Humidity sensor and the mixed signal ASIC in CMOS technology are housed in an organic package. The read-out circuit is connected through wire bonding to the organic substrate. Full moulding is realized using film assist moulding technique.

Figure 17.5 reports an example of the sensor with cap mounted in stacked configuration with the read-out circuit before the moulding step. The hole to expose the sensing element to the environment and the pads and wires to connect sensor to ASIC are clearly visible.

Figure 17.6 shows the top view of the final sensor full moulded with silicon cap exposed. The hole on the top silicon cap is clearly visible: it allows the humidity sensor to be in direct contact with the environment, optimizing responsivity.

The sensor can be used in a several applications like industrial, home appliances as well as wearable devices thanks to the very small dimensions of few mm^3.

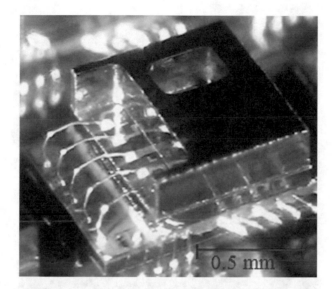

Fig. 17.5 STMicroelectronics humidity sensor bonded in stack up configuration on ASIC chip

Fig. 17.6 STMicroelectronics humidity sensor final package top view. Package size 2 x 2 mm

17.1.5 Interdigitated Capacitive Humidity Sensor

Interdigitated capacitive sensors offer interesting advantages. First, they are based on a polymeric layer deposited at low temperature on top of standard CMOS process

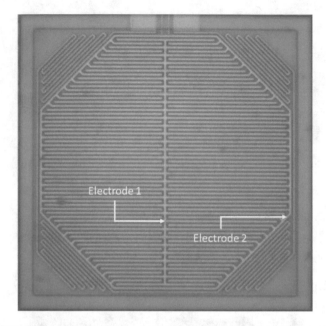

Fig. 17.7 STMicroelectronics interdigitated capacitive structure with planar electrodes configuration

flow. This allows direct integration with the ASIC, processing the sensor signals at the source and decreasing degradation of signal quality.

Moreover, since the capacitors of the sensor are designed as interdigitated structures, the electrode configuration is planar, and this results in a shorter response time of the device. (Fig. 17.7)

In the interdigitated capacitive structure with planar electrodes, the electric field extends into the polymeric material and capacity changes when dielectric constant changes due to water adsorption. Thickness of the polymer and electrodes width and spacing must be properly defined to obtain required sensitivity. Typical values for capacity achievable with the interdigitated structure are in the range of 500 pF, while the sensitivity is in the range of 0.1 fF / %RH. (Fig. 17.8)

In this case, the silicon cap is directly bonded to the ASIC chip through a dry film and the sandwich housed in an organic package with silicon cap exposed.

Some examples of temperature and humidity accuracy of the sensor are reported hereafter. Temperature accuracy is defined as temperature readings of the DUT minus temperature of the reference temperature sensor (NIST certified). Figure 17.9 shows maximum temperature accuracy of the sensor in the temperature range − 40 °C–100 °C.

Figure 17.10 reports max relative humidity %RH accuracy of the sensor at 25 °C. Relative humidity %RH accuracy is defined as relative humidity readings of the DUT minus relative humidity of the reference relative humidity sensor (NIST certified).

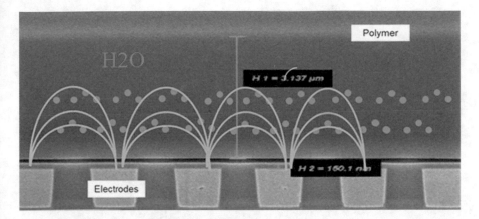

Fig. 17.8 Silicon section of interdigitated electrodes. © 2018 by STMicroelectronics

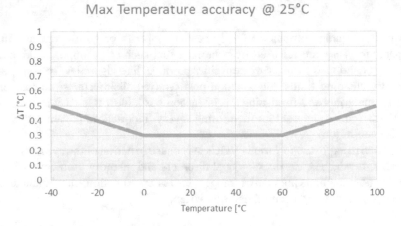

Fig. 17.9 Max Temperature accuracy in the temperature range − 40 °C–100 °C

17.2 VOC GAS Sensor

17.2.1 Air Pollution

Indoor and outdoor air quality is important for healthcare. When air pollutants are delivered to the ambient the air quality decreases. Air pollution has been recognized as one of the ten most death-causing factors, that is, some 500,000 premature deaths worldwide annually.

Substances that affect indoor and outdoor air quality can be solid particles, liquid droplets or gasses. The most important and dangerous substances emitted in the air by human activity are carbon dioxide (CO_2), sulphur oxides (SOx), nitrogen oxides (NOx), carbon monoxide (CO), volatile organic compounds (VOC) particulates.

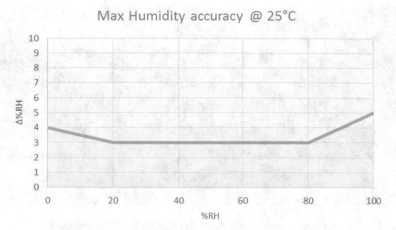

Fig. 17.10 Relative Humidity accuracy of 5 DUTs from 20%RH to 80%RH and T $= 25°C$

Volatile organic compounds easily become vapours or gases even at ambient temperature. They are released from burning fuel such as gasoline, wood, coal or natural gas and also from many consumer products like cigarettes, solvents, adhesives, dry cleaning fluids, glues, wood preservatives, disinfectants, air fresheners, building materials and furnishings, printers and pesticides. It is easy to understand that VOCs indoor and outdoor monitoring is very important to prevent exposure to high concentration.

Short-term exposure to various VOCs may cause irritation of the eyes, respiratory problems, headaches, dizziness, visual disorders and memory problems, while the main long-term effects are nausea, fatigue, loss of coordination, cancer and damage to the liver, kidneys and central nervous system.

17.2.2 Sensors for VOC Detection

Correct evaluation of outdoor and indoor air quality is not an easy task. Air is a quite complicated system subjected to changes even in a short period. Analytical methods and analytical instruments have progressed recently to obtain reliable information on indoor and outdoor air condition and quality, but these methods and instruments imply an high cost of monitoring, therefore limiting their widespread application.

Alternative cheaper methods are based on sensor techniques and especially on chemical sensors. A chemical sensor transforms chemical information from the environment into an analytically useful signal. Main advantages of chemical sensors are low cost of manufacturing and possibility of miniaturization, as well as relatively good metrological parameters, such as sensitivity, selectivity, measurement range, linearity and response time.

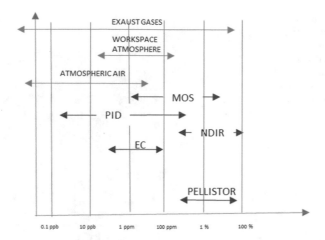

Fig. 17.11 Concentration range of VOCs present in ambient air, indoor air at workplace and exhaust gases and main commercially available sensors for detection and measurement of VOCs. MOS (metal oxide semiconductor), PID (photoionization detectors), NDIR (non-dispersive infrared sensors), EC (electrochemical) and Pellistor (thermal sensors)

Chemical sensors include "electrochemical sensors", semiconductors with solid electrolyte and PID-type sensors (photoionization detectors).

Figure 17.11 reports commercially available sensors designed for detection of VOCs in ambient air, indoor air and exhaust gases [4].

17.2.3 Metal Oxide Semiconductor Sensors

In these sensors, analyte particles diffuse towards the receptor surface, which is a metal oxide normally maintained at suitable temperature using a heater, where they undergo chemisorption. This interaction results in a change of resistance of the receptor element.

Different type of receptor element material can be used like ZnO or SnO2 (type N) – where the resistance changes in case of reducing gas presence – or NiO or CoO (type P), where the resistance changes in case of oxidizing gases presence.

When metal oxide sensor material (typically tin dioxide SnO2-x) is heated at high temperature (e.g. 400 °C) and is ideally exposed to an ambient with oxygen concentration 0%, free electrons flow through the grain boundary of tin dioxide crystals. In clean air (approximately 21% O2), oxygen is adsorbed on the metal oxide surface. With its high electron affinity, adsorbed oxygen attracts free electrons inside the metal oxide, forming a potential barrier (of the order of some electron Volts (eVs) in air) at the grain boundaries. This potential barrier prevents electron flow, causing high sensor resistance in clean air. (Fig. 17.12)

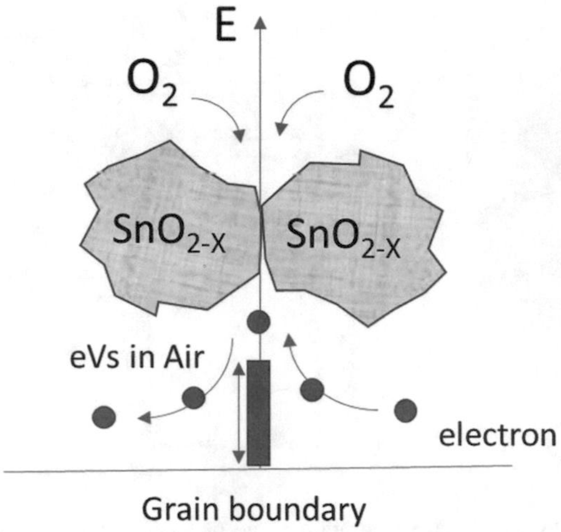

Fig. 17.12 At high temperature, the metal oxide adsorbs oxygen that attracts free electrons inside the SNO$_2$-x forming a potential barrier that prevent electron flow causing high sensor resistance in clean air

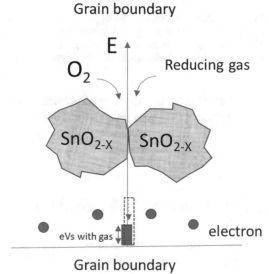

Fig. 17.13 Density of adsorbed oxygen on the tin dioxide surface decreases and the potential barrier is lowered in the presence of a reducing gas like carbon monoxide

When the sensor is exposed to combustible or reducing gas such as carbon monoxide, the oxidation reaction of such gas with adsorbed oxygen occurs at the surface of tin oxide. As a result, the density of adsorbed oxygen on the tin dioxide surface decreases and the potential barrier is reduced, so electrons easily flow through the potential barrier of reduced height and the sensor resistance decreases. (Fig. 17.13)

Gas concentration in air can be detected by measuring the resistance change of MOS-type gas sensors. The chemical reaction of gases and adsorbed oxygen on tin dioxide varies depending on the reactivity of sensing materials and working temperature of the sensor.

In metal oxide sensors, tin dioxide has to be locally heated to several hundred Celsius degrees. Therefore, in order to integrate these gas sensors with electronic

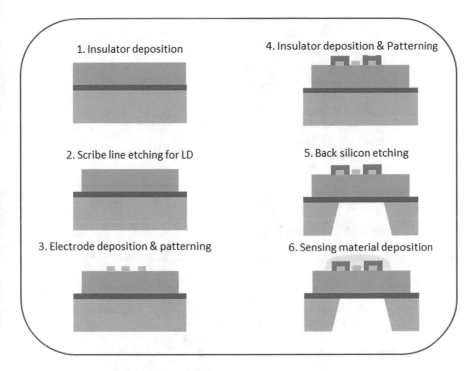

1. Insulator deposition

4. Insulator deposition & Patterning

2. Scribe line etching for LD

5. Back silicon etching

3. Electrode deposition & patterning

6. Sensing material deposition

Fig. 17.14 Schematic process flow description for the MHP

circuits, a micro-hotplate (MHP, which is a MEMS-based heating structure) is required to thermally isolate the sensor from the circuits. Platinum is traditionally used as heater due to linear thermal response up to 600 °C, alternatively TaAl has smaller temperature coefficient.

The resistance R at Temperature T of the heater is

$$R = \text{Rref} \ [1 + \alpha \ (\text{T} - \text{Tref})] \tag{17.7}$$

with a the temperature coefficient of resistance (TCR) of the material.

Schematic process flow for the formation of a closed membrane gas sensor is described in Fig. 17.14.

Figure 17.15 shows a real picture of a MHP. The heater is powered forcing current or voltage through its pads. Metal oxide material is just over the heater and his resistance is sensed through the sensor pads.

Platinum thickness has to be adjusted to reach the suitable resistance value. Normally, for the read out circuit, a low-voltage CMOS technology is used in order to decrease the total power consumption. Therefore, few tens of Ohm is a typical target for resistance value.

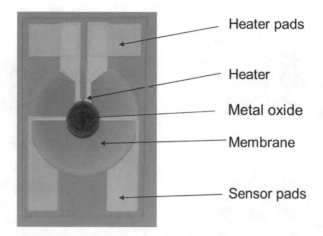

Fig. 17.15 Micro hot plate with screen printed metal oxide on the heater

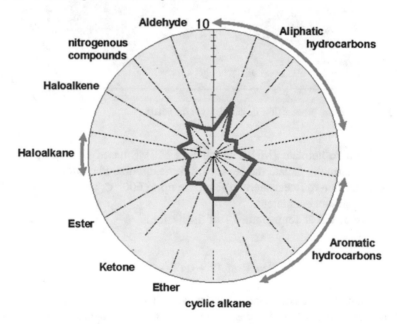

Fig. 17.16 Normalized sensitivity to different gases of a generic air contamination detector using Pd-SnO$_2$ as sensitive material

On top of the heater, a Pd-SnO$_2$ is used as sensitive metal dioxide. A general air contamination detector can be designed adopting Pd-SnO$_2$ at proper operation temperature, with target gases like hydrocarbons, odorous gases and combustible (odourless) gases such as hydrogen. In the Fig. 17.16 is reported the normalized sensitivity to different gas families.

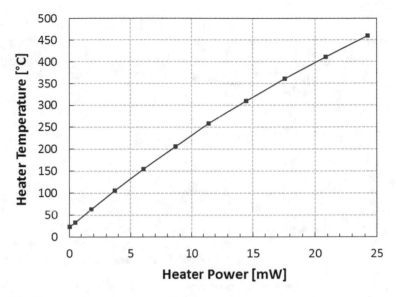

Fig. 17.17 Heater temperature versus applied power

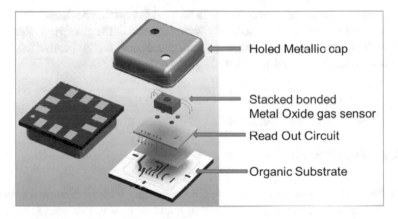

Fig. 17.18 Metal oxide gas sensor and read out circuit housed in a metallic holed cavity organic package. Full dimension is 2.5 mm x 2.5 mm x 0.9 mm

The temperature reached by the heater at different power is hereafter reported. (Fig. 17.17)

In state-of-art sensors, MHP chip is mounted with mixed signal ROC in the same package. ROC drives the MHP at proper temperature and performs resistance readings, converting the value in the digital world for further manipulation. The package uses a holed metallic cap to guarantee good ventilation inside.

Figure 17.18 shows a complete device housed in a cavity package with metallic holed cap.

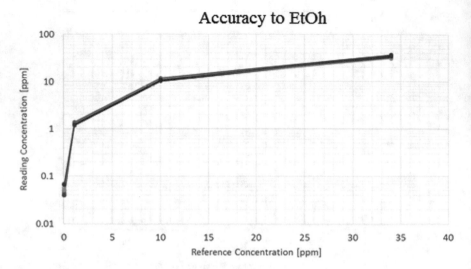

Fig. 17.19 Accuracy vs ethanol of the gas sensor

The form factor enables such device to be used in portable devices.

In Fig. 17.19 is reported the accuracy versus ethanol of the gas sensor. Resistance value of the metal oxide at different gas concentration is red and converted to ppm gas concentration by the ASIC circuit.

17.3 Temperature Sensors

17.3.1 Introduction

While the reader might prefer an introduction starting with an historical overview of the various temperature sensors built to quantify how hot or cold is an object compared to another one (an entire book would not be enough) and the various definitions adopted, a flash back to Lord Kelvin is a necessary starting point in our journey. The concept of temperature is well known and established in daily life: weather forecast, body temperature and sometimes as relative temperature between objects. It is less common and more powerful the absolute or thermodynamic temperature defined by the physicist Lord Kelvin (1824–1907) who defined an absolute zero so called the "infinite cold" [5].

Absolute temperature governs the thermodynamic equilibrium in gases, ideal or not; it defines the radiated power of object at a certain kelvin temperature and defines the energy levels on material regulating the laws behind minority and majority carriers in doped semiconductors.

As consequence, all devices characteristics are affected by absolute temperature and the need of a temperature sensor becomes relevant. Let us focus our interest on the following main items:

1. Temperature Compensation
2. Temperature Monitoring

17.3.2 Temperature Compensation

The first, temperature compensation, falls on the request of a general IC device to maintain, within a specified range of temperature variation, features the device is supposed to have.

In other terms, it is used in a control system feedback to sense the temperature variation and activate a strategy to compensate the variation that otherwise, without compensation, would not respect the specification.

Environmental sensors discussed in the previous paragraphs are good examples of devices, like the pressure sensors, that need temperature information to compensate the temperature-dependent device characteristics. In this case, requirements for the temperature sensor are the monotonicity, time latency, linearity or a well-known non-linearity and noise.

The time latency between the primary sensor and temperature must be minimized, as well as monotonicity is required to design a compensation as a function of the temperature.

It is not intended to be an accurate temperature sensor since its use is merely limited to provide and sense temperature variations to activate the compensation needed from the sensor. Several techniques and approaches are used to design temperature sensor on the IC with advantages and drawback. The following paragraph will summarize the most common and widely used.

17.3.3 Temperature Monitoring

In this case, the temperature sensing is the primary feature of the device. The device will provide the temperature for monitoring purposes or to compensate at higher level the customer application.

In principle, any device, since temperature-dependent, can be used to sense the temperature of the silicon die where it has been realized. Since easier to read out and commonly available in all CMOS technologies, we will limit to introduce an example of resistive temperature sensor.

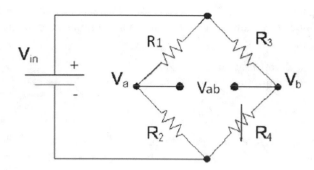

Fig. 17.20 Wheatstone bridge schematic

17.3.4 Resistive Temperature Sensor

Generally, a silicon resistor can be expressed by the formula given below:

$$R(T) = R(T_o) \cdot \left(1 + \Delta T \cdot TC1 + \Delta T^2 + \dots + \Delta T^n \cdot TCN\right)$$

where TCN represents the thermal coefficients of n-order, T_o the reference temperature at which the value of the resistor is $R(T_o)$. In this view, a voltage divider with a resistor sensitive to temperature and a resistor insensitive to temperature can provide a voltage function of the temperature. Unfortunately, resistors dividers have several drawbacks: they provide single-ended measurements sensitive to noise injected by the voltage supply and high-precision instruments are needed to quantify the amount of change of the resistance under measurements.

These limitations can be overcome by using the Wheatstone bridge (Fig. 17.20). Introduced by Sir Charles Wheatstone in 1843, it became popular thanks to his nulling measure nature: Resistor R4 can be found, for example, by varying resistor R2 in order to get a null Vab. In this case, no current flows between the two branches of the bridge opening the possibility to use a non-calibrated voltage or galvanometric instrument.

If the value of resistor R4 is chosen in a way to null the Vab voltage, the Wheatstone bridge can provide high-accuracy measurements, thanks to its differential topology.

Wheatstone bridges are commonly used to read out a voltage (Vab) proportional to the voltage applied to the bridge (Vin) and to the change of the sensing resistor, assuming a small resistance change compared to the value of the resistor itself.

An example of temperature sensor realized by using a Wheatstone bridge with silicon diffused resistors is the temperature sensor used on the pressure sensor.

In this case, two sensing resistors (like R1 and R4 of Fig. 17.21) are implemented in order to double the sensitivity of the bridge. It has been designed in the right bottom corner of the sensor while most of the area is used for the pressure sensor membrane.

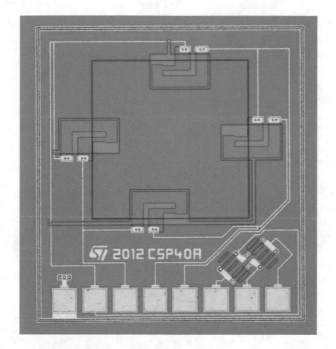

Fig. 17.21 Pressure and Temperature Sensor MEMS picture

As previously explained, in this case, the temperature sensor can be used as temperature sensor monitor of the membrane exposed to the environment and for compensating the pressure sensor itself. Implementation on the same die is a key factor to reduce the time latency between the pressure and temperature channel and optimize the pressure compensation during fast and slow temperature transitions. Main drawback of this kind of temperature sensors is the limited linearity range due to the inherent characteristics of the bridge response in case of large sensing resistor change and to the non-linear coefficient that affect the silicon resistors used to implement the bridge.

References

1. Dubinin, M.M., & Astakhov, V.A. Izv. Akad. Nauk SSR Ser. Khim. 1971 No 1,5,11.
2. Dubinin, M. M. (1966). In P. L. Walker Jr. (Ed.), *Chemistry and physics of carbon* (Vol. 2, pp. 51–20). New York: Dekker.
3. Dubinin, M. M., & Stoeckli, H. F. J. (1980). Colloid interface. *Science, 75*, 35–42.
4. Szulczynski, B., & Gebicki, J. (2017). Currently commercially available chemical sensors employed for detection of volatile organic compounds in outdoor and indoor air. *MDPI Environments, 4*(21), 2–3.
5. Kelvin, L., & William (October 1848). On an Absolute Thermometric Scale. *Philosophical Magazine*. Archived from the original on 1 February 2008. Retrieved 6 February 2008.

Part IV
MEMS Actuators

Chapter 18
Micromirrors

Roberto Carminati and Sonia Costantini

18.1 Application Fields

Micromirrors are among the oldest kind of MEMS solutions which were investigated: the first publication about micromirror devices is traced back to the 1980 [1]. Among the different applications which drove the earlier interest in this type of technology, we can find confocal microscopy, laser printing, bar code reading, switching and interconnection in optical communications.

Micromirror devices were also between the first MEMS to reach high-volume production in consumer market with Texas Instruments' Digital Micromirror Device (DMD) [2]. This device found wide application into light projectors, where the MEMS mirror is used to decide which pixel of the projected image to be on or off. In this case, the micromirror device is composed by a rectangular array of very small (~10 μm $\times$ 10 μm) aluminium mirrors which can stay at rest (ON-state) or be selectively activated to tilt in a position where light does not come out anymore from the projector (OFF-state), as illustrated in Fig. 18.1. The name *Digital* Micromirror Device is related to this binary operation. Another peculiarity of this solution is that the whole image is projected by the MEMS mirror and the size of its mirror elements determines the maximum projection resolution because it corresponds to the image pixel.

R. Carminati (✉)
ST Microelectronics, Analog MEMS and Sensors Group, MEMS Technology and Design R&D, Cornaredo, Italy
e-mail: roberto.carminati@st.com

S. Costantini
ST Microelectronics, Analog MEMS and Sensors Group, MEMS Technology and Design R&D, Agrate Brianza, Monza Brianza, Italy
e-mail: sonia.costantini@st.com

© Springer Nature Switzerland AG 2022
B. Vigna et al. (eds.), *Silicon Sensors and Actuators*,
https://doi.org/10.1007/978-3-030-80135-9_18

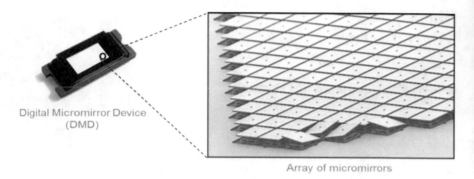

Fig. 18.1 Digital micromirror device with close-up of mirror array [3]. (Courtesy Texas Instruments)

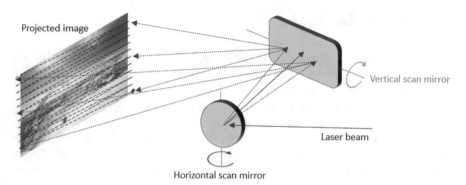

Fig. 18.2 Laser beam scanning display scheme

Such projection concept shows some limitations in scaling into smaller portable applications due to its intrinsically bulky components (light lamp, large micromirror chip to support large resolution). An alternative projector solution has been found in laser beam scanning displays [4], which are based on the generation of an image pixel by pixel through a laser spot which is moved sequentially into different positions. Also, this kind of solution rely on micromirror devices, where in this case only one moving mirror is needed because only one pixel is projected at a time, but where the dimensions of the moving mirror must be quite large (≥ 1 mm) to allow sufficiently small imaged spots on the projection screen and where mechanical scanning angles must be quite large ($\geq 8°$–$10°$) to support an acceptable Field of View (FOV).

A schematic representation of a laser beam scanning display is reported in Fig. 18.2, where a system composed by two mirrors is represented, one for the horizontal scanning and the other for the vertical scanning.

Because in this case the micromirror need to project a laser spot at each pixel position, many hundreds of different positions must be accessible so that practically a continuous scanning micromirror is required: we could then call this kind of design an *analog* micromirror, in contrast to the digital micromirror design previously described.

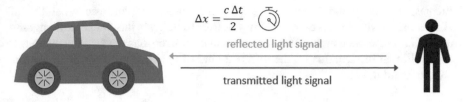

$$\Delta x = \frac{c \, \Delta t}{2}$$

reflected light signal

transmitted light signal

Fig. 18.3 LiDAR working principle. The distance Δx to an object is estimated measuring the time interval Δt between the emission of the light beam and its return, knowing light velocity c

The typical laser beam scanning projector application has a horizontal resolution N_H, which is then related to the mirror scanning angle θ and limited by the diffraction effect on the mirror, whereas the vertical resolution N_V is limited by the scanning rate ratio between the two axes of movement required to project an image [4]:

$$N_H \propto \frac{\theta D}{\lambda}$$

$$N_V \propto \frac{f_H}{f_V}.$$

where D is the mirror diameter, λ is the wavelength of the reflected light, f_H is the scanning frequency for the horizontal image direction and f_V is the frequency of scanning of the vertical image direction.

The advantage of this solution is that it can allow to reach more compact projector dimensions thanks to its smaller components and so it is particularly suited for handheld and wearable devices. Laser beam scanning found commercial application into portable projectors [5] and recently attracts lots of interest in the augmented reality glasses market [6].

In recent years, the research for micromirror application into automotive markets has also increased, driven by the push for autonomous driving and the growing interest for LiDAR (light detection and ranging) sensors. LiDAR is the ranging technology working in the infrared light spectrum and complementary to RADAR. As the RADAR, LiDAR sends off an electromagnetic wave (in this case in the infrared light wavelength range) and estimates distance of an object by measuring the time passing between the wave firing and the reflected echo return. A scheme of this working principle is reported in Fig. 18.3.

LiDAR offers some advantages in terms of range and immunity to atmospheric conditions, and it is considered a mandatory component for the sensing requirements needed to reach full autonomous driving. In such regards, currently available LiDAR adopt macro-mechanical scanning mirrors which are anyway costly, bulky and not always meeting the automotive market reliability standards. Therefore, a lot of companies are looking into inexpensive substitute components to these macro-mechanical mirrors and MEMS mirrors appear as the most natural candidates, thanks to their low cost, small dimensions and intrinsic reliability of silicon-based MEMS. As of today, anyway, R&D activity is this field is still ongoing and still no reference architecture based on micromirrors has reached the market.

Additional information on the different kind of applications for MEMS mirrors can be found in [7].

18.2   Types of Micromirrors

In the following, the focus will be on the *analog micromirrors*, composed by a single moving mirror able to rotate in a controllable way along one or more directions. In analogy to motors, the rotating mass is typically called *rotor*, whereas the static part is called *stator*. The latter is physically separated from the rotor and can incorporate the actuator element.

The peculiarity of MEMS mirrors with respect to other kind of devices is that they need to support rotational movement and not a translational movement like most of MEMS. This is needed because most of micromirror applications require a controllable angular scanning of a light source, so that it can be pointed in different positions. The first consequence of this requirement is that micromirrors need to have enough room around them to allow their movement. For example, if we consider a 1 mm diameter mirror rotating by $10°$, its edges will travel outwards the rest plane by more than 170 μm. This large movement requires, for example, that a deep enough cavity is realized on the back of the mirror so that no mechanical obstruction is present to the mirror movement. A second consequence of this rotational movement is that all the silicon flexures must be designed to work efficiently under torsion, so that the required scanning frequencies are achieved, and no excessive mechanical stress is built up in the flexure itself. In these regards, the most simple and efficient solution is the straight bar flexure, which can work quite efficiently in torsion and that can be made sufficiently unstressed by making it longer.

The first distinction we must do when talking about micromirrors is about how many scanning directions they can support: typically, micromirrors can rotate around one axis (*monoaxial* micromirror, Fig. 18.4a) or around two axes (*biaxial* micromirror, Fig. 18.4b). There is also some literature about what could be called triaxial micromirrors, adding a third movement to the mechanical structure which is not anyway a rotation (because the third orthogonal rotation would be useless for projection purposes), but an out-of-plane translation (tip-tilt-piston micromirrors) [8] useful for phase control, or a mirror deformation, like a controllable curvature of the mirror for focus or aberration correction of the optical system [9]. Considering anyway their specific interest only in academic research, we will not treat them further in this dissertation.

The easiest monoaxial design can be achieved by attaching to the mirror body a couple of torsional flexures, as reported in Fig. 18.4a. Considering the specific kind of actuation principle pursued, a suitable actuator will be then attached to the mirror body with the purpose of forcing the mirror body into motion.

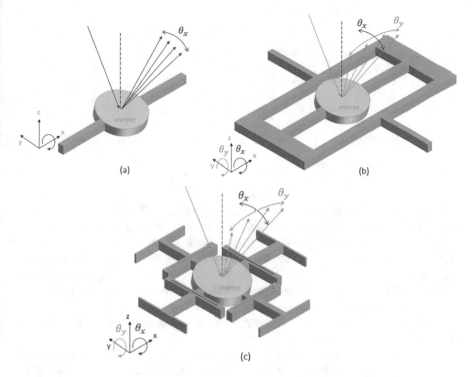

Fig. 18.4 Types of micromirrors: (**a**) monoaxial, (**b**) gimbal biaxial, (**c**) gimbal-less biaxial

On the opposite, a biaxial design needs to sustain rotation of the mirror body around two orthogonal axes. In order to reach this result, two main design solutions are typically used:

- The gimbal design (reported in Fig. 18.4b) is based on the addition of a decoupling rigid frame at which the mirror body and a first couple of torsional flexures are connected, so that rotation is possible around this first axis relative to the frame itself. Such frame is then connected to the fixed substrate by a second set of torsional flexures which allows rotation along a second orthogonal axis. This is the typical design choice for a biaxial micromirror solution, and it is particularly suitable for scanning solutions which require two different and distant scanning frequencies for the two axis, due to the very different inertia of the mirror and frame bodies.
- The gimbal-less design (reported in Fig. 18.4c) avoids instead the presence of a decoupling frame element and it is based on the design of a set of silicon hinges able to support the rotation along the two orthogonal axes. This kind of design solution is more complex due to the plurality of mechanical elements which must work in cooperation, but it is well suited when scanning with similar or equal frequency is needed.

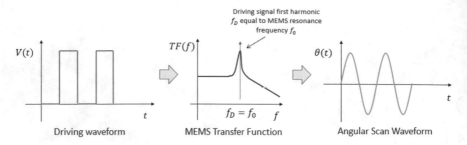

Fig. 18.5 Resonant operation

Most applications require a biaxial scanning to support their scanning requirements (let's consider for example the laser beam scanning projection previously described), so biaxial micromirrors would seem the best choice in most cases. A biaxial scanning can anyway be supported also relying on two monoaxial micromirrors in cascade, and in practice, this solution is preferred in many cases, thanks to its better performances, which comes from the better design optimization achievable for single axis MEMS, whereas biaxial design needs particular care for the additional design trade-offs coming from the integration of two axes scanning into the same MEMS structure.

An additional distinction in micromirrors is the one related to the mode of operation that they need to support. The modes of operation are related to the angular scanning frequency and waveform which are required by the target application:

- *Resonant* operation is a mode of operation in which the micromirror is driven at the frequency of its natural rotational vibration mode, as represented in Fig. 18.5. In this condition, the mechanical structure itself responds with an amplified movement thanks to a mechanical resonance, and larger scanning angles can be achieved with less input electrical power. This kind of operation can anyway only support sinusoidal waveform scanning, because the resonance itself acts as a narrow-band filter. As narrower the resonance is, as larger is the displacement amplification factor. The figure of merit of the resonance is the Q-factor, which represents such amplification factor. To properly work, a resonant micromirror must be designed to have its mechanical eigenfrequency, comprising its manufacturing tolerances, within an acceptable frequency range.

- *Quasi-static* operation is instead a mode of operation where the micromirror can be driven with an arbitrary waveform (see Fig. 18.6), taking into consideration anyway that the micromirror mechanical response will have a bandwidth determined from where its torsional eigenfrequency is located. In such regard, typical quasi-static micromirrors are operated at few tens of hertzs (e.g. 60 Hz), well below their resonance frequencies, which are typically in the range 500 Hz–800 Hz. This mode of operation enables, for example, constant velocity scanning (also called linear scanning) and static operation, that is, angular displacement

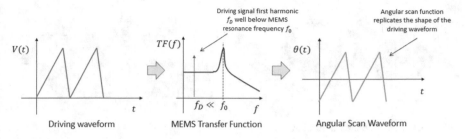

Fig. 18.6 Quasi-static operation

under a DC bias. Quasi-static micromirrors are typically indispensable in most applications thanks to the possibility of handling low frequency and linear scanning required to reach good projection quality without image artefacts.

Finally, micromirrors can be divided based on the wavelength of light radiation they have to reflect. Two main families are typically considered:

- Micromirrors for visible projection must project light in the visible range, typically a triplet of red, green and blue (RGB) laser spots. In this case, the mirror material typically used is aluminium, thanks to its flat reflectance bandwidth of about 90%.
- Micromirrors for infrared projection must project light in the near infrared (NIR) range. In this case, gold is the preferred material to realize the mirror because of its high natural reflectance of about 97%.

18.3 Actuation Principles

Actuators like micromirrors are very much determined by the type of actuation principle exploited to obtain the required motion. Being each actuation principle based on very different physical effects to allow the required electromechanical force conversion, the design principles can vary significantly in the different cases.

In the following, three main actuation principles will be described in detail, which are the most widely adopted in the literature and in the industry. The advantages and disadvantages of these three principles will then be analysed and compared.

18.3.1 Capacitive Actuation and Sensing

Capacitive actuation is likely the first kind of actuation principle adopted for micromirror devices, because of its relative simplicity and inexpensiveness which comes from the fact that the actuator can be realized with the same manufacturing

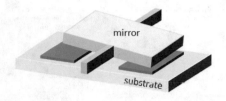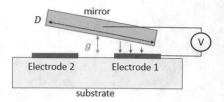

Fig. 18.7 Parallel plate micromirror

process steps which are commonly used to realize other kind of MEMS, for example inertial sensors.

Capacitive actuation is based on the electrostatic force, which is the force built up between the two faces of a capacitor when a potential difference is applied at its two terminals. Electrostatic force arises from the attraction between the electric charges of different sign which accumulate on the faces of the capacitor when it is charged. As a result, the outcome force is always attractive, fact which needs to be taken appropriately into account in order to reach the required actuator operation.

Two main capacitive actuator architectures are typically used.

The first one is the parallel plate architecture: the two faces of the capacitor are realized respectively by the bottom surface of the moving mechanical structure and by a counter-electrode realized on the MEMS substrate. To realize a bidirectional steering of the structure, two separated electrodes are needed, placed to face opposing sides of mechanical structure with respect to its rotational axis. The typical configuration is reported in Fig. 18.7.

The advantage of this architecture is that it can support both resonant and quasi-static operation without additional manufacturing complexity. Its biggest drawback is that the gap between the moving structure and the driving electrodes limits the maximum steering angle both mechanically than electrostatically. The mechanical limitation is trivial to understand referring to Fig. 18.7. Because of the tilting motion of the mechanical structure, its edges will touch the substrate and the driving electrodes when the critical angle θ_{crit} is reached, related to the geometrical dimensions of the MEMS in the following way:

$$\theta_{crit} = \text{atan}\left(\frac{2g}{D}\right) \tag{18.1}$$

where g is the gap between moving structure and driving electrodes and D is the micromirror width, typically equal to its reflecting area diameter.

In addition, pull-in effects could further limit the maximum achievable working angle. Pull-in is an effect for which the mechanical structure collapses over the driving electrode because the position-dependent component of the electrostatic force cancels out the restoring force of the torsional spring. For this kind of MEMS, anyway it has been demonstrated that pull-in voltage can be pushed to be larger

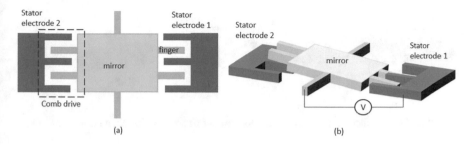

Fig. 18.8 Planar comb drive: (**a**) top view, (**b**) lateral view

than the voltage needed to reach the critical angle θ_{crit} [10], making the mechanical constraint the only limitation to the maximum achievable angle.

Using Eq. 18.1, it can be estimated that for a typical mirror diameter of 1 mm, a gap of about 9 μm would be needed to reach a tilting angle of 1 degree. Such gap dimensions could become technologically complex to realize so that only limited opening angles are achievable, impractical for most applications.

Nevertheless, this actuation scheme has found wide adoption mainly in array micromirrors [11], where the single mirror dimension (few micron large) is no more a limiting factor for the mechanical angle and where the fact that the actuator is hidden below the mirror itself has been found highly advantageous to increase the mirror fill factor.

The second architecture for capacitive actuators is the comb drive. A comb drive is composed by a set of closely packed cantilevers (called fingers) attached to the rotor structure and interdigitated to a specular set of cantilevers attached to a fixed structure, that is, the stator (Fig. 18.8). The capacitor is then obtained by the facing lateral surfaces of the fingers and can be easily increased by adding more fingers to the actuator or by changing their geometrical dimensions.

This architecture is very attractive thanks to its integration simplicity because it can be realized with the same process step used to pattern the mechanical structure. Its main limitation is related to the fact that, being the moving comb fingers and the fixed ones realized in the same plane, the attraction force at rest condition is almost null, so they cannot be used for quasi-static actuation. Resonant operation is anyway feasible thanks to small force imbalances from manufacturing non-idealities, which are able to put in motion the MEMS.

Therefore, a more complex structure is required to obtain quasi-static operation in which an out-of-plane offset is provided by construction between moving fingers and fixed fingers (see Fig. 18.9) and tight alignment between the two layers patterns are needed to keep similar performances to the in-plane comb fingers. Such solution is called staggered comb fingers actuator.

As a first step, let's focus on in-plane comb fingers case and let's derive the electrostatic torque T_{el} relation to the voltage V. Let us start considering the electrostatic energy U in the comb capacitor, which will assume the form:

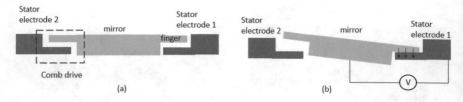

Fig. 18.9 Staggered comb fingers cross section: (**a**) in rest position, (**b**) tilted position under bias

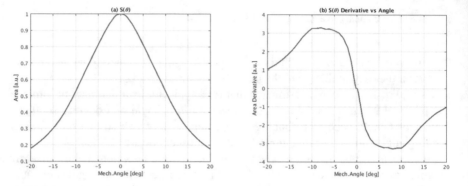

Fig. 18.10 Facing area versus tilt angle (a) and its derivative (b)

$$U = \frac{1}{2}C(\theta)V^2 \tag{18.2}$$

where $C(\theta)$ is the comb capacitance, which is a function of the moving structure tilting angle θ due to the changing facing area of the electrodes during tilt.

The capacitance relation to the comb fingers geometrical dimensions can be then reported as

$$C(\theta) = 2N_f \frac{S(\theta)}{g} \tag{18.3}$$

where N_f is the number of comb fingers, g is the gap between the comb fingers and $S(\theta)$ is a function which represents the comb finger facing area value versus the rotor tilting angle. The function $S(\theta)$ cannot be analytically determined but it can be estimated, for example, through Finite Element Modelling (FEM). An example of $S(\theta)$ function and its derivative are reported in Fig. 18.10.

The electrostatic torque can be obtained then by differentiating Eq. 18.2 with respect to the tilting angle:

$$T_{el} = \frac{\partial U}{\partial \theta} = N_f \frac{1}{g} \frac{\partial S(\theta)}{\partial \theta} V^2 \tag{18.4}$$

It can be observed that the electrostatic torque is directly proportional to the number of fingers and to the square of the driving voltage and inversely proportional to the gap between fingers. It is also to be noted that the torque depends on the derivative of the single finger facing surface with respect to tilting angle.

The same exact considerations can be done for the staggered comb fingers case obtaining the same result reported in Eq. 18.4 except for the fact that the function $\partial S(\theta)/\partial\theta$ will be different due to the different comb finger geometry. We will indicate the facing area derivative function for the staggered comb drive with $\partial S'(\theta)/\partial\theta$.

From a practical standpoint, the easiest implementation of a staggered comb drive requires two structural levels in which the comb fingers are patterned and with an offset equal to the full thickness of one layer. This translates into the fact that the derivative function for the staggered comb is practically equivalent in shape to the in-plane derivative function with an angular offset θ_{off}:

$$\frac{\partial S'(\theta)}{\partial\theta} \cong \frac{\partial S\left(\theta - \theta_{off}\right)}{\partial\theta} \tag{18.5}$$

It can be seen from Fig. 18.10 that the obtained electrostatic torque is linear only on a small angular portion, meaning that a linearization is needed through driving waveform shaping if a linear scan is required.

In addition, being the electrostatic force only attractive, two comb fingers set with opposite out-of-plane offset are needed to provide a scanning in both angular directions.

In both cases, to have torques large enough to reach the typical required opening angle in the range of $7°-10°$, a driving voltage between 100 V and 200 V is needed.

Most micromirror applications require strict projected spot position accuracy which are not possible to be achieved by open loop driving due to effects like self-resonance excitation in quasi-static scanners or change of frequency due to temperature in resonant scanners.

To allow tighter position accuracies by closed loop control, a measurement of the tilt position of the scanner is needed. Also in this case, capacitive effects come to help by a position measurement method based on the same comb fingers used for actuation, without added complexity to the MEMS process.

As a matter of fact, let us consider a comb drive biased at a constant voltage V_{sense}: due to the tilting movement, the capacitance will change in time. As a result, a current I_{sense} will be generated at the ends of the capacitor given by:

$$I_{sense} = \frac{dQ}{dt} = \frac{d\left(C\left(\theta(t)\right)V_{sense}\right)}{dt} = N_f \frac{1}{g}\frac{\partial S\left(\theta(t)\right)}{\partial\theta}\frac{d\theta}{dt}V_{sense} \tag{18.6}$$

Such current is proportional to the velocity of the mirror, so also relevant information on the position can be extrapolated. The presence of the derivative function $\partial S(\theta)/\partial\theta$ in the relation breaks down anyway the linearity of the relation so

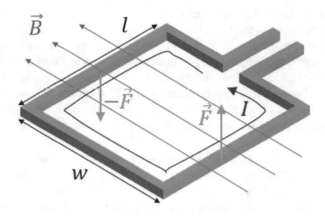

Fig. 18.11 Lorentz's force on a wire loop

that additional considerations will be needed to design an effective control, which anyway will not be treated here.

18.3.2 Magnetic Actuation

An alternative actuation solution to electrostatic force, which has been widely investigated, is magnetic actuation.

Magnetic actuation exploits Lorentz's force to create an electrically controllable torque to drive the MEMS. Lorentz's force is the force created on a wire of length L immersed in an external magnetic field $\vec{B}$ in which it is flowing a current I:

$$\vec{F}_L = IL\,\hat{t} \times \vec{B}$$

where $\hat{t}$ represents the versor directed along the wire.

As we can see, Lorentz's force depends on the vector product between the current carrying wire direction and the external magnetic field. As a result, a wire directed along the magnetic field will not suffer any force, whereas a wire directed perpendicularly to the magnetic field will be subject to the maximum force value.

Let's now consider a wire loop as the one reported in Fig. 18.11. As per what we have described before, the two wire portions directed along the field will not provide any force, whereas the other two wire portions perpendicular to the field will be subject to a force equal in modulus but opposite in direction, that is, a force couple.

The torqueing moment T_L associated to this force couple is

$$T_L = BILw = BIS$$

where w is the width of the wire loop, so the moment is proportional to the area of the wire loop. This relation between moment and loop area has been demonstrated for a rectangular wire loop, but it is valid in general for each loop shape.

In the case we are dealing with a coil composed by N_c wire loops, the torqueing moment of each wire loop will sum up to the others and the total coil moment will become

$$T_{coil} = N_c B I S = K_t I$$

where we have defined a motor constant $K_t \doteq N_c BS$, similarly to what typically done for electric motors. It is easy to understand that, in case the magnetic field B is not homogeneous and dependent from the MEMS angular position θ, as in most of the practical applications, the motor constant will not be a constant anymore and will assume a more general definition

$$K_t (\theta) \doteq \oint \vec{r} \times d\vec{F_L} = \oint \vec{r} \times \left(d\vec{l} \times \vec{B} \right)$$

where $\vec{r}$ is the vector distance from the rotation axis and $d\vec{l}$ is a vector having the local direction of the coil wire and a modulus equal to an infinitesimal length of the wire itself.

The motor constant K_t describes then all the intrinsic properties of the actuator related to its geometry and the magnetic field in which it is moving.

Then Lorentz's force provides an effective way to have a torque controllable through the electrical current flowing into the wire. Differently from the electrostatic case, the torque is linearly dependent from the current, so a bidirectional driving is possible because the actuation forces can be attractive or repulsive. This is particularly helpful for quasi-static micromirrors because a bidirectional scanning is achievable without the need to duplicate the actuator as in the electrostatic case.

We can then exploit the Lorentz's force for actuating a MEMS mirror just by integrating a coil onto the moving mass and enclosing it between permanent magnets which can provide inexpensively a large magnetic field. This is the so-called *moving-coil* architecture (Fig. 18.12a).

It exists also a converse architecture, the so-called *moving-magnet* design, which is based on having magnetic material deposited over the MEMS moving structure and on moving them through the magnetic field generated by the current flowing into an external coil or electromagnet [12]. This architecture is anyway providing an intrinsically weaker actuation torque due to many reasons (e.g. small magnet volume translates into small magnetic fields, not mature MEMS permanent magnet material integration technologies, very small magnetic field generated by external coil). Therefore, we will focus our attention only on the moving-coil architecture.

As we have seen previously, the physical quantity providing the actuation force is the electric current flowing into the coil. Being the coil a resistive load for the driving generator, a direct average power dissipation P_{diss} is present on the MEMS given by the Joule effect:

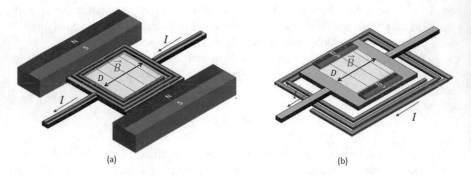

Fig. 18.12 Magnetic mirror architectures: (a) moving coil, (b) moving magnet

$$P_{diss} = R_{coil}I_{rms}^2$$

where R_{coil} is representing the resistance value of the coil and I_{rms} the RMS value of the driving current.

As a result, in a magnetically actuated micromirror, it is critical to minimize both the coil resistance and the driving current RMS value, otherwise a significant power dissipation up to some hundreds of milliwatts could be easily reached. A large power dissipation would be detrimental not only for the final application, where such power losses are very unattractive for battery-based devices, but also for the device reliability when high temperatures could be reached locally due to self-heating coming from the transformation of electric power into heat by the Joule effect.

To minimize coil resistance, the preferred approach is choosing a material with very low resistivity and to optimize coil wire parameters (thickness, width, pitch, number of windings) in order to have the maximum actuation torque for the minimum power consumption. In such regards, it is important to consider that the self-heating effect previously described will contribute to increase the load resistance since most conductors show an increase of resistivity with temperature T, following the law:

$$R(T) = R(T_0)(1 + \alpha_T(T - T_0))$$

Here T_0 is representing the reference temperature and α_T is the coefficient of thermal variation of resistivity in temperature.

As a result, metals are preferred with respect to silicon for manufacturing the integrated coils, and in detail, copper is the material of choice for these devices, thanks to its demonstrated integration capability into MEMS processes and its high conductivity.

Instead, to minimize the driving current keeping the torque level unchanged, the typical approach is maximizing the external magnetic field provided by the permanent magnets. This can be achieved first by an appropriate choice of permanent magnets material. In such regard, the solution of choice are the neodymium–iron–boron magnets, which are inexpensive, and which have proven to provide among the largest magnetic fields achievable among permanent magnetic materials.

Second, a good magnetic assembly design is required, minimizing the distance between the MEMS coil and magnet itself. In such a way, magnetic field values around 0.2–0.3 T are typically achieved in these applications. The volume of the permanent magnet is anyway typically much larger than that of the MEMS chip and it is not possible to scale it significantly without losing in actuation performance, making it hard to reach very small package form factors.

Considering these magnetic field levels and the low resistor values needed to minimize power consumption, currents up to few hundreds of milliamperes can be sustained. The voltage needed for actuation can then stay below 5 V, allowing the use of standard CMOS electronics technologies for driving.

Magnetic micromirrors also present an additional reliability requirement: we have seen that the driving coil is realized with a metal wire that needs to be patterned over the moving structure. Therefore, a metal connection needs to pass onto the silicon flexures around which the mirror rotates and so it is subject to a cyclic mechanical stress. It is then required to use metal materials which are fatigue resistant, to grant the electrical contact over the whole lifetime of the device.

Magnetic actuation is often used also when biaxial scanning micromirrors are needed. In fact, biaxial actuation is more easily implemented with magnetic mirrors, whereas electrostatic micromirrors would need complex processes with vertical electrical insulation to be realized over the moving structure [13].

On the opposite, in magnetically actuated mirrors, it is possible to design two different driving coils, one for each axis, with the same metal layers. One coil will couple with the rotation along a first axis, another one will couple with the rotation along a second perpendicular axis (see Fig. 18.13). A diagonal magnetic field will be required to provide an effective Lorentz's force along both axes.

This design is anyway quite complex to handle: two separate coils need to be designed requiring additional space; additional metal layers are passing over stressed flexure increasing risks of metal trace fatigue.

A more compact and elegant solution have been identified in [14], where a single coil is patterned onto the frame of a gimbal structure. In the coil, two different driving current waveforms are summed up and injected at the same moment. The mechanical structure will then act as a mechanical filter and each axis will react respectively to only one of these frequencies.

In order for this solution to properly work, the two driving frequencies need to be well separated in order to reduce as much as possible cross-axis actuation. This typically translates into having a first axis with a resonant frequency below 1 kHz and a second axis with resonant frequency higher than 20 kHz.

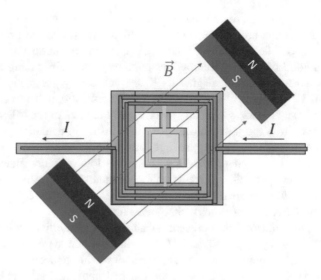

Fig. 18.13 Double coil magnetic mirror

18.3.3 Thin-Film Piezoelectric Actuation

Recently, the emerging of mature industrial processes for thin-film piezoelectric layers integration into MEMS technologies has generated lots of opportunities for the exploration of piezoelectric actuation into MEMS, including micromirrors.

Piezoelectric actuation is attractive thanks to the possibility of integrating a fully working actuator directly on the MEMS chip, which allows complete electromechanical wafer level testing, differently from magnetic micromirrors where the need of external permanent magnets make possible to measure the actuation characteristics only on the final assembled component, shifting potential yield losses later into the production chain and increasing production cost.

In addition, this kind of actuation exploits a physical effect allowing larger actuation forces than electrostatic-based solution enabling better scanner characteristics (larger scanning angle, larger diameter etc.).

Micromirror actuators make use of the converse piezoelectric effect, which consists of the fact that a piezoelectric material shows a change of its mechanical dimensions when subject to an external electric field (see Fig. 18.14b).

In the typical MEMS application, a thin-film piezoelectric material is deposited over a silicon substrate and patterned onto a mechanical structure. Upon voltage biasing, the piezoelectric material would like to change its dimensions, but it cannot do it freely because constrained by the silicon structure so that it builds up internal stresses. As a result, a mechanical force is transferred to the mechanical structure which is consequently deformed.

Among the piezoelectric materials, the most widely used for actuation purposes is the lead zirconate titanate, typically known as PZT, thanks to its demonstrated

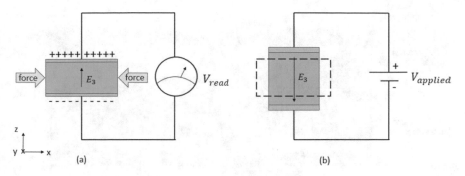

Fig. 18.14 Piezoelectric effect: (**a**) direct, (**b**) converse

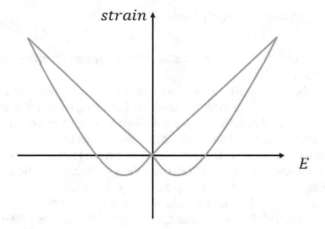

Fig. 18.15 Typical strain versus voltage characteristics of thin-film PZT material

integration capability into MEMS processes and its large piezoelectric parameters compared to other thin-film piezoelectric material.

The piezoelectric constitutive equations, which have been reported in Chap. 10, would lead readers to think that the piezoelectric effect is linearly dependent from the electric field and that a perfect actuation force reversal would be possible upon reversal of the electric field.

This is generally not true for PZT and in particular for its thin films because of its ferroelectric and electrostrictive nature. In fact, in PZT thin films, operating electric fields do easily exceed the coercive field of the material and the material piezoelectric response assumes significant nonlinear fashion, as reported in Fig. 18.15. In particular, PZT shows deformation in the same direction upon complete reversal of the electric field direction, making it unpractical to work with bipolar bias.

It must also be noted that bipolar cycling of the PZT is not suggested in standard operation due to fast insurgence of ferroelectric and piezoelectric fatigue of the

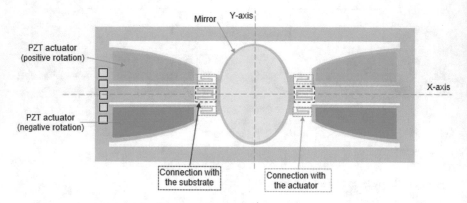

Fig. 18.16 Exemplificative scheme of piezoelectric scanner design

material [15], which causes a degradation of the piezoelectric properties and so of the actuator itself.

Consequently, PZT-based micromirrors need to work in unipolar biasing conditions and rely on an actuation force which has always the same direction. In this regard, PZT-based micromirrors are subject to similar design considerations as the electrostatic micromirrors which must rely only on attractive force. It is also noted that staying in unipolar regime, a linearized model of the piezoelectric force is typically sufficiently accurate to keep using the constitutive equations in Chap. 10 without needing to develop more complex, nonlinear models.

Generating a torsional movement with a piezoelectric actuator, naturally providing an out-of-plane translation movement, is not a trivial task. The working principle of one of the most efficient design solutions [16] will be described in the following, exemplificative of the typical micromirror scanner design.

The piezoelectric micromirror structure comprises a couple of torsional flexures connected to the mirror body on one side and at the fixed substrate on the other side. Two sets of cantilevers with piezoelectric patches on top extend then from the fixed frame up to the edges of the mirror body and connects to it through folded springs, which have the task of efficiently transferring the out-of-plane force to the mirror and allowing the mirror body rotation at the same time. This structure is illustrated in Fig. 18.16. Such sets are put on opposite sides with respect to the rotation axis so that their alternate actuation can create opposite rotation directions of the mirror body. Appropriate tuning of the dimensional characteristics of the structure allows then to support quasi-static actuation or resonant actuation.

An analytic description of a piezoelectric micromirror would be very complex and very much related to the specific design of interest due to the complicated interplay between different materials and multiple mechanical elements. As a result, FEM is considered an essential tool for the design of piezoelectric based micromirrors.

Anyway, meaningful physical insight can be gained by relying on appropriate lumped equivalent models, where an equivalence between mechanical elements and electrical components is drawn [17].

In particular, the electromechanical transformation factor η can be defined as

$$\eta = \frac{dQ}{d\theta}$$

where Q is the charge accumulated on the capacitor and θ the mirror tilt angle. Such electromechanical transformation factor can be found by FEM and used also to have an estimate of the effective piezoelectric torque T_{pze} by using the following equivalence relation:

$$T_{pze} = \eta V$$

The input actuation torque is then dependent only from the driving voltage and the electromechanical transformation factor.

It can be demonstrated [17] that such transformation factor is proportional to the d_{31} piezoelectric coefficient introduced in Chap. 10 and dependent on the stress in the piezoelectric material induced by a rotation by an angle θ, so a piezoelectric actuator is more efficient not only if its piezoelectric properties are better but also if it is stressed sufficiently by the target displacement, which determines the need of appropriate design optimization of the mechanical structure.

18.3.4 Comparison Between Actuation Principles

Now that we have seen the basic properties of the main actuation technologies for micromirrors, let's compare their benefits and the limitations.

The summary of the comparison is reported in Table 18.1:

In terms of pure maximum mechanical torque achievable, electromagnetic and piezoelectric actuations are able to provide larger forces, whereas electrostatic force is limited by the intrinsic weakness of electrostatic attraction and by the maximum

Table 18.1 Comparison of the actuation technologies

	Electrostatic	Electromagnetic	Piezoelectric
Mechanical torque	Low (< 1uNm)	High (>1uNm)	High (>1uNm)
Force direction	Monodirectional	Bidirectional	Monodirectional
Electromechanical efficiency	High	Mid	High
Driving voltage	≤ 200 V	≤ 5 V	≤ 50 V
Power consumption	Low	High	Low
Volume occupation	Mid	Large	Small

driving voltage safely achievable. As a result, electromagnetic and piezoelectric actuations can achieve better scan performances.

To provide good performances, electrostatic actuation needs also large driving voltages, up to 200 V, whereas piezoelectric actuation based on PZT can work with voltages below 50 V and electromagnetic mirrors can stay below 5 V if very small coil resistances are achieved.

Electromagnetic solution has then an intrinsically linear force, which simplifies the design especially when dealing with quasi-static actuation, but on the other side, it is affected by a low-actuation efficiency since a direct power dissipation is happening on the MEMS coil due to Joule effect. Instead, electrostatic and piezoelectric actuations are both having a capacitive drive load which is intrinsically less subject to power dissipation.

In terms of final scanning module size, magnetic solution requires large room to accommodate the permanent magnets needed to create the actuation forces, whereas both electrostatic and piezoelectric solutions can integrate their actuator directly at MEMS process level. In terms of size, then, piezoelectric actuation has an additional advantage related to its intrinsically higher force, so that smaller actuators can be designed with respect to the electrostatic case.

Therefore, electrostatic is a good solution when limited scan performances are sufficient thanks to its efficiency and small dimensions, at the cost of a higher driving electronics complexity required to generate the high voltage driving signals.

Electromagnetic actuation is a good solution when large scan performances or very large mirrors are required at the expense of module size and power consumption.

Finally, piezoelectric actuation offers good compromises in terms of actuation force, size and power consumptions which make this technology very appealing for the commercial application at the price of a higher design complexity.

18.4 Technology Platform

In this section, the technology architectures to realize a Raster scanning mirror are shown and discussed. The technology platform is strictly related to the actuation principle, to packaging constraint and reliability target.

18.4.1 Comb Finger Structure Realization for Capacitive micromirror

In this subsection it is analysed the technology to realize a capacitive micromirror with parallel comb finger. Such technology must integrate a mobile mass, a reflective

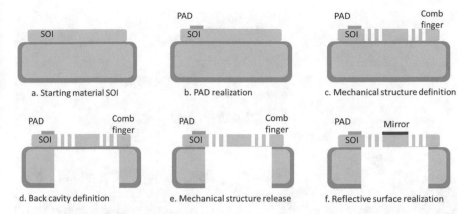

a. Starting material SOI   b. PAD realization   c. Mechanical structure definition

d. Back cavity definition   e. Mechanical structure release   f. Reflective surface realization

Fig. 18.17 schematic process flow for Electrostatically actuated Resonant Mirror: (**a**) starting material, (**b**) pad realization, (**c**) mechanical structure definition, (**d**) back cavity definition, (**e**) mechanical structure release, (**f**) reflective surface realization

surface and a comb finger array that works as actuator and as sensing structure for the mirror position.

A wide part of the published literature is on electrostatic MEMS mirrors, since their fabrication requires the availability of few technology steps, as it will explained in this section.

The simplest process architecture is the one that realizes a resonant, mono-axial, electrostatic mirror. The process schematic is the one explained in [18] and summarized in Fig. 18.17. An SOI wafer is chosen as starting material (for applications with 720p resolution, a 40-60 μm is a suitable SOI thickness), with a low resistivity (1–15 mOhm cm). A first metal layer is deposited on silicon and patterned through photolithography and etch, to realize the pads (Fig. 18.17b). The composition of the metal and its stack depends on the constraint relative to wire bonding and on the desired ohmic characteristic of the contact. After the Pad definition, the mirror mechanical structure is realized through photolithography and dry silicon etch (Fig. 18.17c). During this step, the in-plane comb fingers, the torsional flexure, the electrical domain (stator and rotor) and the mirror area are defined by removing the silicon of the SOI layer and by landing on buried oxide. After these operations, the wafer is reversed and, by using a front to back aligner, the back-cavity mask is realized and the following dry silicon etch helps to remove the silicon under the mirror, stopping on buried oxide again (Fig. 18.17d). The mechanical structure complete release is obtained through hydrofluoric acid (HF) etch of the buried oxide (Fig. 18.17e). The reflective surface is realized through proper metal deposition by using shadow mask technique (Fig. 18.17f).

This simple process scheme has one drawback if it must be manufactured in high volume: during the wafer reverse, the front of the wafer can be damaged by the handling of the equipment and a sort of front protection (sticky tape, temporary

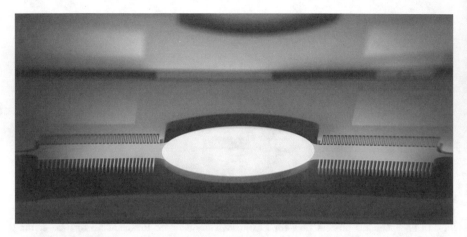

Fig. 18.18 SEM picture of electrostatic resonant mirror

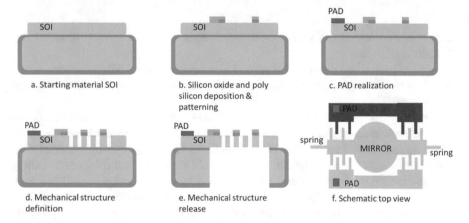

Fig. 18.19 Schematic process flow for electrostatically actuated resonant mirror with deterministic oscillation start: (**a**) starting material, (**b**) silicon oxide and polysilicon deposition and patterning, (**c**) pad realization, (**d**) mechanical structure definition, (**e**) mechanical structure release, (**f**) final device schematic top view

bonding and polymeric film) is necessary. In Fig. 18.18, an example of finished resonant micromirror is shown.

Even if the structure looks perfectly symmetric and the force fully balanced, small asymmetry due to wafer processing allows the start of oscillation. To have a deterministic oscillation start, that is mandatory in pico-projection application, an asymmetry in wafer process can be created in the structure as described in Fig. 18.19. Starting from previously described process flow, a Silicon dioxide and polysilicon deposition is performed before pad realization. Each layer deposition is followed by a patterning step, to short-circuit the polysilicon structure to the SOI (as shown in Fig. 18.19b). Thanks to the presence of Silicon dioxide, the etch of polysilicon does not damage the starting SOI surface. As shown in Fig. 18.19d, this small capacitor is added on stator fingers in asymmetric layout, to create an asymmetry in the electric field. A schematic device top view is reported in Fig. 18.19f, where in red, it is highlighted the region where the polysilicon is left on the top of stator finger. A simpler process approach is to realize a local thinning of stator finger, by using a silicon dry etch at fixed time.

The process scheme for the realization of the electrostatic quasi-static micromirror is slightly more complex since a staggered comb finger structure is needed.

To realize the staggered structure, a double structural layer (double SOI, or SOI and Epi-poly) is necessary, and the comb fingers will be defined in both structural layers. This fact adds a further complexity: in order to avoid the pull-in of the electrodes during mirror operation, the alignment between the staggered fingers must be very tight, or other solution must be adopted on process side. Two different process approaches to realize the staggered solution will be explained.

The first process scheme uses a direct lithographic alignment between the comb fingers. The main process steps are summarized in Fig. 18.20 and the good result on the final device depends on the level of performance of the lithographic alignment in infrared. The starting material is an SOI wafer, where an insulating layer of silicon dioxide is realized by oxidation or deposition. The aim of such dielectric is two-fold: the first is to be the landing layer for the silicon etch, the second is to create a separation between the two electrical domain of the rotor and the stator. In order to realize the electrical connection to the SOI layer, the oxide can be patterned (Fig. 18.20b) or a more complex system of oxide, polysilicon as interconnection and oxide can be realized as explained in [19]. After the realization of the electrical

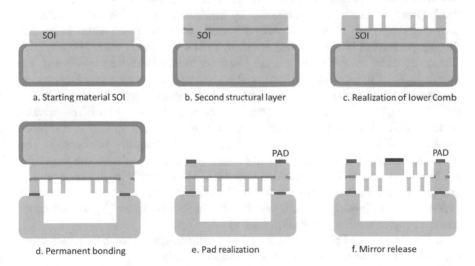

Fig. 18.20 Schematic process flow for electrostatically actuated linear mirror with staggered comb-finger: (**a**) starting material, (**b**) second structural layer deposition, (**c**) realization of lower comb, (**d**) permanent bonding, (**e**) pad realization, (**f**) mirror release

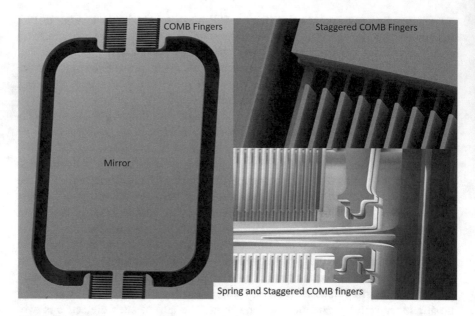

Fig. 18.21 Electrostatic Linear Mirror. Detail of spring and staggered comb fingers

separation of the domain, an Epitaxial silicon layer is grown and through chemical mechanical polishing the same thickness of the starting SOI is reached (for 720p resolution, 40 μm is a suitable thickness). The lower comb fingers array is then realized by using photolithography and dry silicon etch and by stopping on buried oxide (Fig. 18.20c); if it is requested by application, a reinforcement under the mirror structure can be realized with this mask. Such patterned wafer is bonded, through permanent W2W bonding technique to a cavity silicon wafer (Fig. 18.20d). The cavity under the mirror is necessary in order to allow structure movement. The handle layer of the starting SOI material is removed, by using a grinder wheel or chemical solution/gas (Fig. 18.20e). Metal layer for PAD is deposited and patterned (Fig. 18.20e), with similar consideration done on resonant mirror flow. The movable structure definition (mirror structure, upper comb finger and spring) is completed by the photolithography of the last mask and the following dry silicon etch. This mask must be aligned with tight tolerance with respect to the lower comb finger. For a 720p linear mirror that is actuated with 180–200 V bias voltage, the maximum acceptable misalignment between the comb finger is 0.7 μm. The whole structure is released by HF and, at the end, the reflective surface is realized by depositing the metal through a shadow mask (Fig. 18.20f).

In Fig. 18.21, SEM picture of mirror device with staggered comb finger structure, realized with the process scheme described above, is shown.

The second process scheme realized a well control staggered structure by using a self-aligned silicon dry etch between the two comb structures. This solution is cited in [19] and [20]. The idea is to define both upper and lower comb fingers with

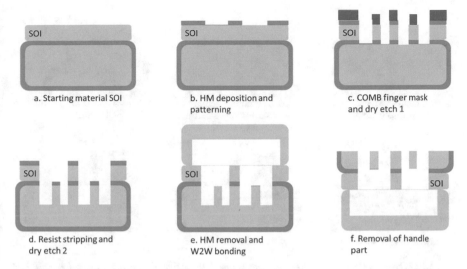

Fig. 18.22 Schematic process flow for Electrostatically actuated linear Mirror with staggered comb-finger (self-aligned approach): (**a**) starting material, (**b**) hard mask (HM) deposition and patterning, (**c**) comb finger mask and first dry etch, (**d**) resist stripping and second dry etch, (**e**) HM removal and W2W bonding, (**f**) removal of handle part

a single dry silicon etch, by using Hard mask and photoresist. A possible flow is the one described in [20] and shown in Fig. 18.22.

The starting material is an SOI wafer, where a Hard Mask (e.g. silicon dioxide layer) is deposited and then selectively removed, in comb finger region (Fig. 18.22b). A photoresist mask is realized to define comb fingers. As shown in Fig. 18.22 c, the photo resist (blue shape) protect both silicon surface and pre-patterned hard mask. After the mask exposure and resist development, a dry etch is performed. The dry etch is composed by three different steps: a first hard mask etch, since the mask is enclosed in previous hard mask patterning, a silicon dry etch, that define the lower comb, and a final buried oxide dry etch. The resist is then removed, and the wafer process continues with a silicon dry etch. During this step (Fig. 18.22d), the upper comb finger is defined using hard mask and buried oxide as silicon etch stop, while the lower fingers, not protected by the hard mask, are etched away. Such wafer is bonded to another one, by using W2W permanent bonding technique, and the process ends by removing the silicon and discovering the upper comb fingers. The simplified process flow here described can be improved by starting from a wafer with two structural layers (SOI and epipoly, or a double epipoly layer), in order to have a better thickness control of the upper comb finger and an etch stop for the last silicon removal step. The other consideration regards the actuation: such approach creates a staggered comb-finger structure slightly different respect to the one described in Fig. 18.22, resulting in a less effective electrostatic force, due to a typically smaller facing area compared to fingers with same thickness.

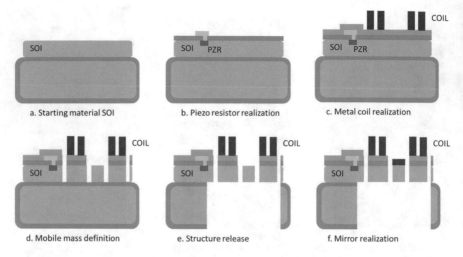

a. Starting material SOI

b. Piezo resistor realization

c. Metal coil realization

d. Mobile mass definition

e. Structure release

f. Mirror realization

Fig. 18.23 Schematic process flow for Electromagnetic actuated uniaxial Mirror (moving coil approach): (**a**) starting material, (**b**) piezo resistor realization, (**c**) coil metal realization, (**d**) mobile mass definition, (**e**) structure release, (**f**) mirror realization

18.4.2 Electromagnetic micromirror Technology

The electromagnetic mirror technology consists on the realization of a reflective surface, the mobile structure, the metal wire necessary for the application of the Lorentz force and the sensing of the mirror position. The permanent magnet is usually placed at die level, in the MEMS package. This approach is called "moving coil"; there is a second possible configuration that uses a fixed coil and a moving magnet. In this section, it will be explained the technology needed for the realization of the first, while for the second, the suggested publication is [21].

In Fig. 18.23, it is shown the process sequence for the realization of electromagnetic actuated micromirror.

The moving structure is realized using an SOI layer; to reach a 720p resolution a suitable thickness is 40–50 μm. While the sensing in an electrostatically actuated mirror is the comb finger capacitance itself and no dedicated process steps are necessary, in electromagnetic mirror, four piezo resistors connected in Wheatstone bridge configuration have to be realized on silicon to verify the mirror movement during actuation (Fig. 18.23b). A powerful sensitive system can be achieved using implanted diffused resistor. The usage of boron doped starting SOI layer, a good lithographic alignment with silicon crystal direction, and a smart layout allow to reach a high sensitivity value (e.g. 5 mV/V/deg).The realization of piezo-resistors bridge (in Fig. 18.23, indicated as PZR) imposes the deposition of an insulant layer (like silicon dioxide) and the realization of a local interconnection layer in metal (in Fig. 18.23b, the orange layer). The choice of the right metal material will be done by considering the desired series resistance and the ohmic behaviour of the contact. After the local interconnection realization, a passivation layer (like Silicon Nitride or Silicon Oxi-nitride) is deposited on top of it (Fig. 18.23c, cyan layer). Vias in passivation layer to reach the local interconnection can be opened by using a dry etch or a wet etch.

The following step is the deposition of a seed metal layer for the coil growth. In order to have the highest force, and to reduce consumption related to power dissipation, the coil has to be long with low resistance. To match these requirements, the electrochemical deposition is the preferred technique to realize such metalliza- tion. On metal seed layer, a mask is realized by using lithography, then the ECD growth is performed. After the growth, the photoresist is stripped, and the seed layer is removed by wet chemical etch. The seed material must match the wettability requirement for the electrochemical growth, while the coil material has to have the lowest resistance value. The preferred material for coil and seed are copper or gold.

In Fig. 18.23c, it is shown the sequence of the described steps. Since the longest the coil is, the stronger is the force, a very compact wire layout must be realized on silicon. To match 720p resolution target, the coil thickness must be greater than 25 μm and the coil pitch around 30 μm, where the space is reduced to a half respect the metal width. In such situation, the photoresist during the growth is stressed, and a smart layout and a robust process window is needed to avoid line collapse, which can have a detrimental effect on force.

In Fig. 18.24, the completed mirror device is shown.

If copper is used as metal material for coil, sometimes a top finishing with nickel is introduced to protect the wire top from oxidation, as shown in Fig. 18.25.

By depending on mirror design, Bi-axial or mono-axial, a second thick metal layer (realized by ECD) with low resistance can be necessary to manage the coil current signal. For Bi-axial design, this second interconnection has to pass over one

Fig. 18.24 Electromagnetic mirror. SEM picture of the final device

Fig. 18.25 SEM cross section of the COIL of electromagnetic Mirror

of the axes springs. In this arrangement, such metallization must be robust enough to sustain the stress coming from mirror movement in both directions. Gold and copper are ductile material, and they are not suitable to ensure the desired lifetime. Other metals and alloys with better resistance to damage and breakage under movement must be used.

After thick metal realization, the mirror surface can be cleaned from dielectric and the mobile mass is defined through lithography and silicon dry etch (Fig. 18.23 d).

In order to match high-resolution target, the mirror mechanical angle has to be as high as possible especially for horizontal scanning, as result the operative stress in silicon can be close to 1GPa. To sustain such stress level in operating condition, the silicon of the spring has to be free of crystal or other defects as explained in [22]. Since dry silicon etch can create some crystal damages on the sidewall, after the mobile mass definition, it is suggested to remove the damaged area, by using an isotropic silicon etch [23]. Dry (SF6 or CF4) or wet (TMAH or KOH) etch can be used for this purpose [24].

The mirror release (Fig. 18.23 e) is performed by using a backside mask and dry silicon etch. As previously commented, the front to back operation can cause severe damage of the wafer frontside and a front side protection is necessary. Respect to electrostatic case, in this process scheme, at this step, the topography present on the wafer front can be important and the chosen front side protection has to be able to manage up to 60 μm of height difference between the top of the coil and the buried oxide. Since the Lorenz force is stronger as the closer is the permanent magnet to the coil, depending on the magnet arrangement in the final package, the handle silicon can be reduced in thickness down to 120 μm. To manage such thin-holed wafer, a carrier wafer is mandatory. With backside mask and etch, a reinforcement under the mirror or coil can be realized, if the design requires it.

The reflective surface is realized as the final step of the flow, by metal deposition through a shadow mask.

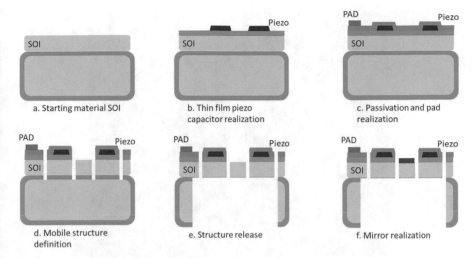

Fig. 18.26 Schematic process flow for thin film piezo actuated Mirror: (**a**) starting material, (**b**) thin film piezo capacitor realization, (**c**) passivation and pad realization, (**d**) mobile structure definition, (**e**) structure release, (**f**) mirror realization

18.4.3 Thin-Film Piezo Technology

Thanks to the availability in semiconductor facilities of equipment able to perform a thin-film piezo deposition on silicon wafer (by sputtering or solgel coating) some devices, like micromirrors, can benefit of this strong actuation principle.

The thin-film piezo mirror technology consists of the realization of a reflective surface, the mobile structure and the thin-film piezo actuator over a thin silicon cantilever.

In Fig. 18.26, a schematic sequence of the major process flow steps necessary to mirror realization is shown.

Like electromagnetic mirror, the sensing of the mirror position has to be performed by using a piezo-resistance bridge or by using the thin-film piezo itself, if its current versus voltage characteristic is linear and stable enough. In Fig. 18.26, the realization of the sensing part is not reported. Starting from an SOI wafer, an isolation layer like silicon dioxide is deposited on silicon surface in order to insulate piezo capacitor from substrate. After such dielectric deposition, the thin-film piezo stack (bottom electrode, piezo material and top electrode) is realized by using standard deposition technique (sputtering PVD or solgel).

The actuators are then defined through photo lithography and etch (Fig. 18.26b): dry or wet chemistry can be used. After capacitor patterning, a dielectric passivation layer is deposited on top. If lead–titanate–zirconate material is used as actuator, the passivation material or passivation stack has to avoid the usage of Hydrogen close to piezo material, since it has a detrimental effect on piezo performance [25]. After passivation deposition, vias are opened in dielectric material with photolithography

Fig. 18.27 PZT actuated quasi-static mirror. SEM picture

and etch, to reach capacitor electrode. A metal layer is deposited to realize the electrical contact to piezo, the interconnection and the pad (Fig. 18.26c).

The dielectric over the mirror surface can be removed using Via mask or after metal patterning, however before mobile structure definition with lithography and dry etch (Fig. 18.26d). Dry silicon etch is used to keep the sharp profile on the spring and the removal of the damaged silicon sidewall is suggested also for this technology if the device performance and reliability require it.

The mirror release and the reflective surface realization follow the steps described in the previous paragraphs. To have an effective actuation, during back-side silicon etch, the silicon handle layer has to be removed also from the back of the piezo cantilever (as shown in Fig. 18.26e), in order to have the maximum structure deflection and mirror rotation. In Fig. 18.27, the SEM picture of PZT actuated mirror is shown. The four thin-film piezo capacitors are disposed in a symmetric way respect to the central mirror, as explained in Fig. 18.16. In Fig. 18.28, other examples of piezo-actuated micromirrors are reported.

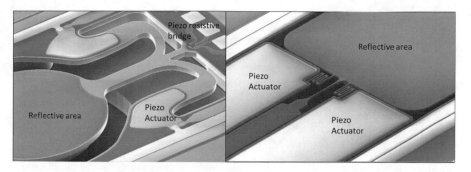

Fig. 18.28 Details of thin-film Piezo actuated mirrors (SEM picture). Resonant (on the left side) and quasi-static mirror (on the right side) are shown

18.4.4 Reflective Surface: Requirements and Realization

In the previous paragraphs, the realization of the mobile structure is widely described, but the reflective surface has been treated in a marginal way. The aim of this section is to describe more in detail that part. In portable projection system, a great effort is spent to minimize consumption: the final image brightness and contrast are directly related to laser power and the loss/light adsorption in optical path has to be avoided as much as possible. To have a more effective system, the mirror reflectance has to be the highest possible (>90%). With such constraint, the silicon itself, even with a high doping level, cannot be a powerful reflective surface and a metal layer is mandatory.

Let' s split the reflective surface versus the application: if the final target is projection, that means visible (VIS) region (400–635 nm), silver or aluminium is the best choice, while for gesture recognition, that means infrared (800–1000 nm), the gold is the preferred material. The reason is that aluminium and silver have a flat reflectivity response along all the visible range, whereas gold has very high reflectivity but only for wavelengths higher than 600 nm.

Aluminium, gold and silver are material that can be deposited in thin film by evaporation or sputtering PVD and are present in a semiconductor factory, with a high purity grade.

In Fig. 18.29, the reflectance characteristics of pure aluminium, silver alloy and pure gold thin film (thickness less than 300 nm), collected with a double beam reflectometer at quasi normal incidence (9 degree), is shown. The data are well aligned to the ones available in literature [26] .

To reach the desired optical performance, the roughness of the metal layer must be the lowest possible (Rms < 10 nm). This imposes the usage of polished grade surface as starting material for reflective surface realization: for example, mono-silicon is preferred to epipoly silicon even after CMP. The silicon surface must be free of defects (like scratches), and oxide wet etch is preferable to dry oxide etch to clean silicon surface before metal deposition.

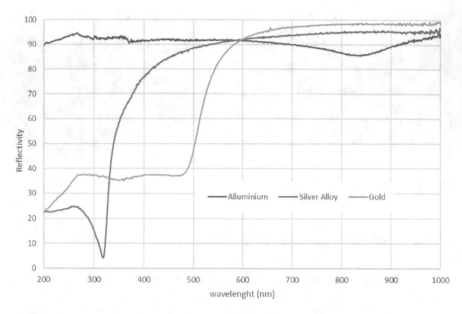

Fig. 18.29 Metal thin-film reflectance spectra

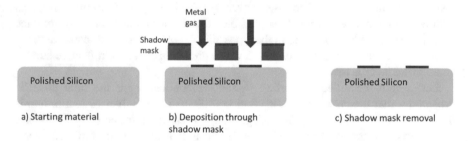

a) Starting material

b) Deposition through shadow mask

c) Shadow mask removal

Fig. 18.30 Shadow mask approach. Mirror realization: (**a**) starting material, (**b**) deposition through shadow mask, (**c**) shadow mask removal

Between metal and silicon, a barrier or an adhesion layer can be present in order to guarantee metal adhesion and to avoid diffusion of metal in silicon, that can cause defects.

There are two possible ways to realize the metal of the mirror: the first is the one suggested in the process flows described before, the deposition through a shadow mask, and the second is the full sheet metal deposition followed by a patterning through photolithography and etch. The two methods are explained in Figs. 18.30 and 18.31.

Shadow mask is a technique widely used in research centres and universities, where it is often performed in a manual arrangement: it is cheap, it has an high flexibility – since it can be applied to released structure – but the drawback is the poor thickness uniformity of the deposited material due to the shadow effect. As

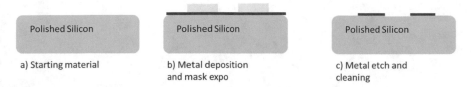

a) Starting material

b) Metal deposition
and mask expo

c) Metal etch and
cleaning

Fig. 18.31 Mirror realization with full sheet approach: (**a**) starting material, (**b**) metal deposition and mask expo, (**c**) metal etch and cleaning

described in [27], the shadow mask wall blocks a part of metal deposition and creates a "dome-shaped" metal profile. Such dome profile can be minimized, by reducing shadow mask thickness, but it can never be completely suppressed.

The "dome profile" contributes to the static curvature of the mirror: by increasing system resolution, this time zero distortion must be limited as much as possible.

Full sheet approach is more expensive and imposes a more complex integration. The plus respect to shadow mask is that the mirror static curvature is related only to mechanical design and to metal film stress control.

Since metal reflectivity is directly related to image brightness and contrast, it has to be as higher as possible in order to reduce laser power consumption. For such reason, in visible range, silver is preferable to aluminium, due to superior optical performance, but the usage of silver usually imposes the adoption of protective coating since the material is prone to tarnish. The adoption of an optical coating for protection and for reflectivity enhancement is something widely used in optics. The thickness and the material choice (in terms of refractive index) is optimized in order to minimize or to improve the impact on optical response. This technique can be also applied in MEMS micromirror as described in [28, 29]. For protection coating, in the visible range, the best choice is the mono-silicon oxide (SiO), and for enhancement coating, an alternate sandwich of low refraction index and high refraction index material is usually applied, as shown in Fig. 18.32. As low refraction index silicon dioxide (n = 1.46 at 632 nm [30]) is a good choice, while for high refraction index, Al_2O_3 (n = 1.77 at 632 nm [30]), HfO_2 (n = 1.91 at 632 nm [30]) or Nb_2O_5 (2.32 at 632 nm [30]) can be good alternatives. All these materials can be easily found in a semiconductor front end fab.

The thickness of the material is a fraction of used wavelength (usually $\lambda/4$), in order to take advantage from constructive interference. The exercise of enhancement reflectivity, widely used in optics, is more complex in scanning micromirror, since the optical path of the incident light is a function also of scanning angle.

In pico-projection application, sometimes there is another optical requirement: to have all the "non-reflective surfaces" as dark as possible, to suppress all the unwanted reflection coming from MEMS fixed part. This kind of defect is called "stray light" and it is related to wide tolerance of alignment of laser beam inside the optical module. Two ways are possible to realize a non-reflective surface on the MEMS from technology point of view: the first is the deposition of a dark coating upon the fixed surface, and the second is to realize a very rough surface that creates

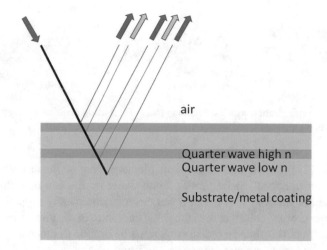

Fig. 18.32 Schematic quarter wave stack for enhanced reflectivity

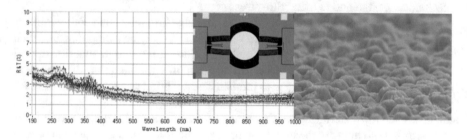

Fig. 18.33 Resonant Electrostatic Mirror with black cavity (reflectivity spectra and SEM picture details of the black surface)

a high level of absorption, like what happens in moth's eyes. In Fig. 18.33, it is shown the realization of a black surface in the cavity where the mirror moves with the reflectivity spectra.

18.4.5 Process Specificities: Vacuum Usage for Improved Performance and Reliability

To suppress air damping effects, the usage of hermetic vacuum package is suggested. The reduced atmosphere improves mirror performance and reduces consumption since higher aperture can be reached with lower power. For electrostatic mirror, it is possible to reduce the extension of comb-drive area since the force in vacuum is more effective [31]. Fraunhofer Institute worked on this approach at wafer level for electrostatic biaxial mirror [19], and showed that 15 degree of

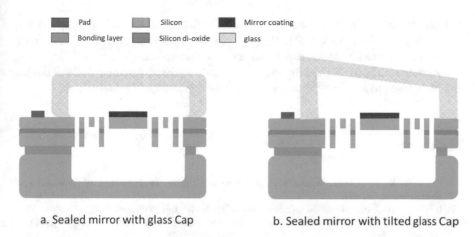

a. Sealed mirror with glass Cap   b. Sealed mirror with tilted glass Cap

Fig. 18.34 Hermetic electrostatic mirror cross section: (**a**) sealed mirror with glass cap, (**b**) sealed mirror with tilted glass cap

mechanical angle on fast resonant axis can be reached with a pulse of 60 V instead of 200 V. The solution presented by Fraunhofer used Titanium Getter to reduce inner pressure down to 0.1 Pa.

The solution of vacuum sealing at wafer level is a technology widely used for MEMS gyroscope, even if in the case of scanning micromirror is more complex, since it requires the usage of three wafers instead of two, and the usage of glass wafer to allow the light transmission.

Regarding to device layout, the scanning mirror has to be surrounded by a closed frame, with minimum topography in order to allow the hermetic W2W bonding. For electrostatic mirror, the connection to stator and rotor electrical domain has to be performed with buried polysilicon connection, as explained in [19] and the process flow can be further complicated by using vertical separation trenches filled by insulator as described in [32]. In Fig. 18.34, a schematic cross view of final device is shown.

The sensor wafer, the one where the mechanical mirror structure is realized, is bonded to a support silicon wafer where a deep cavity is realized by using W2W bonding technique like glassfrit or metallic bonding (e.g. Au-Au). The top cap wafer is made up with glass with several cavities to allow the mirror rotation and it is bonded to the sensor wafer by using glassfrit or anodic bonding. The bonding sequence can be reversed with respect to the flow here described and a getter can be deposited on support wafer through lift-off or shadow mask, if it is required to reach the desired vacuum level and device lifetime.

The usage of the glass in the architecture proposed in Fig. 18.34a has one drawback: the formation of a white spot at the centre of the projected field due to the glass surface back reflection, as explained in [19]. This spurious effect can be avoided by using an antireflective coating on the glass surface or using a tilted glass in order to shift the reflection out of the projection path, as shown in Fig. 18.34b.

Fraunhofer Institute suggested 15 degree of tilting good enough to suppress the white spot [19]. The method of realization of a glass wafer with such characteristics can involve micromachining technique and it is described in [33].

Vacuum and hermetic sealing guarantee better lifetime, since the MEMS device is not exposed to humidity, dust and aggressive contaminants present in air.

References

1. Petersen, K. E. (1980). Silicon torsional scanning mirror. *IBM Journal of Research and Development*, 631–637.
2. [Online]. Available: https://en.wikipedia.org/wiki/Digital_micromirror_device
3. Nelson, P. White Paper – DLP technology for spectroscopy, February 2014. [Online]. Available: https://www.ti.com/lit/wp/dlpa048a/dlpa048a.pdf
4. Urey, H., et al. (2000). Optical performance requirements for MEMS-scanner based microdisplays. *Conference on MOEMS and Miniaturized Systems, SPIE, 4178*, 176–185.
5. [Online]. Available: microvision.com/sony-mp-cl1-pico-projector-announcement/
6. [Online]. Available: https://www.theverge.com/2019/2/24/18235460/microsoft-hololens-2-price-specs-mixed-reality-ar-vr-business-work-features-mwc-2019
7. Holmstrom, S., et al. (2014). MEMS laser scanners: A review. *Journal of Microelectromechanical Systems*, 259–275.
8. Wu, L., & al. (2010). A tip-tilt-piston micromirror array for optical phased Array applications. *Journal of Microelectromechanical Systems, 19*(6), 1450–1461.
9. Inagaki, S., et al. (2019). High resolution piezoelectric MEMS scanner fully integrated with focus-tuning and driving actuators. In *Transducers 2019 – EUROSENSORS XXXIII, Berlin, Germany*.
10. Zhang, X., et al. (2001). A study of the static characteristics of a torsional micromirror. *Sensors and Actuators A*, 73–81.
11. Song, Y., et al. (2018). A review of micromirror arrays. *Precision Engineering, 51*, 729–761.
12. Yalcinkaya, A., & al. (2007). NiFe plated biaxial MEMS scanner for 2-D imaging. *IEEE Photonics Technology Letters, 19*(5), 330–332.
13. Hsu, S., et al. (2008). Two dimensional microscanners with large horizontal-vertical scannng frequency ration for high resolution laser projectors. In *Proceedings of SPIE*.
14. Yalcinkaya, A., et al. (2006). Two-Axis electromagnetic microscanner for high resolution displays. *Journal of Microelectromechanical Systems, 15*(4), 786–794.
15. Genenko, Y. A., et al. (2015). Mechanisms of aging and fatigue in ferroelectrics. *Materials Science and Engineering B*, 52–82.
16. Boni, N., Carminati, R., Mendicino, G., & Merli, M. (2021). Quasi-static PZT actuated MEMS mirror with 4x3mm2 reflective area and high robustness. In *Proceedings of SPIE 11697, MOEMS and Miniaturized Systems XX, 1169708*.
17. Schneider, R. (2016). *High-Q AlN contour mode resonators with unattached, voltage-actuated electrodes*, Graduate thesis in Electrical Engineering and Comupter Sciences, University of California at Berkeley.
18. Arslan, A., Brown, D., Davis, W. O., & al. (2010). Comb-actuated resonant torsional microscanner. *Journal of Microelectromechanical Systems, 19*(4), 936–943.
19. Hoffman, U., Eisermann, C., & Quenzer, H.-J. (2011). MEMS scanning laser projection based on high-Q vacuum packaged 2D-resonator. In *SPIE – The International Society for Optical Engineering 7930*.
20. Kim, M. (2009). High fill-factor micromirror array using a self-aligned vertical comb drive actuator with two rotational axes. *Journal of Micromechanics and Microengineering, 19*, 035014 (9pp).

21. Ataman, C., et al. (2013). A dual-axis pointing mirror with moving-magnet actuation. *Journal of Micromechanics and Microengineering, 23*, 025002 (13pp).
22. Gaither, M. S., Gates, R. S., Kirkpatrick, R., Cook, R. F., & DelRio, F. (2013). Etching process effects on surface structure, fracture strength, and reliability of single-crystal silicon theta-like specimens. *Journal of Microelectromechanical Systems, 22*(3), 589–602.
23. Noonan, E. E., et al. (2008). The scaling of strength in the design of high-power MEMS structures. *Scripta Materialia, 59*(9), 927–930.
24. Tilli, M., Motooka, T., Airaksinen, V. M., Franssila, S., Paulasto-Krockel, M., & Lindroos, V. (2015). Chapter 21 – Deep reactive Ion etching. In *Handbook of silicon based MEMS materials and technologies-second edition* (pp. 464–465). William Andrew – Elsevier.
25. Seo, S., Yoon, J., Song, T. K., Kang, B. S., & Noh, T. W. (2002). Mechanism of low temperature hydrogen-annealing-induced degradation in Pb(Zr 0.4 Ti 0.6)O3 Capacitors. *Applied Physics Letters, 81*(4), 697–699.
26. Paquin, R. A. (1995). Chapter 35: Properties of metals. In *Handbooks of optics-volume II* (pp. 35.30–35.39). New York: McGraw-Hill.
27. Yao, S. K. (1979). Theoretical model of thin-film deposition profile with shadow effect. *Journal of Applied Physics, 50*, 3390–3395.
28. Zannuccoli, M., et al. (2017). Simulation of micro-mirrors for optical MEMS. In *SISPAD Proceedings*.
29. Cianci, E., et al. (2018). Advanced protective coatings for reflectivity enhancement by low temperature atomic layer deposition of HfO2 on Al surfaces for micromirror applications. *Sensor and Actuators A: Physical, 282*, 124–131.
30. FILMETRICS: Refractive index database. Filmetrics, [Online]. Available: https://www.filmetrics.com/refractive-index-database. Accessed 29 Mar 2020.
31. Manh, C. H., & Hane, K. (2009). Vacuum operation of comb-drive micro display mirrors. *Journal of Micromechanics and Microengineering, 19*, 105018–105026.
32. Schenk, H., Durr, P., Kunze, D., Lanker, H., & Kuck, H. (2001). A resonantly excited 2D-micro-scanning-mirror with large deflection. *Sensors and Actuators A, 89*, 104–111.
33. Merz, P., Quenzer, H. J., Bernt, H., Wagner, B., & Zoberbier, M. (2003). A novel micro-machining technology for structuring borosilicate glass substrates. In *The 12th international conference on solid state sensors*. Boston: Actuators and Microsystems.

Chapter 19
Inkjet Printhead

Domenico Giusti, Sonia Costantini, Stefano Brovelli, and Marco Ferrera

19.1 Continuous Jet

Continuous inkjet, abbreviated CIJ, is one of the oldest and well-established printing technologies. In 1867, Lord Kelvin patented the syphon recorder, which recorded telegraph signals on paper using an inkjet nozzle deflected by a magnetic coil. The first commercial device, a medical strip chart recorder, was introduced in 1951 by Siemens. In 1989, Iris Graphics 3407 was widely recognized in the market as a high-definition color printer for wide formats. This technology was transferred to Scitex and Kodak and was commercialized under the name of Prosper, a very high jetting frequency CIJ (jetting frequency 360 kHz) [1, 4, 5, 22].

Figure 19.1 is an example of a continuous jet system. The pumping system, lower part of the image, provides the ink feed to the actuating device, at the right pressure to sustain the selected jetting frequency. The actuator generates the perturbation to allow the jet breakup. The generated drops are selected by airflow or charged plates and only the filtered ones are hitting the substrate. The remaining drops are collected by a gutter into the recirculation system, allowing the ink reuse with no

D. Giusti
ST Microelectronics, Analog MEMS and Sensors Group, MEMS Technology and Design R&D, Milano, Italy
e-mail: domenico.giusti@st.com

S. Costantini
ST Microelectronics, Analog MEMS and Sensors Group, MEMS Technology and Design R&D, Agrate Brianza, Monza Brianza, Italy
e-mail: sonia.costantini@st.com

S. Brovelli · M. Ferrera (✉)
ST Microelectronics, Analog MEMS and Sensors Group, Microfluidic & Microactuators Division, Milano, Italy
e-mail: stefano.brovelli@st.com; marco.ferrera@st.com

© Springer Nature Switzerland AG 2022
B. Vigna et al. (eds.), *Silicon Sensors and Actuators*,
https://doi.org/10.1007/978-3-030-80135-9_19

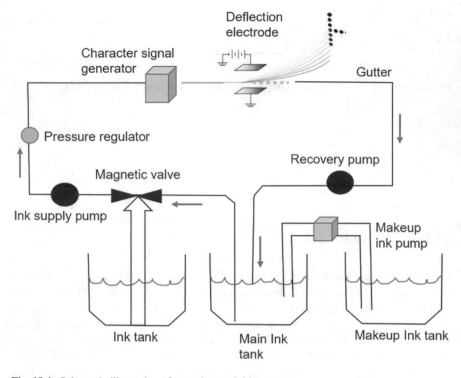

Fig. 19.1 Schematic illustration of a continuous inkjet system

wastes. The device controlling the generation of the droplets is the piezoelectric actuator, exploiting the Rayleigh-Plateau jet breakup at the desired distance. The main idea is to create a perturbation in the fluid jet, in the form of vibrations or heat waves. This perturbation causes the jet to breakup at a certain distance, creating the wanted drops. This is controlled by precisely varying the driving waveform fed to the piezoelectric transducer, which in turn generates the pressure perturbations needed for the jet breakup. Another way to induce these perturbations is by using heat waves, generated by heaters around the nozzle: in this system there is no need for a piezoelectric transducer. This concept is exploited by Kodak with their patented Ultra-Stream technology, fourth CIJ generation.

Figure 19.2 shows a Kodak Ultra-Stream nozzle. The whole silicon nozzle plate consists of an array of 2560 nozzles with an approximate diameter of 9 microns positioned linearly. Ultra-Stream printheads can reach speeds up to 150 meters per minute at 600x1800 dot-per-inch resolution, firing more than 400,000 drops per second per nozzle. Each nozzle has an annular heater positioned at the edge of the orifice. The frequency of heater activation is higher than 400 kHz and provides enough energy to weaken the jet filament by reducing viscosity and surface tension. Drops are then formed individually at velocities of $20m/s$. CIJ generally produces faster drops compared with drop on demand systems, in the order of 20 to $50m/s$. This allows higher distance between the printhead and the paper plane. CIJ systems can also differ according to the way in which drops are selected. In Fig. 19.3, the

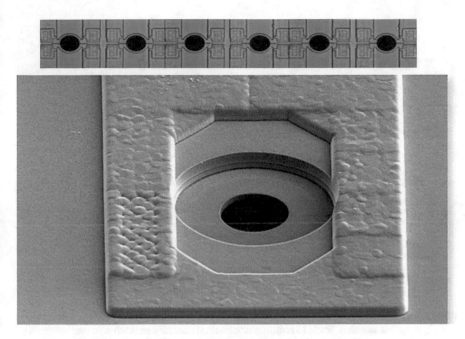

Fig. 19.2 Top image: Kodak Ultra-Stream nozzle array. Each nozzle has an annular heater positioned at the edge of the orifice. Bottom image: magnified nozzle with the annular heater around [2, 3]

two most common drop selection processes are illustrated. In the first schematic, called binary deflection mode, unwanted drops are deflected toward the gutter while others have a straight line till the matrix. Each nozzle can therefore jet on a single position on the matrix. The second schematic shows the multiple deflection mode, in which a single nozzle can print to multiple position on the matrix, reaching higher speeds compared to the binary mode.

The drop deflection is based on a fixed amount of charge transferred to the drops by the charging electrodes, leading to a drift at the desired distance. In earlier CIJ systems, the charging and deflection plates were not used, in favors of a simpler approach, based on a controlled airflow. This is the case of the Kodak Prosper CIJ line-up (Fig. 19.4), predecessor of the current Ultra-Stream, but still used commercially. In that configuration, a precise airflow is deflecting the small drops towards the gutter.

This technique is very flexible achieving a high printing throughput. It is therefore used where high printing speeds are required, in the order of 150 meters per minute. Some applications are printing of newspapers, marketing material, books, and packaging. It is a very reliable system, but due to its higher complexity compared to drop on demand, it could have increased maintenance costs and thus mostly used in industrial environments. For example, charge and deflection require complex electrical equipment and positional displacement may occur because of slight charge noise. In contrast, Prosper's drop selection using airflow is an extremely simple, cheaper, and safe system.

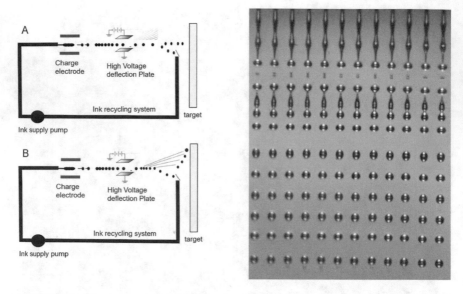

Fig. 19.3 CIJ deflection modes: (**a**) binary and multiple (**b**). In binary mode each nozzle can print to a single matrix position, while in multiple mode a single nozzle can reach different positions on the substrate

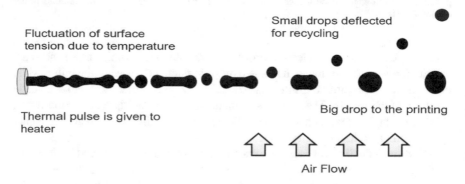

Fig. 19.4 Kodak Prosper airflow schematic. A precise and controlled airflow is used to deflect the unwanted drops toward the gutter

19.2 Drop on Demand (DOD)

Drop on demand, abbreviated DOD, is a more recent market compared to continuous jetting. Its main players are thermal jetting (TJ) and piezoelectric jetting (PJ). The main characteristic is the ability to jet drops of liquid on demand when needed, both technologies have their own strengths and weaknesses, and, depending on the fluid, one might be preferable over the other.

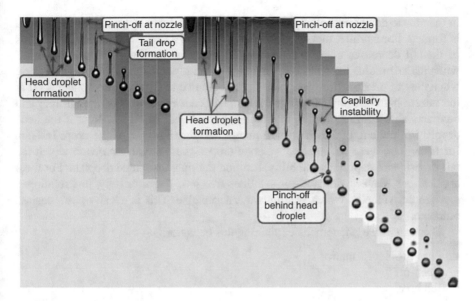

Fig. 19.5 Two droplets' streams are presented: the one on the left has a lower driving waveform amplitude resulting in slower drops, while the one on the right presents a higher amplitude, obtaining faster drops. The first case shows how the tail is joining the main drop after pinching off. Instead, in the right case, the main drop is too fast to be joined by the tail: capillary Rayleigh instability creates satellites

In the field of inkjet printing, companies in the DOD inkjet market mainly focused on a specific kind of technology: Canon, Hewlett-Packard, and Lexmark adopted the thermal actuation route as base for their products while Epson and Fujifilm Dimatix targeted the mechanical actuation.

In this section, after a description of the drop formation, the reader will find a description of both technologies with real examples taken from commercial printheads.

Drop-on-demand inkjet is based on the generation of different droplets, characterized by their volume, velocity, and trajectory, when required. The advantage over a continuous approach is the miniaturization possibility, a simpler system design and better control over used ink. Each single application will have its own preferential drop shapes, according to the resolution, speed, and precision needed. The printhead system with its geometry and acoustics along with the driving waveform play a major role in defining the characteristics of the droplets. The different behaviors can be summarized in the photo below (Fig. 19.5), where all the important jet properties are depicted: main drop, tail formation, capillary instability, and satellites.

We focus on the formation of a single droplet. The first volume of ink that flows out from the nozzle is soon joined by faster moving ink coming from the orifice, resulting in a droplet acceleration, and thus moving away from the nozzle. The drop enlarges due to incoming faster ink, joining the droplet. When the velocity of the ink in the nozzle decreases below the velocity of the droplet, it stops growing and a tail is formed. For a while, this tail connects the droplet and the nozzle until the radius of the tail decreases under a certain value, due to elongation. There is pinch off when the tail radius becomes zero at a single point, while tail breakup is occurring when the tail breaks at several points. Let us assume that the tail first pinches off at the nozzle before it pinches off at the droplet head. For slow jets, the tail rear end contracts and forms a droplet, in which the whole tail collapses. Now if the head droplet is slow enough, the two will merge, forming a bigger single drop. Instead, for faster jets, there will be two separate drops. There are also cases in which the tail experiences capillary instability, forming multiple unwanted droplets. For most applications, there is demand for satellites free jets. Unfortunately, this requires a low droplet velocity, which is not always acceptable. This is usually not a desired outcome.

There are three phenomena explaining this behavior:

- Head droplet formation
- Pinch-off
- Tail breakup

19.2.1 Head Droplet Formation

The main drop is characterized by its volume and velocity. By integrating the mass and momentum flux into the drop over time, an accurate estimate of the volume and velocity of the main drop is obtained. The mass flux is defined as $\rho A_n u_n$, where ρ is the ink density and u_n is the ink velocity exiting from the nozzle with area A_n. The momentum flux is then defined as $\rho A_n u_n^2$. The drop volume, V_d, is then obtained by integrating the flow rate passing through the nozzle $A_n u_n$ over time, between two-time limits: t_0 and t_e. The former coincides with the time when the meniscus emerges from the nozzle, while the latter is the time at which the ink velocity in the nozzle decreases below the drop velocity. Until t_e is reached, the drop continues to grow.

$$V_d = \int A_n u_n dt [t_e t_0] \tag{19.1}$$

The viscous friction causes a momentum transfer, and this is calculated under the slender jet approximation. The viscous component of the tension is so defined:

$$\tau_\eta = 3\eta A \partial_x u \tag{19.2}$$

depending only on the deformation rate and cross-section area.

The continuity equation is linking the stretching rate with the change in the cross section:

$$dA/dt = -A\partial_x u = -\tau_\eta/3\eta \qquad (19.3)$$

The overall momentum transferred from droplet due to viscosity tension is obtained by integrating the above equation between two different time instants and thus positions:

$$A(t_2) - A(t_1) = \int -\tau\eta/3\eta dt[t_2 t_1] \qquad (19.4)$$

This allows to define the momentum associated with a change from cross section A_n to zero: $p_\eta = -3\eta A_\eta$. The minus sign is there to express the droplet slowed down during the pinch-off process, just before detaching from the fluid present in the nozzle. Viscosity is not the only player and surface tension can also transfer momentum when the main drop is ejected. This influences the capillary force created as soon as the fluid exits the nozzle. The change in momentum for the main drop due to capillary forces is so defined:

$$p_c = (t_e - t_0)\pi \, \sigma \, R_n \qquad (19.5)$$

being R_n the jet radius (assumed equal to the nozzle radius) and $\phi\sigma R_n$ the capillary force. An approximated equation for the drop velocity ud is obtained by considering the previous contributes (viscosity, capillary force) along with the fluid advection:

$$ud = 1/\rho V_d(\int \rho A_n u_n{}^2 dt - -3\eta A_\eta \quad p_c)[t_e t_0] \qquad (19.6)$$

The drop velocity so defined is valid when the velocity un, at which the ink is ejected from the nozzle, becomes slower than this droplet velocity. According to this equation, the drop velocity could also be negative. The corresponding jet velocity at which the drop speed crosses zero is defined as the jetting threshold. If the drop speed is much lower than this threshold, drop formation is unlikely to happen.

19.2.2   Pinch-off, Breakup, and Satellites

To effectively predict whether the tail will merge or not with the head droplet, the tail drop velocity is calculated. Without approximation, this velocity also depends on the tail pinch-off time and tail breakup time. Determination of the pinch-off time is not an easy task as it may be caused by many different effects hard to be separated. In the slender jet approximation with neglected radial momentum, the stretching rate tends to infinity at the nozzle when the fluid ejection velocity decreases through zero: this means an instantaneous pinch-off. In real cases, radial inertia has also to be considered and would limit the stretching rate, so that pinch-off will occur later, being not anymore instantaneous. In Fig. 19.5 both pinch-off at the nozzle and at the main drop can be observed. All fluids ligaments, influenced by viscosity, surface, and capillary tension, generate a retraction force, trying to minimize the available surface into spherical drops. Due to Rayleigh-Plateau instability, it may happen that the ligament splits into multiple drops. There is also the relative motion between the meniscus, tail, and the head droplet with different velocities. The meniscus motion is further influenced by the ink channel backpressure and the signal of the driving waveform: when it retracts, it draws a thin amount of tail liquid. Due to these different phenomena, there are portions of the tail with very thin sections, leading to a pinch-off. The acoustics of the printhead also play a main role, possibly introducing Rayleigh-Plateau like perturbations. After the tail has pinched off at the meniscus, the imbalance in the capillary tension at the end of the tail causes the formation of the tail droplet, still linked through a fluid ligament with the head droplet. If velocities are comparable, the capillary tension, providing enough force, pulls the tail droplet toward the head droplet. At the same time, the tail droplet mass increases sweeping up the ink, which slows down the droplet. In this complex process, the jet radius R and its relationship with the pinch-off time t_0 are key players and a power-law relationship is generally used to describe it:

$$R = k(t_0 - t)^m \tag{19.7}$$

being t_0 the pinch-off time and k, m constants. During the last stage before pinch-off, the three forces (inertia, surface tension, and viscous force) are of the same order, which defines the inertial-viscous flow regime, characterized by a $m=1$ [7, 8]:

$$R = 0.0304 L_\eta / t_\eta (t_0 - t) \tag{19.8}$$

with L_η and t_η characteristic viscosity length and time and $k=0.0304$ addressing the most stable solution. This is not the only solution possible for the inertial-viscous flow regime, as others are also possible, but it is the most stable. For example, another solution for low viscosity fluids implies that nozzle radius Rn and tail radius R are way large compared to viscous characteristic length L_η, so viscous forces play even less role. In this case the fluid behaves like an inviscid fluid (with negligible viscosity) and can be described by a simplified version of the Navier-Stokes equations, the Euler equation. According to this, the minimum tail radius can be defined as [9]:

$$R = 0.7(\gamma/\rho)^{2/3}/(t_0 - t)^{2/3} \tag{19.9}$$

with $m=2/3$. As time flows and the tail radius approaches the characteristic viscosity length L_η, viscous forces become more important leading to deviations from the inviscid fluid approximation, just before the pinch-off. With very high viscous forces, this time inertia plays no role. The tail will be very long compared to previous cases with a minimum radius somewhere in the middle. In this case, characteristic lengths such as nozzle and tail radii (R_n and R) are small compared to viscosity

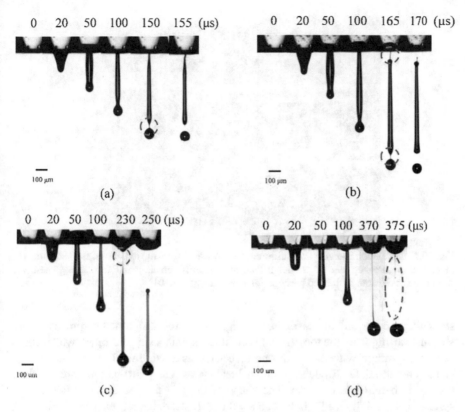

0   20   50   100   150   155 (µs)

(a)

100 µm

0   20   50   100   165   170 (µs)

(b)

100 µm

0   20   50   100   230   250 (µs)

(c)

100 um

0   20   50   100   370   375 (µs)

(d)

100 um

Fig. 19.6 Illustration showing four pinch-off types in DOD printing. The solution used is a sodium-alginate water-based fluid with increasing viscosities and driving waveform amplitudes. (a) 0.30% 35 V, (b) 0.30% 42 V, (c) 1% 50 V, (d) 2% 50 V. The tail pinch-off is highlighted with a dashed blue line [11]

length L_η. In a viscous flow regime, the balance between viscous and surface tension forces results in $m=1$ [7]:

$$R = 0.0709 L_\eta / t_\eta (t_0 - t) = 0.0709 \gamma / \rho (t_0 - t) \qquad (19.10)$$

The ratio defined by $L_\eta / t_\eta = \gamma / \rho$ is known as characteristic capillary velocity of the fluid. In real cases, the flow regimes are not totally separated and there may be some crossover, leading to more complex flow regimes [10] (Fig. 19.6).

Tail breakup is another process, parallel to pinch-off, which can occur during jetting. It may happen that the tail, instead of merging with the head drop, splits in multiple smaller drops, called satellites. Jetting and fluid parameters could generate that different satellites remain separated or eventually join into a bigger droplet, generating a single satellite. The tail breakup cause may be correlated to Rayleigh-Plateau instability, like in CIJ, even if its probability decreases as the viscosity and

Fig. 19.7 Simulation showing three different tail processes, from top to bottom: stable contraction to one drop, breakup due to Rayleigh-Plateau instability, breakup tail formation. These are simulation results ran with fixed perturbation amplitude of $\epsilon = 0.01$

stretching of the tail increase. In the Fig. 19.7 three different tail processes are shown. Starting from the top, the tail is contracting into a single droplet, which could eventually merge with the main one or behave as a satellite. The second process is the characteristic Rayleigh-Plateau breakup, where perturbations are splitting the tail into multiple droplets. The shape of the just formed drops is not always spherical, especially for higher viscosity fluids. Instead, the last example is showing a just formed tail, after meniscus and main drop pinch-off [5, 6, 12].

Satellites formation may be due to the just seen processes or because the driving system is providing too much kinetic energy to the fluid in the channel. Based on ink properties, the provided energy could be too much and win against the surface tension, leading to a drop split. Other satellites may come from the tail splitting. In any case these phenomena are unwanted for the inkjet process and lower the overall quality of the printing, especially in industrial applications.

19.3   Thermal Jetting Actuation

Thermal inkjet was invented and developed concurrently by Canon and Hewlett-Packard in the late 1970s and early 1980s. This method is still used in a large portion of home and office inkjet printers. Inside a cavity behind the nozzle, there is a small resistive heater in good thermal contact with the fluid. Rapid heating, within a few micro-seconds, causes superheating and vaporization of a thin layer of fluid in contact with the heater: a vapor bubble quickly expands, producing the fluid displacement and energy necessary to force a drop from the nozzle. Once the drop is ejected, the heat source is switched off and the bubble rapidly collapses,

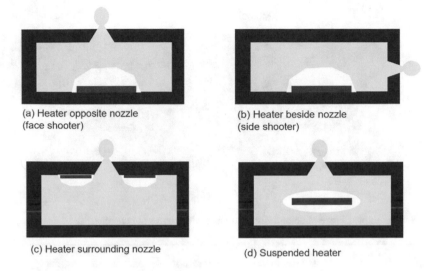

(a) Heater opposite nozzle
(face shooter)

(b) Heater beside nozzle
(side shooter)

(c) Heater surrounding nozzle

(d) Suspended heater

Fig. 19.8 Possible configuration of DOD thermal inkjet printheads

drawing in fresh ink for the next actuation. The second pulse starts again to generate
another droplet. One main advantage of this technique is that the actuator is simply
a resistive track which can be fabricated according with multilayer fabrication
processes such as standard microlithography processes, to guarantee precision,
repeatability, and quality of the product in mass production. The resolution realized
on the different products ranges from 150 to 600dpi and the most recent up to
1200dpi.

The ink formulation plays a key role for TIJ since it should be able to withstand
heating and vaporization processes, limiting deposited components on the heater
surface or within the nozzle (kogation). It also requires certain amount of water or
other easy to boil fluid to promote the vaporization.

The main difference between TIJ designs is the position of the heater relative to
the nozzle, as can be seen in Fig. 19.8 [6].

The most common configurations are the face shooter (also called roof shooter)
and the side shooter. The former has been patented and used by Hewlett-Packard,
while the latter has been invented and adopted by Canon. HP introduced in 1984 the
world's first mass market TIJ personal printer, named Think-Jet. Canon, one year
after, introduced the Bubble-Jet BJ-80 printer.

It is interesting to add that this design can also be used for other applications such
as a generator for small drops for hydroponics application seen in Fig. 19.9.

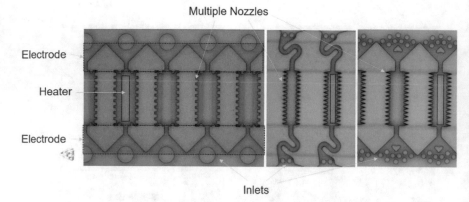

Fig. 19.9 Example of face-shooter design with multiple nozzles per chamber

Name	Roof shooter		Side shooter	Back shooter	Floating
Heater(s) configuration	Single rectangular	Double rectangular	Single rectangular	Double surrounding	Single narrow, long
Employing company	HP Canon third generation Lexmark Kodak	Sony	Canon first, second generation Xerox	BenQ	Silverbrook
Head structure					

Fig. 19.10 TIJ printhead configurations

19.3.1 Thermal Inkjet Printing

An introduction to thermal inkjet printing has been given in previous paragraph. In summary, the picture below presents the common printhead configurations. (Fig. 19.10)

The TIJ technology is based on the rapid and sudden boiling of the ink in contact with the heater. This generates a pressure peak inside the firing chamber, leading to the jetting of fluid from the nozzle. The working principle is focused on the bubble growth process.

19.3.2 *Bubble Growth*

Generally, when there is a heat transfer surface in contact with a liquid, the bubble formation can occur by spontaneous nucleation, either heterogeneous or homogeneous. The former refers to bubble creation by nucleation sites, gas pits, or defects, on the heater surface, while in the latter heat perturbations in the liquid itself are the driving force. In TIJ technology very rapid heating in the order of 100 $°C/\mu s$ is employed. Under this condition, the ink becomes a superheated fluid, well beyond its normal boiling point, and explosive and rapid vaporization is known to occur. Most TIJ inks are commonly water based, thanks to the high heat transfer capacity and detonation of the water. The boiling temperature of water is 100 °C at 1 atm; however, if rapid heating occurs, the boiling temperature of water rises to nearly more than 300 °C. The bubbles that are generated under this condition are the driving force for the ink jetting. If the driving conditions are appropriate for the heater configuration and the selected fluid, then the surface is instantly covered by many small nuclei, and they combine into a thin vapor film, before they become large. Bubbles so formed are growing by inertia, increasing the internal pressure, till an equilibrium is reached with the external ambient pressure and the bubbles shrink. Only when a specific minimum amount of energy, called the turn on energy, is provided by the driving signal, the bubbles grow, and the generated pressure is able to eject the ink. This is summarized in the following pictures. (Figs. 19.11 and 19.12)

Left part of Fig. 19.13 shows the bubbles evolution with time. The boiling incipience is around 250 °C and corresponds to the surface covered by many small bubbles. The driving conditions are providing enough energy and the bubbles grow and merge, providing optimal jetting pressure. In case, too small turn on energy, TOE is provided to the heater, resulting in noneffective bubble growth.

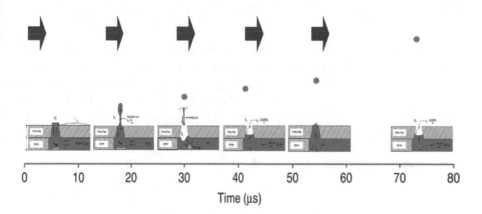

Fig. 19.11 Bubble growth, drop ejection and refill of a TIJ nozzle

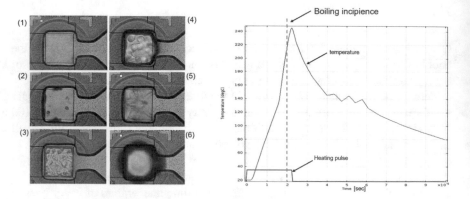

Fig. 19.12 Left: thin-film heater with dimensions of 50×50 μm made of TaSin. Pulse power $Q = 15$ W, pulse width $\tau = 2.5$ μs, turn on energy TOE = 0.0000375 J. The temporal evolution with bubble growth is presented. Right: TIJ jetting mechanism in relation with pulse heating, temperature, pressure, and bubble volume

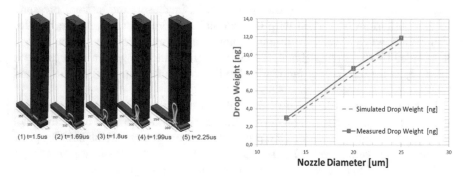

Fig. 19.13 Left: TIJ bubble evolution with time. Right: change of drop volume for several nozzle diameter for the same configuration and driving signal introduced on the left

Since bubbles are generated from a "superheated liquid layer," the size of the bubbles depends on the temperature of the liquid before boiling incipience, and this is reflected on the formed drops. It is possible to increase the amount of ink that is jetted by preheating the liquid layer with multiple pulses before the main one. Figure 19.14 shows an example of a pulse waveform that carries out preheating and a boiling experiment conducted using a commercially available HP51604 with dimensions of 100×110 μm. The boiled ink volume in the (c) case is way higher with respect to a single driving pulse [5, 14, 15].

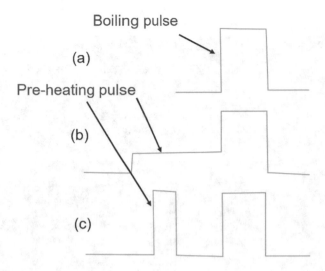

Fig. 19.14 (**a**) boiling incipience, (**b**) bubble formed by a standard single pulse, (**c**) bubble formed by a preheating pulse

19.3.3 Main Jetting Issues

TIJ heaters are thin-film resistive elements dissipating heat by Joule effect. Resistances are usually in the order of hundreds of Ohms with common driving voltages lower than twenty Volts. Dissipated currents are in the order of hundreds of mA to even few Ampere in the case of very small heaters or large arrays. On top of this, high jetting frequencies make the driving signal resembling to a DC current, and thus increasing the adsorbed current. Because of these facts, the design of the printhead and driving board needs to fulfill such power requirements with a careful handling of the generated heat. Stored heat inside the printhead is another critic point, since the materials used are not so heat conductive, especially because heat is affecting the bubble size and thus the droplet volume. Heat dissipation needs to be accounted to guarantee precise performances over prolonged printing periods. One of the common ways to achieve that is to modulate the driving pulses based on the temperature gradient in the printhead. Moreover, crosstalk occurs between the jets because they are closely spaced apart by tens of micrometers. It is composed of an acoustic effect, due to pressure oscillations from boiling, as well as a fluidic effect, where each nozzle competes against the others to draw the ink. Mechanical problems arise when the pressure that is generated from the heater is too large. Pressure is generated twice with the production and collapse of bubbles on the heater surface and could reach several MPa. In particular, the cavitation is concentrated in the center of the heater during bubbles collapse, increasing the mechanical damage and the risk of heater breakage. Tantalum could be deposited to prevent that, but this lowers the efficiency because it increases the heat capacity of the heater. Ink

Top view

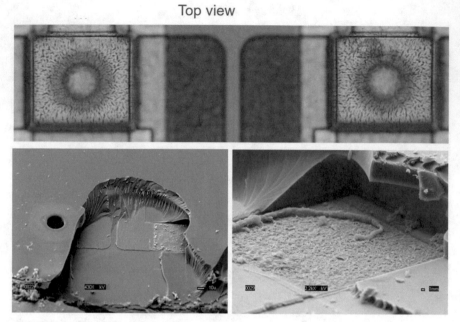

Chamber partially removed

Fig. 19.15 Ink residuals (kogation) deposited on the heater surface after high printing cycles

residuals deposited on the heater surface is another TIJ critic point. Called kogation in the industry, it is usually manifested after high printing cycles, lowering the heater efficiency (Fig. 19.15) [5, 13].

19.4   Piezoelectric Actuation

The first pioneering work in the piezo inkjet was performed in the late 1940s by Hansell of the Radio Corporation of America (RCA), who invented the first drop-on-demand device. By means of a piezoelectric disc, coaxially arranged with an ink-filled conical nozzle, pressure waves were generated, and this led to a spray of ink. However, this invention, designed to serve as a writing mechanism, was never developed into a commercial product [16]. Generally, the basis of piezoelectric inkjet printers is attributed to three patents in the 1970s [17]. The common idea was that the piezoelectric device could be used to convert an electrical signal into a mechanical deformation of the ink chamber, generating the required pressure for the drop formation. The first came from Zoltan of the Clevite company in 1972, who proposed a squeeze mode of operation served by a hollow piezo tube [18]. The second patent came from Stemme of Chalmers University in 1973 who used the piezoelectric bend mode operation. In this mode, the bending of a wall of the ink chamber is used to eject a drop: therefore, the wall is made of a diaphragm with a piezoceramic bonded onto it [19]. The third patent came from Kyser and Sears of the Silonics company in 1976 who also used the bend mode operation [20]. The first PIJ printer to reach the market was the Siemens PT-80 in 1977, which used the squeeze mode. The PT-80 had a resolution of 120 dpi with an array of 12 nozzles, firing drops at a maximum frequency of 2.5 kHz. Silonics PIJ printer, Quietype, present in the market in 1978, requiring a 150 V driving amplitude to fire drops at a maximum frequency of 3 kHz. The shear mode design was present in the market in 1984 with Spectra and their Galaxy and Nova printheads. Spectra was acquired by Markem, later by Dimatix and finally by Fujifilm. A special version of the shear mode is the shared wall design of Xaar. Epson emerged as the PIJ leader with the introduction of the 12 nozzle SQ-2000 in 1984 and the big success of the Stylus 800, introduced in 1993. The printhead had 4 × 12 nozzles at a resolution of 90 npi. The principle behind PIJ printheads is to exploit the anisotropic piezoelectric effect of some ceramics, namely, the lead zirconate titanate (PZT). Depending on the direction of the actuation field, the material can be excited in different ways, each one characterized by a response due to its anisotropic nature, leading to multiple working modes [5, 6, 10, 21] (Fig. 19.16).

The working concept is that when the piezoceramic actuator is deforming, a pressure variation in the chamber is created, able to eject a drop though the nozzle. The squeeze mode design is conceived with two piezoelectric elements squeezing an ink channel, thus creating the needed pressure for the drop ejection. It never evolved further towards a multi-nozzle printhead for industrial printing markets. The electric field is parallel to the polarization direction of the material. Single nozzle devices or heads with only a few operating nozzles were developed in several laboratories. In bend mode, the electrical field is applied in the polarization direction of the piezo material. Since the piezoceramic is bonded onto a passive membrane, the actuator will bend, reducing the ink channel width and thus creating an internal pressure ejecting a drop from the nozzle. The passive membrane is a stack of metal oxides including the electrodes, which are commonly made of platinum Pt, TiW, or LaNiO3. With the push mode, also referred to as bump mode, a piezoelectric element pushes against the chamber wall to later deform it. The electrical field is applied perpendicular with respect to the deformation response of the ceramic material. In the shear mode, the strong shear deformation component is used to deform an ink chamber wall. The electric field is perpendicular to the polarization direction of the material. Each company, dealing with PIJ printheads, concentrated into developing solutions based on a chosen working mode among the mentioned ones (Fig. 19.17).

Epson is not the only one following the PJI market but other main players in the field of commercial printers are Xaar, Fujifilm Dimatix, and Brother. Regarding the first two, they share the common actuation mode: shear mode.

In Fig. 19.17, a special version of the shear mode is shown, using the shared-wall principle, where the piezoceramic element is also the channel plate [23, 24].

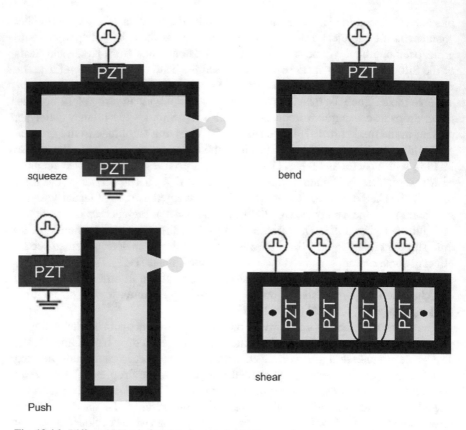

Fig. 19.16 Different PIJ printhead designs and actuation modes

19.4.1 Piezoelectric Inkjet Printing

This chapter is focused on describing the structure dynamics and its working principle with regard to key parameters used to predict the overall performances. As a recall, the printhead configurations can be reassumed in Fig. 19.18. For simplicity, the reference structure will be the bend mode actuation principle. The general concept is to create a large pressure inside the ink channel, also referred to as chamber, used to jet a drop from the nozzle.

19.4.2 Working Principle

A commonly used bend mode configuration is presented in Fig. 19.19. There could be different designs with, for example, the ink supply duct perpendicular to the chamber or with the nozzle positioned just below the actuator. The working mode

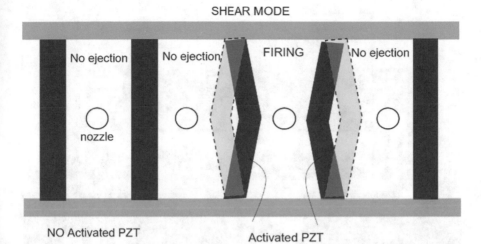

Fig. 19.17 Shared-wall shear mode actuation schematic in a multiple-nozzle drop-on-demand (DOD) printhead

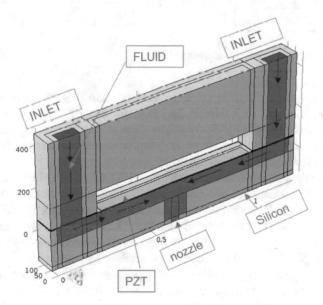

Fig. 19.18 3D bend mode structure with key features and internal geometries

and principle remain the same: a large pressure gradient is generated, able to jet the fluid from the nozzle. The ink supply duct, or inlet, provides the needed fluid flow. It is usually connected to the pumping chamber via a throttle (a relatively long duct with a small cross section). The channel is finally connected to the ambient via a nozzle (a duct with a short length and a small-sized exit opening) from which

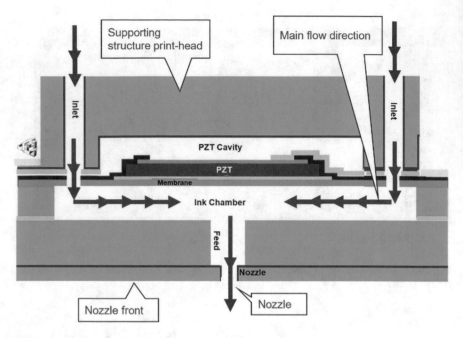

Fig. 19.19 Bend mode Helmholtz resonator chamber

the drops are ejected. Bend mode usually adopts thin-film PZT membranes with
typical dimensions up to 2 μm of thickness, to create highly packed structures
reaching the needed printing density. The term membrane usually addresses the
whole PZT stack, comprehensive of the electrodes and also the insulating layers.
The large pressure gradient is obtained by applying an electric field in direction 3,
relative to the coordinate system introduced in PZT Chap. 10. Membranes actuated
in this way rely on the high $d31$ coefficient to achieve high strain in the direction
1. This turns into a deflection of the membrane itself and consequently of the ink
chamber wall, generating the wanted pressure. The system can be actuated by two
different ways: push-pull and pull-push (Fig. 19.20). Standby mode in push-pull is
the zero voltage (Fig. 19.21). Droplets are produced by applying a voltage pulse
to the actuator, generating a sudden positive pressure in the chamber, able to eject
the ink. The maximum positive pressure is created after the leading pulse edge, by
which the membrane has reduced the channel section. The subsequent trailing edge
restores the membrane to its original shape. Instead, in pull-push mode, standby
voltage is different than zero (Fig. 19.22). To produce ink droplets, the amplitude is
returned to ground removing the strain on the membrane. This movement, during the
leading pulse edge, enlarges the channel cross section, effectively creating a negative
pressure in the chamber and the meniscus is pulled backward. The negative pressure
wave travels through the ink channel towards the inlet, and it is reflected at the ink
reservoir, where the acoustic impedance of the channel changes. After the reflection
at the reservoir, it becomes a positive-pressure wave. The positive-pressure wave is

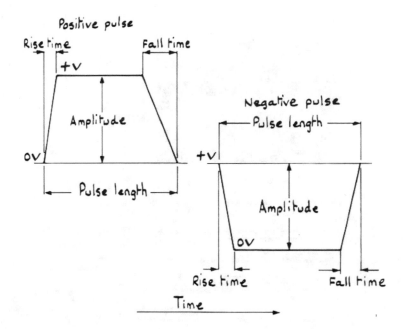

Fig. 19.20 Left: positive pulse used in push-pull mode. Right: negative pulse adopted in pull-push mode

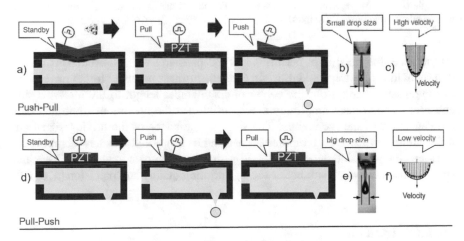

Fig. 19.21 (**a**) Piezoelectric element driving method, (**b**) droplet photograph, and (**c**) schematic diagram of velocity distribution inside inkjet head, for pull-push method. (**d**) Piezoelectric element driving method, (**e**) droplet photograph, and (**f**) schematic diagram of velocity distribution inside inkjet head, for push-pull method

amplified by the trailing edge of the driving waveform, obtaining a large positive-pressure peak at the nozzle, which fires a drop. That is why a perfect timing between the two positive pressure waves is required: only if they generate a constructive

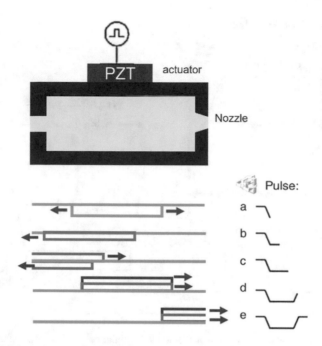

Fig. 19.22 Bend mode actuation scheme along with the pull-push signal on the right. Blue rectangle is the first negative pressure wave, while the red one is the second positive pressure wave, undergoing constructive interference

interference, the maximum performances and pressures are achieved. This means that the width of a trapezoidal driving waveform must be tuned to the travel time of the pressure wave from the center of the chamber length to the reservoir side and back. The voltage amplitude is responsible for the maximum strain and thus the deflection of the membrane. The slope of the edges of the driving signal are influencing how fast the membrane is changing its shape and thus how abrupt is the pressure change inside the channel. Pull-push mode is known to generate faster and smaller droplets, which suit better for high printing quality and precision [5, 6, 10].

19.4.3 Acoustics

The acoustic properties of the ink channels play a central role in the printhead operation. The dynamics of the printhead structure are coupled through the acoustics to the drop formation process in the nozzle. The negative pressure wave generated in a pull-push actuation design and needs to be reflected as much as possible at the inlet or ink reservoir, to minimize the acoustic crosstalk between adjacent chambers. This ensures even and homogeneous performances across the nozzles in highly packed arrays. Moreover, higher performances and nozzle pressures are generated if the whole wave is reflected. Pressure waves propagate in the channel and will be reflected when the characteristic acoustic impedance Z of the channel changes. The acoustic impedance depends on the size of the channel cross section A, the density of the ink ρ, and the effective speed of sound $ceff$ as:

$$Z = (\rho ceff)/A \tag{19.11}$$

The speed of sound is influenced and typically reduced by the compliance β of the channel cross section, reason for which the effective speed of sound $ceff$ is used and defined as:

$$c_{eff} = \sqrt{c_0^2/(1 + \rho c_0^2 \beta)} \tag{19.12}$$

with β being the compliance and $c0$ the speed of sound of the ink.

The effective speed of sound can also be derived from the width of the most efficient driving pulse, the one producing droplets of highest kinetic energy:

$$c_{eff} = l_{ch}/t_{travel} \tag{19.13}$$

being L_{ch} the channel length and t_{travel} the width of the pulse. The reflection and transmission coefficients at the interface between inlet channel (Z_1) and chamber channel (Z_2) are:

$$R = (Z_2 - Z_1)/(Z_1 + Z_2) \quad T = 2Z_2/(Z_1 + Z_2) \tag{19.14}$$

When the compliance does not change, the following relationship holds:

$$R = (A_1 - A_2)/(A_1 + A_2) \quad T = 2A_1/(A_1 + A_2) \tag{19.15}$$

At the large reservoir ($A2 \gg A1$), the transmission coefficient is zero and the pressure wave will be completely reflected. This is an important aspect for both minimizing the acoustic crosstalk effect and achieving high firing performances in pull-push mode. There are different approaches to achieve this but the best one from multiple points of view (production, efficiency, system frequency) is having smaller section inlets in contact with the pressure chamber. The aim is to increase the acoustic impedance Z to have reflected waves. With an inlet located at the beginning of the pressure channel longer than 10% of the chamber length and smaller than 25% of the channel cross section, less than 1% of the pressure waves from the channel are transmitted. In addition to these, there are also other constraints in the design of the system. For example, the inlets must provide the correct ink feed to refill the chamber after each firing cycle and this is hindered by the increasing viscosity of the ink. That is why a proper fluid back pressure needs to be maintained and scaled accordingly with the actuation frequency, since the time to refill the chambers will be way reduced. Another constraint is imposed by the desired resonant system frequency which impacts on the maximum actuation frequency (how many drops per second are achievable) and is reflected on the width of the most efficient pulse (OPW = optimum pulse width). They are linked together by:

$$T = 1/f \, OPW = T/2 \qquad (19.16)$$

If the driving waveform is tuned around the OPW, then the maximum efficiency and energy transfer is achieved, resulting in higher kinetic energy of the drops. The acoustic design of the structure, along with the channel cross sections, play a key role for the frequency response and crosstalk of the whole system [5, 6, 10].

19.4.4 Frequency Response

The Helmholtz resonance is the phenomenon describing the fluid resonance in a cavity. It explains why, when you blow in a bottle, a sound is generated or the principle behind how a flute emits a particular tone. It is applied to a very broad range of fields, some of them are music (instruments, subwoofers), engines (especially two-strokes in exhaust systems), and architecture. The Helmholtz resonator is comparable to a classical mass-spring system, which has a characteristic resonant frequency. The relatively large volume of the actuation channel can act as a spring, with the fluid acting as vibrating masses in the inlets and in the nozzle. The mass of the nozzle or restriction is equal to ρAL, with A_n or A_r the cross section and L_n or L_r the length of the nozzle or restriction. The spring constant k can be derived from the bulk modulus of the liquid and the pressure acting on the cross section of the nozzle or restriction. C is the compliance while the acoustic inertance or inductance is M:

$$m = \rho ALk = pc^2 A^2/VC = V/\rho c^2 M = \rho L/A \qquad (19.17)$$

The Helmholtz frequency is then equal to:

$$f_H = c/2\pi \sqrt{1/V_c(A_n/L_n + A_n/L_r)} \qquad (19.18)$$

In a design composed of an inlet, actuator, pressure chamber, descender, and nozzle, by inserting the compliance C and inductance M defined before, one obtains:

$$f_H = 1/2\pi \sqrt{1/(C_{in} + C_{ch} + C_a + C_d + C_n)(\Sigma_i 1/M_i)} \qquad (19.19)$$

being *in* the inlet, *ch* the channel, *a* the actuator, *d* the descender, and *n* the nozzle. The sum over the inductance is made on the same elements. Some contributes may be neglected, like the compliance of the nozzle. This formula could also be refined by scaling the relevant factors depending on the design or the fluid used. For example, in fluids with higher viscosity, the inductance of small cross sections plays a bigger role. The so defined Helmholtz frequency represents the frequency and tone at which the whole system is in resonance, providing the maximum efficiency [5, 6, 10, 25, 26].

19.4.5   Lumped Elements Modelling

A lumped element modelling (LEM) approach is generally used, when describing such systems, to define the characteristic properties. The main assumption employed in LEM is that the characteristic length scales of the governing physical phenomena are much larger than the largest geometric dimension. For example, in an acoustic system, the acoustic wavelength must be significantly larger than the device itself. If this assumption holds, then the governing partial differential equations for the distributed system can be "lumped" into a set of coupled ordinary differential equations. This approach provides a simple and relatively quick method to estimate the dynamic response of a piezo system. This is possible by the means of an electric-fluid analogy: electrical current analogous to net volume flow rate, pressure analogous to voltage drop, electrical capacitance analogous to fluid capacitance, and fluid inertance analogous to electric inductance. The ink flow at microscale can be considered, with good approximation, to be a viscosity dominated flow with constant fluid density and viscosity. In LEM analysis the individual components of a piezo system are broken into segments of an equivalent electrical circuit, like the one described in Figs. 19.23 and 19.24 [35].

The volume of a channel is represented by its compliance C or in electrical terms by an equivalent capacitor. The resistance to flow is modelized with a resistor while the mass and its inertia by an inductance. This way, the whole system is turned into an equivalent electrical circuit, which can be solved by many simulators and key parameters like the resonant frequency extracted. For a rectangular cross-section design with circular nozzle, the following equations hold [26, 27]:

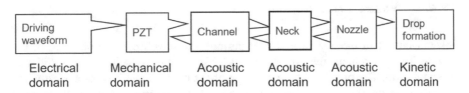

Fig. 19.23 Model structure representing the different domains found in a microfluidic piezoelectric system

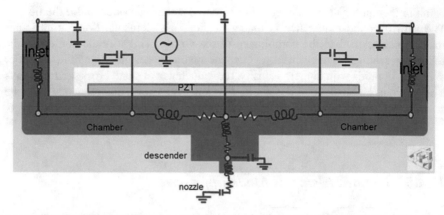

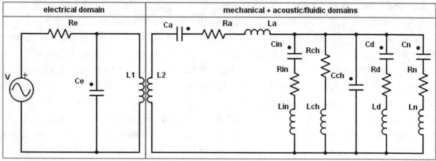

Fig. 19.24 Equivalent electric circuit to simulate the dynamic behavior of an inkjet channel with a small inlet, a cavity with a compliant actuator, a descender, and a nozzle

$$C_{rectangular} = V/\rho c^2 = whl/\rho c^2 \quad L_{rectangular} = M = \rho l/wh \qquad (19.20)$$

$$C_{nozzle} = \pi r^4/3\sigma \quad L_{nozzle} = \rho l/\pi r^2$$

$$R_{rectangular} = (12\eta l/hw^3)/(1 - 192w/\pi^5 h \Sigma_i \tanh(i\pi h/2w)/i^5) \quad R_{nozzle} = 8\eta l/\pi r^4$$

19.4.6 Nozzle Pressure

Ink movement in the nozzle is driven through the pressure at the entrance of the nozzle. To fire droplets, the average meniscus speed u must be of the same order as the final drop. The pressure at the entrance of the nozzle must overcome steady and unsteady inertia, the viscous forces, and forces resulting from the surface tension of the ink. It is possible to have a quick approximate evaluation of the pressure inside the nozzle by considering each governing parameter.

An estimation for the required pressure to overcome inertia is given by the Bernoulli pressure p_b. The unsteady pressure p_i given by the inertia of the mass is calculated by Newton's second law. Surface tension also generates a pressure, and it is accounted with p_c. In designs with sharp nozzle edges, the angle θ is equal to $0°$. The viscous resistance can be estimated assuming a laminar flow (Poiseuille flow profile), which results in a pressure term of p_v.

$$p_b = \rho u^2/2 \, p_i = \rho l_n du/dt \, p_c = 2\gamma\cos\theta/r_n \, p_v = 8\pi\eta l_n u/A_n \qquad (19.21)$$

with a nozzle diameter of 35 μm, length 10 μm, ink viscosity 10 cP, and a final drop speed of 11 m/s, the calculated pressure is 1.3 MPa which is a reasonable result according to simulations [10].

19.4.7 Efficiency

Electromechanical coupling coefficients such as k_{33} and k_{31} describe the conversion of energy by the ceramic element from electrical to mechanical form or vice versa. The ratio of the stored converted energy of one kind (mechanical or electrical) to the input energy of the second kind (electrical or mechanical) is defined as the square of the coupling coefficient k:

$$k^2 = stored\ mechanical\ energy/input\ electrical\ energy$$

$$k^2 = stored\ electrical\ energy/input\ mechanical\ energy \qquad (19.22)$$

A way to calculate k is to express it as function of material coefficients such as d, s, and ε, being the piezo, strain, and permittivity coefficients, respectively. In case of longitudinal vibration through the piezoelectric effect (d_{31}) in a rectangular plate, the sound velocity in the ceramic material, the f_r resonant frequency and k_{31} are given by:

$$v = 1/\sqrt{\rho s_{11}} \, f_A = v/2l = 1/2l\sqrt{\rho s_{11}} \, k_{31} = d_{31}\sqrt{s_{11}\varepsilon_{31}} \qquad (19.23)$$

with l being the lateral length of the ceramic element. A practical method to estimate the coupling coefficient is to use an instrument capable of doing impedance spectroscopy and actuating the piezo element.

This allows to plot the behavior around resonance with respect to the impedance Z or phase, effectively finding the resonance and antiresonance peaks. For a lossless resonator with ignored electrodes, the electromechanical coupling coefficient can be calculated by the equation defined by IEEE Standard on Piezoelectricity [36]:

$$k^2 = \pi f_s/2f_p \tan(\pi(f_p - f_s)/2f_p)k^2 = \pi^2 f s(f_p - f_s)4f_p f_p \qquad (19.24)$$

where the standard defines the series resonant frequency f_s as the frequency corresponding to the minimum impedance, and the parallel resonant frequency f_p as the frequency corresponding to the maximum impedance. The equation on the right is the approximated form in case of low coupling coefficients. The antiresonance frequency, f_p, is also known in literature as f_a, while fs is also referenced as f_r, resonance frequency. Regarding the longitudinal vibration through the piezoelectric effect (d_{31}) in a rectangular plate, the relationships between k and the frequency characteristics are defined as:

$$k_{31}^2/(1 - k_{31}^2) = \pi^2 fr(f_a - f_r)/4f_a f_a k_{eff}^2 = (f_a^2 - f_r^2)/f_a^2 \qquad (19.25)$$

When the effect of the electrode layers and the loss of the piezoelectric thin film cannot be ignored, the values of the critical frequencies usually differ from the ideal ones by a small amount. For this reason, the fundamental resonance of a thin-film resonator is usually quantified by the effective electromechanical coupling coefficient, k_{eff}^2 [10, 28, 29, 30, 31] (Fig. 19.25).

19.4.8 Pulse Design

The piezo actuation typically consists of a positive or negative trapezoidal pulse, depending on the system and the chosen working mode. Residual oscillations of the membrane are generally resulting after each actuation, leading to unwanted side effects such as satellites or, in some cases, to an increased drop volume. Therefore, the piezo driving signal should consist of an additional pulse to damp the residual oscillations. In the literature, there are two categories of actuation pulses, namely, unipolar and bipolar (Fig. 19.26). The unipolar pulse consists of the jetting pulse

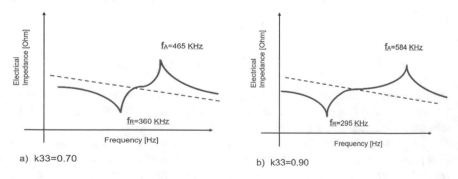

Fig. 19.25 Impedance curves for a good k PZT (k33 = 0.70) and a high k material (single crystal, k33 = 0.90)

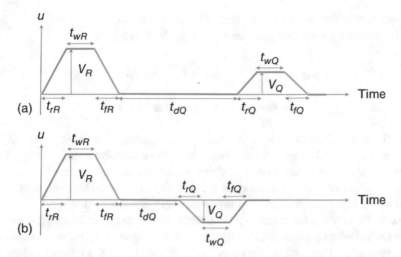

Fig. 19.26 Parametrization of (**a**) unipolar pulse and (**b**) bipolar pulse. tr is the jetting pulse while tq is the quenching or damping pulse

for the droplet and an additional trapezoidal pulse of the same polarity to damp the residual oscillations. The bipolar pulse consists of the jetting pulse for the ink droplet, followed by an opposite polarity damping pulse. From the driving electronics point of view, the unipolar pulse is beneficial as the required driving voltage range is smaller due to the same polarity. On the contrary, the bipolar actuation pulse allows earlier damping of the residual oscillations, improving the maximum jetting frequency. A multi pulses approach is also used to better control the drop volume, either by increasing it or achieving very fine drops.

The set of wave parameters so introduced play a key role into determining the properties of the droplets. They may be tuned by exhaustive experiments when users do not have information about printhead acoustics. However, with a proper use of physical insight, it is possible to obtain the actuation pulses with very few experiments.

The main information for the design of the pulses is the fundamental mode frequency of the inkjet printhead, obtained from the Helmholtz frequency as discussed in paragraph 19.4.4. It is possible to measure the period T $(T=1/F_h)$ of an ink channel by measuring the channel pressure with piezo self-sensing or by observing the displacement of the meniscus. A quicker way of measuring the period of the ink channel is by conducting a series of experiments where the droplet velocity is measured as a function of the pulse width. The droplet velocity will be maximum when the optimal pulse width $(t_{rR}+t_{wR}+t_{fR})$ is approximately equal to $OPW=T/2$. Once T is known, the pulse characteristics can be defined in terms of the period. For common inks used in document printing, rise time and fall time have a negligible influence on the jetting properties, although very short rise/fall times may introduce satellite drops due to the abrupt pressure change in the chamber. For ink viscosity in the range of 3–15 mPa·s, rise/fall times equal to $T8/$ is a fairly good choice. The other actuation parameters of Fig. 19.26 can be set to:

$$t_{rR} = t_{fR} = t_{rQ} = t_{fQ} = T/8 t_{wR} = t_{wQ} = T/2 - (t_r + t_f) \qquad (19.26)$$

$$\text{Bipolar} \, t_{dQ} = T/2 \, \text{Unipolar} \, t_{dQ} = T$$

As introduced before, with a multi pulse approach, it is possible to both achieve smaller and bigger drops, effectively broadening the volume range. To jet tinier drops, the two pulses need to create a partial constructive interference, thus decreasing the amount of the liquid ejected. On the other side, reaching higher volumes of several picolitres requires a different approach: the multi-drop firing protocol. The main idea behind this concept is to let many drops merge during the flight towards the substrate, ending up with a single bigger drop. This is achieved by using multiple jetting pulses, tuned at the OPW, arranged in a train of pulses. The driving voltage needs to increase for each pulse to jet faster drop each actuation. This allows the merging of the droplets during the flight, before reaching the substrate. This technique relies on a good and reliable driving circuit, able to generate very close pulses at increasing voltages. By exploiting the multi-drop firing protocol, drops of about 85pL have been jetted from a nozzle of 35 μm with a solution of ethylene glycol 10 cP and a six pulses train [5, 32].

19.4.9 Jetting Issues

To work as expected the pressure channel of a piezo printhead has to be full of ink between one actuation and the other. Problems arise when an air bubble enters the channel, leading to a very poor performance or even non-firing chambers. This is because air acts as a compressible fluid, vanishing the internal pressure created by the membrane. Impurities in fluid create the same problem. To avoid such issues, proper filters and the correct ink backpressure are employed: the former prevents air bubbles or impurities entering from the feeding channel, while the latter ensures the pressure chamber to be always full, avoiding air entering from the nozzle plate. Other problems are generated from various forms of crosstalk: acoustic, electric, and mechanic. Crosstalk is an issue where adjacent chambers are influencing each other while firing, affecting the velocity and volume of the drops. It could even happen, usually at higher frequencies, that an adjacent chamber starts to fire even if it is not actuated. This is because acoustic and mechanical waves propagate through the ink and close walls, reaching nearby chambers. Electrical crosstalk is given by the circuitry and cables feeding the driving waveform to the printhead and chambers. This could be minimized with a proper connection layout and minimizing the length of the cables. Acoustic and mechanical crosstalk can be controlled by a careful

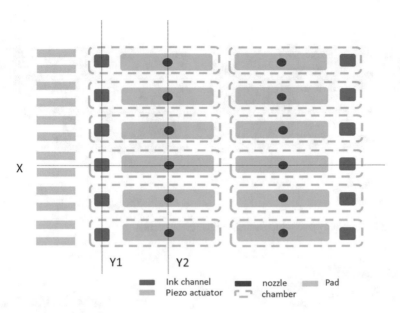

Fig. 19.27 Simplified layout of thin-film piezo printhead

design of the chambers as described in paragraph 19.4.2, as well as adding specific designed cavities near the pressure channels.

19.5   Thin-Film Piezo Printhead: Technology Description

To realize an Inkjet printhead, it is necessary to create some fluidic channels, where the ink can flow starting from the reservoir, some chambers, where the ink can be collected to be blown in an efficient way, and some nozzles, which allows the ink drop jetting.

The chamber has fixed wall and a flexible membrane, where on top of it is placed the piezoelectric actuator.

In Fig. 19.27 an exemplificative layout of thin-film piezo printhead is shown, where the relevant parts are drawn. In Fig. 19.28 the partial die cross section, referred to Fig. 19.27 layout, is reported.

It is clear from Figure 19.28b, the technology needed to realize it has to be easily scalable in Y direction, and since the number of nozzles increases, the drop size control and the jetting speed demands a high fluidic channel density and well-controlled tolerances.

In this paragraph a full silicon micro-machining technology process flow to realize a piezo thin-film printhead is presented. This can be one of the possible solutions, since there are some printhead realized by using a mix of MEMS based

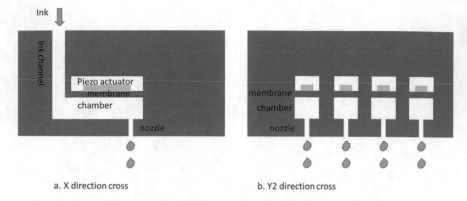

Fig. 19.28 Thin-film piezo printhead. Simplified cross section

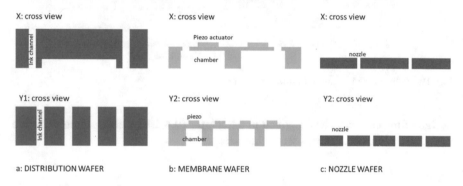

Fig. 19.29 Thin-film piezo-based printhead. Simplified wafer cross section

technology (hybrid MEMS technology) and mechanical part (e.g., metallic nozzle plate).

As described above the printhead die is composed by three main parts: the ink channel, the chamber with the actuator, and the nozzle. The simplest way to realize the final device is to use three different silicon wafers for each part and merge together with wafer-to-wafer (W2W) bonding technique.

By following this approach, three wafers must be realized:

- The first wafer is the one dedicated to the formation of the ink channel. Since it is the one that helps the distribution of the ink from the reservoir to the chamber, it is referred to as "distribution wafer."
- The second wafer is the one dedicated to the active part of printhead, the one where the actuator is placed, and the mobile membrane is realized. It is referred to as "membrane wafer."
- The third wafer is the one where the nozzles are realized. It is referred to as "nozzle wafer."

In Figs. 19.29 and 19.30, a picture of the final desired wafer structure is shown.

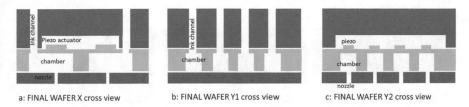

a: FINAL WAFER X cross view b: FINAL WAFER Y1 cross view c: FINAL WAFER Y2 cross view

Fig. 19.30 Thin-film piezo-based printhead. Final wafer cross section: the three different wafers are highlighted by the different usage of color (blue, DISTRIBUTION wafer; green, MEMBRANE wafer; brown, NOZZLE wafer)

The *distribution wafer* has to realize three features: the fluidic channel, the opening of pad region to allow the electrical access to the actuators and the testing, and the cavity where the actuators can move without any blocking. Two of these features are passing hole through the wafer, and to manage it, mainly two technology solutions can be possible. In figure 19.31 a schematic process flow for distribution wafer is proposed. The presence of different layout dimension on wafer (the ink channel has a diameter of 70–100um, while cavity and pad opening have length of millimeters and width of some hundreds of microns) and the different height (the ink channel is a pass-through hole, while the PZT cavity has to be as smaller as possible to make the wafer less weak) impose the twofold usages of hard mask and photoresist mask to reach all the targets.

A starting wafer thickness choice very close to the final ink-channel length can be 400um, which matches the fluidic design and the process industrialization. The first step is the deposition of hard mask material and its patterning as it is shown in Fig. 19.31 a. The hard mask must have a very high resistance to dry silicon etch chemicals (SF6), and a suitable choice can be, for example, silicon dioxide. After this step, a photo resist mask is used to open the first part of ink channel and the pad opening region, by using dry silicon etch. As show in Fig. 19.31 b, due to the different layout between ink channel and pad opening region, the dry silicon etch plasma suffers from loading effects. As a result, the pad opening will be completely open, while the ink channel will be partially etched. Then the photo resist is completely removed, and the deep silicon etch is continued by using the pre-patterned hard mask as etch stop on the front of the wafer. In Fig. 19.31c, the result is shown. The fluidic consideration on ink channel diameter control imposes that dry silicon etch is the only possible solution for channel realization, to keep a well-controlled straight shape. For this reason, the rear side of the wafers must be protected to avoid the equipment chuck damage during the over-etch phase. Different solution can be used: temporary bonding or an appropriate sandwich of layer that reinforces the silicon dioxide membrane and avoid breakage. These protections can be removed at the end of distribution wafer process or after W2W bonding with other wafers. To ensure a good printing quality, the fluidic channel must be very clean, without defectiveness that can compromise printhead

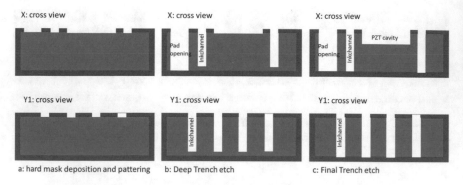

a: hard mask deposition and pattering   b: Deep Trench etch   c: Final Trench etch

Fig. 19.31 DISTRIBUTION wafer. Schematic process flow

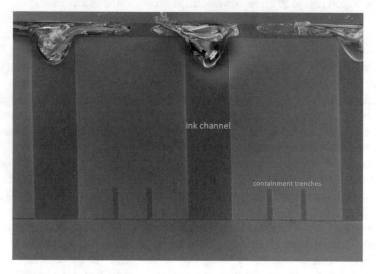

Fig. 19.32 Final ink-jet wafer. Ink channel SEM cross section performed along Y1 direction (see Fig. 19.27 for layout)

performance. Effective polymer cleaning plays a fundamental role in wafer quality. In Figs. 19.32 and 19.33, SEM picture of the ink channel is shown.

The *membrane wafer* schematic process flow is described in Fig. 19.34. The wafer cross sections are taken along the schematic layout reported in Fig. 19.27. The starting wafer material (SOI or bare silicon) choice depends on actuators membrane design requirements. If the silicon is the best material to fit membrane mechanical target, the starting material can be an SOI wafer, or an epipoly layer can be grown starting from oxidized silicon wafer. If a dielectric is preferred as membrane material to reach a higher deflection, a mix of silicon dioxide/silicon nitride can be a suitable arrangement. From integration point of view, since the tolerance in membrane height is fundamental to contain the resonance frequency spread, the starting spread of the as depo/as grown material is important, but a key aspect is to have a sandwich

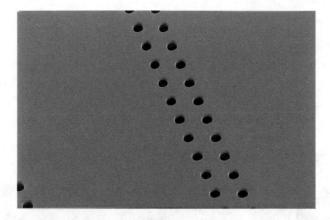

Fig. 19.33 Final ink-jet wafer. SEM top view of ink-channel

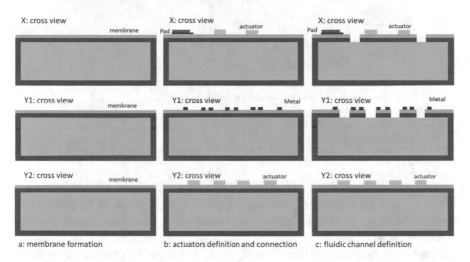

Fig. 19.34 MEMBRANE wafer. Schematic process flow

of material compatible with the selectivity of the following etch chemistry. The membrane formation step is shown in Fig. 19.34a.

After membrane formation, the actuator stack is deposited on top of the wafer and then defined with photolithography and etch (Fig. 19.34b). One of the fundamental printhead parameters, which expresses the quality of the printing, is the *dpi*, the number of drops per inch, because it expresses the resolution of the printed images. The goal is to have the highest resolution, so the highest dpi, and this has a direct impact on technology requirements. From a qualitative point of view to increase the number of drops per inch, the number of fluidic channels per inch must be increased, and this means that the pitch in Y direction (Fig. 19.27) must be decreased. Despite to other VLSI technology (like memory and DRAM), in piezo printhead, there

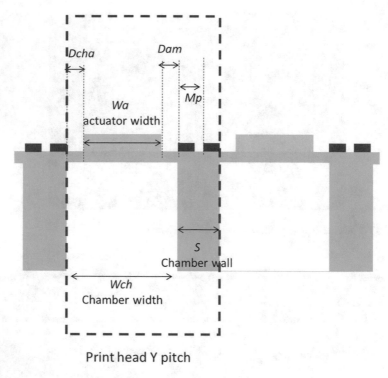

Fig. 19.35 Printhead die simplified cross section (Y direction), to highlight the main technology design rules

is not a universal relationship between dpi and technology pitch in Y direction, because the usage of different color complicated the picture. Below is an example: if the printhead resolution is 1200dpi and if two colors are used in such printhead, the resolution becomes 600dpi for each color, which means that each inch of the printhead realizes 600 drops. So, by assuming that the printhead chip has a Y dimension of 1 inch, with a pitch between actuators of 41.5um, all the 600 nozzles can be placed on the same row. On the other side, it has to be considered that in the pitch some design rules, tolerances and design considerations play a role, as it is shown in Fig. 19.35, where a magnification of the membrane/actuator structure along the Y direction is reported. The pitch in Y direction is given by:

$$Pitch_y = 2 \cdot Dcha + Wa + S$$

where *Dcha* is the distance between actuator and chamber wall, *Wa* is the actuator width, and *S* is the wall that divides the two adjacent chambers. To maximize the membrane deflection, the chamber width (*Wch*) and the actuator width (*Wa*) have to be as wider as possible, and the overlap between the chamber and the actuator (*Dcha*) guarantees that the actuator is always inside the membrane and no force

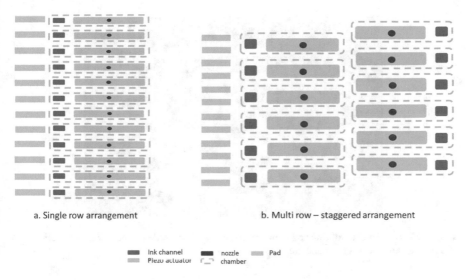

a. Single row arrangement   b. Multi row – staggered arrangement

■ Ink channel   ■ nozzle   ▦ Pad
▦ Piezo actuator   ⌐‾⌐ chamber

Fig. 19.36 Different nozzle arrangement. Simplified layout comparison

is lost. On the other side, the chamber wall (S) must be rigid enough to minimize the mechanical crosstalk between adjacent chambers. By reducing the pitch in Y direction, the membrane has to be designed flexible enough to reach the desired deflection, even with a smaller width and the alignment tolerances between the actuator definition and the chamber wall position has to be kept well under control, in order to avoid force loss. Rather than compress all the nozzles inside a row, it can be more convenient to create a multi-row design, by relaxing the Y pitch from technological point of view and by reducing the complexity of mechanical design. In Fig. 19.36 a comparison between single and multi-row is shown with a simplified layout. In the previous example, by using 4 rows, the pitch Y can increase from 41.5um to 166um, even if the more relaxed Y direction goes to the detriment of the extension in X direction of the chip.

For this reason, it is not possible to define a tight relationship between the printhead dpi and the technology pitch, as is done on DRAM memory or CMOS technology.

After the actuators patterning, some passivation layer can be deposited on top and vias connections are opened to reach the top and bottom electrode. A metal layer, like aluminum, is deposited and patterned to realize the connection to PAD (in Figure 19.34b the metal layer is the red square) . In the "multi-row staggered" solution, the metal tracks coming from the farthest row from pad must run across the chamber wall of the other rows. The number of the metal connections that run between two actuators depends on the number of rows of the printhead. The metal pitch (Mp) of the technology must satisfy the minimum dimension of chamber wall (S) and the maximum number of metal trace, which come from row arrangement. For this reason, a dry metal etch is a good choice for an ink-jet technology.

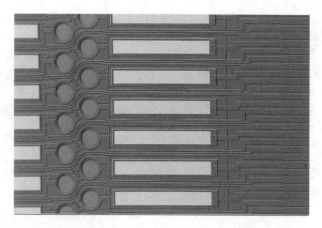

Fig. 19.37 SEM picture of finished membrane wafer. Actuators, dense metal interconnection, and fluidic channel across membrane are visible

After metal patterning, passivation on top of each actuators can be removed, in order to increase the membrane deflection. Of course, this trick must satisfy the reliability requirements of the final application. A dedicated pad finishing (e.g., gold bump grown by ECD) can be realized if the printhead ASSY requires it. The last step of membrane flow is the membrane opening in the correspondence of the ink channel of the distribution wafer. In Fig. 19.37 a SEM top view of finished membrane wafer is shown.

The *nozzle wafer* realization is quite simple, since it is necessary a well-controlled structural silicon layer, where the nozzles are opened. In Fig. 19.38 the simplified process flow is summarized.

The starting silicon wafer is oxidized to create the etch stop layer for the following etch step. After oxide growth, a polysilicon epitaxial layer is done, and a chemical mechanical polishing is applied to the layer to remove roughness and to reach the desired final thickness (typical thickness range is 15–40um). After the surface finishing, the nozzles are opened by using photolithography and dry silicon etch. The etch has to stop on silicon dioxide. The cleaning of polymer is fundamental to reach the desired quality of the drop. This approach, where the nozzle is opened at the beginning of nozzle wafer flow, is called "nozzle first"; there is another possible solution, called "nozzle last," which will be explained after. A well-controlled nozzle dimension inside the die, inside the wafer, and wafer to wafer is fundamental to guarantee the printhead performance. The maximum CD variation allowed has a range below 4%. To reduce the pressure inside the nozzle and to augment the printing effectiveness, tapered nozzle shape is necessary, and, for example, by moving some parameter of dry etch, it is possible to reach tapering below 10 deg. Other solution can be the usage of TMAH etch, starting from a square nozzle layout. In Figs. 19.39 and 19.40, examples of tapered and straight nozzle SEM cross section are shown. In Fig. 19.41 a picture of the final wafer of TFP inkjet taken from nozzle side is shown.

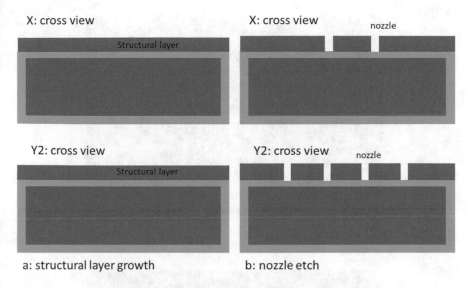

Fig. 19.38 NOZZLE wafer. Schematic cross section of process flow

Fig. 19.39 Tapered nozzle, SEM cross section

When all the three wafers have completed the process flow, they must be merged with W2W technique. The first step is to merge the distribution wafer with membrane wafer. Regarding the choice of the best W2W bonding technique, some clarifications must be done. As outlined before, the dimension control is one of the key aspects of printhead technology since it impacts the functionality. The requirement on W2W bonding alignment reaches on advanced platform 5um of tolerance, and the control of the bonding material deposition respect to the present features (dimension and topography) is a key aspect. There is another strong requirement for the bonding material that is the sealing of the all the fluidic channel. Voids and weak interface must be avoided since it can create ink

Fig. 19.40 Straight nozzle, SEM cross section

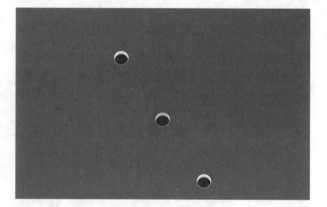

Fig. 19.41 SEM top view of nozzle side of finished inkjet wafer

leak and delamination during printhead functioning. Scanning acoustic microscopy automatic inspection is widely used in printhead production to monitor the device quality.

A good bonding material used in printhead production is the BCB polymer (bisbenzocyclobutene), which can be deposited by stamping and spinning and by spray coat and for some specific composition can be exposed. The stamping is a smart solution since the material is self-aligned to the topography present on wafer, but the thickness dimension control is poor. 3D printing can be an effective solution to BCB dispensing since it improves the material quantity control and its position over the wafer.

During bonding phase, the material must be heated up to 200C and its viscosity drop [37]. To contain the material excess redistribution during bonding phase (that can cause the clogging of the fluidic channel), the usage of containment trenches is

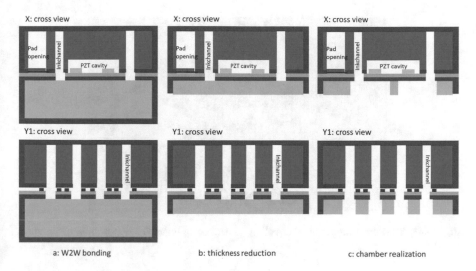

Fig. 19.42 DISTRIBUTON plus MEMBRANE wafer. Schematic process flow

suggested, as visible in Fig. 19.32. To improve the alignment performance during the W2W bonding, some hard stop structure between the two wafers has to be inserted in not active device area.

In Fig. 19.42 it is shown the wafer process flow, after the merge of distribution and sensor wafer. If the stamping is the used technique, the BCB material is deposited on distribution wafer before bonding. After the W2W bonding, the membrane wafer has been reduced in thickness to reach the desired chamber height (a reasonable value can be 70–90um). After thickness reduction, by using infrared alignment, the mask for chamber patterning is exposed. As explained before (Fig. 19.35), the overlap of the chamber versus the actuator (*Dcha*) is a key aspect for the printhead functionality and scalability. Superior alignment performance (<2um) is required on most advanced technology node. The chamber is opened with dry silicon etch, which stops on the rear side of the membrane.

After the chamber definition, BCB layer is applied to the rear side of the membrane wafer and the merge with nozzle wafer can be realized. The excess of silicon present on nozzle side is removed with grinding. The fluidic channel is then opened by removing the silicon dioxide. If the "nozzle first" approach is not used and the structural layer is intact at this stage, a mask can be applied on this wafer structure to realize the nozzle. Such approach is called "nozzle last."

After these steps, the printhead wafer can be considered completed. In Fig. 19.43 and Fig. 19.44, the SEM cross section of the finished printhead wafer is shown.

Two other aspects related to printhead lifetime have a direct impact on silicon technology. The first is the interaction between ink and silicon, and other material present inside fluidic channel. The second is the cleanliness of the nozzle surface.

Regarding to the first item, some inks have PH very low and can etch/damage the structural material of the fluidic channel, thus reducing printhead lifetime. To

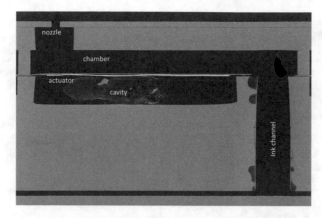

Fig. 19.43 SEM cross section in X direction (as defined in Figure 19.28a) of finished printhead wafer

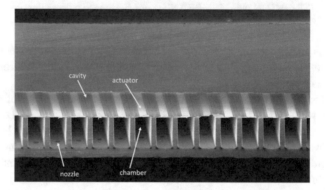

Fig. 19.44 SEM cross section in Y2 direction (as defined in Fig. 19.28b) of finished printhead wafer

improve this aspect, at the end of the process flow, the deposition inside the channel of a material robust versus ink is necessary. Atomic layer deposition is a suitable technique for this purpose, thanks to the perfect step coverage. The chosen material to be deposited is strictly related to the ink composition.

Regarding the second item, to avoid flooding, the nozzle outside surface of the printhead should be hydrophobic. This is usually achieved by depositing hydrophobic coatings in these areas. Also, for such layer, the chemical compatibility with ink is fundamental and the best material is chosen in collaboration with printer makers.

19.6   Inkjet Applications

19.6.1   Graphics

Professional and industrial printing of graphic material, especially adopted in the wide-format market, require high precision and speed, flexibility over the chosen inks, and reliability over high printing cycles. In this scenario, despite the slightly greater cost of the piezoelectric technology compared to the thermal one, the former can fulfill the required constraints and finds wide usage in the market. Most of the companies providing solutions for the wide-format market have their products based on the piezoelectric inkjet technology. The CIJ technology has been nowadays superseded by the PIJ, since the products are more compacts, and the maintenance of the whole system is easier. As an example, Fig. 19.45 shows Canon's Colorado-1640 piezoelectric wide-format printer. This product can print up to 159 $m^2hr/$ with the provided UV inks. The deposited inks are UV sensitive, and, thanks to UV LEDs, the printer directly polymerizes the surface, fixing the image on the substrate.

19.6.2   Electronics

Examples of applications in the electronics field include the microfabrication, pcb manufacturing, and the energy harvesting. Inkjet technology finds a good market into which expanding due to the accurate dispensing of precious material, otherwise

Fig. 19.45 Canon's Colorado-1640 piezoelectric wide-format printer. Printing speed up to 159 $m2hr/$, UV inks for direct polymerization thanks to UV LEDs [courtesy of Canon website]

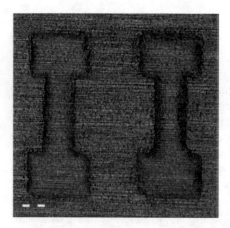 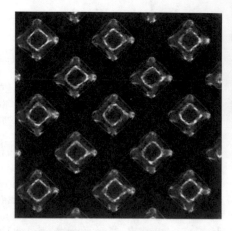

Fig. 19.46 Left: optical image of a SU-8 pattern printed with step reticulation on Cu. Right: optical image of SU-8 printed bumps with step reticulation [33]

wasted by the usually employed processes. SU-8 is a commonly used epoxy-based negative photoresist. It is widely used in microfabrication and chip manufacturing processes to build patterns on the surfaces (Fig. 19.46). Packaging and bonding, especially in medical devices and MEMS components, require noncontact adhesive dispensing to prevent damaging and/or contaminating the surfaces, and this can be provided by the inkjet technology.

Among the pcb manufacturing uses, two examples are chip soldering and direct printing of organic (PEDOT:PSS based) passive elements such as resistors. (Fig. 19.47)

Finally, there is the energy market, especially regarding organic photovoltaics (OPV), in which precious and expensive materials are used. Some of them are PEDOT:PSS (high grade, hole transport layer), ZnO (electron transport layer), P3HT, and derivatives. The inkjet technology provides many benefits, such as low waste of precious materials, low-cost deposition process, noncontact printing, and no masks required process.

19.6.3 Displays

Inkjet DOD technology finds large research and application in the display market, especially since it is focused on the spreading of OLED devices for mass production. This is because OLED displays are conventionally produced with vacuum (PVD) deposition techniques, requiring complex and expensive systems. The main requirements for OLED production are high positional accuracy, critical repeatability of the drops, homogeneous polymer deposits, inert atmosphere, and contaminant-free environment. The last two constraints are explained since the polymers and

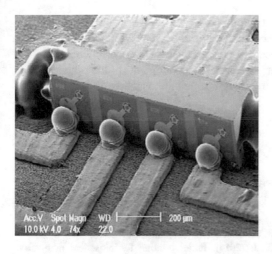

Fig. 19.47 Left: chip soldering, 45° 100um deposits. Right: organic (PEDOT:PSS) resistors [courtesy of Microfab website]

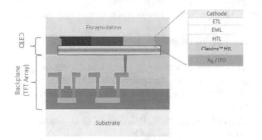

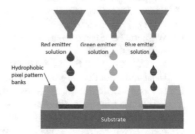

Fig. 19.48 Left: OLED simplified stack structure. Electron transport later (ETL), emissive layer (EML), hole transport layer (HTL), hole injection layer (HIL), transparent electrode (ITO) where the light passes through. Thanks to inkjet printing, both the EML and HIL layers can be printed. Right: direct printing of EML polymer on substrate [Heraeus website, Radiant Vision Systems]

materials used are very sensitive to oxygen and moisture, both typical of the air, and their performance is degraded without an inert atmosphere. To support the high density and resolution displays, with the most challenging market being the small UHD and flexible displays, fine-sized drops with perfect repeatability are needed [34] (Fig. 19.48).

Despite the challenges presented by this market, in the recent years, many different small scale/R&D facilities, along with the first lines close to mass production, have seen the light. J-OLED, in the late 2018, presented their ready-to-be-commercialized 21.6" 4k 204ppi display, along with a 55" prototype. Au Optronics, in 2019, presented a 17.3" 4 k 225ppi display. Research is also focused on the refining of ink formulations, to support the increasing demand for this market. One of the main players in the market, LG, acquired in April 2019 DuPont's soluble OLED technologies, under the name of LG Chem. Kateeva is commercializing the YIELDjet® Explore family products, an all-in-one equipment for the inkjet printing of OLED displays, in a controlled and inert nitrogen atmosphere.

Bibliography

1. Hadimioglu, B., Stearns, R., & Ellson, R. (2016). Moving liquids with sound: The physics of acoustic droplet ejection for Robust laboratory automation in life sciences. *Journal of Laboratory Automation, 21*, 4–18.
2. Kodak, Kodak Ultrastream – Fourth generation continuous inkjet technology, 2016.
3. PrintWeekIndia, What really is Kodak UltraStream inkjet technology?, 2016.
4. Li, J., Rossignol, F., & Macdonald, J. (2015). Inkjet printing for biosensor fabrication: Combining chemistry and technology for advanced manufacturing. *Lab on a Chip, 15*, 4.
5. Hoath, S. D., Fundamentals of inkjet printing, Wiley-VCH Verlag GmbH & Co. KGaA, 2016.
6. Hutchings, I. M., & Martin, G. D. (Eds.). (2012). *Inkjet technology for digital fabrication*. John Wiley & Sons, Ltd.
7. Eggers, J. (1997). Nonlinear dynamics and breakup of free-surface flows. *Rev. Mod. Phys, 69*(3), 865–930, 7.
8. Kowalewski, T. A., On the separation of droplets from a liquid jet, 1996.
9. Eggers, J. (1993). Universal pinching of 3D axisymmetric free-surface flow. *Phys. Rev. Lett, 71*(21), 3458–3460.
10. Wijshoff, H., Structure- and fluid-dynamics in piezo inkjet printheads, University of Twente, Netherlands, 2008.
11. Xu, C., Zhang, Z., Fu, J., & Huang, Y. (2017). Study of Pinch-Off Locations during Drop-on-Demand Inkjet Printing of Viscoelastic Alginate Solutions. *Langmuir, 33*, 5037–5045.
12. Driessen, T., Jeurissen, R., Wijshoff, H., Toschi, F., & Lohse, D. (2013). Stability of viscous long liquid filaments. *Physics of Fluids, 25*, 062109.
13. P. D. J. G. K. D. P. J. S. D. D. Shin. (2012). In J. G. Korvink, P. J. Smith, & D. Shin (Eds.), *Inkjet-based micromanufacturing*. Wiley-VCH Verlag GmbH & Co. KGaA.
14. Iida, Y., Okuyama, K., & Sakurai, K. (1993). Peculiar bubble generation on a film heater submerged in ethyl alcohol and imposed a high heating rate over 107 K s1. *International Journal of Heat and Mass Transfer, 36*, 2699–2701, 7.
15. Okuyama, K., & Yoshida, K. (2018). Dynamic behavior with rapid evaporation of an inkjet water droplet upon collision with a high-temperature solid above the limit of liquid superheat. *International Journal of Heat and Mass Transfer, 116*, 994–1002, 1.
16. Pond, S. F., 1977 Inkjet Technology and Product Development Strategies, Torrey Pines Research, 2000.
17. Hou, S., "Ink Jet Technology," Industry Applications, IEEE Transactions on, Vols. IA-13, pp. 95–104, 2.
18. Bugdayci, N., Bogy, D. B., & Talke, F. E. (1983). Axisymmetric Motion of Radially Polarized Piezoelectric Cylinders Used in Ink Jet Printing. *IBM Journal of Research and Development, 27*, 171–180, 3.
19. Stemme, E., & Sg, L., The Piezoelectric Capillary Injector. A New Hydrodynamic Method For Dot Pattern Generation, 1973.
20. El, K., C. Lf and N. Herbert, Design Of An Impulse Ink Jet, J. Appl. Photogr. Eng.; Issn 0098–7298; Usa; Da. 1981; Vol. 7; No 3; Pp. 73–79; Bibl. 9 Ref., 1981.
21. Zapka, W., Handbook of Industrial Inkjet Printing, Wiley-VCH Verlag GmbH & Co. KGaA, 2017.

22. Le, H. P., Progress and trends in ink-jet printing technology, 1998.
23. Mimura, S. S. T., "Micro-Piezoelectric Head Technology of Color Inkjet Printer," International Conference on Digital Production Printing and Industrial Applications, 2001.
24. The, R., Yamaguchi, S., Ueno, A., Akiyama, Y., & Morishima, K., Piezoelectric Inkjet-based Single-cells Printing By Image Processing For High Efficiency And Automatic Cell Printing, 17th International Conference on Miniaturized Systems for Chemistry and Life Sciences 27–31 October 2013, Freiburg, Germany, 2013.
25. Shah, M., Lee, & Hur, S. (2019). Design and Characteristic Analysis of a MEMS Piezo-Driven Recirculating Inkjet Printhead Using Lumped Element Modeling. *Micromachines, 10*, 757.
26. Yoshida, Y., K. Izumi & S. Tokito, "A push-mode piezo inkjet equivalent circuit model enhanced by diaphragm displacement measurements," AIP Advances, vol. 9, p. 025319, 2019.
27. He, M., L. Sun, K. Hu, Y. Zhu, L. Ma and H. Chen, "Drop-on-Demand Inkjet Printhead Performance Enhancement by Dynamic Lumped Element Modeling for Printable Electronics Fabrication," Mathematical Problems in Engineering, vol. 2014, pp. 1–16, 2014.
28. Chen, Q. & Wang, Q.-M., (2005). "The effective electromechanical coupling coefficient of piezoelectric thin-film resonators," Applied Physics Letters, vol. 86, p. 022904.
29. Sherrit, S., H. D. Wiederick, B. K. Mukherjee and M. Sayer, "Determination of the piezoelectric electromechanical coupling constant k13 using a frequency ratio method," Ferroelectrics, vol. 193, pp. 89–94, 1997.
30. Kriss, M., Handbook of digital imaging, Chichester, West Sussex, United Kingdom: John Wiley & Sons, Inc, 2015.
31. Uchino, K., "Introduction to Piezoelectric Actuators and Transducers," p. 41, 6 2003.
32. Gan, H. Y., X. Shan, T. Eriksson, B. K. Lok and Y. C. Lam, (2009). "Reduction of droplet volume by controlling actuating waveforms in inkjet printing for micro-pattern formation," Journal of Micromechanics and Microengineering, vol. 19, p. 055010.
33. Bernasconi, R., M. C. Angeli, F. Mantica, D. Carniani, & L. Magagnin, "SU-8 inkjet patterning for microfabrication," Polymer, vol. 185, p. 121933, 2019.
34. Chin, B. D., "Solution Deposition: Inkjet-Printed OLED," in Handbook of Organic Light-Emitting Diodes, C. Adachi, R. Hattori, H. Kaji and T. Tsujimura, Eds., Tokyo, Springer Japan, 2019, pp. 1–20.
35. Onoe, M., H. F. Tiersten and A. H. Meitzler, "Shift in the Location of Resonant Frequencies Caused by Large Electromechanical Coupling in Thickness-Mode Resonators," The Journal of the Acoustical Society of America, vol. 35, pp. 36–42, 1963.
36. IEEE Standard on Piezoelectricity - ANSI/IEEE Std 176-1987.
37. CYCLOTENE* 3000 Series: CYCLOTENE Advanced Electronic Resins, DOW, Revised February 2005.

Chapter 20
Micro Speakers

Andrea Rusconi, Sonia Costantini, and Carlo Prelini

20.1 Micro Loudspeakers Applications and Requirements

The global demand for portable audio devices has increased significantly in recent years for smartphones, headphones, and hearables; headphones and headsets are the largest growth drivers in the global market for audio devices with total global available market of around 14 billion euros in 2019. The functionalities of modern devices go far beyond basic audio applications and will continue to increase in the coming years. While there was already added value in the integration of Bluetooth and active noise canceling (ANC), the development of the future headphones revolved around additional functions such as voice command control, and selective hearing (Smart ANC) and automatic sound equalization, but also simultaneous translators, navigation systems, and intelligent personal assistants (e.g., Alexa, Siri, Cortana). Medical hearing aids are undergoing similar developments, with the aim to enable the wearer to connect to additional health-related functions.

The vast majority of loudspeakers are based on a coil-magnet system for generating sound, and this includes all the tiny transducers used today in consumer portable applications like smartphones, headphones, and wearable devices. The original concept is known since more than a century and its applications in audio transducers are mature, being fully characterized, optimized, and industrialized. This resulted in commoditization of the components and consolidation of the market. The aforementioned aspects represent the biggest barriers to entry for

A. Rusconi (✉)
USound GmbH, Graz, Austria
e-mail: andrea.rusconi@usound.com

S. Costantini · C. Prelini
ST Microelectronics, Analog MEMS and Sensors Group, MEMS Technology and Design R&D, Agrate Brianza, Monza Brianza, Italy
e-mail: sonia.costantini@st.com; carlo.prelini@st.com

© Springer Nature Switzerland AG 2022
B. Vigna et al. (eds.), *Silicon Sensors and Actuators*,
https://doi.org/10.1007/978-3-030-80135-9_20

MEMS technology, which is challenging the status quo with a new approach with a great potential yet still at the beginning of its roadmap.

Electrodynamic loudspeakers based on voice coil (coil-magnet system) are widely adopted by the industry. The working principle is well known and fully characterized, and its applications in audio transducers are mature, optimized, and fully industrialized. This results in commoditization of product families and consolidation of the market. The aforementioned aspects represent the biggest barriers to entry for MEMS speakers, which is challenging the status quo with a new technology with great potential yet still at the beginning of its roadmap. But ever-increasing requirements for modern devices pose new challenges for components and systems where traditional loudspeakers offer limited incremental improvements; in particular the speakers must become smaller and effectively integrate into digital systems without impairing performance such as sound quality or battery life.

20.1.1 Voice Coil Transducers for In-ear Applications

The loudspeaker of an in-ear headphone must generate a certain sound pressure level at the eardrum. The ear canal is closed at one side by the headphone, and on the other side by the eardrum, the movement of the loudspeaker membrane generates pressure waves in the ear canal. Assuming an ear canal perfectly sealed by the headphones and a loudspeaker surface of 10 mm^2, the membrane deflection required for a sound pressure level of 85 dB at low frequency is approx. 1.5 μm, for 91 dB approx. 3 μm and for 140 dB approx. 800 μm [1]. Although by removing the perfect sealing hypothesis, these values can get up to 10 times larger. Still, compared to the required diaphragm deflection of a hi-fi speaker playing in free field, these values are very small. The headphone speaker benefits from its close proximity to the eardrum and the almost complete sound radiation into the ear canal. In addition to the sound pressure level, further requirements are placed on a loudspeaker for headphones or hearing aids. Table 20.1 shows examples of typical specifications for the respective area of application. The acoustic parameters relate to the measurement in an ear simulator, e.g., DIN EN 60318-4, which is modeled on the human ear canal.

Table 20.1 Typical specification for in-ear headphone and hearing aid application

Parameter	In-ear headphone	Hearing aid
Max sound pressure level (SPL)	>/=110 dB	>/=120 dB
Total harmonic distortion (THD)@max SPL	<3%	<5%
Bandwidth (-20 dB)	20 Hz to 20 kHz	100 Hz to 10 kHz
Footprint	⌀ 5 to 13 mm	10 to 50 mm2
Thickness	3 to 6 mm	2 to 5 mm
Weight	</= 4 g	</= 2 g

20.2 "More than Moore" Technologies: Background for Audio MEMS

In the past three decades, microelectronics R&D progress has been focused in integrated circuit (IC) miniaturization down to nanometer dimension and system on chip (SoC) integration which led to outstanding results in products and services for users. The best way to summarize this roadmap is with the Moore's Law, which is the observation that the number of transistors in a dense integrated circuit doubles about every 18 months, driving up computing power and driving down costs. In parallel to this trend, there are ever-increasing awareness, R&D efforts, and business drivers to push the development and application of "More than Moore" (MtM) concepts that are derived from silicon technologies but do not scale with Moore's law [2, 3]. This emerging trend is triggered by the increasing need to pair computing power with analog sensory functionalities able to interact with the environment but still fully integrated into a digital system; the goal is to realize systems that are sensitive and responsive to the environment and to the presence and needs of people. A key enabler for such applications is the ability to integrate diverse microfabrication technologies with complementary functions in a system up to a direct interface to the human body or environment; MEMS play a fundamental role in this trend with a number of successful applications in consumer, automotive, medical, and industrial fields. This is the environment in which audio MEMS have been developed, with MEMS microphones now widely adopted in many fields of application; looking at smartphones and hearables, MEMS microphones are enablers for advanced system functionalities and have been successfully integrated, thanks to their small form factors and for being reflow solderable thus compliant with SMT manufacturing processes. Very similar key advantages are true for MEMS micro speakers, with the goal to enable new product roadmaps outperforming traditional voice coil transducers.

20.3 MEMS Speaker Architecture and Performances

There are different ways to generate sound waves based on silicon transducers, including analog and digital approaches. In this chapter we will describe the concept developed by USound GmbH which uses piezoelectric benders for generating the force necessary to move an acoustic membrane.

In principle, there are four main MEMS technology roadmaps that make a new generation of MEMS speakers and ultrasound transducers possible. Those technology and design innovations are:

1. Microfabrication technology: piezoelectric materials thin film deposition on silicon, maximized material properties and performances, high deposition rates at low costs, availability at silicon foundries

2. Assembly techniques: enable heterogeneous technology integration into modules
3. Device design: advanced design and F.E.M. tools, design for manufacturing and testing
4. System design: efficient driving of piezo actuators, enable new integration concepts and applications

The convergence of these innovations makes MEMS devices able to produce high acoustic pressure levels and sensitivity per unit area; enables precise manufacturing miniaturized transducers; makes possible radical innovation in smart system designs by integrating electronics circuits (ASIC), transducers, and passives in very small and thin form factors; leverages a low-cost, scalable manufacturing infrastructure; and enables roadmaps in both device performances and cost reduction.

The biggest challenge for generating high sound pressure levels in a tiny form factor is to be able to generate large forces and elongations in a linear way. This has always been too much of a problem for MEMS technologies and one of the reasons why silicon transducers have been mostly limited to sensing. This has changed in the last couple of years with the advent of new materials, in particular thin-film piezo ceramics with high piezo coefficients.

The MEMS micro speaker is based upon three building blocks: a MEMS linear motor realized on silicon with piezo actuation; an integrated package with acoustic emitting surface (a membrane and a stiffening plate); and an ASIC power amplifier for efficient driving.

PZT (lead-zirconate-titanate) is the material among piezo ceramics that can generate enough energy per unit area thus mechanical force and elongation to compete with standard voice coil transducers. The MEMS motor is realized on silicon using multiple piezoelectric actuators connected together to realize a precise vertical out-of-plane movement. These trends are clearly pushed by technology development and expressed in more than a decade research on thin-film deposition of piezoelectric materials, in particular PZT (lead-zirconate-titanate) and ScAlN (scandium-aluminum-nitride). But the technological breakthrough alone is not enough to produce a radical innovation; it is necessary to marry it with new concepts of device design and application.

A key technical innovation is to integrate the MEMS actuator chip with an acoustic membrane at package level. The main reason is that the emitting surface must be as large as possible for maximizing sound pressure levels, which is in contrast with the cost structure of MEMS manufacturing where the area used on wafer is the main driver; to the contrary typical polymeric membranes used in acoustics are inexpensive with costs barely sensitive to their surface area. Also, by decoupling MEMS and membrane, one can optimize the design characteristics of both without trade-off on performance. In particular the MEMS have to have an open structure to generate large forces and elongations with high linearity performances. In Fig. 20.1, the exploded view of the main components is shown: MEMS, membrane, substrate, and bottom lid.

In Fig. 20.2 the main acoustic performances of a MEMS loudspeaker measured in a 711 coupler (2 cm^2 hear simulator) are shown.

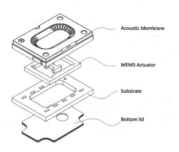

Fig. 20.1 Main components of the transducer: MEMS chip, membrane, substrate, and bottom lid

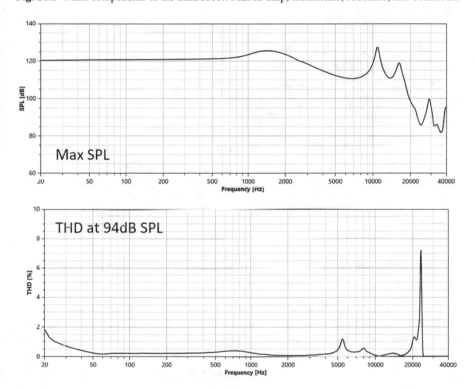

Fig. 20.2 Max SPL of MEMS speaker up to 40 kHz (top) and THD@94 dB SPL (bottom)

The amplifier is a closed-loop system with digital loop control. The input for the loop control is the digital audio input signal and the digital value of the current output voltage, provided by the feedback ADC. The digital control loop sends the input signal to a pulse width modulation (PWM) stage. The required output voltage is defined by changing the pulse width. The output power stage provides the high voltage required at the speaker (up to 30 V$_{PP}$).

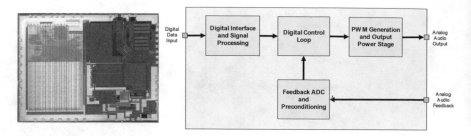

Fig. 20.3 Image of the audio amplifier and corresponding basic block diagram

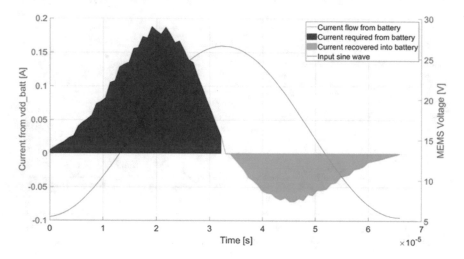

Fig. 20.4 Current flow during one sine wave period showing energy recovery functionality [8]

For typical audio applications, there is the possibility to configure all important settings like digital filter settings, predistortion coefficients, DC offset voltages, and gain settings. A hybrid can also be implemented, for example, one where the configuration is stored outside the audio module and is transferred to the module during start-up. In Fig. 20.3, an image of the mixed signal silicon chip and the basic block diagram of the audio amplifier is shown.

A key feature of the amplifier is the energy recovery functionality to reduce overall power consumption, especially at high amplitudes. A novel energy recovery architecture is included inside the output driving stage, and Fig. 20.4 shows the current flow based on a schematic-based system simulation during one sine wave period. During the rising phase of the sine wave, energy is transferred to the MEMS speaker; during the falling edge, the energy is recovered and brought back to the system. The maximum peak efficiency during charging is up to 80% and close to 70% during discharging. More detailed information on the power stage and energy recovery can be found under ref. [8].

There are other MEMS solutions in development for micro speakers, which are not discussed in detail in this chapter:

1. The DRS (digital sound reconstruction) approach, based on an array of MEMS loudspeaker to produce time-varying pressure waveform. A flexible membrane is moved between two electrode with electrostatic force, and the possible technology solutions are an array of CMOS-MEMS loudspeaker like the one proposed by MEMS Laboratory of Carnegie Mellon University [4, 5], or a MEMS array of small speakers, like the one proposed by AudioPixel (https://www.audiopixels.com.au/).

2. The NED (Nanoscopic Electrostatic Drive) from Fraunhofer institute for Photonic Microsystems (IPMS), described by Conrad et al. [6] and applied by Kaiser et al. [7]. This is a novel micro speaker concept based on in-plane electrostatic bending actuators exploiting chip volume rather than surface for sound generation and using bulk micro machining fully CMOS compatible for the realization.

3. Silicon monolithic integration, xMEMS micro speaker technology is based on the inverse piezoelectric effect using piezoMEMS material. The inverse piezoelectric effect is created by applying electrical voltage to make the piezoMEMS contract or expand, converting electrical energy to mechanical cncrgy. This energy excites an integrated silicon membrane to move air and generate acoustic sound waves.

20.4 PZT Thin-Film MEMS Speaker Modeling and Design

By definition, a piezoelectric material will show a change in physical volume when an electrical field is applied and generates a charge when deformed. PZT (lead-zirconate-titanate) is the most widely used piezoelectric material in MEMS actuator, for example, it is used in inkjet head printers, micromirror scanners, and autofocus lenses for smartphones, Thanks to high piezoelectric coefficients which enable designs with large actuating forces and displacements for its adoption.

An electric field is applied to a piezoelectric cantilever, which consists of two layers: an active layer and a passive layer. One side of the structure is fixed which results in a displacement of the opposite side. There is an even number of cantilevers (2, 4, or 6) arranged in a symmetric way and connected in the middle to a central structure via spring elements. The resulting mechanical movement is perfectly vertical ("piston mode") and is used to move a membrane and to generate the acoustic output of the speaker. The magnitude of the movement depends on the polarization of the material as well as the applied voltage. (Fig. 20.5)

To design the speaker, the engineering task is to maximize three main performance parameters: actuator elongation, force, and linearity.

The dynamic behavior of a piston-like acoustic transducer can be modeled as a damped mass-spring-resonator with air load; the forces acting on the dynamic elements are described in the following model (Fig. 20.6):

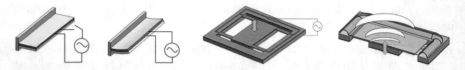

Fig. 20.5 Schematic concept of a MEMS speaker using cantilever principle for driving an acoustic membrane

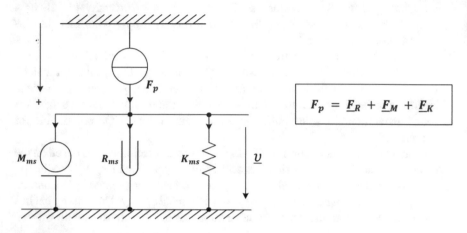

$$F_p = \underline{F}_R + \underline{F}_M + \underline{F}_K$$

Fig. 20.6 Design iterations for optimizing key transducer parameters

- The total moving masses are summarized in a point mass M_{ms} and generate an inertia F_M proportional to the acceleration a. The total mass includes the mass of all moving parts and the air mass moved by the transducer.
- The mechanical resistance R_{ms} causes a damping force F_R proportional to the velocity v. The total resistance includes the mechanical losses in the materials and the air load at the emitting surface.
- The stiffness of the system K_{ms} generates a restoring force F_K proportional to the displacement x. The total stiffness includes the mechanical stiffness of the actuator and suspension and the stiffness of the acoustic cavity (if present).

Table 20.2 summarizes those forces and establishes a relation to the emitting surface velocity by making use of the mechanical theory of vibrating systems oscillating in one dimension.

The *equation of motion* of the system is:

$$F_P = M_{ms}\frac{d^2x}{dt^2} + R_{ms}\frac{dx}{dt} + K_{ms}x$$

where x is the displacement of the emitting surface. In case of a harmonic oscillator, the same equation can be written as follows:

Table 20.2 Mechanical forces in relation to membrane velocity

Forces	Physical parts of the transducer	Basic relationship
Total driving force FP	Piezo actuator system	$F_P = {}^F M + {}^F R + {}^F K$
Inertial force $_{FM}$	Moving mass: cantilevers, membrane, acoustic plate	$F_M = M_{ms}*a = M_{ms} * j2\pi f * v$ Moving mass M_{ms}, acceleration a, frequency f, complex operator j
Dumping force $_{FR}$	Friction losses in the mechanical and acoustical elements; air load at the emitting surface	$F_D = R_{ms}*v + Z_R*v$ Velocity v, mechanical loss R_{ms}, radiation impedance Z_R,
Potential (restoring) force F_K	Deformation of cantilevers, springs, membrane; air compression in acoustic enclosure	$F_K = K_{ms} * x + K_e * x = $ $= K_{ms} * \frac{v}{j2\pi f} + K_e * \frac{v}{j2\pi f}$ Displacement x, system stiffness K_{ms}, enclosure stiffness K_e

Table 20.3 Relationship between frequency and mechanical admittance

Frequency range	Mechanical impedance	Dominant force
$f \ll f_0$	$H_{ms}(f) \approx j2\pi f/K_{ms}$ Stiffness is dominant	$F_P \approx F_K$ Driving force equals potential force
$f = f_0$	$H_{ms}(f) = 1/R_{ms}$ Mechanical loss is dominant	$F_P = F_R$ Driving force equals losses
$f \gg f_0$	$H_{ms}(f) \approx 1/j2\pi f M_{ms}$ Moving mass is dominant	$F_P \approx F_M$ Driving force equals inertia

$$F_P = M_{ms} j\omega v + R_{ms} v + K_{ms} \frac{v}{j\omega}$$

where $\omega = 2\pi f$ is the angular frequency of the oscillatory motion, related to the general frequency f. The same equation yields:

$$F_P = \left(M_{ms} j2\pi f + R_{ms} + K_{ms} \frac{1}{j2\pi f} \right) v$$

where $\left(M_{ms} j2\pi f + R_{ms} + K_{ms} \frac{1}{j2\pi f} \right) = Z_{ms}$ is the mechanical impedance of the system.

The *mechanical admittance* $H_{ms} = {}^1\!/_{Z_{ms}} = v/F_P$ describes the complex ratio of velocity v and total driving force F_P. It is a function of frequency and becomes maximal at the resonance frequency where the sum of the restoring force and inertia force is zero. Table 20.3

The *accumulated acceleration* is defined as the integral over the area of the absolute value of the acceleration of the emitting surface [10]:

$$p_{aa}\left(\mathbf{r_a}\right) = \int_0^{S_c} W \left| a\left(\mathbf{r_c}\right) \right| dS_c; \ W = \frac{\rho_0}{2\pi \left| \mathbf{r_a} - \mathbf{r_c} \right|}$$

$$AAL\left(\mathbf{r_a}\right) = 20\log\left(\frac{p_{aa}\left(\mathbf{r_a}\right)}{\sqrt{2}p_o}\right) dB$$

where a is the acceleration, S_c is the total emitting surface, r_c is the radial position on the emitting surface, r_a is the distance from the emitting surface on its perpendicular symmetry axis, and p_o is the reference sound pressure typically $20\mu Pa$ in the case of sound radiated in air. The acceleration is weighted by the distance $\left| \mathbf{r_a} - \mathbf{r_c} \right|$.

AAL describes total mechanical vibration of the acoustic transducer; it is very useful because it can be modeled with distributed parameters (FEM) or analytically (given some important simplifications); it is comparable with the SPL curve at point r_a; it predicts the ideal potential acoustic output as it is never smaller than SPL(r_a); it is identical with SPL(r_a) for a rigid body mode before resonance frequency; it is higher than SPL(r_a) after resonance frequency because it does not take in account acoustical cancellation effects due to rocking (circular) modes and other losses. In case of a micro speaker, the dimension $\mathbf{r_c}$ is negligible compared to $\mathbf{r_a}$; therefore $\left| \mathbf{r_a} - \mathbf{r_c} \right| \cong r_a$.

The MEMS micro speaker has an inherent piston movement where the rocking modes are damped by the actuation system; by making a strong simplification, we assume that the transducer surface S_c is equal to the effective radiation area S_d; therefore the acceleration is constant over the whole emitting surface:

$$AAL\left(f, r_a\right) = 20\log\left(\frac{\rho_0 S_c a(f)}{2\pi\sqrt{2}r_a p_o}\right) dB$$

or in order to highlight the velocity $v(f)$:

$$AAL\left(f, r_a\right) = 20\log\left(\frac{\rho_0 S_c jf v(f)}{\sqrt{2}r_a p_o}\right) dB$$

By using the relationship $v(f) = F_P/Z_{ms}(f)$, we obtain *AAL expressed by the main transducer design parameters:*

$$AAL\left(f, r_a\right) = 20\log\left(\frac{\rho_0 S_c jf F_P}{\sqrt{2}r_a p_o Z_{ms}}\right) dB$$

AAL depends on the emitting surface area, the force, and the mechanical impedance of the actuator. The following graphs show AAL using piezo MEMS

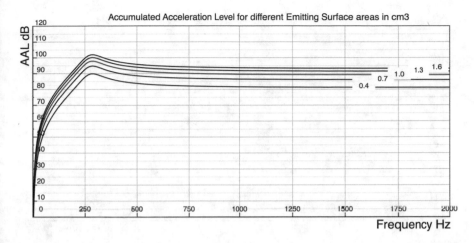

Fig. 20.7 AAL in dB in relation to emitting surface variations. A larger emitting surface increase AAL in all conditions, in particular a doubling in area contributes for around +6 dB after resonance

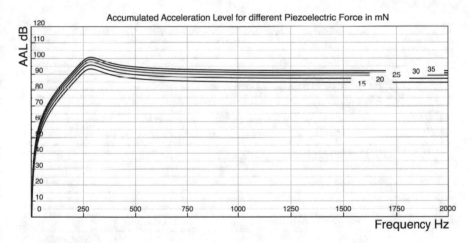

Fig. 20.8 AAL in dB in relation to piezoelectric force variations. More piezo force improves AAL especially after resonance, the bandwidth range dominated by the moving mass of the transducer

values for different design parameters; the table summarizes the fundamental parameters with their variations. (Figs. 20.7, 20.8, 20.9, 20.10, 20.11, 20.12 and Table 20.4)

The system can be also modeled with lumped parameters; in particular we can make the following extensions to the acoustic theories:

- A loudspeaker drive unit can be approximated as a series connection of a zero-impedance voltage source, a resistor, a capacitor, and an inductor. The values depend on the specifications of the unit and the wavelength of interest.

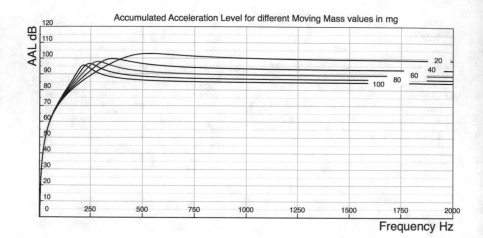

Fig. 20.9 AAL in dB in relation to moving mass variations. Reducing the amount of moving mass moves the resonance frequency toward higher values and substantially increases AAL after resonance

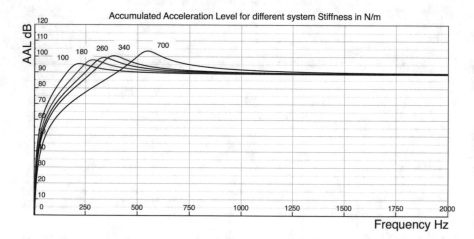

Fig. 20.10 AAL in dB in relation to system stiffness variations. Lower total stiffness of the system moves the resonance frequency toward lower values and substantially increases AAL before resonance

- Damping materials or meshes can be approximated as a resistor. The value depends on the properties and dimensions of the material. The approximation relies in the wavelengths being long enough and on the properties of the material itself.
- A rigid-walled cavity containing air can be approximated as a capacitor whose value is proportional to the volume of the cavity. The validity of this approximation relies on the shortest wavelength of interest being significantly (much) larger than the longest dimension of the cavity.

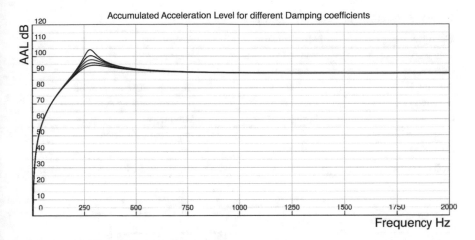

Fig. 20.11   AAL in dB in relation to damping coefficient variations. In a linear analysis, the system damping only affects the Q factor of the resonance

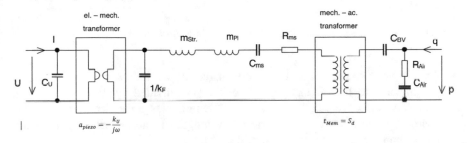

Fig. 20.12   Model of speaker radiating in free field [9]

Table 20.4   Main design parameters and variations

Parameter	Center value	Variations	Comment
Emitting surface S_c	$1\ cm^3$	0.4; 0.7; 1; 1.3; 1.6	Variation values from receiver to micro speaker
Piezoelectric force F_P	$25\ mN$	15; 20; 25; 30; 35	Total actuator force as from FEM simulation
Moving mass M_{ms}	$60\ mg$	20; 40; 60; 80; 100	Including silicon actuator frame, plate, and air load
Stiffness K_{ms}	$180\ N/m$	100; 180; 260; 340; 700	From FEM simulation $700 = 180 +$ closed cavity
Damping R_{ms}	0.038	0.038 +/− 20% +/−40%	Reference value
Distance r_a	$10\ cm$	–	Industry standard for micro speakers
Air density ρ_0	$1.18\ g/cm^3$	–	Standard value at sea level

Table 20.5 Electrical-mechanical-acoustical analogies

Electrical	Mechanical	Acoustical
Voltage, V	Force, F	Pressure, P
Current, i	Velocity v	Volume velocity, vv [a]
Resistor, R	Dashpot, b	Dumping material or mesh
Inductor, L	Mass, m	Air mass moving in a channel
Capacitor, C	Spring, k	Volume, V
Impedance, Ze [Ω]	–	Impedance, Za [Rayl]
Charge, Q	Displacement, x	–

[a] Volume velocity $v_v = v_m A_m$ where v_m is the membrane velocity and A_m is the membrane emitting area

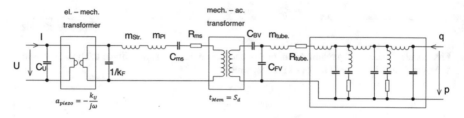

Fig. 20.13 Model of speaker radiating in coupler (ear simulator) [9]

- An air vent channel can be approximated as an inductor whose value is proportional to the effective length of the port divided by its cross-sectional area. The effective length is the actual length plus an end correction. This approximation relies on the shortest wavelength of interest being significantly larger than the longest dimension of the port.

Table 20.5 shows the analogies between the three domains and the corresponding conjugated variables.

In Fig. 20.12 the assembled system model is drawn with the speaker radiating into free air. Starting left with the electrical domain going through the piezo transformer into the mechanical domain, via the MEMS actuator connected to the membrane into the acoustical domain.

In Fig. 20.13, the assembled system model is drawn with the speaker directly connected to the coupler (ear simulator). (Table 20.6)

20.4.1 PZT Nonlinearities Effects on MEMS Speakers

Linear models are valid only in first approximation as ferroelectric piezo materials like PZT display nonlinear behaviors in polarization and strain. If the driving signal is bipolar and therefore the electric field between electrodes changes its sign, the material exhibits a polarization switch. The electric field needed for the PZT to switch its polarization when the signal is increasing is different from the electric

Table 20.6 Description of model parameters

Parameter	Coefficients [units]
Relative permittivity	ε_{33}
Length of the piezo layer	l_{PZT} [mm]
Width of the piezo layer	w_{PZT} [mm]
Thickness of the piezo layer	h_p [μm]
Thickness of the passive layer	h_s [μm]
Piezo electric coefficient/constant	d_{31} [C/N]
Elastic constant (with short circuited) piezo	s_{11}^P [m^2/N]
Elastic constant of the poly-Si layer	s_{11}^s [m^2/N]
Volume of the MEMS structure in the middle	V_{Str} [mm^3]
Mass of plate	m_{Pl} [kg]
Compliance of the membrane	C_{ms} [mm/N]
Friction of the mechanical system	R_{ms} [Ns/m]
Membrane size	S_d [mm^2]
Back volume	V_{BV} [mm^3]
Front volume	V_{FV} [mm^3]
Length of the tube	l_{tube} [mm]
Diameter of the tube	d_{tube} [mm]
Permittivity of vacuum	ε_0
Density of the structure (silicon)	ρ_{Str} [kg/m^3]
Velocity of sound in air	c [m/s]
Density of air	ρ_{air} [kg/m^3]

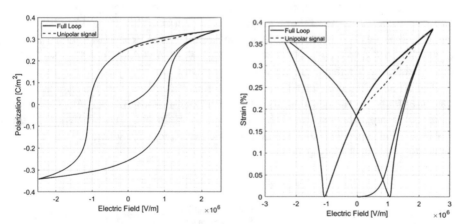

Fig. 20.14 Polarization loop (left) and strain loop (right) for a typical PZT thin film

field needed when the signal is decreasing. This means that the polarization behavior is hysteretic. The consequence on the strain, which is to a first approximation proportional to the square of the polarization, is shown in Fig. 20.14, with a cycle that, because of its peculiar shape, is often referred to as a butterfly loop.

However, by applying a DC offset to the signal, it is possible to avoid the polarization switch. In this case, as can be seen from the small loops in Fig. 20.14, the system presents only small hysteretic behavior and the nonlinearities are greatly reduced. Therefore, thin-film PZT loudspeakers are driven with a DC offset combined with the desired acoustic signal. Still, the piezoelectric coefficients are proportional to the state of polarization of the material; therefore the polarization directly influences the electromechanical coupling, and the remaining nonlinearities in the material will cause a nonlinear response of the transducer. Such nonlinearities are acoustically measurable in terms of THD (total harmonic distortion) and have to be added to other mechanical nonlinearities inherent of the MEMS actuator design and membrane. For reference, in traditional voice coil loudspeakers, the main causes of harmonic distortion are the nonlinear Bl factor (force factor), the nonlinear inductance of the coil, and the nonlinear stiffening of the membrane.

To link the electrical and mechanical nonlinearities, one can make the approximation that the elongation of the transducer is proportional to the tensor of the piezoelectric moduli e, which is nonlinear with respect to the applied electric field and inversely proportional to the stiffness of the device, which is nonlinear with respect to the elongation itself. This can be written as:

$$x \propto -\frac{e_{31}(V)}{K_x} V$$

This relation can be used to model, within a good approximation, the nonlinear behavior of the speaker and to compute a linearization of the system. The nonlinear factors are modeled approximating the real behavior with polynomial fits and the compensation can be fitted into a polynomial expansion. This method is convenient because it is easily implementable in the digital signal processing (DSP) block. Figure 20.15 shows the signal flowcharts of the system with the main blocks being the equalization filters and the polynomial predistortion.

The effect of the compensation can be seen in Fig. 20.16, which shows acoustic measurements of the speaker in an ear simulator (G.R.A.S 43 AC Coupler), at a sound pressure level of 118.5 dB SPL at 1000 Hz. The efficiency of the compensation is particularly good below resonance frequency where the mechanical behavior is dominated by stiffness.

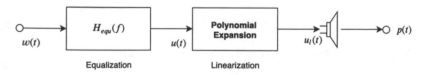

Fig. 20.15 Signal flowchart of the system with polynomial predistortion

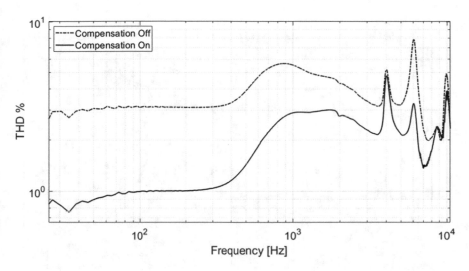

Fig. 20.16 Acoustic THD with and without polynomial predistortion at 118.5 dB SPL

20.5   MEMS Speakers Compared to Electrodynamic Micro Speakers

Compared to micro speakers based on voice coil, piezo MEMS are superior in three categories: sound quality, form factor, and system integration.

Sound Quality
The piezo linear motor is faster and more precise than a voice coil; for in-ear headphones, this means precise bass and crisp and clear treble; covering both aspects within one transducer results in wider bandwidth and using on single element instead of multiple drivers. THD distortions are determined by the linearity of the MEMS design, with the adoption of a closed-loop control, THD has the potential to go down to virtually zero.

Mechanical Integration
MEMS and microelectronics are planar technologies, and the MEMS micro speakers achieve a thickness of 1 mm compared to 2–5 mm voice coil transducers; the lack of heavy metals and use of low density materials enable a 10x lighter design; contrary to a voice coil, piezo does not generate any heat thus improving reliability and durability; devices can be connected directly to PCB as SMD components.

System Integration
PZT consumes 95% of reactive power, and an energy recovery loop can transfer more than a half back to the system making these devices very efficient. The ASIC driver works purely with digital inputs eliminating the need of a DAC. Thanks to the system architecture, electromechanical overdrive is impossible, and this eliminates

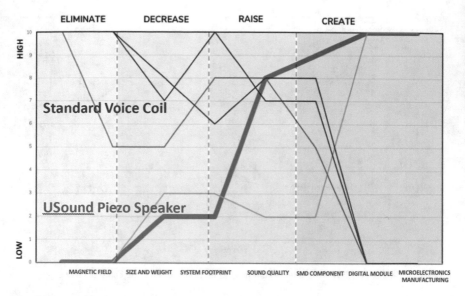

Fig. 20.17 ERRC grid explaining the value proposition of MEMS speakers. USound technology value proposition in dark blue and voice coil speaker solutions in red

the need of smart power amplifiers; tight specs coming out of the fab avoid the need of calibration and enable array configurations.

In particular the goal is to realize a fully digital speaker module which only needs a power supply and a digital audio stream to function. In Fig. 20.17, the comparison to standard electrodynamic speaker technology is summarized: elimination of magnetic components and magnetic fields, reduction of system footprint and volume, and creation of a digital component manufactured with microelectronics production equipment.

20.6 MEMS Speaker Applications

MEMS speakers can be integrated in a variety of audio products; the most relevant applications for demonstrating the impact of MEMS into audio products are described in the next paragraphs.

20.6.1 In-ear Headphones: USB-C and Audio Module for TWS

A MEMS driver fits very well for in-ear headphones as it is the smallest transducer available to effectively cover the whole hearable bandwidth form 20 Hz to 20 kHz.

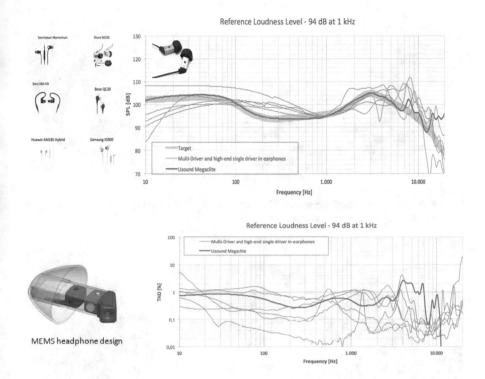

Fig. 20.18 Benchmark between MEMS and ED/BA occluded hear earphones

The acoustic design including volumes and channels can be optimized for its characteristic performances to produce a crisp and clear sound signature. In Fig. 20.18 we report a benchmark of a MEMS speaker headphone reference design compared to several commercially available products based on voice coil speakers.

The audio signal chain is purely digital, from the Bluetooth or USB-C digital interface to DSP up until the power stages of the amplifier; this makes the system very easily integrable into modern systems like TWS headphones (True Wireless Stereo).

Most widely adopted TWS speaker is a 6 mm voice coil driver which is typically 4–5 mm thick. The MEMS component needs 5 times less volume (30 mm3 vs 140 mm3), is 10 times lighter (50 mg vs 500 mg), produces 5 times less vibrations (0,05 mg/s2Pa vs 0,1 mg/s2Pa), and generates no heat. Figure 20.19 compares the standard technology with the MEMS solution; the additional free space made available can be used for designing a more compact product with more advanced digital and sensors integration.

The MEMS speaker enables an audio module that sits directly in the ear canal, leading to an ideal electronics integration and fitting standardized headphone ear tips. Figure 20.20 shows the audio module elements.

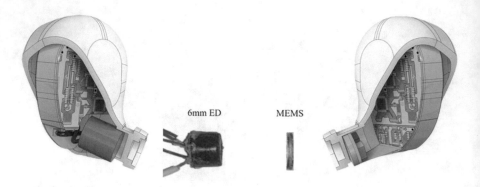

Fig. 20.19 Study for integrating a MEMS speaker into TWS earphones; the MEMS speaker uses much less volume that a voice coil speaker

Fig. 20.20 Audio module shell (left); rigid-flex PCB containing MEMS speaker, amplifier, and ANC feedback microphone (middle); module integration with Bluetooth connectivity (right)

The module includes all the electronics required for the audio signal chain including Bluetooth low energy connectivity. The microphone and speaker are integrated as a sub-assembly connected via rigid-flex PCB, a digital bottom-port MEMS microphone is used to keep the electronics package as small as possible. The shell creates a closed front volume for the speaker and microphone. This is ideal for rich, full-bodied sound reproduction and for high-quality ANC capability. The module incorporates front venting to allow for pressure adjustment of the TWS in the ear canal and includes a mesh for dust protection, along with a waterproof membrane.

This concept greatly improves the manufacturing processes of TWS headphones; with MEMS loudspeakers, all components are assembled using fully automated SMT manufacturing lines; this is a key aspect for mass manufacturing TWS at low costs.

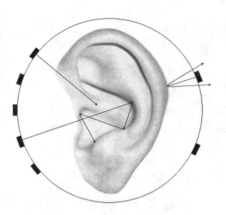

Fig. 20.21 Configuration of MEMS speakers placed around the ear (left) and design of an open headphone for 3D audio (right)

20.6.2   Virtual Reality and Gaming

MEMS enable multi-driver headphones that can improve virtual reality and gaming headphones to the next level. The system comprises of an electrodynamic woofer and several MEMS tweeters positioned around the ear. By placing the speakers in specific positions, it is possible to improve the sound source localization in space by having the listener using its own ear pinna to localize the source. The hardware configuration is supported with HRTF (head-related transfer function) and room model algorithms; compared to standard HRTF headphones, the MEMS cue preserving headphone significantly increases the externalization perceived sound and doesn't need an individualized configuration procedure. Figure 20.21 is an image of a configuration of 7 MEMS speakers placed around the ear and the product design of an open headphone for 3D audio.

20.6.3   Smart Glasses and Augmented Reality

MEMS speakers are used in an audio system for smart glasses and augmented reality applications. The speakers are mounted in the temples delivering great audio quality with the smallest form factor. A two-way system combines a woofer and a tweeter to get the maximum out of the acoustic performances of both transducers: the tweeter is MEMS with crisp and clear treble, and the woofer is electrodynamic with best in class bass performances. The system has a specific acoustic architecture in order to damp and reduce the noise radiated to the ambient and maximize privacy: an acoustic dipole is helping cancelling out leaking sound by providing a 180° inverted

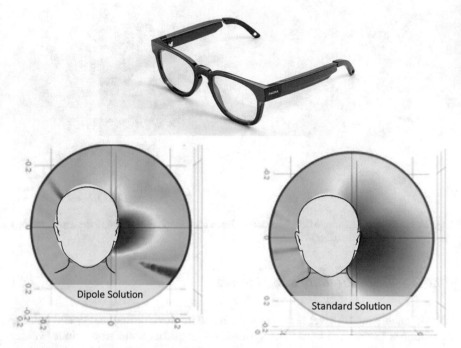

Fig. 20.22 Smart glasses with MEMS speakers (top) and effect on SPL of using acoustic dipoles for improving audio privacy

phase signal from a back port, In Fig. 20.22 an image of the final product is shown and the effect on SPL of using acoustic dipoles for improving audio privacy.

20.7 MEMS Speaker Manufacturing Process

USound concept replaces the voice coil engine with a piezoelectric MEMS driver to move a diaphragm and produce sound. This transducer is designed to convert an electric signal into movement using the piezoelectric properties of a lead-zirconate-titanate crystals. We saw in the past paragraph that the movement is generated by the deflection of two or more suspended cantilever connected to a piston that is glued to the membrane. In the following part, the fabrication process of the MEMS actuator will be discussed, trying to highlight the most peculiar passages.

The piezo MEMS device has a dimension of 3.5 mm x 4.4 mm and is composed by:

1. *Two cantilevers actuated by Thin film PZT* connected with
2. *One piston* having the full wafer final thickness by
3. *Several springs* (depending on design).

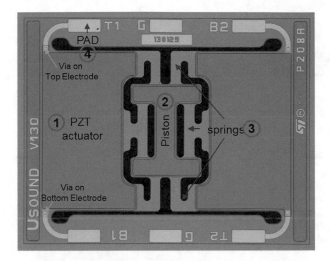

Fig. 20.23 Optical microscope picture of MEMS thin-film piezo loudspeaker

4. Electrodes of the 2 cantilevers are connected to external by *4 gold PADS* (2 for top electrode and 2 for bottom electrode).
5. Release of cantilever is obtained by a *back cavity* deep silicon etch after thickness reduction of the whole wafer to 400um.

In Fig. 20.23, the optical image of realized MEMS device is shown. The 3D micromachining of the MEMS speaker can be divided mainly in two part: the frontside, with the patterning of the piezoelectric material, the related electrical connections and the cantilever with the springs connecting mechanically with the piston, and the backside opening to release the cantilevers.

The very first step of the process is to create the structural layer for the cantilever, and this is done with a silicon epitaxial growth on a thermal oxide grown on the substrate. The Epi-poly layer is one of the key players on the final MEMS response since its dimensions and mechanical properties will influence the device performances; for this reason thickness and residual stress must be tightly controlled. The former is obtained by controlled growth and tuned by CMP (chemical mechanical polishing); the latter is mainly dependent from the epitaxial growth parameter and could be further modified with a thermal annealing. Let's see more in depth.

Since for the considered architecture the actuated cantilevers fold only upward (in the direction of the PZT itself), due to the tensile stress induced by the polarization of the material domains, for the correct functioning of the device, the cantilever without any electric field applied to PZT must be naturally bended downward (in the direction of the cavity side). This allows the up and down movement of the piston driving with an AC input superimposed to a DC voltage that brings the cantilever in a planar position (see Fig. 20.24).

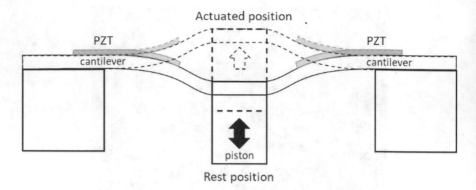

Fig. 20.24 Scheme of piston movement with cantilever bending during PZT actuation when voltage is applied

The cantilever initial bending is the result of the sum of the device design and of the mechanical properties of the thin-film layer deposed on it (other than its own as said before). In our case, fixing some input as materials Young modules and some boundary conditions of design, it has been studied experimentally and by mechanical simulations the possible variables that can be modified to tune the final position. The final result was that the best way to achieve the right bending was to introduce an annealing in oxidizing atmosphere to create a stress gradient in the Epi-poly layer.

After the realization of the structural cantilever layer, the process proceeds with the deposition of the piezoelectric material and its electrodes, namely, a barrier layer of TiO2, a bottom electrode of Pt, a 2um PZT by Sol-Gel deposition, and a top electrode. After these process steps, the patterning of the device starts with alignment and etch of the top electrode and PZT layer, then alignment and etch of the bottom electrode. In both cases, dedicated dry etch steps and cleaning are used.

At this point the actuators have been defined, and the following steps are intended to make the metal interconnections to the I/O pads, the intermetal dielectrics, and the passivation. The PZT passivation and the intermetal dielectric are deposed and vias to top and bottom electrodes opened. Standard AlCu metal interconnections obtained by PVD sputtering and dry etching allow to bring external electrical signal to the actuators and gold pads are realized after nitride CVD passivation to grant correct connection with the speaker package.

The last step of the frontside of wafer is the patterning of the EPI-poly to define the cantilevers and the piston shapes with springs bridging them. These structures are very important to the final performances of the device so the design must be carefully evaluated, and the realization needs to be strictly controlled in terms of critical dimension and profile.

The frontside of the device is completed; a parametric testing is done to verify the good electrical functioning of the single structures and their connections. To have the final MEMS speaker, it only remains to release the cantilever structures

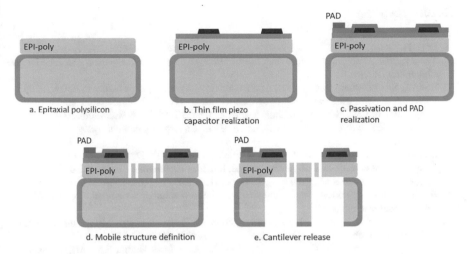

a. Epitaxial polysilicon

b. Thin film piezo
capacitor realization

c. Passivation and PAD
realization

d. Mobile structure definition

e. Cantilever release

Fig. 20.25 Schematic process flow for MEMS speaker realization

removing the bulk silicon through the realization of a cavity from the back of the wafer. To do such step, some precautions need to be taken, in particular the frontside wafer protection and the landing of the cavity etch on the previously realized holes between the springs (Fig. 20.25).

Different temporary frontside protection scheme can be applied to the silicon wafer in order to preserve the frontside, during backside pattering: temporary bonding scheme, protection tape or polymeric thick film, deposited with lamination, and spinning are the possible solutions. The choice of the best method depends on topography present on wafer, equipment used in the following process step, and chemical and thermal compatibility of the layer present on wafer and the process itself. After protection application to frontside, the rear side of the device wafer can be grinded to the desired thickness (400 um) and the backside patterned.

Back cavity dimension accuracy is another important aspect to take care of because this deep silicon etch (obtained via BOSCH etch) determines the length and the symmetry of the cantilevers that obviously must be as much as possible aligned to target on every dice with respect to position or wafer batch.

Finally, the protection is removed and usually a final cleaning of the thin holed wafer is necessary to remove all the residuals of the temporary material used. At this stage, O2 plasma in barrel system or wet bench is used, depending on wafer structure a material used for protection.

After last cleanings, the thin, through holed wafer, is ready to be measured (piston deflection screening) and quality checked. The wafers are fragile and not easily handy, so they are mounted on tape with flat frame to be tested in EWS, inspected by a final automatic inspection and shipped to BE fab for stealth dicing.

References

1. Beer, D. (2020). MEMS-Lautsprecher – Ein Paradigmenwechsel. *DEGA Journal Jan.*
2. The Rational Paradigm of "More than Moore" in *56th Electronic Component and Technology Conference*, 2006.
3. ENIAC, MtM Domain Summary in Applications specify technologies section, ENIAC, 2005.
4. Diamond, B. M., Neumann, J. J., & Gabriel, K. J. (2002). Digital sound reconstruction using arrays of CMOS-MEMS microspeakers, in *MEMS 2002 IEEE International Conference. Fifteenth IEEE International Conference on Micro Electro Mechanical Systems*, Las Vegas, NV, USA
5. Diamond, B. M., Neumann, J. J., & Gabriel, K. J. (2003). Digital sound reconstruction using arrays of CMOS-MEMS microspeakers, in *12th International Conference on Solid-State Sensors, Actuators and Microsystems. Digest of Technical Papers*, Boston, MA, USA
6. H. e. a. Conrad. (2015). A small-gap electrostatic micro-actuator for large deflections. *Nature Communications, 6*(10078).
7. B. L. S. E. L. e. a. Kaiser. (2019). Concept and proof for an all-silicon MEMS micro speaker utilizing air chambers. *Microsystems & Nanoengineering, 5*(43).
8. Hänsler, M., & Auer, M. (2019). Design Considerations for a Digital Input MEMS Speaker Audio Amplifier with Energy Recovery, in *15th Conference on Ph.D Research in Microelectronics and Electronics (PRIME)*, Lausanne, Switzerland.
9. Rusconi, A. (2019). MEMS Micro Loudspeaker: Modeling, Technology, and Applications, in *7th International Symposium on ElectroAcustic Technologies (ISEAT)*, Shenzen, China.
10. Klippel, W. Schlechter, J. Distributed Mechanical Parameters Describing Vibration and Sound Radiation of Loudspeaker Drive Units. https://www.klippel.de/know-how/literature/papers.html

Chapter 21
Tunable Lenses for Autofocus

Sonia Costantini, Irene Martini, Dario Paci, Adriana Cozma,
Gjermund Kittilsland, and Pierre Craen

21.1 Tunable Lenses and Voice Coil Motor

Autofocus technologies are nowadays fundamental for mobile camera and having compact, reliable, and high performance tunable optical systems is mandatory for the success of any smartphone.

Voice coil motor (VCM) technology is the main technology in the market for such kind of systems [1]. VCMs are moving the entire lens stack with respect to the sensor to create the focusing mechanism as described in Figs. 21.1 and 21.2.

Figure 21.1 shows the position of the imaging lens stack for an object located at far distance from the first lens #1 (far distance from the camera module or the phone in example), while Fig. 21.2 is representing the position of the imaging lens stack for an object at close distance (10 cm) from the camera.

The VCM has been introduced in the mobile camera as early as the resolution of the sensor started to increase over 3 megapixels, and the F number has reached values as low as F/2.8. Indeed, a focusing mechanism was needed to increase the imaging quality of pictures taken from infinity to close distances.

S. Costantini · I. Martini
ST Microelectronics, Analog MEMS and Sensors Group, MEMS Technology and Design R&D,
Agrate Brianza, Monza Brianza, Italy
e-mail: sonia.costantini@st.com; irene.martini@st.com

D. Paci
ST Microelectronics, Analog MEMS and Sensors Group, MEMS Technology and Design R&D,
Cornaredo, Italy
e-mail: dario.paci@st.com

A. Cozma · G. Kittilsland · P. Craen (✉)
poLight ASA, Skoppum, Norway
e-mail: adriana.cozma@polight.com; gjermund.kittilsland@polight.com;
pierre.craen@polight.com

© Springer Nature Switzerland AG 2022
B. Vigna et al. (eds.), *Silicon Sensors and Actuators*,
https://doi.org/10.1007/978-3-030-80135-9_21

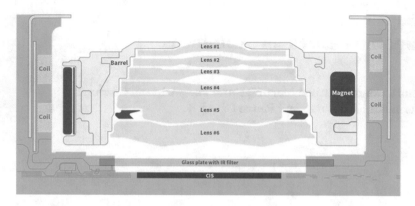

Fig. 21.1 Illustration of VCM when object is far away from the camera

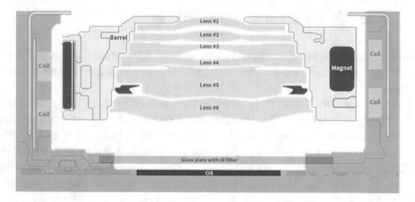

Fig. 21.2 Illustration of VCM when object is at close distance from the camera

The VCM technology was first used in the CD player head, where focusing of the optical reader head was needed to keep focus on the CD and secure reading the data accurately. Introducing it to mobile camera was an obvious and natural step. Despite the challenges it has created to the camera module supply chain, the VCM has been introduced and gradually improved to successfully take more and more market share. The more recent introduction of the phase detection (PD) capability has helped reducing the limitations related to the focusing speed inherent to the VCMs [2, 3]. The earlier autofocus (AF) algorithms were based on a full scan search and contrast hill climbing method, which for VCM required at least 15–20 frames/image acquisitions to reach the focus, resulting in a time to focus in the best case of around 500 ms. With the PD sensor, the time to focus using the same VCM has been reduced to a few frames/image acquisitions and capable to focus in less than 200 ms, which makes a quasi-instant focus for the human eye.

Despite many attempts to introduce alternative solutions, the VCM technology has established itself as predominant and most widely used in the mobile phone cameras, providing the best compromise between maturity, reliability, performance, and cost [1]. Technologies like liquid lens based on electrowetting, developed originally by Varicotic, the liquid crystal technology developed by LensVector, as well as non-tunable lens technologies like the one developed by Siimple, have not succeeded to compete with VCM for different reasons, despite some clear performances or cost advantages [1]. The liquid crystal technology had the potential to offer a cost competitive solution, but the thermal behavior and image quality could not reach the always raising performance requirement in terms of image quality. The electrowetting technology had hard time to compete with VCM from the point of view of the cost/performance's ratio. Simple was based on a MEMS actuator for moving only the first plastic lens element of the entire lens stack (as opposed to VCM that moves the entire lens stack made of 4–6 lens elements) and had therefore a potential to be cost competitive, fast, and with low power consumption. But the technology did not come through because the required image quality could not be reached. The concentricity of the moving lens with the other lenses in the stack needed to be around 1–2 micrometers, which turned out to be very challenging.

poLight® has developed a novel technology and a products family (TLens®), based on polymer optical material and piezoelectric PZT MEMS actuators, which offers a cost competitive solution that can compete with the VCM by offering higher performances, in terms of autofocus speed, power consumption, compactness, non-magnetic interference, as well as a good robustness and long lifetime (number of autofocus cycles). poLight® is a registered trademark in European Union, India, Japan,, Norway, South Korea (KR), Taiwan, and the United States, and TLens® is a registered trademark in China, European Union, India, Japan, Norway, South Korea (KR), and the United States.

The TLens® response time is 2–3 ms (about 5 time less than typical VCM), which enables instant focus even without PD (phase detection) imaging sensor or other telemeter technology.

TLens® low power consumption when focusing has been evaluated to around 5 to 6 mW, compared to 200–250 mW for VCM. The low power consumption is key for wearable devices and video applications.

The constant field of view associated with the non-magnetic interference as well as no field of view change while changing focus (unlike VCM) make the TLens® one of the most promising options for multi-camera solutions where image stability is important to secure image stacking and image fusion.

21.2 TLens® Concept

The current poLight® TLens® is based on a thin-film piezoelectric MOEMS element, which is placed on a thin-glass membrane, and plays the role of an actuator. A patented soft polymer is sandwiched between this thin glass and another high-quality rigid window lens. Under voltage application, the piezoelectric material changes shape depending on the applied voltage, forcing the thin-glass membrane to bend and, consequently, to change the shape of the polymer. This leads to the

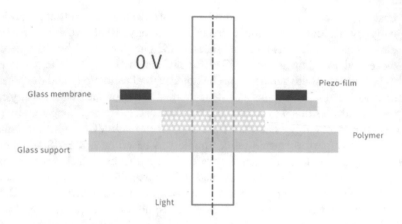

Fig. 21.3 Working principle with no voltage applied to Piezo of the TLens® (0 V)

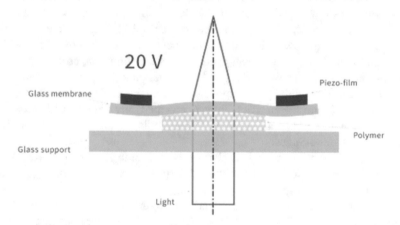

Fig. 21.4 Working principle with voltage applied to Piezo of the TLens® (20 V)

variation of the optical power (OP) of the lens, allowing focusing at a therefore voltage-dependent distance. As better detailed in Sect. 21.3.3, the optical power is the inverse of the distance of focus from the lens, and it is measured in diopters (dpt, 1/m): e.g., if focus is far from lens OP~0dpt, if is at 10 cm from the lens OP = 10dpt.

The working principle is given as follows; simplification has been made for easiness and clarity. When the piezo actuator is in standby mode, no force is applied to the thin glass, and light passes through the two elements of glass and the polymer without deviation (Fig. 21.3). When a voltage is applied, the piezo actuator will immediately force the thin-glass membrane to bend accordingly (Figs. 21.4 and 21.5).

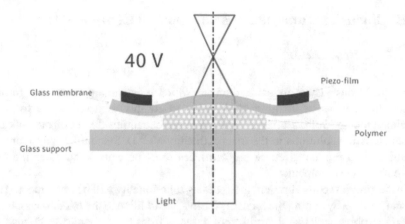

Fig. 21.5 Working principle with voltage applied to Piezo of the TLens® (40 V)

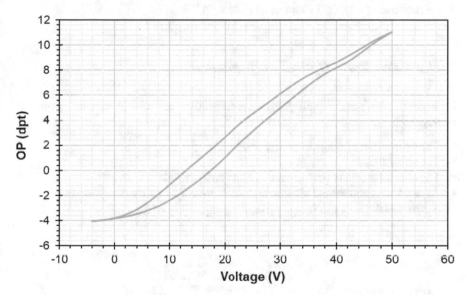

Fig. 21.6 Typical behavior of poLight® tunable lens optical power versus voltage applied to PZT material OP(V)

Membrane bending results in a change of curvature radius of one side of the lens. The following graph represents the typical behavior of the optical power versus the voltage measured at ambient temperature. There is a hysteresis in the OP(V) curve mainly coming from hysteresis of the piezoelectric material used for the membrane actuation. (Fig. 21.6)

21.3   Focusing Tunable Lens and Optical Components

21.3.1   Introduction

Since the TLens® is an optical component which is integrated in addition to other lenses (fixed focus or imaging lens), it has to comply with optical performance requirements, as well as offer a focusing range capability that competes with the already existing solutions in the market (mainly VCM). Standard requirements for reliability and product lifetime performances must be considered since the early phase of the development.

In the simplest configuration, the focusing functionality will be implemented on a camera module by adding the focusing TLens® in addition to the fixed focus camera. When the object is located at different distance from the camera, the TLens® can refocus and secure a sharp image on the sensor. (Fig. 21.7a shows a schematic representation of a typical Camera module with TLens®.)

21.3.2   The Basic Optical Performances

In general, the image quality is very subjective but still it can be quantified through:

- The estimation of contrast of the image
- The color fidelity
- The overall artifacts performances like ghost, or flair
- As well as geometrical distortion which usually can be compensated up to a certain level, thanks to image processing capability of the platform

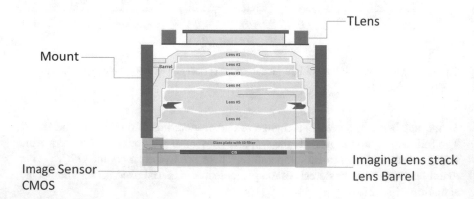

Fig. 21.7a   Camera module with TLens®

The following paragraph provides simple explanations about the way TLens® requirements have been established. The reader is recommended to refer to optical engineering books for more detailed explanations and better understanding [4, 5].

The image contrast is the result of combined effects coming from different contributors. Therefore, to secure a good image contrast, the lens will have to be compliant with specification in terms of:

(a) Optical cosmetic: Scattering due to inclusions/bubbles in optical material and scratches on glass surface over the useful aperture in the optical path can affect the contrast and degrade the optical quality. The optical cosmetic needs to be specified and tested such that the lens will minimize its impact.
(b) Wavefront error (WFE): The sharpness of the image is quantified in terms of MTF (Modulation Transfer Function of the lens stack), and it is related to WFE of single optical components and accuracy/performances of the focusing actuator. Please refer to reference below for theory and correlation between MTF and WFE.
(c) Scattering: The scattering can reduce the contrast of the image and therefore should be specified for any optical components like TLens®.

The color fidelity is related to the stability of the spectral transmittance as well as reflectance of the optical component that the light is propagating into. The TLens® will not affect the color transmittance when its own transmittance over the spectral range is constant and over 96%. The color fidelity is also of importance when creating an imaging system, and any component added shall not affect the spectral transmittance of the system over the visible spectral range.

In general, the below parameters will also affect the global image quality due to the lack of good transmittance:

(a) Ghost and stray light: The image artifacts (ghost and stray light) can be generated by the lack of lens stack baffling and lack of antireflective coating on every single lens component used in the lens stack.
(b) Scattering/optical cosmetic: The scattering by small inclusion will generate scattered light. Since scattering of light is dependent on $1/\lambda^4$ (Mie scattering mode [4]), the spectral transmittance may be affected by scattering). Since blue will be more scattered than red when small particles and inclusions are present, it may affect the color fidelity of the image. This is a requirement to consider for the product design.
(c) Antireflective coating (ARC): When adding an optical component in an optical stack, special care must be taken to avoid creating parasitic reflection when light is transiting from air to the optical material. Antireflective coating is in general a stack of multiple layers whose thicknesses and reflective indexes are chosen in order to minimize reflection of the surface in a certain range of wavelength. Indeed, without ARC, 4% of the light will be reflected by every glass to air and air to glass interface. Thus, it is mandatory to apply ARC on each of these interfaces. The ARC requirement shall also be part of the basic requirements for the TLens®.

Also the geometrical distortion has to be controlled and must be considered at system level, rather than at lens level. Among the sources of geometrical distortion, there is the optical axis stability, which includes the gravity effect, which is a well-known weakness of VCM. The tilt generated by VCM is wide enough to require very specific processing and calibration, which could lead to non-robust image solution/low-quality image when it comes to image stacking/computational photography applications. TLens® shows significant advantage with respect VCM in terms of optical axis stability.

21.3.3 Main Mechanical Features for an Autofocus System

In the previous paragraph, we have mentioned the optical performances that a tunable lens must guarantee. It is also important to guarantee focusing performances of the tunable lens.

Equation 21.1 expresses the relationship between the equivalent stroke of VCM for a given optical power variation for a typical camera module that has a lens with a given effective focal length EFL or field of view and the optical power variation OP of a tunable lens to perform the equivalent refocus or defocus.

$$\Delta VCM = EFL^2 \times \Delta OP \tag{21.1}$$

where ΔVCM is the stroke of VCM (μm), ΔOP is the optical power variation (in diopter or 1/m), and EFL is the effective focal length of the imaging lens (mm).

Considering $d1$ as the minimum distance of an object from the camera that system can focus, while $d2$ the maximum distance, ΔOP can be written as:

$$\Delta OP = \frac{1}{d1} - \frac{1}{d2} \tag{21.2}$$

Considering a typical condition where EFL is 4 mm, $d1 = 10cm$ and $d2 = \infty$, the equivalent stroke of the VCM is 160 μm.

On the other hand, in the case of poLight® TLens®, the OP change is given by the change of curvature of the glass membrane R_c, as shown in Figs. 21.2, 21.3, 21.4, and 21.5. Following the well-known Lens Maker's [5], it is possible to write the optical power OP as:

$$OP = \frac{1}{d} = (n_l - 1) \left(\frac{1}{R_c} - \frac{1}{R_0} \right) = \frac{(n_l - 1)}{R_c} \tag{21.3}$$

where d is the distance of the object to be focused, n_l is the refractive index of the lens, and R_0 is the curvature radius of the glass support which is infinite.

The aperture diameter of a tunable lens is important too, since it could affect the so-called F-number FN of the system:

$$FN = \frac{EFL}{\emptyset} \tag{21.4}$$

where EFL is the effective focal length of the lens stack and $\emptyset$ is the minimum useful diameter of the imaging lens aperture.

Allowing small F-number, i.e., having large lens aperture, is fundamental to get picture in dark condition, since it allows to capture more light given a certain focal length.

Finally, another important parameter for a camera is the field of view α, which is defined in Fig. 21.7b and is fundamental to guarantee the capture of a scene as wide as possible. Field of view can be written in terms of focal length and image sensor diagonal D, assuming of course that lens aperture/lens stop does not limit it:

$$Tan\left(\frac{\alpha}{2}\right) = \frac{D/2}{EFL} \tag{21.5}$$

Thus, allowing large diameter lenses helps also to keep high field of view.

21.4 TLens® Product Architecture

21.4.1 TLens® Components

The autofocus system developed by poLight®, i.e., the TLens®, is made of three components, a MOEMS actuator, the poLight® polymer, and the back window, as illustrated in Fig. 21.8.

The final product is the result of the assembly of the three components, which are manufactured by different suppliers and poLight®. The manufacturing of the MOEMS part, called front-end operation, is done by ST Microelectronics. The final assembly of the different components, called the back-end operation, is made by a specialized company.

The MEMS actuator is made of a piezoelectric layer deposited on a thin-glass membrane that is attached to a silicon wall. The polymer is attached to the membrane as well as the glass window at back-end.

The simplified assembly process flow of the TLens® is shown in Fig. 21.9.

After the assembly, a customized final test is performed on all lenses. The final test has been designed specifically to guarantee the optical and focusing performances of the TLens®. As the TLens® is a new product, with new functionalities, no standard equipment was available when the development started. poLight® has developed dedicated test processes and equipment together with commercial equipment vendors.

poLight® has developed the TLens® with a package (Fig. 21.10) to facilitate the integration of the TLens® on a camera module, with minimized mechanical stress

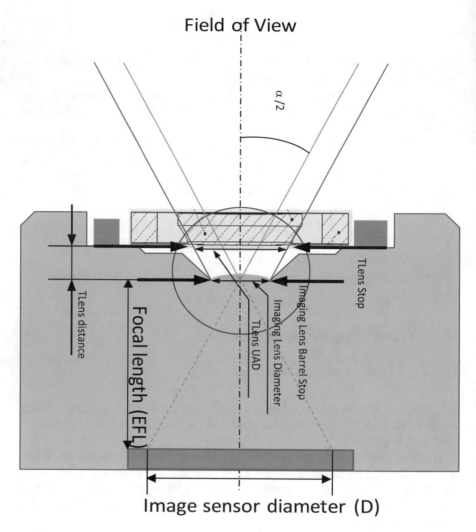

Fig. 21.7b Simplified sketch of a camera module showing the relation between EFL, FOV, and lens and TLens® Stop

impact on the MEMS actuator from external forces. It guarantees optimum TLens® performance over the operating temperature range and storage conditions, as well as mechanical robustness to withstand tests according to the most demanding mobile phone requirements [6].

It goes without mentioning that the performances of the MEMS actuator will be fundamental to enable high product performances. Indeed, the high optical quality, low power consumption, fast focus speed, high optical axis stability, and low hysteresis are key performances enabled by the MOEMS design and the materials used.

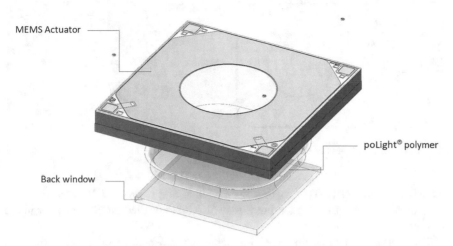

Fig. 21.8 Exploded view of a TLens®

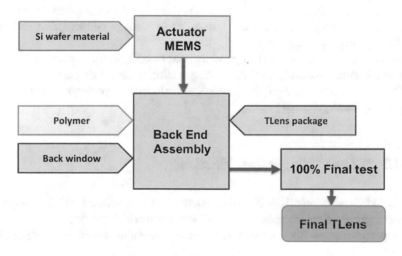

Fig. 21.9 Simplified process flow of the TLens® assembly

To secure the optical performances, the TLens® has been designed such that the actuation-induced deformation of the glass membrane is as spherical as possible, generating a high-quality plano-concave or convex lens (low WFE), while providing an optimum optical power variation range.

The optical index of refraction matching between the glass membrane, the polymer, the supporting glass (referred as back-window), and the antireflective coating ensures that the optical transmittance is optimized for the visible spectrum. If needed, the antireflective coating could be tuned to other wavelength range, for

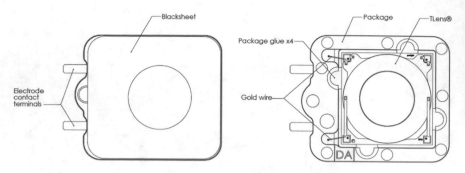

Fig. 21.10 Essential parts of packaged TLens® (example)

example, infrared (IR). This coating is embedded and part of the MEMS structure, as described later in the chapter, but it may also be deposited on the lens later in the process.

The characteristics of the piezo element enable a change of focus in the lens within a millisecond (ms). As the piezo element is controlled by voltage and is electrically equivalent to a small capacitor, the power consumption is extremely low, compared with existing VCM, since there is only consumption related to charge/discharge of a capacitor.

Such an architecture exhibits a good thermal stability, which is needed to enable a stable focusing mechanism over a wide operating temperature range.

The basic architecture of the TLens® and different important features of the device have been patented [7].

21.4.2   TLens® Target Specifications

The table below summarizes the list of parameters to be considered for design and process optimizations to create a robust and competitive product family. Typical target performances for an example TLens® are summarized in Table 21.1 as example.

The TLens® has been designed and developed to comply with demands of a mission profile specific to applications like camera for mobile phones. Table 21.2 gives a summary of the minimum environmental tests that any optical component shall comply with for the mobile phone application.

Table 21.1 Typical target performances

Description			
1. Mechanical specification	*Symb*	*Unit'*	**Target**
1	Length of TLens®	*TLL*	*mm*	<6
2	Width of TLens®	*TLW*	*mm*	<6
3	Potential added thickness on camera module	*TLT*	*mm*	<0.5
2. Optical specification	*Symb*	*Unit'*	**Target**
1	Useful aperture (UA) diameter – front window	*FUA*	*mm*	1.55
2	Useful aperture (UA) diameter – back window	*BUA*	*mm*	2.05
3	Wavefront error (RMS) over useful aperture	*WFE*	*nm*	<50
4	Average transmittance (spectral range 400–650 nm)	*T*	*%*	>95%
3. Actuator specification	*Symb*	*Unit'*	**Target**
1	Optical power @ min voltage (offset)	*PL*	m^{-1}	<−1
2	Optical power @ max voltage	*PH*	m^{-1}	> 12
3	Response time	*RT*	*ms*	<5

Table 21.2 Summary of typical reliability test (according to typical mobile phone mission profile)

Test	Test full name	T (°C)	RH (%)	Biased	Duration	Relevant standard
ELFR	Early life failure rate	85	–	Yes	48 h	IEC 60068-2-2
LTS	Low temperature storage	−40	–	–	240 h	JESD22-A119A IEC 60068-2-1
HTS	High temperature storage	85	–	–	240 h	JESD22-A103E IEC 60068-2-2
THS	High temperature and humidity storage	85	85	–	240 h	IEC 60068-2-67
TC	Thermal cycling	−40 to 90	–	–	96 cycles	JESD22-A104E IEC 60068-2-14
LTO	Low temperature, operating	−20	–	Yes	96 h	JESD22-A108D IEC 60068-2-1
HTO	High temperature, operating	85	–	Yes	96 h	JESD22-A108D IEC 60068-2-2
THB	High temperature and humidity, operating	85	85	Yes	96 h	JESD22-A101D
ESD	Electrostatic discharge	RT	–	–	–	IEC 61000-4-2
FF	Free fall drop test	RT	–	–		
TBL	Tumble test	RT	–	–	–	IEC 60068-2-31
SR	Solar radiation	25 to 55	–	–	10 cycles	IEC 60068-2-5
VIB	Vibration	RT	–	–	–	JESD22-B103B IEC 60068-2-64
TS	Thermal shock	−40 to 85	–	–	96 cycles	IEC 60068-2-14

21.5   Thin-Film PZT MEMS Technology for Tunable Lens

21.5.1   The MEMS Actuator: General Considerations

Several different actuator principles and technologies have been investigated at the early stage of the product development. Electromagnetic actuators (like VCM) have the advantage of large displacements and are therefore widely used in the mobile industry. They have however important limitations related to size, power consumption, and speed. A few linear actuators based on conventional electro-magnetic or piezoelectric actuation have been considered but were discarded as too bulky. Thermal actuators are usually slow and have high power consumption (more than 100 mW) ([8, 9]). Microscale hydraulic and pneumatic actuators have also been demonstrated ([9, 10]), but they have serious limitations in terms of reliability, low speed, and high thermal sensitivity. Electrostatic actuators, widely used in microdevices, have been tested at early stage ([11, 12]), but discarded very quickly by poLight® because the achievable forces were too weak, which constitutes a serious limitation for the TLens®.

Based on experiments and early testing, the thin-film piezoelectric actuation has been identified as the best option for the TLens® actuator, due to its advantage in terms of speed and power consumption. This choice has driven the design of the product architecture since the early stage of the development.

The need of a transparent membrane was also a clear requirement. Furthermore, mechanical properties of the membrane are fundamental since its stiffness affects the actuation range and it will secure the required optical surface sphericity that the polymer alone could not have secured. This has been a key constraint. Considerable effort has been made to find a material that could offer the correct mechanical and optical performances, as well as a good compatibility with the MEMS processes in general, as well as the specifics of the thin-film piezoelectric material and processes, which usually involve relatively high temperatures. More information about this topic can be found in [13].

Many iterations have been required to eventually converge to a material and structure that could guarantee the lifetime of the product over the operating condition of mobile devices.

Indeed, the stability, over the operating temperature range, as well as extreme humidity conditions, must be guaranteed. The evolution of the mobile phone design, increasingly more compact, thinner, stiffer, and harder, has driven the need of increasingly high shock resistance for all mobile phone components. The accelerations applied to the mobile phone component has raised from a 5000G in the 2008 to currently over 40000G, which creates significant challenge for mechanical robustness. poLight® has implemented solutions that are patent pending. The reader will find the description of the feature that has been added to improve the drop test resistance in [14].

21.5.2 MEMS Actuator Architecture

The selected architecture of the MEMS actuator is shown schematically in Figs. 21.10 and 21.11, and described later in this paragraph. All features and layers have been implemented to guarantee performances and robustness of the manufacturing processes. poLight® has patented the key MEMS architecture [14].

Because of the mechanical stresses of the different layers and because of drift phenomena and temperature-induced effects, the glass membrane is usually not entirely flat for zero actuation voltage. When the membrane is bent upwards, the TLens® presents a positive optical power (OP), and when the membrane is bent downwards, the optical power of the lens is negative. The amplitude and the sign of the membrane deflection in non-actuated stage are mainly related to the stresses induced by the different materials and processes used to manufacture the MEMS actuator.

At system level (camera module level), it is important that focus on infinity (i.e., zero optical power) is always achievable over the entire operating temperature range and over the lifetime of the device. The architecture of the MEMS design is therefore made such that it is always possible to achieve this condition by actuation within the voltage range available. Since the piezoelectric actuation using a single piezoelectric layer as shown in Fig. 21.12 is basically monodirectional, bending the membrane upwards, the room temperature optical power at zero voltage (offset) for the as-fabricated actuator shall be negative, i.e., downward bent membrane. Consider that also contributions from the camera module (not illustrated in the figure) must be considered when the whole system is analyzed.

Since the actuator has to be used under voltage, electrical connection and pad areas have been included in the layout. The reader will note the presence of redundant contact pads, since only two electrodes are needed, and that the contact pads are not located on top of the active piezoelectric material. Such implementation allows robust electrical connection for final testing as well as electrical connection in any device, without altering the reliability under extreme operating and humidity conditions.

As described earlier, an antireflective coating (ARC) shall be applied on any glass/air interface, to minimize the reflections over the visible wavelength range. The ARC layer is deposited by common wafer fab methods. It consists of a multiple stack of alternatively high and low index of refraction materials for which the

Fig. 21.11 TLens® actuator layout and cross-section line

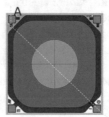

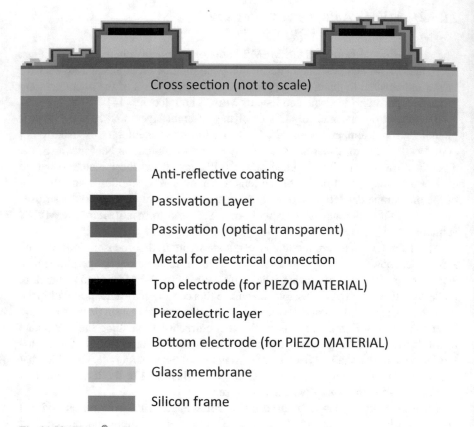

Fig. 21.12 TLens® actuator cross section and color code

thickness has been optimized to create an antireflective function. More information about the ARC coating is given in [15].

From an optical point of view, the antireflective coating deposited on top of the actuator chip is one of the most important parts, designed to avoid ghost images. The choice of materials was determined by the materials and processes available at the fabrication site. The layers thicknesses have been chosen considering the process capability, for stable performance.

The top ARC covers as much of the chip as possible. It is only removed from the pads, where electrical contact to the metal is needed, and at the chip edges, in the dicing area.

The capability to manage the stress in each material and layer is usually a key point for a MEMS process and this is even more true for the TLens®, since any stress and thickness variation in the material stack of the basic elements, such as the piezo, the glass membrane, and the silicon frame, will be translated in variability of the glass membrane bending, meaning variation of the optical power of the non-actuated TLens® eventually. Furthermore, intrinsic stresses, especially the one of the membrane can affect also the structure stiffness and thus the actuation range. Consequently, poLight® and ST engineers have optimized the TLens® architecture to minimize the influence of stress variations.

To obtain the target specification in terms of typical optical power OP of the non-actuated lens (i.e., at 0 V), thickness and stress of the passivation layer in Fig. 21.12 have been properly chosen, so that this layer has both passivation and compensation function. In particular, to guarantee a negative OP at 0 V, the layer stress is compressive, so that it drives downwards the membrane.

Since this layer also performs a passivation function, it covers as much of the chip as possible. It is removed from the aperture area, out of optical considerations (since it is not transparent), from the pads area where electrical contact to the metal is needed, and at the chip edges, in the dicing area. In this way, the membrane deflection at zero applied voltage is downward and the room temperature offset is negative.

The deposition conditions of the layers have been optimized to improve the stability of the film with respect to humidity (i.e., the stability of the stress in the layer and consequently the stability of the offset).

21.5.3 Thin-Film PZT MEMS: The Front-End Processing

The front-end processing of a tunable lens MEMS can be divided in four main process steps performed on a standard silicon wafer substrate:

– Thin-glass membrane deposition
– PZT stack fabrication
– Passivation and antireflective coating of the lens
– Cavity patterning and membrane release

21.5.3.1 Thin-Glass Membrane Deposition

The so-called glass membrane is designed to have a thickness in the range of tens of microns and it is composed of different dielectrics layers. The whole membrane stack is part of the tunable lens itself and therefore needs to fulfill demanding optical requirements, such as high transmittance in the visible range and high-level optical quality (no micro defectiveness, low roughness). A dielectric material was chosen as the main component of the stack. Its thickness is higher than standard dielectrics deposited by CVD (chemical vapor deposition) techniques commonly used in MEMS technology, and it is not difficult to understand that its uniformity and quality are of crucial importance. The dielectric thickness, intrinsic stress, and its stability strongly influence the final lens performance: in particular, the deflection

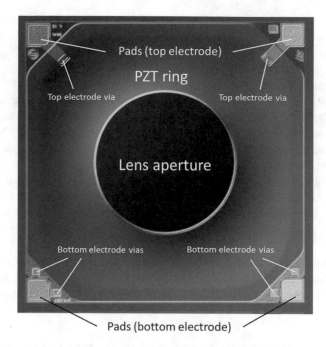

Fig. 21.13 Top view of tunable lens: main components are highlighted

of the non-actuated lens, the optical power range as well as the wavefront error WFE (please refer to the design chapter for more details).

Considering the peculiarity of the application, this process needed a long optimization to match both the design requirements and to find a stable working device.

The material needs to fulfill strict average stress requirements together with extreme thickness uniformity and a very low defectiveness. The gas flows during the CVD growth must be closely controlled. Such a deposition step must be supported by a subsequent thermal procedure to stabilize the layer itself. The integration of additional dielectric films is another important point. The introduction of a moisture barrier is mandatory to guarantee the lens membrane film stress stability over time. The sequence of dielectrics layer involves different deposition techniques together with annealing procedure. The combination of such dielectrics films is able not only to prevent moisture to diffuse inside the glass membrane but also to guarantee a combination of thickness and stress that matches the lens deflection requirement, directly related to the optical performances. The integration of this specific sequence required a particular care in order not to impact the lens optical quality (refer to Sect. 21.3.2). As a matter of fact, the substrate surface and the wafer bevel status can strongly impact the layer growth quality inducing defectiveness and nonuniformity.

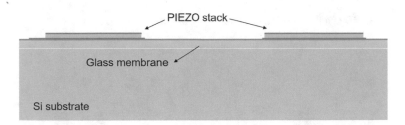

Fig. 21.14 Tunable lens cross section after PZT stack fabrication

21.5.3.2 PZT Stack Fabrication

Thin-film piezo technology is the chosen platform for the tunable lens actuator, considering its easy integration on silicon wafer and the low power consumption. The current piezo stack is made of two metallic layers (Pt as bottom electrode and TiW as top electrode) with a PZT layer deposited by sol-gel between them (Fig. 21.14). The details of deposition and patterning of such stack are illustrated in the dedicated chapter as well as the piezo figure of merit and the physical/chemical characterization techniques.

In the tunable lens process, the PZT stack is integrated on top of the glass membrane. Being fabricated via sol gel deposition, the defectivity of the layer beneath needs to be closely controlled to guarantee the good quality of piezo film.

As mentioned in the design chapter, the stress and thickness of such stack impacts the deflection of the non-actuated membrane and the OP(0 V) eventually. The intra-wafer stress and the wafer-to-wafer variability impact the device performance and the tunable lens design must be robust enough to allow all these tolerances.

The PZT layer is grown on the bottom electrode and the piezo crystallographic orientation is closely linked to the electrode properties. As explained in the dedicated chapter, the sol-gel layer needs to undergo several annealing steps to properly crystallize. The PZT layer stress is crosslinked to the bottom electrode, as the intrinsic stress is generated by the mismatch of the deformation of the two different materials during the previously mentioned annealing steps. The top electrode intrinsic stress is also influenced by the substrate on which is deposited, i.e., the PZT stack or a dielectric layer. Following such considerations, the PZT actuator stack may be considered as a unique complex system made of three different interconnected materials.

21.5.3.3 Passivation and Antireflective Coating of the Lens

The passivation layer has multiple purposes. As in all MEMS actuators based on PZT technology, the piezo actuator needs to be passivated with a pre-metal deposition (PMD) stack which acts as a protection during the following process steps and as an electrical insulator between electrodes and metal routing (see Fig.

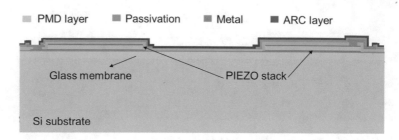

Fig. 21.15 Tunable lens cross section after passivation and ARC deposition

21.15). The passivation layer includes dielectric materials either as single layers or a plurality of layers stacked. The metallic stack for electrical connection is deposited by PVD sputtering (physical vapor deposition) and subsequently patterned via dry etching. The dry etching recipe is optimized to have a good selectivity over the passivation layer and to guarantee no residues and low final roughness, especially in the lens area, for a good image contrast and color fidelity.

As explained in Sect. 21.4.2, the final product has also a specific mission profile, and a barrier against potential detrimental effects of the atmosphere, in particular due to humidity, is in general mandatory. For tunable lenses, those protective films, deposited on top of the metal routing, play an important role in the device functioning, due to its compensating role of the static deflection, as shown in Sect. 21.5.2. A retuning of target thickness of about 20% in the moisture barrier, for example, can impact the optical power as much as more than 1 diopter. The device functioning is even more sensitive to other layer characteristics: the same optical performance variation can be obtained with less than 15% of intrinsic stress variation. A specific equipment management procedure was implemented to better control the "as depo" properties of such material. As explained in Sect. 21.5.2, the deep knowledge of the product sensitivity to this passivation layer has allowed to adapt its thickness and layout to properly tune the OP (0 V). Such dielectric material, which acts as moisture barrier as well as compensation layer, is removed by dry etching from the lens area, from the scribe lines to facilitate the dicing process and from the metal pads for electrical connection.

Another fundamental goal of machining a MEMS lens is to increase the transmittance spectrum intensity in the wavelength range required by application (visible light in this case). Therefrom a sequence of dielectrics layers deposited by PECVD (plasma-enhanced chemical vapor deposition) was developed to match the desired optical requirements at glass/air interface. The films sequence includes oxide layers based on silicon and nitrogen that were specifically chosen (i.e., with proper refractive index mismatch) to reduce the light reflection. The thickness of such oxides needs to be strictly controlled (tolerance +/− 10%) to guarantee the performance. The ARC (antireflective coating) stack covers the whole die, apart from metal pads for electrical connection.

21.5.3.4 Cavity Patterning and Membrane Release

To allow a very robust optical path and a simple implementation of the tunable lens, the final device thickness is limited to few hundreds of microns. After the actuator is machined, the fabrication flow involves wafer thinning. The silicon substrate undergoes a silicon grinding procedure followed by a final stress release step. Thickness measurements are performed to check that the target value is matched. Further process steps are required to pattern the cavity and release the dielectrics membrane. Figure 21.16 shows a view of the device after cavity etch.

To carry out such fabrication, a sequence of lithography, deep silicon dry etching (hundreds of microns), and cleaning steps is performed. The previously mentioned procedures involve reverse-side wafer processing. To preserve the integrity of the top layers, particularly focusing on the lens surface optical quality, the wafer frontside needs to be protected during the patterning of the backside. The integration of a protective layer in the final steps of the fabrication of a tunable lens device is sophisticated and challenging.

The selected procedure needs to fulfill many requirements. First, the protective layer needs to be robust in order to prevent the frontside surface from scratches and from particles that could be eventually embedded in the functional layers during the reversed wafer processing. Such a layer will be the contact surface of many handling tools; therefore its impact on the different chucks needs to be carefully evaluated as well as its influence on wafer manufacturability. The protective layer needs to withstand high temperature procedure, remaining as stable as possible to not impact the wafer handling feasibility. The bow of the wafer after thinning can be strongly impacted by any additional layer and should be carefully managed. Finally, this support layer is removed with a sequence of cleaning steps properly selected to

Fig. 21.16 Backside view of the tunable lens device after cavity patterning (optical microscope)

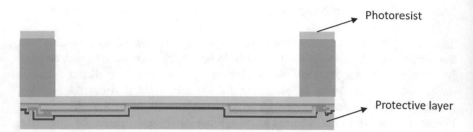

Photoresist

Protective layer

Fig. 21.17 Simplified sketch of a die after backside etching (drawing not to scale)

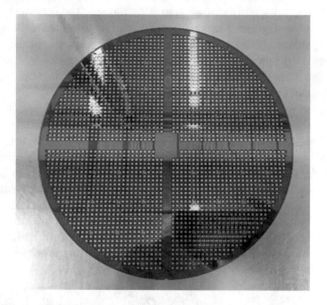

Fig. 21.18 Wafer backside after cavity patterning with the typical thick silicon cross

address both the chemical compatibility with the layers on the wafer and the final cleanliness. The surface needs to be clean and with the same optical performances as designed, particularly no residues that can cause light scattering should be present, and the ARC layer must be preserved.

The backside etching step is a delicate phase of the process and it has a huge impact on many features of the final device. During this machining, a significant part of the silicon substrate is removed, leaving a fragile wafer where the main part of the device is the thin membrane (<20 um thickness) as shown in Fig. 21.17. The backside mask layout was properly modified to improve the robustness of the etched wafer (see Fig. 21.18, with the "cross" where silicon is not removed), which needs to be handled during the following procedures (resist stripping, cleanings, metrology steps, automatic optical inspection). Firstly, a fine-tuning process was needed to guarantee high uniformity of the cavity dimensions within the wafer. It should be noted that the discontinuous pattern (thick silicon cross), introduced to reduce the wafer fragility, strongly influences the etching uniformity. The process recipe was

Fig. 21.19 Cavity array appearance at scanning electron microscope (SEM)

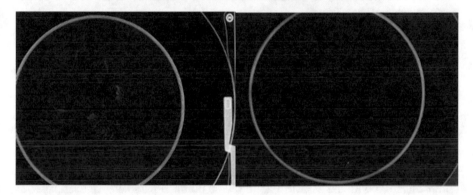

Fig. 21.20 Dark field optical microscope inspection of lens area: example of residues due to not optimized cleaning sequence (left), optimized cleaning sequence (right)

optimized to guarantee the cavity dimensions specification. Another crucial point is the oxide landing layer of the silicon etching. As a matter of fact, the residual oxide thickness impacts the lens performance, and it should be uniform from die to die. Many efforts are put in place from an industrial point of view: equipment control procedures, preventive maintenance routine, logic test management, and process control strategies. A detail of the typical appearance of the back of the wafer after cavity patterning can be seen in Fig. 21.19.

Furthermore, the cavity inside status, both silicon sidewalls and oxide upper surface, has an impact on the final delicate procedure of lens assembly. Silicon sidewall defects created during a non-optimized etching step or any undesired residues (polymers, photoresist, particles) can negatively influence the back-end process during which the polymer is encapsulated between the back window and the actuator itself. Examples of residues due to not optimized cleaning sequence are shown in Fig. 21.20.

Surface wettability, as determined by the final cleanings, may induce from one side polymer overflow or, on the opposite, a totally inefficient coupling with the dielectric membrane. The consequence would be a lens nonconformity or may eventually evolve in a die failure. Following the previous considerations, some options of final resist and polymer removal steps had to be discharged. The final cleanings were therefrom optimized including a combination of oxygen plasma and organic cleaning and excluding water-based solution.

21.6   Design and Simulations of the Actuator

Finite element modelling and physics simulations by COMSOL software have been used to model the entire TLens®. Detailed layers geometry and physical properties have been implemented to simulate the actuator performances as well as optical performances of the TLens®.

Main parameters that could be extracted by simulations are *OP(V)*, i.e., optical power as a function of *V* and *WFE(V)*, i.e., wavefront error as a function of *V*. *OP(V)* can be calculated by using Eq. (21.3) where the curvature radius is extracted by deflected shape from FEM simulations. *WFE* can be extracted also from simulations, by calculating the rms error of deflected shape of the lens with respect to ideal spherical deformation, which leads to a nonspherical wavefront of the focused light. Figure 21.21 shows a typical deflected shape of the lens (complete of polymer and back-window).

At a first stage, the model was used to predict the nominal performances of the TLens® and to tune/optimize the TLens® actuators. This tuning was done in communication with the MEMS Fab to set the nominal parameter of each layer, as well as the target tolerances for each of the critical parameters to control. The model has also been used to widen the TLens® product family, targeting larger aperture products. The chip size and the membrane thickness were the main parameters to play with, for an optimum size/performance trade-off. The graph in Fig. 21.22 shows

a)   b)

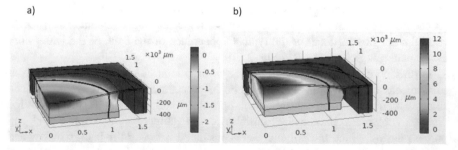

Fig. 21.21 Simulated deflected shape of TLens® a) at 0 V actuation b) at 50 V actuation. One quart of the geometry only is simulated exploiting geometry symmetry axis

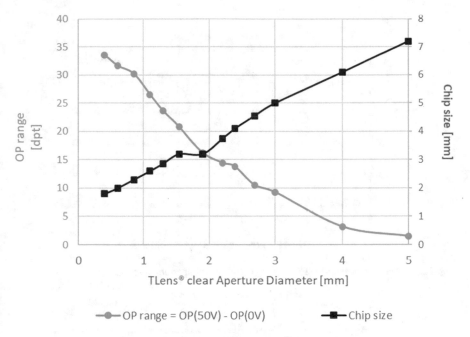

Fig. 21.22 Optical power range and chip size versus TLens® diameter

the simulated maximum optical power range as a function of the diameter of the clear aperture, for a set of parameters that are in line with the nominal processes and material parameters in use at the MEMS fab. Obviously, the die size increases with the clear aperture size. At the same time, the maximum optical power (obtainable for a given maximum actuation voltage) decreases (Fig. 21.22) and the WFE RMS increases (Fig. 21.23). It means that the available focusing range increases and the WFE RMS decreases when the useful aperture decreases. As a rule of thumb, the actuator size is about 1.5 mm larger than the diameter of the useful aperture. Note also that the optical power of the non-actuated TLens® approaches zero as the aperture size increases (Fig. 21.24).

For big apertures, the focusing range is mainly limited by the maximum WFE acceptable (larger as the membrane deflection increases). If optical quality requirements can be relaxed, TLens® can be optimized differently and exhibit larger optical power range.

Expected behavior has been validated by measurements, and impact of temperature has been evaluated as well. Figure 21.25 shows typical measured curves describing the *OP* of the TLens® over the actuation voltage range at different temperatures, while Fig. 21.26 shows behavior of *WFE* over actuation voltage and temperature.

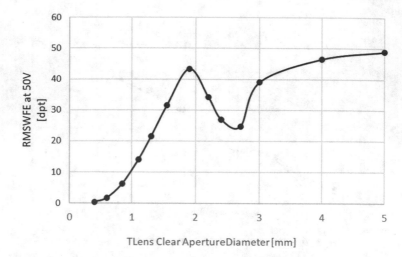

Fig. 21.23 Actuator size and WFE at 50 V versus useful aperture diameter

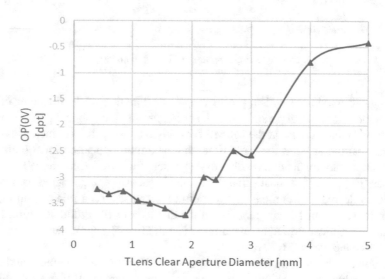

Fig. 21.24 Optical power in non-actuated state (@0 V) versus the useful aperture size of the TLens®

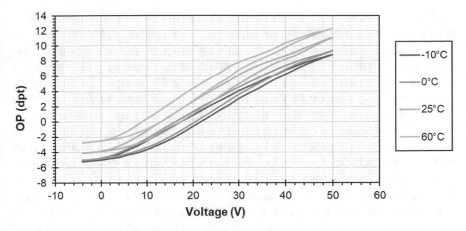

Fig. 21.25 Typical optical power versus voltage behavior for TLens® over temperature

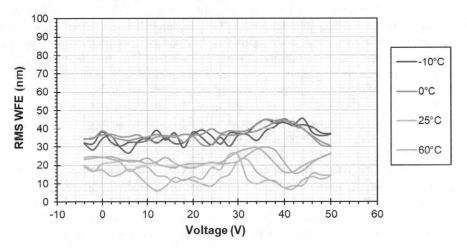

Fig. 21.26 Typical WFE over voltage range for TLens® over temperature

21.7 TLens® Real Performances Versus Simulations: Tolerance Analysis

The mechanical properties of the different layers with their tolerances must be considered to estimate the spread of the product performances in production. Based on preliminary sensitivity analysis, some parameters have not been considered in the tolerance analysis, as their effect is negligible compared to the other parameters.

For most layers, the thickness variation is within $+/-10\%$ of the nominal thickness, according to inline measurements.

The nominal values of the mechanical stress in the different layers were calculated from wafer bow measurements on test wafers, representing the average values across a wafer. The given tolerances windows cover mostly wafer-to-wafer and batch-to-batch variations. The stress variations within a wafer have been estimated by measuring the deflection of test structures specifically designed for this purpose (i.e., TLens®-like devices with only a few of the constituent layers), which are distributed across the wafer. Still this method provides just a first assessment, since the test structures contain more than just one layer, so that some assumptions should be done to extract parameters. By comparing the results of this tolerance analysis with statistical data from production, first assessment has been verified and refined.

Among the process parameters, stress variations of piezo material and glass membrane are expected to have a significant impact on the variations of lens optical performance. Making a pareto analysis, the glass membrane stress variation is the main contributor to the *OP* range variation, while piezoelectric material stress variability plays a significant to role in *OP(0)* spread.

A Monte Carlo analysis has been performed using *COMSOL*, considering, as input for parameters distribution, inline measurements for layer thickness variability, while layer stress distribution was derived from a combination of wafer-to-wafer average value variability coming from inline bow measurements and in-wafer variation coming from test structure. Piezo coefficient distribution has been estimated by inline measurements on some dedicated test structures.

Figure 21.27 shows a comparison between the simulated and the measured distributions of optical power at the minimum and the maximum actuation voltages, $OP(V_{min})$ and $OP(V_{max})$, respectively. Production data from almost 63,000 lenses have been used for comparing to simulation data.

In general, the tolerance analysis results are relatively well aligned with the production data. It is also important to note that the included measured data are coming only from the dies that have passed all final test criteria. The only significant difference between the simulated and the measured values is the wider distribution of the simulated OP(0 V). On the other hand, the fit is good for OP(Vmax).

Table 21.3 presents some statistical values from the simulated and measured distributions.

The tolerance analysis has helped to confirm the following:

For *OP(0 V)* (defined by the deflection of the non-actuated membrane), the parameters with the largest effect on it are *the passivation layer thickness and stress* (larger offset for thicker layer and for larger compressive stress), the *stress and thickness of the glass membrane* (smaller offset for thicker membrane and larger tensile stress), and the stress in the *piezo material* (smaller offset for larger tensile stress).

The optical power range is given by the achievable membrane deflection when actuated compared to its non-actuated position. It is mostly affected by *the glass membrane stress and thickness* (range decreases for higher tensile stress, i.e., stiffer membrane and for thicker membrane) and by the *piezo coefficient*.

The parameters with the largest impact on the WFE are the *glass membrane thickness and stress*. This is mostly due to the effect on the *spherical aberration*.

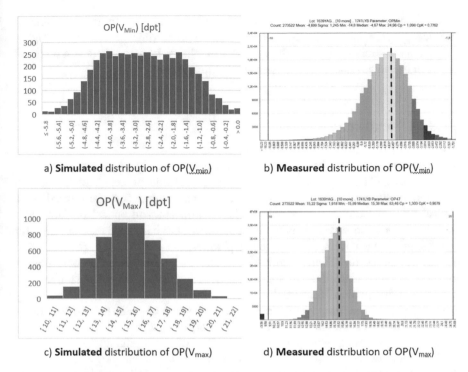

a) **Simulated** distribution of OP(V_{min})

b) **Measured** distribution of OP(V_{min})

c) **Simulated** distribution of OP(V_{max})

d) **Measured** distribution of OP(V_{max})

Fig. 21.27 Comparison between simulated and measured distributions of optical power. (**a**) is representing the distribution of optical power at Vmin (V = 0) result of the Mote Carlo simulation. (**b**) is representing the distribution of optical power at Vmin (V = 0) measured on 60K parts TLens®. (**c**) is representing the distribution of optical power at Vmax (V = 50 V) result of the Mote Carlo simulation. (**d**) is representing the distribution of optical power at Vmax (V = 50 V) measured on 60K parts TLens®

Table 21.3 Statistical values for simulated and measured optical power and WFE

Parameter		Median	St.dev
Simulated	OP(V_{min}) (1/m)	−2.8	1.2
	OP(V_{max})(1/m)	15.4	1.9
	Range (1/m)OP(V_{max})-OP(V_{min})	18.1	1.9
	RMSWFE (V_{max})(nm)	25	5
Measured	OP(V_{min}) (1/m)	−4.8	1.2
	OP(V_{max}) (1/m)	15.4	2
	Range (1/m)	21.6	2.5
	RMSWFE max. (nm)	25	24

Aberration components can be extracted by writing the wavefront as the sum of Zernike polynomials [16]. The stiffer the membrane (thicker membrane and/or larger tensile stress), the lower the optical aberrations. It means there is a trade-off between the achievable optical power and the image quality.

21.8 Conclusions

We have presented the basic principle of work and technology for a poLight® Tunable lens (TLens®) with the associated fabrication process for the MEMS actuator. A system-level optical performance analysis has been performed for a typical camera module to establish the key parameters and optical performances to be achieved. Those system level target performances have been used as guide for the design and the development of such a product.

The design of the TLens® architecture and of its fabrication process flow is based on detailed finite element analyses (FEA/FEM), using, as input data, process parameters distributions (essentially layer thicknesses and stresses) coming partly from inline measurements, partly from characterization of dedicated test structures.

The developed simulation models were subsequently also used for extending the TLens® product family with tunable lenses with different useful apertures.

The effect of the MEMS process tolerances on the optical performance of the TLens® has been simulated. Overall, the analysis shows good fit to production data.

It is also important to note that the FEM and the tolerances analysis will be key for potential further process improvement activities and an important tool to improve production yield.

The selected MEMS architecture and material choice leads to several advantages:

(i) Instant focus – enabling recording of several images, to be combined into one for all-in-focus image, featuring instant focus above and beyond competing technologies. The TLens® is up to 10 times faster than competing technologies.

(ii) Constant field of view – provides smooth and accurate focusing as the image or video is recorded, so that the zoom effect is not visible as focus changes. Constant field of view significantly reduces the complexity of image stitching and bracketing, which is relevant for both single- and multi-camera systems.

(iii) Extremely low power consumption – is an important quality for new applications in smartphones and wearables. TLens® power consumption is in the range of a few mW, where others have up to 500 mW power consumption. TLens® will therefore affect less the image sensor's temperature when focusing, which leads to better image quality.

(iv) No magnetic interference – as the TLens® does not have coils or springs, the magnetic interference is negligible.

Despite all the listed TLens® advantages, the VCM is still the most used technology and there are applications where TLens® will not be able to compete with the VCM. This is true when it comes to cost/performances balance, where the very mature VCM technology is still far ahead, benefitting from the economy of high volumes production, in which TLens® does not have yet.

In 2019 poLight® entered mass production with the two TLens® products and reached the consumer market in January 2020 (camera for a smartwatch phone), as well as industrial market in September 2020. Continuing successful market penetration will improve the competitiveness towards VCM.

21.9   Potential Future and Improvement

The piezo material and process are in constant evolution, and the use of new or improved piezo material enabling larger actuation range, lower hysteresis, and higher breakdown voltage is obviously a path to improve performance of the TLens®.

The recent development of camera module add in concepts, for which the TLens is placed inside the lens stack, enables a very compact solution. Such solution is a good match for the current front facing smartphone cameras, under the screen camera, screen size trends and WFOV back-camera.

In combination with lens design optimization, further piezo material improvement and process will be an interesting path to generate new TLens® structure, it will enable bigger Tlens® for a wide variety of mobile camera.

It is also interesting to observe that there are other applications, such as Augmented Reality, medical, industrial application, where small size, low power consumption, no magnetic interference and high focus speed are key performances, and as such TLens® is a promising solution and has the capability to take the lead.

In addition, poLight® key technology platform, combining optical polymers and actuators, could be enriched by other functionalities like optical image stabilization (OIS) and beam steering, which could give further opportunities to improve and enable better performances cameras.

References

1. Galstain, T. (2013). *Smart mini cameras*. CRC Press. ISBN 978-1-4665-1293-1.
2. Chan, C., & Huang, S., & Chen, H. H. (2017). Enhancement of phase detection for autofocus, *2017 IEEE International Conference on Image Processing (ICIP)*, Beijing, China, 2017, pp. 41–45.
3. Hamada, M. Imaging device for phase difference detection. Patent US 9,525,833, 2016.
4. Waren, J. S. (2008). *Modern optical engineering* (4th ed.). MacGrayHill.
5. DiMarzio, C. A. (2012). *Optics for engineers*. CRC Press. ISBN 978-1-4398-0725-5.
6. Ole-Morten Ruud, V. K. (2016). Packaging of a Tuneable Lens. In J. Kutilainen (Ed.), IMAPS-International Microelectronics and Packaging Society, Nordic. Tonsberg, Norway.
7. poLight® Patents, 1. 1. (2020). Patent No. 8724198, ZL2007800346046, 2074465, 5581053, 5323704, 101496157, 078347507, 8199410, 037943602, 6897995, 2313798, ZL2016108066627, 2313798, 2313798, 5580819, 101372042, 2313798.

8. Ashtiani, A. O., & Jiang, H. "Thermally actuated liquid tunable microlens with embedded thermoelectric driver and sub-second response time," 2013 Transducers & Eurosensors XXVII: The 17th International Conference on Solid-State Sensors, Actuators and Microsystems (TRANSDUCERS & EUROSENSORS XXVII), Barcelona, Spain, 2013, pp. 2604–2607.

9. Lee, J.-K., Park, K.-W., Lim, G.-B., Kim, H.-R., & Kong, S.-H. (2012). Variable-focus liquid lens based on a laterally-integrated thermopneumatic actuator. *J. Opt. Soc. Korea, 16*, 22–28.

10. Hoshino, K., & Shimoyama, I. An elastic thin-film microlens array with a pneumatic actuator," Technical Digest. MEMS 2001. 14th IEEE International Conference on Micro Electro Mechanical Systems (Cat. No.01CH37090), Interlaken, Switzerland, 2001, pp. 321–324.

11. Koga, A., Suzumori, K., Sudo, H., Iikura, S., & Kimura, M. (1999 Jan). Electrostatic linear microactuator mechanism for focusing a CCD camera. *Journal of Lightwave Technology, 17*(1), 43–47.

12. Pouydebasque, A. et al., Process optimization and performance analysis of an electrostatically actuated varifocal liquid lens, 2011 16th International Solid-State Sensors, Actuators and Microsystems Conference, Beijing, China, 2011, pp. 578–581.

13. Vladislav Vasilyev, G. K. (n.d.). Review—Mechanical Stress in Chemically Vapor Phase Deposited Boron- and Phosphorus-Contained Silicate Glass Thin Films: A Review. ECS Journal of Solid State Science and Technology, 2020 9 043003.

14. Henriksen, L., Kartashov, V., & Kilpinen, J. (2018). Europe Patent No. WO2019224367A1.

15. Nicolas Tallaron, L. H. (2018). Europe Patent No. EP3170037A1 PIEZOELECTRICALLY ACTUATED OPTICAL LENS.

16. Wyant, J., & Creath, K. (1992). Basic wavefront aberration theory for optical metrology. *Appl Optics Optical Eng, 11*.

Part V
Electronic Interfaces

Chapter 22
Electronic Sensors Front-End

Tommaso Ungaretti, Sergio Pernici, Daniele De Pascalis, Deyou Fang, Mario Maiore, and Giovanni Pelligra

22.1 Integrated Interfaces for MEMS Microphones

22.1.1 Introduction

Over the last decade, MEMS microphones have emerged over the traditional microphones' techniques becoming the most used solution in portable, consumer, automotive, and industrial applications.

The success of MEMS microphones is strongly related to the advances in the integrated interfaces which are continuously improving in term of S/N, dynamic range, and power consumption. Moreover, they evolved from simple amplification stages to mixed signal circuits including A/D converters and digital signal processing. MEMS microphones can deliver an analog output or a digital audio bitstream which opens new horizons for microphone applications.

MEMS microphone can be based on capacitive or piezoelectric transducers. The first type is largely the most used, thanks to the better performances in term of noise and sensitivity and more compatibility with mass production. The piezoelectric types are gaining interest since they do not require a bias voltage for the sensor, resulting in lower power consumption, and have better immunity to humidity and dust.

T. Ungaretti (✉) · S. Pernici · D. De Pascalis · D. Fang · M. Maiore · G. Pelligra
ST Microelectronics, Analog MEMS and Sensors Group, MEMS Sensors Division, Cornaredo, Italy
e-mail: tommaso.ungaretti@st.com; sergio.pernici@st.com; daniele.depascalis@st.com; deyou.fang@st.com; mario.maiore@st.com; giovanni.pelligra@st.com

© Springer Nature Switzerland AG 2022
B. Vigna et al. (eds.), *Silicon Sensors and Actuators*,
https://doi.org/10.1007/978-3-030-80135-9_22

22.1.2   Interface Circuits for Capacitive MEMS Microphones

A single-ended MEMS capacitive microphone has a displaceable membrane (M) and a fixed backplate (B) and can be modelled as a two terminals variable capacitor whose capacitance value is related to the acoustic pressure applied in this way:

$$C_{MEMS} = \frac{\varepsilon_0 A}{d} = \frac{\varepsilon_0 A}{d_0 - kP_S} \tag{22.1}$$

d_0 is the membrane to backplate distance under the voltage bias V_b and without sound pressure P_S and is related to V_b in this way:

$$d_0 = aV_b^3 + bV_b^2 + cV_b + const$$

kP_S is the membrane displacement due to the sound pressure.

The "constant charge" reading configuration is the most used and is shown in Fig. 22.1.

This configuration is realized with a first resistor connecting the membrane to a high fixed voltage bias V_{CP} and a second resistor connecting the backplate to another fixed voltage bias V_{cm} having a level compatible with the ASIC input preamplifier A. The impedances of this second resistor and that of the of the preamplifier's input must be very high to guarantee that the charge stored in the microphone capacitance is kept constant. MOSFET transistors are used at the input for this reason.

The charge when $P_S = 0$ is $Q = C_0 V_b = V_b \varepsilon_0 A/d_0$. Since the charge is constant, the membrane movements due to the air pressure translate into a voltage signal following the charge conservation law:

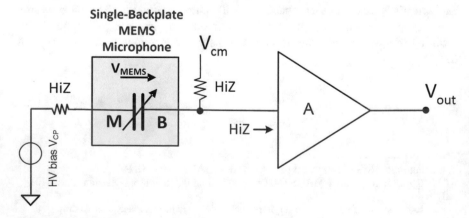

Fig. 22.1 Constant charge readout circuit

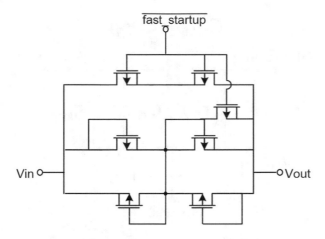

Fig. 22.2 HiZ bias resistor implementation

$$V(t) = \frac{Q}{C(t)}$$

So the capacitance variation due to a sound pressure variation PS leads to a voltage signal $V_{MEMS} = \Delta V$ given by

$$\Delta V = \frac{Q}{C_{MEMS}} - \frac{Q}{C_0} = \frac{Q\Delta d}{\varepsilon_0 A} = \frac{\Delta d}{d_0} V_b = -\frac{kP_S C_0 V_b}{\varepsilon_0 A} = -MS \cdot P_S \quad (22.2)$$

where $MS = \frac{kC_0 V_b}{\varepsilon_0 A}$ is the voltage sensitivity of the microphone, proportional to the bias voltage V_b.

To increase the sensitivity and the SNR, the membrane bias voltage V_{CP} must be quite high, typically in the range from 5 V to 19 V, and is realized with a charge pump circuit starting from the standard CMOS power supply voltage (1.8 V÷3.3 V). The high impedance resistor used to connect V_{CP} to the membrane is needed to filter the charge pump noise.

The most common implementation of the high impedance resistors in Fig. 22.1 is based on series of diode connected MOSFET transistors in back-to-back configuration (Fig. 22.2). These MOSFET must remain in sub-threshold for the whole signal swing across them so that they exhibit a very high impedance (hundreds of GΩ). Reset switches for fast startup are included in Fig. 22.2.

In practice the MEMS microphone is not just a variable capacitor, but some parasitic components should be included in a more accurate model. The equivalent circuit of an actual MEMS microphone is shown in Fig. 22.3. Besides the variable capacitance $C_{MEMS} = C(P_S)$, the equivalent circuit includes three parasitic capacitances, C_{P2} and C_{P3}, connected between each plate of the MEMS microphone and the MEMS substrate, and C_{P1} between the MEMS substrate and the package substrate, which are separated by an insulating layer. The value of these parasitic

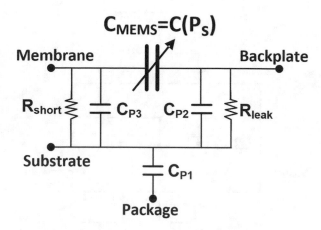

Fig. 22.3 Equivalent circuit of a MEMS microphone

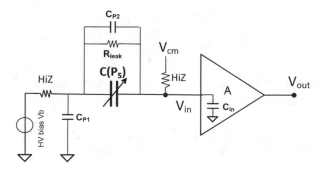

Fig. 22.4 Readout circuit with parasitic components

components depends on the specific implementation of the microphone, but typically C_{P1} and C_{P3} are of the order of few pF while C_{P2} is a fraction of pF. A leakage is present between the backplate and the substrate and is modelled with the resistor R_{leak} that is $>10^{15}$ Ω. The membrane is usually shorted to the substrate through a resistor R_{short} with a typical value <100 Ω. The package is connected to the microphone ground contact.

If R_{short} is considered a real short, the readout circuit of Fig. 22.1 can be modified using the MEMS equivalent circuit, as shown in Fig. 22.4. The parasitic capacitance C_{P1} and C_{in} produce an attenuation of the microphone voltage ΔV, given by (22.2), leading to:

$$V_{in} = \Delta V \frac{C_0}{C_0 + C_{P1} + C_{in}} = -M_S \cdot P_S \frac{C_0}{C_0 + C_{P1} + C_{in}} \qquad (22.3)$$

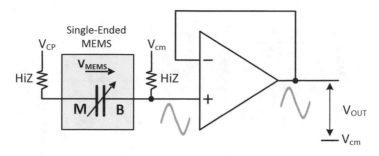

Fig. 22.5 Single-ended output readout circuit

This attenuation may lead to a degradation of the actual microphone sensitivity and, hence, increase too much the contribution to the overall SNR of the electronic noise of the preamplifier which could become dominant. A typical value of V_{in}/P_S at the preamplifier input of a well-designed microphone and readout circuit, including the attenuation of the parasitic capacitances, is in the order of -38dBV/Pa corresponding to about 12.6 mVrms/Pa.

22.1.3 Preamplifier

The function of the preamplifier in a MEMS microphone interface is to buffer the output of the MEMS microphone, providing in some cases a proper amplification in the band of interest (audio band) and delivering an output voltage with low impedance, capable of driving the output pads and their loads, in case of analog microphones, or the input of an ADC in digital microphones. To limit the attenuation due to the preamplifier input capacitance, the size of the input MOSFET cannot be increased too much, as shown in (22.3), even if this may help to lower the noise. The simplest case is that of analog microphones with a single-ended output. In this microphone, an operational amplifier in non-inverting unity gain configuration can be used, as shown in Fig. 22.5.

In case of differential output analog microphones, a buffer connected differential-difference-amplifier is commonly used for the readout interface as shown in Fig. 22.6. C_{dummy} is a capacitor with a value as close as possible to C_0. This solution is not optimized when it comes to linearity and to input dynamic range. Some circuit solutions based on a resistive degeneration of the differential input pairs partially overcome these limitations [1]. A better solution is shown in Fig. 22.7. It employs a differential sensor, having one single membrane and 2 backplates. It allows the use of two single-ended buffers and can achieve better performances in terms of linearity since the input stage of each buffer operates with a lower level with respect to the previous solution, since the signal of the sensor is not doubled but is increased of 3–4 dB. The drawback is that the differential MEMS microphones are subject to both

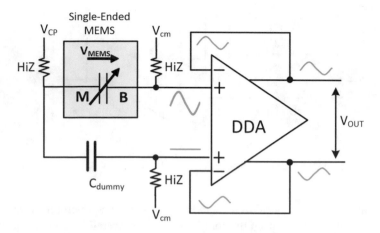

Fig. 22.6 Differential output readout circuit with single-ended MEMS

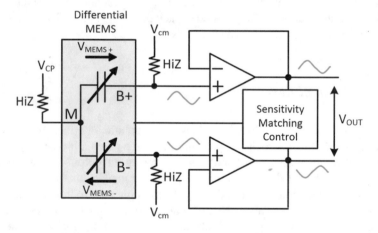

Fig. 22.7 Differential output readout circuit with differential MEMS

reliability weaknesses and technological complications. In addition, they usually require an auxiliary controlling circuit to correct potential capacitive mismatches caused by different distances of the membrane from the backplates [2].

The solution, shown in Fig. 22.8, merges the advantages of single-ended MEMS microphones with those of differential topologies [3]. The basic idea is to implement a feedback path that senses the output common mode voltage and through a proper amplification factor returns to the membrane half the differential output signal.

The relation between R_1, R_2, and R_{FB} is:

$$R_1 = R_2 // (1 - \alpha) R_{FB}$$

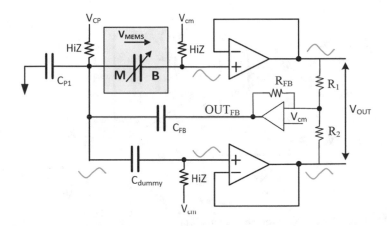

Fig. 22.8 Improved differential output readout circuit with single-ended MEMS

Since usually $C_{P1} \gg C_{in}$, as shown in [3], the attenuation coefficient α can be written as:

$$\alpha \cong 1 - \frac{C_{FB}}{C_{FB} + C_{P1}} \cdot \frac{C_0}{C_0 + C_{in}}$$

This condition can almost be reached by reducing the main parasitic capacitances, thanks to an accurate package and layout strategy, and by choosing C_{FB} as high as possible without exceeding in area occupation.

This topology can reach a better linearity than the differential MEMS approach shown in Fig. 22.7, with similar complexity and area occupation. It has also the same C_{in}, hence the same attenuation coefficient caused by the input capacitive partition. The preamplifiers' noise is similar in the two topologies, while the MEMS noise is almost 3 dB lower in this solution, thanks to the single backplate used, which compensates almost completely the 3–4 dB lower signal of the single-ended MEMS.

In the analog microphones, the preamplifier usually has a unitary gain, but quite often an amplification is required in case of digital microphones. A proper amplification can be provided using a non-inverting topology for the operational amplifiers, rather than a buffer configuration, as shown in Fig. 22.9. The gain can be adjusted changing the ratio between the feedback and the input capacitor, to adapt the MEMS microphone output signal, which can change depending on the MEMS type and the fabrication tolerances, to the ADC input signal range.

Nowadays, microphones have increased their capability to capture pressure levels up to an acoustic overload level AOL = 130dBSPL or more, with distortion lower than 4%. This level corresponds to 36dBPa since 94dBSPL is the sound pressure level at 1 Pa. Since 12.6mVrms is the typical voltage level from the MEMS at 1 Pa, an AOL of 36dBPa results in nearly 0.8Vrms or 1.13Vp at the preamplifier input. The supply voltage of the microphone can be as low as 1.6 V and an LDO is usually needed, at least for the opamps' input stages, to guarantee a good power supply

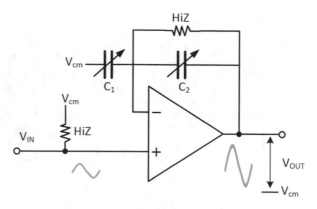

Fig. 22.9 Amplifier with gain for the readout circuits

rejection. So, to be compliant with such high input signal levels, the amplifiers in the readout circuit must have a rail-to-rail complementary input stage.

The AOL and THD behaviors of a MEMS microphone depend mainly on the preamplifier, while the SNR performance is usually dominated by the sensor noise.

22.1.4   A/D Converter

In case of digital microphones, the readout circuit is followed by a sigma-delta ($\Sigma\Delta$) ADC. This type of ADC is the standard choice for audio applications in view of its inherent linearity and low-power consumption. A discrete time (DT) 4th order topology with a single-bit pulse density modulation (PDM) output is usually chosen, allowing a single line digital connection with the digital processing companion chip. The quantization noise is kept well below the sensor noise in the audio band (20 Hz–20 kHz), using a sampling frequency FS in the range 1.4–3.3 MHz, and in the voice band (100 Hz–8 kHz) using an Fs in the range 500 Hz–1.4 MHz.

The block diagram of a converter based on cascade of four integrators with weighted feedforward summation (CIFF) is shown in Fig. 22.10. It is implemented using switched-capacitor (SC) circuits. Even if not shown, each amplifier is equipped with a circuit that senses when its output signal crosses a threshold and progressively reduces the order of the loop filter from 4th to 2nd to 1st by inserting SC dumping on the 2nd, 3rd and 4th integrator feedbacks, respectively. A 1st order loop has no stability problems and can guarantee a good dynamic behavior for input levels close to the reference one. Moreover, due to the pseudo random nature of voice and audio signals, this progressive adaption of the order does not introduce spurious tones.

Alternatively, a DT multi-bit $\Sigma\Delta$ converter can be used. An architecture providing the proper SNR is a second-order 5-bit modulator, whose block diagram

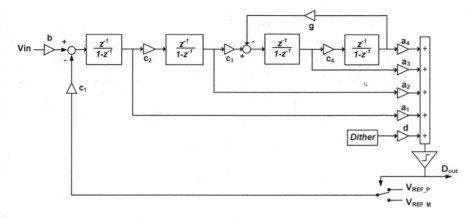

Fig. 22.10 4th order 1-bit $\Sigma\Delta$ ADC

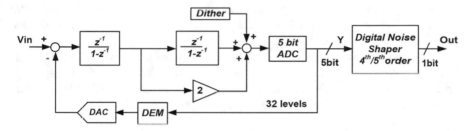

Fig. 22.11 2nd order 5-bit $\Sigma\Delta$ ADC followed by a single-bit digital noise shaper

is shown in Fig. 22.11. It has not the stability issues of single-bit higher-order modulators but does not provide a single-bit output stream. This drawback can be solved by connecting the output of the multi-bit ADC to a single-bit, fourth or fifth order, digital $\Sigma\Delta$ modulator, operated at the same sampling frequency SF, which truncates the multi-bit output down to a single bit and shapes the resulting truncation error with a fourth-/fifth-order transfer function. The digital, $\Sigma\Delta$ modulator, is less critical than its analog counterpart, since it can be easily verified under any operating conditions. Using sufficiently large word length in the integrators, a suitable noise transfer function, and the order degeneration already illustrated for the single-bit analog modulator, instability can be avoided.

The quantizer (flash ADC) consists of 31 comparators and 32 level output code. Dithering and DAC dynamic element matching (DEM) are needed to improve the linearity and avoid spurious tones.

To save area and consumption keeping the same performances, a modified topology based on a tracking ADC quantizer with only 6 comparators (7 quantization levels) can be used as described in [4] and shown in Fig. 22.12.

An alternative to the previous DT ADC consists in a hybrid structure (CT-DT) that can be obtained from the architecture shown in Fig. 22.12 substituting the

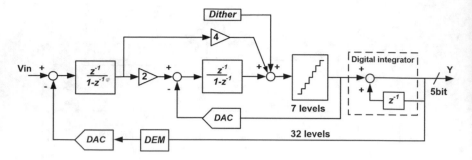

Fig. 22.12 2nd order 5-bit $\Sigma\Delta$ ADC with reduced number of comparators

first stage with a continuous-time one, including an active RC first integrator and a current steering DAC, while keeping a discrete-time SC second stage, as described in [5]. Since the digital microphones require a wide range of clock frequency, on-chip RC automatic tuning is necessary to ensure stability and to maintain the modulator performance under all process conditions.

This approach may result in lower noise, thanks to the absence of the KT/C noise in the first stage. It is also possible to increase the maximum sampling frequency up to 4.8 MHz with a limited consumption increase. This is needed in case the microphone should include the ultra-sound processing capability.

22.1.5 Digital Microphones with Extended Dynamic Range

The capability to process sound over 130dBSPL, avoiding signal swing limitation in the analog blocks, can be obtained supplying the MEMS readout circuit with a higher supply generated by an on-chip voltage booster [6].

Another approach is based on an automatic gain controller (AGC) digital block [7]. The AGC detects the level of the output digital signal, and when it crosses appropriate signal levels, it switches the gain of the preamplifier (for instance from 0 to 19 dB in 1 dB steps) and at the same time that of an output digital multiplier in the opposite way (from 0 to -19 dB in -1 dB steps), so that the overall microphone gain remains constant but the linearity is increased at high signal levels. Also the noise is increased at high signal levels but this is not relevant since it is masked by the signal. A block diagram of an architecture of this type is shown in Fig. 22.13.

22.1.6 Analog and Digital Microphones ASICs

The photographs of the readout ASICs of an analog and a digital microphone are shown in Figs. 22.14 and 22.15. They include the main blocks described before

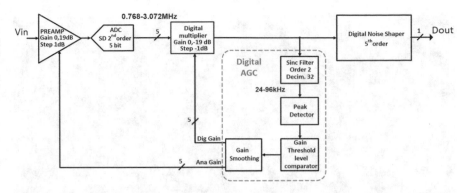

Fig. 22.13 Digital microphone with extended dynamic range, based on an AGC

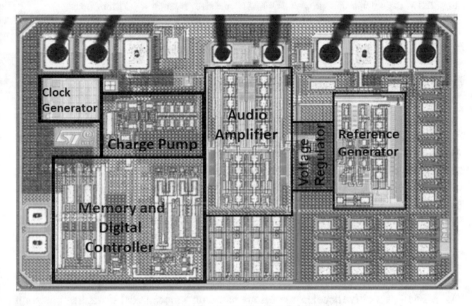

Fig. 22.14 Analog microphone ASIC (die size: 1.19mm × 0.85 mm)

plus some service blocks such as current and voltage references generators, voltage regulators, charge pump for high voltage generation, clocks generators, memory and digital blocks for processing, and programmability.

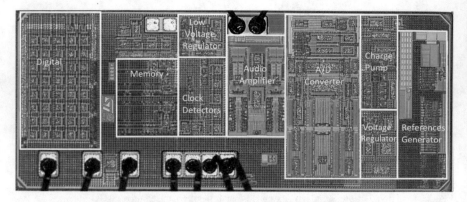

Fig. 22.15 Digital microphone with AGC ASIC (die size: 1.72mm × 0.71 mm)

22.2 Interfaces for Capacitive Accelerometer

22.2.1 Introduction

In the last two decades, MEMS capacitive accelerometers have conquered accelerometers sensor market, thanks to their low temperature sensitivity, good linearity, and low cost, features that make them suitable for a lot of applications, from consumer to biomedical, from automotive to industrial.

In this section, an overview of the main capacitive accelerometer interfaces is presented.

22.2.2 MEMS Basic Concepts

All MEMS capacitive accelerometers are based on a mechanical structure consisting of an inertial mass suspended by a spring as in Fig. 22.16 for a single-axis device.

The proof mass motion equation is

$$m\ddot{x} + b\dot{x} + kx = ma \qquad (22.4)$$

where k is the spring's constant and b is the air damping coefficient.

In order to access to the proof mass, two electrodes (noted as stators) capacitively coupled to the proof mass (noted as rotor) are used, so a more realistic picture of the structure is shown in Fig. 22.17.

The motion equation becomes

$$m\ddot{x} + b\dot{x} + kx = ma + Fel \qquad (22.5)$$

Fig. 22.16 Mechanical
element

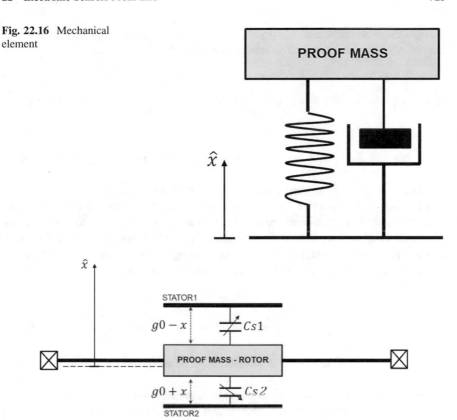

Fig. 22.17 Electromechanical model of parallel plates sensing in single bridge configuration

where *Fel* is the electrostatic force caused by the applied bias voltage between rotor
and stators

$$Fel = \frac{1}{2}(Vs1 - Vr)^2 \, \varepsilon \, \frac{A}{(g0 - x)^2} - \frac{1}{2}(Vs2 - Vr)^2 \, \varepsilon \, \frac{A}{(g0 + x)^2} \qquad (22.6)$$

for $x \ll g0$

$$Fel \simeq \frac{1}{2}(Vs1 - Vr)^2 \, \varepsilon \, \frac{A}{g0^2} - \frac{1}{2}(Vs2 - Vr)^2 \, \varepsilon \, \frac{A}{g0^2} +$$

$$+ (Vs1 - Vr)^2 \, \varepsilon \, \frac{A}{g0^3} \, x + (Vs2 - Vr)^2 \, \varepsilon \, \frac{A}{g0^3} \, x \qquad (22.7)$$

The first two terms of *Fel* give the net electrostatic force applied on the proof
mass by stator1 and stator2, while the other two terms, proportional to *x*, can be

thought as an electrostatic contributor *kel* (always negative) to spring's constant *k*, known as spring softening effect, so, the formula can be written as

$$m\,\ddot{x} + b\,\dot{x} + (k + kel)\,x = m\,a + \frac{1}{2}\,(Vs1 - Vr)^2\,\varepsilon\,\frac{A}{g0^2} - \frac{1}{2}\,(Vs2 - Vr)^2\,\varepsilon\,\frac{A}{g0^2}$$

$$(22.8)$$

Stators and Rotor can be used to sense the acceleration signal or to force an electrostatic deflection.

In open-loop accelerometers, stators and rotor are mainly used to sense the acceleration and $|Vs1\text{-}Vr| = |Vs2\text{-}Vr|$ any time during accelerometer normal operating mode; instead, in closed-loop architectures, stators and rotor are used to sense the acceleration, as well as to force a deflection in order to close the loop, so *Vs1*, *Vs2*, and *Vr* voltage profiles change between the sensing phase and the forcing phase.

The main parameters that define mechanical element are

$$fres = \frac{1}{2\pi}\,\sqrt{\frac{(k + kel)}{m}} \quad \text{and} \quad Q = \frac{1}{b}\,\sqrt{(k + kel)\,m} \tag{22.9}$$

while the acceleration signal is extracted by sensing the capacitance variation $\Delta Cs = Cs1 - Cs2$.

Finally, proof mass thermal noise (noted as Brownian noise) completes the picture of mechanical element performances, and it's expressed by the following formula [8]:

$$An^2_{brw} = \frac{4\,K\,T\,b}{9.8^2\,m^2} = \frac{8\,K\,T\,\pi}{9.8^2}\,\frac{fres}{Q\,m} \tag{22.10}$$

Brownian noise defines the lower limit to the noise performances of any accelerometer, and in the following part of this chapter, it's not treated; instead, the electrical noise will be discussed; moreover Flicker and thermal noise of the specific AFE used to extract ΔCs signal, as well as the thermal noise due to parasitic series resistances of interconnections between mechanical element and AFE will be treated.

22.2.3 Reading Interfaces Overview

Once introduced the proof mass motion equation, let's discuss the various circuital solutions to read the MEMS.

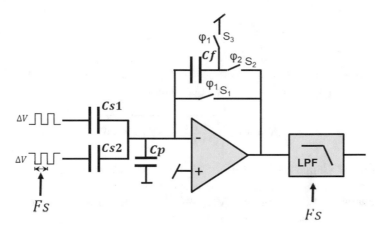

Fig. 22.18 Switched capacitors DT AFE

22.2.4   Open-Loop Architectures

Open-loop architectures are widely used because, despite their relative simplicity, they fit the main applications in the market, where low-medium requirements in terms of bandwidth, stability, linearity, and noise performances must be addressed.

Basically, there are two main approaches to read the proof mass, noted as DT approach and a CT one.

22.2.4.1   DT Approach

In Fig. 22.18, a conceptual implementation of DT AFE is shown, where switched capacitors technique is used.

The operational amplifier (OA), the additional circuitry placed in feedback (capacitors and switches) and the driver (not drawn) that generates ΔV voltage stimuli define the CVC. The signal is extracted from the MEMS by a square waveform (of amplitude ΔV and frequency Fs) applied on the two stators, converted by CVC, finally sampled at Fs by LPF block. For an effective sampling of MEMS signal, the sampling frequency should be $Fs \sim 10\, fres$.

OA offset and its variation, as well as its low frequency noise (Flicker noise), heavily affect 0 g level sensor stability, so they must be reduced by some circuital technique; in DT architectures, CDS is the election technique, here implemented by S1, S2, and S3 switches directly placed on the CVC. To make CDS effective, sampling frequency Fs must be well higher than OA Flicker noise corner frequency fk.

To calculate SNR, considering only the OA noise source, we can write (in case of single-pole OA and φ_1 and φ_2 of the same duration $Ts/2$)

$$S = \Delta V \frac{\Delta Cs}{Cf} \frac{1}{\sqrt{2}} \left(1 - e^{-\frac{\pi\,fcvc}{Fs}}\right) \tag{22.11}$$

for a sinusoidal input signal (where $fcvc$ is CVC bandwidth and $e^{-\frac{\pi\,fcvc}{Fs}}$ is CVC output sampling error at the end of φ_2 phase) and

$$Nrms \simeq \sqrt{\frac{\pi\,fcvc}{Fs}\,SVnout\,BW} \tag{22.12}$$

for rms noise (supposed that $Cp + Cs1 + Cs2 \sim 10\ Cf$, so making negligible the thermal noise transfer in φ_1 phase), where $SVnout$ is the thermal noise power spectral density at the output of CVC, $\frac{\pi\,fcvc}{Fs}$ is the undersampling factor [9], and BW is the signal bandwidth, so having

$$SNR \simeq \sqrt{\frac{Fs}{\pi\,fcvc}}\,\frac{\Delta V \frac{\Delta Cs}{Cf} \frac{1}{\sqrt{2}}}{\sqrt{SVnout\,BW}}\left(1 - e^{-\frac{\pi\,fcvc}{Fs}}\right) \tag{22.13}$$

SNR can be increased working with high voltage stimulus (e.g., using a charge pump to generate ΔV), increasing the bias current of CVC OA (so reducing $SVnout$ keeping $fcvc$ fixed) and finally working on $Fs/fcvc$ coefficient.

$Fs/fcvc$ coefficient must be defined by a trade-off between SNR maximization and sensor scale factor stability; in practice $Fs/fcvc < 1$ (at least $fcvc \sim 2\ Fs$) and the formula (22.13) can be simplified and rearranged as

$$SNR \simeq \sqrt{\frac{4\,Fs}{\pi\,fcvc}}\,\frac{\Delta V \frac{\Delta Cs}{Cf} \frac{1}{\sqrt{2}}}{\sqrt{4\,SVnout\,BW}} \tag{22.14}$$

Moreover, SNR is further reduced by KT/C noise due to the S1, S2, and S3 switches added to implement CDS and by thermal noise of parasitic series resistance of the rotor connection at MEMS side (typically realized with poly routing, so in the $k\Omega$ range, not drawn in the Fig. 22.18).

22.2.4.2 CT Approach

If low-power consumption is a key requirement, a CT approach should be considered. Figure 22.19 shows a conceptual implementation.

The signal is extracted from MEMS by a modulation of amplitude ΔV and frequency $Fchop$ applied on the two stators, converted by CVC and demodulated in base band, ready to be processed then by LPF block.

CVC OA offset and its variation, as well as OA Flicker noise, are reduced by chopper technique, consisting into offset and Flicker noise modulation at $Fchop$

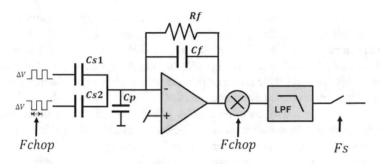

Fig. 22.19 CT AFE

frequency out of LPF bandwidth; to have an effective Flicker noise filtering by a low order LP filter (so easy to integrate), *Fchop* should be well higher than *BW* + *fk* (typically *Fchop* − *fk*~10 *BW*).

After LP filtering, the signal is sampled at sampling frequency *Fs*.

In order to calculate *SNR*, considering only OA noise source, we can write (in case of single-pole OA [10])

$$S \simeq \frac{\Delta V}{2} \frac{\Delta Cs}{Cf} \frac{1}{\sqrt{2}} \left(1 - 4\frac{Fchop}{2\pi\, fcvc}\right) \qquad (22.15)$$

for a sinusoidal input signal and

$$Nrms = \sqrt{SVnout\, BW} \qquad (22.16)$$

for rms noise, so having

$$SNR \simeq \frac{\Delta V \frac{\Delta Cs}{Cf} \frac{1}{\sqrt{2}}}{\sqrt{4\, SVnout\, BW}} \left(1 - 4\frac{Fchop}{2\pi\, fcvc}\right) \qquad (22.17)$$

The *SNR* depends on *ΔV*, *SVnout*, and *Fchop/fcvc*; in practice *fcvc*~10 *Fchop* (to guarantee a good scale factor stability), so (22.17) can be simplified as

$$SNR \simeq \frac{\Delta V \frac{\Delta Cs}{Cf} \frac{1}{\sqrt{2}}}{\sqrt{4\, SVnout\, BW}} \qquad (22.18)$$

showing that the *SNR* in CT solution is not directly dependent from CVC bandwidth *fcvc*.

Comparing the *SNR* formulas (22.14) and (22.18) and *fcvc* constrains for DT and CT solutions (i.e., fixing *fres*~3*kHz*, *BW*~1*kHz* and *fk*~1*kHz*, it should be *Fs*~30*kHz* in case of DT architecture and *Fchop*~11*kHz* in case of CT architecture, so *fcvc*~60*kHz* for DT and *fcvc*~110*kHz* for CT), we can say that it would be possible to hit *SNR* and scale factor stability targets with similar OA bias current, so with similar current consumption between the DT and CT approach.

Actually, as previously mentioned, OA is not the only noise source and CT approach turns out to be more power efficient than DT one. In fact, in case of CT approach, no switch is necessary to implement electrical offset cancellation, so no additional *KT/C* switches noise must be considered and managed (in DT approach *KT/C* noise is reduced by increasing *Cf* value and sampling frequency, so increasing the system power consumption). Furthermore, in case of CT architecture, the thermal noise due to rotor parasitic series resistance is filtered over the signal bandwidth *BW*, while in case of DT architecture, it's integrated over CVC bandwidth *fcvc* and folder over *Fs/2*.

22.2.4.3 DC Level Control of High Impedance Node

DC level of high impedance node of OA (in the previous description the rotor node) must be managed. In case of DT architecture, the node is continuously refreshed during ϕ1 phase of CDS technique; in case of CT architecture, instead, the node is typically controlled putting a *Rf* resistor in feedback to the OA (so in parallel to integration capacitor *Cf*).

In order to have a suitable band pass behavior of CVC, the pole *fp* due to *Cf* and *Rf* must be

$$fp = \frac{1}{2\,\pi\,Rf\,Cf} \ll Fchop \qquad (22.19)$$

From (22.19), *Rf* must be ~ *GΩ*, so a solution requiring wide silicon area in case a passive resistor is integrated on-chip; *Rf* can be more easily implemented by using MOS transistors in series in diode configuration as described in Sect. 22.1.2.

Thanks to the modulation and filtering performed by chopper technique, *Rf* thermal noise is heavily filtered, so it only marginally affects the noise performances of CT AFE [11].

22.2.4.4 Circuital Implementations

Typically, in order to improve power supply rejection ratio and the substrate noise rejection capability, as well as to simplify the circuital implementation of the demodulator (in CT AFE approach), the two previous architectures are implemented in a fully differential configuration, so, the MEMS is actually stimulated by the rotor

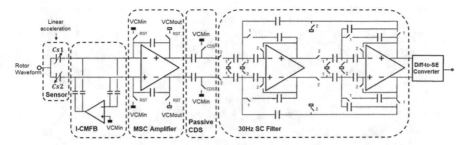

Fig. 22.20 Fully differential DT approach: example of circuital implementation

connection, while the two stators are connected to the high impedance nodes of fully differential OA. To avoid any scale factor drift, the AC level of stators (the OA inputs) must be managed.

An example of fully differential switched capacitors DT approach is reported in Fig. 22.20 [12]. Here the AC level of stator lines is controlled by an additional input common mode feedback circuitry (I-CMFB in the figure), while the CDS is implemented at the output of main SC (MSC) amplifier.

Figure 22.21 shows a micrograph of the fully differential AFE integrated in one of the first analog accelerometers launched on the consumer market.

An example of fully differential CT approach is reported in Fig. 22.22 [13]. In this case the AC level of stators is self-controlled adopting a full bridge MEMS solution, so no additional (passive or active) circuitry is needed, reducing at minimum the capacitive load on the stator lines, and so the CVC transferred noise as well as the current consumption.

Furthermore, fully differential approach is key to simplify the demodulator, which can be easily implemented with simple transmission gates.

Figure 22.23 shows a micrograph of a fully differential CT AFE, where the main blocks are highlighted.

22.2.5   Closed-Loop Architectures

When high linearity in combination with wide dynamic range performances are required, as well as in case of high bandwidth is needed, closed-loop approach should be evaluated.

Closed-loop accelerometer uses electrostatic force to balance inertial force, so, a feedback loop measures the position x of the proof mass and applies a force in the opposite direction to keep the proof mass in a stable state and centered ($x \simeq 0$); this means that the motion Eq. (22.8) can be simplified as follows

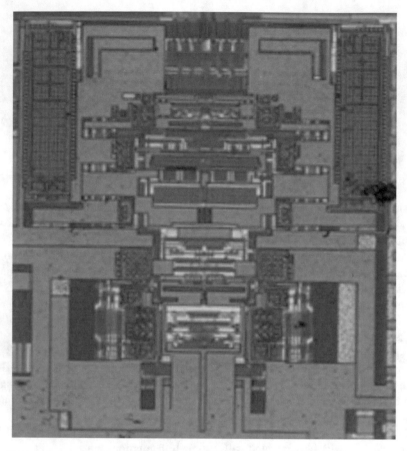

Fig. 22.21 Micrograph of fully differential DT AFE

$$ma = \frac{1}{2}(Vs2 - Vr)^2\, \varepsilon\, \frac{A}{g0^2} - \frac{1}{2}(Vs1 - Vr)^2\, \varepsilon\, \frac{A}{g0^2} \tag{22.20}$$

and considering that $Vs1$ and $Vs2$ are typically driven with squared voltage stimulus of the same frequency, opposite polarity and ΔV value with respect to the rotor voltage, and that the feedback voltage is applied to the rotor, we can write [14]

$$Vout = Vr = \frac{m\, g0^2}{2\, \varepsilon\, A \Delta V}\, a \tag{22.21}$$

that is the output of feedback loop and the scale factor of the sensor.

The formula (22.21) says that in closed-loop architectures the output doesn't depend from spring constant and its nonlinearity; furthermore, the nonlinearity due to parallel plates sensing elements results minimized, thanks to the limited excursion

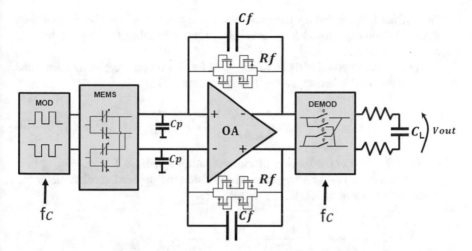

Fig. 22.22 Fully differential CT approach: example of circuital implementation

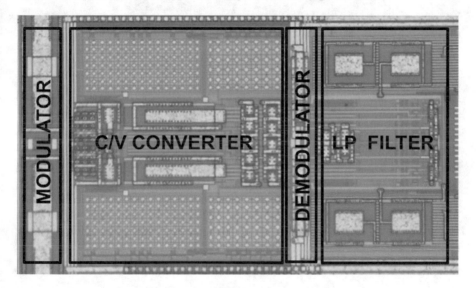

Fig. 22.23 Micrograph of fully differential CT AFE

of the proof mass ($x \simeq 0$). These features make these architectures very suitable in applications where wide dynamic range and good linearity performances must be addressed.

Furthermore, the force feedback performs a positive shift of MEMS resonance frequency *fres*, which becomes *fres* $\sqrt{1 + Gloop}$ (where *Gloop* is the system closed-loop gain), extending the system bandwidth.

Finally, the loop, reducing the proof mass excursion, performs an electrostatic Q damping, making closed-loop architecture very suitable to manage high Q mechanical elements.

The access to the proof mass in sensing and forcing mode is typically performed in time division multiplexing, so Eq. (22.21) becomes [14]

$$Vout = \frac{m\, g0^2}{2\, \varepsilon\, A \Delta V\, \eta} a, \eta = \frac{Tf}{T0} \tag{22.22}$$

where Tf is the time of the forcing phase over sampling period $T0$.

The system model of a closed-loop time division multiplexed accelerometer can be described as shown in Fig. 22.24, where

$$Hsens = \frac{\frac{1}{m}}{s^2 + \frac{b}{m}s + \frac{k+kel}{m}}, Ks = \frac{2\,\varepsilon\, A}{g0^2}, Kcv = \frac{\Delta V}{Cf},$$

$$A = A0, \beta(s) = \frac{1 + \frac{s}{\omega z}}{1 + \frac{s}{\omega p}}, KF = \frac{2\,\varepsilon\, A \Delta V}{g0^2}\eta \tag{22.23}$$

and *Gloop* in the signal bandwidth range ($s \simeq 0$) is

$$Gloop = \frac{1}{k + kel} \frac{2\,\varepsilon\, A}{g0^2} \frac{\Delta V}{Cf} A0 \frac{2\,\varepsilon\, A \Delta V}{g0^2} \eta \tag{22.24}$$

Once considered the mechanical parameters fixed, formula (22.22) and (22.24) say that it's possible to increase together full-scale range *a* and *Gloop*, so the sensor linearity and bandwidth, working with high voltage stimulus ΔV and high η, but decreasing the scale factor.

Furthermore, in order to assure the correct working mode and stability of the force loop, the sampling frequency $F0 = 1/T0$ must be much higher than force loop unity gain bandwidth, thus making the force feedback approach very power consuming.

22.2.5.1 Circuital Implementations

The system model in Fig. 22.24 can be implemented, as shown in Fig. 22.25, by using a fully analog single-ended circuital solution [15].

In sense phase *Ts*, acceleration signal is extracted from MEMS, thanks to *S6* switch connection to CVC; following the sense path, the signal is first amplified (by *OA2*, *R1*, and *R2*), therefore cleaned from offset and Flicker noise, thanks to CDS circuitry (C5 and S8). After that, the signal passes through the S&H circuit, providing the output of closed-loop system.

The force feedback path instead is implemented by a PID circuit (buffered at the input and output); the proportional and derivative part is needed to put a

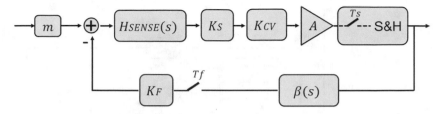

Fig. 22.24 System model of a closed-loop DT AFE

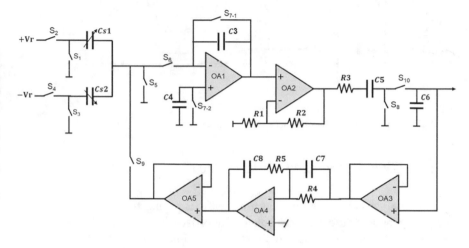

Fig. 22.25 Closed-loop fully analog accelerometer interface

compensation zero in the loop (to compensate MEMS's two poles transfer function), while the integral part significantly increases the loop gain at low frequency, thus improving the frequency response performances. $S5$ switch performs a rotor voltage reset between sense and force phase.

When a direct high-resolution digital output is required, the loop can be closed by digitizing the feedback signal using a sigma-delta A/D converter, obtaining a very compact solution.

Furthermore, high resolution implies low electrical noise as well as low Brownian noise, so, as per formula (22.10), high Q mechanical element is needed, making force loop compensation very challenging.

A design example of fifth-order sigma-delta accelerometer is reported in Fig. 22.26 [16].

The system consists of a MEMS, an SC charge amplifier, the CDS and hold circuits, a lead compensator in the direct path (due to high Q of MEMS, a heavy loop compensation is necessary), a third-order sigma-delta modulator (that, in combination with the MEMS, performs as a fifth-order system), and a 1-bit DAC to close the loop on the stators in time division multiplexing.

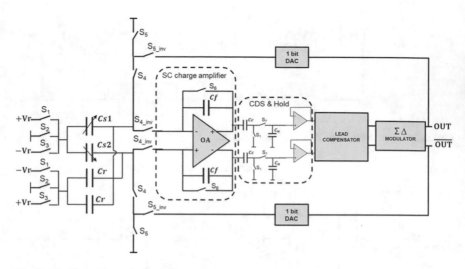

Fig. 22.26 Closed-loop sigma-delta accelerometer interface

22.2.6 Charge Balanced Architectures

When the target is to have high linearity without paying for the complexity of a closed-loop architecture, a charge balanced approach should be considered.

In a parallel plate sensing scheme, the major linearity-limiting factors are the nonlinear displacement to capacitance conversion and the nonsymmetrical and nonlinear electrostatic forces; the self-balancing bride (SBB) approach presented in Fig. 22.27 solves both [17].

In fact, typically, in open-loop interfaces (as presented in 22.2.4), the two parallel-plate sensing capacitors are biased by acceleration independent voltage profiles, so the output of the parallel-plate sensing capacitors is inversely proportional to the displacement; furthermore, the electrostatic force (due to the voltage profiles applied) changes under acceleration, causing signal distortion.

Instead in Fig. 22.27, a charge balanced AFE is shown, where the rotor voltage dynamically changes under acceleration, thanks to the voltage feedback provided by the integrator output, making equal the charges in the two sensor capacitors, so providing a sort of linearization of parallel plates output, as well as a balancing of the electrostatic forces, theoretically nulling the spring softening effect [18].

At low frequency, the scale factor is

$$Vout = \text{Vref} \frac{Cs1 - Cs2}{Cs1 + Cs2} \sim \frac{x}{g0} \tag{22.25}$$

and the formula (22.25) clearly shows the linear behavior of SBB approach.

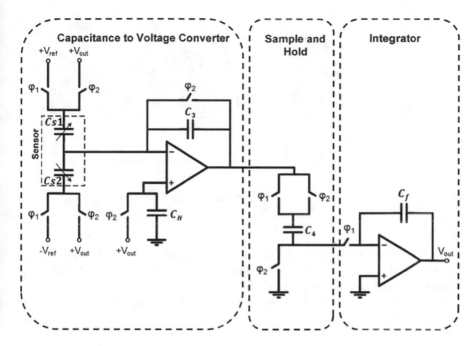

Fig. 22.27 Charge balanced AFE

Despite the use of feedback, SBB architectures are still considered open loop since the feedback is not used to cancel the movement of the proof mass (no force feedback applied), so no gain in terms of dynamic range, bandwidth, and electrostatic damping (typical of force feedback architectures) is obtained.

22.2.7 Interfaces for High-End Applications

When extreme bias stability must be addressed, all contributors to 0 g level drift must be managed by electronic interfaces, so not only the electrical and mechanical offset drift (the first typically addressed by CDS or chopper techniques, the second by force feedback), but even any offset drift due to parasitic capacitance changing between MEMS and ASIC bonding wires.

In fact, the bonding wire parasitic capacitances can appreciably drift due to deformations created by thermal and mechanical stresses as well as due to the package moisture absorption, which changes the permittivity of the dielectric between the bonding wires. The literature reports 0 g level drift in the order of tens mg for 1 μm of displacement variations between two bonding wires [19].

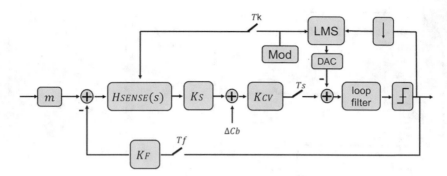

Fig. 22.28 System model of a closed-loop accelerometer with K modulation technique for very high 0 g level stability

In order to separate the contributor of parasitic capacitances of bonding wires from ΔCs mechanical signal, an interesting solution based on k modulation concept is now presented through a system model (Fig. 22.28).

As already discussed, when ΔCs signal is extracted from the mechanical element, stiffness constant k is lowered by a factor kel due to the previously described spring softening effect.

Now looking into the system model, in the range of signal bandwidth, the transfer function for the acceleration signal a and bonding wire capacitance ΔCb can be simplified into

$$Out \simeq \frac{1}{KF} \, m \, a + \frac{1}{KF \, KS} \, (k + kel) \, \Delta Cb \tag{22.26}$$

suggesting that it is possible to reduce the transfer of ΔCb by the modulation of kel parameter.

In fact, thanks to kel modulation at frequency $Fmod$ (higher than signal bandwidth and generated by Mod block in Fig. 22.28), a spectral replica of ΔCb at $Fmod$ is generated at the loop output and continuously background-correlated against $Fmod$ in least mean square block, which injects a corrective offset $\sim \Delta Cb$ (via DAC block) into the loop filter to suppress ΔCb itself.

The discussed solution has been declared able to reduce the ΔCb contribution of about 41 dB, so reporting a residual 0 g level drift post moisture stress of few mg [19].

22.2.8 Conclusions

Capacitive interfaces for MEMS accelerometers have been presented, from easier open-loop solutions to a more complex closed-loop ones, highlighting the main features for each solution, starting from the key equations and going through the details of the circuital implementations.

22.3 Electronic Interfaces for Coriolis Gyroscope

22.3.1 Introduction

Due to their capability to measure angular rate with a high accuracy and at a very low cost, MEMS gyroscopes have a wide range of applications in motion detection, stability control, platform stabilization, and navigation. In the past two decades, they have been deployed in a variety of products in consumer, automotive, and industrial sectors.

The working principle of MEMS gyroscopes is based on the Coriolis force acting on objects which move in a non-inertial frame of reference. A representative implementation of MEMS gyroscope is the tuning fork structure, which contains a pair of masses that are driven to oscillate with equal amplitude, but in opposite directions. An angular rate applied to a plane orthogonal to the drive plane gives rise to Coriolis acceleration, resulting in a differential vibration of the masses, orthogonal to the drive oscillation (and commonly called the sense direction). The resulting differential signal, which is directly proportional to the angular velocity, is amplitude modulated at the drive frequency. Capacitive gyroscopes use comb-type structures to electrostatically actuate the drive resonators, while the vibrations of both drive and sense directions are detected with parallel-plate capacitive electrodes through motion-induced capacitance change.

In real fabricated MEMS gyroscopes, mechanical imperfections inflict parasitic coupling of oscillation between resonators. The resulting parasitic component, which is 90-degree phase shifted compared to the Coriolis signal, is commonly referred to as the quadrature signal. One of the most significant challenges in MEMS gyroscopes is the detection of a very small capacitive angular velocity signal, in the presence of several mechanical and electrical non-idealities such as quadrature signal and capacitive cross-coupling.

The electronic interfaces for MEMS Coriolis Vibrating Gyroscope (CVG) play a key role in meeting more and more demanding specifications in the industry, which mainly include angle random walk (also referred to as noise), zero-rate-output (ZRO), scale factor accuracy (SF), full-scale range, and bandwidth [20]. Compared with other transducing mechanisms for MEMS sensors, capacitive interfaces have notable advantages in both sensing and actuation. On the sensing side, capacitive MEMS gyroscopes have been capable of achieving a resolution down to zF (10^{-21}F) [21]. On the actuation side, comb-finger capacitive actuation allows a linear motion with large distance up to 10 μm. In addition to sensing and actuation, the interface circuits (IC) can also provide calibration, temperature compensation, self-test, and

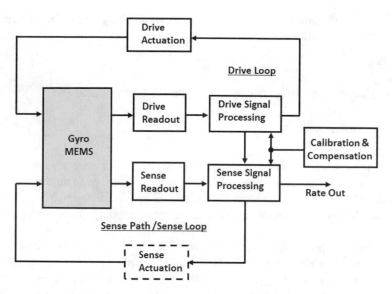

Fig. 22.29 Diagram of functional blocks in gyro interface circuit

conversion of sensed signals into digital and provide digital communication to the output [22].

This section starts with a description of the functionalities of CVG interface circuits, followed by a discussion of major circuit blocks and requirements on their performance. Then design concerns at system and architecture level are covered, illustrating the trade-off between achievable performance and architectural complexity. Finally, this section is concluded with a brief discussion on IC technology selection for gyroscope interface circuits.

22.3.2   Functionalities of Gyroscope Interface Circuit

The key functional blocks in a CVG IC are shown in Fig. 22.29. At the interfaces between MEMS sensing element and IC, first, the drive and sense readout circuits are utilized to convert the variable capacitance signals, provided by the MEMS drive pickup electrodes and sense pickup electrodes, into electrical voltage signals. Second, the IC needs to provide high-precision voltages to bias all MEMS electrodes. Third, it needs to provide high voltage signals to actuate the drive proof mass into oscillation at its resonant frequency (Fd). The drive loop, which includes drive readout, drive signal processing, and drive actuation, is utilized to keep the MEMS drive proof mass in a self-sustained oscillation at Fd, preferably with a constant displacement amplitude. On the sense side, in an open-loop operation, the Coriolis signal is coherently demodulated into a rate signal; in a closed-loop operation, the sense actuation block provides a feedback force to nullify the Coriolis force induced by an input rotation signal. Non-idealities from MEMS sensing element and readout circuits, especially those that are sensitive to variations in process, voltage, and temperature, are trimmed and compensated in the calibration and compensation block. Calibration and compensation in CVG ICs are of great importance to meet specifications in a large volume production. In addition, advanced CVG ICs can convert the sensed rate signal into a digital output and provide digital communication to the outside world through standard protocols such as SPI and I^2C.

22.3.3 Main Circuit Blocks and Requirements

A variety of circuit blocks are used in MEMS CVGs IC, including but not limited to capacitance-to-voltage converters (C2V), precise amplifiers (PreAmp), analog and digital filters, analog-to-digital converts (ADC), digital-to-analog converters (DAC), phase-locked-loop (PLL), and high-voltage charge pump (CP). Among them, the performance of the gyroscope is highly determined by the performance and quality of the C2V, PreAmp, PLL, ADC, and CP. Requirements and design considerations of these major blocks are described below.

22.3.3.1 Capacitance-to-Voltage Converters (C2V)

Together with the bias voltage on MEMS proof mass, the C2Vs of drive readout and sense readout convert the capacitive signals of drive displacement and sense displacement into voltage signals. Depending on how the proof mass is biased, the C2Vs can be implemented with either switched-cap C2Vs (SC-C2V) or a continuous-time (CT) C2Vs, the latter can be further categorized into trans-impedance amplifiers (TIA) and trans-capacitance amplifiers. Figure 22.30 shows topologies of these three C2Vs at a conceptual level. Since C2V is the first analog-front-end circuit (AFE) in the drive path and sense path, minimizing its noise is always the first design priority. In the CVG system, the Coriolis signal is amplitude modulated by the drive velocity, while the unwanted quadrature is amplitude modulated by drive displacement, so a coherent demodulation is required to minimize the leakage from quadrature to rate. To obtain the best performance from coherent demodulation, both amplitude noise and phase noise in the sensed signals need to be minimized. In both drive path and sense path, C2V noise is always the dominant contributor.

When the MEMS proof mass is excited with a square-wave pulse, normally the SC-C2V topology is chosen for the implementation of both drive C2V and sense C2V. In SC-C2V, correlated double sampling (CDS) is generally used to cancel flicker noise and KT/C noise, with the penalty of folding of thermal noise [23].

Since the mechanical variable capacitance signals are AC at frequency Fd, the MEMS proof mass can also be biased with a DC voltage, and then a continuous-time C2V can be utilized as the readout AFE. In the literature, both TIA and TCA have

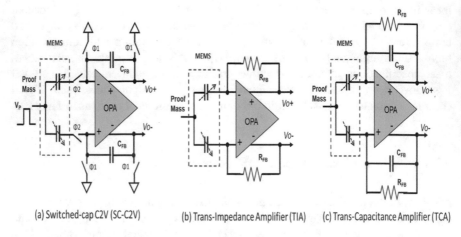

(a) Switched-cap C2V (SC-C2V)   (b) Trans-Impedance Amplifier (TIA)   (c) Trans-Capacitance Amplifier (TCA)

Fig. 22.30 Topologies of C2V

been utilized as gyroscope C2Vs. Due to the large motion resistance in CVGs, TIA normally needs a feedback resistor (R_{FB}) with a resistance in the range of several to tens of mega ohms, which inevitably results in large noise, a big disadvantage against the low-noise requirements of gyroscope C2Vs. Nevertheless, when TIA is used as drive C2V, a 90° phase shifter is not needed in the drive loop, because the signal at TIA output is already in phase with drive velocity.

TCA can intrinsically achieve very low noise, at a cost of more complexity than TIA. Therefore, it is widely used as gyroscope C2V. The design challenge of TCA is to provide a robust and stable DC biasing path for the sensing electrodes. Furthermore, to minimize its noise, TCA generally needs a huge feedback resistance R_{FB}, typically in the range of hundreds of mega ohms to giga ohms [21]. For all the C2V topologies, when the supply voltage is high enough, a telescopic operational amplifier (OPA) is recommended to minimize noise contribution from OPA.

22.3.3.2 Precise-Amplifier (PreAmp)

When C2Vs are targeted to minimize noise, usually they are followed by a PreAmp to maximize signal-to-noise ratio (SNR) in the signal path. In the drive path, when the sensed drive sinusoidal signal is converted to a clock at Fd, maximizing SNR of the sine signal helps to reduce phase noise of clock Fd. In the sense path, when the C2V is designed to minimize noise, a PreAmp following C2V can help to (a) adjust the full scale range, (b) reduce noise contribution of any signal processing circuits after the PreAmp, and (c) reduce equivalent ZRO contributed by the analog DC offset of signal processing circuits after the PreAmp. With a proper gain distribution between C2V and PreAmp, noise contribution of PreAmp will become negligible.

22.3.3.3 High-Voltage Charge Pump (CP)

Availability of high voltage can greatly improve the performance of CVGs. First, the drive force is proportional to the voltage across proof mass and drive actuation electrodes, when a differential actuation mechanism is applied. The drive force is proportional to the square of the voltage across proof mass and drive actuation electrode when a single-ended actuation mechanism is applied. Second, in the pickup side, the total charge going into the drive and sense C2Vs is proportional to the voltage across proof mass and the drive and sense pickup electrodes. Therefore, high voltage helps to increase both signal gain and drive force. Furthermore, for a given drive motion displacement, using high DC voltage reduces amplitude of the drive-forcing AC components, thus minimizing the capacitive couplings and the related ZRO drifts. The high voltages normally are much higher than the supply voltage, and fortunately they can be provided by a high-voltage charge pump (CP) circuit. Robustness and reliability over process variation, operating temperature range, and lifetime are of great importance of CP. In the industry, both Dickson-type and voltage-doubler-based CPs have been developed for MEMS CVGs [24, 25].

22.3.3.4 Analog-to-Digital Converter (ADC)

Analog-to-digital converters (ADC) are widely used in gyroscope ICs, and they play a key role in determining overall system architecture. In the drive loop, ADCs can be utilized to digitize drive displacement signal from drive C2V output or to digitize the amplitude signal of drive displacement. In the sense path, ADCs can be used to digitize the Coriolis force signal from sense C2V output or to digitize the demodulated rate signal.

When ADCs are used to digitize the amplitude signal of drive displacement or rate signal, since the input signal is in baseband, offset and noise of these ADCs must be minimized. Linearity and full-scale range are also important requirements. Both discrete-time and continuous-time sigma-delta ($\sum - \Delta$) low-pass ADCs have been used for this application, and their power consumption normally is very low due to the low frequency and low bandwidth of the analog input signals.

In high-performance CVGs, ADCs are used to digitize drive displacement signal and the Coriolis force signal, which has a frequency of Fd. In order to guarantee high SNR for both amplitude and phase signals in the digital domain, these ADCs generally require very high resolution in the frequency range of Fd, preferably above 16bit ENOB. If a low-pass ADC is used, its signal bandwidth is required to be more than 10Fd to minimize phase delay in both drive and sense loop, which is mandatory to minimize ZRO in final output. To reduce power consumption, these ADCs can also be implemented with band-pass ADCs with central frequency locked at Fd. The clocks of band-pass ADC need to have frequencies at a multiple of Fd, which normally can be provided by a phase-locked loop (PLL) with clock Fd as its reference. In the industry, both discrete-time and continuous-time ADCs have been used for this application [23, 26].

22.3.3.5 Phase-Locked Loop (PLL)

Phase-locked loop (PLL) has been widely used in CVG ICs. Firstly, when TCA is used as drive C2V for its low noise, PLL can be used in the drive loop to provide the 90° phase shift, which is necessary to form a positive feedback loop. Compared with self-oscillating drive loop [27], with a multiplier-based phase detector (PD), PLL within drive loop can help avoiding drive oscillation locked to spurious modes [22].

Secondly, due to the high drive Q of CVGs, the drive C2V output can be used to generate a clean reference clock at Fd. A PLL with clock Fd as reference can provide clocks to all blocks in the interface circuits; thus the whole system becomes self-clocked. Utilizing a PLL offers the advantage that MEMS drive Fd is constantly tracked and higher multiplies of the Fd are provided as a reference clock signal for the sense ADC and for post-processing of the Coriolis signal. Since poles and zeros of digital filters scale with the sampling frequency, therefore, all sense and drive loop filtering tracks the variations of Fd [22, 26, 28].

Thirdly, in a closed-loop forced-feedback sense system, the continuous-time nature of the mechanical filter makes the performance of the electromechanical $\sum - \Delta$ modulator sensitive to the exact shape of the feedback pulse. Low-jitter clock from the PLL can guarantee the best performance of force feedback, leading to a low noise floor at the system output [26].

22.3.4 System and Architecture Design

System design of CVGs is strongly impacted by the electromechanical interfaces between MEMS sensing element and interface circuits. At this level, major design considerations include MEMS operation mode (the relationship between drive Fd and sense resonance frequency, Fs), proof mass bias, drive actuation method, quadrature cancellation method, and rate sensing operation mode. The design options for these considerations, together with their advantages and disadvantages, are summarized in Table 22.1.

Typically, the IC architecture is chosen for the balance of performance, power, and size to meet the market requirements. In order to choose the right architecture, the requirements imposed by final applications on the gyroscope specifications need to be well understood. System architecture and circuit design are the dominant factors determining the noise performance. They also play a significant role in determining ZRO and bias instability (BI). CVGs for different applications can have significant differences in their IC architectures. Critical performance requirements, such as noise, ZRO, robustness to vibration, and power consumption, are the

Table 22.1 System design options

Design considerations	Option	Advantages	Disadvantages
MEMS Operation	Matched-Mode (Fd = Fs)	Boosted MEMS sensitivity	Complex MEMS and system design
		Intrinsically lower noise, higher SNR	Need frequency tuning
			limited bandwidth in open loop sense
	Split-Mode (Fd ≠Fs)	Simpler MEMS and System design	Low MEMS sensitivity
		Wide bandwidth in open loop sense	Intrinsically higher noise, lower SNR
Proof-Mass Bias	DC Bias	Can apply high voltage to boost signal gain	Crosstalk from drive actuation to sense readout is in band, contributing to ZRL
		Low noise, simpler readout Circuit	
	Clocked	Crosstalk from drive actuation to sense readout is out of band, no ZRL contribution	Limited voltage level, low signal gain
			More complex readout Circuit
Drive Actuation	Sine-Wave	Low harmonic frequency components	Lower forcing efficiency
		MEMS Spurious modes will not be excited	Needs higher voltage for same force
	Square-Wave	Higher forcing efficiency	High harmonic frequency components
		Needs lower voltage for same force	MEMS Spurious modes could be excited
Quad cancellation	Electrical/Open-loop	Simple and accomplished only electrically	Quadrature is not removed at its origin
		Can cancel very large quadrature	Demodulation phase error will result in ZRL
		No extra electrodes in MEMS	Sensitive to temp and mechanical stress
	Electromechanical/ Closed-loop	Quadrature is removed at its origin	Need extra electrodes and control loop
		Robust to temperature and mechanical stress	Quad. Cancellation is limited by voltage level

(continued)

Table 22.1 (continued)

Design considerations	Option	Advantages	Disadvantages
Rate Sensing	Open-loop	Simpler MEMS, ASIC and system design	trade-off between bandwidth and sensitivity
		Robustness against vibration	sensitive to variations of MEMS parameters
	Closed-loop/Force-feedback	Reduced sensitivity to MEMS nonlinearity	Intrinsically more MEMS spurious modes
		Increased bandwidth	Vulnerable to vibration when spurious modes not well controlled by the loop
		Achieve lower noise with Mode-matching	

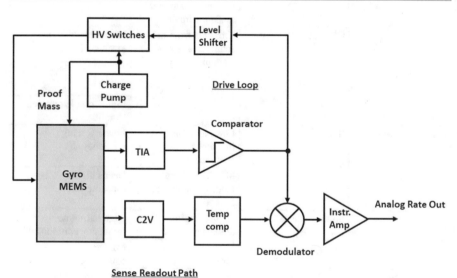

Fig. 22.31 Block diagram of purely analog architecture

dominant differentiators in choosing different architectures. In addition, how to distribute signal processing tasks into analog domain and digital domain is another important factor. The diagrams of three representative IC architectures, from simple purely analog to mixed signal and to highly digital intensive, are shown in Figs. 22.31, 22.32, and 22.33, respectively.

The simple analog architecture shown in Fig. 22.31 includes almost the minimum number of circuit blocks for a gyroscope IC [21]. It uses high-voltage square-wave drive actuation to provide the highest drive force for a low-Q MEMS CVG,

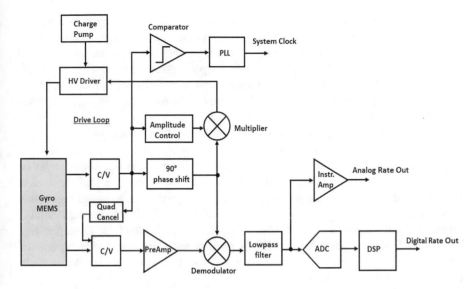

Fig. 22.32 Diagram of mix-signal IC architecture

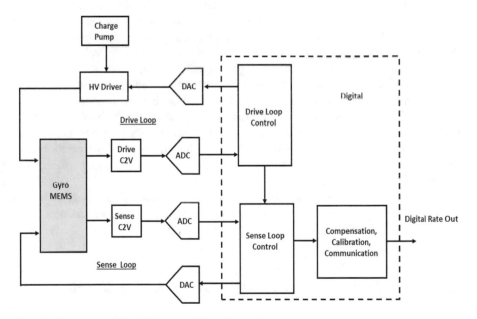

Fig. 22.33 Advanced digital-intensive IC architecture

in which the drive oscillation displacement is too small to hit stoppers in normal operation. Both drive amplitude control and quadrature cancellation circuits are not used. MEMS sensitivity variation over temperature due to a changing drive Q is compensated in the Coriolis signal sensing path. An obvious drawback of this architecture is that even critical performance factors, such as noise, ZRO, and scale factor are susceptible to MEMS and ASIC non-idealities. Therefore, it is only used for very low-cost consumer applications.

The mixed-signal architecture in Fig. 22.32 is widely used in large-volume consumer products. Compared with its purely analog counterpart, it has more functionalities. First, drive oscillation amplitude is maintained constant with an automatic-gain-control (AGC) block in the drive loop. Second, a PLL is used to track drive frequency Fd and provide clocks at multiples of Fd to signal processing blocks including ADCs and DSP. Third, the sensed rate signal is converted to digital, so that a digital output can be provided. Last, quadrature is electrically compensated so that ZRO is less sensitive to MEMS process variation and over temperature. These added circuit blocks are necessary to provide a better performance at the cost of more complexity [29].

To meet high-performance requirements that are needed in automotive applications or in tactical-grade gyroscopes, more and more of the control and signal processing are shifted from analog domain into the digital domain, leaving only the readout circuits and high-voltage actuation circuits in the analog domain [22, 26]. Figure 22.33 shows a digital-intensive architecture as an example. After the drive oscillation signal and Coriolis signal are converted into voltage signals, they are immediately converted into digital signals through high-speed and high-resolution ADCs. All the drive loop control functions including Fd tracking, AGC, clock generation with a digital PLL, and sense loop control functions, such as electromechanical closed-loop cancellation of quadrature, force feedback loop to nullify Coriolis force, and demodulation of Coriolis signal into rate signal, are all accomplished in digital domain. With this architecture, very-high-performance MEMS gyroscope can be achieved at the cost of relatively high-power consumption and interface circuit complexity.

Figure 22.34 shows a micrograph of a low-power readout IC for 3-axis digital-output MEMS gyroscope, which has been widely used in consumer products [30]. In the micrograph, all major functional blocks are highlighted.

22.3.5 IC Technology Selection for Gyro Interface Circuits

For MEMS CVGs with large-volume applications, the interface circuits are usually implemented in dedicated application-specific integrated circuits (ASIC) chips. IC technology selection for these products is of critical importance to their success in the market. First, since high voltage is critical to the function and performance of CVGs, high-voltage option and its matureness in IC technology are of a high priority. The second requirement is the capability to provide very-low-flicker noise

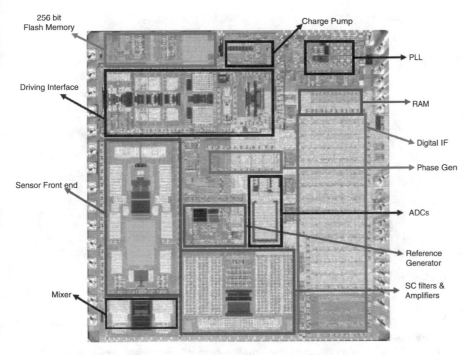

Fig. 22.34 ASIC die micrograph of a 3-axis digital-output gyroscope (die size: 2.5mmx2.5 mm)

transistors, since the working frequencies of MEMS CVGs are not high enough to be in a thermal-noise-only frequency range. Last but not least is the capability to provide high-density digital processing, since high-performance CVG ASICs more and more rely on complicated digital signal processing and digital control, which will have advantages in both size and power consumption with high-density digital technology.

In conclusion, both design techniques and IC technology have great impact on the successful introduction to market of high-performance MEMS gyroscopes in large volume.

22.4   Interfaces for Pressure Sensors

22.4.1   Introduction

The physical output of a pressure sensor, generally, is a differential output voltage of a resistive bridge for the piezo sensor (PPS) or a small capacitance variation for the capacitive sensor (CPS). For the piezo sensor, generally, the output voltage can vary from few mV to few tens of mV, depending on the sensitivity of the sensor and the

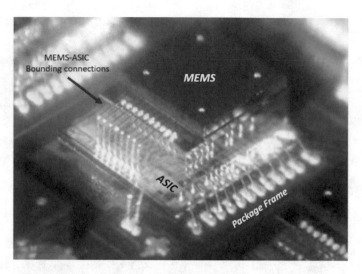

Fig. 22.35 Example stacked assembly in package of a MEMS pressure sensor on ASIC

pressure to which it is subjected. For the capacitive pressure sensor, the capacitance variations are of the order of few fF.

An ASIC allows to manage the small output of the sensor and converts it into a signal more appropriate (analog or digital) to the final application for which the sensor is intended. Therefore, an ASIC for pressure sensor is implemented by means of the combination between an analog front-end, to which the sensor is connected, and a digital processing system that allows to process, store, and carry out data compatible with the most common communication protocols that the applications required today.

The first evident aspect is the importance of the role played by interconnections between MEMS and ASIC. For this reason, in the pressure sensor device, there is a large focus on continue studies of possible package solutions aimed at reducing any parasitic effect between the two dices housed together inside the same package. Minimum distances, materials with minimum sensitivity to external stresses, and the optimization of the dice size are only among the main aspects that are considered in the manufacture of a pressure sensor. The example of a stacked solution, as shown in Fig. 22.35, is one that best meets the design criteria mentioned above.

Obviously, the final application will define the type of package to use. Full molded solutions, such as in Fig. 22.36 (a), are widely used for barometric applications, where the sensors must provide pressure indications without being particularly influenced by the surrounding environment. Otherwise, if the sensors have to work in contact with substances, which can strongly influence the physical characteristics of the sensor itself, cavity solutions are preferred where both the sensor and the ASIC are immersed in particular gels that protect the device from external corrosive effects without altering its performance.

Fig. 22.36 (a) Plastic full molded package; (b) metal cap cavity package

The pressure sensor output characteristic is a function depended on temperature (T) and process (Proc). It can be expressed as

$$Xout_{psens}(T, Proc) = Xoff(T, Proc) + S(T, Proc) \cdot (P - Po, int)$$

where $S(T, Proc)$ is the sensitivity of the sensor (mV/V/bar for Piezo-PS or aF/Pa for the CPS) and $Xoff(T, Proc)$ is the output offset of the sensor. This offset is defined as the output voltage or output capacitance variation of the sensor when the environmental pressure is equal to the pressure inside the cavity under the membrane *Po, int* and the diaphragm displacement is a rest condition, respectively, for the piezo-PS and the CPS.

22.4.2   Pressure Sensor Design Consideration

The most common applications in the consumer or industrial sector operate in a temperature range that can vary from −40 °C to 125 °C. Since both sensitivity and offset voltage have a strict and not linear dependence on temperature, a digital compensation of the pressure in temperature is often necessary, in order to achieve good performance of absolute pressure accuracy in all the temperature ranges. For this reason, normally, the pressure sensors work in pair with temperature sensors and use the same interface to alternately acquire pressure and temperature data.

Since the offset and sensitivity of the pressure sensor are even dependent from the process variation, the $Xout_{psens}$ can be represented from a family curve. Therefore, the interface has to be, also, versatile enough to cover all possible sensor variations with the process and the temperature. About that, some techniques of calibration

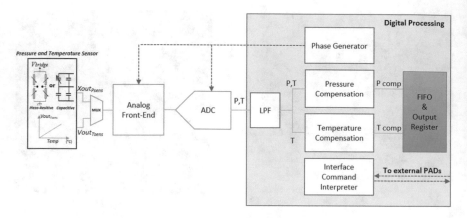

Fig. 22.37 Pressure sensor interface (ASIC) block diagram

for the analog and digital parameters are normally implemented into the interface to overcome these variations (Fig. 22.37).

The first block of the interface is the analog front-end (AFE). Since the electrical inputs to be monitored have a small value, an amplifier with one or more gain stages can be implemented in inverting or not-inverting configuration, which represents a key building block of the AFE.

Generally, fully input-output differential operational amplifiers are preferred to have a high common mode rejection ratio (CMRR). These topologies allow to reject some possible common mode disturbance that can propagate, through the sensor, at the input of the interface from the voltage ground or supply paths.

A high input impedance is necessary. In fact, the interface has to detect a small signal but must not load the output of the sensor, especially for the piezo-resistive sensor. Therefore, a very high input impedance is required to the input of the front-end. For this reason, CMOS technology is preferred for the implementation of the operational amplifier used in the input stage.

After the technology to use is defined, the complexity of the design consists, in fact, in finding the correct trade-off among some key features:

- Low intrinsic noise
- Low current consumption
- Low input offset voltage
- Small area occupation

Obviously, to use PMOS or NMOS devices as input pair of a differential amplifier with width (W) and length (L) such as to have a big W/L ratio improve the noise, and big W·L to improve the input offset performance. On the other hand, the die size of the whole interface is often constrained by the dimension of the final package used. This is a limit for the occupation of area reserved for each block, including the analog front-end.

The importance to have a very low noise system with a fine resolution in the measurement of pressure is strongly linked, today, to the consumer application. In fact, the pressure sensors are used, also, as absolute barometer and to support the indoor or external GPS navigation, as altimeter.

The relation between absolute pressure read and the altitude is

$$P = Po \cdot \left(1 - \frac{Altitude}{44330} \right)^{5.255}$$

where Po is ambient pressure of 1013.25 hPa if it is referred to sea level. For small altitude the relation can be linearly approximate to

$$P - Po - 0.119 \cdot Altitude$$

A variation of altitude of 10 cm with respect to an initial position, where the initial absolute pressure is Po, correspond at about 1.2 Pa of pressure variation. The height variation of 10 cm is a significant specification because it identifies the variation in height between two steps of a staircase. For this reason, the input pressure noise specification required for the pressure sensor interface (in particular, in the GPS system for indoor navigation) is lower than 1 Pa that corresponds to a magnitude of input signal on the order of few microvolt (μV).

Another way to improve the noise performance is to design an input stage of an operational amplifier with a high level of bias current, but this strongly increases the power consumption with consequent reduction of the battery operating time. Luckily, noise reduction and input offset voltage cancellation techniques, such as chopper or CDS (correlated double sampling) technique, could be implemented in the analog front-end to prevent high level of current and achieve good performance in terms of pressure noise and pressure accuracy.

By their nature, capacitive sensors have the great advantage of having a sensor with zero current consumption; this allows to concentrate the current budget available mainly for optimizing front-end noise performance. However, the possibility of integrating digital interfaces that can accurately manage conversion times, mediation and filtering operations, and, especially, the weighted selection of key blocks needed during the acquisition has ensured that, today, the current and noise performance, between capacitive and piezo-resistive, are equivalent.

From the point of view of the front-end design, the main difference between the two types of sensors is the implementation of the input stage. Unlike piezo-resistive sensors that use low noise voltage amplifiers, the capacitive must convert and amplify the variation of input capacitance into an output voltage that can be managed by the analog to digital converter. Since the discussion of interfaces that manage variation of small capacitances has already been addressed in the paragraph of capacitive accelerometers, this section will be mainly focused on piezo-resistive AFE. About that, AFEs for pressure sensor devices could be implemented either with switched capacitor discrete time or with continuous time architecture. A brief discussion of the two topologies, with advantages and disadvantages, is reported below.

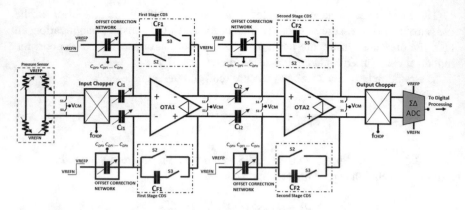

Fig. 22.38 Switched capacitor AFE and ADC

22.4.3 Switched-Capacitor Amplifier Front-End

Figure 22.38 shows a basic schematic of an amplifier chain designed with the cascade of two switched capacitor gain stages.

To save the die dimension, the same amplifier channel can be used both for pressure acquisition and, with appropriate gain and offset scale factor, for temperature. Normally, chopper technique could be used to modulate the input rms noise contribute of the front-end to the chopping frequency and subsequently filtered by the finite bandwidth of ADC.

A good design strategy could be to use the same band-gap voltage reference to drive the sensor, the offset correction network, and the ADC. The advantage of this approach allows to remove the low frequency noise contribution coming from the reference and bias circuitry. Moreover, it guarantees a high immunity to the power supply variation and to noise injection from the voltage ground path.

Offset correction networks could be used to correct the offset voltage contribution of the sensor for all the process and temperature variations. These circuitries are connected to the input virtual ground of both the stages. Figure 22.39 shows an example of an offset correction network that uses a digitally programmable capacitor arrays to have a charge partitioning with a basic capacitance module.

A reference voltage drives the capacitive network. A correction offset binary word [C_{OFn}, C_{OFn-1}, ..., C_{OF1}, C_{OF0}] is used to open or close the switches, in Fig. 22.39, and force the same capacitance variation but with opposite sign between the positive and negative virtual ground node of the operational amplifier. As consequence, a differential voltage is generated to input of both the stages to compensate the intrinsic offset voltage of the sensor. The factors k and n define, respectively, the resolution step and the maximum voltage correction. Their values are the trade-off between the available area and the level of accuracy and resolution desired.

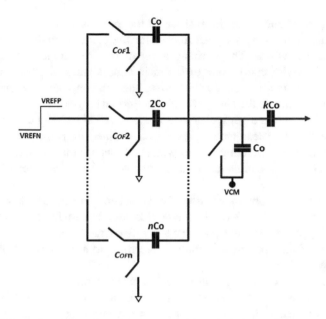

Fig. 22.39 Offset voltage correction with programmable switched capacitor arrays network

OTA1 uses a fully differential topology with a trimmable high closed-loop gain given by the ratio of the input capacitance $Ci1$ and the feedback capacitance $CF1$.

$$H_{OTA1}(z) = -\frac{Ci1}{CF1} \text{ with } Ci1 = C_{F1} \cdot \sum_{i=1}^{m} 2^i \cdot b_i$$

where m is the number of bits used for the gain value trimming. A typical gain value higher than 10 is normally used to reduce the noise contribute, reported to the input of the front-end, from the second stage and of the ADC.

Also, the second stage is a switched capacitor fully differential operational amplifier (OTA2). The input capacitances $Ci1$ and $Ci2$ could be implemented by a digitally programmable capacitor array. Therefore, by means of binary words, the value of the capacitances is fixed and hence even the first and second closed-loop gain. The capacitive value range is defined, during a feasibility study of the system, to cover all the process variation of the $S(T, Process)$. A typical range of gain values, for the second stage, can get from 1 to 4 or 5.

$$H_{OTA2}(z) = -\frac{Ci2}{CF2} \text{ with} Ci2 = C_{F2} \cdot \sum_{i=1}^{n} 2^i \cdot b_i$$

Even in this case n is the number which defines the resolution step for the closed-loop gain.

A closed-loop gain value insensitive to the process or temperature variation allows to have good accuracy performance for the entire operating pressure range where the sensor works. The intrinsic offset of the operational amplifier and its temperature behavior can be one of the main causes that can affect the accuracy performance. For this reason, further offset cancellation technique such as CDS technique, in addition to the chopper technique, could be used. The schematic example in Fig. 22.38 shows a particular case where the CDS technique is implemented for both the stages. These techniques are constrained from the generation of an opportune number of operating phases with the drawback that the size of the digital phase generator block can have an important role in the budget area of the digital side.

In fact, three main phases are used for the scheme in Fig. 22.38, synchronized with the chopping clock period. The phase S1 is a reset phase for the amplifier chain. During the S1 phase, all the capacitances are discharged, and the input and output node voltage of the operational amplifier are fixed to a common mode reference voltage VCM.

The phase S2 forces the second stage in a buffer configuration. In this way, the input offset voltage of the second stage is reported and sampled to the output. When the S3 phase is active, the second stage is in not-inverting configuration, and the input signal is amplified while the offset is deleted.

The big advantage to use a discrete time implementation such this one is that with an opportune synchronization between the chopping phase and the CDS phases is possible to have a boost of the signal-to-noise ratio (SNR) parameter of about 3 dB. On the other hand, the time discrete nature of the system sample and hold leads some drawback such as the thermal noise of the switched capacitor $\left(\frac{K \cdot T}{C}\right)$ and the folding effect of all contributions of noise in baseband with the consequence of a higher white-noise value in-band. The folding noise factor is strongly dependent from ratio between the bandwidth of the front-end and the sampling frequency (Fs) of the ADC. To reduce this factor is one of the main challenges of the design of discrete time architectures, to achieve the noise performance compatibly with the market demand.

22.4.4   Continuous Time Amplifier Front-End

Unlike discrete time implementation, the continuous time architecture has no noise contributions due to the in-band folding effect or capacitive thermal noise. Therefore, with a fixed current consumption and front-end gain-bandwidth, it achieves a better noise specification than a discrete time architecture. Furthermore, with a same noise specification is possible to achieve a lower power consumption and smaller conversion time acquisition (higher ODR).

On the other hand, however, there are no front-end control phases, synchronized with the clock period of the chopper, so there is no boosting effect of the signal. This means that, with the same gain-bandwidth, is possible to have the same SNR, despite of higher current consumption. For the continuous time architecture, the same general considerations used for the discrete time are still valid, in terms of high input impedance, low noise, offset, and power consumption requirement. Normally, architecture such as instrumentation amplifier or capacitive closed-loop amplifier offers a good solution to achieve the desired performances. The second architecture allows to combine the characteristics of the CMOS operational amplifiers together with the use of input capacitors in order to achieve very high input impedance, for the very low operating frequencies of the pressure sensor, normally in the order of few Hz.

In particular, Fig. 22.40 shows a capacitive closed-loop amplifier. The operational amplifier could be a high gain fully differential OTA. The design considerations of the OTA must be focused to achieve the good trade-off between power consumption, noise, and closed-loop bandwidth specifications. Pressure and temperature acquisitions could use the same front-end, whereby a gain scale factor should be implemented on the feedback capacitance CF by using a digitally programmable capacitor network.

The OTA offset cancellation and noise reduction could be implemented by using the chopping technique where the demodulation of the signal can be embedded inside the output stage of the OTA or inside the $\Sigma\Delta$ ADC architecture.

The DC continuity needed for the closed-loop stability is guaranteed by the RFEED blocks. The implementation of this RFEED blocks has been discussed in Sect. 22.1.2. Typical values, included from tens of hundreds of MOhm to few GOhm, are often required and digitally calibrated with trimming techniques.

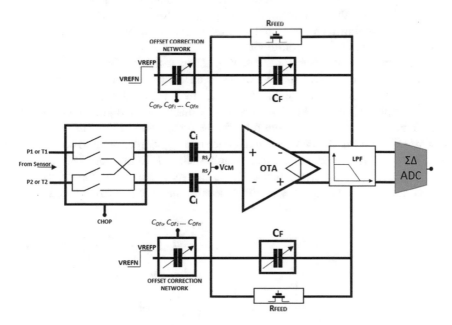

Fig. 22.40 Capacitive inverting closed-loop front-end and ADC architecture

Even for continuous times, offset correction networks could be used to correct the offset voltage contribute of the sensor. The basic principle of operation is the same described for the scheme in Fig. 22.39.

In the architecture of Fig. 22.40, only a reset phase RS is used to discharge all the capacitances and to reset the input and output common node voltage of the amplifier to a common mode voltage reference, and before that the device starts a pressure or temperature acquisition. To use only one phase is an advantage in terms of digital phase generator complexity and area occupation.

Once the signal is amplified and corrected, by the front-end, it will be converted by the analog-digital converter. Although, as described previously, the higher the sampling frequency, the lower the folding factor with a defined front-end bandwidth, to define the architecture of the ADC, it is necessary to find the right compromise, also, with other key parameters such us the conversion speed, linearity, output resolution, and power consumption.

The analog-digital converter, working in a border area between the analog and digital domain, lends itself well to the implementation for both discrete and continuous time architectures or hybrid solution which exploits the advantage of both the previous one. Successive approximation ADC (SAR) or sigma-delta converters ($\Sigma \Delta$ ADC) could be good solutions to get the right trade-off between the specifications mentioned above.

The choice between one topology and another one is often dictated by the process technology available, by the design times often imposed by the market and by the possibility of having or not IP that can be reused and adapted to the specific requests. For example, Fig. 22.41 shows blocks scheme of second-order $\Sigma \Delta$ ADC.

Regardless of the architectural strategy used for the ADC implementation, normally, chopping technique could be used embedded in the first integrator to reduce his noise and offset contribute.

Typically, the first integrator the ADC presents a signal transfer function (STF) with a low-pass characteristic that works as an embedded anti-alias filter. In this way, the in-band folding noise that contributes from the front-end is cancelled.

The output of the $\Sigma \Delta$ ADC is processed by means of the digital interface which include key blocks such us digital filter for the decimation of the data, volatile and non-volatile and FIFO memory, and interface command interpreter to communicate with the standard protocol of communication such as I2C, I3C, and SPI.

A digital phase generator block allows the management of the pressure and temperature acquisition phase, by the generation and the synchronization of a set

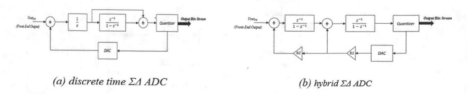

(a) discrete time $\Sigma \Delta$ ADC   (b) hybrid $\Sigma \Delta$ ADC

Fig. 22.41 Discrete time (**a**) and hybrid structure (**b**) second-order $\Sigma \Delta$ ADC

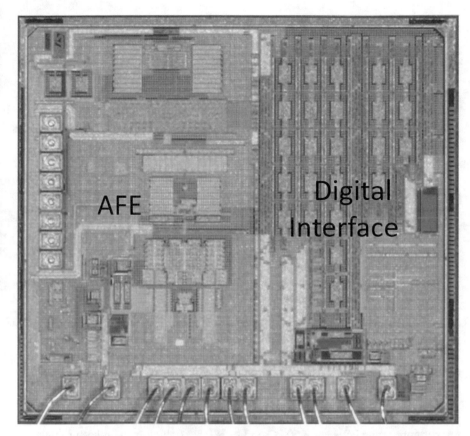

Fig. 22.42 Micrograph of a pressure sensor ASIC top view (die size:1.44mm × 1.55 mm)

of signals used to drive the front-end, the ADC, and the digital operations of the pressure and temperature compensation. All the signals are synchronized by an internal clock used to set; also the output data rate (ODR) is used to update the data of pressure and temperature on dedicated user reading volatile registers.

All the design and technical considerations described above are implemented during the realization of the layout of the ASIC, as shown in Fig. 22.42, where an obsessive attention is paid to the interconnections of each single path and to the circuit symmetries of the devices in the key blocks.

Figure 22.42 shows an ASIC photo with analog path shown on the left side, and the digital interface on the right side. This separation allows to avoid that high rate digital signals do not influence the correct analog processing of the input signal.

22.5 Interface for PTAT Temperature Sensors

22.5.1 Introduction

In most of today's temperature sensors, both voltages V_{BE} and the difference ΔV_{BE} between the base-emitter voltages of two bipolar transistors having different collector current densities are used as a measure for temperature. The sensor essentially digitizes the thermal voltage kT/q (PTAT), using the bandgap voltage as reference V_{REF}.

In CMOS core-based technologies, substrate bipolar transistors are often used as temperature-sensitive devices and are combined with a precision interface circuitry in an analog front-end to extract two voltages that can play the role of the signals required for the temperature measurement:

1. The difference in base-emitter voltages ΔV_{BE}, used to generate a voltage V_{PTAT}, proportional to absolute temperature
2. The silicon bandgap voltage, which is the basis for generating a temperature-independent reference voltage V_{REF}

Their main advantages are that they are suited to be integrated in standard CMOS process, because of their high sensitivity and acceptable long-term stability, their low cost, and small size. The favorable properties of transistors for temperature sensor applications are due to the predictable and accurate way in which the base-emitter voltage V_{BE} is related to the temperature.

In smart temperature sensor, the sensing elements and sensor-interface electronics are combined in a single chip. Such temperature sensors can work in the temperature range between $-50\ °C$ and $+180\ °C$ (which is the relevant range for most applications), with high accuracy and high reliability.

Precision interface electronics, as shown in Fig. 22.43, are used to make the most of bipolar transistors temperature characteristics. Two main parts can be considered:

- The analog front-end whose main blocks are the temperature sensor core and the ADC
- The digital front-end for temperature digital processing

The sensor provides a digital output signal that can be directly interpreted by a microcontroller. Communication between sensor and microcontroller requires converting the analog signals into digital ones. For this purpose, an analog-to-digital converter (ADC) is implemented and it determines a digital representation of the ratio between the two voltages V_{PTAT} and V_{REF}. This ratio is a measure of the chip's temperature and is communicated to the outside world (e.g., a microcontroller) by means of a digital interface (like the I^2C serial interface).

In order to optimize the accuracy, various trimming techniques have been introduced to correct for temperature errors resulting from process spread of

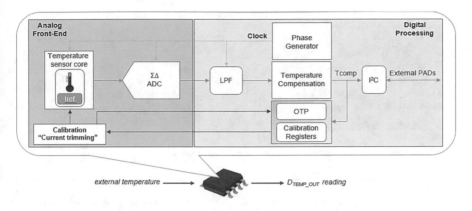

Fig. 22.43 Smart temperature sensor interface

the base-emitter voltage V_{BE}. These techniques require that the error in V_{BE} is determined by means of a calibration.

The temperature sensor can be simply trimmed at one temperature, calibrating the PTAT current I_{REF} in the analog front-end, but when targeting high accuracy over the entire temperature range, additional digital compensation could be adopted so to compensate, after calibration process (current trimming), second-order effect of temperature such as offset and gain compensation. This compensation can be also considered a "fine calibration" in addition to the "coarse calibration" performed by means I_{REF} current trimming.

22.5.2   Principle of Operation and Accuracy Requirements for the Interface Electronics

Figure 22.44 illustrates, in more detail, the operating principle of an analog front-end of a temperature sensor. To obtain, therefore, a digital temperature reading, a ratiometric measurement has to be performed: a temperature-dependent signal has to be compared to a reference signal. The PTAT voltage ΔV_{BE}, generated from the difference in base-emitter voltage between two transistors Q1 and Q2 biased at different current densities (I_{PTAT}, $n{\cdot}I_{PTAT}$), is amplified of a factor α to get a useful voltage V_{PTAT}. The voltage reference V_{REF} is based on the V_{BE} voltage of a bipolar transistor to which is added the voltage $\alpha{\cdot}\Delta V_{BE}$ resulting in a voltage that is essentially temperature independent.

Temperature dependency of the key voltages in the sensor is shown in Fig. 22.45. The graph illustrates the trend of the following voltages:

- V_{REF}, temperature-independent reference voltage
- V_{BE}, base-emitter voltage with a negative temperature coefficient of about 2 mV/°C

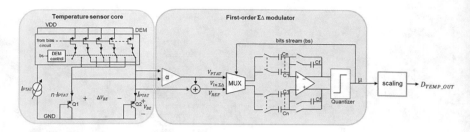

Fig. 22.44 Principle of operation of a smart temperature sensor

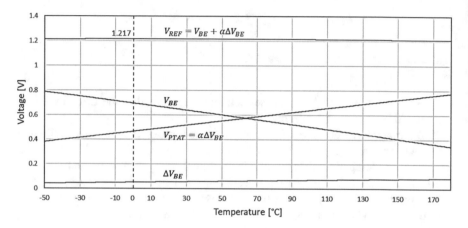

Fig. 22.45 Temperature dependency of the various voltages: V_{REF}, V_{PAT}, V_{BE}, ΔV_{BE}

- ΔV_{BE}, difference in base-emitter voltage between two matched bipolar transistors proportional to absolute temperature (PTAT)
- V_{PTAT}, ΔV_{BE} voltage amplified of a factor α

The digital temperature reading is obtained using an analog-to-digital converter (ADC) that takes V_{PTAT} as input and V_{REF} as reference. A first-order $\Sigma\Delta$ ADC (Fig. 22.44) will provide an output μ (bitstream) given by the following equation:

$$\mu = \frac{V_{PTAT}}{V_{REF}} = \frac{\alpha \cdot \Delta V_{BE}}{V_{BE} + \alpha \cdot \Delta V_{BE}} \tag{22.27}$$

In this ratio, the numerator is proportional to absolute temperature (PTAT), while the denominator is temperature-independent (ignoring curvature effects for simplicity), so that the ratio μ is PTAT.

The output D_{TEMP_OUT} obtained with appropriate digital scaling produces a temperature reading in degrees Celsius.

The accuracy reached by the temperature sensor is ultimately determined by the accuracy of the temperature characteristics of the bipolar transistors. Dynamic offset cancellation and dynamic element matching techniques are used to eliminate the non-idealities of the transistors, such as offset and mismatch.

The component mismatch of standard CMOS processes does not permit to achieve this level of accuracy, even with precision layout techniques, unless dynamic techniques are used.

Dynamic element matching (DEM), used in temperature sensor core block, allows to obtain an accurate 1:n bias current ratio and, then, to reduce mismatch errors. The implementation of a set of switches commutating nominally equal current sources I_{PTAT}, with one of them connected to Q2, while the others $n \cdot I_{PTAT}$ connected to Q1, leads to n possible connections each of which results in an inaccurate ΔV_{BE} due to mismatches between the current sources. The average, however, is accurate, because the mismatch errors cancel out and the inaccuracy induced by the mismatch between n-current sources can significantly be reduced to the second order.

When applying DEM technique in a continuous time circuit, two potential problems have to be taken into account. First of all, DEM introduces extra switching transients. If these transients are integrated, they are likely to introduce errors. Such errors can be avoided if the output in question is disconnected from the integrator during the switching. The second problem is related to the non-zero on-resistances of the switches used to interchange the unit elements. The switches have to be designed in such a way that voltage drop across their on-resistance does not introduce significant errors. This can be realized by using wide switches, which have a low resistance, but such switches, therefore, require a large chip area.

Another significant source of error is the offset. Offsets in the μV range are required, while CMOS amplifiers typically have offsets in the mV range. Trimming could be used to reduce the offset, but it cannot compensate for offset drift and temperature dependency. The better solution is to use dynamic offset cancellation techniques (DOC) such as chopping, which remove the offset during operation and have the added advantage that they eliminate the Flicker noise too [31].

Overall better accuracy and noise performances usually come at the cost of a higher power consumption. Power consumption, however, cannot be increased indefinitely, as the associated self-heating will eventually limit the accuracy. A way to maintain accuracy, noise, and power consumption performances, for example, is to operate the sensor in one shot mode. In this operating mode, the measuring chain is put in power-down condition once it has completed the temperature processing.

22.5.3 Analog-to-Digital Conversion

The ADC is a key building block in a smart temperature sensor, and it serves to digitize the ratio between the voltages V_{PTAT} and V_{REF}. The output μ (22.27) is obtained by connecting the bipolar core, from which the ΔV_{BE} and V_{BE} voltages are generated, to the charge-balancing converter. Figure 22.46 shows a block diagram of a first-order $\Sigma \Delta$ modulator that consists of an integrator and a clocked comparator.

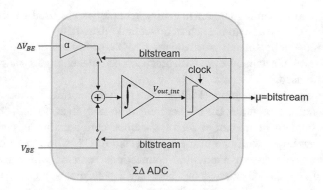

Fig. 22.46 Block diagram of a sigma-delta that produces a bitstream whose average is PTAT

The resulting voltages ΔV_{BE} and V_{BE} constitute the inputs of the modulator. Every clock cycle, the comparator produces a bit of the bitstream (μ) in the following way: if the bitstream in a given clock cycle is zero, the ADC integrates α $\cdot\Delta V_{BE}$, while if the bitstream is one, the ADC integrates $-V_{BE}$.

As a result of the feedback in the modulator, the average input to the integrator is zero. In other words, the charge added by $\alpha\cdot\Delta V_{BE}$ is balanced by the charge removed by $-V_{BE}$. This charge balancing can be expressed as:

$$(1 - \mu) \cdot \alpha \cdot \Delta V_{BE} = \mu \cdot V_{BE} \tag{22.28}$$

where the bitstream average is μ.

The required scale factor α is established by appropriately sizing the sampling capacitors at the input of the modulator.

Solving for μ we can obtain Eq. (22.27).

A final digital output D_{TEMP_OUT} in degrees Celsius can be obtained by linear scaling:

$$D_{TEMP\_OUT} = A \cdot \mu + B \tag{22.29}$$

where A and B are the extremes of the operating range that corresponds to a certain temperature range in which the sensor functionally works [32].

22.5.4   Calibration Techniques

A smart temperature sensor needs to be calibrated and optimized according to the accuracy target. Sensor is calibrated by comparing it with a reference thermometer of known accuracy. To save production costs, this is typically done at only one temperature (one-point calibration, OPC).

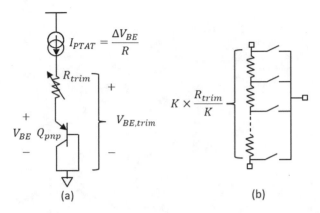

Fig. 22.47 (a) Voltage trimming by means of a programmable resistor in series with the emitter (b) Implementation of the R_{trim} resistor

The trimming of the I_{REF} current, as mentioned before, is a way of trimming V_{BE} voltage and it is referred to as a "calibration procedure" to compensate its spread: it can be implemented by means of a digitally programmable resistor, as shown in Fig. 22.47, in series with the emitter of the diode-connected substrate PNP transistor, biased, accordingly, at a programmable PTAT/R_{trim} current. By adjusting the size of R_{trim}, the magnitude of the PTAT current can be adjusted with the aim of optimizing the accuracy at one temperature.

In addition to I_{REF} current trimming, other possible calibration procedures like

- Using an integrated bipolar transistor (Q_{CAL})
- Using an external voltage (V_{ext})

are also used to correct the spread of the base-emitter voltage V_{BE}, which is, in principle, the only reason why CMOS temperature sensors need to be trimmed.

Figure 22.48 shows how the Q_{CAL} calibration can be implemented. A PNP dedicated transistor Q_{CAL} is integrated in the temperature sensor die.

Three externally generated bias currents I_{ext1}, I_{ext2}, and I_{ext3} are successively applied to the emitter of this transistor to generate three different base-emitter voltages (V_{BE1}, V_{BE2}, and V_{BE3}) that are then measured using an accurate external voltmeter. From the measured base-emitter voltages, the chip's temperature can be calculated. Two pins are needed in order to access Q_{CAL} from outside the chip. These pins, however, only need to be connected to Q_{CAL} during calibration. During normal operation, they can be reused, for instance, for a digital bus interface, so that the total number of pins required on the sensor's package is not increased.

Kelvin connections are used to reduce the effect of series resistances. However, the base and emitter resistances of the transistor itself still affect the measurement. To cancel the effect of Rs, V_{BE} and IC are measured for three different values of the emitter current.

This gives two independent equations in ΔV_{BE}:

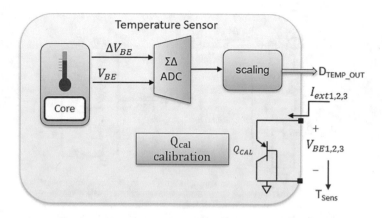

Fig. 22.48 The temperature of the sensor, T_{Sens}, derived from the base-emitter voltages of the calibration transistor Q_{CAL}

$$\begin{cases} \Delta V_{BE2,1} = (I_{E2} - I_{E1}) R_S + V_T \ln \left(\frac{I_{C2}}{I_{C1}} \right) \\ \Delta V_{BE3,2} = (I_{E3} - I_{E2}) R_S + V_T \ln \left(\frac{I_{C3}}{I_{C2}} \right) \end{cases} \tag{22.30}$$

where $V_T = \frac{nkT}{q}$. These equations can be solved for V_T, and R_s.

The absolute temperature can be so derived as follows:

$$Temp \left({}^\circ C \right) = \left(\frac{q}{k \cdot N_F} V_T \right) - 273.15 \tag{22.31}$$

A trade-off between trimming and accuracy consists in I_{REF} current trimming until the device supplies the same temperature as the Q_{CAL} temperature value. An important advantage of calibration based on Q_{CAL} measurements is that its accuracy is independent from the one of the sensor circuitry. Therefore, a calibration transistor can be easily added to an existing design to reduce calibration costs with minimum design effort.

Calibration by V_{ext} is shown in Fig. 22.49:

This calibration can be considered as the implementation of the voltage reference calibration indirectly. An external voltage V_{ext}, which can be switched in place of V_{BE}, is applied to the sensor. Thus, a temperature reading $D_{TEMP\_OUT,cal}$ is obtained. The corresponding ratio μ determined by the ADC can be calculated. This ratio is equal to:

$$\mu_{cal} = \frac{\alpha \Delta V_{BE}}{V_{ext} + \alpha \Delta V_{BE}} \tag{22.32}$$

Since V_{ext} and α are known, ΔV_{BE} can be calculated:

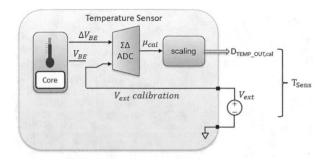

Fig. 22.49 Voltage calibration by replacing V_{BE} with an external voltage V_{ext}

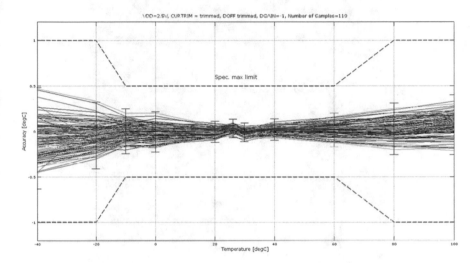

Fig. 22.50 Smart temperature sensor accuracy of 119 DUTs: one-point calibration @30 °C

$$\Delta V_{BE} = \frac{\mu_{cal}}{(1 - \mu_{cal}) \cdot \alpha} V_{ext} \qquad (22.33)$$

Thus, ΔV_{BE} has been measured indirectly using the external voltage as a reference voltage. From the value of ΔV_{BE} thus found, the sensor's temperature T_{Sens} can be calculated. Finally, the sensor can be trimmed to ensure that its output D_{TEMP_OUT} during normal operation equals the calculated temperature.

An example accuracy compensated of a smart temperature sensor is reported hereafter.

Figure 22.50 shows maximum temperature accuracy of an ultra-low-power, high accuracy, digital temperature sensor offering high performance over the entire operating temperature range from −40 °C to 100 °C.

The DUTs are trimmed using a single-point calibration method at 30 °C and a digital calibration to compensate the second order effects. The 119 samples after

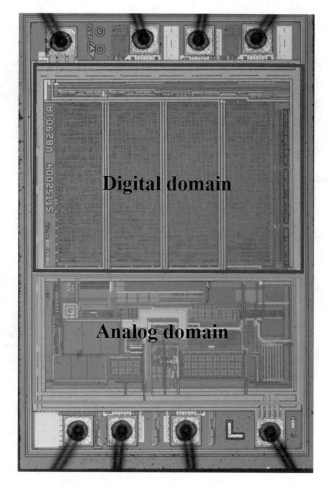

Fig. 22.51 Micrograph of the smart temperature sensor chip (die size: 1.454mmx0.934 mm)

the calibration can realize an inaccuracy ($\pm 3\sigma$) within ± 0.5 °C in the temperature range from -40 °C to 100 °C.

Overall, the design philosophy is to compensate all temperature errors, resulting from circuit non-idealities, down to an accuracy target based on the specifications. The usage of design techniques, such as dynamic offset cancellation and dynamic element matching, allows to reduce errors due to offset variations in the current gain of the bipolar transistors and other mismatch sources.

Spread of the base-emitter voltage of the bipolar transistors is then the only remaining significant error source which is trimmed out using calibration techniques with the aim of optimizing the accuracy at a given temperature.

A chip micrograph which includes digital interface, analog circuitry, and bond-pads is shown in Fig. 22.51.

References

1. Barbieri, A., & Pernici, S. (2016, September). A differential difference amplifier with dynamic resistive degeneration for MEMS microphones. In *Proceedings of ESSCIRC* (pp. 285–288). Losanne, Switzerland.
2. Gaggl, R., & Buffa, C. (2017). Silicon microphones: from concept to design. *Topics on Microelectronics, XI*, 95.
3. Barbieri, F., Barbieri, A., & Nicollini, G. (2018, April). Pseudo-differential analog readout circuit for MEMS microphones performing 135dBSPL AOP and 66dBA SNR @1Pa. In *Proceedings of CICC, San Diego, CA, USA*.
4. Pinna, C. (2014, November 18). *Multi-Level Sigma-Delta ADC with reduced quantization levels*. Patent No. US 8890735B2.
5. Nguyen, K., Adams, R., Sweetland, K., & Chen, H. (2005, December). A 106-dB SNR Hybrid Oversampling Analog-to-Digital Converter for Digital Audio. *IEEE JSCC, 40*(12).
6. Bach, E., et al. (2017, Febraury). A 1.8 V true-differential 140 dB SPL full-scale standard CMOS MEMS digital microphone exhibiting 67 dB SNR. In *IEEE ISSCC Dig. Tech. Papers* (pp. 166–167).
7. Pinna, C., Mecchia, A., & Pesenti, P. (2011, December 22). *Digital microphone device with extended dynamic range*. Patent No. EP2608569B1.
8. Lemkin, M. A., & Boser, B. E. (1999). A three axis micromachined accelerometer with a CMOS position-sense interface and digital offset-trim electronics. *IEEE Journal of Solid-State Circuits, 34*(4), 456–468.
9. Enz, C. C., & Temes, G. C. (1996). Circuit techniques for reducing the effects of op-amp imperfections: Autozeroing, correlated double sampling and chopper stabilization. *Proceedings of the IEEE, 84*(11), 1584–1614.
10. Makinwa, K. *Dynamic offset-cancellation techniques*. [Online] Available: https://ieee-sensors.org/wp-content/uploads/2011/archive/Tutorials/makinwa.pdf
11. Kevin, T., Chai, C., Han, D., Singh, R. P., Pham, D. D., Pang, C. Y., Luo, J. W., Nuttman, D., & Je, M. 118-dB dynamic range, continuous-time, opened-loop capacitance to voltage converter readout for capacitive MEMS accelerometer. In *IEEE Asian Solid-State Circuits Conference November 8–10,2010/Beijing, China*.
12. Andrea, B., Gola, A., Chiesa, E., Lasalandra, E., Pasolini, F., Tronconi, M., & Ungaretti, T. (2003). A 1-g dual-Axis linear accelerometer in a standard 0.5-μm CMOS technology for high-sensitivity applications. *IEEE Journal of Solid-State Circuits, 38*(7), 1292–1297.
13. Nag, D. (2016). *A 6 nV/√Hz high precision front-end with sub-μV input offset for MEMS accelerometer*. IEEE 978-1-4673-9019-4/16.
14. Huang, J., Zhang, T., Zhao, M., Hong, L., Zhang, Y., Lu, W., & Chen, Z. (2012). *Yilong Hao "time divided architecture for closed loop MEMS capacitive accelerometer*. IEEE 978-1-4673-2475-5/12.
15. Liu, Y., Liu, X., Liang, Y., Chen, W., & Wu, Q. (2009). CMOS interface circuitry for a closed-loop capacitive MEMS accelerometer. In *Proceedings of the 2009 4th IEEE International Conference on Nano/Micro Engineered and Molecular Systems January 5–8, 2009, Shenzhen, China*.
16. Xu, H., Liu, X., & Liang, Y. (2015). A closed-loop interface for a high-Q micromechanical capacitive accelerometer with 200 ng/Hz input noise density. *IEEE Journal of Solid-State Circuits, 50*(9), 2101–2112.
17. Amini, S., & Johns, D. A. (2015). A flexible charge-balanced ratiometric open-loop readout system for capacitive inertial sensors. *IEEE Transactions on Circuits and Systems – II: Express Briefs, 62*(4), 317–321.
18. Langfelder, G., Frizzi, T., Longoni, A., Tocchio, A., Manelli, D., & Lasalandra, E. (2011). Readout of MEMS capacitive sensors beyond the condition of pull-in instability. *Sensors and Actuators A*.

19. Lajevardi, P., Petkov, V., & Murmann, B. A $\Delta\Sigma$ interface for MEMS accelerometers using electrostatic spring-constant modulation for cancellation of bondwire capacitance drift. In *2012 IEEE International Solid-State Circuits Conference*.
20. Yazdi, N., Ayazi, F., & Najafi, K. (1998). Micromachined inertial sensors. *Proceedings of the IEEE, 86*(8), 1640–1659.
21. Geen, J. A., et al. (2002). Single-chip surface micromachined integrated gyroscope with 50 =h allan deviation. *IEEE Journal of Solid-State Circuits, 37*(12), 1860–1866.
22. Balachandran, G. K., Petkov, V. P., Mayer, T., & Balslink, T. (2016). A 3-axis gyroscope for electronic stability control with continuous self-test. *IEEE Journal of Solid-State Circuits, 51*(1), 177–186.
23. Boser, B. E. (1997, June). Electronics for micromachined inertial sensors. In *Tech. Dig. 9th Int. Conf. Solid-state sensors and actuators (Transducers'97), Chicago, IL, USA* (pp. 1169–1172).
24. Dickson, J. F. (1976). On-chip high-voltage generation in MNOS integrated circuits using an improved voltage multiplier technique. *IEEE Journal of Solid-State Circuits, SC-11*(3), 374–378.
25. Favrat, P., Deval, P., & Declercq, M. J. (1998). A high-efficiency CMOS voltage doubler. *IEEE Journal of Solid-State Circuits, 33*(3), 410–416.
26. Omar, A., Elshennawy, A., Ismail, A., Nagib, M., Elmala, M., & Elsayed, A. (2015). A new versatile hardware platform for closed-loop gyro evaluation. In *Proceedings of the Inertial Sensors and Systems 2015 (DGON ISS), Karlsruhe, Germany, 22–23 September 2015*.
27. Eminoglu, B., Alper, S., & Akin, T. (2011). Novel, simple, and q-independent self-oscillation loop designed for vibratory MEMS gyroscopes. In *Proceedings of 16th International Solid-State Sensors, Actuators and Microsyst. Conf.* (pp. 1440–1443).
28. Rombach, S., Marx, M., Nessler, S., De Dorigo, D., Maurer, M., & Manoli, Y. (2016). An interface ASIC for MEMS vibratory gyroscopes with a power of 1.6 mW, 92 dB DR and 0.007 °/s/$\sqrt{Hz}$ noise floor over a 40 Hz band. *IEEE Journal of Solid-State Circuits, 51*(8), 1915–1927.
29. Seeger, J., et al. (2010). Development of high-performance, high-volume consumer MEMS gyroscopes. *Solid-State Sensors, Actuators and Microsystems Workshop*, 61–64.
30. Prandi, L., et al. (2011). A low-power 3-Axis digital-output MEMS gyroscope with single drive and multiplexed angular rate readout. *ISSCC Dig. Tech. Papers*, 104–105.
31. Pertijs, M. A. P., Makinwa, K. A. A., & Huijsing, J. H. (2005). A CMOS smart temperature sensor with a 3σ inaccuracy of ±0.1 °C from −55 °C to 125°C. *IEEE Journal of Solid-State Circuits, 40*(12), 2805–2815.
32. Pertijs, M. A. P., & Huijsing, J. H. (2005). Precision interface for a CMOS smart temperature sensor. *IEEE*, 943–946.

Chapter 23
Electronic Interfaces for Actuators

Andrea Barbieri, Luca Molinari, Mauro Pasetti, and Marco Zamprogno

23.1 MEMS Mirror Driving

Two distinguished family of MEMS mirror have been developed, allowing in one case a linear-like mechanical movement and, in the other case, an oscillating (resonant one). A combination of two of them, also from the same family, is used at application level to implement a 2-dimensional scanning.

Behaviors of the MEMS micro-mirrors, in terms of equivalent transfer function load, actuation voltage, driving bandwidth, and performances to be fulfilled, differ a lot between linear and resonant solutions.

This leads to driver design approaches well different and defined.

23.1.1 Linear Driver Architectures

In general, quasi-static (linear) micro-mirrors are intended to allow a slow and linear scan into the spatial domain [1]. As an example, in a very basic raster scan pico-projection solution, linear mirror is tailored to move in a direction (horizontal or vertical, depending on specific approach) to cover frame definition while a resonant mirror scans the other spatial direction.

According to that, the drive signal to be provided to a linear micro-mirror is in general a very low-frequency signal; 60 Hz or 120 Hz triangular waveform is a typical example for that.

A. Barbieri · L. Molinari (✉) · M. Pasetti · M. Zamprogno
ST Microelectronics, Analog MEMS and Sensors Group, Microfluidic & Micro Actuators Division, Cornaredo, Italy
e-mail: andrea.barbieri@st.com; luca.molinari@st.com; mauro.pasetti@st.com; marco.zamprogno@st.com

© Springer Nature Switzerland AG 2022
B. Vigna et al. (eds.), *Silicon Sensors and Actuators*,
https://doi.org/10.1007/978-3-030-80135-9_23

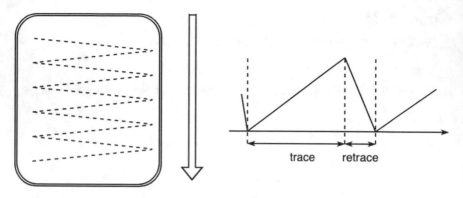

Fig. 23.1 2D raster scanning example. Linear micro-mirror is devoted to the frame scanning (trace). During retrace micro-mirror recovers the initial position in order to start a new frame scanning

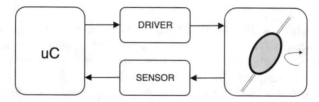

Fig. 23.2 Control loop for a micro-mirror linear driver path

Figure 23.1 shows an example of raster scan approach and a consequent reference signal for linear micro-mirror, where two different phases (trace and re-trace) can be easily highlighted.

Often two actuators must be driven through two separated single-ended drivers while, for some other solutions, more reliable purely differential approach can also be used. Figure 23.2 shows a block diagram for a linear driver chain, where also the control loop can be identified.

For what concern in detail the driving section, digital data provided by a generic micro-processor are translated through a DAC in the analog domain to input the linear driver itself, topology of which depends on specific micro-mirror solution.

Moreover, if on one side bandwidth performances are not so challenging, output noise requirements and reduced power consumption lead to non-trivial designs. Linearity (such as THD performance) is often not critical due to control loop adjustment; nevertheless, to guarantee proper driving performances and to simplify the control loop implementation, at least 15b equivalent absolute resolution should be addressed [2].

Historically, first MEMS micro-mirror solutions were relying to electrostatic activation. From electrical point of view, they behave like a simple capacitive load with a capacitance value in the order of tens of pF to be driven by a high-voltage

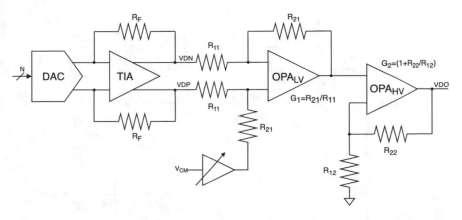

Fig. 23.3 Linear driver block diagram. DAC converts N-bit input word in analog domain. Low-voltage signal is amplified, filtered (filter capacitors not shown in the figure for simplicity), and then inputted to the high-voltage driver .

signal. To get acceptable force to the actuator (for instance to get proper opening angle for the micro-mirror), a maximum voltage close to 200 V can also be required.

Considering all the above-mentioned requirements, a specific driving architecture for electrostatic micro-mirrors is presented in Fig. 23.3.

Digital data defining the driving signal input a steering current DAC. Both a Nyquist and a $\Sigma\Delta$ approach can be followed for DAC design.

Considering the desired resolution, $\Sigma\Delta$ implementation allows to strongly relax the design by exploiting oversampling to reduce quantization noise. The only recommendation in case of oversampled DAC implementation is to ensure a proper quantization noise filtering.

Nevertheless, this is quite easy in such kind of chain, considering that the output load (the mirror) offers a 2nd-order filter. Additional filter orders can be implemented into the intermediated stages taking advantage to the fact that the chain itself must manage only low-frequency signals.

$\Sigma\Delta$ DAC are generally designed with a N-bits thermometric topology, and it can be driven or directly through related N-bits bus or through a single-bit signal implementing well-known FIR additional filtering.

Independently on the chosen DAC architecture, noise (mainly low frequency) is a task to be faced with into the steering current generators design. On top of usual hints (very low pass frequency filter for current seed, large area devices, etc.), again $\Sigma\Delta$ DAC implementation allow to directly integrate some approaches, such as large-scale excitation (LSE) and scrambling, quite useful as additional means to contain the low-frequency noise sources.

In Fig. 23.4, an example of some DAC blocks is represented.

DAC output current is then translated into a voltage signal through a transimpedance amplifier (TIA) and then filtered and pre-amplified before to be

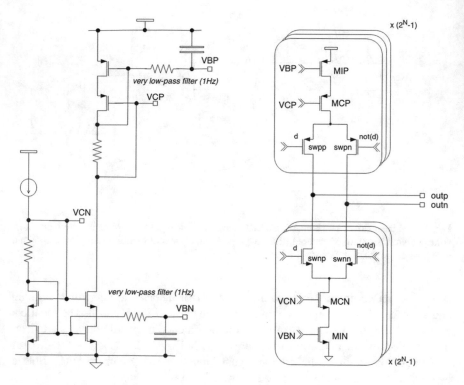

Fig. 23.4 Main analog blocks for fully differential steering current DAC. Cascode arrays both for N and P sides are designed according to a pure thermometric topology (2^N-1 unity modules)

presented as an input for the high-voltage linear driver aimed to amplify the signal up to required output voltage.

In this implementation the high-voltage linear driver has a single-ended topology. This because not all the micro-mirror pairs of actuators must be driven in a pure differential way (differential signal defined with respect to a fixed common mode and superimposed to that), but in some cases two independent (similar) single-ended driving signals must be used.

To keep all the benefits from differential approach, driving chain starts just resorting differential blocks (both for DAC and TIA) while last pre-amplification stage and high-voltage linear driver move to a single-ended (pseudo-differential) topology.

In detail, pre-amplification stage is aimed to provide a gain G_1 to exploit all the available dynamic range for the low-voltage supplied stage so that high-voltage stage can be consequently designed with the lowest possible gain G_2 required to get the desired output voltage. In addition, pre-amplification stage adds to the driving signal a (in the case) programmable common mode voltage V_{CM}.

Linear driver output voltage will be so defined by the following formula:

$$VDO = G_1 \bullet G_2 \bullet (VDP - VDN) + G_2 \cdot V_{CM}$$

being (VDP-VDN) the differential input voltage to the pre-amplification stage.

While the design of the low-voltage blocks relies on well-known architecture, high-voltage driver requires some dedicated solutions to overcome limits and difficulties sometimes related to the high-voltage technology.

A first task to deal with is related to the resistive gain ladder. It is evident that, playing at high voltages, a main contributor for the overall power consumption will be represented by the current consumption across feedback resistance, directly connected to the output voltage.

To reduce as much as possible this contribution, very large resistor value (above 1 MΩ and more) must be implemented. In this context, the choice for the specific technological solution sets a trade-off between several parameters. An important effect to be considered is related to the fact that resistance value has a dependency on voltage across resistor itself and on voltage between terminals and well pocket just below.

Modular approach, already required by basic matching approaches, allows to fix or reduce the first dependency.

For the second one, a possible solution comes from some technological options, such as a metal resistor, where, however, the specific resistance is very low with relevant area penalties, or a shielded resistor, where a shielding layer interposed between resistor and well allows to get an equivalent decoupling strongly reducing the terminal-well voltage dependency. The cost of this solution is, in any case, not negligible: a complex distributed R-C ladder net will be so defined, and, considering the high value for the resistors, several singularities (poles-zeros) will be added in the loop transfer function making critical the stability of the loop itself.

An interesting solution has been found using more traditional poly resistance. As a first step, well pocket can be biased through a direct connection with one terminal resistor so zeroing the voltage drop responsible for the resistor nonlinearity. Nevertheless, similarly to shielded solution, this will add non-negligible parasitic (well-substrate) capacitances across the feedback resistor. A parallel, highly resistive dummy ladder can be inserted in parallel of the effective one, aimed only to provide proper bias to the wells.

Figure 23.5 shows schematically this solution. Dummy ladder is a load for the operational amplifier so not affecting gain and linearity and only increasing a little bit the power consumption.

Feedback capacitor across gain resistor connected between output and inverting input is usually a well-known solution to improve stability by adding a zero into the loop transfer function. Again, working at high voltages, only very low specific capacitance devices are available unless to use large silicon area.

High-voltage operational amplifier topologies are limited to quite basic solution. Two-stage architectures are suitable to have large enough gain (exploiting the possibility to have multiple cascode) and allow to keep moderated noise and power consumption.

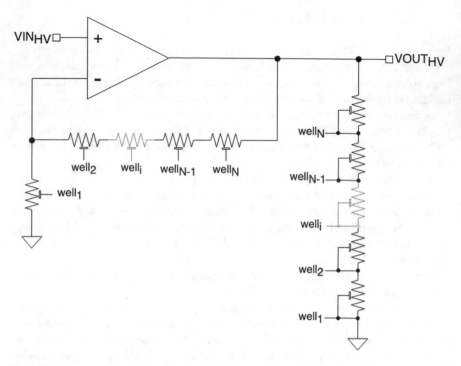

Fig. 23.5 Dummy resistor ladder, designed with very high resistor modules, allows to generate bias voltage for well resistors in order to compensate for terminal to well voltage dependency. In the feedback network, modular topology is mandatory

In Fig. 23.6, a schematic for a high-voltage operational amplifier is depicted.

Single-ended configuration leads to the necessity to manage an input voltage almost rail to rail and, in any case, variable in a range wide enough to require a complementary input stage with obvious disadvantages in terms of noise and distortion.

Relying to a low-voltage domain, well-known solutions such as g_m transconductance equalization can be easily implemented to mitigate non-linearity effects due to input stage.

Since signal bandwidth is limited to few hundreds of Hz, flicker noise is the main contributor for the output noise. Also consider the load can be modeled as second-order transfer function having a resonance not so far from driving signal fundamental and so behaving, at medium-high-frequency like a -40 dB/decade low-pass filter: thermal noise is reduced enough to be less relevant than 1/f noise contribution.

Chopping to filter out low-frequency noise is generally mandatory unless increasing CMOS area and/or using bipolar where possible allows to get required performances. By the way chopping implementation is quite easy and risk-less: moreover, output load ensures by itself to proper filter the modulated noise.

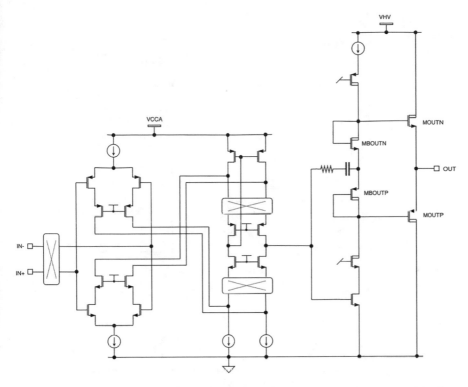

Fig. 23.6 High-voltage output operational amplifier with complementary input stages (circuit to compensate for g_m variation not depicted). Topology is for a class A amplifier with output push-pull buffer to offer low output impedance

One of the preferred solutions for high-voltage operational amplifier output stage is a simple class A architecture. The second stage is then followed by a push-pull unity gain buffer: it is aimed to offer output current capability and decouple output load capacitance so relaxing stage compensation [3],[4].

Miller compensation with RHP zero control must be implemented always keeping in mind not so large capacitor can be implemented. To improve stability without to resort to too much area expensive capacitor, an interesting strategy is to move the 2nd pole at high frequencies sizing large enough the transconductance g_m of the 2nd stage itself. This can be mainly done increasing its bias current: but, instead of to directly increase the current provided by the high-voltage supply referred current generator, a dedicated current bias is provided by a low-voltage supply one with evident benefits on power consumption.

Thanks to this trick, 2nd stage bias current provided by high-voltage supply can be strongly reduced relying of multiplication factor between diode-connected devices MBOUTP and MBOUTN and related output devices MOUTP and MOUTN to set the needed bias current in the output stage.

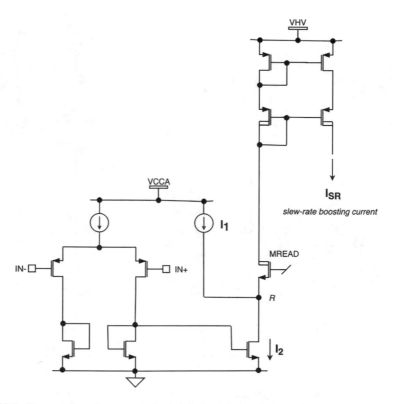

Fig. 23.7 Slew-rate boosting current circuit (single-ended implementation)

A so reduced bias current for the 2nd stage leads however to a limited slew-rate capability for this stage that becomes the bottleneck for the overall slew-rate performances.

This limitation can be exceeded through a slew-rate booster circuit; a simplified principle schematic is presented in Fig. 23.7. The core of the circuit is represented by the subtraction of two currents, the first one I_1 is a fixed bias current while the second one I_2 is a mirror of one current reading an unbalancing of a replica of the input stage. Normally $I_1 > I_2$, net R is clamped one threshold voltage below supply voltage VCCA and cascode device MREAD is off. If the input stage is significantly unbalanced reaching a slewing condition, its replica reads at the same time this unbalancing, I_2 becomes larger than I_1, and the excess current (I_2-I_1) is sinked through the cascode device MREAD to bias an auxiliary mirror that, with an appropriate mirror factor, boosts the bias current of the second stage.

A dedicated attention must be used to manage the turn-off of the high-voltage side. Current-driven level shifters (Fig. 23.8), biased (in DC) with few nA currents, must be inserted to properly ensure the pull-up of critical nets. Additional switched currents can be introduced to speed-up, if required, the switching of the level-shifter output.

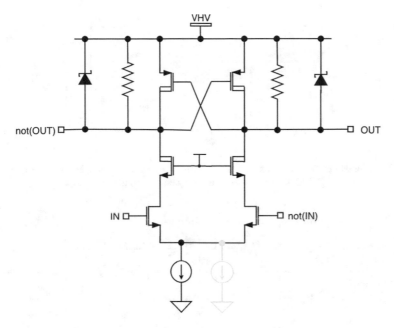

Fig. 23.8 Current-driven level-shifter for high-voltage supply operation

Completely different solution for MEMS micro-mirror led to an electromagnetic actuation. From electrical point of view, a quasi-perfect dualism can be described by comparing electrostatic versus electromagnetic drivers.

So, if on one side, as previously described, electrostatic drivers must be intended to drive an equivalent load capacitor up to very high voltages, and electromagnetic ones must be designed considering the load is practically a coil so equivalently described by a resistor load (in the order of few ohms) and a series inductance, while the resulting output voltage can be kept reasonably low (less of 5 V, usually).

Non-negligible (peak and average) currents must be provided, but, in terms of power consumption, very-high-efficiency drivers can be design by exploiting class D architectures.

Just as an example, Fig. 23.9 shows a full bridge class D driver with a generic coil load.

In fact, from driving point of view, electromagnetic actuators are not so far from a classical audio speaker load, and so, all the several years past developments on PWM driving can be efficiently reused with very few specific technical precautions.

Maybe the main difference that must be considered with respect to the class D design for audio purposes relies again to the flicker noise, here to be properly managed since effectively seen in-bandwidth.

Both open-loop and closed-loop class D architecture can be case-by-case evaluated with related pros and contra.

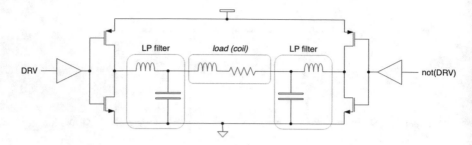

Fig. 23.9 Class-D driver for electromagnetic mirror. Load can be modeled as a series resistor-inductor. Low-pass LC filter suppresses for high-frequency modulation tones

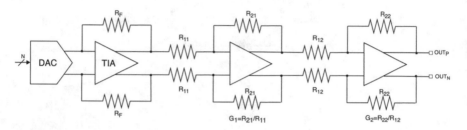

Fig. 23.10 Fully differential driving chain for piezoelectric linear micro-mirrors

More recent and new micro-mirrors solutions refer to piezoelectric actuation. While piezoelectric physics is completely different from electrostatic one, for what concern the electrical driving, a lot of the considerations previously done in that domain can be here recovered [5].

In fact, for the driver, piezoelectric load behaves again like a capacitor, and, to be more precise, it is also here possible to model the load like a 2nd-order resonant filter.

The main differences are related to the capacitance value, more than 10 nF, usually, for linear micro-mirrors and to the driving voltage now limited to maximum values in the order of 40 V.

On top of that, fully differential driving solutions have also been implemented with relevant advantages in the high-voltage linear driver design.

Figure 23.10 shows an example for a piezoelectric load linear driving chain (control loop is not here depicted). A fully differential topology is kept for all the paths.

As usual, to optimize the chain noise performances, all the possible gain compatible with the operational amplified headroom must be in the first stages so that the high-voltage linear driver will have the minimum gain needed to fulfill the output voltage swing required.

Some gain programmability can be introduced in one or more of the pre-amplification stages. This will help avoiding any gain programmability in the

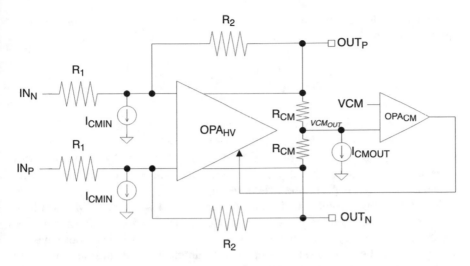

Fig. 23.11 Input/output common mode setting for high-voltage operational amplifier

high-voltage stage since it is not so trivial and often requiring (expensive) high-voltage switches to deal with safe operating area (SOA) constraints.

To properly fix output common mode, a dedicated common mode feedback must be introduced. Well-known solution to get common mode voltage through a simple resistive partitioning (R_{CM}) must be adapted as shown in Fig. 23.11 to develop this control loop in low-voltage supply domain both for simplicity and power consumption optimization. A current generator sinks a current I_{CMOUT} from the intermediate node VCM_{OUT}. When the common mode loop is properly closed, this intermediate node will set to VCM (a low-voltage reference) and the output common mode will be defined by the formula:

$$VCM_{OUT} = \frac{1}{2} \cdot R_{CM} \cdot I_{CMOUT} + VCM$$

As a good design practice, common mode current I_{CMOUT} can be got through a simple voltage-to-current circuit starting from the same VCM reference voltage and based on a matched fraction of the common mode resistance R_{CM}. By this way, output common voltage VCM_{OUT} will be precisely defined (in the limits of matching performances) as a (programmable) multiple of the reference voltage VCM itself.

Always in Fig. 23.11, two additional current generators are present at the operational amplifier inputs. They are aimed to adjust the input common mode with respect to the selected output common mode.

Their value I_{CMIN} can be calculated to keep the input common mode equal to output common mode of the last pre-amplification stage. Assuming for instance that the previous stage has an output common mode equal to VCM, it is enough to set those currents to

$$I_{CMIN} = \frac{VCM_{OUT} - VCM}{R_2}$$

Implicitly assuming all the resistors (R_1, R_2, and R_{CM}) are designed starting from the same unity module to get good matching results and remembering that for what previously described VCM_{OUT} is defined as a multiple of VCM, it results that I_{CMIN} can also be advantageously got from the same voltage-to-current circuit (with proper and eventually programmable mirroring).

To set the input common mode equal or as close as possible to VCM has at least two benefits. It reduces the power consumption of the previous stage (zeroing the DC current it must provide to the next one), and moreover, avoiding requiring a full swing signal to be managed by the operational amplifier input stage also allows to have a single-type input stage itself.

Figure 23.12 shows the schematic of a fully differential high-voltage operational amplifier suitable for piezoelectric linear mirror driving.

On top of similar design considerations done for the electrostatic case, the most relevant task in that case is related to the large capacitive load to be sustained. Even increasing a lot output stage bias current (with unacceptable power consumption effect) to reduce the output impedance of the operational amplifier, the pole given by

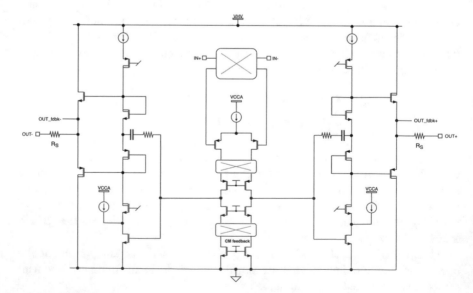

Fig. 23.12 High-voltage fully differential operational amplifier. If allowed by low-voltage supply VCCA, telescopic cascode can be designed for 1st stage, while folded cascode will be suitable in case of reduced headroom

this load and the output impedance itself will be inevitably located into the closed-loop bandwidth, strongly affecting its stability.

Immediate solution to face with this task is to compensate with an output series resistor R_S large enough (100 Ω is a reasonable order of magnitude) to decouple the load capacitance. Said R_{OUT} the output impedance of the operational amplifier and C_L the load capacitance, it possible to calculate that the closed gain $G_{LOOP}(s)$ will be modified accordingly to the formula:

$$G^*_{LOOP}(s) = G_{LOOP}(s) \cdot \frac{1 + s \cdot C_L \cdot R_S}{1 + s \cdot C_L \cdot (R_S + R_{OUT})}$$

So, a zero is also introduced recovering part of the phase margin lost for the pole.

Series resistance compensation emphasized the fact, even if a low-frequency signal must be still managed, strongly limited bandwidth leads to relevant distortion effects on driving signal harmonic components.

This is not a problem in general, relying on a control loop system to minimize any nonlinear component of the system.

One of the disadvantages of that solution is related to headroom losses both to ground and high-voltage supply. In fact, output saturation will start to occur about a threshold voltage (of a high-voltage device) before to reach ground or supply.

This translated into an efficiency loss requiring more supply voltage that strictly needed to provide proper differential voltage to the load.

Improved output stage keeps actual topology putting in parallel an equivalent AB output stage able to move the operational amplifier saturation closest to ground and/or supply.

For MEMS micro-mirror actuation, both electrostatic and piezoelectric, linear driver solution based on high-voltage operation amplifier is today an easy and quite reliable solution. Power consumption $C \cdot V^2 \cdot f$ due to the load is not so relevant mainly due to very low-frequency driving signal: quiescent driver consumption dominates the overall power budget.

Even without evidence that the frequency can be increased soon, the research is quite active to move towards different kind of driver where some charge recovery approach can be implemented.

This is mainly driven by a different kind of actuators used to implement MEMS speakers. They are like electrostatic load, to be driven into a quasi-linear way, but, due to MEMS topology, equivalent output load can be in the order of 1 nF and actuation voltage close to 40 V or more. In addition, on that, bandwidth to be managed here is obviously defined by audio requirements (so up to 20 kHz, theoretically). Load $C \cdot V^2 \cdot f$ power consumption becomes consequently the main contribution of the overall consumption.

Recent research on large bandwidth boost converter could open an important field of development for such kind of driver. Weber circuit-based solution starts to be developed.

An example is shown in Fig. 23.13.

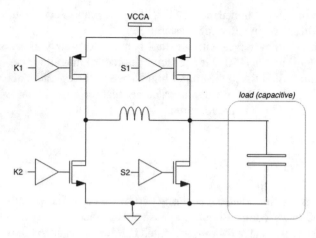

Fig. 23.13 Inductor based driver for piezoelectric actuators

23.1.2 Resonant Driver Architectures

So-called resonant micro-mirrors are intended to be operated to exactly exploit their resonance frequency f_R where properly sized quality factor Q allows to optimize the energy transfer efficiency from electrical to mechanical domain.

The same high-level block diagram previously presented in Fig. 23.2 is also valid for a resonant micro-mirror system.

Resonant driver must generally provide a square (or better trapezoidal) waveform at the exactly micro-mirror resonance frequency to sustain the resonance itself against mechanical losses that would damp the oscillations [6, 7].

Control loop must provide both a control for the driving frequency that can slowly vary a little bit due to environmental condition (such as external pressure) and a control for the mechanical opening, mainly adjusting the amplitude of the driving signal (directly linked to the actuation energy provided to the micro-mirror).

According to that, resonant driver is basically a buffer for an input digital signal generated by control micro-processor: buffer output amplitude must then be (low frequency) updated according to opening angle control mechanism (often named OAC).

Amplitude control requires an accuracy dependent on the precision according to which mechanical angle must be controlled. For typical actual application, it means something in the order of 14–15 bits across the whole full-scale range.

For what concern the performance required to such kind of driving signal, jitter is for sure the most important one, since the equivalent noise will affect mechanical movement. Clearly, spatial definition accuracy will be defined by absolute jitter performances.

Assuming the input driving signal clean enough in terms of noise (precise fractional PLL can be at the basis of that signal), resonant driver design must be addressed to minimize both the effective switching jitter and the noise of the reference voltage used to set the output signal amplitude.

Resonant frequency depends on micro-mirror design according to specific application to be tailored. Considering for instance generic raster scan application, to get adequate video resolution, resonant micro-mirror deals with resonance frequencies in the order of 20–30 kHz, while, for different application, even very low frequencies (below 1 kHz) or higher ones (up to 70 kHz) can be addressed.

Similarly, to their linear counterpart, resonant electrostatic micro-mirror can be modeled as a 2nd-order resonant transfer function, and, for what concern the load to be driven, it is no more than a few tens pF capacitor. Driving voltage again is quite high, up to 180 V. Due to quite load value of the load capacitance, $C \cdot V^2 \cdot f$ power consumption related to the load itself is not so high, allowing to address the design for a classical high-voltage buffer. Additional request is to allow rising and falling edges slew rate. MEMS micro-mirror in fact, in addition to their fundamental resonance mode, has several other resonance modes practically in all the XYZ mechanical directions.

Energy spectrum provided to the micro-mirror by the driving signal must be optimized to excite as less as possible any spurious mode. Wide range energy is located on fast edges so that slewing control is of relevant utility to mitigate unwanted excitation effects.

Figure 23.14 illustrates an example topology for a high-voltage buffer.

Circuit is basically a high-voltage supplied inverter configured into an integrator-like topology. High-voltage inverter is built in an HV N-type device M_N and an

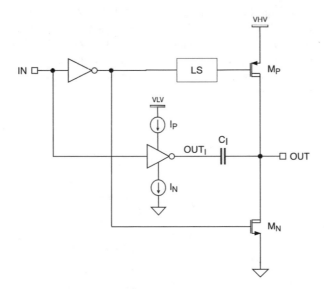

Fig. 23.14 Slew-rate programmable driver for resonant micro-mirrors. LS stands for a level shifter able to properly shift up to VHV domain the driving signal

HV P-type one M_P. M_N and M_P gate are obviously not directly connected but a level shifter (LS) is needed to translate the digital input to high-voltage domain for M_P driving. So, the digital input drives (through an inverter/buffer) the high-voltage output inverter directly for the ground referred device (M_N) and properly translated for the VHV referred one. A feedback integration capacitor C_I is connected between the output of the high-voltage inverter and an internal (OUT_I, low-voltage net) that is charged/discharged through fixed currents (I_P and I_N) because of the same digital input.

At the beginning of a digital input rising edge (IN = 0), M_N is turned on, M_P is turned off, and output voltage (OUT) is tied to ground, while the current generator I_P is keeping internal net OUT_I close to the low-voltage supply (VLV). When IN signal switches to high logical state (IN→1), M_N is switched off, M_P is switched on, and current generator I_P is un-connected form the internal net OUT_I, while its counterpart I_N starts to discharge that internal net. Output voltage OUT will rise up to VHV with a slope (I_N/C_I) limited (and definable) by integration capacitor charge.

Dually, falling edge will be managed at the same way, reversing the role of switches and current generator between upper and lower side.

Current value I for the generators (I_P, I_N) can be made programmable to have the possibility to vary positive and negative slew rate.

Current generators (I_P, I_N) and output devices (M_N, M_P) must be sized to get required noise performances leading to acceptable jitter.

To control the amplitude of the driving signal, resonant driver output buffer is supplied through a programmable reference voltage.

General topology (see Fig. 23.15) includes a low-voltage DAC providing a reference and then amplified by a high-voltage (fixed gain) operational amplifier.

Since the main parameter to be satisfied refers again to the noise, and since for the noise the low-frequency contributions are often dominant, resistive DACs are a good choice. Also, it must be considered amplitude update is managed by the control loop to follow very low-frequency effects, so limited bandwidth is required to amplitude control blocks.

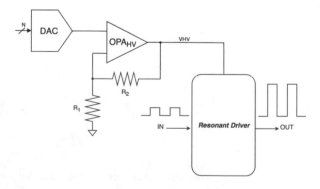

Fig. 23.15 Resonant driver amplitude regulation conceptual schematic

Trade-off between area, matching, and consumption from one side and noise on the other side has to be considered in any case for DAC topology choice: even steering current (both Nyquist and $\Sigma\Delta$) can be evaluated as reliable solutions for such kind of required resolutions.

Reference voltage amplifier OPA_{HV} can be designed according to the same single-ended topology already discussed for linear driver. A pure LDO topology is generally not suitable for this purpose because of too much low constant time to reduce output voltage.

Electromagnetic actuation for resonant micro-mirrors starts from the same considerations done for linear ones. From electrical point of view, no more than a load resistor in series with an inductor must be driven at quite low voltages. Simple digital-like buffer with appropriate current capability is the reference design for those applications.

As for piezoelectric linear micro-mirrors, also the resonant ones require moderate actuation voltages (up to 40 V) with an equivalent load capacitance to be driven in the order of 1 nF or more.

So, for these new solutions, load $C \cdot V^2 \cdot f$ power consumption becomes relevant.

This aimed the research to move towards different topologies, trying to exploit adiabatic mechanism to implement charge recovery approach [8, 9].

The key point that leads to the $C \cdot V^2 \cdot f$ expression is that classical pulse driver will discharge, each half-period, the load charge to ground, completely losing it.

Basic idea is to charge/discharge the actuator (modeled through a pure load capacitor for simplicity) moving across N voltage steps from ground to voltage supply VHV (and vice versa), each i-point of the step defined by a tank capacitor storing a voltage $(i/N) \cdot VHV$, capacitor large enough to behave like a voltage generator.

Figure 23.16 can help to understand the mechanism.

During the rising edge, for instance, actuator capacitor C_A, resonant micro mirror equivalent capacitor moves from ground to voltage supply VHV being connected, at each step, to a tank capacitor C_T (C_1 or C_2 in the Fig. 23.16). Assuming this tank capacitor is bigger than C_A, the charge sharing occurring during their connection will allow to charge C_A quite close to the voltage value (previously) stored into the tank capacitor. Charge sharing formula is the following:

$$V_T^* = V_i = \frac{C_T \cdot V_T + C_A \cdot V_{i-1}}{C_T + C_A}$$

where V_T and V_{i-1} are the initial voltages, respectively, on the tank and actuator capacitors, while V_T^* and V_i are the related final voltages: assuming the charge sharing transient fully completed, it will be $V_T^* = V_i$.

Tank capacitor will lose (small) part of its charge to charge up the actuator capacitor.

During the falling edge, when the actuator comes back to the same tank capacitor, it will give back (approximatively) the charge previously taken from that.

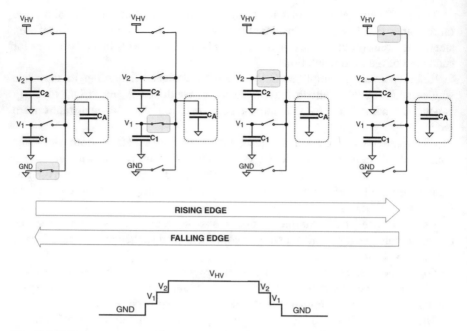

Fig. 23.16 Charge recovery mechanism

So, for each step of the rising from a generic tap *(i-1)-th* to the following *(i)-th*, the *(i)-th* tank capacitor will provide a charge to the actuator that will be given back during the falling edge at the step from *(i + 1)-th* to *(i)-th*. Only on the step to ground the system will lose the charge on the actuator and it will be recovered, dually, when the actuator is connected to the voltage supply VHV at the end of the rising edge.

For what purely concern the load power consumption, considering the charge lost on the last falling edge step and the power lost during each charge sharing, it's possible to demonstrate a reduction by a factor N with respect to the value expected for a classical pulse driver:

$$P^* = \frac{C_A \cdot VHV^2 \cdot f}{N}$$

Previous relationship is valid only if, during each step, charge sharing transient will settle within at least 4 constant times. Incomplete transient leads to a progressive degradation of the charge recovery factor, so reducing the attenuation factor below the value N.

Design rule asks to satisfy the condition:

$$R \cdot C_A < \frac{1}{4} \cdot \frac{\alpha}{2 \cdot N \cdot f}$$

where R is the resistance of the switch to connect actuator to tank capacitor and α is the percentage of each half-period devoted to rise (or fall) time.

Rise and fall time can be programmed playing, for a specific N, on the step allocated time, always keeping previously highlighted rule. Time step can be defined as number of periods of a higher frequency clock.

To exploit (1/N) reduction, N must be chosen properly large, but this requires the usage of a related number of tank capacitors, the value of which (as a rule of thumb, let us assume 100 times bigger than actuator one) does not allow silicon integration.

Since usually for micro-mirrors two driving signals in phase opposition must be provided, intermediate level at value (VHV/2) can be got for free without any extra-tank capacitor only connecting the two actuators together. On top of that, having to be N even for obvious symmetry reason, number T of tank capacitor is given by:

$$T = N - 2 \ (N \geq 2)$$

Increasing the number N of levels will lead also to require more and more a small value or the switch resistance. Therefore, consumption to drive the switches themselves will increase till to represent a contribution that frustrates (1/N) reduction.

Switch design must be compatible with high-voltage operation at both terminals. Dedicated floating switch (Fig. 23.17), built with the usage of appropriate high-voltage class devices, allows to deal with SOA.

Driving on of the switch is done providing and sinking a current I through the generators (I_P, I_N) to activate the Zener diode D_Z: it sets V_{GS} voltage for the series devices M_{SW1} and M_{SW2}.

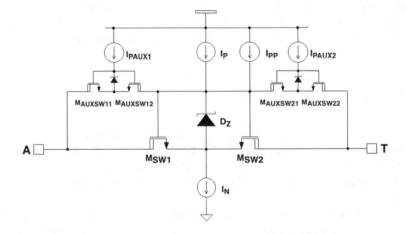

Fig. 23.17 High-voltage bidirectional switch for charge recovery resonant driver

Series devices must be sized, as anticipated, to figure out a low enough on-resistance. In addition of working at the maximum V_{GS} compatible with technology operating voltage, increasing size factor (W/L) remains the only way to reduce on-resistance.

Nevertheless, going in that direction, time to charge up gate switches parasitic capacitances through current generators (I_P, I_N) takes relevant portion of the settling time so frustrating the wanted on-resistance reduction effect.

To increase bias current is not a reliable solution, both for power consumption and also for SOA constraints.

A pre-charge solution is so implemented to speed up the turning-on time of the switches. Two additional auxiliary switches ($M_{SWAUX11}$, $M_{SWAUX12}$) and ($M_{SWAUX21}$, $M_{SWAUX22}$), with the same topology of the main one, are activated as soon as the switch must be turned on: they allow to quickly charge the gate capacitance using the low impedance provided by the tank and actuator capacitors. Auxiliary switches are kept on for a monostable period just enough to complete gate parasitic capacitance charging. In parallel, always to help a main switch fast turn-on, a current pulse I_{PP} is also provided in parallel to the bias one I_P.

Charge recovery operation is self-sustaining and self-stabilizing. However, a long start-up will be required if tank capacitors must be charged to their respective values through the charge recovery operation itself. A voltage divider, kept on for a limited time and partitioning the supply voltage in the same (i/N) •VHV levels, quickly charges the tank capacitors close to their proper values. Switches like the main switch above described are used to connect tank capacitors to related voltage divider taps.

Starting from the charge recovery method explained, several different options can be evaluated.

Once the working principle that is based on simple charge sharing management is understood, interesting solutions where tank capacitors are not referred to ground but "floating" between two levels have been implemented. A trade-off between tank capacitor number and main switches number to manage all the possible connection must be considered for area/external components/power consumption evaluation.

23.2 MEMS Mirror Sensing

Pure actuation of a MEMS mirror is not enough to provide reliable mechanical movement. As already highlighted and better discussed in the next session, both for linear and resonant solutions, a control loop must be implemented to get required output overcoming specific non-linearities and non-idealities of the loads themselves.

For that, a feedback from the mechanical movement must be extracted from the structure to the electrical domain, converted and elaborated to monitor how the movement itself is happening.

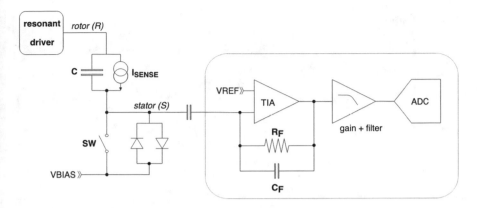

Fig. 23.18 Readout circuit for electrostatic resonant micro-mirror system

Different solutions have been implemented as sensing solution, each one with its peculiarities.

23.2.1   Capacitive Sensing

Electrostatic MEMS micro-mirrors have been suitable to implement capacitive sensing approaches.

Figure 23.18 shows an example of the equivalent circuit for a possible solution to readout information related to the movement of a resonant micro-mirror.

From the same capacitor previously considered as load for the resonant driver, the information is extracted. This capacitor appears connected on one terminal (rotor, R) to the resonant driver output and on the other terminal (stator, S) to a generic voltage reference V_{BIAS} through a switch SW and a couple of back-to-back diodes and also (AC coupled) to the input of the sensing circuit.

According to MEMS micro-mirror mechanical design, capacitor value depends (slightly) on movement. This means that capacitance variation (dC/dt) will induce an equivalent current I_{SENSE} also including micro-mirror movement information:

$$I_{SENSE} \propto \frac{dC}{dt} \cdot V_C$$

where V_C is the (assumed fixed) voltage across the capacitor.

Turning off the switch SW, node S will be high impedance biased to (about) the same reference voltage as before and the current I_{SENSE} can flow into the sensing circuit.

To get required information on resonant movement, this periodic (not continuous) sensing approach is often enough.

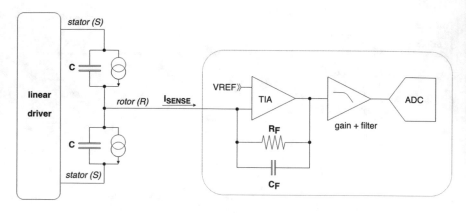

Fig. 23.19 Readout circuit for electrostatic linear micro-mirror system

For the linear micro-mirror, the sensing configuration that can be used is shown in the circuit of Fig. 23.19.

Equivalent load capacitors are here connected from linear driver output (stators, S1 and S2) and a common floating node (rotor, R) directly connected to the sensing circuit input: input that so also provides a low impedance to bias the node R itself.

Here, a continuous sensing of the linear movement is made by reading the current I_{SENSE} flowing into the actuation capacitor:

$$I_{SENSE} \propto C \cdot \frac{dV_C}{dt} + V_C \cdot \frac{dC}{dt}$$

As for resonant case, capacitance variation (dC/dt) contains the effective movement information to be read, while an additional useless current $C \cdot (dV_C/dt)$ is sent to the sensing circuit.

Independently on different equivalent input circuit architectures, from electrical point of view, capacitive sensing deals with the readout of a (small) input current.

Transimpedance amplifier (TIA) architecture, shown in Figs. 23.18 and 23.19 in a single-ended configuration, is a well-known solution for this kind of task [10].

A first operational amplifier stage reads the input current on a virtual ground and converts it into a voltage output through the gain feedback resistance R_F. Following stages are aimed to provide additional gain and/or filter before to present output voltage to an ADC for analog-to-digital conversion. Acquired digital data can then be processed extracting the movement information.

Since a major part of the gain is generally provided by the first stage, following stages do not have critical design requirements.

On the contrary, first stage requires to be carefully designed with respect to several items [11]. A challenge is often represented by the fact that a non-negligible capacitance (the actuator one and PCB, MEMS, additional parasitic) results to be connected to the input virtual ground.

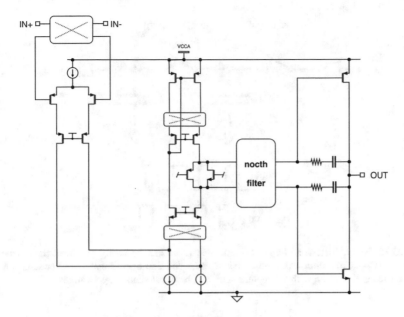

Fig. 23.20 2-stage operational amplifier (AB output class)

At first approximation, gain loop G_{LOOP} is calculated by the formula:

$$G_{LOOP} = -A(s) \cdot \frac{1 + s \cdot C_F \cdot R_F}{1 + s \cdot (C_F + C_I) \cdot R_F}$$

where C_I is the overall capacitance seen at the TIA input virtual ground; C_F is additional feedback capacitance, useful for G_{LOOP} stabilization; and A(s) is the operational amplifier transfer function.

Values and variability to be considered both for C_F and R_F lead to require large enough bandwidth for the operational amplifier: let's consider also R_F has quite large value. Operational amplifier DC gain can be generally moderate.

Two (or three)-stage class AB well-known topology operational amplifier (i.e., Figure 23.20) can be chosen [12].

Noise performances of the first stage are quite challenging, considering input (current) signal has small dynamic range (few μA). In case of sensing circuit tailored for linear mirror where the input signal has low-frequency, chopper amplifier is mandatory to filter out flicker and shot noise contributions [13].

Unless to use high-frequency f_S for chopper amplifier (not suitable for operational amplifier design and power consumption), noise modulation (seen for instance as output ripple) can be quite large, leading to an output dynamic issue.

Introduction of related notch filter for modulated noise is an interesting option to minimize this effect. Practically, simple switched-capacitor circuit allows to

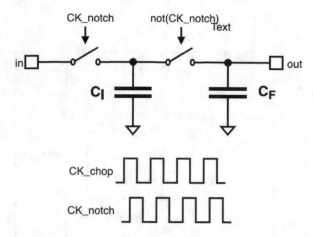

Fig. 23.21 Notch filter for chopping modulated noise can be easily got through a switched-capacitor ($C_I \ll C_F$) solution interposed between 1st and 2nd operational amplifier stages. Notch filter switches are driven with 90 degrees shift with respect to chopping switches

introduce a zero at the same chopping frequency f_S, reducing output effects. Figure 23.21 shows an example for implementation of notch filter in the previously presented operational amplifier.

Due to frequency signal to be managed, both for linear and resonant options, following stages can only provide moderate filter function, unless to introduce expensive external components.

Nevertheless, adequate filtering and signal processing can be easily moved in digital domain, as soon as analog path output is converted through an ADC.

Again, input signal low-frequency domain and resolution to be provided lead to oversampled $\Sigma\Delta$ ADC as preferred solution for analog-to-digital conversion. Continuous-time (CT) implementation allows to avoid alias effect and to reduce power consumption.

Mixed solution, in which only the first integrator stage keeps continuous-time topology while the following integrators are designed through switched capacitors architectures, gets the benefits of both the implementations.

Figure 23.22 illustrates a block diagram for a 2nd-order mixed CT + SC $\Sigma\Delta$ ADC: tracking comparators architectures reduces more and more consumption with small complexity increasing. Converted output is finally processed in digital domain both to provide efficient filtering (i.e., bi-quad (BQ) filters can follow decimator) and also to directly extract data required for system control loop.

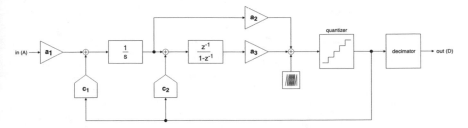

Fig. 23.22 Block diagram for a 2nd-order (mixed) $\Sigma\Delta$. While 1st integrator and related feedback DAC (c_1) refer to a continuous-time topology, 2nd stage and DAC (c_2) follow a sampled (discrete) architecture

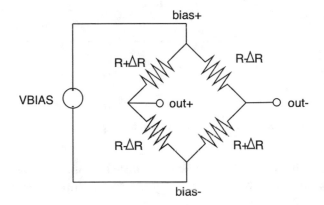

Fig. 23.23 Whetstone bridge translates mechanical stress in a resistor mismatch ($+/-\Delta R$) and, when properly biased, it leads to a differential voltage between output terminals

23.2.2 Resistive Sensing

Electromagnetic and piezoelectric micro-mirror architectures led to different sensing approaches, where, instead of a capacitive sensing architecture, a resistive sensing is in general implemented.

A Wheatstone resistive bridge can be implemented in the micro-mirror structure: bridge resistors experience piezoelectric effect, showing variation in their value according to mechanical stress.

A soon as properly biased, an appropriate layout of the Wheatstone bridge in the micro-mirror allows to get an output differential signal (see Fig. 23.23) representative of the mechanical stress.

The key parameter of such kind of resistive bridge is its sensitivity S. Sensitivity is defined by geometrical parameters: in practice, it is the "capability" to have a mismatch ($\Delta R/R$) in the modules of the Wheatstone bridge consequently to a mechanical stress.

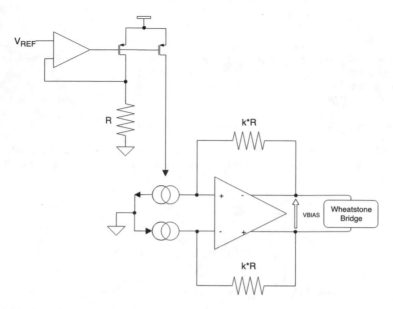

Fig. 23.24 Fully differential bias circuit for Wheatstone bridge

Said VBIAS the bias voltage applied to bridge and ϑ the mechanical movement (i.e., the micro-mirror opening angle), voltage signal V_{OUT} at the output of the bridge itself is given by the following relationship:

$$V\,OUT = V_{BIAS} \cdot S \cdot \vartheta$$

V_{BIAS} must be sized as large as possible, with respect to available supplies headroom, to maximize overall input signal.

Fully differential topology both for bridge bias and sensing circuit is a good approach for noise immunity, PSRR, and any other well-known single-ended-driven worsening effects.

In Fig. 23.24 an example for bridge bias circuit is presented: it's based on a current-to-voltage feedback. Starting from a (precise) reference voltage V_{REF}, a differential current is generated from a voltage-to-current circuit (V_{REF}/R) and then feeds to the same type of feedback resistors: playing with modularity, a differential voltage proportional to the reference one is output to bias the load bridge.

VBIAS voltage noise must obviously be reduced, being directly transferred at the output through the sensitivity gain. Low-frequency contributions are the most critical ones, leading to usual approaches to face with (i.e., chopper amplifier).

As for capacitive sensing, readout circuit is based on an analog fronted followed by ADC and digital processing, both for linear and resonant options.

A preferred solution for analog front-end is based on instrumentation amplifier (INA) as illustrated in Fig. 23.25. The 1st stage is based on two single-ended

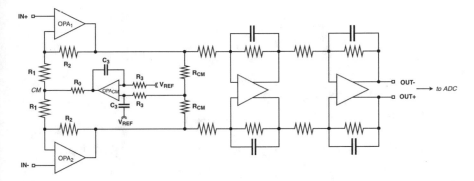

Fig. 23.25 Instrumentation amplifier (INA) architecture for piezoresistive bridge readout

operational amplifiers (OPA$_1$ and OPA$_2$) and provides a gain $G_1 = (1 + R_2/R_1)$. Following stages are introduced to provide additional gain, filter, and drive the ADC.

Basic two/three-stage operational amplifier topologies can be used.

Wheatstone bridge differential bias circuit has, depending on bridge resistors matching, a quite well-defined input common mode. Single-type input stage can be designed for the 1st stage operational amplifiers, also considering that, due to limited sensitivity value, input signal dynamic range is very small.

To control the 1st stage output common mode, a dedicated loop can be introduced through the operational amplifier OPA$_{CM}$ and related local feedback network. This loop is aimed to provide a DC/low-frequency control of the output common mode, usually extracted through a simple voltage divider; loop controls the bias voltage of the intermediate node CM between gain resistors R$_1$.

Wheatstone bridge has often poor matching performances with consequent offset and common mode shift of the output signal: for instance, DC offset can be of the same order of magnitude of the effective input signal full scale.

Although offset can be canceled or managed in digital domain, a too large value would limit operational amplifier dynamic range. Calibration is so required, and this can be accomplished for instance through a differential current DAC by injecting/sinking (differentially) a current, I_{OS}, to/from the output of the bridge (out+ and out- in Fig. 23.23). It is easy to see that a correction voltage ΔV_{OS} can be applied:

$$\Delta V_{OS} = R \cdot I_{OS}$$

where R is value of resistor bridge module.

To compensate for process effect and keep as constant as possible offset cancellation, correction current I_{OS} must be generated starting from a seed based on the same bridge resistor type.

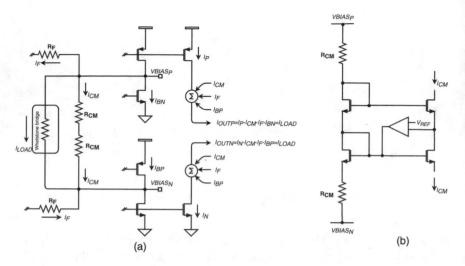

Fig. 23.26 Readout circuit for Whetstone bridge (bias) current (**a**) and simple solution to mirror common mode resistor bias current (**b**) R_F is the feedback resistor for the bias circuit (Fig. 23.24) while R_{CM} is its common mode feedback resistor. Resulting current I_{OUTP} and I_{OUTN} are the seed for the offset current (IOS) to be sunk/injected in the Wheatstone bridge for its offset cancellation

Once compensated for bias, feedback and common mode contributions (Fig. 23.26 (a)), current flowing in the output devices of the bias circuit is exactly representative for the ratio V_{REF}/R and so they can be mirrored providing the bias reference (seed) for I_{OS}.

Feedback I_F and bias I_B currents are known, and they can be directly subtracted. A simple circuit (Fig. 23.26b) allows to also extract a copy of the current flowing in the output common mode voltage divider.

Offset DAC low-frequency noise must be obviously addressed an important design parameter to be reduced. Large scale excitation (LSE), scrambling, and very low-pass frequency filter for seed current, such as, if available, the usage for bipolar devices, can be evaluated.

While it is important for dynamic range reasons to have a coarse offset cancellation directly applied to the resistive bridge, additional fine correction can be implemented in the last operational amplifier stage of the analog chain.

Wheatstone bridge resistors mismatch also gives a shift of the output common mode. This common mode is only mathematically defined in the middle point of the differential output voltage, but it is dependent on resistor matching. Also, common mode shift must be reduced to simplify 1st stage operational amplifier design.

Similar calibration can be implemented through a common mode cancellation DAC.

A switched capacitor (SC) analog front-end is an option for design (Fig. 23.27) even if anti-alias filter is needed and area efficiency on capacitor implementation

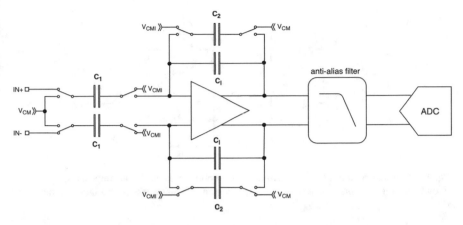

Fig. 23.27 Switched capacitor (SC) analog front-end. Any continuous time functional block, like offset cancellation, gain programmability, and son on can be designed accordingly to its SC counterpart

(i.e., considering gain programmability often required) can be a drawback to be managed.

On top of that, design considerations are basically the same as for continuous time solution: only gain/filter network is "translated" on switched capacitor counterparts.

For low noise filter, moreover, correlated double sampling (CDS) is an additional resource, to be implemented for free, for improving performances.

Resistive sensing suffers from a critical issue related to the sensitivity dependency on temperature. At least a second-order dependency is highlighted:

$$S(T) = S_0 \cdot \left[1 + \beta_1 \cdot \Delta T + \beta_2 \cdot \Delta T^2\right]$$

where S0 is the sensitivity value at reference temperature, ΔT is the difference with respect to reference temperature, and $\beta 1$ and $\beta 2$ are the linear and quadratic coefficients describing the dependency.

Calibration allows to extract $\beta 1$ and $\beta 2$ parameters to compensate for temperature variation. Nevertheless, resistor bridge temperature must be known to complete the equation.

Different solutions have been explored to fulfill this task.

More classical one requires to have the micro-mirror design, as close as possible to the mechanical bridge, another resistive bridge with at least one resistor with a temperature coefficient (to be calibrated too) for thermal measurements.

In that case, two (couple of) signals will come from MEMS device, one for mechanical output and one for thermal one. To save area and consumption, the same analog front-ended can be multiplexed (and eventually properly rearranged).

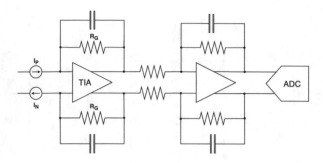

Fig. 23.28 Readout circuit for Wheatstone bridge replica currents. A transimpedance amplifier is followed by (at least one, as in the figure) filtering/amplification stages and then inputted to an ADC

Complexity in MEMS design and precision on temperature measurement are the main limits of this approaches.

A different solution has been found using only mechanical Wheatstone bridge. It is based on the fact also that the resistance of the bridge shows a temperature dependency like the one highlighted for the sensitivity:

$$R(T) = R_0 \cdot \left[1 + a_1 \cdot \Delta T + a_2 \cdot \Delta T^2\right]$$

where R0 is the module resistor value at reference temperature, ΔT is the difference with respect to reference temperature, and $\alpha 1$ and $\alpha 2$ are the linear and quadratic coefficients describing the dependency.

Since the bridge is biased to a fixed VBIAS voltage, extracting the value of the related bias current (flowing into the bridge) will lead to know (after digital processing) the resistance value. So, monitoring that current and knowing (from calibration) the parameters $\alpha 1$ and $\alpha 2$, the temperature shift ΔT can be calculated.

Similar to what has been discussed for offset and common mode cancellation, current flowing into the bridge is already available from the bias circuit.

A dedicated front-end can be designed to read out this current (Fig. 23.28). It is based on a transimpedance amplifier (TIA). A mirrored version of the bridge current is inputted, keeping a differential approach, to the virtual grounds of the TIA, translated to a voltage through a gain resistance RG, amplified and filtered by additional stages and converted through an ADC.

Complexity is so moved from MEMS design to read out circuit, basically requiring a dedicated path even if sharing (with reconfiguration) of analog front-end and ADC is possible.

To get enough resolution on temperature measurement, quite high (at least 16-bits) accuracy is requested to this front-end. Oversampled ADC is again the best solution, considering, in addition, that temperature can be considered a DC-like signal.

By the way, even SAR ADC can be used to save power consumption: 12-bits 14-bits design is not so challenging. Additional resolution can be got through dithering and averaging techniques in digital domain.

About digital filtering, an important aspect to be highlighted is related to the fact that current output will include spectral components of the effective sensing input. For instance, in case of sensing circuit for a resonant mirror, a tone at the resonance frequency will appear in the current spectrum. This must be considered in the design of the digital filter implemented to extract (only) the DC/low-frequency component related to the temperature variation.

Sensitivity compensation against temperature variation is a critical task. MEMS micro-mirror in fact changes its performances (mainly quality factor Q) depending on system temperature. So, sensing circuit cannot be affected by its own temperature drifts unless to confuse an actuation drift (to be compensated in the control loop since related to the effective movement) with a sensing one.

If on one side sensitivity compensation is the first task to face with, on the other side, all the sensing circuit must have performances temperature independent.

For instance, the problem is quite critical for resonant micro-mirror sensing. That sensing must be able to provide a temperature-stable information related to the phase of the micro-mirror oscillation. Sensing must in fact track the oscillation to provide feedback, to be managed by the control loop, about an increase or decrease of the resonance frequency.

It is clear that the transfer function of the analog front-end must have gain and phase as independent as possible from temperature.

Phase shift leads to not so trivial consequences. Assuming for simplicity that the transfer function of the analog frond-end can be modeled as 1st-order one with dominant pole frequency f0:

$$A(s) = \frac{A_0}{1 + \frac{s}{2 \cdot \pi \cdot f_o}}$$

Transfer function phase is given by

$$\Phi(f) = arctg\left(\frac{f}{f_0}\right)$$

In the time domain, output waveform will have a shift driven by a phase shift in the transfer function given by

$$\Delta t = \frac{1}{2 \cdot \pi \cdot f} \cdot \frac{d\Phi(f_0)}{df_0} \propto \frac{1}{f_0} \cdot \frac{1}{1 + \left(\frac{f}{f_0}\right)^2}$$

In the previous formula only the phase shift related to the shift of the dominant pole frequency f0 has been considered. Real situation is a little bit more complex since 1st-order transfer function model can be too much approximated, but general trend can be brought back to this model.

Apparently small variation in f0 frequency determines important shift in the time domain with an order of magnitude comparable with the spatial resolution required in the projection system.

While absolute value of the phase shift is not issue since it is usually calibrated, its variation, the main cause of which is related to temperature dependencies, is not and must be minimized.

To get this goal, several design recommendations must be followed.

First, a wide-bandwidth operational amplifier is required, then, since dominant pole in the transfer function is generally defined by a RC constant time related to the feedback network, this constant time must be as temperature independent as possible. While several options for integrated capacitor without temperature coefficient are usually available, this is not for resistors. Some flavors with reduced temperature coefficient exist, while possibility to mix positive and negative coefficients flavors mitigate the overall drift even if, across process spread, this is often not resolutive.

Local control loop for temperature drift self-tracking is an elaborate but effective solution.

One simple solution is to have a (simplified) dummy version of the analog front-ended with overall behavior similar, in terms of temperature dependency, to the main one. This dummy cell can be managed like a delay cell in a delay-locked loop (DLL) architecture, based on a properly divided version of a fixed and stable system clock (PLL). Equivalent analog delay of the delay cell will be adjusted in the DLL to keep the loop locked. Adjustment can be done in different ways, by changing the operational amplifier current (so that its bandwidth is adjusted) of changing RC constant time. Assuming this replica cell with the main analog front-end is well matched, the same adjustment can be transferred to this one.

The same approach can be applied directly to the main analog front-end at the cost to un-connect it from the sensing inputs for a time periodically allocated to evaluate the drift in a DLL-based configuration.

As an alternative, a solution to directly use the main analog front-end without online has also been developed. A test signal to a frequency f_T high enough to do not impact input signal can be injected as additional input: its output amplitude can be measured (through the ADC, properly set to its frequency). Always in a 1st-order approximation, a drift on f_0 frequency will lead to a change in the output amplitude so allowing to understand how f_0 is changing. Following adjustment can be accomplished as previously described.

Although well optimized, this solution faces with limited dynamic range for test signals. It cannot be too much large to do not compromise operational amplifier available bias headroom. Then, attenuation variation due to dominant pole temperature drift is quite small, requiring strong filtering (i.e., averaging) to extract reliable information against noise.

On the other side, this approach has the advantage to have the possibility to be applied starting from the Wheatstone bridge bias circuit. Test signal can be inputted in addition to VBIAS DC voltage. By this way also, the phase shift associated with the RC filter made by the bridge itself and the parasitic capacitances (package, PCB, analog front-end input capacitance) will be considered in the local control loop. Just considering bridge resistors are significantly variable in temperature, this low-pass frequency variation, even if located at frequency quite high, can be the major actor for the phase drift.

A backup way to face with this task is to insert a series resistance between bridge outputs and analog front-end input, with a temperature coefficient opposite than the one characterizing the Wheatstone bridge. Obviously, such kind of compensation is only moderate with respect to the performances to be guaranteed.

23.2.3 Piezoelectric Sensing

Resistive sensing suffers from reduced sensitivity values and its own temperature dependency, aspects that lead to challenging design to get accuracy needed for system control loop purposes.

In addition, implementation of resistor bridge requires additional mask and process step in MEMS realization.

Piezoelectric effect-based micro-mirror opens the possibility to exploit the same piezoelectric mechanism not only for actuation but also for sensing.

It is a dual effect. For actuation, a voltage signal applied to piezoelectric structure makes a movement on that. For sensing, a movement can be transferred into an electrical signal. No additional masks are so required. Moreover, sensitivity can be designed very large and with negligible temperature dependency.

Sensitivity variation due to aging could be the only drawback.

Several options can be implemented for a readout circuit in the piezoelectric domain. Two main strategies are known: readout in charge or in voltage.

Figure 23.29 shows an example for a fully differential sensing circuit in charge. Piezoelectric sensor is modeled like a capacitor CI with, in parallel, a charge generator and a resistor (to define for instance leakage effects, assumed large enough to be neglected in a first approximation).

Sensor is connected between a bias voltage VBIAS and virtual ground input of the sensing circuit.

Piezoelectric sensor has a sensitivity that depends on bias voltage VBIAS and geometrical design and sizing. Generally, its value has an order of magnitude considerably larger than piezoresistive counterpart: this can avoid using high-voltage value for VBIAS.

Sensing circuit starts with a charge amplifier OPA with a feedback network made by a gain capacitor CF and a resistor RF.

It is possible to calculate the transfer function of this 1st stage:

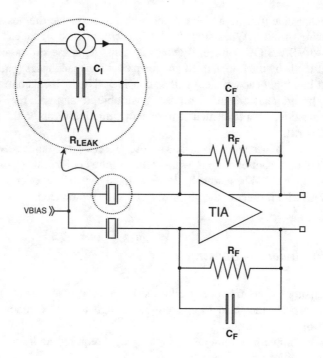

Fig. 23.29 Readout circuit in charge for piezoelectric sensor

$$A_1(s) = \frac{s \cdot C_I \cdot R_F}{1 + s \cdot C_F \cdot R_F}$$

First point to be highlighted, with respect to resistive sensing, piezoelectric one behaves like a high-pass filter, so, it is not able to readout a DC (that means a fixed spatial position).

Constant time CF·RF defines the lower limit for available bandwidth, where the gain is given by the ratio CI/CF.

As usual, after this 1st stage, additional stages are present to further amplify and filter the signal and drive it to the following ADC.

Design recommendations for operational amplifier remain the same previously discussed for the other sensing methodologies.

Figure 23.30 shows an example for a sensing circuit in voltage. Piezoelectric sensor is biased to VBIAS through a (very) high impedance and decoupled in DC to the input of the sensor circuit. Similar high impedance load allows to bias the input of the sensing circuit.

Such kind of high-impedance coupling can be designed or through a pure resistance or, in the case not large enough specific resistance is available, through a reversed biased diode.

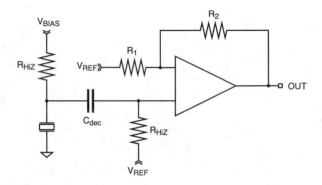

Fig. 23.30 Readout circuit in voltage for piezoelectric sensor

Pseudo-differential instrumentation amplifier (INA) topology can also be easily implemented: in that case, on top of DC decoupling in any case present, all the considerations done for a resistive sensing can be recovered.

23.2.4   ASIC Implementation Examples (Figs. 23.31 and 23.32)

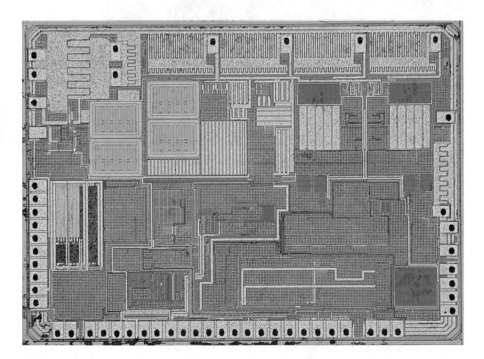

Fig. 23.31 Dual-axis (resonant and quasistatic) ES mirror driver and analog front-end in SOIBCD8s technology (200 V) 0.16um. 8mm^2

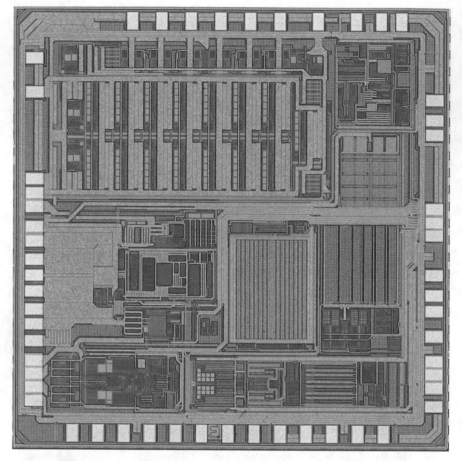

Fig. 23.32 Dual-axis (resonant and quasistatic) PZT mirror driver in BCD8sp technology (40 V) 0.16um. 11mm^2

23.3   MEMS Mirror Control Loop

MEMS mirror are indeed a mechanical structure that have limited bandwidth, non-linearities and environmental dependencies that may prevent the mirror to move as per its applied control voltage.

The variables that affect the mirror movement are as follows:

- Actuation nonlinearity

    Electrostatic mirrors: The relationship between voltage and force depends on the electrostatic forces acting between rotor and stator. Since those forces depend on complex and position-dependent geometries, the applied force is usually far to be linear with respect the angular position.

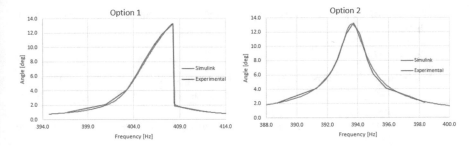

Fig. 23.33 Nonlinear (harden) resonant mirror vs linear resonant mirror

PZT mirrors: Here the linearity is higher than electrostatic mirrors since the dependence from geometry is quite reduced. Anyway, some hysteresis and voltage to force conversion is present, letting some nonlinear behavior still affecting the movement.

- Mechanical nonlinearities

    The simpler resonant mirror can be represented by a second-order differential equation like

$$\ddot{x} + \delta\dot{x} + \alpha x = A \cdot \cos(\omega t)$$

This equation is represented by a transfer function of a low-pass filter with a resonance frequency $\omega 0$ and a certain Q.

Some kind of resonant mirrors are often behaving like a nonlinear oscillator. In a nonlinear oscillator, a phase discontinuity is present close to the resonance. This discontinuity may prevent the mirror to have a working point close to the resonance, since any small variation from the equilibrium may bring the mirror to cross the discontinuity and so to close. Such mirrors are following the Duffing equation:

$$\ddot{x} + \delta\dot{x} + \alpha x + \beta * x^3 = A \cdot \cos(\omega t)$$

In those mirrors, a transfer function cannot be defined and the mirror opening angle may be different in case the operative point is reached moving from low frequency or from high frequencies (Fig. 23.33).

- Mechanical resonances

    Resonant mirrors, as per their operation, require a mechanical resonance with the highest Q as possible to reduce the energy needed to open the mirror at a certain FOV.

    In quasistatic mirrors (so-called linear mirrors), the mechanical resonance is an unwanted behavior, and since it is not possible to eliminate it completely, to minimize the Q is an important design target; otherwise a so-called angular ripple will be always overimposed to the mirror trajectory generated by the drive signal.

- Temperature

  Mechanical structures change their behavior at different temperature since springs may soften or harden depending on the temperature (typical values are 30 ppm/C).
  PZR sensors change their sensitivity over temperature (typical coefficients are 0.3%/C).

- Pressure

  The air drag may also change depending on the ambient pressure level, so the Q of the mirrors may be affected by that.

23.3.1 Resonant Control Loop

Linear resonant mirrors are controlled driving them exactly at their resonance frequency. This will bring the maximum opening with the minimum required energy. A simple approach to drive such mirrors requires to measure the resonance of the specific mirror (to remove the variable of the process spread) and to drive it at this specific frequency. This condition is then not stable since the resonance of the mirror may change in temperature as said before. That means that the resonance frequency must be tracked to adapt continuously to the variations.

The approach to track the resonance requires to measure the phase difference between the mirror drive and the mirror position obtained from the sensor. The resonance condition is achieved when the phase difference is 90Deg. For mirrors that have position sensors (like PZR), any phase or delay measurement approach is valid, measuring time (e.g., using counters or TDC) as far as quadrature approaches. In case of electrostatic mirror, the sensor is not present, so the phase measurement needs some additional consideration. As described in Section 23.2.1, what is measured in electrostatic mirrors is the variation of the rotor-stator capacitance over time. To do that the mirror is driven at 50% of the period and sensed for the other 50%. The capacitance is roughly related to the mirror position, so the derivative of the capacitance is related to the mirror velocity. The velocity is so shifted 90Deg with respect to the position, that's phase shift should be added to the phase shift related to the mirror resonance to control the mirror drive frequency. Since the sense is done only in the 50% of the period and also because the capacitance variation is not linear, the sensing waveform is a signal like the one in Fig. 23.34. From this signal, the zero-crossing information can be obtained. The zero cross is then used to measure the delay with respect to the drive and then using the delay as an estimation of the phase.

A general control loop for resonant mirrors works in this way:

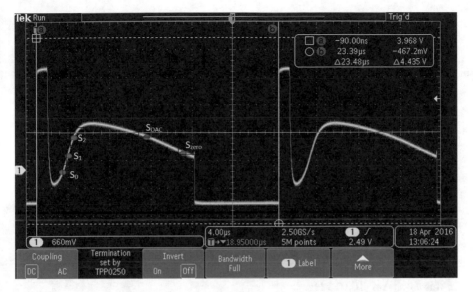

Fig. 23.34 Electrostatic resonant mirror sensing waveform

- Measure the phase between drive and sense.
- If the phase is greater than the target, it decreases the drive frequency.
- If the phase is smaller than the target, it increases the drive frequency.

This algorithm considers a generic phase target so that the mirror may be also driven close to resonance just changing the target phase. This is useful for nonlinear resonant mirrors where the resonance region is presenting the phase discontinuity described in the previous section.

Despite this, control may be also realized in analog domain; generally a digital implementation is more flexible and easier to be implemented. The drive generator can be realized by a fractional PLL or a DDS + PLL structure able to change the drive frequency with good resolution. The phase between drive and sense in a generic resonant mirror (linear) is

$$\alpha = \text{atan}\left(\frac{\omega * \omega_0}{Q * (\omega^2 - \omega_0^2)}\right)$$

In this case, if we are controlling the mirror perfectly at resonance and linearizing in this point the phase relation, we have the following equation:

$$\Delta\alpha = \frac{2 \cdot Q}{\omega_0} \cdot \Delta\omega$$

In the laser beam scanning applications, the resonant mirror is used to project a line with a certain resolution. The resolution is determined by the number of distinct pixels that can be projected. Usually, the projected area is a subinterval of the complete mirror opening angle, so if the mirror moves as A*sin(2*pi*f*t), usually around the 90% of the opening is used (0.9*A). That means the phase of the sinus will span from asin(0.9) to -asin(0.9). Defining a pixel in angle, if X is the number of pixel, each pixel (in angle unit) will be 2*asin(0.9)/X.

That means that the frequency step (i.e., the minimum frequency adjustment step that we can apply to the drive waveform) should be

$$\text{asin}(0.9) \cdot \frac{\omega_0}{X \cdot Q} > \Delta\omega$$

The above formula is indeed an approximation (in particular the 90% ratio and the pixel definition may differ in different systems), but it could be useful to define the PLL specification. Let us try to apply it to a real case:

$$\omega_0 = 27 \text{ kHz}; X = 2000; Q = 1000$$

in this case, $\Delta\omega < 0.015$ Hz.

Also the phase accuracy can be derived from the above consideration, any phase measurement should be more accurate than 2*asin(0.9)/X.

To close the loop, a classical PID approach is suitable. Since the set point is defined as a constant phase value, a single integrator in the loop is enough to guarantee the error will be zeroed.

The phase lock is then not enough to control a resonant mirror. In case of pressure changes, the Q factor may change, which means that the drive amplitude must be also controlled.

The OAC (opening angle control) requires the measurement (in case of position sensor is available) or the estimation of the opening angle of the mirror.

In case the peak opening angle can be measured directly, a classical PID approach may be used to adjust the driving waveform amplitude to keep constant this angle. In case a position sensor is not available (like in the electrostatic mirror), the control may be trickier because the amplitude must be estimated from a signal that looks like the one in Fig. 23.34. A possible approach is considering some geometrical properties of such waveform, so taking some specific points on it (S_0, S_1, S_2, S_{OAC}, S_{zero}) and considering their value, it may be possible to keep the amplitude constant.

23.3.2 Linear Control Loop

For linear (or quasistatic) mirror, any mechanical resonance should be avoided. The effect of the resonance in such mirrors is depicted in Fig. 23.35. The ripple is excited by the drive signal discontinuities and it is generating a dumped sine wave overimposed to the main mirror trajectory. Depending on the quality factor of the resonance, the amplitude as far as the decay of the ripple may be larger or smaller.

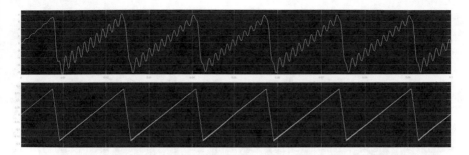

Fig. 23.35 Ripple on the quasi-static mirror trajectory (top) with respect the drive signal (bottom)

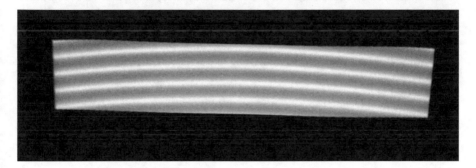

Fig. 23.36 Ripple effect on the projected image in a raster LBS

The final effect of the ripple is the change of mirror speed that will not be always constant during the trace period. This is generating light modulation in the projected images, and the image will be brighter when the mirror speed is smaller and darker when the speed will be faster (see Fig. 23.36).

Since it is not possible to remove completely the mechanical resonances, an active mirror control is needed. In case a sensor (e.g., PZR sensor) is available to measure the mirror position, a classical control approach may be used. The complexities of the control are mainly due to the frequency of the resonance with respect the bandwidth of the control. The larger the closed loop bandwidth, the more critical will be the stability of the system.

Normally triangular or sawtooth wave are used to move linear mirrors, at frequencies that can be from 10 Hz up to 120 Hz. To have good accuracy, the system bandwidth should be able to accommodate at least 10 harmonics. Considering that the resonance frequency for such mirrors span normally from 500 Hz up to few kHz, it is clear that the singularity of the controller must be well controlled, especially for mirrors with low resonance frequency.

A very basic control for linear mirrors relies on the mechanical relationship of (Eq. 23.1). Here the quality factor of the resonance is related to the coefficient of the mirror velocity δ:

$$Q = \sqrt{\alpha}/\delta$$

The basic idea is that if we can apply a derivative to the mirror sensing, scaling it with the right coefficient, this signal can be added back to the input so that it may act as an additional (electrical) dumping. In this way the Q of the mirror can be lowered from some hundreds down to small numbers (around 0.5). This approach will avoid any resonance-related ripple, but in this way the bandwidth of the closed-loop system is still limited by the mirror, and moreover, since the derivative is magnifying the high frequencies, it may bring to higher noise if used in this simple form. While with a more sophisticated controller, the bandwidth of the closed-loop mirror system can be also enlarged above the resonance, being able so to reproduce well 120 Hz triangular waves even with 700 ~ 800 Hz mirrors.

In case that no good sensor is available to measure the mirror position, a classical control cannot be used. An example of such situation is the electrostatic mirrors case, where only a pure capacitive sensing is possible. That sense approach cannot directly measure the position but a mix of the drive signal and the velocity of the mirror. What is important anyway is that using band pass filters in some condition, it is possible to extract an information related to the ripple. The electrostatic sensing is highly nonlinear so that it is not possible to measure the ripple amplitude, but the control loop goal in this case will be to minimize it, whatever will be the absolute value.

Once the mirror ripple is measured, it cannot be fed directly in a loop, but it will be used to adjust in a feedforward approach the drive signal waveform shape. The drive waveform of the mirror will be so the target sawtooth with added some compensation waveform that will be able to minimize the ripple. The shape of the compensation waveform will be updated iteratively at each N mirror cycles from the ripple estimated parameters.

Obviously, the advantage of having a sensor is that a closed-loop control can be implemented. This is something more robust from the environmental condition changes and from mechanical vibration rejection point of view. Moreover, a feedforward approach needs a calibration procedure able to compensate for any mechanical nonlinearity. The calibration purpose is mainly to measure the voltage to angle mirror characteristic in the application. This characteristic is measured in quasistatic condition, i.e., applying a very slow ramp to the mirror avoiding as much as possible to stimulate the resonance. With this information it is possible to define the reference waveform shape (a sort of pre-distorted sawtooth) and scaling factors to drive the mirror.

23.4   Piezo Inkjet Printers

23.4.1   Nozzle Drive Waveform

As described in <19.4> section, to fire a single PZT nozzle, a driving pulse must be applied to be able to create a pressure difference between the chamber and the external atmosphere so that a drop will be ejected. This pulse must satisfy to several constraints to be able not only to eject the drop but also to let the drop have stable trajectory, speed, and size.

The pulse will therefore be characterized by a voltage value, V (that controls the droplet mass), by a duration in time T (that controls the repetition rate or firing frequency), and by edges with a determined slew rate SR (that controls the droplet velocity). Given that in the design phase of the print heads, the characteristics of the piezo used may be different or that due to spread the PZT of the same head may differ slightly due to project spreads, a hypothetical driver will have to guarantee sufficient flexibility and configurability to manage the different situations avoiding having to lose yield due to excessive MEMS screening.

An additional degree of freedom of the PZT printers not enabled by the thermal inkjet technology is to allow with a more complex waveform to control the size of the drop in a discrete way. Starting from a base "small" drop, it is possible to merge more drops together enlarging the overall drop size. This is achieved by using a pulse train of trapezoidal waveforms like in Fig. 23.37. Here the different pulse amplitudes allow multiple drop ejection at different speed so that at a certain point a single drop will be created.

At the end of any fire sequence, the membrane will stop with a speed related to the mechanical damping of the fluidic system. To avoid unwanted liquid/drop ejection, sometimes a damp pulse may be needed where the first pulse is used for jetting and the second pulse (typically 1/3 amplitude of the first pulse) is used to suppress the unwanted damping.

The generic nozzle drive waveform is illustrated in Fig. 23.37. The period of the pulses depends on the Helmholtz frequency (about 200 KHz). This is the resonant frequency of a chamber closed with an exit hole and filled with a fluid. The value of this frequency depends on the volume of the chamber, the size of the nozzle hole, and the density of the fluid. The amplitude of the pulses will depend on the piezo characteristic of the actuator and of the fluidic characteristics of the system. Any driver to be used to drive a PZT printer head must be able to generate this kind of waveform on a capacitive load (the nozzle) up to 40 V amplitudes. Rise and fall time of the pulses should be fast enough (30 V/us to 100 V/us) or (250 to 400 ns) and possibly well repeatable to be able to guarantee uniformity over time of the drop's parameters.

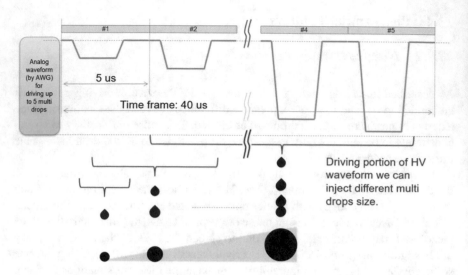

Fig. 23.37 Generic driving waveform for a PZT nozzle

23.4.2 Driver Architectures

A PZT printer head is normally built from one or more PZT MEMS element that comprises several hundreds of nozzles. That is clear that one of the most challenging issues in designing a driver for a PZT printer is to find a good solution to manage arrays of actuators.

With reference to what has been said before, driver systems must be designed with an amplifier for each PZT to be controlled, with its own programmable pulse generator. This solution certainly allows to optimize the impulse for each single nozzle to obtain drops of constant shape, size, and speed as conditions change. Could change, for example, the fluid characteristics, temperature and any other parameter including also the process spread. Of course, it is complex and not very convenient to have as many signal generators and related amplifiers as there are nozzles to be controlled, considering thousand channels that are tricky to manage in terms of space and costs on a print head. The complexity can be reduced by defining only a few different waveforms to be selected for each nozzle and then sending them to the single channel for D/A conversion and the necessary amplification.

Figure 23.38 shows an example of realization of an optimized structure with 12 waveforms to be selected and sent on the nozzles according to the selected configuration. This is a simple solution for a silicon implementation that maintains a certain flexibility on driving the nozzles.

Many different approaches exist to generate the waveforms, RAM/ROM tables, and parametric generators (e.g., polynomial), with different tradeoff between area, consumption, and flexibility. Nevertheless, even minimizing the number of waveforms at the minimum, the number of the drivers is still a bottleneck.

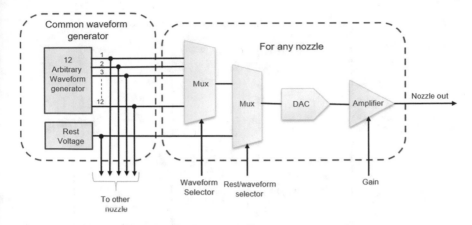

Fig. 23.38 Driver architecture example

Considering 1000 nozzle driven by its own driver, it may require areas in the range of 350 mm^2. Also, in case the area/cost may be a questionable constraint, the power dissipation maybe remains a critical point. 1000 nozzles of 270 pF (typical value for the PZT nozzle actuators) each driven at 40 V at 200 kHz will require 80 W to be dissipated. This is normally a value compatible with power packages and/or active cooling, so this point must be well considered from assembly and application point of view.

Another approach may be to think about a single programmable signal generator with a single signal amplifier that drives many nozzles simultaneously. Of course, to drive each single nozzle independently, a switch matrix must be interposed to make the fire signal to reach the desired nozzles. In this way the amplifier will have to supply all the current necessary to activate all the nozzles provided and the waveform will be the same for everyone.

These systems with single amplifier + switches have the advantage of facilitating the system integration, since the complexity and the power dissipated are now shared between the matrix and the amplifier. The counterparty to be paid is less flexibility of the system, and it will no longer be possible to choose the waveform parameters independently for each nozzle. The amplifier will also have to manage a highly variable load depending on the current print data. In fact, it will have to drive a capacitive load equal to 1 Pzt and up to the maximum number of PZT in parallel without changing the pulse slew rate, its accuracy while managing the high currents involved. The introduction of a switch matrix in between the driver and the MEMS allows indeed a better flexibility in terms of assy. The driver can be moved far from the MEMS, while the switch matrix may be integrated on the MEMS flex (or even in a rigid flex approach) reducing the number of connections between the main printer control board and the printer head. Having a driver on the main printer board, it can be optimized reducing the power dissipated by using inductive or capacitive adiabatic current recovery approaches. The matrix approach could be

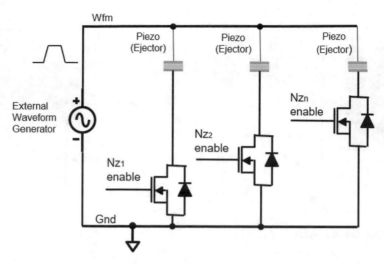

Fig. 23.39 Matrix architecture example

quite flexible, depending on the spread information of the MEMS (the periphery nozzle may have normally different characteristics from the ones on the center) the number of independent drivers can be defined, and then the MEMS nozzle can be grouped, and using a certain number of switch ASICS, those nozzle group can be connected to a driver that can be correctly sized in terms of driving capability.

23.4.2.1 Switch Connected to Ground

A simple configuration could be the one shown in the Fig. 23.39. Where a single high-side amplifier drives all the positive poles Pzt together, the negative ones are closed to ground controlled by low-side Dmos switch. In this way only the nozzles that have the Dmos turned on will shoot the drop of ink on the paper. The advantages are given by the simplicity of the structure, one-way switch connected to ground. The main disadvantage is that the negative pole of the unused PZT remains floating and subject to unwanted couplings with the other adjacent signals.

23.4.2.2 Switch Floating

By changing the type and position of the switches, other configurations can be obtained; the following is an example (Fig. 23.40):

This structure is more complicated than the previous one as it requires bidirectional, floating switches, but allows to have the external waveform signal applied only to the PZT to be activated by reducing the parasitic couplings. This structure allows to drive the PZT with negative voltage because the signal, Wfm, refers to

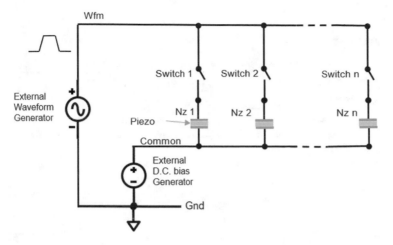

Fig. 23.40 Matrix floating switch architecture example

ground, while the negative pole of the PZT refers to a positive voltage with respect to ground. It also allows the implementation of more complex structures such as the one described in the next section.

23.4.2.3   Multiswitch Floating

The matrix of the switches can also be designed in a more complex way, that is, switches can be added to connect the PZT of the MEMS of each individual channel to different voltages or to test terminals to test its operation and characteristics. An example can be the one in Fig. 23.41, where three switches are used to connect the nozzle's PZT, Nzn, to three different lines, a command signal line, Wfm; a rest voltage line, rest; and a test line, test can be used as input to supply a signal to the PZT or can be used as a monitor to verify the correct system operation.

The main advantage is given by the flexibility of the structure which allows to have under control the polarization of the two poles of the nozzle PZT. The PZT can be left inactive on a low impedance so that it is not disturbed by the firing signal, Wfm, when it must not be activated. It also allows you to test the individual PZT after assembling the fluidic MEMS with the switch matrix.

Furthermore, using the rest lines, two main configurations can be created, the first with PZT always at rest voltage (pre-charged) and the second with PZT without at rest voltage.

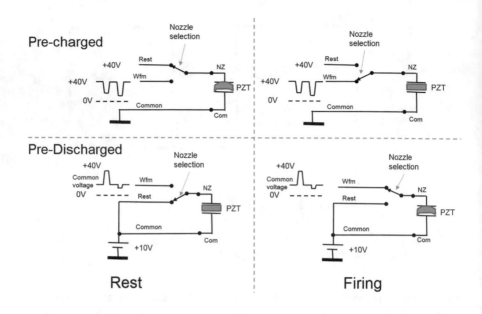

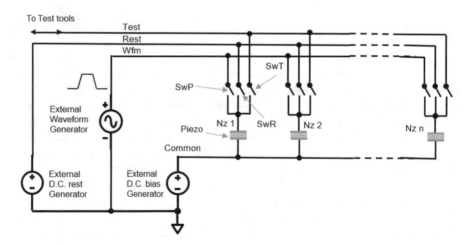

Fig. 23.41 Matrix multiswitch architecture example

The aim of the first method, named pre-charged, is to have the piezo actuator always supplied with a DC voltage, rest voltage, and consequently the piezo will be normally stressed. In the firing state, the piezo actuator will be connected to an external waveform generator (activation signal) that with a proper waveform will discharge the piezo relaxing it to eject the drop. To avoid extra current and extra

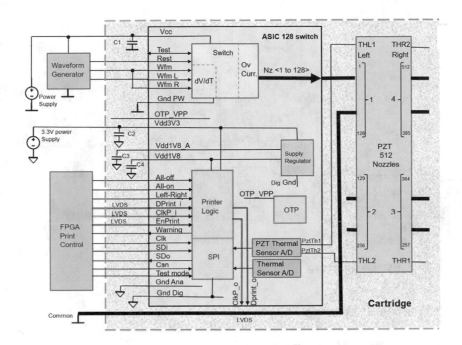

Fig. 23.42 Example of matrix and MEMS connections

drop at the first turn on the rest voltage, it must be applied with slow rising edge (Slew rate < 1 V/μs) with the nozzle selection switch closed on rest voltage.

The second method, pre-discharged, instead has as a rest condition the null voltage applied to the piezo referred to a common voltage. In the firing state, the piezo is connected to an external voltage that have a proper waveform that will stress the piezo ejecting the drop.

In both cases, for rest condition, it will be also possible to open the switches to leave the NZ terminal in high impedance. The pre-charged implementation is not possible with the switch connected to ground structure.

23.4.2.4 Print Head Assembly

For the systems with single amplifier + switches the idea is to assemble the print head with fluidic MEMS and switch matrix, leaving the signal amplification outside of the head print. The possible circuit configuration is shown in Fig. 23.42 as an example. The integrated switch matrix controls 128 nozzles and the fluidic MEMS is designed with 512 nozzles, so, it will take 4 I.C. matrix to drive the full MEMS.

With this configuration a print head can be assembled by putting the 4 matrix I.C. on a flexy as shown in Fig. 23.43.

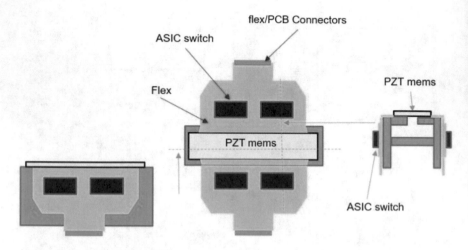

Fig. 23.43 Example of PH assy

23.5 PMUT Sensing

23.5.1 Sensor Modeling

A piezoelectric micromachined ultrasonic transducer (PMUT) is a device that, different from bulk piezoelectric transducers, uses the motion of a flexible membrane covered with a thin film of piezoelectric material to convert energy from the mechanical to the electrical domain and vice versa [14],[15],[16].

Assuming the flexion of the membrane as a piston-like movement of a rigid plate of mass m, from a mechanical point of view, the device can be modeled as a mass-spring-damper element as in Fig. 23.44. When an electrostatic force is applied to the membrane, it flexes with an opposite force and restores the equilibrium; this effect can be modeled by a spring with elastic coefficient k. All the energy losses, including the acoustic-mechanic coupling losses (membrane-medium), can be modeled as a damper with damping coefficient b.

To describe how the PMUT interacts with the electric world, it is of convenience to use the Mason's linear model in Fig. 23.45 that includes, in addition to the mechanical parameters m, b, and k, the parameters Z_m, A_{eff}, and h. Z_m represents the acoustic impedance of the surrounding fluid, A_{eff} is the effective area of the PZT membrane, and h is the conversion coefficient between the mechanical and the electrical domain and is mostly associated with the piezoelectric properties of the material. Even if it does not consider nonlinear effects, the Mason's model includes electrical and mechanical losses, the transducer-medium acoustic interactions, and provides a strong basis for the definition of the PMUT performance [17].

From the Mason's model, it is possible to obtain an equivalent circuit completely referred to the electric domain, as in Fig. 23.46, by gradually eliminating the

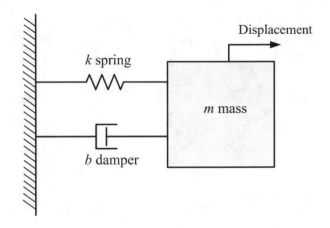

Fig. 23.44 Mass-spring-damper model

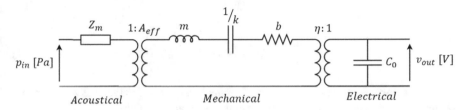

Fig. 23.45 RX Mason's model

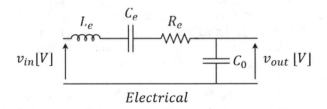

Fig. 23.46 Electrical Mason's model

conversion transformers multiplying the elements by the conversion factor and restoring the branch impedance.

After the transformation, each element will be defined as

$$v_{in} = p_{in}\frac{A_{eff}}{\eta} \ [V]; \tag{23.1}$$

$$L_e = \frac{m}{\eta^2} \ [H]; \tag{23.2}$$

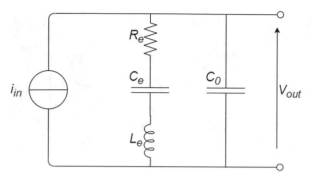

Fig. 23.47 pMUT Norton equivalent model

$$C_e = \frac{1}{k}\eta^2 \ [F]; \tag{23.3}$$

$$R_e = \frac{b}{\eta^2} + Z_m\frac{A_{eff}^2}{\eta^2} \simeq Z_m\frac{A_{eff}^2}{\eta^2} \ [\Omega]; \tag{23.4}$$

From a pure electric point of view, the transducer, when analyzed as a sensor, can be furtherly considered as a current generator with an associated Brownian noise, as shown in Fig. 23.47.

C_0 is the intrinsic capacitance of the transducer and is related to the film thickness of the piezoelectric material and its dielectric constant. All the other parameters are related to the same thickness and to the transducer shape, dimensions, and conversion factor. Depending on the value of these parameters and on the application field, the sensing architecture is usually chosen between the charge reading mode and the voltage reading mode.

Since in most of cases the impedance associated with the branch formed by elements R_e, C_e, and L_e is negligible, from now on the simplified model shown in Fig. 23.48 will be adopted.

A high-frequency analysis, excluding eventual bias resistors, of both the solutions will be carried out to facilitate the comparison and highlight the advantages of one respect to the other [18, 19].

23.5.2 Charge Reading Mode

In this solution the pMUT output current is applied to the input of a charge amplifier as shown in Fig. 23.49.

The current produced by the transducer is transferred into the feedback capacitor C_F and converted to an output voltage v_{outc} where

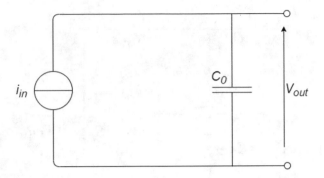

Fig. 23.48 pMUT simplified model

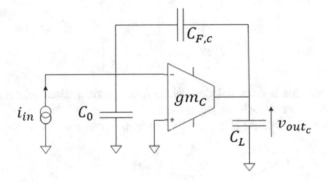

Fig. 23.49 Charge amplifier

- $i_{in} = p_{in} * H(s) * \frac{A_{eff}}{\eta}$ is the output current provided by the sensor
- C_0 is the sensor output capacitance
- $C_{F,c}$ is the feedback capacitance
- C_L is the charge amplifier load capacitance

Applying the Kirchhoff Current Law at the negative input and output nodes, and solving the system of equations, it is possible to find the input/output transfer function defined as

$$\frac{v_{out_c}}{i_{in}} = -\frac{1}{sC_{F,c}} * \frac{\left(1 - s\frac{C_{F,c}}{gm_C}\right)}{\left(1 + s\frac{C_{TOT_C}}{gm_C}\right)} \left[\frac{V}{A}\right] \tag{23.5}$$

The low-frequency transimpedance gain is defined as

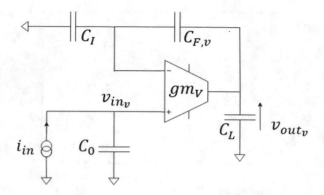

Fig. 23.50 Voltage amplifier

$$\frac{v_{out_c}}{i_{in}} = -\frac{1}{s C_{F,c}} \left[\frac{V}{A} \right] \tag{23.6}$$

This means that it does not depend, in first approximation, on C_0 and on the transimpedance amplifier input capacitance.

The frequency response is mainly limited by the pole at

$$\omega_{p,c} = \frac{gm_C}{C_{TOT_C}} \tag{23.7}$$

where

$$C_{TOT_C} = C_0 + C_0 \frac{C_L}{C_{F,c}} + C_L \tag{23.8}$$

In this case, the bandwidth is strongly affected by C_0, hence, for sensors with a high output capacitance used for high-frequency applications, this reading architecture is not recommended.

23.5.3 Voltage Reading Mode

Instead of a charge amplifier, a voltage amplifier with high input impedance can be used as a signal conditioning circuit for the PMUT (Fig. 23.50).

The current produced by the transducer is converted into an input voltage v_{in_v} and amplified by the voltage amplifier where

- $i_{in} = p_{in} * \frac{1}{R_e} * \frac{A_{eff}}{\eta}$ is the output current provided by the sensor
- C_0 is the sensor output capacitance

- $C_{F,v}$ and C_I are the voltage amplifier gain capacitances
- C_L is the voltage amplifier load capacitance

If we consider the input of the voltage amplifier as an infinite impedance, the input voltage can be expressed as $v_{in_v} = \frac{i_{in}}{sC_0}$.

In the same way we did for the charge amplifier, by solving Kirchhoff equations, it is possible to find the input/output transfer function:

$$\frac{v_{out_v}}{i_{in}} = \frac{1}{sC_0} * \frac{\left(1 + \frac{C_I}{C_{F,v}}\right)}{\left(1 + s\frac{C_{TOT_V}}{gm_V}\right)} \left[\frac{V}{A}\right] \tag{23.9}$$

The low-frequency transimpedance gain is defined as

$$\frac{v_{out_c}}{i_{in}} = -\frac{1 + \frac{C_I}{C_{F,v}}}{sC_0} \left[\frac{V}{A}\right] \tag{23.10}$$

This means that it depends on C_0, on the transimpedance amplifier input capacitance, and on the input parasitic; hence, for applications with long connection between the sensor and the readout circuit, this solution is not recommended.

The frequency response is mainly limited by the pole at

$$\omega_{p,v} = \frac{gm_V}{C_{TOT_V}} \tag{23.11}$$

where the capacitance C_{TOT_V} is expressed as

$$C_{TOT_V} = C_I + C_L\frac{C_I}{C_{F,v}} + C_L \tag{23.12}$$

In this case, by choosing a low value of C_I, the bandwidth is only affected by C_L; hence, for sensors with a high output capacitance and for high-frequency applications, this reading architecture is to be preferred.

23.5.4 Linear Array Sense Architecture

In many applications, for instance, when ultrasound waves are used to reproduce multidimensional images, the reading probes are composed by an array of sensing elements where each element is composed by several pMUTs. The number of transducers per each element depends on the dimensions required: for this reason,

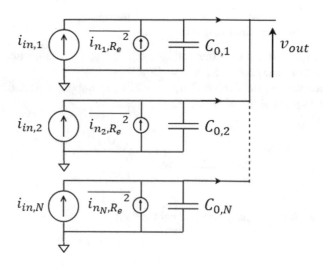

Fig. 23.51 Parallel membranes

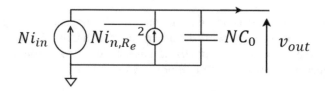

Fig. 23.52 Parallel membranes equivalent circuit

both in the literature and on the market, there are transducers made up of elements of various shapes and sizes.

We will now investigate how the electrical behavior of the transducer, in terms of output voltage and signal-to-noise ratio, varies if the transducer is composed by a single membrane or by an element, i.e., a set of many PMUT membranes, considering also the two possible electrical connections: parallel connection and series connection [20, 21].

23.5.4.1 Parallel Membranes Connection

In the parallel connection, shown in Fig. 23.51, N equivalent PMUT electrical models are placed in parallel.

This easily leads to the equivalent circuit in Fig. 23.52.

The characteristics of the circuit in Fig. 23.52 are the following:

– The equivalent input current generator is the result of the linear sum of the individual current generators, obtaining Ni_{in}.

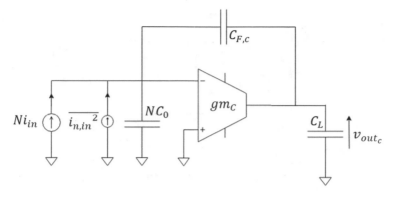

Fig. 23.53 Charge amplifier with N parallel membranes

- The total output noise spectral density is the result of the quadratic sum of the individual contributors, obtaining $\overline{Ni_{n,Re}^2}$.
- The total output capacitance is the result of the parallel of the N output capacitances C_0, obtaining $N * C_0$.

The best-suited amplifier to manage this type of connection is the charge amplifier. If the parallel membranes had been connected to the voltage amplifier, there would have been no benefit since the voltage generated by the element is equal to the voltage generated by the single membrane.

Figure 23.53 shows a simplified schematic of the charge amplifier connected to N membranes in parallel. The current generator $\overline{i_{n,in}}^2$ represents the input-referred noise current including the values of the N parallel PMUTs and the input-referred noise contribution of the charge amplifier.

Recalling Eq. 23.5, the transfer functions (voltage/input current) and output voltage noise/input-referred current noise of the circuit in Fig. 23.53 are given by

$$\frac{v_{out_c}}{i_{in}} = -\frac{N}{sC_{F,c}} * \frac{\left(1 - s\frac{C_{F,c}}{gm_C}\right)}{\left(1 + s\frac{C_{TOT_C}}{gm_C}\right)} \quad \left[\frac{V}{A}\right] \qquad (23.13)$$

$$\frac{v_{n,out}}{i_{n,in}} = -\frac{\sqrt{N}}{sC_{F,c}} * \frac{\left(1 - s\frac{C_{F,c}}{gm_C}\right)}{\left(1 + s\frac{C_{TOT_C}}{gm_C}\right)} \quad \left[\frac{V}{A}\right] \qquad (23.14)$$

At first glance, from Eq. 23.13, it is noticeable that the output voltage increases by a factor N as compared to the case when a single membrane is used.

On the other hand, observing Eq. 23.14, it is possible to notice how the total output-referred noise increases by a factor $\sqrt{N}$. This means that the effect of a

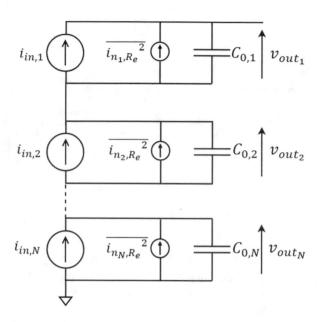

Fig. 23.54 Series membranes

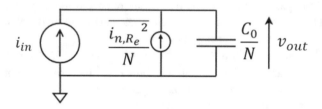

Fig. 23.55 Series membranes equivalent circuit

parallel connection of N pMUTs membranes in conjunction with a charge mode reading architecture is to increase the sensitivity of a factor N and the signal-to-noise ratio of a factor $\sqrt{N}$. The main drawback of this solution is that, due to the huge increase of the sensor equivalent output capacitance, the system bandwidth is strongly impacted.

23.5.4.2   Series Membranes Connection

In the series connection, shown in Fig. 23.54, N equivalent PMUT electrical models are placed in series.

This easily leads to the equivalent circuit in Fig. 23.55.

The characteristics of the circuit in Fig. 23.55 are the following:

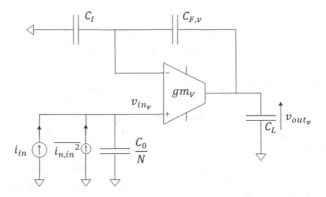

Fig. 23.56 Voltage amplifier connected to N series membranes

- The equivalent input current generator is the same of the individual current generator i_{in}.
- The total output noise spectral density is the result of the quadratic sum of the individual contributors divided by N, obtaining $\overline{i_{n,in}^2} = \frac{i_{n,Re}^2}{N}$.
- The total output capacitance is the result of the series of the N output capacitances C_0, obtaining $C_0\big/N$.
- The total output voltage v_{out} is the result of the current generator value i_{in} multiplied by the output capacitance: $v_{out} = \frac{i_{in}}{s\frac{C_0}{N}}$.

The best-suited amplifier to manage this type of connection is the voltage amplifier. If the series membranes had been connected to the charge amplifier, there would have been no benefit since the charge generated by the element is equal to the charge generated by the single membrane.

Figure 23.56 shows a simplified schematic of the voltage amplifier connected to N membranes in series.

The current generator $\overline{i_{n,in}}^2$ represents the input-referred noise current including the values of the N series PMUTs and the input-referred noise contribution of the voltage amplifier.

Recalling Eq. 23.9, neglecting the voltage amplifier noise, the transfer functions (voltage/input current) and output voltage noise/input-referred current noise of the circuit in Fig. 23.56 are given by

$$\frac{v_{out_v}}{i_{in}} = -\frac{N}{sC_0} * \frac{\left(1 + \frac{C_I}{C_{F,v}}\right)}{\left(1 + s\frac{C_{TOT_v}}{gm_v}\right)} \quad \left[\frac{V}{A}\right] \qquad (23.15)$$

$$\frac{v_{n,out}}{i_{n,in}} = -\frac{\sqrt{N}}{sC_0} * \frac{\left(1 + \frac{C_I}{C_{F,v}}\right)}{\left(1 + s\frac{C_{TOT_v}}{gm_v}\right)} \left[\frac{V}{A}\right] \tag{23.16}$$

At first glance, from Eq. 23.15, it is noticeable that, as for the parallel connection used with the charge amplifier, the output voltage increases by a factor N as compared to the case when a single membrane is used.

On the other hand, observing Eq. 23.16, it is possible to notice how the total output-referred noise increases by a factor $\sqrt{N}$. This means that the effect of a series connection of N pMUTs membranes in conjunction with a voltage mode reading architecture is to increase the sensitivity of a factor N and the signal-to-noise ratio of a factor $\sqrt{N}$. The main advantage respect to the parallel connection is that the system bandwidth is not affected by the number of membranes. The only limitation to the number of membranes N is when the element equivalent output capacitance C_0/N becomes comparable with the transconductance amplifier input capacitance. When this occurs, increasing the number of membranes does not produce any further increasing of the output sensitivity and of the signal-to-noise ratio.

References

1. Pham, D. D., Singh, R. P., Yan, D. L., Tiew, K. T., Bernal, O. D., Langer, T., Hirshberg, A., & Je, M. (2011). *Position sensing and electrostatic actuation circuits for 2-D scanning MEMS micromirror*. Defense Science Research Conf. and Expo.
2. Pierco, R., Torfs, G., Verbrugghe, J., Bakeroot, B., & Bauwelinck, J. (2015). A 16 channel high-voltage driver with 14-bit resolution for driving piezoelectric actuators. *IEEE Transactions on Circuits and Systems I: Regular Papers, 62*(7).
3. Guo, Z., Xu, R., Geng, L., & Li, B. (2019). *High-Voltage driver for fully-integrated piezoelectric inkjet printhead module*. ICICDT.
4. Main, J. A., Newton, D. V., Massengill, L., & Garcia, E. Efficient power amplifiers for piezoelectric applications. *Smart Materials and Structures, 5*(6).
5. Chaput, S., Brooks, D., & Wei, G. *A 3-to-5V input 100Vpp output 57.7mW 0.42% THD+N highly integrated piezoelectric actuator driver*. ISSCC 2017 – Session 21 – Smart SOCS for Innovative Applications.
6. Pashmineh, S., & Killat, D. (2015). *Design of a high-voltage driver based on low-voltage CMOS with an adapted level shifter optimized for a wide range of supply voltage*. ICECS.
7. Chen, Y., Lee, F. C., Amoroso, L., & Wu, H.-P. (2003). *A low voltage to high voltage level shifter circuit for MEMS application*. UGIM.
8. Nakata, S., Honda, R., Makino, H., Mutoh, S., Miyama, M., & Matsuda, Y. (2012). General stability of stepwise waveform of an adiabatic charge recycling circuit with any circuit topology. *IEEE Transactions on Circuits and Systems, 59*(10).
9. Khorami, A., & Sharifkhani, M. (2016). *An efficient fast switching procedure for stepwise capacitor chargers*. IEEE Transactions on Very Large-Scale Integration (VLSI) Systems.
10. Eschauzier, R., & Huijsing, J. H. (1995). *Frequency compensation techniques for low-power operational amplifiers*. Kluwer Academic Publishers.
11. Enz, C. C., & Temes, G. C. (1996). Circuit techniques for reducing the effect of op-amp imperfections: autozeroing, correlated double sampling, and chopper stabilization. *Proceedings of IEEE, 84*(9).

12. Burt, R., & Zhang, J. (2006). A micropower chopper-stabilized operational amplifier using a SC notch filter with synchronous integration inside the continuous-time signal path. *IEEE Journal of Solid-State Circuits, 41*(12).
13. Gray, P., Hurst, P. J., Lewis, S. H., & Meyer, R. G. *Analysis and design of analog integrated circuits* (5th ed.). Wiley.
14. Qiu, Y., Gigliotti, J. V., Wallace, M., Griggio, F., Demore, C. E. M., Cochran, S., & Trolier-McKinstry, S. (2015). Piezoelectric Micromachined Ultrasound Transducer (PMUT) arrays for integrated sensing, actuation and imaging. *Sensors, 15*, 8020–8041.
15. Akasheh, F., Myers, T., Bose, F. J. D. S., & Bandyopadhyay, A. (2004). Development of piezoelectric micromachined ultrasonic transducers. *Sensors and Actuators A, 111*, 275–287.
16. Wang, Sawada, & Lee. (2015). A Piezoelectric micromachined ultrasonic transducer using piston-like membrane motion. *IEEE Electron Device Letters, 36*.
17. Gabrielson. (1993). Mechanical-thermal noise in micromachined acoustic and vibration sensors. *IEEE Transactions on Electron Devices, 40*, 903–909.
18. Liu, W. Q., Feng, Z. H., Liu, R. B., & Zhang, J. (2007). The influence of preamplifiers on the piezoelectric sensor's dynamic property. *Review of Scientific Instruments, 78*, 1–4.
19. Hopkins, M. B., & Lee, P. (2015). High frequency amplifiers for piezoelectric sensors noise analysis and reduction techniques. In *2015 IEEE international instrumentation and measurement technology conference (I2MTC) proceedings*. Pisa, Italy.
20. Gurun, Hochmann, Hasler, & Degertekin. (2012). Thermal-mechanical-noise-based CMUT characterization and sensing. *IEEE Transactions on Ultrasonics, Ferroelectrics, and Frequency Control, 59*(6), 1267–1275.
21. Bozkurt, A., & Yaralioglu, G. G. (2016). Receive-noise analysis of capacitive micromachined ultrasonic transducers. *IEEE Transactions on Ultrasonic, Ferroelectrics, and Frequency Control, 63*(11), 1980–1987.

Part VI
MEMS Back-end

Chapter 24
MEMS Package Design and Technology

Marco Del Sarto

24.1 Introduction

The first scope of an electronic package is to transmit the electrical signal from IC to the user. This means a kind of "translation" of the physical dimension from the tens of micrometer of the IC pad to the hundreds of micrometers (it was millimeter just 25 years ago) needed to match the technology scale of PCB (printed circuit board) manufacturing and SMT (surface mount technology) process capability. This interconnection and routing function of the package can be defined and will be referenced in this chapter, as fan-out.

The second scope is to protect the sensitive IC from the environment. This is obtained by an enclosure that can vary from metallic can to ceramic cavity up to molded resin around the component and its routing substrate. The latest is the most used solution nowadays due to parallel producibility that leads to high volumes and massive throughput at low cost.

The sensor world changed a bit the paradigm and roadmap of classical IC package since its early adoption [1–3]. The reason is that sensor package has another main role and scope: transmit the physical signal from the environment to be sensed to the actual part of the system that is sensing the stimuli. This part can be called in this chapter as "sensing element." The transmission path created on the sensor package has finally to not interfere with the signal itself also keeping the protection purpose as much as possible [4, 5]. As an example, Fig. 24.1 is showing a section of the typical pressure sensor used in industrial and automotive application since the 0mid-1980s where MEMS is a membrane that in case is also packaged without control ASIC for peculiar applications. The package structure is a premolded SO

M. Del Sarto (✉)
ST Microelectronics, Analog MEMS and Sensors Group, MEMS Package Design R&D, Agrate Brianza, Monza Brianza, Italy
e-mail: marco.delsarto@st.com

© Springer Nature Switzerland AG 2022
B. Vigna et al. (eds.), *Silicon Sensors and Actuators*,
https://doi.org/10.1007/978-3-030-80135-9_24

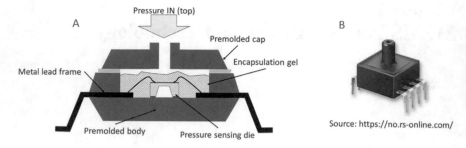

Fig. 24.1 Premolded SO package section used for pressure sensor. Sketch (**a**) and actual example (**b**)

Table 24.1 Feature summary for sensor family

Sensor family	Fan-out	Protection and housing	Sensing signal path
Motion sensor	Copper frame, organic substrate, ceramic	Top cap, full mold, ceramic housing	Die attach and housing material, die placement
Pressure sensor	Copper frame, organic substrate, ceramic	Top cap, full mold, pin assisted mold, ceramic housing	Pressure port, silicon trenches, die attach, and housing material
Waterproof pressure sensor	Ceramic	Metal cap, gel coating	Metal cap, gel coating, die attach material, housing material
Microphones	Organic substrate	Metal can, plastic can	Die attach material, accurate die placement, filtering elements
Other environmental sensor	Copper frame, organic substrate, ceramic	Metal can, plastic can	Die attach material, filtering elements, material contamination

(small outline) where in its cavity, the die (or dice) is attached then wire bonded. Cavity is then filled with an encapsulation and protection material (silicone-based gel as example). Finally, a cap with the pressure inlet is glued on the cavity to add a protection structure and eventually a hydraulic pipelike connection.

Table 24.1 is summarizing the main features of the package based on the typical sensing families; here it is reported the technologies that are mostly used in the market for routing of electrical signals, protection method, and the sensing signal path.

24.2 Sensor Package Design Structures

Package structure can be defined as the overall technology used to contain and protect the dice, route the signal, and allow sensing.

Sensor packages and generally packages can be divided in two categories:

– Full molded.
– Cavity.

The two can then split again based on the substrate used for signal routing and for material used for cap/can in case of cavity packages. Figure 24.2 is showing a diagram that exploits some of the most used combination.

Full mold package is the most used standard in semiconductor industry. On this process the dice mounted and wire bonded on a substrate or frame are covered by a dedicated epoxy resin. This structure is obviously suitable only for MEMS and sensor that have the sensing element protected by a cap (e.g., by wafer to wafer level bonding process). This is mostly used for motion MEMS sensor (accelerometer, gyroscope, magnetometer, and their combinations) and in peculiar case for pressure sensor as based on STMicroelectronics Bastille process.

MEMS wafer level bonding is often considered a first level package on sensor. Due to the manufacturing environment used for the W2W bonding process and

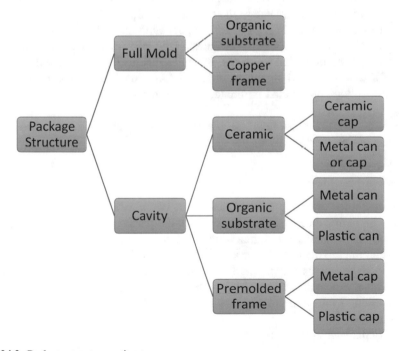

Fig. 24.2 Package structure option tree

Fig. 24.3 Full mold package most used structure. SO package (left), LGA package (center), and QFN (right)

involved processes, more correct classification for this process family is indeed at front-end process. It must be noted that in peculiar case it acts as a relevant package feature like in STMicroelectronics pressure sensor.

Full mold packages are directly derived from standard semiconductor package and have the main advantages of high-volume production capability at very low cost, so this technology is the mainstream for consumer and mobile application. Main drawback for this solution is that the thermomechanical stress exhibited at MEMS die level by the mold and substrate material can impact the performances of the sensor itself [4–6]. In addition to dedicated sensing element design, some package design rules and material selection hints have to be considered to minimize the effect. These will be treated in the dedicated sections.

Most used structures (Fig. 24.3) for the package used in MEMS/sensor package are small outline (SO) as leaded package and land grid array (LGA) or quad flat no-lead (QFN) for non-leaded packages. SO and QFN packages are based on copper frame for die attach and signal routing, and LGA are instead based on laminate substrate (called copper clad laminate, CCL) composed by resin core and copper foils.

In terms of die placement on the routing substrate or frame, two design architectures are the most used:

– Side-by-side placement.
– Stacked placement.

Figure 24.4 is reporting these structures. A combination of the two is also used when more than 2 dice must be placed inside the package. On the stacked version, the solutions of standard die attach and film over wire is depicted. The latest is used when the top die has similar size as the bottom die and becomes a need when control ASIC becomes smarter (and thus bigger) with the addition of more and more digital signal processing. Figure 24.5 is showing a 9-axis inertial measurement unit (IMU) where 7 dice and passives are mounted in the package using all described configurations.

Cavity packages [7] are used for different needs:

• Reduce the stress transferred by package material to the sensing element.

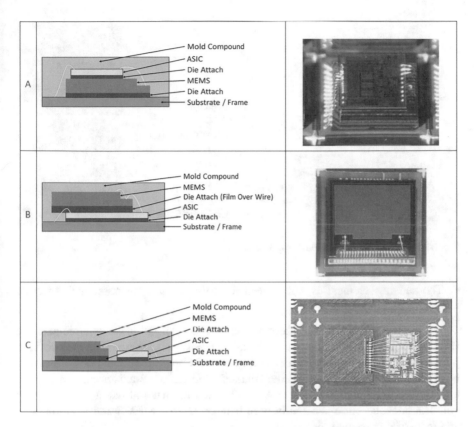

Fig. 24.4 Full mold package most used schematic internal structures. (**a**) Original stacked structure with Mechanical element on bottom and ASIC on top; (**b**) Reversed stacked structure. When ASIC is similar size or larger than MEMS, it can be placed below it; (**c**) Side-by-side structure where ASIC and mechanical element are placed at same level on substrate

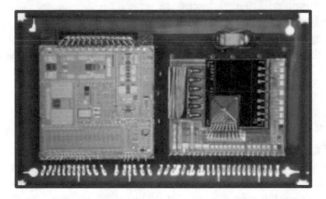

Fig. 24.5 Combined side-by-side and stacked structure for a 9-axis MEMS-based IMU

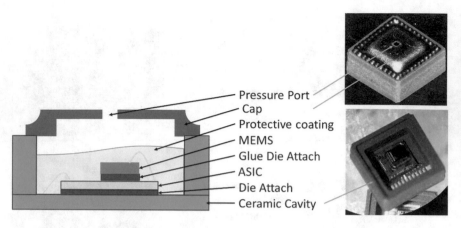

Fig. 24.6 Cavity ceramic-based MEMS sensor

- The sensing element is not protected and/or need connection with external environment.

For the low stress package, the historically used concept is the ceramic package (Fig. 24.6). Concept for this can be a ceramic cavity that host the dice then covered by a cap or a flat substrate then closed by a can. Cap sealing could be hermetic as in the first accelerometers (like the "classic" ADXL202, 2-axis accelerometer) or having vent holes and die then protected by coating material like a silicone-based gel. The latest solution is actually used both on motion MEMS and in automotive and industrial pressure sensor.

More recent examples of successful and high volume cavity package are the ones used for consumer waterproof pressure sensor and microphones.

The market success of wearable devices like smart watches leads to the transformation of industrial flavored pressure sensor on a consumer market. In fact, the so-called sport watches need to be waterproof objects that can be used also for swimming in pools and seawater, and dedicated pressure sensor is introduced to measure the barometric pressure variation. To build a real waterproof pressure sensor, there is the need to work on three aspects [8]:

- No leaking materials with the use of ceramic substrate and metal can.
- Design a custom can (or cap) capable to fit O-ring.
- Dice protection obtained by gel coverage.

Figure 24.7 is showing the typical construction parts of a waterproof pressure sensor aimed for consumer/wearable market. Base substrate is built with HTCC (high-temperature cofired ceramic) material that ensures no absorption of moisture or water. The ceramic substrate can be designed on different layers so to allow routing of electrical signals (see Sect. 24.3). The peculiar shape of the metal can define a bore designed to fit an O-ring that is contained in position by the top lid and base of the cap. ASIC die is film attached on the ceramic base and sensing element

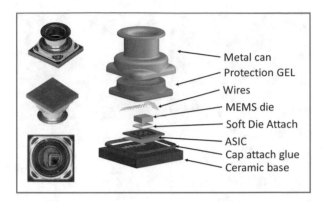

Fig. 24.7 Waterproof pressure sensor construction

(pressure membrane) is attached on top of ASIC with a low modulus glue. The soft glue is needed to decouple the thermomechanical stress from the environment to the sensing membrane itself. All dice and wire bond volumes are then covered with special potting gel material that has the purpose to protect from water, moisture, and chemicals the sensitive parts of the circuits and mechanics.

Microphones are one of the most successful MEMS in the market; they can be seen as kind of pressure sensors capable to measure the fast pressure variation (audio frequency range between 30 Hz and 10KHz typically), thus the structure of this package is a cavity package with few peculiarities that allows high volume production and good accuracy on the sound sensing. The most used microphone has the so-called bottom port structure where the sound inlet (sound port) is in the backside of the package, on the same side of the soldering land (Fig. 24.8). The substrate to hold the mechanical element and the control ASIC is based on organic substrate to allow design flexibility, hole capability, and integration of passive component (see dedicated section). ASIC is attached with standard glue or film while the membrane die must be attached with a soft material that decouples the package stress to the membrane itself. The cap is instead soldered by use of standard solder paste through mass reflow process derived by the flip chip package technology. Figure 24.9 shows the typical process flow used for the assembly of microphones.

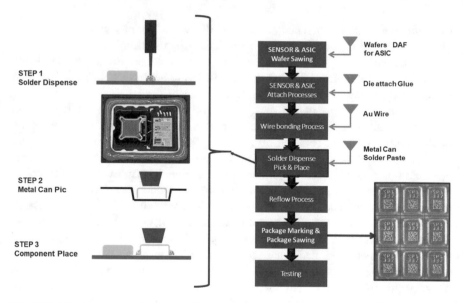

Fig. 24.8 Bottom port and top port microphone packages

Fig. 24.9 Metal can microphone assembly process flow

24.3 Basic Technology and Processes for Sensor Package

24.3.1 Fan-out Technology

As seen in previous section, the structure of the package is composed of a base that is holding the dice and allowing electrical signal routing (fan-out). At project startup, it is then a must to understand which is the most suitable technology for driving the electrical signal out of the die to the external of the package.

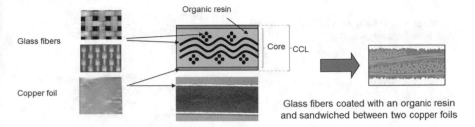

Fig. 24.10 Copper clad laminate construction; it's the base for organic substrate processing

Most used technologies for MEMS and sensor are:

- Ceramic substrate.
- Organic laminate substrate.
- Copper frame.

Few other technologies are emerging (molded interconnect substrate (MIS), wafer level fan-out) and will probably lead for next wave of innovation on sensor package.

Ceramic substrates (flat or cavities) are mainly used for their intrinsic stability capability due to hardness of material itself, providing a good insulation to external environment, and thermal expansion coefficient that is close to the silicon die one. This is resulting in low stress transfer from package housing to the sensing die that then is exhibiting highest performances also versus temperature changes and environmental stress. Attention should be placed to the use of ceramic package for large area package (greater than 8 mm on one side) due to the high mismatch of CTE (coefficient of thermal expansion) between ceramic itself and circuit board (PCB, printed circuit board) where component is mounted.

Organic laminate substrates are based on a copper clad laminate material where two copper foils are laminated around a glass fiber cloth material. Typical material used is organic resin (e.g., BT, bismaleimide-triazine) immersed in warp-weft of glass wires (Fig. 24.10). Starting from this core, material copper can be etched and plated, while core material can be drilled and plated to connect electrically top and bottom copper layers. Following a typical process flow as per Fig. 24.11, a two-layer connection circuit substrate can be realized. Multilayer structures are possible by subsequent lamination of starting core material with so called "pre-preg" material that has similar content as CCL but only one side is copper laminated.

Evolution of substrate technology is today following two parallel paths:

1. In the computing area, the need of more electrical lines, signal integrity concern, and high pinout count is driving the "in plane" technology advancement.
2. On out-of-plane/3D integration, the biggest innovation has been driven by MEMS microphone industry.

In the latest case, the need of reduce the output electrical noise on analog microphone leaded to the need of using RC filters on supply voltage and output

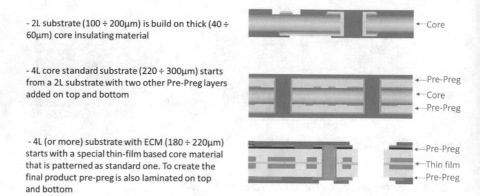

- 2L substrate (100 ÷ 200μm) is build on thick (40 ÷ 60μm) core insulating material

- 4L core standard substrate (220 ÷ 300μm) starts from a 2L substrate with two other Pre-Preg layers added on top and bottom

- 4L (or more) substrate with ECM (180 ÷ 220μm) starts with a special thin-film based core material that is patterned as standard one. To create the final product pre-preg is also laminated on top and bottom

Fig. 24.11 Construction of standard 2 or 4 layers of substrate versus the thin-film/embedded capacitance one

Table 24.2 Summary of advantages and drawbacks for any of the solution

	Routing flexibility	Size	Development cost	Unit cost
Ceramic	Mid	Mid-high	High	High
Copper frame	Low	Low-mid	Mid-high	Low
Organic laminate	High	Low	Low	Mid

itself. To integrate these in the package, the substrate stack is built starting from a core material that has very thin thickness and dedicated material (embedded capacitor material, ECM) to obtain high capacitance density (around 5 nF / cm^2) as reported in Fig. 24.11. With dedicated design and process, a capacitor can then be designed in the package. To realize resistor, a stack of copper and nickel/phosphorous is used for the connector. By means of masking and material selective etching, the Ni/P layer is left with no copper on top on the region where resistor must be created (Fig. 24.12).

Copper frame is also used as substrate on MEMS sensor. With the main advantage of very low cost of material, the copper frame has not the capability to reroute the signal, and a "direct" fan out from electronic component to the package pad is realized. The results of this limitation are typically a larger package size and less flexibility in design. The usage is in any case spread especially in automotive market where the possibility to have or lead (like in SO, small outline package) or wettable land side is appreciated for extended solder joint reliability (Figs. 24.13, 24.14 and Table 24.2).

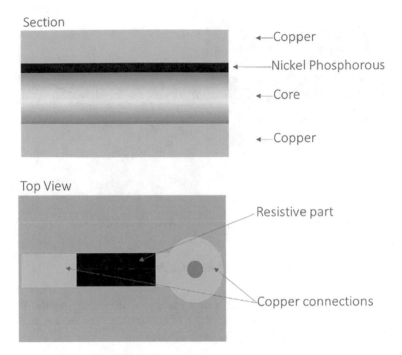

Section

←—Copper

←——Nickel Phosphorous

←—Core

←—Copper

Top View

Resistive part

Copper connections

Fig. 24.12 Starting from a laminate construction where thin nickel-phosphorous (NiP) layer is below the copper, a resistor can be designed and fabricated by means of masking and selective etching processes

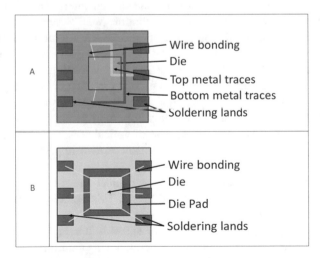

A

Wire bonding
Die
Top metal traces
Bottom metal traces
Soldering lands

B

Wire bonding
Die
Die Pad
Soldering lands

Fig. 24.13 Difference between a routable substrate and standard copper frame. (**a**) On a two (or more)-layer substrate signal out of die can be routed to any of soldering lands. (**b**) On a copper leadframe, there is no routing capability and the output pad must be directly facing the die pad

Fig. 24.14 A QFN package (**a**) and a SO package (**b**) soldered on PCB

24.3.2   Die Attach Processes and Materials

On the MEMS sensor, the method and accuracy of die attach to the substrate determine performances and characteristics of the sensor. The reason is different for different sensor type and generally these are due to:

- Geometrical factors, like placement accuracy of the sensing die.
- Thermomechanical aspect as stress transfer.

Most important geometrical factor for die placement is the rotation angle of the die with respect to package edge. In motion MEMS sensor (accelerometer, gyroscope, magnetometer), the output relation between a rotation respect to input signal and cross axis can be considered linear for small angles. In this case, a rotation of mechanical sensing element of 1° inside a package is leading to a 1% of cross axis signal specification. Research is progressing in the direction of better equipment and stable process due to the need of more and more accurate sensor.

Displacement accuracy is relatively less important as is not impacting one-to-one on the device performances. In the case of optical sensor, also the translational tolerances play instead a role on the performances as it impacts the alignment of light signal path.

Reducing the package-induced thermomechanical stress is key to increase the sensor performances for all MEMS-based sensor. This is achieved in different way but always die attach plays a role.

For full mold package, the reduction of stress is related to the uniformity of stress transfer to the sensing element. In general, having a uniform stress level helps the designer to counteract by peculiar structure or eventually by calibration during testing phase (Table 24.3).

Epoxy glues are the first choice for die attach as used on standard semiconductor packages. Due to overall spread on processing, from dispensing to cure, the thickness of the material below the die (BLT, bond line thickness) and its uniformity

Table 24.3 Die attach medium and their primary characteristics

	Suitable package type	Size reduc-tion	Stress transfer	Uniformity	Geometrical impact	Further process-ability
Epoxy glue	All	Low	High	Low	High	Standard
Die attach film *	All	High	High	High	Low	Standard
Low modulus glue	Cavity	Low	Low	Low	High	Difficult

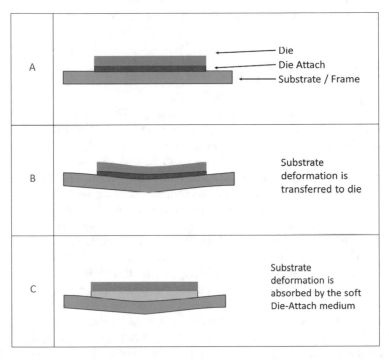

Fig. 24.15 From unstressed situation (**a**) if the substrate tends to warp for an external stress, if a rigid die attach is used (**b**), the deformation is transferred to the die. Usage of low Young's modulus die attach (**c**) decouples the deformation of the substrate to the die

are not well controllable. The Young's modulus of these materials is in the range of 100's to thousands of MPa, leading to a high stress transfer by the material below to the die itself (Fig. 24.15).

To overcome the issues related to geometrical uniformity, new materials have been developed in the form of adhesive films. These are called die attach film (DAF) and film over wire (FOW). These materials are prelaminated on dicing tape and then results directly attached to wafer backside after the dice singulation. The die attach process is then directly from singulation frame to substrate without the need of glue dispensing. The nature of the tape is then leading to an almost perfect uniformity in thickness variation and also to a very low thickness variation from lot to lot. Typical range of thickness used is in the range between 10um and 20um. The drawback of these materials is the high Young's modulus (GPa range) that in any case transfer strongly the stress from substrate to the die.

Low modulus glue (Young's modulus in the range of 1 MPa to 10 MPa) is also widely used for stress decoupling especially in pressure sensor and microphone in cavity packages. The main drawback of the use of these materials is processability. In fact, both control of BLT (bond line thickness) and uniformity is critical; at same time the subsequent wire bonding process can be tricky due to the low mechanical resistance that the material itself offers at ultrasonic power used during the process.

24.3.3   Electrical Interconnection Methods

Electrical interconnection between IC controlling ASIC and mechanical element and between ASIC and substrate is typically done with standard thermo-sonic wire bonding where heat, pressure, and ultrasonic wave is applied to form the joint between wire and pad (see Fig. 24.16). Gold wire is still the first choice when developing a new package for sensor due to relatively short wire length used for sensor. Aluminum wire is instead used on ceramic packages when glass frit sealing is used as this is the process known to survive the high bonding temperatures.

The wire diameter can go down to 0.6 mils (15 um) for full mold package, while for cavity packages it is recommended a slight higher diameter (e.g., 0.7 mil = 17,5 um) to have a higher mechanical stiffness and robustness being the wires not embedded and fixed on a rigid (e.g., mold) material.

Ultralow loop profile and reverse bonding are also a must-have on the portfolio of sensor package development. The first is used on the ASIC on top stack-up to keep overall low package height, the second when the total height of the dice stack is so high that a standard loop will create difficulties on the approach of second bond to the receiving pad or finger. Figure 24.17 is showing the basic process flow and typical shapes where an ultralow loop profile can be set around 50um and a reverse profile can be set at 75um.

Flip chip technology is not widely used in sensor environment except few cases where ASIC die is mounted facedown on substrate for pressure sensor or in peculiar three-axes compass. Magnetometer sensing elements, used for compass application, are relatively easy to be built for in-plane sensing. To achieve a sensor capable to measure also out-of-plane magnetic field, the flip chip technology is used. In this case a dedicated design of sense element and wettable (typical Ni/Au finish) metallization on the chip is needed. The die is then mounted rotated at 90° (images/sketches), and dots of solder paste (or balls) are placed at the corresponding pad on substrate (or ASIC) and MEMS. Figure 24.18 is showing the principle of the process and the results on a case where the sense element is mounted on substrate and when it is mounted directly on the controlling ASIC via a dedicated RDL.

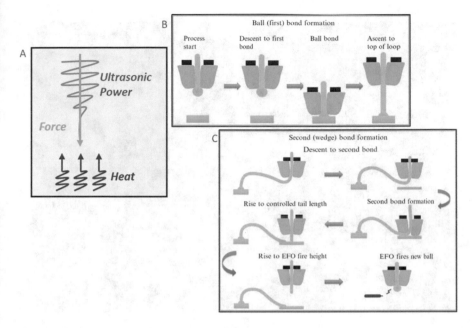

Fig. 24.16 Basic wire bonding technology explanation. (**a**) Thermo-compression wire bond implies energy transferred by force, hear, and ultrasound at same time. (**b**) To perform a wire bond, first a ball is formed by EFO (electro-formation) and then bonded on the pad. (**c**) The other size of the wire is then "stitched" on the other pad or land

24.3.4 Molding Processes

Molding is the process where the dice mounted on the substrate and connected electrically is encapsulated inside a plastic component. Material used is the so-called molding compound, typically abbreviated in EMC (epoxy molding compound), which is a thermoset material specially manufactured for semiconductor industry and used to encapsulate the assembled device. The resin is injected at high temperature (170° C to 185° C) and high pressure (around 1 Ton) in liquid format inside a cavity (mold chase). Subsequent step of curing is then solidifying the material and completing the polymer cross linking (see Fig. 24.19).

For sensor performance optimization, low stress EMC must be considered in material selection while the choice between transfer or compression molding is more devoted to the stress that can act on the dice itself during the molding process.

Stress generated by a material over another (like in case on molded package) can be defined as the integral result of the multiplication of coefficient of expansion versus the Young's modulus as in Eq. 24.1 and summarized in Fig. 24.20:

$$s(T) = \int E(T) \cdot \alpha(T) \cdot dT, \qquad (24.1)$$

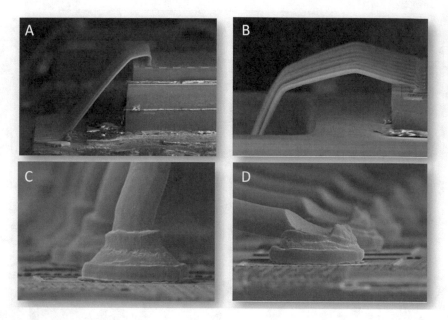

Fig. 24.17 Wire bonding typical shapes. Standard forward loop (**a**) where ball is formed and placed on the silicon pad (**c**) and stitch on substrate. Reverse loop (**b**) has a stitch formed on a pre-placed ball (**d**) on the silicon

- $s(T)$ is the stress versus temperature,
- $\alpha(T)$ is the expansion curve (CTE).
- $E(T)$ is Young's modulus.

CTE can be varied and tuned by suitable filler loading and resin selection while additives and catalyst are the ingredients for low modulus material. Wide material selection is available at material supplier to choose as function of technology to be developed.

Molding process exerts relatively high pressure on the dice during the process. In the process flow, the resin is first injected (for "transfer mold tools") on the mold chase (cavity). The resin is then pressurized with the force defined by the chase itself, called "clamp force." This latter is the key feature to reduce the mechanical stress on the dice that can lead, especially in case of MEMS with a cap, a crack on the cap itself or on the bonding ring. Typical values for the clamp force may vary from 15 to 35 Tons. It should be noticed that the lower the pressure, the higher the risk of voids in the mold compound itself.

Another variant of mold process used in sensor package is the film-assisted molding (FAM) to produce cavity packages or exposure of silicon on top of the package. Figure 24.21 is showing the basic principle and Fig. 24.22 is showing the

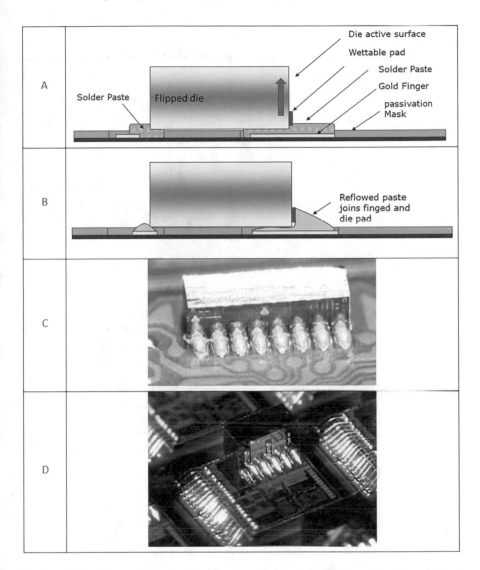

Fig. 24.18 Flip chip assembly for vertical axis magnetometer. (**a**) After solder dispense, die is placed flipped by 90° on the substrate. (**b**) After reflow the solder material joints the substrate land to the die pad. (**c**) Z axis magnetometer die mounted on organic substrate. (D) Die mounted vertically direct on control die

STMicroelectronics pressure sensor that uses the FAM process to expose the silicon cap of the mechanical.

Fig. 24.19 (1) Strip is inserted inside cavity. (2) Cavity is closed. (3) Molding compound is transferred inside cavity by means of a pressure difference

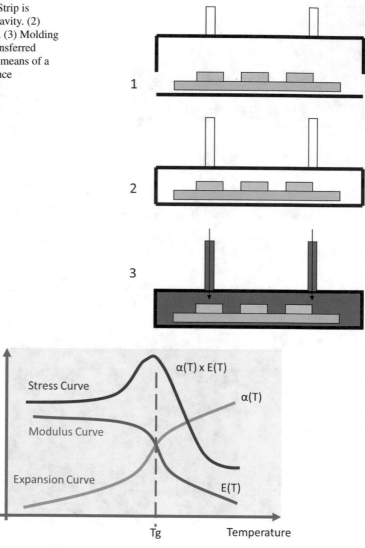

Fig. 24.20 Graphical representation of stress curve versus temperature

24.3.5 *Cap/Cans Material and Attach Processes*

As introduced in Sect. 24.2, many sensors are built in a so-called cavity package for sensing microphones, gas sensor, pressure sensor, etc. or for peculiar needs like stress decoupling in high accuracy inertial measurement units (IMU) [9].

For the design of cavity package, there are two starting choice:

1. Use a substrate that is a cavity and put a flat lid on top.

Fig. 24.21 1. Strp inserted inside cavity. Top part ot mold chase is supporting a tape. 2. when mold chase is closed the tape get in contact to silicon surface. 3. Tape presence blocks the resin flow on top of the die so exposing the surface

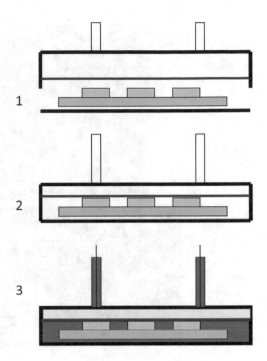

2. Use a flat substrate and a modeled cap.

The option (1) is mostly used on ceramic packages and on premolded packages. Ceramic cavities are obtained by lamination of different layers where base ones are flat and subsequent ones are "rings" that create the cavity at the end (Fig. 24.23). Premolded package are obtained by combination of copper frame and film-assisted molding (Fig. 24.24).

The cap/can/lid can be produced in various materials depending on actual needs, from ceramic itself to various metal (typical use is stainless steel, Kovar, and brass) up to different types of plastic caps (Fig. 24.25). Cross-reference Table 24.4 can be used to verify the suitable cap material family versus the used base.

Sealing of cavity packages plays then a key role to define the characteristics of the product in combination with the material. At first point, another discrimination can be done if the sealing is leading to a hermetic package or not. With the present technologies, real hermetic sealing is possible in ceramic-ceramic or ceramic-metal packages.

Hermetic sealing is using mainly three technologies:

– Low-temperature glass sealing (glass-frit).
– Seam-welding.
– Solder paste/mass reflowable material.

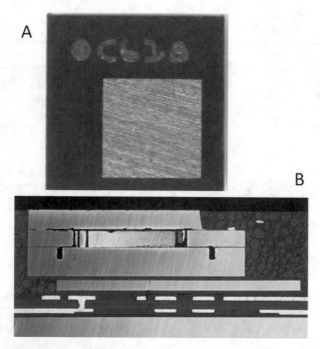

Fig. 24.22 Pressure sensor with exposed silicon cap by STMicroelectronics. The top view (**a**) is showing the exposed silicon with the pressure ports. Cross section (**b**) shows the silicon stack with the evidence of cap silicon that is the same top level as mold resin

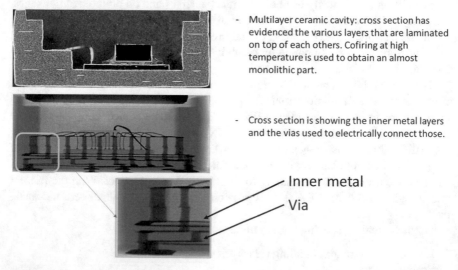

- Multilayer ceramic cavity: cross section has evidenced the various layers that are laminated on top of each others. Cofiring at high temperature is used to obtain an almost monolithic part.

- Cross section is showing the inner metal layers and the vias used to electrically connect those.

Inner metal

Via

Fig. 24.23 Ceramic cavity structure. Several layers of ceramic can be stacked and fired together. Metal-based vias are used to electrically interconnect the various routing layers

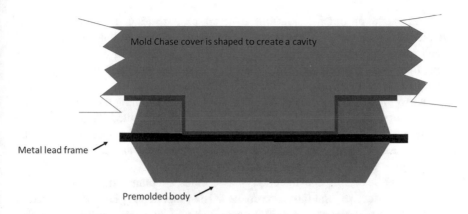

Fig. 24.24 Premolded cavity can be obtained via PIN-molding. The mold chase top cover is shaped to get in contact with the support frame/substrate during the mold flow. The mold material flow so around to create a cavity around the pin

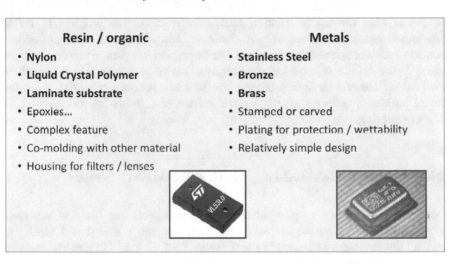

Resin / organic	Metals
• **Nylon**	• **Stainless Steel**
• **Liquid Crystal Polymer**	• **Bronze**
• **Laminate substrate**	• **Brass**
• Epoxies...	• Stamped or carved
• Complex feature	• Plating for protection / wettability
• Co-molding with other material	• Relatively simple design
• Housing for filters / lenses	

Fig. 24.25 Cap/can materials and their main features

Table 24.4 Cross-reference scheme between substrate material and cap material family

	Cap material		
Substrate material	Ceramic	Metal	Organic/plastic
Ceramic	Yes	Yes	No
Copper premolded	No	Yes	Yes
Organic/laminate	No	Yes	Yes

Glass-frit sealing is suitable for ceramic-ceramic joint and it's well known and stable process. As processing temperatures are over 400° C, the electrical connection has to be done with aluminum wires.

Solder paste or other reflowable material process (e.g., eutectic Au/Sn) is suitable for ceramic-metal joining and organic-metal joining. Metal cap surface and metallization on the ceramic has to be wettable by the material with, as example, Ni/Au finishing. Temperature is driven by the material used and generally in the 240 deg. C range so compatible with gold wire bond process.

Seam welding is using localized high temperature for joining the metals and the bulk of the package as well as the dice are at low temperature. This is the main advantage for this process that has the drawback on material selection (metal cap must be in Kovar steel), and also ceramic has to have a Kovar steel brazed ring to allow the joint thus resulting in higher cost of materials.

Non-hermetic sealing is using epoxy-based glues, eventually filled with metallic material if conductive joint has to be created. The proper epoxy has to be studied as function of the material couple to be joined. It should be noticed however that also the material used for hermetic sealing can be designed to be non-hermetic. Peculiar case is the one of microphones where the soldering process is used to glue the metal can to the organic substrate while package is clearly non-hermetic.

Design-wise the cavity solution must counteract for the problem of inside pressure increase if these are hermetic or in any case not opened. Due to that low bonding pressure and robust design are mandatory. On this path in recent years, the epoxy-based solution evolved in the "semi-hermetic" packaging capability where material choice and dedicated processing (like low pressure bonding) allow the cavity package to survive to temperature stresses (e.g., soldering reflow) with no crack or bulging.

24.4 Computed-Aided Engineering for MEMS Sensor

Design of a sensor is a contribution of several aspects and often is vertically integrated starting from wafer technology development to design and ideally in parallel development of packaging technology [10]. Several CAE methodologies and tools have been developed during the years to couple the different physics behind (almost) any aspect of the sensor design, trying to predict the system performance.

Several points peculiar to sensor package design can be considered, as examples:

– Package to silicon interaction where the critical point is the scaling of FEM modeling from millimeter size packages to micrometer scale of MEMS element.
– Novel characterization techniques are needed to analyze stress and strain on small packages.
– Different materials coupling needs to be studied where environmental and optical sensor need metal or plastic cap and sensing window.

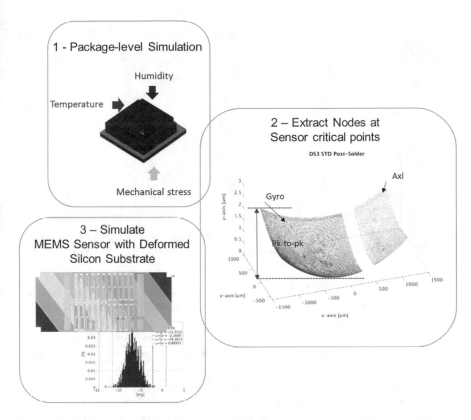

Fig. 24.26 Three-step analysis used for typical MEMS package vs sensor FEM simulation

Creating a single FEM model that exploits both package and sensor feature is practically impossible due to exploding number of simulation elements, for this reason a box-in-box approach can be used.

Starting from a package modeling, where the silicon features are not designed (Fig. 24.26, block 1), the analysis is performed to extract a deformed shape on the MEMS critical points (Fig. 24.26, block 2). The critical point is, for example, the anchoring points of the mechanical structure. These deformed shapes can be used both for a more detailed analysis and as comparative analysis between different designs and material choices. In Fig. 24.26, block 3, it is shown that the deformed shapes are used as input for FEM models that start from and already stressed-strained structure thus giving more accurate results.

Sensor packages that mostly feed the consumer market are getting smaller and smaller. At present the standard for a 3-axis accelerometer is to have a 2 mm x 2 mm package footprint while a 6-axis sensor, containing also a 3-axis gyroscope is smaller than 3 mm x 3 mm. At these dimension levels, the standard deformation analysis techniques (like Shadow Moire) start to show some limitations, and the use

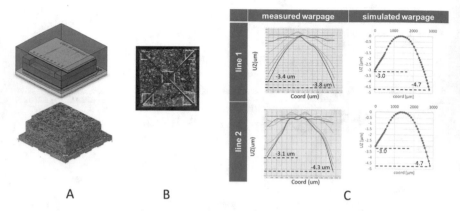

Fig. 24.27 (**a**) 3D model of accelerometer in LGA package. (**b**) Diagonal lines on top of package where the deformation is measured. (**c**) Comparative results between measurement (left) and FEM model (right)

of a micro digital image correlation (uDIC) instrumentation [11] can be ideal to measure small deformation on small scales.

Analysis of a full mold LGA accelerometer package of 2 mm x 2 mm x 1 mm though package FEM and measured through DIC is reported in Fig. 24.27. Applied stress was aimed to simulate the reflow profile with standard lead-free peak at 260 °C. Data extracted by the uDIC analysis are the displacement along the two diagonal lines on top of the packages (Fig. 24.27b). Correlation results (Fig. 24.27c) are impressive showing an almost perfect matching on the extracted displacement lines.

24.5 Advanced Design and Processes

In the field of MEMS sensor, the progresses are going in different directions [12]. The IEEE (Institute of Electrical and Electronics Engineers) is working, through its Electronic Package Society (EPS), to develop the so-called Heterogeneous Integration Roadmap. MEMS and sensor are part of this roadmap. First analysis of the dedicated working group is pointing three main challenges that can be summarized as follows:

1. Evolutionary part: improve sensor performances, stability to address more and more stringent condition for both automotive and consumer market.
2. Revolutionary part: increase digital signal and radio-frequency transmission and detection capability as system in package/heterogeneous integration.
3. Simplification/standardization: define platform commonality for signal chains.

On first point the research is working on material selection and new structure definition. Combination of process bricks developed for different sensor and package structure can be joined to reach the best or desired performance. As described in [9],

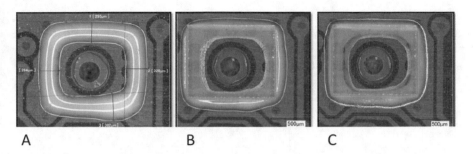

A   B   C

Fig. 24.28 B-stage type soft glue maintains the aspect ratio obtained at dispensing (**a**), after B-staging and die placement (**b**), and after full curing (**c**)

an organic substrate with metal can structure, typical of MEMS microphones, can be used for motion MEMS (inertial measurement units, accelerometer, gyroscope) with the eventual adoption of a "mechanical interposer" that can be designed to "mix and match" the different material characteristics to obtain a high performance sensor under different and frequently contrasting aspect (like solder joint reliability versus parametric stability).

On material side the low modulus die attach (often silicone based) material is getting into commercial on film-based structure thus easing the processing steps. At the same time, interesting development is coming from "B-stage" material that can be dispensed and kept in shape with a first quick cure. Full curing will then be applied after die/cap attach (Fig. 24.28).

On structure and process integration standpoint (point 2 of EPS roadmap), one path is going to wafer level package and more in general on the integration of typical wafer processing for MEMS in the package processing. In [13] it is found the activity done by Infineon on the fabrication of a microphone at wafer level package integration. In this case, to handle a fragile membrane in the packaging process, the front-end (intended as processing the silicon wafer) process itself was not completed, but the sacrificial layer was kept and finally removed at the end of assembly process.

On similar path, Fraunhofer institute published [14] the activities done to manage sensitive surfaces of MEMS wafer during the WLP (wafer level package) and the use of completely novel material for the wafer-level or panel-level packaging.

Standardization in MEMS consumer and personal electronic industry, the point 3 of EPS roadmap, was avoided in the first era of introduction of this technology as each manufacture was leveraging on peculiarity of own production process also for business penetration perspective. In the middle of 2010s, the differences on the technologies used for the fabrication of sensitive elements have been overcome and package technologies converged in a couple of main building structures. As shown in [15], the motion MEMS field focused on full mold LGA packages (consumer electronic), and microphone arena is dominated and somehow standardized on metal can on organic substrate structure.

Different is situation on automotive market where, especially on advanced IMUs used for vehicle control, the customer-driven design and specification lead to a wide differentiation on package structures.

On the same page but on opposite direction, the pressure sensor arena, if standardized in industrial and automotive marked, still differentiates on consumer one for the main reason of different sensing technologies still used for fabrication of sensitive membrane.

24.6 Conclusion

This chapter should have illustrated the basic principle of the structures and processes useful to build a successful MEMS sensor package.

Package design and processes for MEMS sensor follow the basic rules of semiconductor package adding, on top as complexity and in parallel as study, the need of having a sensing path that do not corrupt the element to be sensed. Peculiar attention needs then to be paid on the aspects of the package that influence the sensing path as well as the aspect of protection and electrical signal integrity.

The physical/sensing signal integrity can be managed by different aspects of package design and technology. On design base the first choice is if a full molded or a cavity package is the more suitable choice for both stress relief and signal path. The substrate on which the dice are then mounted is playing a role and correct trade-off must be chosen between cost, signal integrity, and again stress transfer.

The three selections should then proceed for each process from die attach to wire bonding and encapsulation in the above-mentioned framework.

The following paragraph shows then some examples on how computer-aided engineering can help in package design and selection of best suitable bill of material.

The chapter concludes with some references to the roadmap of sensor packaging from industrial and research center point of view.

References

1. Southwest Center for Microsystems Education, *History of MEMS PK*, http://scme-nm.org
2. Fujita, H. (1997). A decade of MEMS and its future, 0-7803-3744-1/97/$5.00 0 1997 IEEE.
3. Fujita, H. (2007). Two decades of MEMS – From surprise to enterprise", MEMS 2007, Kobe, Japan, 21–25 January 2007.
4. Krondorfer, R. H., & Kim, Y. K.. (2007). Packaging effect on MEMS pressure sensor performance, 1521–3331/$25.00 © 2007 IEEE.
5. Wen, J., Sarihan, V., Myers, B., & Li, G. (2011). Multidisciplinary approach for robust package design of MEMS accelerometers. *IEEE Transactions on Components, Packaging and Manufacturing Technology*, Vol. 1, No. 12, December 2011.
6. Zhang, X. (2007). *Accurate assessment of packaging stress effects on MEMS sensors by measurement and sensor–package interaction simulations*, 1057–7157/$25.00 © 2007 IEEE.

7. Lee, J. S., & Faheem, F. F.. (2009). *A cost-effective MEMS cavity packaging technology for mass production*, 1521–3323/$25.00 © 2009 IEEE.
8. Duca, R., & Ghidoni, M. O. (2019, May). *A comprehensive methodology for design for package miniaturization*. Minapad 2019, Grenoble.
9. Del Sarto, M., et al. (2020). *Hybrid package for high performance Inertial Measurement Units*. 70th Electronic Components and Technology Conference (ECTC).
10. Del Sarto, M., et al. (2019). *Multiphysic simulations for MEMS sensor package*. 21st Electronics Packaging Technology Conference (EPTC).
11. https://www.dantecdynamics.com/measurement-principles-of-dic
12. Vigna, B. (2019). *Sensor integration: Feynman or Moore*. IMAPS Device Packaging Conference.
13. Theuss, H. et al.. (2019). A MEMS microphone in a FOWLP. IEEE 69th Electronic Components and Technology Conference (ECTC).
14. Braun, T. et al. (2019). *Fan-out wafer level packaging - A platform for advanced sensor packaging*. IEEE 69th Electronic Components and Technology Conference (ECTC).
15. SystemPlus consulting. (2017, October). *MEMS packaging reverse technology review*, Version 1.

Links

www.st.com
https://invensense.tdk.com/
www.Infineon.com
https://amkor.com/
http://www.cea.fr/
www.fraunhofer.de
https://eps.ieee.org/hir
https://www.systemplus.fr/
www.st.com
https://invensense.tdk.com/
www.Infineon.com
https://amkor.com/
http://www.cea.fr/
www.fraunhofer.de
https://eps.ieee.org/hir
https://www.systemplus.fr/

Chapter 25
MEMS Testing: Sensors Testing and Calibration Impact on Product Performances

Michele Tronconi, Paolo Aranzulla, Giacomo Calcaterra, Matteo Catalano, Paolo Cerri, Fabiano Frigoli, Valeria Montuori, Dario Premi, Marco Rossi, and Giancarlo Spoldi

25.1 Introduction

25.1.1 Testing Purposes: Screening, Calibration, Compensation

The fabrication process of a silicon device in general and of a sensor in particular is impacted by non-idealities and testing is intended primarily to eliminate parts that due to these non-idealities of production process could present physical defects that are preventing correct functionality of the device, like in case of open and short contacts, or that are impacting the performances at different severity levels, like power consumption, voltages, and linearity. This capability of testing to remove parts showing defective behavior is commonly called "screening" and has to be applied of course to sensors; it is common to embed dedicated blocks during design phase to facilitate the implementation of screening and to make it effective. This kind of blocks are part of design for testability and design for productivity concepts.

A MEMS sensor is typically a system made of electronic integrated circuitry and of a dedicated electromechanical part that is the transducer of a physical quantity to electric signal. Hence the defects can come from electronic interface;

M. Tronconi (✉) · P. Aranzulla · P. Cerri · F. Frigoli · G. Spoldi
ST Microelectronics, Analog MEMS and Sensors Group, MEMS Sensor Division, Cornaredo, Italy
e-mail: michele.tronconi@st.com; paolo.aranzulla@st.com; paolo.cerri@st.com; fabiano.frigoli@st.com; giancarlo.spoldi@st.com

G. Calcaterra · M. Catalano · V. Montuori · D. Premi · M. Rossi
ST Microelectronics, Analog MEMS and Sensors Group, MEMS Technology and Design R&D, Agrate Brianza, Monza Brianza, Italy
e-mail: giacomo.calcaterra@st.com; matteo.catalano@st.com; valeria.montuori@st.com; dario.premi@st.com; marco.rossi-apm@st.com

© Springer Nature Switzerland AG 2022
B. Vigna et al. (eds.), *Silicon Sensors and Actuators*,
https://doi.org/10.1007/978-3-030-80135-9_25

from mechanical parts like moving masses, suspended membranes, micro tubes and caps in silicon, micro actuators, and dedicated layers; and from the way these parts interact with each other like bonding and packaging. Defects can be catastrophic, so that the impacted parts are not functional and cannot be recovered; testing strategy has to cover efficiently all these possible faulty behavior to screen out that parts possibly in an early stage of the process.

In addition to catastrophic defects, since the production steps are not ideal, all physical and electrical parameters vary inside process window and they can produce, alone or in combination with others, different behavior part to part impacting heavily on specification. In cases where parameter accuracy is a sensitive element, like it is in all sensors, the standard control level of the process could be inadequate to reach target specifications as it is; in that case, it is possible to use strategies intended to internally calibrate each device based on its particular behavior. Normally this techniques make use of a controlled, repeatable, and well-positioned "reference" and all the parts are compared to this reference and adjusted to reach the expected specification target when a stimulus is applied. In sensors this normally means that a known physical stimulus (acceleration, angles, angular velocity, magnetic field, pressure, etc.) is applied to the sensor under calibration and to the reference at the same time measuring the output and storing in a dedicated programmable memory the difference between the output read from device and the expected, reference value. This allows to reconstruct for each single device the correct output and finally to reduce devices difference one from the other inside defined limits (resolution) and to reach target performance (accuracy). Reconstruction of accurate output using these data can be done by the device itself with dedicated embedded circuitry or by external components, typically a microcontroller; in any case accuracy and resolution at testing stage are differentiating element since they highly impact final performances of the product and they can enable applications where specific precision and good repeatability are necessary. This step is called "calibration."

Testing engineers duty is to assure that equipments chosen for production have the right stimulus and reference to be applied and to guarantee that in all test condition, and during all the testing steps, the reference is stable enough and there are no other effects that could continuously or temporarily modify the value applied to device under test (DUT) preventing to calibrate correctly and within desired accuracy. Dealing with new family of sensors or with new applications that require an improvement in specification or a significant increase in market volumes of known sensors, it is often needed to develop specific setup since equipments are not yet available or they are not suitable to manage the target production volumes and efficiency. In this case, the challenge is to obtain the most accurate calibration measurement and flow in very large scale integration (VLSI) devices respecting time to market, target costs, and volume of production depending on application defined by business model.

A further step to be considered to reach desired performances is to avoid that external, environmental conditions like temperature could impact device output and modify its worsening overall accuracy. In this case, it is possible to find a correlation between device output and the external conditions impacting it to infer the correct value. For example, if a pressure sensor changes its own output response not only as a consequence of pressure but also of temperature due to second order effects on the reading chain, it is possible to "compensate" this effect inferring for each part the characteristic equation and the coefficients that modify the value of the output under measurement. If coefficients and equation are known for each part, it is possible to read pressure and temperature and reconstruct correctly the output removing undesired effect of the latter. This step is called "compensation" and it allows a stable, reliable, and precise output for final application.

In accurate MEMS sensors, all the three steps described are executed during test flow using dedicated equipment. The success of the device and the possibility to penetrate and address defined segment of the market depends largely from performances reachable in accuracy and cost of the full process in terms of equipment, complexity and efficiency of the flow, lead time and total throughput.

25.2   Wafer Level Testing

This session describes the main features of the MEMS wafer level test. In particular it will focus on the test of the mechanical part of the MEMS and on the introduction of stimuli which are the most distinctive aspects compared to the usual wafer level test.

The wafer level test setup consists of a tester and a prober. The tester deals with the generation and acquisition of signals, data processing, and saving of fundamental parameters. The prober instead manages the movement of the wafers and allows to carry the signals from the tester to the pads of the device under test. Regarding MEMS wafer level testing, the instrumentation contained in the tester is pure analog instrumentation. The necessary instruments are arbitrary waveform generators, voltage and current generators, and analog-to-digital converters to acquire and convert from analog to digital the output coming from the device under test. The characteristic parameters of a MEMS device have extremely small electrical signals to be measured, and therefore aspects such as the layout of the electrical connections and the management and processing of the generated and acquired signal become fundamental for obtaining valid measurement and a decent signal-to-noise ratio.

For example, a typical sensor like an accelerometer is characterized by very small capacitance values to be read during wafer level testing to guarantee correct functionality. The order of magnitude of picoFarad is common for these sensors and equally small current values, of the order of few nanoAmpere. Under these conditions, the electrical connections, the hardware, and the environment can easily influence the measurement.

The choice of components and an easily scalable layout are key factors for a successful development of MEMS wafer-level testing. The characterization of each element of testing equipment normally requires working in close collaboration with suppliers of the various components of the test setup. The concept of parallelism is at the basis of the calculation of the testing capacity of a system and is equivalent to the number of devices that can be tested in the same time frame. High parallelism values are essential in high volume productions. Currently on MEMS, a parallelism up to 64 is common.

In most of the cases, it is not proper to apply at wafer level the physical stimulus for which the sensor was designed. For example, the need to maintain good contact between the device pads and the tester resources is in contrast with the application of an acceleration or rotation; for this reason systems capable of doing it are normally less efficient from a production point of view. The fundamental characteristics can be tested by applying electrical signals instead of the physical stimulus. Mechanical parts of inertial products, for example, can be moved applying appropriate electrical test signals, and checking at the same time the characteristic parameters of the MEMS, it is possible to identify in an early stage devices that do not comply with the specifications.

25.2.1 Calibration of the Measurement Setup

MEMS measurement requires capability to detect very small current and capacitance values, thus making necessary to adopt a correct calibration system to check the operating status of the hardware and to eliminate the contribution due to parasites. A certified component is used to periodically certify equivalent components suitably positioned inside the tester. Through the components inside the tester, it is possible to regularly perform measurements without connecting any device to an extent. The values obtained with this measurement will then be subtracted during the measurement with the devices to be tested, and this allows to correctly eliminate all the contribution due to the parasites.

25.2.2 Specific Electromechanical Screening (Capacitive, Magnetic): Offset, Sensitivity, Quadrature, Freq Response

In the next paragraphs, the main parameters that can be tested and verified at wafer level on inertial MEMS devices will be explained, with references mainly to accelerometers and gyroscopes.

MEMS inertial sensor can be represented as a differential capacitor; basic structures with two stators and one rotor per axis can be considered as described in Chap. 16.

The four main measures that are performed on these devices are the capacitance measure, sensitivity and offset included, the resonance frequency, the quality factor measure, and the quadrature error estimation.

All these measures are useful to avoid yield loss at final test and to provide rapid feedback to the technology development team.

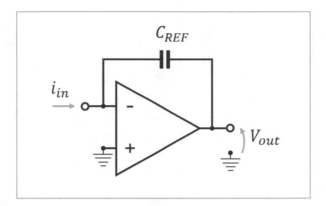

Fig. 25.1 Analog front-end circuit schematics for capacitance measurement

25.2.2.1   Capacitance Measure

Capacitance values in inertial sensor are typically in the range of few hundreds of femtofarad (fF) up to tenth of picofarads (pF); to perform a measure with a good repeatability, a setup such as calibration is used.

Using a waveform generator, a sinusoidal signal is forced to the stator (refer to Chap. 13 for stator-rotor definition) under measurement while rotor is connected to the measure chain. The signal acquired from rotor is processed by a charge amplifier with different possible sets of range and accuracy; the range is decided based on the capacitance values, and the lower range has the highest repeatability (Fig. 25.1).

After the acquisition, the signal is processed to find the amplitude applying well-known mathematical transform. Finally using the calibration values, it is possible to obtain the value of the stator capacity (C0) that has been measured. This measurement will be repeated on all stators to verify that the device complies with the design specifications.

Moreover sensitivity and offset values can also be evaluated, thanks to the same measurement. Typically, sensitivity is represented by the capacitance variation of the electrodes with acceleration stimulus applied, at wafer level if the device cannot be rotated to apply gravity force, an electrical stimulus is used to emulate acceleration since it can move the rotor to estimate variation of capacitance. Knowing relation of electrostatic force applied versus acceleration, it is possible to determine if capacitance variation measured is correct and aligned with target sensitivity.

25.2.2.2   Sensitivity Measure

To perform sensitivity measure at wafer level, a DC voltage is added to the sinusoidal signal; thanks to the DC voltage, the mass will move towards the stator

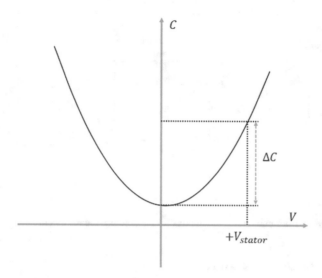

Fig. 25.2 Sensitivity measure

and the capacitance measure will increase (Fig. 25.2). Sensitivity values expressed as dC/dV is the difference between this measure and C0 measure.

This measure allows to verify that the mass was correctly released during the related process step and that there are no anomalies in the stiffness of the springs that anchor the rotor.

25.2.2.3 Offset Measure

Offset measure is the difference between C0 values of the two stators of the same axis (Fig. 25.3). These values should be zero by design but, due to process variability, the devices show a Gaussian distribution on wafer centered on zero values.

After the assembly in a package, the ASIC can trim (calibrate) the MEMS device applying a voltage on stator to align stator 1 and stator 2 values. Screening at wafer level of this parameter is useful to avoid to assembly and test not trimmable devices.

25.2.2.4 Resonance Frequency and Q Factor

Quality factor is a dimensionless damping parameter of an oscillation, which summarizes two fundamental parameters of the oscillation: damping and natural resonance frequency.

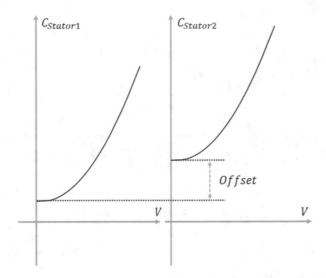

Fig. 25.3 Offset measure

Quality factor measurements are mainly performed in two different methods depending on the value of Q factor and finally from the kind of device being tested:

1. Decay resulting from a stimulus is used for high Q-factor (above 50) components like in gyroscopes.
2. Bandwidth at half power is used for low Q-factor (below 50) devices like in accelerometers.

25.2.2.5 Exponential Decay Method

The stimulus for Q factor test can be either mechanical or electrical. Mechanical stimulus is the best choice, but the device's response could be too weak or cannot be a viable solution if the vibration mode of interest is not directly excited by external accelerations (as for drive mode of common gyroscopes). In this cases it is more reliable to force an electrical sine wave stimulus swapping frequency in the range of interest around the natural frequency of device using as an example voltage step obtained by chirp function as in the following formula:

$$x(t) = \sin\left[\phi_0 + 2\pi\left(f_0 t + \frac{k}{2}t^2\right)\right]$$

where f_0 is the start frequency and k the so-called chirp rate necessary to increase the frequency.

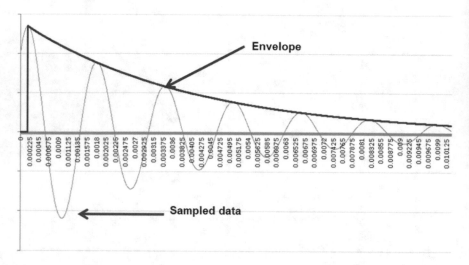

Fig. 25.4 Output signal during oscillation phase

Applying a pulsed stimulus, the device oscillates with a damping ratio ξ. The damping ratio is an important parameter to evaluate the Q factor. It's obtained from the envelope of the signal vibration intensity, in time domain.

In Fig. 25.4 the oscillation frequency of sampled data is the resonance frequency of the device and the envelope can be fit with a negative exponential decay function $y(t) = e^{-\frac{t}{\tau}}$.

The envelope fitting the time constant (τ) can be calculated.

Knowing the frequency (f_0) and the time constant (τ), it is possible to estimate Q factor from relation:

$$\tau = \frac{1}{\xi \omega_0} = \frac{Q}{\pi f_0}$$

To perform this measure, the stimulus signal is forced on one stator, and a DC voltage signal is applied on the rotor and the output signal is acquired from the other stator on the axis under test (Fig. 25.5).

25.2.2.6 Bandwidth Amplitude at Half Power Method

The exponential decay method is the right solution when τ is high enough, and it is possible to acquire a sufficient number of samples for the fitting algorithm. In an accelerometer, typically the resonance frequency is not as high as in gyroscope and Q is very low (normally lower than 10) so it is not possible to use the exponential decay fitting.

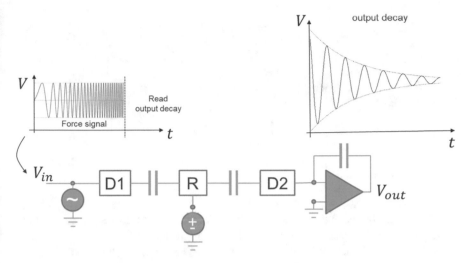

Fig. 25.5 Measure setting

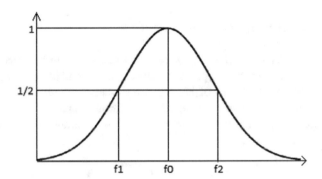

Fig. 25.6 DUT frequency response

The quality factor can be also calculated from the relation $Q = \frac{f_0}{\Delta f}$ where f_0 is the peak frequency and Δf is the full width at half maximum (FWHM) Fig. 25.6). This parameter is the bandwidth over which the power of vibration is greater than half the power at the resonant frequency: $\Delta f = f_2 - f_1$.

This measurement is not calibrated; therefore, two measurements are needed. In the first measurement, a 0V signal is forced to the rotor (measure OFF), so only the $C_{\text{feedthrough}}$ (Fig. 25.7) is considered. During the second measurement, a constant DC voltage is forced to the rotor (measure ON). The difference between the second and the first result is the DUT frequency response without feedthrough signal. During both these measurements, a chirp signal is forced on one stator and the other is connected to measurement chain.

For both signals, the transfer function normalized on input signal is calculated. After this a Fourier transform (FFT) is performed to calculate amplitude and phase.

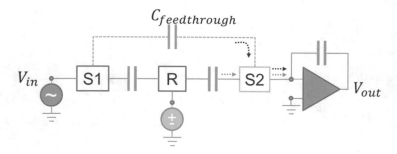

Fig. 25.7 Measure setting

The difference between the second and the first measurement in frequencies domain is the DUT frequency response without feedthrough signal.

Knowing the frequency response of the device, and applying the equation above, peak resonance frequency and Q factor can be estimated.

25.2.2.7 Quadrature Error Measurement

Quadrature measurement is the only wafer level measurement on inertial sensor in which the gyroscope sensor is operated with drive movement at target amplitude, emulating the normal behavior of MEMS after assembly with ASIC. During this measurement on one drive stator, a square waveform is forced. This waveform has a forcing frequency aligned to the resonance, and an amplitude that can be derived from the equation:

$$A = 2V_{RD} + \sqrt{V_{RD}^2 - \frac{kf_D^2}{Q_D}}$$

where

- V_{RD} is the difference between rotor and drive voltage.
- f_D and Q_D are the drive resonance frequency and the quality factor.
- k is a constant linked to the design of the device.

To measure quadrature error, while the square waveform is forced on drive stator, two couple of measurements are performed: two on the drive sense stator and another couple on the sense stators. The two measurements for each stator are necessary (as for Q measurement) to compensate feedthrough signal. The measurement with 0V applied on rotor is called measure OFF and the other is called measure ON.

Signals acquired on sense axes are normalized with respect to the signal on sense drive axis. Amplitude and phase are obtained from the FFT signals, so it is possible to estimate the quadrature error (refer to design chapter for definition) using the following formula:

$$quad_{Sense} = \frac{quad\_amp_{S1} \cdot \cos\left(\frac{quad\_ph_{S1}}{180\pi}\right) - quad\_amp_{S2} \cdot \cos\left(\frac{quad\_ph_{S2}}{180\pi}\right)}{2}$$

where

$$quad\_amp_{sx} = \frac{amp_{Sx}\,(ON - OFF)}{amp_D\,(ON - OFF)} const\_quad_{Sx}$$

And $const\_quad_{Sx}$ is a parameter defined by sensor's design.

25.2.3   Physical Stimuli Applied at Wafer Level for Screening and Calibration

In this section examples of wafer level testing applying physical stimulus are presented.

In Sect. 25.2.3.1, the case of typical resistive pressure sensor is described, followed by humidity (Sect. 25.2.3.2) and magnetometer (Sect. 25.2.3.3).

25.2.3.1   Pressure-Temperature Sensor

The piezoresistive pressure sensor is typically composed by a Wheatstone bridge and a suspended membrane called diaphragm which deflects under a pressure stimulus. The sensing element consists of four resistors in a Wheatstone bridge configuration: when a pressure is applied on the membrane, two opposite resistors will see longitudinal strain and the others two resistor will see transversal strain. The voltage between the output terminals is proportional to the applied pressure (Fig. 25.8).

Piezoresistive pressure sensor converts a physical stimulus into an electrical output signal, and in order to test this kind of sensor, a controlled air blow is required to modify the rest condition. The physical parameter which describes the response of the sensor to the pressure stimulus applied is called sensitivity, and it is an indirect measure calculated as

$$\text{Pressure Sensitivity} = \frac{(\text{Output@ForcedPressure} - \text{Output@AmbientPressure})}{(\text{ForcedPressure} - \text{AmbientPressure})}$$

where Output@ForcedPressure is the output of the Wheatstone bridge measured during the forced stimulus and Output@AmbientPressure is the output of the Wheatstone bridge measured at ambient pressure.

Since pressure depends also on temperature, usually the pressure sensor is coupled with a temperature sensor, which is also a Wheatstone bridge with

Fig. 25.8 Schematic diagram
of pressure sensor structure

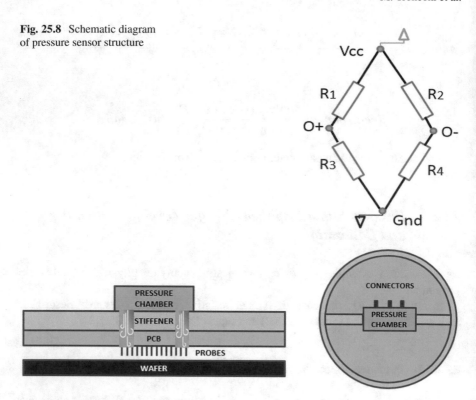

Fig. 25.9 Section and front view representation of the pressure sensor probe card

thermoresistor components. The working principle is similar to the one described for pressure sensor: when a temperature stimulus is applied, the opposite resistors will see a deformation due to their expansion or contraction. For this reason, two different testing temperatures are required to calculate the temperature sensitivity parameter:

$$\text{Temperature Sensitivity} = \frac{(\text{Output@HighTemp} - \text{Output@LowTemp})}{(\text{HighTemp} - \text{LowTemp})}$$

where Output@HighTemp is the output of the Wheatstone bridge measured during the test performed at higher temperature, while Output@LowTemp is the output of the Wheatstone bridge measured at ambient temperature.

At wafer level, pressure-temperature sensor test is executed with a dedicated probe card with a pressure chamber mounted on the probes in order to create a controlled environment during test (Fig. 25.9).

An external pressure stimulus applied with two instruments is necessary to implement the sensitivity test: a pressure regulator which forces air proportional to the voltage applied and a reference barometer which measures pressure inside the

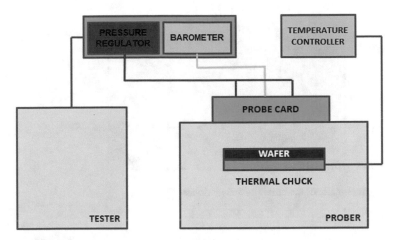

Fig. 25.10 Schematic description of testing setup

chamber itself. Both instruments are connected to the pressure chamber using small vacuum pipes and the pressure regulator is connected to the facility line. Pressure regulator and reference are also connected to the tester in order to be controlled during the electrical test, as shown in Fig. 25.10. Moreover, in order to control the temperature of the device under test, the setup is composed by an accurate thermal chuck and an accurate temperature controller.

25.2.3.2   Humidity Sensor

The humidity sensor is the result of the process integration of a structure sensitive to humidity embedded on the ASIC circuit (Chap. 17). In this case, since ASIC and sensor are integrated on the same die, it is possible to perform screening and complete calibration of the device under test at wafer level, and it is possible to store on the memory the calibration data, and at the end of the wafer level testing process, the device is fully calibrated. With this approach, it is possible to simplify the testing flow at package level and it is possible to increase the productivity (Fig. 25.11).

At wafer level, humidity sensor test is executed with a dedicated probe card with a chamber mounted on the probes in order to control the humidity level on the device under test (Fig. 25.12).

In addition, an external dew point generator is necessary to apply the stimulus and a reference which control the humidity inside the chamber itself. Both instruments are connected to the pressurized chamber using small vacuum pipes, and all the connectors are integrated with automatic probe card change mechanism of the prober equipment. The dew point generator and the reference are also connected to the tester in order to be controlled during electrical test, as shown in Fig. 25.13.

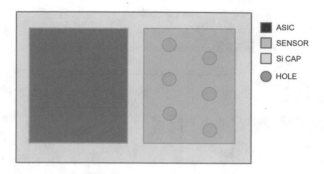

Fig. 25.11 Schematic of the humidity sensor device

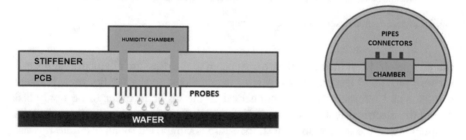

Fig. 25.12 Section and front view representation of the humidity sensor probe card

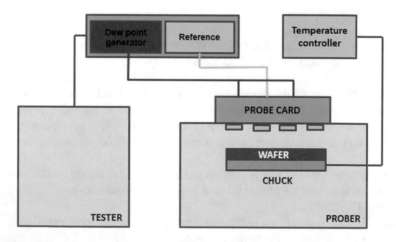

Fig. 25.13 Schematic description of testing setup

On the PCB (wafer side), there are calibrated humidity sensors in order to increase as much as possible the humidity measurement accuracy on the wafer side.

25.2.3.3 Magnetometer

A magnetometer is a sensor capable of measuring magnetic field amplitude and it can be implemented using different physical effects. Based on the way the field is measured, they can be piezo, AMR, GMR, TMR, and Hall between others. In this chapter, AMR magnetic sensors are used to describe the typical testing and calibration flow, and in any case all concepts can be extended to other magnetometer families.

A magnetoresistive sensor is a resistor that changes its resistance when a magnetic field is present. Four of these resistors can be connected in a Wheatstone bridge configuration to allow the measurement of the magnetic field magnitude. One bridge for each sense axis is needed and devices up to 3 embedded sensitivity axes are available. In addition to the bridge structure, the AMR sensors could have an extra resistance called Set/Reset coil. Set/Reset coil is normally driven by a current pulse, in the range of hundreds of mA to generate a strong magnetic field with the purpose to align the magnetic domains of the magnetoresistive sensors into one direction. Realigning the magnetic domain is mandatory to ensure a high sensitivity, low noise, and repeatable reading, and based on the current pulse direction, positive (SET) or negative (RESET), the polarity of the output response of the bridge can be flipped. It is possible to take advantage of this behavior by measuring the bridge output twice, one after positive current pulse and one after negative current pulse, to improve linearity and reduce offset and temperature effects.

Usually the AMR sensor is coupled with an ASIC that drives the MEMS, providing all the needed signal. In order to test the sensors alone at wafer level, it is needed to give the necessary supply voltage to the bridges, to measure the bridge output with a dedicated signal conditioning circuit, to drive the Set/Reset coil forcing a specific current pulse and to apply a proper magnetic field in the sensitive axis direction; the effects on characteristics of tester and external circuitry are treated in next paragraph.

On automatic test equipment, power supply and per pin measurement units are usually available, so it is possible to apply voltage to the bridges and to read their offset after suitable signal conditioning. To drive Set/Reset coil, instead, a specific external circuitry is needed. It is placed in the electrical board located between tester and devices, and it is able to give current pulses with programmable amplitude (in the range of hundreds of mA) and duration. Regarding the magnetic field generation, a dedicated, specific solution is used to apply the field to the devices. In order to reach good field magnitude on the device, current amplitude around one ampere must flow into the magnetic generator structure. Driving properly the magnetic generator structure, a magnetic field into X, Y, and Z directions can be applied at wafer level to stimulate all the magnetic axes of the devices and properly determine the sensor sensitivity.

On the probe card PCB, few magnetic sensors are soldered in order to be used as reference during the testing flow, to verify the goodness of the magnetic field value. A deep characterization has to be done to measure the field value at devices location. In fact, due to high parallelism used to test these devices, at wafer level each device position can be exposed to a slightly different magnetic field stimulus. Using characterization results, the precise value of the magnetic field per each position of the devices under test is known; therefore, a calibration can be performed prior to the testing.

Test flow performed on magnetic devices comprehends standard tests such as continuity check, leakage, resistance, and offset measurements performed to guarantee coverage and magnetic sensitivity of the device under test.

Focusing the magnetic test, the description of the steps that must be performed to reach good results are described hereinafter.

First of all the magnetic generator structure is turned on, forcing the proper current, in order to apply the magnetic field into specified direction on devices:

$$MAG_1 = +x \ [G]$$

With the magnetic field applied, a SET pulse current is forced, to align magnetic domain into SET direction. The bridge offset (V_{SET}) for the axis under tests is therefore measured. To evaluate the cross-axis parameters, axes not under test can be also measured.

The same procedure is repeated for the RESET direction, applying a current in the opposite direction in respect to the SET direction, across the SET/RESET coil. Bridge offset (V_{RESET}) for the axis under test is measured the second time. Also in this case, to evaluate the cross-axis parameters, axes not under test can be measured.

Calculations can be done with the two bridge offset measurements, to calculate the electrical offset referred to the applied magnetic field MAG_1:

$$V_{OFFSET+} = \frac{V_{SET} - V_{RESET}}{2} \ [V]$$

The same procedure is repeated with an external magnetic field of same intensity but inverted direction:

$$MAG_2 = -MAG_1 = -x \ [G]$$

Forcing the SET current will align the magnetic domain into the SET direction. The bridge offset (V_{SET}) for the axis under test is then measured. The Reset current is then forced into the SET/RESET coil to invert the magnetic domain into RESET direction and the bridge offset (V_{RESET}) is measured. As before, to evaluate any cross-axis parameters, axes not under test can be also measured.

The electrical offset referred to the applied magnetic field MAG_2 is calculated:

$$V_{OFFSET-} = \frac{V_{SET} - V_{RESET}}{2} \ [V]$$

Finally, the magnetic sensitivity of the axis under test can be evaluated by applying the following formula:

$$Sensitivity = \frac{V_{OFFSET+} - V_{OFFSET-}}{MAG_1 - MAG_2} \ [V/G]$$

Performing the bridge offset measurement for the two opposite magnetic field direction, any unwanted external magnetic fields and any electrical offset introduced by circuitry or by device itself are compensated.

To evaluate the sensitivity for each magnetic axis embedded in the device, the full procedure must be repeated changing appropriately the magnetic field generator driver parameters.

25.2.4 MEMS Actuators: Testing Characteristics

In this paragraph, the testing of the MEMS actuators will be treated explaining the implications at wafer level testing.

An actuator is a system capable of converting an electrical signal into a physical effect, and it is clear that the testing of MEMS requires both the possibility to apply a proper electrical signal in order to convert it into a physical stimuli and the ability to read back the generated physical stimuli in order to verify that the energy conversion relation is confirmed and acceptable according to the device specifications.

Energy	Parameter	Physical effect	Example actuators
Mechanical	Rotation angle, displacement	Piezoelectric, capacitive, inductive, electromagnetic	Micromirrors, printheads, loadspeakers, microfluidics
Thermal	Temperature, heat flow	Seebeck effect, thermoresistance, pyroelectricity	Printheads

To choose the correct testing setup, the requirements should be expressed in terms of actuator driving capability and physical stimuli readback. According to the different product families, testing setups may therefore differ.

25.2.4.1 Micromirrors

One of the major family of actuator devices is represented by the micromirrors. They can be classified in multiple ways, based on their actuation principle (capacitive, electromagnetic, piezo), their working method (linear, resonance, lissajou), the number of rotational axis (monoaxial or biaxial), and the fabrication technology characteristics (wafer thickness, full or holed wafers, mirror materials).

The technology characteristics will determine the requirements on the handling equipment. Holed wafers require specific handling prober equipment since vacuum cannot be used to manage wafers. If wafers are also very thin (<200 um), wafer bowing and warpage can easily lead to a challenging specification for equipment manufactures. This aspect is not furtherly investigated.

Considering the driving requirements, for capacitance actuation mirrors, the key parameter is the maximum voltage that can be applied between stator and rotor. This value can easily rise up to hundreds Volts. Generally the number of independent sources required are two, one for each stator.

Based on working principle, it is possible to discriminate two main capacitance micromirrors: linear and resonant. For linear capacitance mirrors, the rotation angle varies in a proportional relation with the applied voltage and the rotation direction depends on which stator is driven. In a common testing configuration, a first stator is driven with a positive rising ramp signal in respect to the common-grounded rotor. The second stator is at zero potential. Maximum angle will be achieved at maximum applied voltage. The second phase will consist in lowering the applied voltage on the first stator to zero potential with a descending ramp. Then the procedure will be repeated on the second stator with the same rising voltage ramp to determine the maximum angle in the other direction and a descending ramp to return to stable condition (Fig. 25.14).

In case of resonant capacitance mirrors, the working principle is slightly different. Two voltage sources are needed, one for each stator. The applied voltage signal on the first stator is a positive square wave or a positive sinusoidal wave. On the second stator, the same signal is forced but with a 180deg shift. This will lead to a "swing effect" with the two voltages on the stators always in opposite status. With a fixed maximum voltage, the signal frequency will vary following a rising or descending sweep in a determined operating range. The mirror will start to rotate once the mechanical resonance frequency will match the applied forcing signal frequency. The maximum angle at a fixed voltage is detected at the resonance frequency (Fig. 25.15).

Since the MEMS mirrors have a nonlinear behavior, maximum opening angle and the resonance frequency will also depend on the maximum applied voltage and the direction of the frequency sweep.

At wafer level, the testing conditions must be identified and a tradeoff must be taken according to what parameters must be guaranteed. To reduce both the complexity of the test and the overall testing time, usually both the maximum voltage and the frequency sweep direction are fixed, and only the forced frequency is changed, guaranteeing a minimum angle at the specific test conditions.

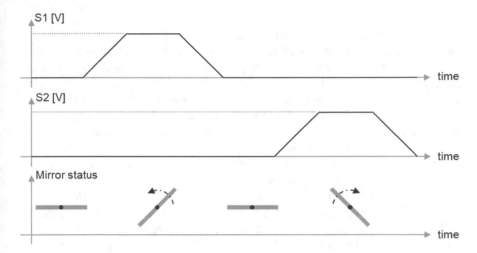

Fig. 25.14 Linear capacitance mirror typical rotation angle dependency with applied voltage on stators [S1 and S2]

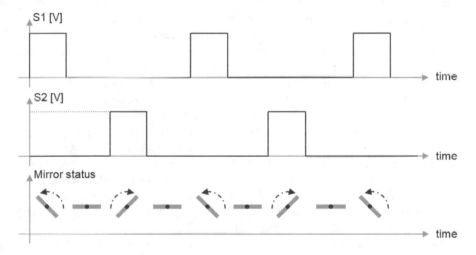

Fig. 25.15 Resonant capacitance mirror typical rotation angle dependency with applied voltage on stators [S1 and S2]

The thin-film piezo mirror actuation principle is similar to the abovementioned for both linear and resonance mirrors since the piezo structure can be modeled, at first order, as a capacitance. The difference is within the value range of the structure capacitances. For capacitive mirrors, the range is in the units of tens of picofarad, while for the piezo mirrors, the range is in the units of tens or hundreds of nanofarads. The different capacitance range will certainly need to be treated properly while designing the driver stage for the testing equipment.

The last family of micromirrors is represented by the electromagnetic mirrors. The driving principle can be reduced to the well-known Lorentz force law (or Laplace law): if a current flows into a wire placed inside a constant magnetic field, a mechanical force will develop moving the wire in an orthogonal direction to the current flow. As seen in the previous chapters, the MEMS electromagnetic mirrors are made of a solenoid or coil, in which the current will flow, which is placed on a movement element solidal to a fixed structure by the means of springs. The generated Lorentz force will result in a torsional force around a rotational axis.

From testing point of view, two are the key elements to consider for the requirements to the supplier: the maximum driving current and the magnetic field necessary to develop the torsional force.

The current driver must be dimensioned to force a current in the range of hundreds of milliamps, enabling current signal profiles such as sinusoids or ramps.

For linear electromagnetic mirrors, the flow direction of the current inside the coil will determine the direction of the rotation. The maximum angle will be related to a combination of the current absolute value and the magnetic field absolute value.

For resonant electromagnetic mirrors, a sine signal is required, leading to a periodical invertion of the Lorentz force and consequently of the direction of the rotation of the mirror. Maximum rotational angle will be reached when the forcing sinusoidal frequency matches the mechanical resonance frequency.

The main difficulty in testing this particular family is the requirement to have a strong magnetic field. In the final application developed by customers, this can be possible by placing magnet close to the edge of MEMS mirrors inside the final optical module. A different approach must be chosen for the wafer level testing since there are some constrains in the geometry of the testing. First of all, dice are not singulated and still close to each other on the wafer. This means that magnets cannot be placed on the edge. Secondly, the testing solution must consider the required space for the contact needle tips and the wanted testing parallelism. An effective and costless solution to these problems has been patented by ST (ref: US2018/0237293 A1, AUGUST 23 2018), using a custom probe card with a particular configuration of strong permanent magnets inside a special housing on top of the wafer, reaching magnetic fields on the dut of the order of $0.1 \div 0.3$ T.

In the previous paragraphs the actuating driving requirements for all mirror families have been described. For a complete wafer level, testing is also important to determine how much the mirrors respond to the same driving actuation signal. Different techniques have been considered both at dut level and at equipment level.

Rotational angles can be easily read in terms of the unbalance of a piezoresistive Wheatstone bridge, linked to the wanted rotational axis. According to the bridge sensitivity, the higher the bridge output absolute value, the higher the angle. For piezo and electromagnetic mirrors, this opportunity is embedded with the technology and is present on the mirror also enabling in application, a run-time monitoring of the angle aperture.

Wheatstone bridge sensitivity might also be affected by process spread; therefore in order to determine the overall relation between actuation force and the rotational angle, a direct angle measurement must be considered at equipment level. For the

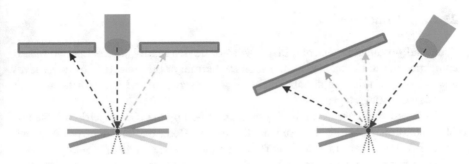

Fig. 25.16 Laser beam illustration

same reason, this kind of measurement can be effective also in capacitive mirrors. The angle rotation can be read by a capacity readout of the position only in a closed-loop architecture, by reading a periodic signal through an ASIC. This technique cannot be taken into account at wafer level where a single shot measurement is preferrable and a direct optical measurement is a better choice.

A direct measurement of the angle can be obtained directing a laser beam on the mirror and reading the deflection angle using arrays of photodiodes or camera sensors like a charge-coupled device (CCD). Different geometrical laser configurations can be considered according to the maximum rotational angles to read and the available equipment and probing spaces.

A first configuration is obtained placing the laser beam perpendicular to the mirror surface, with two optical detectors near each side, one for each rotation direction. The advantage of this configuration is that it can detect wider angles and the geometrics is quite easy. The disadvantage is the impossibility to read the deflected beam when the mirror is not moving, since the return beam path will collide with the inlet laser beam.

In order to be able to detect also the rest position of the mirror, the laser must be placed aside with a direct beam deflecting with a certain incident angle on the mirror. The optical detector must be placed according to the deflected beam in rest position and sufficiently big to detect the reflective beams at maximum rotational angles. Due to geometrical considerations, this solution can be used for smaller opening angles (Fig. 25.16).

In both the setups, the accuracy of the detection will have to take into account the variation of the offplane position of the mirror due to probing specifications and the laser beam spot size on the detector. An optical calibration procedure must be considered.

When testing linear mirrors, since the angle is controlled, the optical detectors will read a single spot. In case of resonance mirrors, the detectors will likely detect a line, due to the fact that usually the angle sweeping frequency is higher than the detector readout frequency. For angle detection this is not a problem.

25.2.4.2 Printheads

Printhead products represent a major actuator device family, divided in two macro categories based on the actuation principle: thermal or piezoelectric. At wafer level test, the related microfluidic obviously cannot be screened and it is demanded to the successive testing stages.

An interesting case of study can be represented by the piezoelectric printheads. The actuation principle is based on the piezo effect, explained in the previous chapter. When a certain voltage is forced on the opposite ends of a piezo material, coherently to a specific crystallographic orientation, the material will contract or stretch, leading to a deformation of the MEMS structure. In a piezoelectric printhead, this physical principle is used to squeeze an ink chamber and eject the drops of ink through the nozzle.

The deformation of the structure is enhanced by applying a stress at the mechanical resonance frequency of the MEMS element. Wafer level testing is therefore important to screen the devices with an unwanted behavior, by measuring the resonance frequency and some electrical parameters that models the piezoelectric structure, e.g., the impedance.

To perform such task, the measurement of this device is done applying a polarization voltage overimposed by a small test sine signal, sweeping in frequency in the range of the expected resonance frequency.

Two readout circuits have been evaluated: a capacitive partitioner and a transimpedance amplifier. The first circuit is easier to implement but is more affected by parasitic capacitances that may lead to unwanted accuracies or inability to measure. The second circuit has the advantage to be more reliant and less effected by parasitic effects. Consequently, the design of the reading stage can be more complex due to the presence of the high polarization voltage.

In any case, for both circuits, an important task to consider for proper measurements is the calibration of the readout stages and the correct determination of the transfer function. This non-trivial procedure is required to correctly convert the measured test signal to the device's impedance value.

All electrical parameters may then be extracted by correctly fitting the relation of the impedance value trend with the frequency, according to the equivalent electrical model of the MEMS structure. In Fig. 25.17 an example of a typical fitted printhead resonant curve is reported.

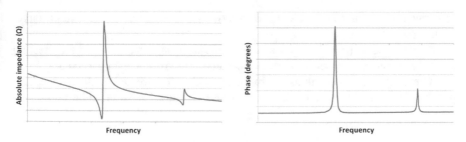

Fig. 25.17 Typical fitted printhead resonant curve

Finally, from testing point of view, particular attention must be paid to the device measurements since crosstalk effects can occur between the piezo structures of each nozzle that share one same electrode. Forcing the resonance of a single structure, while reading the impedance over the frequency sweep, may excite a close structure that is mechanically coupled, generating a disturbing return unwanted signal back to the device under test.

Particular attention must be paid to the needle tips manufacturing since generally the distance between close pads (pitch) and the pad dimensions are very small, in the order of few tens of microns. To relax the specifications, a design for testability study is generally performed at early stages. A possible solution is to use a staggered layout disposition of the probe tips to increase the pitch.

25.2.4.3 Loudspeakers

Loudspeakers transform electrical signals into a mechanical effect. An interesting case study is represented by the piezo loudspeakers, where a piezo capacitance is used to move a mechanical element responsible of the displacement of the membrane.

The wafer level testing has the possibility to screen the dice that show an insufficient displacement. Such task can be performed by measuring the piezo capacitance at different polarizations of the capacitance, since the deformation is in a proportional relation with the applied voltage.

The capacitance measure is performed in the same way as previously described for the other piezo actuators.

25.3 System in Package Calibration

This paragraph treats more specifically items related to test, calibration, and compensation of complete system where it is possible to store in dedicated non-volatile memories and apply on the output data all calibration and compensation coefficient, different and measured separately part to part, allowing to reach specifications target of the product.

In Sect. 25.3.1, there is a description of the aspects to be considered on testing and calibrating inertial sensors, accelerometer, and gyroscopes in particular.

In Sect. 25.3.2, temperature and humidity sensors are treated, while Sect. 25.3.3 provides information on microphones and passive optical sensors calibration.

25.3.1 Physical Stimulus: Characteristics, Main Requirements, Limits

As it commonly happens in all measurement systems, MEMS sensor characteristics need to be stated with enough accuracy for the target application. From an industrial point of view, such accuracy is defined in terms of specifications which usually has the form of a confidence interval for the given quantities. Stating the level of confidence, together with the condition of measurement, is of utmost importance both for manufacturer and user, as it defines the extent of reliability.

Coping with increasing demand of tighter and tighter specification is a task to be addressed both during design phase and at test. The spread that originates from micromachining affects the system in such a way that finished-goods compliancy with datasheet could not be guaranteed if a proper calibration process did not take place. While the calibration procedure is directly connected to ASIC design, which sets the highest attainable accuracy, the measurement process on the production line is by far the most accuracy-killing factor. This is due to essentially two reasons: environmental noise and thermo/mechanical stress. In principle, noise effects could be mitigated by taking a great number of samples; in practice however, some limitations arise: in a production line the test time available is limited, the specific nature of the physical stimulus to be provided.

The stress-inducing boundary conditions (thermal/mechanical) play, in turn, a great role in reducing the measurement accuracy, because they represent *spurious* inputs of the same order of magnitude of the real inputs to be transduced. This is mostly true for capacitive motion MEMS.

In the remainder of this paragraph, the most critical parameters considered will be offset and sensitivity. For those parameters it is required that the physical stimuli must be known with high accuracy and with fixed boundary condition (temperature, humidity).

25.3.1.1 Gravity Stimulus Used for Low-G Accelerometer

To calibrate an accelerometer, one should measure and study the sensor output in response to a physical stimulus. For the calibration of low-G accelerometers, the input acceleration is usually given by Earth's gravitational field.

In order to have a proper description of the calibration process, it is important to consider non-idealities affecting the output signal generated by the sensor. The main effects that must be kept into account are a possible non-orthogonality among the accelerometer axes, and the misalignment of the sensor inside the cavity where the DUT is placed. The former is mainly due to intrinsic misalignment resulting from design and fabrication process; the latter implies that the actual reference frame of the accelerometer is different from the socket's one, which in turn is directly relatable to the "laboratory" frame where the sensor is tested. In fact, in a production line it is not possible to keep under control all mechanical tolerances, and, for pick-and-place systems in particular, the size of the cavity that houses the DUT is a source of misalignment between the stimulus frame of reference and the sensor. Thus, a transformation is needed to link the real output signal with the physical stimulus entering the sensor.

Let $\underline{o}$ and $\underline{u}$ be, respectively, the output signal and the input stimulus. It can be shown [1, 2, 3] that the output signal may be written using the following transformation:

$$\underline{o} = M\underline{u} + \underline{b}, \quad \underline{o}, \underline{u}, \underline{b} \in R^3, \tag{25.1}$$

where the offset $\underline{b}$ is a characteristic bias which is strictly dependent on the device being tested. Notice that, for the time being, we are working in the approximation where the background noise coming from the external ambient is neglected; this approximation will be removed at the end of the section. The M matrix is called sensitivity matrix, and it translates the input acceleration in the output signal of the sensor. It can be shown [1] that this matrix is actually given by the product of three other matrices:

$$M = SCR, \tag{25.2}$$

where S is a scaling matrix, C is a lower-triangular matrix, and R is a rotation matrix.

The S matrix is diagonal, and it allows the system to have different magnitudes along the three accelerometer axes:

$$S = \begin{pmatrix} S_x & 0 & 0 \\ 0 & S_y & 0 \\ 0 & 0 & S_z \end{pmatrix}. \tag{25.3}$$

The effect of S is therefore to scale the amount of input signal u_i along the i-axis transferred to output signal o_i along the same axis. Thus, this matrix is required to keep into account deviations from the ideal situation where the accelerometer responds to equal inputs along each axis with equal outputs.

The C matrix is responsible for the first major non-ideality, i.e., non-orthogonality among accelerometer axes. To do so, one should transform the Euclidean orthogonal reference system into a non-orthogonal system. Thus, C

should be a lower triangular matrix, which, in the approximation where α, β, and γ are approximately equal to 90°, reduces to

$$C = \begin{pmatrix} 1 & 0 & 0 \\ \cos\alpha & 1 & 0 \\ \cos\beta & \cos\gamma & 1 \end{pmatrix}. \tag{25.4}$$

With reference to Fig. 25.18, we can see that the transformation induced by this matrix maps the z-axis in itself, it rotates the y-axis in the yz-plane with angle α and the x-axis towards the z- and y-directions with angles γ and β, respectively. To see this, one may compute the transformation via matrix multiplication

$$\begin{pmatrix} x' \\ y' \\ z' \end{pmatrix} = \begin{pmatrix} 1 & 0 & 0 \\ \cos\alpha & 1 & 0 \\ \cos\beta & \cos\gamma & 1 \end{pmatrix} \begin{pmatrix} x \\ y \\ z \end{pmatrix}, \tag{25.5}$$

using the Euclidean basis for R^3 in place of the non-transformed vector (x, y, z).

By computing the transformation in Eq. (25.5), one realizes that the multiplication by C introduces mixed terms between the different accelerometer axes: for a non-ideal device, an output signal along one axis is the result not only of an input signal along the same axis but also of a small leak signal along the other axes. This can be seen by interpreting the primed terms on the left-hand side of Eq. (25.5) as the output signal of the sensor, and the non-primed terms on the right-hand side as the physical stimuli: as an example, if the misalignment between the axes is set as $\alpha = 89°$, $\beta = 90°$, $\gamma = 90°$, the output signal along the y-axis reads as

$$y' = y + \cos\alpha x, \tag{25.6}$$

meaning that the sensor output y' derives from accelerations along the y-direction, plus a small contribution coming from physical inputs along the x-axis.

Going back to Eq. (25.2), the R matrix keeps into account the second non-ideality: the physical misalignment deriving from the placing of the sensor inside the socket. This also means that there is a misalignment between the physical-stimulus reference frame, which is the same of the sensor package, and the sensor reference system. To keep this into account, one should apply a rotation of the DUT reference frame to align its axes to the accelerometer frame.

Notice that there is no conflict with the non-orthogonality introduced by the C matrix, as Eq. (25.1) may be written as

$$\underline{o} = SCR\underline{u} + \underline{b}, \tag{25.7}$$

which means that the reference frame of the sensor is rotated first using the multiplication of the input $\underline{u}$ with the matrix R, and only then each axis is

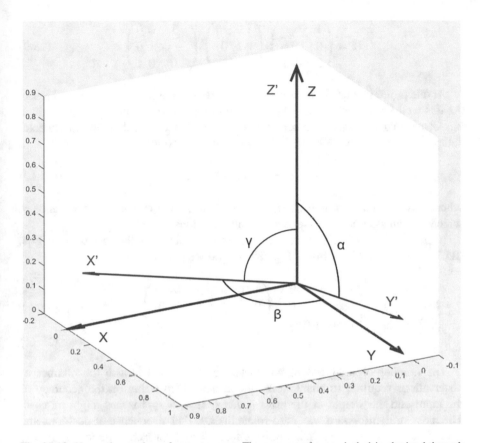

Fig. 25.18 Non-orthogonal accelerometer axes. The new set of axes x',y',z' is obtained through the application of the C matrix on the orthogonal set x,y,z

independently transformed to obtain the non-orthogonality through the matrix multiplication $C(R\underline{u})$.

Thus, R is a rotation matrix around the three Euclidean axes, which can be written as the product of three independent rotations around each axis

$$R = \begin{pmatrix} 1 & 0 & 0 \\ 0 & \cos\phi & \sin\phi \\ 0 & -\sin\phi & \cos\phi \end{pmatrix} \begin{pmatrix} \cos\theta & 0 & -\sin\theta \\ 0 & 1 & 0 \\ \sin\theta & 0 & \cos\theta \end{pmatrix} \begin{pmatrix} \cos\psi & \sin\psi & 0 \\ -\sin\psi & \cos\psi & 0 \\ 0 & 0 & 1 \end{pmatrix}. \tag{25.8}$$

By taking a first-order Taylor expansion around small angles, which is always the case in practical applications, the rotation matrix simplifies to

$$R = \begin{pmatrix} 1 & 0 & 0 \\ 0 & 1 & \phi \\ 0 & -\phi & 1 \end{pmatrix} \begin{pmatrix} 1 & 0 & -\theta \\ 0 & 1 & 0 \\ \theta & 0 & 1 \end{pmatrix} \begin{pmatrix} 1 & \psi & 0 \\ -\psi & 1 & 0 \\ 0 & 0 & 1 \end{pmatrix}. \qquad (25.9)$$

At this point, it is useful to remove the approximation of zero background noise. As it is well known, white noise is usually modelled with an additional term which has Gaussian distribution with zero mean, and a variance typical of the experimental setup. Thus, we can write Eq. (25.1) in its final expression

$$\underline{o} = SCR\underline{u} + \underline{b} + \underline{\delta}, \qquad (25.10)$$

where $\underline{\delta}$ is the Gaussian noise term with zero mean and characteristic variance σ, which we can assume to be the same for all directions.

In order to write the full expression, we can substitute the results from Eqs. (25.3), (25.4), and (25.9) inside Eq. (25.10) and we get

$$\underline{o} = \begin{pmatrix} S_x & 0 & 0 \\ 0 & S_y & 0 \\ 0 & 0 & S_z \end{pmatrix} \begin{pmatrix} 1 & 0 & 0 \\ \cos\alpha & 1 & 0 \\ \cos\beta & \cos\gamma & 1 \end{pmatrix} \begin{pmatrix} 1 & 0 & 0 \\ 0 & 1 & \phi \\ 0 & -\phi & 1 \end{pmatrix} \begin{pmatrix} 1 & 0 & -\theta \\ 0 & 1 & 0 \\ \theta & 0 & 1 \end{pmatrix} \begin{pmatrix} 1 & \psi & 0 \\ -\psi & 1 & 0 \\ 0 & 0 & 1 \end{pmatrix} \underline{u} + \underline{b} + \underline{\delta},$$

$$(25.11)$$

The number of independent parameters is 12, which is also the number of observations required to characterize the system. Depending on the accuracy of the input and the weight of the error sources, several approaches might be used. The most straightforward way is to rotate the DUT in a sequence of positions and averaging its output to minimize noise effects. This approach is static, and it only requires the positioning accuracy to be large enough compared to the calibration step for offset and scale factor trimming.

25.3.1.2 Rotational Stimulus for Gyroscope Calibration

The calibration process for a gyroscope consists in studying the response of the sensor when it is rotated around a given axis with constant velocity. Commonly adopted techniques take advantage of rotating systems which move with a trapezoidal speed profile, as in Fig. 25.19, where the required constant velocity is achieved within the flat region.

To describe the calibration process, we are going to derive an equation which relates the sensor scale (i.e., its calibration) to the mean value of the sensor output over the full angular rotation. The result is derived in the approximation where the gyroscope is rotated around a single axis.

Exploiting the general value of Eq. (25.1), which relates the output signal of a sensor placed inside its DUT with its physical stimulus, we may also describe the calibration process of a gyroscope. The only difference is that the sensitivity matrix

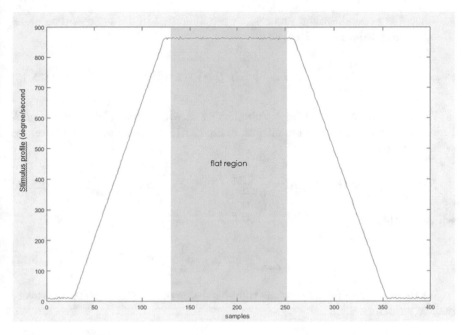

Fig. 25.19 Physical input rotation with trapezoidal profile. The flat region in the middle corresponds to a constant speed rotation applied to the DUT

describes rotation axes, instead of acceleration axes. The analogue of Eq. (25.1) in one dimension may be rearranged to give

$$\frac{1}{S}(o(t) - b) = C R u(t),$$
(25.12)

where the output signal and the physical stimulus have an explicit time dependence. Notice that the right-hand side of Eq. (25.12) contains all the information regarding the rotation of the system, since C, R are rotation matrices and the stimulus $u(t)$ is a rotation.

Thus, we define the infinitesimal rotation of the DUT in the time interval $(t, t + dt)$ as

$$d\theta = C R u(t) dt,$$
(25.13)

and integrating over the rotation time, we get the full angular rotation θ, which is defined as

$$\theta = \int_0^\theta d\theta = \int_0^T dt\, C R u(t) = \frac{1}{S} \int_0^T dt\, (o(t) - b).$$
(25.14)

Since this is a calibration process, we are interested in the scale of the device. Thus, we may write Eq. (25.14) as

$$S = \frac{1}{\theta} \int_0^T dt \ (o(t) - b).$$

(25.15)

We divide the interval T in N subintervals of width $\delta t = T/N$. In the large-N limit, one may approximate the integral in Eq. (25.15) with a discrete sum over the N subintervals

$$S = \frac{1}{\theta} \sum_{i=0}^{N-1} \delta t \ (o(t_i) - b) = \frac{1}{\theta} \sum_{i=0}^{N-1} \delta t o(t_i) - \frac{1}{\theta} b \ \delta t \sum_{i=0}^{N-1} 1,$$

(25.16)

which simplifies to

$$S = \frac{1}{\theta} \sum_{i=0}^{N-1} \delta t o(t_i) - \frac{1}{\theta} bN \delta t.$$

(25.17)

We focus on the first term on the right-hand side of Eq. (25.17). The sum over i may be written as

$$\frac{1}{\theta} \delta t \sum_{i=0}^{N-1} o(t_i) = \frac{1}{\theta} \delta t \ (o(t_0) + o(t_1) + \cdots + o(t_{N-1}))$$

$$= \frac{1}{\theta} \delta t \left(\frac{o(t_0)}{2} + \sum_{i=0}^{N-2} \frac{o(t_i) + o(t_{i+1})}{2} + \frac{o(t_{N-1})}{2} \right).$$

(25.18)

It is now time to exploit the trapezoidal-shaped physical stimulus, with reference to Fig. 25.10: at t_0 and t_{N-1}, i.e., at the extremes of the time interval, the input signal is close to zero. Thus, we can substitute the following approximation

$$u(t_0) \approx u(t_{N-1}) \approx 0$$

(25.19)

inside the one-dimensional analogue of Eq. (25.1) to get

$$o(t_0) \approx o(t_{N-1}) \approx b.$$

(25.20)

Using this approximation, Eq. (25.18) may be written as

$$\frac{1}{\theta} \delta t \sum_{i=0}^{N-1} o(t_i) = \frac{1}{\theta} \delta t \left(\frac{b}{2} + \sum_{i=0}^{N-2} \frac{o(t_i) + o(t_{i+1})}{2} + \frac{b}{2} \right)$$

$$= \frac{1}{\theta} b \delta t + \frac{1}{\theta} \delta t \sum_{i=0}^{N-2} \frac{o(t_i) + o(t_{i+1})}{2}.$$

(25.21)

Notice that each addendum inside the sum on the right-hand side of Eq. (25.21) is nothing but the mean value of the output signal between the time steps t_i and t_{i+1}

$$\langle o\,(t_i)\rangle = \frac{o\,(t_i) + o\,(t_{i+1})}{2}, \tag{25.22}$$

and thus we have

$$\frac{1}{\theta}\delta t \sum_{i=0}^{N-1} o\,(t_i) = \frac{1}{\theta}b\delta t + \frac{1}{\theta}\delta t \sum_{i=0}^{N-2}\langle o\,(t_i)\rangle. \tag{25.23}$$

In order to complete the derivation, we need to massage the last term on the right-hand side of Eq. (25.23).

The mean-value theorem states that the average output of the sensor is given by

$$\langle o\rangle = \frac{1}{T}\int_0^T dt\, o(t), \tag{25.24}$$

and thus we need to switch our description of Eq. (25.23) back to a continuous one.

Exploiting once again the fact that we are working in the large-N limit, we may use the approximation $N - 1 \approx N$ to extend the upper limit of the sum from $(N - 2)$ to $(N - 1)$, and consequently the upper limit of the integral from $(T - \delta t)$ to T.

Thus, using this last approximation and the definition of $\delta t = T/N$, we write Eq. (25.23) as

$$\begin{aligned}
\frac{1}{\theta}\delta t \sum_{i=0}^{N-1} o\,(t_i) &= \frac{1}{\theta}b\delta t + \frac{1}{\theta}\frac{N}{N}\frac{T}{T}\sum_{i=0}^{N-2}\delta t\,\langle o\,(t_i)\rangle \\
&= \frac{1}{\theta}b\delta t + \frac{1}{\theta}N\frac{T}{N}\frac{1}{T}\sum_{i=0}^{N-2}\delta t\,\langle o\,(t_i)\rangle \\
&\approx \frac{1}{\theta}b\delta t + \frac{1}{\theta}N\delta t\frac{1}{T}\int_0^T dt\, o(t) \\
&= \frac{1}{\theta}b\delta t + \frac{1}{\theta}N\delta t\,\langle o\rangle,
\end{aligned} \tag{25.25}$$

where the average $\langle o\rangle$ represents the mean value of the output signal over the entire time-interval T. Substituting this results inside Eq.(25.17), one gets the final expression for the calibration equation

$$S = \frac{1}{\theta}b\delta t + \frac{1}{\theta}N\delta t\,\langle o\rangle - \frac{1}{\theta}bN\delta t = \frac{1}{\theta}N\,\langle o\rangle\,\delta t - \frac{1}{\theta}b\,(N - 1)\,\delta t. \tag{25.26}$$

Summarizing the results from this section: the gyroscope calibration is a process which requires to physically rotate (input stimulus) the sensor around a chosen axis, with a trapezoidal-shaped velocity profile. Furthermore, one must measure the full angular rotation θ, the mean value of the output signal $\langle o\rangle$ coming from the sensor itself, and the bias b characteristic of the analyzed gyroscope. The values of N, δt are linked to the number of measurements performed during the angular rotation to derive the average value $\langle o\rangle$.

The main limitation to this method is the mean signal value, which should be measured through a signal integration. Integrating the DUT output over time necessarily means integrating the noise too, which causes deviations from the correct values at the expense of measure reproducibility.

25.3.2 Temperature and Humidity Sensors

In environmental sensors, the calibration flow is usually simpler to implement compared to inertial sensors since there are less compensation channels, i.e., degrees of freedom. Temperature and humidity MEMS sensors do not have suspended structures, and they can be realized on a single die where sensor is integrated together with the ASIC, so it is possible in principle to apply calibration at wafer level and in general mechanical related parameters impact on device performance is minimized; stress gradients on the output due to mechanical pressure over the device during handling and contacting are normally considered negligible.

The biggest effort of industrialization phase is due to calibration equipment hardware development since the stimulus is a continuous flow of air properly conditioned in temperature and humidity or a chuck changing temperature that has to be kept stable overtime and over the whole socket area. Main difficulties are represented by its intrinsic slowness in changing its conditions due to mass to be conditioned and its intrinsic drift, being based on an equilibrium requiring tight-controlled closed-loop mechanism and accurate reference measurement systems.

Different solutions are appropriate depending on different target accuracy to be reached as explained in following paragraphs.

25.3.2.1 Temperature Sensors

Sensors for temperature measurement are usually small and thin (thicknesses are in the range of 0.5mm) since they must be fast in sensing temperature changes affecting the device volume. They are usually calibrated putting them in mechanical contact with a thermal chuck structure which is able to force temperature on a wide span on its surface during the calibration of the unit using, for example, thermal resistors, liquid or air coolers connected to chillers or Peltier cells.

A simple correlation with an external reference which is put in proximity of the device under test is enough to reach calibration accuracy of 2.5 or 3 C, while to reach higher accuracy it is not a matter of scaling the system, it is a matter of rethinking it.

In particular it is needed:

- Usage of periodically certified gauges
- Runtime compensation of reference measures including drift compensation of active elements used in the acquisition chain
- Layout study of boards in order to remote active elements positioning, thus minimizing their dissipations over the dut area which might have an impact on temperature

- Ad hoc designed air flows or thermal chuck in order to equalize areas and avoid even the smallest generated temperature gradients inside test equipment
- Specific geometry and material choice of the contacting socket, with peculiar thermal properties
- Very accurate studies of asymptotic behaviors of stimuli, monitored and compensated drifts

25.3.2.2 Humidity Sensors

For humidity sensor or a combined device of temperature and humidity sensors, a further complication is that humidity is not an independent parameter, but it is strictly dependent on the temperature and the two parameters have to be considered and controlled together in order to reach an accurate calibration target. These sensors have a window opening on the top of the package, in order to expose the sensitive part (usually a capacitor structure with an oxide-based element between the plates whose dielectric constant will vary with humidity absorption) to the external world. When humidity is forced over a device, using a preconditioned air blow, a dew point generator, or a whole climatic chamber, the temperature of the air and the temperature of the device must be equal; otherwise the relative humidity at the stimulus source will not be the same which is felt by the device. In alternative temperature, difference must be known with very high accuracy, in order to apply proper correlation formulas.

The choice of the stimulus depends on the response time of the sensitive element to humidity change: in case humidity sensors has very long response time constant, in the order of minutes, it is necessary to condition in parallel batches of material for long time, up to hours, before calibrating the device, and every humidity point to be observed requires hours of preparation before test, so a chamber is needed to let the production work by batches. In this case being the whole testing machine inside the chamber, the setup is designed and produced to be able to work in humid and warm environment without impact on its correct functioning and robustness. With the response time of the devices becoming faster and faster, calibration flow could access to more efficient conditioning concepts, like local air blow over duts area, or small chamber with on-the-fly humidity setting change.

Dealing with tighter and tighter accuracy, it is needed to develop dedicated hardware that has been used and applied into many other fields of MEMS testing; the most significant example is the socket closing mechanism without air. In standard electronic devices, sockets are closed mainly with high pressure dry air pushing down a piston. If unwanted air leakage, even very small and unperceptible, is coming out from socket piston, the calibration of a humidity sensor is totally compromised, especially when stimulating with humid air. This brought to the development of sockets where air is used only to lock the device but not to retain it, or completely mechanical closing mechanism concepts. Moreover the idea of remote placing of active and dissipative components, or even the redistribution of layout paths of connections inside pcb, is a concept that has been developed from the observation of their parasitic impact on the correct evaluation of the humidity.

25.3.3 Acoustic Waves Sensors: Microphone

A microphone is indeed a sort of environmental sensor, its goal being to catch and reveal sound signals in the environment. It is a delicate device built with cavity package (molded or metal) where a hole must be present to let the sound in, so it is very sensitive to contamination coming from external world, and it must be handled softly not to press it too much affecting its shape, internal volume, and consequently acoustic behavior.

In order to calibrate a microphone, the stimulus to be applied is an acoustic pressure, i.e., soundwaves, usually divided into frequencies (single or multiple) and single pressure levels.

The main parameter that need to be calibrated is the acoustic sensitivity, defined as the signal output amplitude for an applied acoustic sine wave of 1kHz frequency and 94dBSPL pressure (1Pascal, expressed in absolute sound pressure Level scale, which is referred to the audible threshold pressure, corresponding to 20uPa).

Other parametric characteristics have to be checked, for example, frequency response, harmonic distortion, phase, linearity versus high pressure level signals, noise floor, and SNR.

In general, all the coupling system between the sound source and the device must be accurately designed. The equipment evolution for microphone followed the requests for coverage, accuracy, and performance to be guaranteed on the calibrated device:

- First-generation microphones were using a standard handler equipped with a good performance loudspeaker pointing to the devices under test area, properly mounted with small anechoic walls and propagating sound in a free field configuration. This solution was good enough to check main parameters like sensitivity and total harmonic distortion (THD), and in case they are not so stringent in accuracy, no phase and no noise measurement are guaranteed and maximum pressure signal to be applied is 10 to 20dB more than the calibration target.
- Second-generation microphones require tighter limits and better calibration accuracy for wider frequency response and sound pressure level ranges, THD, and phase measurement, and so the concept of stimulus has to be completely different. It is no longer a single free field sound source, but it requires usage of dedicated pressure field chambers for each device under test where each speaker source emits very stable, runtime compensated signal with closed-loop measurement of electrical speaker stimulus, pure signals with minimum THD (one order of magnitude less than the average THD of MEMS devices). Minimized volumes of space between sound source and microphone acoustic

hole is needed to reach very high pressure levels signal, hermetic closure system of the "dut + speaker" system to extend the frequency response range and signal acquisition synchronization with stimulus electrical source of sine wave to allow phase measurement.

- SNR inline measurement has been introduced later and it had a considerable impact not only on the forcing and measuring chain around the device but on the whole environment around it. Socket area is mounted on suspended platforms to reject vibration noises, and wood/foam-based covers are used to isolate the parts from external sources of noise. The latter have all to be minimized including tester setup remoted far from the device and equipped with silent fans, handler setup with motors of all moving parts stopped during acoustic measurement, and a full isolation from external noises with the construction of small anechoic rooms to allocate each measurement setup, with aircon system which was piloted by the measurement tester, and paused during most sensitive tests.

References

1. Jurman, D., Jankovec, M., Kamnik, R., & Topi, M. (2005). Calibration and data fusion solution for the miniature attitude and heading reference system. *Sensors and Actuators A: Physical,* 411–420.
2. Vcelak, J., et al. (2005). *Sensors and Actuators A: Physical, 123–124,* 122–128.
3. Skog, I., & Handel, P. (2006). *Calibration of aMemsInertial measurement unit, XVIII IMEKO World Congress, metrology for a sustainable development* (pp. 17 22).
4. Deb, N., & Blanton, R. D. (2006). Built-in self-test of MEMS accelerometers. *Journal of Microelectromechanical Systems, 15*(1), 52–68.
5. Deb, N., & Blanton, R. D. (2000). Analysis of failure sources in surface-micromachined MEMS. *Proceedings of IEEE International Test Conference,* 739–749.
6. Lakdawala, H., & Fedder, G. K. (2004). Temperature stabilization of CMOS capacitive accelerometers. *Journal of Micromechanics and Microengineering, 14*(4), 559–566.
7. Fokhrul Islam, Mohamad Ali, M. A. (2006). *On the use of a mixed-mode approach for MEMS testing,* Semiconductor Electronics 2006. IEEE International Conference on ICSE '06, pp. 62–65, 2006.
8. Courtois, B., Mir, S., Charlot, B. & Lubaszewski, B. (2000). *From microelectronics to MEMS testing,* ISRN TIMA-RR-00/1001-FR, TIMA Laboratory, France.
9. Wu, Z. C., Wang, Z. F., & Ge, Y. (2002). Gravity based online calibration for monolithic triaxial accelerometers' gain and offset drift. *Proceedings of the 4th World Congress on Intelligent Control and Automation,* 2171–2175.
10. Krohn, A., Beigl, M., Decker, C., Kochendrfer, U., Robinson, P., & Zimmer, T. (2004). Inexpensive and automatic calibration for acceleration sensors. *Proceedings of UCS,* 245–258.
11. Giansanti, D., Maccioni, G., & Macellari, V. (2005). The development and test of a device for the reconstruction of 3-D position and orientation by means of a kinematic sensor assembly with rate gyroscopes and accelerometers. *IEEE Transactions on Biomedical Engineering, 52*(7), 1271–1277.
12. Avrutov, V. V., Aksonenko, P. M., Bouraou, N. I., Henaff, P. & Ciarletta, L. (2017). *Expanded calibration of the MEMS inertial sensors,* 2017 IEEE First Ukraine Conference on Electrical and Computer Engineering (UKRCON), Kiev, pp. 675–679.
13. Probe card for a magnetically-actuable device and test system including the probe card, Marco Rossi, Sergio Mansueto Reina, Giacomo Calcaterra, IT102017000019437A.

Part VII
Reliability

Chapter 26
Reliability

Alessandro Balzelli Ludovico, Fabio Banfi, Stefano Losa, Francesco Petralia,
Ernesto Fabrizio Speroni, Aldo Ghisi, and Stefano Mariani

26.1 Introduction and Historical Evolution

The last two decades have seen the impressive growth of MEMS devices which
became more and more pervasive in many market segments enabling new different
applications as described in the previous chapters. One of the key factors of this
MEMS booming is the successful penetration in some high-volume consumer
market and in particular the strong link with smartphone revolution. MEMS are
key components in a smartphone; inertial sensor (accelerometer and gyroscope),
MEMS microphones, pressure sensor, and magnetometer cover fundamental and
well-known functions. To serve this type of market with its fast dynamic and huge
involved volumes, MEMS development and manufacturing had to adopt the typical
strategies and methodologies of semiconductor industry. In this scenario quality
and reliability assurance played a key role to enable an effective and sustainable
development of MEMS business, on one hand implementing qualification exercises
able to discover and to address in advance possible reliability weaknesses within a
development time frame compatible with time to market and on the other supporting
the production rump-up by monitoring possible quality and reliability defectiveness
linked also to process spreads which needs to be faced timely to assure the quality
levels required by market and customer in a context of fast learning curve.

Around the 2000s, at the beginning of the MEMS journey in STMicroelectronics,
there was poor knowledge and lack of experience on practical reliability of MEMS

A. Balzelli Ludovico · F. Banfi · S. Losa · F. Petralia · E. F. Speroni (✉)
ST Microelectronics, Analog MEMS and Sensors Group, Quality & Reliability, Cornaredo, Italy
e-mail: alessandro.balzelli@st.com; fabio.banfi@st.com; stefano.losa@st.com;
francesco.petralia@st.com; fabrizio.speroni@st.com

A. Ghisi · S. Mariani
Politecnico di Milano, Dipartimento di Ingegneria Civile e Ambientale, Milano, Italy
e-mail: aldo.ghisi@polimi.it; stefano.mariani@polimi.it

© Springer Nature Switzerland AG 2022   899
B. Vigna et al. (eds.), *Silicon Sensors and Actuators*,
https://doi.org/10.1007/978-3-030-80135-9_26

devices and of their relevant failure modes with a real major impact on the field. Already since that time, the reliability approach was based on the study of dedicated test structures (see Sect. 26.2) but also on reliability stress tests on first MEMS products. Test structures were intended to characterize intrinsic failure mechanisms, mainly in the mechanical domain, defining proper design rules to prevent them; on the other hand, reliability on product allowed to discover and to investigate other failure mechanisms like those that are defects related or coming from the interaction with package or more inherent to the application.

The first successful MEMS sensor developed by STMicroelectronics was the three-axis accelerometers family, consisting in a MEMS sensing element and an IC interface assembled in a same plastic package typical of consumer semiconductor industry. In this contest it was natural to apply the basic reliability test matrix consolidated for ICs standard qualification with some important customization to consider mechanical stresses. This approach allowed to cover the standard reliability stresses necessary to validate the IC interface embedded in accelerometer; meanwhile these same stresses applied to MEMS sensing element highlighted new failure modes specific of MEMS itself or arising from the interaction with plastic package as deeply explained later in this chapter. Another important inertial sensor is the MEMS gyroscope which is a key component for motion MEMS application; in this case, MEMS structures work in resonant mode, and this poses additional reliability challenges as it will be discussed later.

Other types of high-volume MEMS sensor developed in the following years were microphone and pressure sensor. Differently by the inertial sensors, these cannot be protected by a closed plastic package since the sensing element has to be exposed to the atmospheric environment, and this can induce, for instance, failure modes linked to environmental contamination and mishandling.

As said, MEMSs are key components in cellular phones, and hence, they follow the same market dynamic with large volumes and fast ramp-up; therefore, to assure the quality level required in this scenario, it is important to study and to prevent failure modes related to defectiveness and process spreads, in addition to intrinsic failure modes that are usually well covered during qualification exercise. Monitoring of production, especially during ramp-up, with proper methodologies and collecting quality feedback by customer from the field are necessary steps for a fast and effective learning curve.

Different issues and hence different approaches characterize the other MEMS macrofamily, the actuators. These types of devices are often codeveloped between MEMS producer and whole system maker. The MEMS assembly, its full functional testing, and sometimes the last MEMS fabrication steps are done at system manu- facturer level. Differently by MEMS sensors that are basically system-in-package devices, MEMS actuators cannot work alone, but they need proper IC drivers which are coupled to MEMS by system manufacturer. Therefore, full reliability of MEMS actuators can be investigated only at system or subsystem level, and effective qualification can be only done jointly between MEMS and system maker partners. In this scenario special subassembly solutions dedicated to reliability tests are developed, and at MEMS manufacturing side, the effort is mainly on material and technology reliability investigation, while full product qualification is done at system maker side.

First MEMS actuators produced in STMicroelectronics were thermal actuators for inkjet printhead. Other types of actuation based on electrostatic and magnetic working principle have been developed for mirror devices. Piezo-based actuators are presently one of the most important areas of development in the field of mirrors but also for other application like autofocus lens and speakers described in the previous chapters and for last generation inkjet printhead. In all these cases dedicated joint qualification approaches have been developed as already mentioned in some previous chapters and will be further discussed in the following.

26.2 Material Characterization and Basic Failure Modes

Though failure modes of structural parts will be extensively described in the following sections, for each sensor type, we start here by providing a sketch of the links between failure, device geometry, and mechanical properties of silicon as a structural material. Since silicon is a brittle material, stress intensification induced by the geometry of the movable parts can interplay with that provided by morphological details of the structural films, if of polycrystalline type; readers can find thorough discussions, e.g., in [1].

Several round-robin tests have been carried out to fully characterize polycrystalline silicon films, in terms of their effective elastic, strength, and toughness properties; see, e.g., [2]. The large dispersion in the published results can be basically linked to two sources: the production process, in terms of parameters involved in the sequence of crystal growth, masking, oxidation, and etching, and the mechanical characterization itself, as each laboratory carries out experiments in a non-standardized way in terms of equipment, specimen size and shape, and environmental conditions. To provide a further step beyond those studies and try to reduce the level of uncertainties in the characterization outcomes, some testing configurations have been proposed to determine the mechanical properties of ThELMA polysilicon using on-chip testing approach with dedicated test structures integrated on a chip exploiting electrostatic actuation. Such approach has been adopted to avoid any detrimental effect induced by the handling of small specimens in the laboratory, and the design of ad hoc, complex grips and data acquisition systems. Nevertheless, the on-chip approach suffers the same technological limitations associated with MEMS production, induced for example by etching for which the actual sample geometry may differ from the designed one, and to the limited possibility to use different loading conditions to stress the specimens.

On-chip testing devices have been considered, designed with the purpose of characterizing the stiffness and the strength of polysilicon; see Fig. 26.1 [3–5]. Capacitive sensing and actuation have been adopted for the tests, mainly because of the relevant measurement accuracy and efficiency. Both in-plane and out-of-plane actuation mechanisms have been adopted, so that either tensile-dominated

Fig. 26.1 On-chip testing devices for polysilicon mechanical characterization [4, 5]

or bending-dominated loading conditions were induced to let polysilicon fail. A description has been proposed for the failure strength in accordance with a Weibull probability distribution, as eventual defects dominate the mechanism of rupture, and under the two loading conditions, the film region influenced by the maximum stress level is different, with the probability to find a defect resulting higher for larger regions (under tensile loading) than for smaller ones (under bending loading); it was thus important to assess the material response in both cases.

Numerical simulations have been carried out to assess in a virtual environment the mentioned mutual effects of the geometry of the movable structure and of the film morphology; see [6]. Some results are reported in Fig. 26.2, where the failure type is reported for analyses at the device level, wherein details of the film morphology were considered negligible, and at the film level, wherein the micromechanical details of the columnar morphology have been instead fully accounted for. It can be seen that, due to the specific device geometry and the kind of (impact-induced) loading conditions, results prove to be approach-independent.

The fatigue of polysilicon has been also investigated through on-chip devices in, among others, [7]. The obtained fatigue lifetime results are relevant for devices under repeated loads, such as oscillators and resonators. To induce an alternating tension/compression stress state at a notch, a system of levers has been driven through standard capacitive actuation controlled by a set of comb fingers subject to an AC voltage; see Fig. 26.3. Since the system has been dynamically actuated, the shuttle, which pushes/pulls the levers, induced a continuous opening/closing of the surfaces facing at the notch, and finally produced a sharp crack propagating, at least in an initial stage, in a stable way. When the fatigue crack reached a critical value of the length/ligament ratio, an unstable propagation then followed. A number of cycles on the order of 10^9 were typically necessary to attain the end of the fatigue

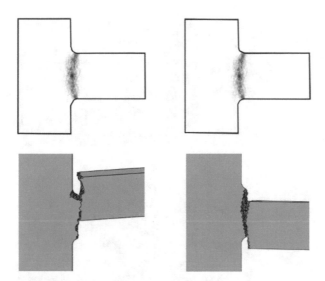

Fig. 26.2 (Left column) tension-dominated versus (right column) bending-dominated failure modes in a suspension spring of a uniaxial accelerometer close to the anchor: comparison between the results of (top row) microstructure-informed and (bottom row) phenomenological approaches [6]

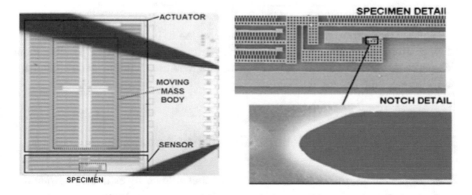

Fig. 26.3 On-chip testing devices for polysilicon fatigue and fracture tests [7]

lifetime; such amount obviously depends on stress level, stress ratio, humidity, and load type (i.e., oscillatory with constant or increasing amplitude). Typically, the AC voltage has been applied close to but not coincident with a resonance peak, to reduce the duration of the experiments.

With the help of microstructure-informed numerical simulations, data reduction has been obtained as shown in Fig. 26.4; see [7]. This test revealed that a sharp crack must be produced, to assure that the fracture toughness is not overestimated; this issue can explain the scattering in the experimental results collected in the

Fig. 26.4 Exemplary crack path in the polysilicon at the notch of the experimental device depicted in Fig. 26.3. A numerical vs experimental comparison is shown during crack propagation (with a stress map at the tip)

literature. The fracture toughness has been reported to be not an isotropic property of polysilicon; see [8]: this microstructural feature, combined with possible local defects at the grain boundaries, can explain the resulting tortuous and difficult-to-predict crack path shown in the figure.

26.3 MEMS Device General Reliability Approach

Reliability can be defined as the discipline that allows to estimate the devices lifetime versus all the critical potential failure modes that could occur during the product useful life. Starting from the sensor mission profile (in use ambient condition, maximum supply voltage, power on-off duty cycles, etc.), the purpose of reliability is to address both *INTRINSIC* and *EXTRINSIC* potential failure modes. Intrinsic failure mechanisms are strictly related to product design and performances; the extrinsic failure mechanisms are instead related to the supply chain steps performances involved in the products manufacturing flow (wafer manufacturing, assembly, testing, etc.)

Reliability plays a crucial role in all the phases of product development, industrialization, and production flow optimization. The three fundamental phases are summarized here below:

1. Phase 1 new products and technology introduction
2. Phase 2 industrialization and volume ramp-up
3. Phase 3 process changes managements (capacity expansion, cost reduction, quality improvements)

Phase 1 starts when first samples are available. In this context the intrinsic potential failure modes have to be investigated in order to explore the product design compatibility with the fabrication process. The devices are submitted to stress tests according to JEDEC (Joint Electron Device Engineering Council) standard specifications, but in some cases, the reliability stresses are customized according to product peculiarities and market requirements (consumer, automotive, medical, industrial).

Phase 2 starts together with the industrialization steps. During this phase both intrinsic and extrinsic potential failure mechanisms are investigated. In general, a large number of devices are submitted to short reliability stress test. The stress test used depends on product peculiarities and eventual marginalities occurred in the previous phase.

Phase 3, in most cases, starts before the final production process is frozen. Also, in this phase both intrinsic and extrinsic potential failure modes are investigated.

This general approach is used mainly for sensors products.

Today sensors are based on well-consolidated technologies and manufacturing flow. They represent in terms of volume the main MEMS devices introduced in the market by our company.

Since few years STMicroelectronics started to invest in the development of actuators devices in order to increase the overall product portfolio. The business model in this case is different compared to the sensors one. While in the sensor market we provide to our customer a complete device assembled inside a dedicated package and ready to be mounted on a dedicated PCB, in case of actuators, STMicroelectronics provides to customer wafers in which only the mechanical element is present. Then our customers produce the final module suitable for their application.

In this specific case, the STMicroelectronics production flow stops after EWS (electrical wafer sorting) test. This implies that the general reliability approach is different from that of sensors and is mainly composed by two phases.

1. Phase 1. Reliability is performed at wafer level to check the compatibility of the devices with the front-end production flow.
2. Phase 2. Reliability is performed on a dedicated set of test vehicle (TV) assembled on a support package. The aim of this activity is to explore the devices behavior under biased and unbiased condition for a long-time stress.

In this context the reliability at TV level is carried out following a dedicated stress test matrix that considers the product/technology potential vulnerabilities. As for the sensor case, the starting point is the JEDEC or other international standards. All the measurements on TV are carried out in manual mode; this implies that the number of samples submitted to reliability stress test is less than in case of sensor.

Different techniques have been developed to have the possibility to investigate the manufacturing process window in this case.

26.4 Sensors Reliability Overview

In the following sections, we deal with the main types of MEMS sensors developed and brought on the market by STMicroelectronics, discussing challenges, reliability strategies, and main failure modes based on the wide experience gained over the years during qualification and industrialization of various sensor families.

26.4.1 Inertial Sensors

Inertial sensors are represented by accelerometer and gyroscope which are produced with very similar technologies and conceived as a system on package, combining the MEMS sensor die with the ASIC (application-specific integrated circuit) into a package which protects them from environment and handling (see Fig. 26.5). The choice of the package, in terms of materials and manufacturing process, plays a key role on the reliability of inertial sensors influencing its functionality since the beginning of its lifetime. In fact, any sensor protected by the package materials (e.g., with epoxy resin, substrate, etc.) can be subjected to thermomechanical stresses arising during the package fabrication or caused by interactions between package and environmental agents such as ambient humidity.

Inertial sensors hereafter considered are those devices encapsulated into plastic package and surrounded by epoxy resin (see Fig. 26.5). Even it is not specified, these devices can be accelerometer or gyroscope stand-alone or combo sensor, which

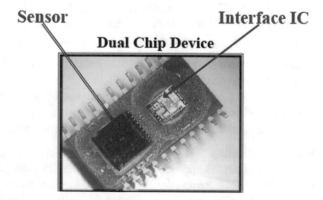

Fig. 26.5 Historic photo showing MEMS accelerometer and interface IC (ASIC) into epoxy resin SO16 package, which was partially de-capped

integrates both MEMS sensors on the same die and represents state of the art from Thelma technology platform.

For this family of sensors, the most relevant failure modes we encountered were the following ones:

- Lack of hermeticity
- Parametric stability
- Mechanical robustness
- Spurious adhesion

Some of them, such as lack of hermeticity, are critic for the device survivability, others can change the electromechanical behavior resulting acceptable or not for final application. These phenomena will be described in detail, where they happen and related countermeasures if applicable, underlying commonalities or differences among accelerometers and gyroscopes.

Lack of Hermeticity
Lack of hermeticity is a serious failure mode for the sensor functionality, and it can be caused by some defects present in the sealing material used to join the MEMS sensor die to cap die (glass-frit or metallic bonding). This defect can be present at time-zero when the sealing ring is formed, appearing as passing through voids or not-complete sealing ring; alternatively, the defect in the sealing can be a crack or a delamination which can be caused by some thermomechanical stresses, for instance, when the package resin is molded into the forming cavity or when the plastic package is exposed to reflow during the on board soldering (see Fig. 26.6). Time zero defects or induced damages, alone or together, can weaken the sealing strength and meaningfully affect electromechanical behavior of the sensor. Besides, for gyroscope, even a small pressure change in the cavity can be catastrophic because this device requires very low cavity pressure to properly oscillate (i.e., to have high-quality factor).

Different solutions, for full-molded plastic packages, were taken to solve hermeticity issue:

- At MEMS sensor and package design level, leading to important design rules formalizations:
  - Sensor: fixing safe margins when designing bonding ring area
  - Package: choosing die attach material and package substrate layout
- At MEMS wafer and package manufacturing steps:
  - Front-end: glass-frit stencil design and wafer to wafer bonding process optimization
  - Back-end: adopting innovative resin molding process to reduce the stress on glass-frit sealing ring

Validation methodology was based on electrical measurements, done both at wafer and package levels, then followed by dedicated reliability stress tests such as the ones reported in Table 26.1).

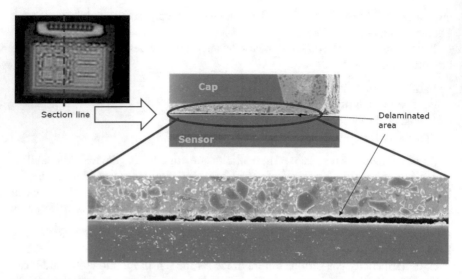

Fig. 26.6 Scanning acoustic microscope (SAM) image revealing delamination on the sealing glass frit (grey zone). The cross section at SEM shows a glass-frit de-attaching at sensor/glass interface

Table 26.1 Reliability stress tests used to monitor hermeticity problem on inertial sensors

Reliability test	Conditions/reference	Product lifecycle	Note
Autoclave	121 °C, 100% RH, 96 h	Qualification phase	Technology-oriented test
	JESD22-A102		
u-HAST (unbiased highly accelerated stress test)	130 °C, 85% RH, 96 h		u-HAST can replace autoclave
	JESD22-A118		
THS (temperature humidity storage)	85 °C / 85% RH, ≥24 h	Qualification phase or during production ramp-up monitoring	
	JESD22-A101		
Reflow	Tpeak @260 °C	Qualification phase or during production ramp-up monitoring	Alone or prior other stress tests
	J-STD-020		

Today hermeticity is well controlled performing electrical measurements combined with reliability stress during early qualification phase and the production ramp-up. The complete flow involves different controls to screen out initial defectiveness, firstly at wafer level by means of automatic acoustic inspection (by SAM equipment) and electrical measurement (it is worth to note that gyroscope hermeticity is well controlled by measuring quality factor parameter) and secondly performing other controls at package level by testing the final product before and after accelerated reliability stress tests, based on moisture application or high temperature excursion (sometimes these stresses can be combined making the screening more selective; see Table 26.1)

Electrical controls combined with environmental stress are most effective on final production line, screening out possible and residual defectiveness induced by assembly process, such as cracked glass-frit sealing and caused by some outgoing stress.

Parametric Stability

Talking about parametric stability, we can distinguish between:

- Intrinsic behavior mainly dependent on product design and its tolerance vs natural process variability
- Extrinsic behavior, triggered by manufacturing process spread or defectivities, typically random and enhanced by some product susceptibility

Key parameters of accelerometer and gyroscope, such as offset (Zero-g Level Offset and Zero Rate Level respectively) and sensitivity are strongly influenced by interactions between MEMS sensor and surrounding package materials. Some important environmental stresses impacting both board and package and – in turn – the sensor parametric stability are hygro-swelling phenomenon which arises during accelerated reliability tests or harsh applicative conditions and fast temperature excursion as that present during on board soldering. From a physical point of view, moisture absorption and temperature cycling can produce package deformation and stress causing submicrometric displacements on sensor moveable masses.

STMicroelectronics developed a dedicated reliability test aimed to evaluate intrinsic susceptibility to hygro-swelling stress. This test, which is typically performed in early qualification phase, consists in submitting the sensor device in a climatic chamber and monitoring their outputs while test temperature and humidity are changed. Besides the devices are soldered on test board thus considering the influence of package-board interactions. Goal of this test method is to estimate sensor output change while humidity is varied passing from a "wet" condition to a "dry" one (from a "soak" to a "post soak" phase; see Fig. 26.7).

Furthermore, sensor outputs monitoring based on soldered parts and automatic control on temperature and humidity variables have brought to develop special reliability board adapter and optimize climatic chamber software (see Fig. 26.8).

Intrinsic sensibility of accelerometer Zero-g Level Offset (ZgL) or gyroscope Zero Rate Level (ZRL) to temperature excursion can be measured carrying out HTOL (High-Temperature Operative Life) reliability test and collecting sensor output during oven heating and cooling. This method to estimate output sensitivity when temperature changes, for instance, from room temperature to 125 °C, implies to sample sensor behavior at different temperature values and hence to acquire at each temperature point few hundreds of raw data in order to cancel mechanical noise (see Table 26.2). This procedure is applied also for abovementioned humidity test which aimed to estimate ZgL or ZRL versus humidity level (see Table 26.3).

Furthermore, new usage of HTOL test to characterize ZgL/ZRL versus temperature does not imply to customize reliability equipment, but it can be done

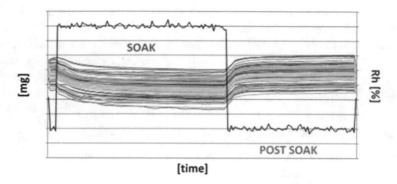

Fig. 26.7 Typical accelerometer Zero-g Level Offset behavior during "soak" phase and "post-soak" phase. Humidity change of the chamber is done keeping the temperature constant. In this manner, sensor output change happening during humidity transition is only the effect of moisture percentage variation

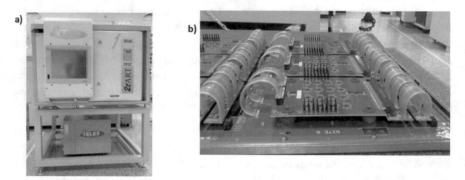

Fig. 26.8 (**a**) Dedicated climatic chamber with the refrigerator compressor put outside to minimize the mechanical noise during the test. (**b**) Reliability board and adapter: there are four soft gimbals and one flex cable for each reliability board, aimed to hold the board and measure the sensor output minimizing external mechanical vibrations

Table 26.2 Approach to characterize accelerometer Zero-g Level Offset (ZgL) or gyroscope Zero Rate Level (ZRL) sensitivity versus temperature during HTOL oven heating and cooling

Reliability test	Conditions/reference	Product lifecycle
HTOL test (high-temperature operative life test)	Device is set at max operative conditions, Tmax = 125 °C, duration from 24 to 1000 h depending on HTOL scope, ZgL/ZRL sampled at different T point. Ref JESD22-A108	Qualification phase or during production ramp-up monitoring

Table 26.3 Approach to characterize accelerometer Zero-g Level Offset (ZgL) or gyroscope Zero Rate Level (ZRL) sensitivity versus humidity

Reliability test	Conditions/reference	Product lifecycle
Humidity test	Device is soldered and set at typ. Operative conditions; temperature and relative humidity are changed; duration is of 168 hrs; ZgL/ZRL is sampled at different T/Rh points. Internal specification	Qualification phase

on standard system; indeed, it requires rather complex test-program development for sensor data acquisition and elaboration, maintaining ambient temperature synchronism. Nevertheless, HTOL potentiality is impressive because it can be extensively used at production plant, collecting meaningful "big data," and doing reduced temperature characterization on very large sample size.

The most effective ways to reduce interactions and make parametric stability as much as possible insensitive to environmental stress are smart choices at sensor and package level. At die level, adopting for instance innovative MEMS sensor architecture and leveraging on anchors positioning or mass and spring design; at package level, adopting cavity package and ceramic based substrate for board connection, avoiding surrounding MEMS with any hygroscopic materials. Some further benefits on offset and sensitivity stability can be achieved at ASIC and application board level, the former by optimizing the signal reading from MEMS and the latter by decoupling its interaction with package.

Parameter stability can be influenced by other factors which are not intrinsic of the device and do not affect most of the devices. In this case we are in front of extrinsic phenomenon triggered by manufacturing process spread or defectivities embedded in the process itself.

One of most important events is the reaction between moisture and epoxy molded resin, triggered by some resin morphological irregularities and happening locally around the MEMS sensor pads: this phenomenon can influence the electrical signals provided by MEMS to ASIC and sometimes cause serious parametric failure of sensor device. This finding is the results of the several analyses of failed sensor devices and of a cross-functional collaboration between STMicroelectronics and Politecnico of Milano [9]. As stated in [9], we have realized that resin material may locally present electrical characteristics different from the typical bulk material, somehow behaving like an electrochemical solution where a so-called double-layer capacitance effect can take place. Indeed, molded resin, which is mainly composed of fillers, can be distributed not homogenously around the MEMS sensor pads leading to formation of some micro-voids; after humidity exposure, condensed water into filler voids can act as an electrochemical solution (H_2O + ions) making the formation of the double-layer capacitance between adjacent pads (see Fig. 26.9).

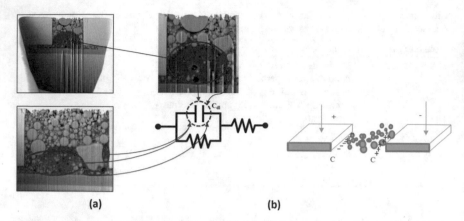

Fig. 26.9 (**a**) FIB micro-section done across MEMS sensor pads and highlighting filler voids (dark area circled in red) acting as an electrochemical solution. (**b**) Double-layer capacitance effect explanation and equivalent electrical circuit modelling of Randles circuit (**b**) (courtesy of the authors of [9])

This fact can be a quality problem for the sensor, and it can be controlled or somehow mitigated by performing reliability test based on moisture absorption on final production line, using the same test flow used to screen out not hermetic parts.

Mechanical Robustness and Spurious Adhesion

Mechanical breakage represents one of most "classical" failure mode of MEMS sensor after a mechanical shock. Indeed, other failure mechanisms can be observed on accelerometer and gyroscope, some of them with no catastrophic effects or simply with intermittent behavior. This miscellaneous scenario is the effect of the widespread application of MEMS sensor on consumer and automotive markets, which has brought to adapt the reliability approach to specific mission profiles, sometimes under the customer impulse. For instance, for sensor embedded in handheld electronics such as smartphone or wearable device, testing is mainly based on high-g level mechanical shock which better simulates the drop event happening in the real life; for inertial sensors used for car motion control, mechanical characterization is mainly based on long-term vibration test and medium g-level shock repeated for thousands of times.

Main procedures used to assess mechanical robustness are based on the following reliability tests:

- Vibration test
- Mechanical shock test
- Drop test

Vibration test is suitable to reproduce the mechanical effects generated during transportation or normal usage in range of some tens of g. Mechanical shock test and drop test instead can produce shock acceleration level up to hundreds of thousands

Table 26.4 Reference MIL-STD 883 method 2002.3

Test condition	g level (peak)	Duration of pulse (ms)
A	500	1.0
B	1500	0.5
C	3000	0.3
D	5000	0.3
E	10,000	0.2
F	20,000	0.2

Table 26.5 Reference JEDEC 22 B110

Peak acceleration (G)	Pulse duration (ms)
2900	0.3
2000	0.4
1500	0.5
900	0.7
500	1.0
340	1.2
200	1.5
100	2.0

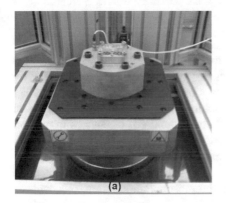
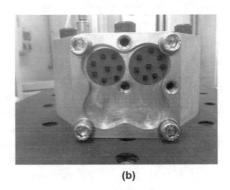

Fig. 26.10 (a) Example of pneumatic shock machine with the carriage and special fixture acting as shock amplifier. (b) Aluminum holder on top of which sensor devices (with package LGA 3 × 3 × 1 mm) is attached to sense produced acceleration; ± x and ± y directions are given by rotating the metallic support

of g, well simulating effect of fallen handheld devices (followed specifications are reported on Table 26.4 and Table 26.5).

Mechanical shock test can be performed using pneumatic machine which combines falling height and pressurized air to produce mono-axial and repeatable shock; adding a special fixture, acceleration level can arrive up to 20,000 g (see Fig. 26.10).

Drop test is a free fall that can be done on various hard surfaces and from different heights. With this test very high acceleration levels up to 200,000 g can be reached.

Fig. 26.11 Tumble equipment is a rotating machine where devices are repeatedly dropped

One example is drop generated with Tumble equipment, which is a rotating machine able to produce repetitive random free fall drops in a controlled repeatable manner (see Fig. 26.11). Starting from one important smartphone maker request, today we use Tumble to drop inertial sensor devices as loose parts (i.e., without a container acting as drop box) from 1 m of height, obtaining acceleration in the range of 100,000 g. Tumble test has the advantage to stress hundreds of devices in parallel and it is simple for execution, so this methodology is widely used to validate new sensor design and to monitor performance variability in production.

After two decades of experience on mechanical stress tests, performed on accelerometers before and gyroscopes after, the most common failure mechanisms encountered are:

• Spring and electrode breakage (see example in Fig. 26.12)
• Stopper and electrode anchoring breakage (see example in Fig. 26.13)
• Stopper chipping (see example in Fig. 26.14)
• In-use adhesion

Various solutions at process or design level were adopted to improve sensor mechanical robustness, then formalized as design guidelines for next sensors. For instance, to solve spring breakage or out of control rotor displacement, most frequent solution was to introduce or redesign stoppers both in-plane and out-of-plane (e.g., for Z displacement); besides, to avoid chipping, big effort was spent to model stopper shape lowering impacting energy density.

As far as adhesion phenomenon is concerned, for instance, between movable mass and stopper, it happens when the surfaces adhesion forces (e.g., electrostatic attraction, van der Walls forces, etc.) are larger than mechanical restoring force of

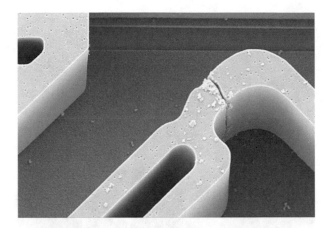

Fig. 26.12 Example of broken springs

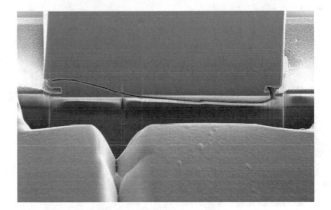

Fig. 26.13 Example anchoring breakage (cross section)

suspended micromechanical element. It may occur during wafer manufacturing or device usage: in the second case, it is known as "in-use adhesion." This condition is activated in the presence of electrical field for instance when the device is powered and shocked at the same time. Detection and characterization are done performing dedicated capacitance versus voltage measurements during wafer electrical testing or monitoring sensor outputs of final product. Test on final product is done applying acceleration while the device is supplied (see Fig. 26.15): this "powered shock" method is a good reproduction of possible in-use adhesion. Finally, "powered shock" test results are used to validate or improve design, for instance, by increasing moveable mass stiffness or reshaping the electrodes.

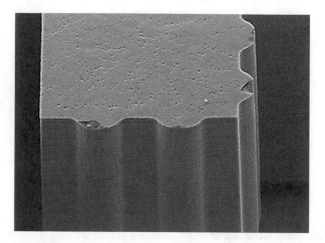

Fig. 26.14 Example of stopper chipping

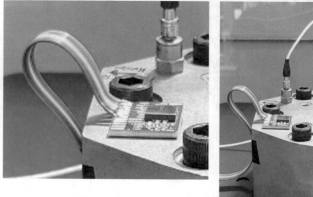

Fig. 26.15 Example of in-use adhesion characterization based on "powered shock" test: sensor device is soldered on a small board which is fixed to metallic fixture by mean of bi-adhesive tape; board and cable wires permit to drive sensor and monitor its output when shock occurs, thus verifying if any adhesion inside the MEMS can happen

26.4.2 Magnetometer

In this section a global overview on magnetometer reliability is presented. STMicroelectronics developed the AMR (anisotropic magneto resistance) process for magnetic element fabrication. This kind of technology is a key pillar for the e-compass devices suitable for navigation application.

The sensor is composed by an acceleration plus a magnetic sensing element.

Entering more in detail, the acceleration sensor is composed of a MEMS transducer (mechanical element that detects acceleration giving as an output a capacitance variation signal proportional to the acceleration stimulus) and a signal conditioning integrated circuit, usually referred to as ASIC (application-specific integrated circuit). The two elements are electrically connected by means of wire bonding. The ASIC input signal came from the transducer element, then it is elaborated and converted (e.g., from capacitance to voltage) so that it can be available for the final user. This architecture is the same of accelerometer standalone devices.

The magnetic field sensor is realized using the same approach of the acceleration one. In this specific case, the magnetic transducer was fabricated by means of AMR technology (no movable masses are present). A dedicated ASIC is connected to the sensor with the same functionalities described in the previous case.

The final e-compass module is composed by two transducers (one accelerometer and one magnetometer) and two ASIC interfaces (one for each transducers). As a result, four silicon dice are assembled inside an LGA (land grid array) package (Fig. 26.16). The typical dimensions are 2mmx2mmx1mm (1 mm height). The e-compass represents an example of a very complex system on package. This aspect is crucial and can impact the product reliability. Therefore, a dedicated reliability activity both during the development and industrialization phases has been done in order to verify the intrinsic and extrinsic device vulnerabilities.

The major potential failure mode that needs to be addressed during product development and industrialization phase can be resumed as follows:

1. Accelerometer and magnetometer parametric stability
2. Wire bonding integrity (wire bonding complexity due to four silicon dice in the same package)
3. Magnetic shocks to verify eventual permanent damage when high magnetic field is applied

The real system complexity can be appreciated considering the figures below.

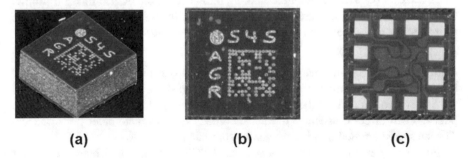

(a)   **(b)**   **(c)**

Fig. 26.16 LGA package images: (**a**) 3D view, (**b**) top view, (**c**) bottom view

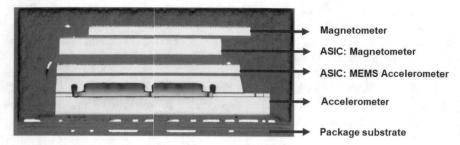

Fig. 26.17 LGA cross section. Stack sequence of the four dice

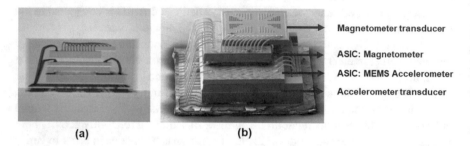

(a)   **(b)**

Fig. 26.18 (a) X-ray side view image, (b) 3D view SEM image after package resin removal

In Fig. 26.17 the die stack sequence inside the package is reported. This sequence plays a fundamental role in the device parametric stability performance vs thermomechanical stress that the devices experience during its lifetime.

In Fig. 26.18 the overall assembly including wire bonding complexity can be appreciated. The right die placement and the tight tolerances together with wire bonding integrity are two of the major challenges to be addressed by reliability.

The assessment on reliability during product development with the purpose to investigate intrinsic potential failure mode starts considering the standard reliability test matrix as per JEDEC reference taking into account the device mission profile. On top of that, dedicated stress tests are applied to investigate the potential device vulnerability vs the potential failure modes already mentioned.

Parametric Stability

Both accelerometer and magnetometer parametric stability has to be investigated because it is highly influenced by the overall silicon stack inside the package.

Any environmental condition that induces thermomechanical stress of the entire package can affect drastically the accelerometer and magnetometer performance. The stress test applied in this case can be summarized as follows.

First, samples are soldered in a dedicated PCB and measured before and after soldering so that the drift due to thermomechanical stress induced by soldering can be evaluated. Then all the PCB and devices are submitted to humidity and temperature stress. Both humidity and temperature induce on soldered devices mechanical stress that generates package deformation (in particular humidity can activate hygro-swelling phenomena). The sensor output is monitored during all the duration of the environmental stress test. In case of any abnormal behavior, a deep-dive investigation is needed to fix it.

Wire Bonding Integrity

As a result of integration of four dice inside the same package, a very complex wire bonding scheme is implemented. Due to package dimension and the quantity of wire, the tolerance for this process step (as well as the tolerance for die placement) is very tight. The potential failure mode associated to this specific topic is the wire misplacement that can occur at wire bonding and molding process steps. Wire can come close to each other and in some cases, they can touch (Fig. 26.19).

In case the wires are in contact, a leakage current occurs among them and the devices does not work anymore. The critical cases to be investigated are those in which the wires become closer after some stresses but at time zero (at the end of ST manufacturing flow), the devices work properly; for example, thermomechanical effect induced by soldering process or temperature variation during customer manufacturing flow or during the life of the devices could induce a package deformation; in this case the wire becomes closer to each other causing device malfunctioning. To address these marginalities solder preconditioning plus temperature and humidity stress plus temperature variation cycles are applied in sequence. At each steps the key parameters of the devices are monitored. DOE (design of experiment) at die attach, wire bonding, and molding steps have been done to fix this marginality and find the appropriate process window at back-end manufacturing side.

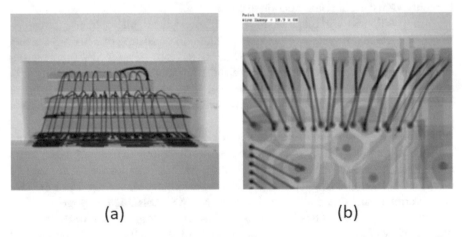

(a)   (b)

Fig. 26.19 (a) X-ray side view showing wire bonding. (b) X-ray top view of wire bonding misplacement

Magnetic Shock

Magnetic transducers are realized by material that can acquire a permanent magnetic polarization if submitted to high magnetic field (magnetic shock).

To verify the product robustness vs this potential failure mode, a dedicate setup was realized to take care of all the magnetic characteristic of its component; it was very important in order to have the real response of the devices (artifacts due to permanent magnetic polarization of the measurement component must be avoided).

This peculiar aspect was investigated both at time zero and after aging induced by reliability stress. The extrinsic vulnerability was then investigated during the industrialization and preproduction phase. In general, very short reliability stress test that can activate the abovementioned potential failure modes was applied to a huge quantity of material. The results and a detailed data analysis review were the basis for the reduction of residual defectivity due to manufacturing process steps. At the end, an estimation of residual ppm at customer manufacturing was released. Any deviation, if present, from a fixed target, agreed with customer, must be addressed before the full production release.

26.4.3 Microphone

In MEMS microphones, the sensing element is directly exposed to the environment to receive and transduce the sound signal. This triggers various threats by reliability viewpoint during manufacturing phase and sometimes on the field; microphone is a key component of mobile phone and hence it is subjected to frequent mechanical shocks on the field caused by the phone drop events; for this reason, mechanical robustness is a key aspect to consider during the microphone design and then to carefully verify during qualification exercise.

In this section we will focus on the most diffused MEMS microphones, the capacitive ones, which are system in package (Fig. 26.20) consisting of a MEMS sensing element transducing sound in an electrical signal and of an ASIC (application-specific integrated circuit) reading and elaborating this signal and translating it in a digital or analog output; package does not just act as protection and electrical connection interface but it also plays an active role in the acoustic performance of the system.

The characteristics and performances that define the quality of microphones are sensitivity, frequency response, total harmonic distortion (THD), signal-to-noise ratio (SNR). As said, the main application of MEMS microphone is in the field of mobile phone, and to guarantee the above performances, scaling down the size poses various reliability challenges involving MEMS, ASIC, and package.

Qualification strategy has to address the different microphone elements covering the reliability of MEMS sensor, ASIC, and package; at this purpose a reliability test matrix typical of MEMS devices is implemented on the microphone as shown in Table 26.6; in this case the tests are done on the stand-alone microphones and the check of the devices under test (DUT) after reliability stress is done by the automatic testing equipment used in production.

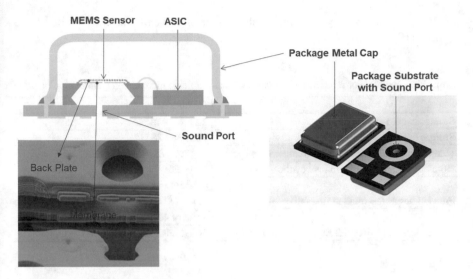

Fig. 26.20 MEMS microphone key elements overview

Table 26.6 Typical reliability test matrix done on microphone component stand-alone

Stress Test	Conditions
Preconditioning	J-STD-020 MSL3; Pb-free reflow profile
High-temperature operating life	Dynamic test: Vdd @ max op. Voltage, T − 125 °C, 1000 h
Temperature humidity storage	85 °C/85% RH, 1000 hours
Temperature cycling	−40 °C to +125 °C, 1000 cycles
Low-temperature storage	−40 °C, 1000 hours
Mechanical shocks	Repetitive half sine shocks for each axes/direction:
	- MS1: 3000 g / 0.3 MS
	- MS2: 10000 g / 0.1 MS
ESD (electrostatic discharge):	
- human body model	- 2 kV HBM
- charge device model	- 500 V CDM
Latch up	+/−100 mA current injection

Moreover, depending also on the requirements of mobile phone customers, additional tests are done with microphone units mounted on dedicate qualification boards (QB), meaningful of real assembly solution; an example of such additional test's matrix is reported in Table 26.7.

Working on parts mounted on such qualification board allows to better study some reliability aspects like:

− Investigate package solder joint reliability.
− Consider the stress induced by the board on microphone during environmental exposure.

– Better reproduce the mechanical stresses during events like the mobile phone drop [10].

Table 26.7 Example of reliability test matrix with microphone mounted on qualification board

Stress Test	Conditions
Storage in cold environment	72 h @ −40 °C
Storage in hot dry environment	72 h @ +85 °C
Temperature change endurance (temperature cycling)	-40 °C/+85 °C, 100 cycles
Humidity endurance (damp heat cyclic)	18 cycles; 1 cy = 8 h
	+30 °C /+65 °C; ~ 93% RH
Free fall durability (guided free fall)	1.5 m (12 + 12 + 16) drops on to concrete
	Total of 40 drops on each unit
Impact durability (free fall repeated, tumbler test)	300 drops, 1 m height

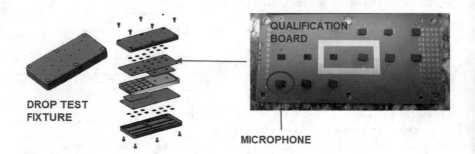

Fig. 26.21 Board and fixture for drop test

In this case the check of microphone after stress test is done at bench with a dedicated setup of characterization which allows to measure key functional parameters with part mounted on QB.

As said mechanical robustness is a key aspect to assure the microphone reliability; for this reason, besides the standard mechanical shock test on stand-alone parts, mobile phone makers have defined specific tests for part mounted on QB with dedicated reliability equipment. Key ones are:

– Guided free fall test
– Tumbler Tests

For these test QB with the microphone DUTs is housed in a fixture intended to simulate the phone as shown in Fig. 26.21.

Guided free fall tests are carried out with a dedicated equipment shown in Fig. 26.22 which allows to control drop direction and impact angle on the concrete basement; side, edge, and corner impacts can be reproduced in controlled way.

Fig. 26.22 Guided free fall test equipment

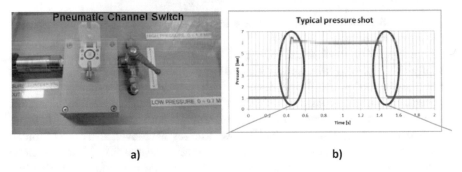

Fig. 26.23 Compressed air test experimental setup (**a**) and pressure shot (**b**)

Another test specific for microphone again aimed to check mechanical robustness is the so-called compressed air test. By a dedicated setup, a shot of pressure is applied through the sound port (Fig. 26.23) causing a mechanical stress on the microphone sound sensing structure.

Main Failure Modes

As said microphone catastrophic failures and malfunctioning are mainly due to insufficient mechanical robustness of MEMS mechanical structures [11] or linked to the feature of the holed cavity package which is intrinsically vulnerable to

Table 26.8 Microphone main failure modes and malfunctioning

Physical failure mode/defect	Malfunctioning symptom/effect	Defect detection/activation
Broken structures: • Membrane • Back plate • Anchoring points	• Catastrophic "mute" response in case of fully broken membrane • Instable sensitivity for partial breakages	• Highlighted by mechanical test on component (GFF, tumbler) in qualification • Also activated by parts handling and shipment at/after manufacturing
Particles	• Low sensitivity • Noise	• Mainly highlighted in humidity tests during qualification • Sometimes detected at cell phone final checks also linked to possible contamination in manufacturing environment
Package cap bonding bad sealing	• Low sensitivity • Bad frequency response	• Mainly activated by soldering reflow in qualification tests and at cell phone manufacturing • Temperature cycling test can also highlight this defect

environment contamination and which can have itself in case of degradation an impact on microphone acoustical performance.

Table 26.8 presents a non-exhaustive list of some key failure modes and their related malfunctioning.

An example of mechanical failure of MEMS sensor is shown in Fig. 26.24. Usually, this type of weakness is detected during qualification investigating intrinsic device failure modes, and proper design guidelines have been developed to improve mechanical robustness. Also, parts picking and handling during manufacturing can play a role in the occurrence of this failure mode, so tools and equipment must be designed to avoid dangerous mechanical overstresses. Moreover, also process spread in general can generate some weakening, and for all these reasons, this failure mode has to be monitored not only in qualification but also during industrialization and volume ramp-up phase to cover the possible extrinsic occurrence of this type of failures.

An example of particles or residuals contamination is shown in Fig. 26.25. This defect affects the functioning of MEMS sensing structures impacting mainly noise and sensitivity performance. Contamination can arise during the assembly of MEMS sensor die in the package but also in the following manufacturing steps at phone maker line; for this reason, a careful environment control of all involved assembly lines is key to assure quality and reliability of microphones. Humidity tests are able to highlight this type of defects since moisture capillarity condensation or absorption around these particles and residuals can activate leakage current and local spurious adhesion phenomena between mobile sensing structures; therefore,

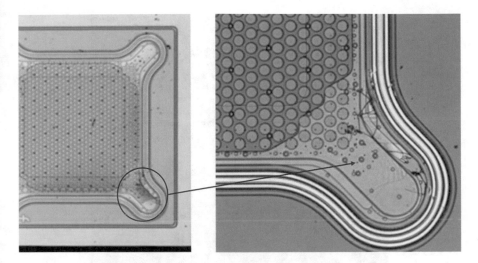

Fig. 26.24 Example of microphone broken membrane

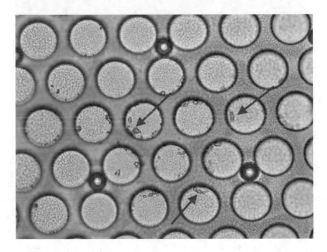

Fig. 26.25 Example of contamination defect between back plate and membrane

periodic humidity tests on sample basis can be an effective tool to monitor possible contamination problem in the assembly line.

As said package has an active impact on microphone acoustical performance; Fig. 26.26 shows degradation examples of package cap soldering with sealing failure impacting the microphone response at low frequency. Careful control of raw material quality and of cap soldering process are key to avoid this failure mode; temperature cycling tests on production can provide a way to monitor this problem and in case even to screen out it.

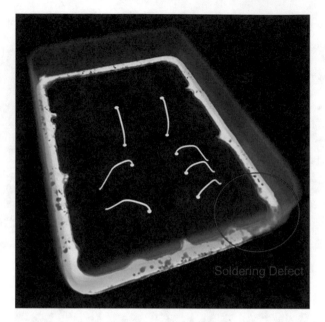

Fig. 26.26 X-ray 3D tomography analysis showing package cap sealing failure

26.4.4 Pressure Sensor

The sensor considered in this section is an ultra-compact piezoresistive absolute pressure sensor with functions of digital output barometer. The device comprises a sensing element and an IC interface which communicates through I^2C or SPI from the sensing element to the application. The sensing element, which detects absolute pressure, consists of a suspended membrane manufactured using a dedicated process developed by ST.

This device is available in a solution (called "Bastille") with a dedicated full-molded package leaving exposed a silicon cap with suitable vent holes to sense the external pressure, or in another version (called "water-resistant") in ceramic package with metal holed lid which keeps the sensing element in contact with the environment; gel inside protects the electrical components from water corrosion.

Typical structure of Bastille version is presented in Fig. 26.27, showing some key features of sensor chip. Picture of the package with exposed silicon cap is shown in Fig. 26.29.

In addition to the usual stress tests (from thermal cycles, temperature storage, humidity exposure, etc.), widely described in the previous sections, further dedicated stresses have been introduced during last years, like:

- The nicknamed *e-juice* test for e-cigarette (secret liquid blend suggested by the customers) or – generally called – chemical stress test. During these tests, the

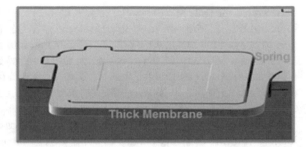

Fig. 26.27 Bastille chip features: sensing membrane, thick suspended membrane with spring

Fig. 26.28 Air leakage test, manual setup; tool for testing multiple DUTs in parallel; overpressure of 10 bar applied for 10 min, by looking for air bubbles formation in the water, as symptom of air leakage

product performances are monitored while samples are subjected to repeated cycles of chemical contaminant and hot-moist environment exposure.

- Air leakage test for water resistant pressure sensor, where the device is submitted to high pressure to check the mechanical assembly robustness and the water-proof capability; Fig. 26.28 shows an example of manual setup developed for air leakage test.

Pressure sensor reliability engineering team approaches the mission profile requirements by starting with a life-use plan: in this case, the product can be used as an absolute digital output barometer, but also like altimeter for portable devices, for sport watches, e-cigarettes and drone's application. Some major considerations can be established, starting from common incorrect events that may occur during the life-use, including drop onto a floor, into a wet sink or, accidentally, being left out in the rain. Finally, the customer requires a standard level of reliability over a period of years (from 2 years, recently requested for a consumer product, to 10 years in industrial applications).

When assessing the reliability of a semiconductor product, there are many elements to be considered. A typical product consists of a semiconductor die (ASIC) attached to a package substrate using either solder or epoxy, with wire bonds between the die pads and the package pins. Concerning the pressure sensor products, in addition to IC die, we need to consider also a "MEMS mechanical transducer" able to convert the barometric pressure dimension into the electrical one.

Problems with any of these elements, either individually or in interaction with other parts of the assembly, may lead to device failure, either catastrophic or parametric. In order to systematically measure the contribution of any of these elements to product reliability, it is often useful to design reliability tests that stress only one element of the entire assembly: this is the case of the above-described e-juice or chlorine or salt exposure to stress the chemical features, or the air leakage test to stress metal lid and sensing membrane resistance under the very high-pressure stress.

The objective of performing reliability investigation and trials is to assure that pressure sensor device manufacturing processes can produce products with acceptable long-term reliability. This objective is typically achieved by performing a series of experiments that stress our pressure sensor manufactured using these processes over a range of environmental conditions. Functional dependence of device degradation to these stresses is monitored during testing. Appropriate stresses are used so the devices either degrade parametrically (e.g., related to the pressure accuracy drift) or fail catastrophically (with an electrical overstress causing open or short circuit, heavy mechanical damage on structure or similar).

From these tests, it is possible to identify and quantify the major failure mechanisms causing the degradation of the electrical performance of the pressure sensor, including the functional degradation dependence caused by the environmental stress. Quality and reliability, process, and design engineers may then use this information to improve the processes and/or the design with respect to the failure mechanisms, resulting in higher reliability.

Typical failure mechanisms, observed on pressure sensor, are:

1. Contamination
2. Mechanical breakages
3. Corrosion (particularly on water resistant products)

Each of the above items are induced/highlighted considering the following reliability trials (some of them are dedicated, other ones belong to the standard JEDEC flow):

1. Reflow, or THS (temperature humidity stress)
2. Mechanical stress test, like dedicated tumble or drop
3. HAST (highly accelerated temperature and humidity stressing test, ref. JESD 22A110) or autoclave (ref. JESD 22A102)

Accuracy drift analysis provides an effective way of problem detection.

Contamination
Particle – variable dimensioned portion of matter – can be generated by process manufacturing flow or penetrate from environment thru the vent hole located on

Fig. 26.29 Pressure sensor device, vent holes (1–6) on exposed silicon cap (magnified on the right)

 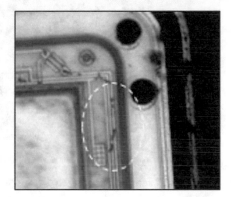

Fig. 26.30 Resin removal (on the left) and IR microscope inspection (on the right) to observe foreign material lying on sensing area

silicon cap of the pressure sensor (Fig. 26.29), potentially causing a pressure accuracy drift reading once lying on sensing area (Fig. 26.30).

Generally, this kind of anomaly is caught just after applying a thermal stress like the reflow trial (which is simulating the soldering step of the device on printed circuit board, at final customer side).

At manufacturing process flow level, this kind of occurrence has been strongly mitigated applying more severe screening and a better handling at back-end assembly steps.

Corrosion

Corrosion phenomenon is very dangerous, especially from reliability viewpoint and lifespan on field; it refers to any chemical alteration of die surface, or wires bonding or pads or metallic elements inside a device. Particularly dangerous in case of

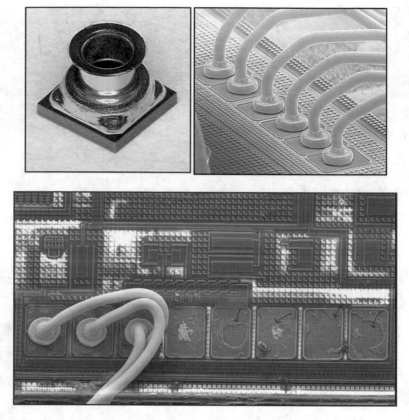

Fig. 26.31 Water-resistant product: good pads (above), corroded pads (below)

humidity (water) presence, to deal with it in some cases, protective gel has been used, like in the case of pressure sensor water-resistant products.

Even in this case the reliability tests help us to investigate these phenomena, using, for example, the HAST test (ref. JESD 22A110) or also the autoclave (ref. JESD 22A102), known also as pressure cooker test (121 °C/100%RH), able to activate any potential leaky path thru the whole structure, causing the contact between humidity (water) with the exposed metal. Corrosion causes damages on die, wires bonding, pads, and metallic finishing (Fig. 26.31) and the damages cause electrical malfunction.

26.5 Actuators Reliability Overview

Actuators are a family of products that spread over a wide range of applications, from the classical ones of inkjet to the new applications in the field of augmented reality, gesture recognition, or camera autofocus and earphone speaker (Fig. 26.32).

Dedicated MEMS design has to be developed to cover the application range of the specific device also selecting the more appropriate actuation mode to reach the required performances; for instance, different type of actuation can be used for the same scope to move a μmirror as shown in Fig. 26.33.

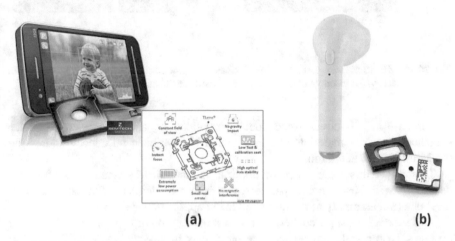

(a)　　　　　　　　　**(b)**

Fig. 26.32 Picture shows two different field of application for the MEMS actuators: (**a**) autofocus, (**b**) speaker

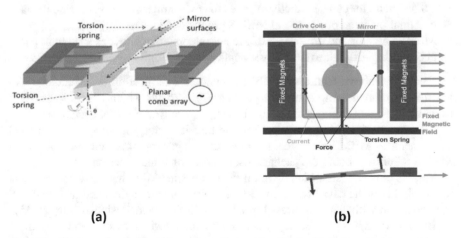

(a)　　　　　　　　　**(b)**

Fig. 26.33 Picture shows two different actuation modes for two MEMS μmirror used in the augmented reality field: (**a**) electrostatic actuation, (**b**) electromagnetic actuation

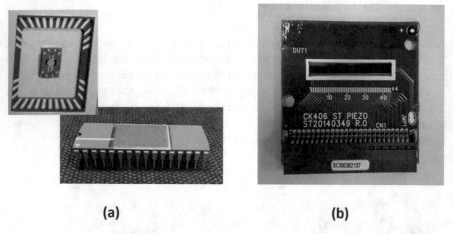

(a)   **(b)**

Fig. 26.34 MEMS actuator assembled in a dedicated support package: (**a**) ceramic package with inside the MEMS; (**b**) example of customized chipboard used to support the reliability activity

Furthermore, to cover some specific applications and comply with device specifications, new materials like PZT and AlN have been introduced addressing the MEMS piezo actuation.

To support this wide range of application, the MEMS actuator is often co-developed with customers or partners, and it is provided as MEMS die at wafer level or as subassembly to be integrated in the application system.

This scenario poses a challenge from a reliability point of view for this type of products. Compared with sensor case previously discussed, some key features of MEMS actuators have to be considered facing their reliability strategy:

- MEMS actuator can be directly exposed to environmental stresses due to the lack of a final package protection at MEMS producer side.
- MEMS actuator and driving ASIC are not integrated together as system in package, making difficult to fully cover application reliability by MEMS producer.

Facing the above features, a MEMS actuators qualification test vehicles (TV) is often developed using a support package (Fig. 26.34) which allows to perform reliability stress test, in particular for the test requiring an electrical actuation.

The drawback of a support package is that it is not possible to have a final assessment for MEMS/package interaction that, in general, it is an important task in MEMS reliability, being the package itself source of some stress on MEMS.

For this reason, tests which do not need an electrical stimulus are done mainly at die level; this is the case of mechanical tests performed to understand if the MEMS structures can withstand the stress level for which they were designed (Fig. 26.35).

By these tests it is possible to assess the maximum level of mechanical robustness that the actuator could reach and to discover possible weaknesses in the design; using the results obtained in terms of breakage, it is possible to study adequate

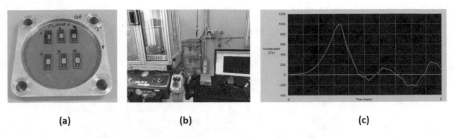

(a) (b) (c)

Fig. 26.35 Setup used to perform the shock test: (**a**) MEMS actuators on the holder used for shock test; (**b**) equipment used for shock; (**c**) signal of acceleration peak (first peak) with following damping

redesign to overcome the problems. Reliability results in terms of maximum shock level that the actuators can sustain are useful also at customer side to better design the package for the final application. Anyway, on the final system side, where at the end actuators will be placed, mechanical tests are repeated by the customer to better understand if the MEMS itself shows the same range of robustness or it is impacted by some package effect.

In general, since final assembly solution and full application environment are not available at MEMS producer side, an exhaustive study of MEMS actuators failure modes can be done also involving customer in a joint qualification approach starting in development phase. An example of such joint qualification test matrix is reported in Table 26.9.

Even if, as said, full investigation of MEMS actuator reliability needs also tests at final system level involving customer, some failure modes more intrinsically inherent to MEMS technology and design can be effectively investigated at MEMS level.

An example of such failure modes is reported in Fig. 26.36 which shows a MEMS mirror where the mirror metal surface is degraded, and its reflectivity decreases over time; this degradation can sometimes appear during exposure to humidity during the reliability tests performed on this device when aluminum surface is directly exposed.

This failure mode can be prevented by covering the mirror surface with a protective layer that do not impact on the reflectivity and inhibit the contact between humidity and aluminum surface and hence the reaction with possible residues, for example, of fluorine which, at some extent, can be present in MEMS wafer fabrication processes or can be introduced later-on in assembly processes.

Another failure mode that is found during reliability test on a different MEMS mirror in humid ambient is illustrated in Fig. 26.37.

In general, cyclic fatigue could cause fracture formation and propagation in materials till their breakage. Other than cycling fatigue itself, also environment could play a role in crack formation and propagation.

As reported in literature (es [12].), bulk silicon does not show fatigue cracking behavior. However, at microscale level such as MEMS structure, the presence of humidity associated to fatigue during actuation could create some cracks that

Table 26.9 Simplified test matrix for a printhead actuator MEMS device (some tests were performed at STMicroelectronics (ST in the table) side and other tests were done at customer side)

Test	Owner	Purpose
Operating life	Customer on final system	To guarantee minimum number of actuation according to mission profile, checking reliability simulating field operative conditions
Ink bias test	Customer on final system	To check ink compatibility under bias condition on final system
Therma cycling(TC)	Customer on final package	To check failure modes activated by thermal excursion
Ink compatibility	Customer	Compatibility of materials with various ink types
Temperature humidity bias (THB)	ST on support package and customer on final system	To check failures modes activated at high humidity and bias conditions
High-temperature reverse bias (HTRB)	ST on support package	To check failures modes activated at high temperature and bias conditions
ESD	ST on MEMS at wafer level	Robustness vs. ESD (electrostatic discharge)
Mechanical shocks (MS)	ST on MEMS at die level	Mechanical robustness of MEMS structures

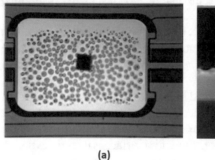
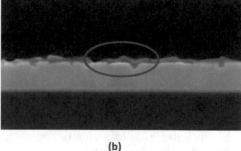

(a)   (b)

Fig. 26.36 Mirror surface degradation after humidity stress test: (**a**) optical microscope top view of mirror. (**b**) SEM cross section of aluminum degraded surface

propagate till the fractures reach certain length that cause the abrupt breakage of the structure. Most accredited mechanism for this failure mode is the so-called stress corrosion cracking (SCC), where micro-cracks in silicon native oxide work as nucleation center for the starting and cracking propagation under cycling fatigue, till structure brakeage [1, 12, 13] (Fig. 26.38).

This type of failure mode is more evident in MEMS actuators than in sensors, where the movements of the mobile structure are lower with respect to the actuator ones. Such a failure mode is not always easy to check during the stress test

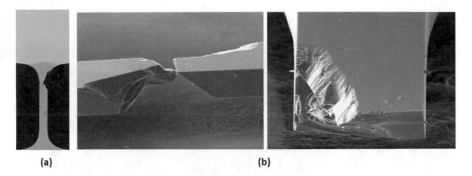

(a)   **(b)**

Fig. 26.37 (**a**) Mirror torsional axis brakeage in humid ambient. (**b**) SEM images of the fracture

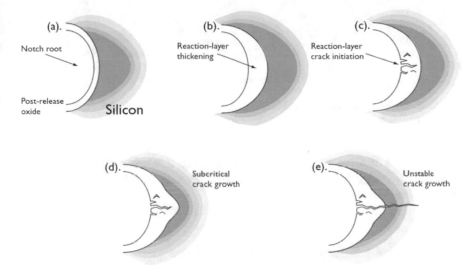

Fig. 26.38 Picture shows stress corrosion cracking mechanism with show crack formation in native oxide and its propagation (courtesy of the authors of [12])

monitoring or during the readout, but it could appear suddenly. Furthermore, this issue is not always easy to solve at design and/or at wafer process level.

Finally, even if stress corrosion cracking is the theory widely accepted to explain such type of failure mode, there is not a complete and general law able to describe the physics and kinetics of the mechanism. This implies that it is not easy to make a general model for the prediction in terms of lifetime in humid environment, and depending on actuator type, application field, and conditions, a series of studies of fatigue vs. humidity could be necessary to investigate and characterize this failure mode. As an example of these studies, Fig. 26.39 reports a graph with failures occurrence depending on fatigue and humidity levels.

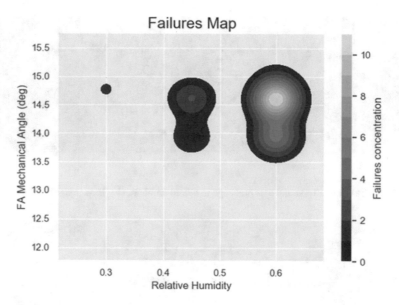

Fig. 26.39 Occurrence tortional mechanical failure vs fatigue level (mechanical angle) and humidity exposure in operating life test

The results of these studies are also useful when it is necessary in some cases to develop package solution at costumer side that can withstand humid ambient without affecting the product lifetime.

26.5.1 Case Study: Electromagnetic Mirror

In this section, the general approach for the reliability of actuators will be illustrated with one example or case study to better explain how to analyze the new failure modes that could be found with these devices. The case study considered is a MEMS mirror magnetically actuated, as described in Chap. 18. For the sake of convenience, a sketch of the structure is also reported in Fig. 26.40.

Mirror has two rotation axes: Y-axis and X-axis as reported in Fig. 26.40; around Y-axis the mirror oscillates at structure resonance frequency while X movement is actuated in a quasi-static mode. Besides X and Y axes torsional spring, a silicon frame is also present on which a copper coil is grown by ECD (electrochemical deposition). The mirror is actuated by means of Lorenz force generated by current flowing in the coil and by magnetic field generated by external fixed magnets (Fig. 26.40).

Driving system and final package are developed by the customer based on final application field. Then, to perform the powered reliability tests and to measure the functional parameters of the device, it was developed inside STMicroelectronics a

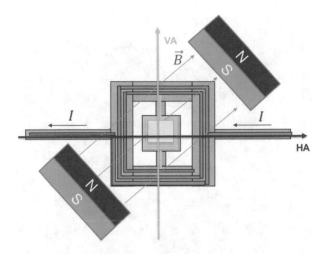

Fig. 26.40 MEMS mirror with magnetic actuation (from Chap. 18 Fig. 18.13): X-axis (HA) and Y-axis (VA) are shown as well as the magnetic field direction; silicon frame with copper coil is also sketched

dedicated driving system to actuate the MEMS; this driving system was designed considering the requirements in terms of opening angles, currents, coil resistance, and frequency defined by the customer applications.

In the present case, MEMS is mounted on a dedicated assembly by customer which is the test vehicle (TV) for reliability test, then this TV is fit by a connector on the driving board (DB) done by STMicroelectronics (Fig. 26.41).

The driving board includes a microprocessor that could actuate the mirror in different ways (both axes together, just one single axis, different opening angles combination etc.) and monitor some parameters during stress test like Y-axis resonance frequency. In this way it is possible to follow the drift in this parameter and try to correlate it to some failure modes, for example, fatigue induced.

Indeed, the final assessment on the reliability of the device is done also in parallel by the customer on the final system with final package and driving system. Using the homemade system of Figure 26.41 allows anyway to highlight and study some intrinsic failure modes that could be due to the design and/or process and others which could happen in application.

To investigate the reliability of this type of MEMS mirror, a suitable stress test matrix has been developed with the purpose to address and study the possible intrinsic failure modes (FM) of the MEMS and related acceleration factors.

Table 26.10 is a sketch of the test matrix elaborated for the reliability activity on this mirror. In principle, matrix can be divided in the following three blocks:

- Mechanical tests (shock and random vibration): they address the mechanical robustness of the mirror determining the maximum level of stress which the device can withstand.

Fig. 26.41 MEMS mirror dedicated driving board to actuate the mirror during reliability stress test

- Powered tests: they are used to targeted FM that could appear during mirror actuation as, for example, life test, or to check for FM that could be induced by current flowing (as Cu oxidation discussed later in this chapter) in different ambient (dry vs humid) and/or fatigue.
- Non-powered tests: they are normally done to address FM related to metal passivation and/or MEMS to package interactions; parametric stability is also investigated by these tests.

From the tests reported above, some failure modes have been detected and then studied; as outcome, solutions have been implemented increasing the reliability of the device. In the following paragraph it is discussed the intrinsic failure mode related to Cu coil oxidation and its modelling by a reliability point of view ([14–16]).

Cu Coil Oxidation Model in Constant Current Test
When driving the mirror, a relatively high current is flowing in the Cu coil. The coil resistance causes a power dissipation (RI^2) which in turn causes a temperature increase. Being the Cu exposed to the ambient, it starts to oxidize, and the coil increases the resistance value and as consequence the power dissipation ($P[RI^2] \rightarrow +T \rightarrow +R \rightarrow +P \ldots$). To avoid a thermal runaway phenomenon that could cause a catastrophic failure of the device, the temperature characterization vs the flowing current was performed by means of an IR camera (Fig. 26.42). By FEM (finite element method) thermal modelling, it is possible to estimate coil temperature and its evolution in the time (Fig. 26.43) and hence to assess the maximum operative

Table 26.10 Sketch of the test matrix used for reliability and failure mode study for this electromagnetic mirror

Test category	Test	Investigated factors
Mechanical	Shocks	MEMS mechanical robustness
	Random vibration	
	Sweep vibration	
Environmental unpowered	Low-temperature storage (LTS)	Interaction between MEMS and package
	Temperature humidity storage (THS)	
	Thermal cycles (TC)	
	Electrostatic discharge (ESD)	ESD induced degradation (leakage current, dielectric breakdown)
Powered	On-off (dry and humid ambient) cycles	Corrosion phenomena
	Operative life test	Fatigue effects
	Constant current test (CCT)	Power dissipation effects

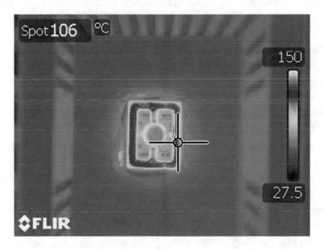

Fig. 26.42 Temperature map by IR camera showing coil self-heating

current that can be used to avoid excessive temperature increase. Of course, this maximum current has to be compliant with the requirements of the application (opening angle) so to match them also a coil resistance decrease by increasing its thickness was considered.

Referring to Table 26.10, to experimentally study the Cu oxidation, a constant current test (CCT) was implemented in order to investigate this phenomenon and find a law governing it.

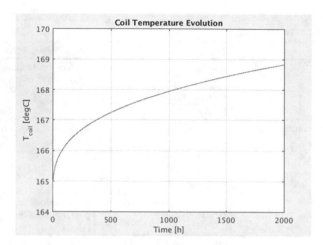

Fig. 26.43 Coil temperature estimated trend at a fixed current accounting Cu oxidation

Fixing a threshold value for the maximum coil resistance (or its maximum drift) as criterion for the parametric failure, therefore the probability in term of time to failure (TTF) can be determined as a function of the driving current and ambient temperature.

Using data collected during the CTT tests, it was found the probabilistic distribution function that better fit the experimental data for all considered current level. Probability charts were built considering the more common distribution used for TTF. Using coefficient R^2 as indicator for the best data fitting, it was found that the log-normal function is the one that better describes the failure distribution for this μmirror failure mode (Figure 26.44).

Now to move from the experimental data related to the set of currents used in the CCT experiments to a more general model, it is necessary to link the Cu oxidation phenomena due to current power dissipation to a general law that could describe it. Formally, by a chemically point of view, the formation of Cu oxide is to be considered as an oxidation reaction, where its rate (oxidation rate) is driven mainly by the temperature. So, based on this assumption, Cu coil oxidation could be viewed as the copper oxidation itself and the law that describes the kinetic of phenomena of this type is the well-known Arrhenius law.

Therefore, to account dependence of time to failure by temperature, an Arrhenius-like equation was used:

$$t = Ae^{\left(-\frac{E_a}{K_B T}\right)}$$

where A is a constant, Ea is the activation energy of the phenomenon, K_B is the Boltzmann constant, and T is the absolute temperature. The temperature to be used

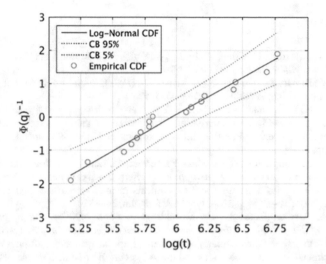

Fig. 26.44 Experimental data fitting with log-normal distribution ($R^2_{Log-Normal} = 0.9756$)

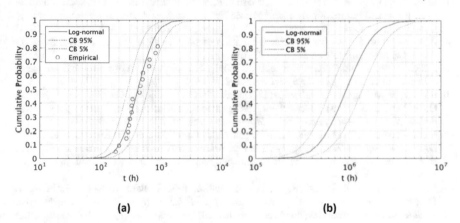

(a)　　　　　　　　　　　　　　**(b)**

Fig. 26.45 (a) Experimental data collected during CCT test; (b) predicted failure distribution in use conditions

in the above equation is the coil temperature, as it was characterized with IR camera and evaluated by thermal model FEM simulation.

Finally, using a model based on log-normal failure distribution and Arrhenius law and finding the parameters which best fit the experimental data, it is possible to obtain a predictive law to describe the coil oxidation in terms of time to failure (Fig. 26.45).

References

1. Corigliano, A., Ardito, R., Comi, C., Frangi, A., Ghisi, A., & Mariani, S. (2018). *Mechanics of microsystems*. Wiley. ISBN: 9781119053804.
2. Sharpe, W., Brown, S., Johnson, G., & Knauss, W. (1998). Round-Robin tests of modulus and strength of polysilicon. *MRS Proceedings, 518*, 57.
3. Cacchione, F., Corigliano, A., Zerbini, S.(2007). *Parametric study of fracture properties in polycrystalline MEMS*. 2007 International conference on thermal, mechanical and multiphysics simulation experiments in microelectronics and micro-systems. EuroSime 2007, London, pp. 1–5.
4. Cacchione, F., Corigliano, A., De Masi, B., Ferrera, M. (2004). *Rupture tests on polysilicon films through on-chip electrostatic actuation*. 5th International conference on thermal, mechanical and multiphysics simulation and experiments in micro-electronics and micro-systems, EuroSimE 2004, 10–12 May 2004, Brussels, Belgium, 347–350.
5. Cacchione, F., Corigliano, A., De Masi, B., Ferrera, M. (2005). *Out of plane flexural behaviour of thin polysilicon films: mechanical characterization and application of the Weibull approach*. 6th International conference on thermal, mechanical and multiphysics simulation and experiments in micro-electronics and micro-systems, EuroSimE 2005, April 2005, Berlin, Germany, 100–104.
6. Mariani, S., Ghisi, A., Corigliano, A., Martini, R., & Simoni, B. (2011). Two-scale simulation of drop-induced failure of polysilicon MEMS sensors. *Sensors, 11*, 4972–4989.
7. Corigliano, A., Ghisi, A., Langfelder, G., Longoni, A., Zaraga, F., & Merassi, A. (2011). A microsystem for the fracture characterization of polysilicon at the micro-scale. *European Journal of Mechanics A: Solids, 30*, 127–136.
8. DelRio, F. W., Cook, R. F., & Boyce, B. L. (2015). Fracture strength of micro- and nano-scale silicon components. *Applied Physics Review, 2*, 021303.
9. Rolandi, P., Magagnin, L., Valzasina, C., De Pascalis, D., Tocchio, A., Garnier, A. B. M., Filoni, G., & Pecchia, F. (2018). Electrochemical capacitance based method applied to epoxy molded devices. *International Symposium on Microelectronics, 2018*(1), 000685–000693.
10. Meng, J., Mattila, T., Dasgupta, A., Sillanpaa, M., Jaakkola, R., Luo, G., Andersson, K. (2012). *Drop qualification MEMS components in handheld electronics at extremely high acceleration*. In 13th InterSociety conference on thermal and thermomechanical phenomena in electronic systems.
11. Li, J., Makkonen, J., Broas, M., Hokka, J., Mattila, T. T., Paulasto-Kröckel, M., Meng, J., Dasgupta, A. (2013). *Reliability assessment of a MEMS microphone under shock impact loading*. In 14th international conference on thermal, mechanical and multi-physics simulation and experiments in microelectronics and microsystems.
12. Alsem, D. H., Pierron, O. N., Stach, E. A., Muhlstein, C. L., & Ritchie, R. O. (2007). *Advanced Engineering Materials, 9*(1–2), 15.
13. Ciccotti, M. (2009). Stress-corrosion mechanisms in silicate glasses. *Journal of Physics D: Applied Physics*.
14. Baiardi, M. (2018-2019). *Reliability of an electromagnetically actuated micromirror*, Master degree thesis at Politicnico di Milano, academic year 2018–2019
15. Beretta, S. (2018). *Affidabilità delle costruzioni meccaniche* (1st ed.). Springer. ISBN: 9788847010789.
16. Tobias, P. A., & Trindade, D. (2012). *Applied reliability* (3rd ed.). CRC Press. ISBN: 978-1584884668.

Part VIII
The Future of Sensor and Actuators

Chapter 27
MEMS: From a Bright Past Towards a Shining Future

Alessandro Morcelli, Simone Ferri, and Anton Hofmeister

27.1   Introduction

"Benedetto, hurry up, come here, come to see how our epitaxial polysilicon accelerometer is moving under the microscope," Paolo cried out excitedly in October 1997, echoing the more famous "And yet it moves" of G. Galilei.

"Your electrostatic mirror is working properly for our Optical Cross connect switches and it is also qualified, but we cannot use it, we are firing all our employees," a VP of a Californian startup told us right after the internet bubble of 2001.

"I do not see any added value in this accelerometer for airbag, we have already another supplier," one big automotive customer of STMicroelectronics replied to us in first half of 2002.

"I need to use a 2-axis accelerometer in our washing machines to reduce water consumption and your product is perfect," a customer based in Newton, USA, told us in second half of 2002, placing the first order of ST accelerometers. Which better place to start a business if not in a city having the same name as of the Great Isaac!

"We need a 3-axis accelerometer to protect the Hard Disk Drive of our Computers, but if you want to win this business, you need to tell me now, I repeat noooow, how to solder without any soldering drift," a General Manager of a big Japanese company asked us in September 2003.

A. Morcelli (✉) · S. Ferri
ST Microelectronics, Analog MEMS and Sensors Group, MEMS Sensor Division, Agrate Brianza, Monza Brianza, Italy
e-mail: alessandro.morcelli@st.com; simone.ferri@st.com

A. Hofmeister
ST Microelectronics, Analog MEMS and Sensors Group, Microfluidic & Micro Actuators Division, Agrate Brianza, Monza Brianza, Italy
e-mail: anton.hofmeister@st.com

© Springer Nature Switzerland AG 2022
B. Vigna et al. (eds.), *Silicon Sensors and Actuators*,
https://doi.org/10.1007/978-3-030-80135-9_27

"I do not believe you will ever succeed in selling your MEMS for mobile phones and game controllers," a veteran of semiconductor company claimed in 2004.

"I understand very well how your 3-axis accelerometer works, and I can use it in our game controllers, but I need a price lower than 2 $, otherwise there is no business," Satoru Iwata, CEO of Nintendo, told us in March 2005.

Jumping to 2020, STMicroelectronics is leading the silicon sensor and actuators market, together with a handful of semiconductor players.

As the reader can appreciate from the quotes listed above, a great, brave, resilient, innovative, and lucky team stands behind this incredible success.

Very often, in our lives, we need to select a possible path forward having alternatives in front of us. Each choice has an unknown probability of success; we can only perceive its borderless vastness. In any case, whatever the choice, we cannot pretend to analyze in detail the complete map of all possible alternatives.

Time is of the essence, and nimbleness is key, as W. Shakespeare wrote "Only Nimble thought can jump both sea and land."

If luck plays an important role in the choice of the most successful alternative, the ability to follow the chosen path forward depends on the talent of the team and its trust in the business vision.

The team of STMicroelectronics started this fantastic journey in silicon sensors and actuators 25 years ago. It has been, and it continues to be, a great journey, a further demonstration of the insightful set of two equations mentioned by the Nobel Prize winner Daniel Kahneman[1]

Success = Talent + Luck

Big Success = Talent + Big Luck

We always believed there were many business opportunities for MEMS. But honestly only now we are in the position to appreciate the positive impact that the "MEMS Consumerization Wave," spurred from the "Secret MEMS Revolution," had on the overall field of sensors and actuators.

Nowadays, small, highly reliable, energy-efficient, and tiny silicon sensors are ready to be coupled on one side with low-cost and low-power radios (sub-GHz, Bluetooth Low Energy, WiFi, SigFox, LoRa, Narrow Band IoT, etc.) and on the other side with cheap ultra-low-power microprocessors to transition the mankind towards the third and fourth wave of artificial intelligence, the sensing and autonomous ones, respectively (ref to Kai Fu Lee Book).

Thus, we apologize in advance with the reader if in this last chapter we do not foresee exactly where the micromachined sensors and actuators will find the most successful commercial use cases.

We strongly believe, however, that we are entering a new era of MEMS. MEMS components are moving successfully from offline era to online era, and this transition resembles, on much bigger scale, the same quantum leaps we will describe in the following paragraph.

---

[1] https://en.wikipedia.org/wiki/Daniel_Kahneman

Last but not the least, we will ask you to be patient for the next 20 years after which, together, we will be able to draw a clear strategic rationale behind the present MEMS innovation adventure!

27.2 The Quantum Leaps of Offline Era of MEMS

Nowadays we live in a world full of MEMS: piezoresistive pressure sensors, capacitive inertial sensors, microphones, thermal inkjet printheads, to mention a few of them. They are used across all the market segments from automotive to personal electronics, from industrial to computers and printers. This has been possible thanks to several technical and commercial quantum leaps that happened in the last 60 years.

In the end of the **1950s**, R. Noyce of Fairchild Semiconductor invented the first monolithic chip (1958), C.S. Smith discovered the piezoresistive effect in silicon and germanium (1954). This discovery showed that silicon and germanium could sense air or water pressure better than metal. Strain gauges began to be developed commercially in 1958 and Kulite was founded in 1959 as the first commercial source of silicon strain gages.

During the **1960s**, many scientists and engineers developed bulk micromachining technologies so that in the 1970s bulk micromachined piezoresistive pressure sensors started to be used in blood pressure monitoring devices and in cars to reduce gasoline consumption. Pressure sensors can be considered as the earliest commercial success of MEMS sensors.

Moreover, in the late **1970s**, Hewlett-Packard and Canon came up independently with an alternative for dot matrix printing, called Thermal Inkjet Technology. The ink droplets were ejected through an array of micromachined nozzles and allowed the rapid creation of an image onto a paper sheet and other media. Still today Canon and HP use silicon micromachining technology to manufacture their inkjet printhead nozzles (IPN). Without any doubt IPNs are the most successful MEMS actuators, so far. Below picture is an example of oxide nozzle array realized by STMicroelectronics (Fig. 27.1).

But it was only in 1982 that the famous paper "Silicon as Mechanical Material," written by the visionary K. Petersen and published on IEEE – a provocation for electronics engineers considering the silicon only for its semiconductor properties – attracted the scientific interest of a wider engineering community.

In the **1980s** many process technologies have been conceived and developed. Among them, the surface micromachining took the commercial lead and in 1993 Analog Devices was the first to produce a surface micromachined accelerometer in high volume. That accelerometer was adopted in cars for airbag deployment system, as mandated by NHTSA.

As already detailed in the previous section, in **1994**, Dr. Franz Laermer and Andrea Schilp from "Robert Bosch GmbH" developed a new process to help manufacturing their MEMS devices. This technique patented and licensed by Bosch

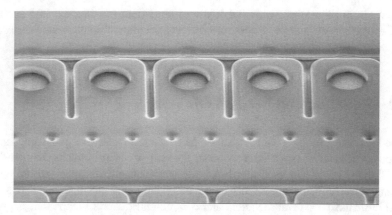

Fig. 27.1 Silicon oxide nozzle array

company is frequently called TMDE (time-multiplexed deep etching), but it is mostly known as "Bosch process," and it combines the effectiveness of a fluorine-rich plasma on silicon surfaces with the etch-inhibitory property of Teflon-like films generated by fluorocarbon discharges without needing to reach cryogenic temperatures. Thanks to its increased aspect ratio in the silicon etching process, it enabled to increase performances while ensuring proper repeatability of MEMS structures during the manufacturing process.

While thermal inkjet printhead adoption was increasing, MEMS started to establish themselves in the automotive market. The adoption of sensors in automotive was driven by regulation.

All modern cars have active (i.e., vehicle dynamic control™) and passive safety systems (airbag) based on micromachined inertial sensors. High-g accelerometers, firstly adopted in the airbag system, can be located either in the central part of the car or in the periphery of the car. Since **1996**, active safety systems such as vehicle dynamic control (VDC™) or electronic stability control (ESC®) started to use high-grade low-g accelerometers (up to 5 g), yaw-rate gyroscopes, and magnetic speed sensors at each wheel. These safety systems help drivers handle a car on icy or rainy roads, by preventing the dangerous situation of under- or over-steering. After wide adoption of sensors for active safety in European vehicles, in April 2006, the NHTSA mandated the use of ESC active safety systems in all US passenger vehicles, beginning in 2012.

After the year **2000**, in which the high failure rate of a specific tire brand was implicated in more than 270 deaths, pressure sensors started to gain momentum for measuring tire pressure and alerting the driver in case of anomalies. All vehicles manufactured in 2008 or later have a tire pressure monitoring system. Pressure sensors are also used to measure the ambient and the intake manifold pressure to optimize the fuel-to-air ratio with the goal to reduce carbon footprint.

Till year 2005, year over year growth of MEMS devices in the market was relatively small, until something exceptional happened in 2006. Nintendo CEO,

Satoru Iwata, bet the future of its company on a new concept of game console, leveraging the inherent benefits of a micromachined accelerometer. Indeed, the three-axis accelerometer of STMicroelectronics together with Wii™, the revolutionary game console of Nintendo, changed dramatically the way users were interacting with the digital world of electronic games.

At that time, STMicroelectronics was the first company to build an 8″ wafer MEMS facility to sustain the sky-rocketing volumes of Nintendo Wii Consoles, as well as the adoption of MEMS in digital still cameras for image stabilization, by hard disk drives for data protection, and by emerging smartphones for user interface.

After Nintendo, Sony's PS3™ started to use MEMS sensors to realize human centric wireless game controller: for the first time it was the technology adapting to people and not vice versa. This triggered a paradigm change in the man-machine interface and the same technology got adopted in TV sets, PCs, and set top box remote controllers.

After this first great success, sensors spread into personal electronics devices such as smartphones, tablets, and wearables. If sensors are used mostly for user interface in game controllers, they began enabling multiple applications in personal electronics devices. For instance, in smartphones:

- Accelerometers for automatic screen rotation
- Gyroscope, measuring angular rate for gaming application
- Magnetometers for magnetic field, necessary for maps and navigation
- Microphones for voice pick-up to exploit low-cost calibrated arrays of sensors

Overall, during the first decade of the twenty-first century, MEMS established themselves as an enabling technology for new products and applications. However, we were still at the beginning and in the "offline era," with the cloud still being an abstract idea, computing power still being limited and expensive, and efficient power management still being an issue.

27.3 MEMS Moving from the Offline Era to the Online Era

Let's now jump straight to the second decade of the twenty-first century (2010–2020). Thanks to an acceleration in technology development, the variety of MEMS devices and their performance kept increasing. Also, MEMS devices became more compact and affordable, making them the most popular sensors in the market.

In this decade there was an important conceptual step ahead of the market: from single component there was a massive deployment of fusion of sensors and SW usage to extract more information and meta-data. It is all about MEMS? For sure no, beside the availability of sensors, a couple of enablers can be identified:

- **MCU**: the only component in this revolution following the Moore law
- **RF**: cheaper than in the past, finally low power and well standardized for easy system integration

From an application perspective, this generated two new trends: the use of multiple types of MEMS device to broaden the range of applications and the increased specification requirements. To achieve the needed accuracy, MEMS technology and design must go hand in hand with improvements in final test and calibration equipment. At this point everything was ready to efficiently replicate a "living system": sensing, elaborating, and wireless transmitting/receiving of data.

2010–2020 has been the decade of the smartphone proliferation, equipped with more and more sensors to support applications such as the indoor navigation, barometer, optical image stabilization, and face ID. But this decade is also doubtlessly the Internet of Things (IoT) era. Millions of affordable devices paired via Bluetooth to our smartphones were launched to offer new services and use cases to consumers.

Because of that, MEMS products also found interest in more conservative markets such as industrial, medical, infrastructure, etc. Sensors are now being adopted in the industrial domain to replace bulky and expensive devices for vibration analysis and tool diagnostics (predictive maintenance had still to come …). Accelerometers also started being used in pacemakers and neuro-stimulators and MEMS sensors were used for glucose monitoring devices and DNA analysis.

During the same timeframe, more and more vehicles got equipped with electronics devices, among which sensors and actuators, to support vehicles dynamics control systems, electrification (EV and hybrid models), body of the car monitoring, infotainment, connectivity, and ADAS (advanced driver-assistance systems). Finally, the last years of this decade have seen the beginning of a revolution in the automotive industry: autonomous driving.

Besides MEMS sensors, there is another product family started to emerge: MEMS actuators. MEMS actuators, as seen previously, are devices that transform a electrical signal into a micro-actuation. First application of actuator was in printheads to eject tiny droplets of ink using thermal actuation; other techniques like electrostatic, electromagnetic, and most recently piezoelectric actuation were used to design MEMS mirrors for tiny laser beam scanning systems. While MEMS mirrors have been around since the late 1990s, it took them quite a while to find its "killer application." Numerous adoptions in diverse areas like optical switches, head-up displays, and portable pico-projectors took place, but none of them did sell in high volumes. With the arrival of smart glasses and augmented reality headsets, this is expected to change, but let us postpone this interesting technology to the next section.

27.3.1 Consumer and IoT

When thinking about personal electronics, we think of the smartphone. On top of being a formidable standalone device, it has become an enabler of a multitude of objects, connected to it, to support people with additional smart features and services during their personal life: so-called IoT devices, such as smartwatches, bracelets, fitness bands, wireless speakers, earbuds, etc.

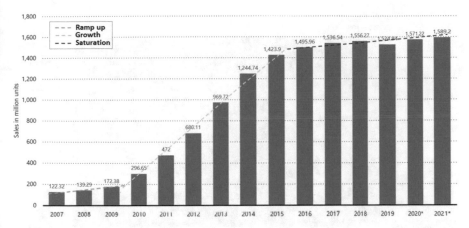

Fig. 27.2 Smartphones sales to end users worldwide from 2007 to 2021. (Source: statista.com)

27.3.1.1 Smartphones Boom and Feature Increase

The smartphone success is typically associated with the launch of the first iPhone in 2007. After that, the market boomed between 2010 and 2015 and started to mature from 2016 onwards (Fig. 27.2).

Sensor-wise, the first phones were equipped with accelerometers to support basic functions such as screen rotation. The first gyroscope in a phone has been adopted by Apple Inc. in 2010 with iPhone4 and used for application that exploits the information of the angular rate. Cupertino designers understood that, since the gyroscope allows the calculation of orientation and rotation, it could be used in combination with accelerometer to achieve a more accurate recognition of movements, through the so-called sensor fusion. One of the most common smartphone applications for gesture recognition is gaming, which requires sensors capable of accurately recognizing device orientation and user inputs to properly support user experience (rotate screen, tap, double tap, etc.). As always, after the pioneers successfully launch new products and generate new trends, competitors come and the challenges increase.

After 2015 the market saturated and global shipment of phones stagnated around 1.5BU/year. In an attempt to differentiate, phone makers started focusing on the quality of the phone's camera. Nokia N95TM using the five-megapixel STMicroelectronics CMOS image sensor started to displace the Japanese digital still cameras. And now 14 years later, digital still cameras have been largely replaced by smartphones with high-resolution cameras, up to the recent 48 megapixels. Smartphones are offering increased resolution and image quality not only thanks to better camera modules but also thanks to multiple cameras and the adoption of motion sensors to implement optical image stabilization. Since in a flat device the optical zoom cannot be supported by a telescopic technique, the only way to support different focuses is to embed different ranges camera optics.

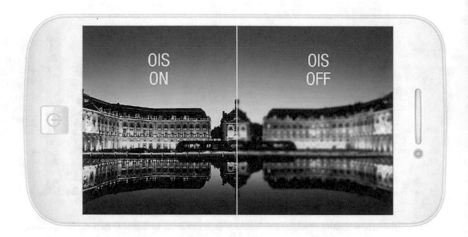

Fig. 27.3 MEMS sensor based – optical image stabilization

The quality of an image, without blurring, would depend on the ability of the user to hold the phone steady while taking a picture. Nevertheless, taking a picture with a smartphone without inducing vibrations is practically impossible. For this reason, a compensation system called optical image stabilization has been developed. The concept is easy, the accelerometer and gyroscope are adopted to sense the linear accelerations and rotation, thus activating a mechanics to compensate the unwanted movement in the opposite direction (Fig. 27.3).

OIS systems were already adopted in digital still camera based on piezoelectric gyroscopes. Finally, the sensor based on this technology, mostly manufactured by Japanese suppliers, lost the race against silicon sensors, thanks to their smaller dimensions and higher performance in terms of sensitivity and stability. Moreover, silicon gyroscopes can measure angular rates along the pitch and roll axes simultaneously, and thanks to their reduced dimensions, they can be integrated easily in the camera module itself. Finally, silicon gyroscopes are sold already calibrated and thus the final product manufacturer does not have to waste time and money to calibrate the final system, i.e., the smartphone (Fig. 27.4).

By the end of 2020, six phone vendors own almost 80% of the global market, and one of the key selling points of brands such as Samsung, OPPO, Vivo, and Huawei is the superior quality for the photographs. For example, Huawei has a partnership with Leica to execute this strategy.

Summarizing, the first MEMS sensors used in smartphones have been the motion sensors, accelerometer, and gyroscope, to support basic function, sensors fusion, gaming, and optical image stabilization. Across this period, sensor vendors were able to offer devices with higher level of integration and miniaturization to save space and cost; for this reason accelerometer and gyroscope standalone left the stage in favor of combo sensors embedding both sensors in a unique package. More

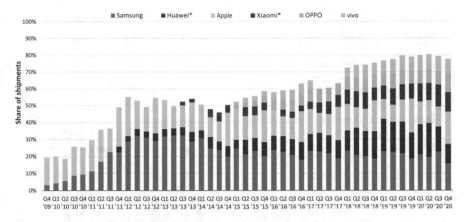

Fig. 27.4 Global market share by vendor – top 6 phone makers. (Source: statista.com)

and more sensors have been added, such as pressure sensors to implement a digital barometer, for absolute and relative altitude measurement. Altitude measurements became a requirement to support both indoor and outdoor navigation, useful not only for navigation but also as accurate person positioning in case of an emergency such as fire in a building. Temperature and humidity sensors to monitor the environment both inside and outside the smartphone got added as well.

Nevertheless, these added features supported by sensors did not really represent new technology challenges, but more of an evolution and optimization of a product during its own life. The disruption in personal electronics and its impact on sensors is underway and will take places in the years to come.

Smartphones were not only the boost for MEMS sensor sales but also image sensors. Once again, the pioneer has been Apple Inc. launching for the first time in the end of 2017, with iPhone X model, the face ID function. A time-of-flight device is adopted by this application to measure the distance from the phone and multiple points of the face to rebuild a 3D model. After Apple's launch, demand for such products increased and having time-of-flight is a must have technology for high-end smartphones.

27.3.1.2   Fitness Bands and Entertainment Devices

The combination of affordable sensing, processing capability, and connectivity, as enablers for innovative products, is what smartwatches, fitness bands, and smart speakers have in common. Starting from 2013, first Fitbit and later other companies like Apple announced products embedding sensors of different types in their watches. Even though Fitbit, with its accurate fitness bracelets, was the first company launching such a device, in China several brands (Xiaomi, Huawei, BBK, Huami, etc.) offer mostly fitness bracelets (from high-end to entry-level solutions)

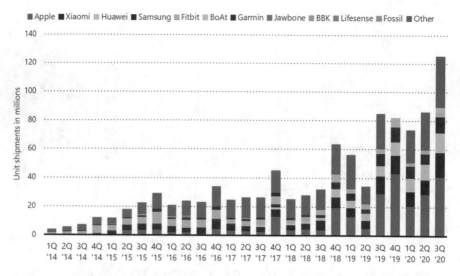

Fig. 27.5 Wearable unit global shipments by vendor. (Source: statista.com)

and Apple and Samsung with their high-quality fashion smartwatches were able in few years to achieve a major market share (Fig. 27.5).

Xiaomi unveiled its first generation of Mi Band and priced it at RMB 79 (USD 11). At that time, prices offered by Mi Band's competitors were several times that of Mi Band's. In 2015, the company sold more than ten million units of Mi Band. In June last year, Xiaomi launched its Mi Band 2 at RMB 149 (USD 23), which holds a heart rate sensor and allows users to instantly read their heart rate. Xiaomi's pricing strategy of Mi Band was successful, as Mi Band saw its 2016 shipments at 15.7 million, making Xiaomi the second largest wearable device manufacturer in the world by shipment. Another big player in the market, offering a wide portfolio of wearables, is Huawei with its series of Huawei Bands and watches for fitness. It was 2015,when the Chinese tech giant claimed the intent to put a huge focus on the wearable device after the great success of the first Huawei watch (Fig. 27.6).

Samsung and Apple instead focused on a high-end product targeting the luxury and fashion sector (Fig. 27.7).

The sensor adopted in such devices depends on features offered by product. Purely sports-related products usually include accelerometer used to measure the physical activity. Or 6-axis combo sensors to support the sensor fusion for a better movement and precise sport recognition. High-end fashion watches also include an accelerometer or the 6X more focused on the gesture recognition.

Affordable smart speakers, artificial intelligence-enabled, have been selling in huge volumes during the past years, showing the full potential of MEMS microphones, once seen only as a mere replacement of old electret microphones in the mobile phones (Fig. 27.8).

Fig. 27.6 Fitness bands

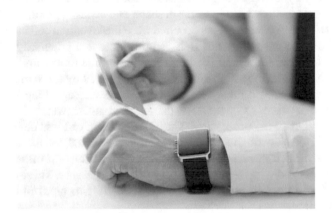

Fig. 27.7 Fashion smartwatches

Fig. 27.8 Smart speaker examples

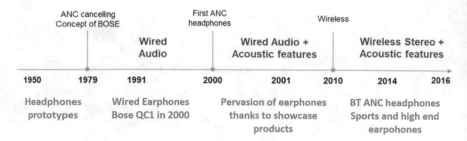

Fig. 27.9 Headphones evolution milestones

27.3.1.3   TWS and Hearables

Acoustic transducers to be worn on human body were invented around the middle of the last century, better known as headphones. Such devices had their evolution across the whole century both in terms of shape and acoustic features (Fig. 27.9).

Up to 2010, the two main events occurred were the invention and industrialization of the active noise-cancelling headsets and the increased popularity of earphones boosted by the great success of the iPod. People became more familiar with an in-ear device characterized by high quality of audio granted by the signal processing.

The use case, requiring a sensor, is the active noise-cancelling headset where a microphone is used to "listen" to the ambient noise which is subsequently cancelled through signal processing. The big revolution of these devices happened in the second decade of the century once again enabled by the MCU and the RF technology. Such devices allowed the deep transformation of headsets, which became wireless and with high capability of signal processing. These are the perfect conditions for sensors to proliferate. In this scenario two types of earphones are established, commercially known as:

1. TWS – True wireless stereo
2. Hearables

Even if they look similar, these two devices deeply differ from each other. As a matter of fact, they are both wireless stereo earphones with capability of local audio processing, but with very different product features (Table 27.1).

In few words, while TWS is focused on a better listening experience, hearables are focused on human biometric signal monitoring. This definition is based on the analysis of the most popular devices available on the market, and it is not unique but generally valid.

Let us observe which trends are valid for both products. The traditional user interface is no longer convenient. The shared features are mostly related to communication protocol and basic functions such as motion sensor adopted for user interface (XL to sense single or double tape to support distinguished functions like pick up a call or trigger voice commands). Let us see a practical example: If you are using wireless headphones while running or working out at the gym, chances

Table 27.1 Features vs. device

Feature type	TWS	Hearables
Dedicated	Active noise cancelling	Passive noise isolation
	Noise filtering/beam forming	Biometrics and environmental sensor
Common	Stereo BT into neckband or NFMI (NFMI stands for near-field magnetic induction, a short range communication system coupling 2 devices through a non-propagating magnetic field)	
	Audio transparency (microphone listening the environment)	
	Ear bone conduction for voice clarity	
	Motion sensor for user interface	

are slim that you are going to be staring at your phone the entire time. This makes it inconvenient for users to rely on their phones to control their hearables. Buttons put directly on the device itself tend to be small and are not visible when they are in the user's ear, making it difficult to locate and press them. A more convenient user interface is gesture detection. With motion tracking, simple gestures can provide instruction for specific controls and actions. For instance, your device could sense a simple "tap" on the earbud to increase the volume. It is much easier to find and tap the entire hearable compared to pressing a specific button on it. This approach makes the use of the accelerometer necessary to support command detection. "In-ear" detection is a gesture that can be used to automatically pause the audio when the user takes their earbud out. Think of how much easier that would be when you run into a friend in between sets at the gym; the audio just stops automatically as you politely take out an earbud and resumes the moment you put it back in your ear. This is not only convenient from a user experience standpoint, but it is also instrumental for an optimized power budget management. Do not forget that all these devices are power battery operated making the battery duration a key selling point of the product. The electronic device adopted to support this feature is typically the IR proximity sensor. Finally, both devices are embedding at least one microphone pear each earbud to support phone calls. Except for the proximity sensor, TWS and hearables selected the MEMS technology for the microphone and accelerometer.

Moving more specifically to analyze the **TWS**, as anticipated, on top of the standard accelerometer to support gesture user interface and proximity sensor for application management, such products include more than one microphone, per earbud, and a wide bandwidth accelerometer with the scope to pick up the signal generated by the head bones and enhance the acoustic experience. So, TWS are typically adopting two microphones, each earphone, enabling an audio processing to support the most advanced functions such as beam forming or active noise cancellation. The wider bandwidth accelerometer is also used in combination with microphones to optimize the audio quality. The sensor fusion between audio and vibration of the bones allows to implement additional functions such as the VAD (voice activity detection) or noise reduction. Those algorithms enhance the user experience supporting an optimized quality of the calls and playback of your favorite music (Fig. 27.10).

Fig. 27.10 True wireless stereo device

The most popular TWS are without doubt AirPods® from Apple. It has been the inspiration and somehow a design reference for many other TWS that followed. A more advanced device can also use a 6-axis IMU, made up of an accelerometer and a gyroscope, to track orientation. With the additional data from the gyroscope, the hearable device can find the user's relative head orientation. After ensuring the proper sensor rates and latency, this enables the accurate head tracking necessary for immersive 3D audio and XR applications. Pairing either the accelerometer or the 6-axis IMU with a proximity sensor increases the robustness of features like in-ear detection. The more information, the better the algorithmic result, and hearables are no exception.

The other family of this wireless stereo device are Hearables. As anticipated, the difference in the architecture and sensors adopted resides into the main scope of the application that is the human biometrics and activity monitoring. The ear is among the best suited body parts to measure the human biometrics. On top of the common sensor for user interfaces, low-power accelerometer, and microphones, to support phones calls and noise reduction algorithms, many other sensors are embedded to support the primary cope of such devices (Fig. 27.11).

This architecture and several sensors make hearables a professional tool for both fitness activity and common life. Fitness tracking through the head has robustness built into it since the range of motion for a head (and ears) is relatively consistent compared to your wrist or pocket. Still, it's possible to fool fitness algorithms, with many false positives and negatives affecting the output data if the motion tracking is not precise. If your hearable can detect and classify activities automatically, it can track full body movement and gain context – are you running? Biking? Standing in line at a cafe? Accurate classification can be integrated with a software library to convert step counts to calorie counts as well, giving a more complete picture of your day. Another important sensor involved in the audio quality is the motion sensor. But when you pair accurate head tracking with hearable technology, however, the

Fig. 27.11 Sport hearable

listening experience becomes an immersive experience. Hearables equipped with spatial audio changes as you turn your head, putting you right in the middle of the music, as if you were there. This life-like experience requires high accuracy head tracking with low latency, to ensure that it moves with you, and without delay. Spatial audio elevates the user experience for gaming or XR applications as well.

Two different products focusing on different user experience features but sharing same technological approach: full of MEMS sensors.

27.3.2 Fourth Industrial Revolution

The phrase Fourth Industrial Revolution was first introduced by Klaus Schwab, executive chairman of the World Economic Forum, in a 2015 article published by Foreign Affairs. The articles quoted,

> "We stand on the brink of a technological revolution that will fundamentally alter the way we live, work, and relate to one another. In its scale, scope, and complexity, the transformation will be unlike anything humankind has experienced before. We do not yet know just how it will unfold, but one thing is clear: the response to it must be integrated and comprehensive, involving all stakeholders of the global polity, from the public and private sectors to academia and civil society. The First Industrial Revolution used water and steam power to mechanize production. The Second used electric power to create mass production. The Third used electronics and information technology to automate production. Now a Fourth Industrial Revolution is building on the Third, the digital revolution that has been occurring since the middle of the last century."[2]

---

[2]The Fourth Industrial Revolution, What It Means and How to Respond. Klaus Schwab, December 12, 2015.

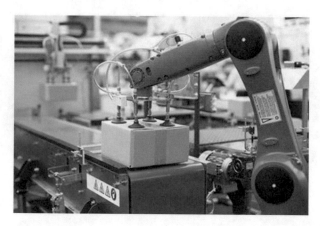

Fig. 27.12 Cobot in warehouse

27.3.2.1 Industrial IoT

MEMS sensors together with the processing and connectivity devices took a relevant part in this industrial revolution. Some examples of application exploiting these technologies are as follows:

- Vibration analysis for equipment, the foundation for most of the predictive maintenance approach.
- Infrastructure tilt monitoring (building, tunnels, bridges, etc.).
- ultrasonic analysis for leakages detection. Predictive maintenance can now be done on power lines and gas pipes.
- Robots and cobots in industry 4.0 factories.
- Smart logistic.
- Home automation.

Micro-fabricated sensors start to be widely utilized in numerous applications both in the industrial and medical market space. Smart factory automation, condition monitoring for preventive machine diagnostics, asset tracking for more efficient logistic services, optimization of water and energy consumption of household appliances, continuous monitoring of people activity, and ambient assisted living are just a few examples of applications in this market. Beyond this growing and essential healthcare market, sensors will leave a sign in the new generation of home and factory robots, including the cobots, enabled by artificial intelligence (Fig. 27.12).

The recent market trends in electrification, digitalization, and automation will require more and more the deployment of wireless sensor networks, which today are still in their infancy. A wireless sensor network results from the combination of different wireless sensor nodes (mote) consisting of some combination of a sensor, a micro-controller, a wireless receiver, an antenna, and a battery, or, even better, an energy scavenger. The potential market for motes is limited only by our imagination, and as soon as overall power consumption of the mote will be lowered (big current technological hurdle), motes will ultimately become a natural part of our lives. Their potential applications range from security and bio-detection to building and home automation, industrial control, pollution monitoring, and agriculture. Moreover, rising concerns for safety, convenience, entertainment, and efficiency factors, coupled with worldwide government regulations in terms of security, could boost sensor usage to unprecedented levels, although not all of them are necessarily silicon micro-fabricated. In fact, motes will have to measure real-world variables such as pressure, temperature, heat, flow, force, vibration, acceleration, shock, torque, humidity, and strain as well as capture and record images in the visible and near-IR range. Suppliers must be ready to integrate different technologies in a modular format as well as to achieve high-volume manufacturability of motes. Most of the transducers, with the exceptions of the CMOS image sensors, are coming from advanced 8″ fabs, while the rest of the active silicon components is fabricated in state-of-the-art 12″ fabs. Then, system in package solutions allows a modular and flexible integration of the different components in a miniaturized and compact module. It is of course not clear which exact application will win in the long run, but different new sensors, currently being developed by many companies and research centers, will enable a lot of new applications, driving new revenue streams.

27.3.2.2 Medical Devices

Another interesting products family belonging to the Fourth Industrial Revolution is represented by medical devices. Smart watches and fitness bracelets launched as personal electronics devices to track body activity have been also equipped with devices like time of the light sensors to measure the body's function through blood oxygen content measurement. Such evolution enabled the transformation of an IoT product to a certified medical device.

Some cardiologists have raised concerns about two-lead ECG: according to them, it is unnecessary for the general population or it could cause problems, including false positives. At the best, they say, that could result in stress for users and unnecessary visits to doctors, helping further burden an increasingly sluggish healthcare system. Even worse, false positives could also lead to unnecessary follow-up tests with the costs and health risks those tests can involve.

The next steps are noninvasive glucose monitoring. Several companies are busy in developing and selling add-on devices from blood pressure monitor to a sleep-aiding headband. Last 2019 Consumer Electronics Show in Las Vegas hosted many companies working on digital health. However, mental health has strong impact on blood pressure and ECG readings and thus on a person's overall well-being. Thus, several cardiologists are questioning the usefulness of continuous monitoring in consumer wearable gadgets. The authors share the same concerns. Recently two-lead electrocardiogram (ECG) appeared in the Watch 4 series. Global health care spending is estimated to reach almost $9 Trillion annually by 2020, and this is the field where smart sensors can make a big difference. We will all discover the future of medical wearable devices in the course of next decade; the authors do not have

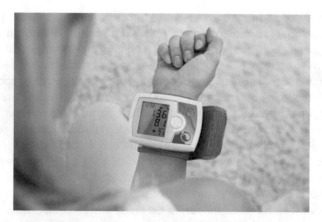

Fig. 27.13 Wireless electrocardiogram

Fig. 27.14 ESC – airbag – navigation

doubts about the successful market adoption of medical-grade wearable devices. Examples of such products manufactured by both traditional big companies and clever startup have already landed into the market (Fig. 27.13).

27.3.3 Automotive

The electronics content of a vehicle is consistently increasing; after safety, the pervasion is happening in navigation, the driver convenience, and now the research is on autonomous driving:

- Accelerometer and gyroscope adopted in the electronic stability control system to control slip and roll over of the car
- Accelerometer and gyroscopes to support dead reckoning for navigation
- Accelerometer and pressure sensor for tire pressure monitor system
- Accelerometers for key for and car alarm (Fig. 27.14).

However, the pervasion of motion sensors in mobile phones did not stop there since they started to be used more and more for pedestrian navigation to enable location-based services, so attractive for network operators in constant search of new revenue streams. When navigating in unknown places, the electromagnetic signals coming from several geostationary satellites and from the cellular network infrastructure can be faded, and thus the GPS function embedded in our phones is not 100% reliable. Tunnels, bridges, and skyscrapers are all obstacles for telecommunications operators who want to sell location-based services such as zone-based advertising and augmented-reality experiences. In this context, micro-fabricated low-power and high stability motion sensors can assist and substitute the GPS signal, so that dead reckoning system continues to track people movements. To implement dead reckoning, it is necessary to know the distance and the direction the phone user travelled. An ideal module for personal navigation comprises a 3-axis accelerometer plus a 3-axis magnetometer plus a 3-axis gyroscope and a pressure sensor, everything coupled with the appropriate application software. Often a module like the previous one is also called 10-axis. The accelerometers help to determine the traveled distance and the average speed of the user, the compass picks the heading direction with respect to the magnetic north, the gyroscope measures the change in the direction of user path, and the pressure sensor measures the inclination of travelled path. Emergency services like the e911 in the USA mandate the integration an altimeter (pressure sensor) in the phone so that in combination with a GPS, it can tell the emergency responders on which floor of the building the caller is. In the last 10 years, many integrated modules have been developed to address this interesting market application, but as of today, we do not have yet a reliable solution. Suppliers are able to delivery highly accurate altimeters and very low noise magnetic sensors, but we still need a couple of years before to see in the market highly precise and stable six-axis inertial modules suitable to address these long-waited applications.

It's also interesting to notice that all the attempts, done by companies such as STMicroelectronics, TDK, and Bosch, to integrate different sensors in the same modules miserably failed because of the high calibration cost of the final solution. The development of high accuracy six-axis inertial sensors will bring benefits also to future autonomous cars who must rely on digital maps and onboard sensors to navigate in the city jungle.

27.4   MEMS: The Journey Has Just Begun – The Online Era

The evolution of the attitudes and characteristics of human generations is a necessary premise to be analyzed before going into details of this section. There is a direct link between the attitudes or the macro trends among the populations and the product definition (Fig. 27.15).

We can read the past 100 years in another way:

– Greatest/silent generation experienced the worst beginning of twentieth century, maybe the worst decades ever and the impact technology can have on people.

Generation	Silent	Baby boomers	Gen X	Millennials	Gen Z
years	1923 -1945	1946-1964	1965-1980	1981-1997	1998-2016
Life events	World Wars Great depression	Cold War Moon landing	Berlin Wall Live Aid	9/11 attacks Iraq war	Great recession Arab springs
Technological breakthrough	Electric appliance	Transistor invented	Electronic scalability	Social media	Rise of AI
Key tech product	Car	TV	PC	Smartphone	AR/VR
Communication style	letters	Telephone	Email	Chat	Images
Mobility style	Public / car	Own car	Bicycle / car	Shared mobility	Rent

Fig. 27.15 Generation vs attitude

- Baby boomers created the web: a way to destroy the barricades between people and nations and to (try) prevent the experience of their fathers and a new cold war.
- Gen X exploited the www bringing him in (almost) every corner of the world.
- Millennials connected people and the objects to the net because they wanted a digital world.
- Gen Z is digital innate, the experience seamlessly reality and digital. Digital is as real as reality.

Here are some ways[3] that American Gen Zers are different from their millennial counterparts:

- They like brick-and-mortar stores: in a sense, members of Generation Z are always shopping, because they are always connected. They buy on any device and in any format or channel. At the same time, they are surprisingly old school. They are much more likely to shop in physical stores than are millennials, who were the first generation to grow up with online shopping and who are more likely to shop that way.
- Gen Zers want to stand out, not fit in, so brands are not as important to them. Rather, they are looking for the next unique product.

---

[3]Source: https://www.mckinsey.com/industries/retail/our-insights/the-young-and-the-restless-generation-z-in-america

- Gen Zers care about experiences, but to them it is more important to spend on those that enrich their everyday lives, such as hobbies and home entertainment. Millennials are more likely to splurge on things like travel and luxury hotels.
- Almost 40 percent of adult Gen Zers (age 18 to 23) say their purchasing decisions are most influenced by social media. For millennials, not so much.

Based on this brief people generation analysis, the product fitting the new generation attitude should be connected to everything, special and offering unique and distinctive features and able to integrate into the everyday life, not limited to professional or educational context. Trying to make a simplified guess, what we can expect in the respective markets as products:

1. Personal electronics pervasion of artificial intelligence

It is supposed, by people belonging to new generation, that the electronic devices not only own personal information but also use those data to return to the user a service or to influence human decision based on artificial intelligence processes. Somehow the people will delegate, on purpose, their actions and decisions to the electronics devices trusting in them as the most optimized decision makers. For this reason, the products that will pervade more and more in our lives will be enabling the augmented reality or the virtual reality.

2. In the industry, the reduction of workers in favor of robots and intelligent automations

The artificial intelligence will also establish more and more in the industry domain both as part of evolution of factory automation and as working tool. The tools offering augmented reality to optimize the efficiency of a worker are already available on the market. Also for healthcare worker, we have example of devices to monitor the worker conditions and automatically send an alert in case of emergency (anomalous heart rate, man down, etc.).

3. In automotive, the establishment of the new mobility

With regard to new generation attitude, the vision of the car of the future is doubtless represented by an electric vehicle, because it copes with the trend of the sustainability and green economy, and self-driving.

27.4.1 Personal Electronics Trends and Future Applications

The companies making personal electronic devices, after having filled of sensor and actuators their devices and having exploited most of the MEMS potential, recognized to own a new gold mine, the DATA, more precisely the personal data, incredibly obtained with the consent of the users. Huge amount of data is generated by the applications using the most popular devices. Just think about three words:

Table 27.2 Example of data path

Device	Application	Service provider	Database
Smartphone	Navigator	Google	Google One Drive

where, who, what, and how (W3H), and now recognize what knows yourself more than you. Your smartphone knows:

- Where you are (geo-localization)
- Who you are (face recognition, human biometric parameters); what you like (from the internet search of your favorite food, goods, interest, etc.) and say (phone call contents, messages, etc.)
- How you behave (when you use your car; if you practice some sport, indoor-outdoor; potentially who you meet, etc.)

And again, all these personal information have been consensually delivered into a database sitting into the servers of the services providers supporting the applications running on the personal electronics devices. Let's see a basic example (Table 27.2):

This is just a simple subset of the measurable parameters related to a person that a simple standalone smartphone can collect. The potential is even higher, in fact the electronics devices are changing and there is a specific technology on which such are relying on: The 5G. The 5G technology, with its increased bandwidth and transmission capacity, emphasizes and amplifies the possible ways of using data. As analyzed by some consultancy companies, the new generations are hunting the revolutionary product, something disruptive capable to make their lives trendy and comfortable. Smartphone will not disappear but also won't be the primary interface. The actual shape, or more simply what will be, of this revolutionary product cannot be clearly envisioned nor anticipated; however, it will be surely linked to artificial intelligence and highly connected. The challenge for semiconductor manufacturers is represented by the comprehension of this market need, and the company that is designing a product to cope with this trend will greatly succeed.

27.4.1.1 Augmented Reality and Virtual Reality Devices

The sensors and micro actuators, all the types we listed in the previous sections, but also high processing engines and connectivity devices are the key devices to collect, process, and transmit the personal data. Smartphones, bracelets, TWS, and IoT already started to do this, but the return that such devices can offer to the user life is limited or even totally absent. There is, instead, a category of products that represents the first attempt of be such a revolutionary product. It is somehow linked to the IoT device, but it goes far beyond: the augmented reality and virtual reality devices, simply known as AR/VR. This product satisfies the requirements of such revolutionary product; in fact it relates to the smartphones, and it is equipped with local capability of sensing processing and connectivity and supports features of augmented reality. So it can be defined as an IoT device but it is seen as a device able to "redefine our relation with technology" as said Mark Zuckerberg, CEO of Facebook.

Smart glasses are at the moment considered the device that will most likely replace the smartphone as primary interface and simultaneously can support user with multiple feedbacks based on what is in the field of view. There will be two distinct categories: all-day wearable smart glasses for all of us and AR/VR headsets for limited professional use during a certain period of the day. Such headset, embedding audio codecs, speakers, and microphones can replace the phone supporting audio playback, audio notifications, and calls. Including high-end cameras can also be adopted to take pictures and record videos. Motion MEMS are also included to support the motion tracking and the user interface replacing buttons. Laser beam scanning modules built around MEMS mirrors are widely considered the most promising projection technology for AR headsets as well as all-day wearable smart glass. Examples of products already in the market are Microsoft's HoloLens 2 headset, using a pair of MEMS mirrors for high-end industrial, professional markets. A different use case in the classical consumer electronics space are North's (now Google) all day wearable glasses named "Focals," built around a different set of MEMS mirrors. So far, the list of sensors/actuators supporting the different features have been enumerated, but what makes this product special is what is projected in field of view. The next picture is representing a possible use case during a working session, the device can observe what you are observing, transmit the data as is, and project you back the aid of an assistant (Fig. 27.16).

But not only the device can elaborate, either locally or in cloud, what is in your field of view and return in your vision useful information, for example, specific parameter of the various components of a motor you are looking at (Fig. 27.17).

Considering instead, some personal life situations, a smart glass can assist you during your shopping or can help you for a pedestrian navigation across the city, in other words relevant data projected on the glass based on the life context (Fig. 27.18).

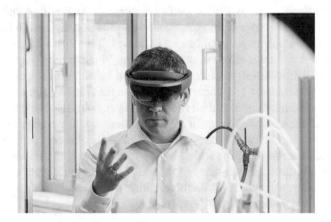

Fig. 27.16 Smart glass – transmitting data

Fig. 27.17 Smart glass – augmented reality

Fig. 27.18 Augmented
reality device

All these simple examples have behind a complex technology and imply the presence of an electronics that both processes data locally and transmits data to the cloud for much more complex calculations. These two calculus locations generate the distinction of edge computing and could computing. Again, these two locations imply a powerful processing component to support the edge computing and powerful transmission components to send into the cloud a huge amount of data. It is worth to remember again that all this data are personal data because monitoring a worker constantly, even if it is for his convenience, still remains personal data covered by strict regulations in every country. Knowing what people like to buy in a shop and extending this on large scale gives an important indication from a business point of view. As direct consequence of the complexity of the system and number of inputs to be managed, such device is representing a big business opportunity for the sensors and actuator components. The semiconductor company who can succeed in

Table 27.3 AR/VR
partitioning

Function block	Product description
Optical engine	Laser pico projection image processor
	MEMS mirror driver
	Laser diode driver
	Horizontal MEMS mirror
	Vertical MEMS mirror
System board	MCU
	6-axis – A + G
	3-axis – magnetic
	Microphone
	Bone conduction accelerometer
	LDO

this challenging scenario is that one able to create an ecosystem. The table below is an example of the multitude of semiconductor components that can be adopted in a smart glass (Table 27.3):

On top of the MEMS mirror chipset, two additional technologies are crucial for the design of small, low-power, and high-resolution AR headsets and smart glasses: laser diodes and waveguide components. To create an ecosystem of component vendors in all these domains, ST has created in October 2020 together with Osram, Dispelix, Applied Material, and Quanta the "LaSAR Alliance." By aligning their respective product roadmaps, the alliance members will boost the adoption of laser beam scanning solutions and help system companies accelerate the path towards all-day wearable glasses.

But that's all about the innovation? For sure no. So far what started from smart glasses and AR/VR headset is representing the first step, but the future will give us more surprises.

27.4.1.2 From an Idea to a Vision

What is anticipated by AR/VR product will be elaborated and expanded for sure. The personal data not only will be used for basic user convenience such as navigation, retail, or notifications but also will be processed with artificial intelligence for a new and more challenging scope. The ultimate purpose of the data usage is to support extensively people along their life, including uncertain conditions, by leveraging the digital context they live.

Let's see which electronic device will be surely in the game:

1. Local data processing, the so-called Edge AI. The idea of sending everything to the cloud has two drawbacks:

   • Amount of data is too high.
   • The delay in between event and reaction.

As anticipated in the previous section, the data processing task in the electronic device has been assigned to the microcontroller so far. Nevertheless, the increased number of components and the high complexity of the algorithm, which elaborated the data coming from sensor, are generating the need of dedicated device. One of the devices perceived as key enabler of innovative products is a component between the sensors and the microcontroller working as interface for the sensor and implementing complex algorithms of artificial intelligence. A kind of, already knows, sensor hub with boosted capability of local processing.

2. Sensing. Sensors must be increased in number and type, calibration-friendly for industrial scale up, on top of inertial and environmental sensor, and it might be useful to have sensors capable to detect also chemicals, gas and presence.
3. System in package: New packaging technology to enhance integration of BOM in tiny and unique form factor embedding actives, passives, connectivity, and power.
4. The software:

   • Edge AI is mandatory as well as cooperative algorithms that, by connecting the nodes, can react, reconfigure, and share the important information to the cloud.
   • This is a new concept, the product is no longer a simple product, and it is an entity capable to provide meaningful data. At this point it will be possible to abstract from the HW and to create a model of what under monitor (building, factory, apartment, etc.) to observe and control it. Or, if we push the boundary, to create digital twins.

To reach the target to obtain a device able to share meaningful data without necessarily knowing the scope, the only way to do this in a compact, reliable, and accurate manner is to use MEMS sensors together with local processing and connectivity. Two virtuous examples of a product offering a tool to pick, elaborate, and return meaningful data are the STMicroelectronics SensorTile.Box and STWIN. Those two devices represent actual proof of concept since according to the great popularity they achieved a large adoption in the market. Those two tools, respectively, focused on personal electronics and industrial markets which make possible the described revolution of moving from a simple product to a complex and highly technology system that enclosed in a black box (user does not even need to know what it is inside) is able to support user with a meaningful information leveraging on multiple sensing, local processing capability, and connectivity (Figs. 27.19 and 27.20).

Encouraged by the success of the SensorTile.Box, ST has started in 2021 to expand the idea towards its Microfluidics MEMS expertise. The result is the "FluidicJet.Box," a reference design built around a microfluidics actuator, which will allow users to develop products and applications which require accurate dispensing of a broad variety of fluids. Typical applications for this reference design are aerosol devices for home care, cosmetics, or medical use. Also, the reference design is being evaluated by companies for more accurate dispensing of water or

Fig. 27.19 STMicroelectronics SensorTile.Box

Fig. 27.20 STMicroelectronics STWIN

nutrients in "Smart Farms," resulting in important environmental benefits for all of us.

So, is it everything about business? We do not think so and we would like to give another fascinating interpretation of the current and future trend. Have you ever heard about the quantum entanglement theory? It explains how quantum particles cannot be described independently, even if separated by a large distance. In a way, an entangled quantum system of more than one particle must be treated as single unit. In another way, entanglement is about being connected, and entanglement is also about efficiency; most researchers believe that entanglement is necessary to realize quantum computing; among the best-known applications of entanglement are superdense coding and quantum teleportation. If we scale up to the world human being can perceive, the society that we are creating with IoT, 5G, etc. is a society where human being, machines, and objects are fully entangled (unconsciously?) emulating what happens in quantum physics, thanks to the share of information between the particles of the system itself.

27.4.2 The Automotive Future

The progressive and natural technological evolution, along with global regulations focused on sustainability, and the impact of economic macrotrends are driving a profound revolutionary change in the concept of vehicles. Market analysis identifies three major trends underway:

- Evolution of propulsion
- Architecture
- Concept of mobility

Regarding the propulsion, global attention to the green economy and a more sustainable world are driving the migration from the standard internal combustion engine (ICE) to the electric vehicle; this trend is also known as powertrain electrification. This migration is not perceived to happen suddenly, but transitionally, over time, with the purchase of hybrid, hybrid-mild, and plug-in hybrid cars. Considering the architecture, regardless of the method of propulsion, the system is increasingly complex. To meet the complexity of these demands, engineers are completely reshuffling the electronics of the traditional car. Since more control units, more sensors, more vision, RF and ultrasonic systems, more connectivity, and more components related to power train are required, the architecture needs to be organized in specific clusters. This approach is referred to as "domain-based architecture." One of the most important clusters taking place for modern cars is the advanced driver-assistance system (ADAS). Regarding the evolution of mobility, the vision of many OEM and car manufacturers, as well as startups, is to develop unmanned vehicles for the transportation of people. New mobility will not be limited to transporting groups of people but will also target individuals: pay per use instead of owning a car, or the combination of different vehicles for one trip (car + eScooter + eBike + passenger drone). In brief, it can be reasonably assumed that the field of innovation for car makers is represented by electric vehicles with highest level of automation. Autonomous driving and electric cars are deeply interlinked. Easier integration and component control make self-driving cars easier to be realized with electric vehicles than with traditional combustion engines.

Even if the future looks defined, strong challenges are still there. The adoption of electric vehicles might be prevented by the lack of infrastructure, e.g., charging stations, limited range, high up-front costs, lack of consumer knowledge, and the

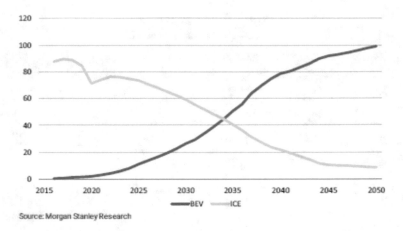

Source: Morgan Stanley Research

Fig. 27.21   BEV vs ICE sales (Mu)

pressure of oil companies and the car manufacturer lobby. Not only are charging stations still very few and far between, but they are also usually manufactured by different suppliers without a standardized charging and payment system. High up-front costs make electrical vehicles still less attractive as compared to traditional internal combustion engine cars. Although a major transformation involves complex technical challenges, the ecosystem is yet to come, and government regulations are not yet defined; the trend is well defined, and sooner or later the traditional combustion engine will be abandoned in favor of the electric one. The picture below shows an estimation of Morgan Stanley forecasting how such conversion will take place (Fig. 27.21).

If electrification represents an opportunity for electronic devices related to the power train, the area in which there is a high potential for innovation is autonomous driving. This automation of a car requires, same as personal electronics products, powerful processing engines, several sensors and actuators, and enhanced transmission capacity. An autonomous vehicle needs to transfer huge amount of data:

- They need to interface between each other.
- Talk with objects in the environment and infrastructures.
- They need to know their actual position in the glove (accurate precise position-ing).

A necessary premise is to understand how the architecture of the case is changing. The complexity of the onboard electronics and the increased number of control units are forcing car makers to change vehicle architecture with that of a more optimized organization of these multiple building blocks. Vehicle architecture is moving from a traditional approach to a domain-based architecture. Basically, the domain-based approach aims to optimize the architecture in terms of efficiency, design complexity, interconnections, and associated costs. The three main categories involved in the

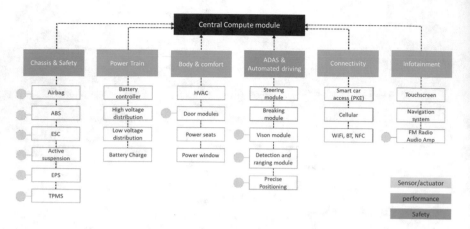

Fig. 27.22 Example of electric car domain-based architecture

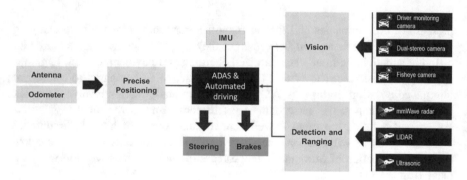

Fig. 27.23 Example of ADAS system

electronics of a car are electronic components to support car performance, safety, and sensor/actuators. The following figure represents an example of a domain-based architecture of an electric car (Fig. 27.22).

ADAS and automated driving domain, on top of the electronics used to control the car like speed and steering, includes several devices to replace human senses. During autonomous driving, the input to the car, steering and velocity, is provided by the system not by the driver. In this scenario the system needs to monitor and process many environmental signals to properly control the vehicle. This type of system needs to have the capability to recognize the position of the car either by global positioning or relative to other objects. According to this distinction, the electronics of the ADAS system with the task of the positioning is assigned to the precise positioning control module and environmental monitoring modules (vision, detection, and ranging) (Fig. 27.23).

The environment monitoring modules are adopted to define the presence and the distance of any object surrounding the car to avoid collisions or any unwanted contact. Due to the complexity and the diversity of the objects to be recognized and detected, the system is further split into two blocks, one managing the optical devices and the other one for the detection and ranging of the objects. Regarding the vision system, the ADAS system includes cameras installed in several locations (rear, front, inside) and with different fields of view (limited or 360°). Regarding the detection and ranging block, the system includes long-range and short-range radars, LiDAR, and ultrasonic sensors. **The precise positioning module** is composed of GNSS (Global Navigation Satellite System), an antenna, and an inertial sensor. The GNSS allows the system to locate the car by global positioning and to control the trajectory of the car from a starting location to a destination according to a specified path (satellite navigation). The satellite navigation system input is the antenna. The **IMU** (inertial measurement unit) is adopted by the ADAS host processor in combination with other elements, like the GNSS module and the vision system, in order to enable the automated navigation of the car. The ADAS control unit implements the navigation system with a fusion of all the signals coming from the different elements.

Summarizing, the described modules (precise positioning and the environment monitoring modules) enable autonomous driving, replacing the driver. The precise positioning module allows the system to drive and locate the car along a specified path to a defined destination, and the environment monitoring modules allows the system to implement spatial controls to avoid collisions. The following image represents an example of the main tasks operated by the different modules and their respective input. Each of the listed detection systems has its own specific function; the list of the most common functions is shown in the following figure (Fig. 27.24).

But it is not all about the adoption of motion sensor in automotive market. As clearly indicated in the Fig. 27.24, a car with high level of automation is full of vision and optical systems. A key element for the correct and optimized working of those systems is the proper installation of the devices. This is true for both first and over life installation. Six-axis sensors are finding an adoption into the radar and LiDAR systems to check proper orientation at the beginning and as condition monitoring over life. The same purpose is shared by light levelling system where sensor can monitor the installation to monitor if the illumination is properly done and corresponding to the expected. As part of autonomous driving system, also those applications are safety critical related making the safety integrity level a requirement for sensors as well. In conclusion, as a matter of fact, the MEMS sensors are strongly pervading across many applications in the car, from traditional to new, and they are required to be developed according to the safety standard.

On top of the various opportunities described for MEMS sensors, autonomous driving will not be possible without a MEMS actuator, in the form of a MEMS mirror for LiDAR systems. Recent advances in MEMS process technology have generated a wave of LiDAR modules to be launched into the market, using MEMS mirrors for LiDAR/3D sensing. In order to address this new opportunity, ST has announced in 2020 the development of piezo-actuated mirror and ASIC to be integrated into the LiDAR reference design, offered by "Leddartech," one of the leaders in the field of LiDAR.

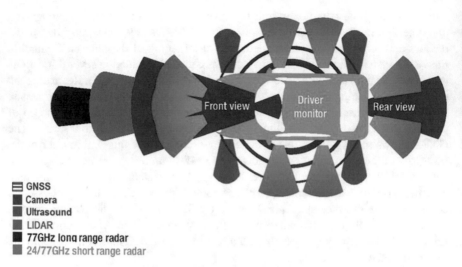

GNSS
Camera
Ultrasound
LIDAR
77GHz long range radar
24/77GHz short range radar

Domain	Module	Sensor	Functions
ADAS and Automated driving	Detection and Range	Long range radar	Adaptive Cruise control
		Mid/Short range radar	Traffic alert
			Rear collision
			Blind spot detection
		Lidar	Emergency braking
			Pedestrian detection
			Collision avoidance
		Ultrasound	Park assist
	Vision	Front camera	Traffic sign recognition
			Lane departure
		Rear camera	Park assist
		Inside camera	Driver monitor
	Precise positioning	GNSS	Global localization
		IMU	Inertial navigation system

Fig. 27.24 Highly automated ADAS system functions

But the autonomous driving will not be enabled only by the sensor fusion implemented by local processing. Safety is representing a critical challenge due to the high liability exposure of the car maker in case of accident or person injury while car is set in self-driving mode, and the way to reinforce the capability of a car to drive safely is to create an ecosystem in which the vehicles communicate to each other. So, a safe self-driving is not only achieved by car able to monitor what is surrounding it, thanks to sensors and cameras but also knowing accurate information related to other cars. Refer to the simple example reported in the following figure (Fig. 27.25).

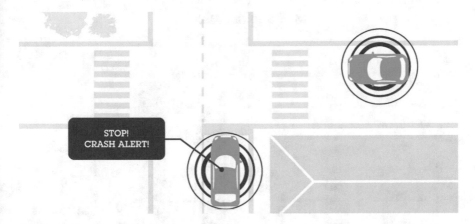

Fig. 27.25 V2X example

In this example, the vision system might have difficulties to detect a coming car due to obstruction. The only way to know that another car is coming is that all the autonomous cars send information about their position. This system provides 360° coverage of surrounding vehicles, works at any environment, and helps in real life to prevent accidents. This electronic system is also known as vehicle to everything (V2X); somehow it is a wider concept of the ADAS system and not only relies on sensing and local processing but also implies a huge capability of data transmission. Once again, the 5G is seen a key enabler for the execution of such challenging vision.

Finally, the artificial intelligence plays here a relevant role because the car needs to autonomously make its own decisions. Is the autonomous car, the unique product that guys belonging to generation Z are looking for? We don't think so, but for sure it is assonant with the trend described at the beginning of this section: based on artificial intelligence and highly connected!

27.5 Conclusion

Sensors, MEMS based, are established and spread in the market and they are expected to still grow much more, thanks to all the applications described in this chapter. It is worth to substantiate this statement with numbers. The following figure gives a clear overview of the size of the market represented by MEMS sensor in the past and the size that is expected to be in the coming years (Fig. 27.26).

What are the takeaways from those numbers:

- MEMS sensor market is still expected to grow, thanks to the multitude of different applications they can enable and support.

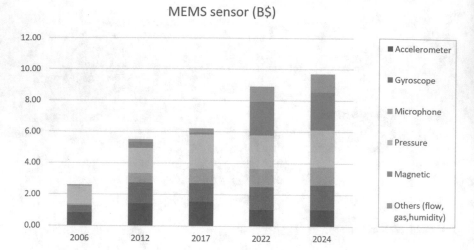

Fig. 27.26 MEMS sensor market size. (Data: courtesy of OMDIA. www.OMDIA.com)

- Automotive electrification and digitalization represents the major segment boosting up the presence of sensors.
- Despite of mobile phone demand is flat and price erosion is ongoing, the market size of accelerometer, gyroscope, and other sensors are not decreasing due to increase of the adoption of sensor in several IoT device, industry 4.0 sensor nodes, and innovative solutions.

We are living the era of MEMS. The synthesis is really in the title of this section: from a bright past towards a shining future.

Correction to: Silicon Sensors and Actuators

Benedetto Vigna, Paolo Ferrari, Flavio Francesco Villa, Ernesto Lasalandra, and Sarah Zerbini

Correction to:
B. Vigna et al. (eds.), *Silicon Sensors and Actuators,*
https://doi.org/10.1007/978-3-030-80135-9

One of the authors' name was incorrectly published as "Paolo Ferrari" in Chapters 4 and 10 and it has been updated now.

The updated version of these chapters can be found at
https://doi.org/10.1007/978-3-030-80135-9_4
https://doi.org/10.1007/978-3-030-80135-9_10

Index

© Springer Nature Switzerland AG 2022
B. Vigna et al. (eds.), *Silicon Sensors and Actuators*,
https://doi.org/10.1007/978-3-030-80135-9